SCHAUM'S SOLVED PROBLEMS SERIES

3000 SOLVED PROBLEMS IN

LINEAR ALGEBRA

by

Seymour Lipschutz, Ph.D.

Temple University

SCHAUM'S OUTLINE SERIES
McGRAW-HILL PUBLISHING COMPANY

*New York   St. Louis   San Francisco   Auckland   Bogotá   Caracas
Hamburg   Lisbon   London   Madrid   Mexico   Milan   Montreal
New Delhi   Oklahoma City   Paris   San Juan   São Paulo
Singapore   Sydney   Tokyo   Toronto*

Seymour Lipschutz, Ph.D., *Professor of Mathematics at Temple University.*

Dr. Lipschutz earned the Ph.D. at the Courant Institute of Mathematical Sciences of New York University. He has authored a number of volumes in the Schaum's Outline Series, including the internationally used LINEAR ALGEBRA.

Other Contributors to This Volume

Frank Ayres, Jr., Ph.D., *Dickinson College*

David Beckwith, Ph.D., *McGraw-Hill Book Company*

Richard Bronson, Ph.D., *Fairleigh Dickinson University*

Orin Chein, Ph.D., *Temple University*

Project supervision was done by The Total Book.

Library of Congress Cataloging-in-Publication Data

Lipschutz, Seymour.
 3000 solved problems in linear algebra.

 (Schaum's solved problems series)
 1. Algebras, Linear—Problems, exercises, etc.
I. Title.   II. Title: Three thousand solved in linear
algebra.   III. Series.
QA184.5.L56 1988   512'.5   88-8944
ISBN 0-07-038023-6

   3 4 5 6 7 8 9 0 SHP/SHP 8 9

0-07-038023-6

CONTENTS

iv ▯ CONTENTS

To the Student

This collection of thousands of solved problems covers almost every type of problem which may appear in any course in linear algebra. Moreover, our collection includes both computational problems and theoretical problems (which involve proofs).

Each section begins with very elementary problems and their difficulty usually increases as the section progresses. Furthermore, the theoretical problems involving proofs normally appear after the computational problems, which can thus preview the theory. (Most students have more difficulty with proofs.)

Normally, students will be assigned a textbook for their linear algebra course. The sequence of our chapters follows the customary order found in most textbooks (although there may be some discrepancies). However, whenever possible, our chapters and sections have been written so that their order can be changed without difficulty and without loss of continuity.

The solution to each problem immediately follows the statement of the problem. However, you may wish to try to solve the problem yourself before reading the given solution. In fact, even after reading the solution, you should try to resolve the problem without consulting the text. Used thus, *3000 Solved Problems in Linear Algebra* can serve as a supplement to any course in linear algebra, or even as an independent refresher course.

CHAPTER 1

Vectors in $\mathbf{R}^n$ and $\mathbf{C}^n$

1.1 VECTORS IN $\mathbf{R}^n$

1.1 A vector u in the vector space $\mathbf{R}^n$ is an ordered set of n real numbers: $u = (a_1, a_2, \ldots, a_n)$. The real number a_k is called the kth *component* or *coordinate* of u. Compare this with the definition of a vector in physics.

▌ Physics defines a vector **u** to be a quantity with magnitude and direction, represented by means of an arrow or directed line segment emanating from a reference point O. In Fig. 1-1 a planar vector **u** is identified with the coordinates of its endpoint, $P(4, 2)$. That is, **u** $= (4, 2)$—in accord with the above definition of a vector in $\mathbf{R}^2$.

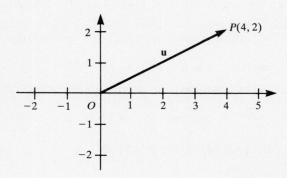

Fig. 1-1

1.2 State the difference between a row vector and a column vector.

▌ A *column vector* u is a vector whose components are arranged vertically:

$$u = \begin{pmatrix} a_1 \\ a_2 \\ \ldots \\ a_n \end{pmatrix}$$

A *row vector* is a vector whose components are arranged horizontally, as in Problem 1.1. [In this chapter, vectors will normally be written as row vectors.]

1.3 To which vector space $\mathbf{R}^n$ does each vector belong?

$$(a) \quad (3, -2, 5, 8) \qquad (b) \quad (3, 6 + 2i) \qquad (c) \quad (\pi, 2, 5\pi)$$

▌ (a) $\mathbf{R}^4$, since there are four components. (b) None, since not all the components are real numbers. (c) $\mathbf{R}^3$ [π and 5π are real numbers].

1.4 For vectors u and v in $\mathbf{R}^n$, when is $u = v$?

▌ The vectors u and v are equal if and only if the corresponding components are equal.

1.5 Let $u_1 = (1, 2, 3)$, $u_2 = (2, 3, 1)$, $u_3 = (1, 3, 2)$, $u_4 = (2, 3, 1)$ be vectors in $\mathbf{R}^3$. Which of the vectors, if any, are equal?

▌ Only u_2 and u_4 are componentwise equal.

1.6 Find x and y if $(x, 3) = (2, x + y)$.

▌ Since the two vectors are equal, the corresponding components are equal to each other: $x = 2$, $3 = x + y$. By subtraction, $y = 1$.

1.7 Define the *zero vector* in $\mathbf{R}^n$.

▌ $\mathbf{0} \equiv \underbrace{(0, 0, \ldots, 0)}_{n \text{ components}}$

1.8 Find x and y if $u = (x + y, x - 3)$ is the zero vector.

▌ First set each component of u equal to 0 to obtain the equations $x + y = 0$ and $x - 3 = 0$. The second equation gives $x = 3$; then the first equation gives $y = -3$.

1.9 Define the *negative*, $-u$, of a vector $u = (a_1, a_2, \ldots, a_n)$ in $\mathbf{R}^n$.

▌ $-u \equiv (-a_1, -a_2, \ldots, -a_n)$

1.10 Find the negatives of (a) $u = (3, -5, -8, 4)$, (b) $v = (-4, 2\pi, 0)$, (c) $\mathbf{0} = (0, 0, 0, 0)$.

▌ Take the negative of each component of the vector. (a) $-u = (-3, -(-5), -(-8), -4) = (-3, 5, 8, -4)$; (b) $-v = (4, -2\pi, 0)$; (c) $-\mathbf{0} = (-0, -0, -0, -0) = (0, 0, 0, 0) = \mathbf{0}$.

1.11 Show that $-(-u) = u$ for any vector u in $\mathbf{R}^n$.

▌ Let $u = (a_1, a_2, \ldots, a_n)$. Then, because $-(-a) = a$ for any scalar a in $\mathbf{R}$,

$$-(-u) = -(-a_1, -a_2, \ldots, -a_n) = (a_1, a_2, \ldots, a_n) = u$$

1.2 VECTOR ADDITION AND SCALAR MULTIPLICATION

1.12 Let $u = (u_1, u_2, \ldots, u_n)$ and $v = (v_1, v_2, \ldots, v_n)$ be any vectors in $\mathbf{R}^n$. Then the *sum*, $u + v$, is the vector

$$u + v \equiv (u_1 + v_1, u_2 + v_2, \ldots, u_n + v_n)$$

Show how this definition conforms to the addition of vectors in physics.

▌ In physics, vector $\mathbf{u} + \mathbf{v}$ is the diagonal of the parallelogram formed by vectors $\mathbf{u}$ and $\mathbf{v}$, as pictured in Fig. 1-2(a). Now let O be chosen as the origin of a coordinate system (of $\mathbf{R}^2$), and suppose (a, b) and (c, d) are the endpoints of $\mathbf{u}$ and $\mathbf{v}$, respectively, as pictured in Fig. 1-2(b). Then, using geometry, one can show that $(a + c, b + d)$ will be the endpoint of $\mathbf{u} + \mathbf{v}$. On the other hand, adding corresponding components, we obtain

$$(a, b) + (c, d) = (a + c, b + d)$$

Both definitions of addition give the same result.

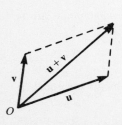

(a)

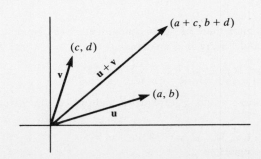

(b)

Fig. 1-2

1.13   Compute: (*a*) $(3, -4, 5, -6) + (1, 1, -2, 4)$,   (*b*) $(1, 2, -3) + (4, -5)$.

  ▮   (*a*) Add corresponding components:

$$(3, -4, 5, -6) + (1, 1, -2, 4) = (3 + 1, -4 + 1, 5 - 2, -6 + 4) = (4, -3, 3, -2)$$

(*b*) The sum is not defined, since the vectors have different numbers of components.

1.14   Compute: (*a*) $\begin{pmatrix} 7 \\ -4 \\ 2 \end{pmatrix} + \begin{pmatrix} -3 \\ -1 \\ 5 \end{pmatrix}$,   (*b*) $\begin{pmatrix} 1 \\ 3 \\ 5 \end{pmatrix} + \begin{pmatrix} -2 \\ 4 \end{pmatrix}$.

  ▮   (*a*) Add corresponding components:

$$\begin{pmatrix} 7 \\ -4 \\ 2 \end{pmatrix} + \begin{pmatrix} -3 \\ -1 \\ 5 \end{pmatrix} = \begin{pmatrix} 7 - 3 \\ -4 - 1 \\ 2 + 5 \end{pmatrix} = \begin{pmatrix} 4 \\ -5 \\ 7 \end{pmatrix}$$

(*b*) The sum is not defined, since the vectors have different numbers of components.

1.15   Let $u = (u_1, u_2, \ldots, u_n)$ be any vector in $\mathbf{R}^n$ and let k be any scalar [real number] in $\mathbf{R}$. Then the [scalar] *product* ku is the vector

$$ku \equiv (ku_1, ku_2, \ldots, ku_n)$$

Show how this definition conforms to the scalar multiplication of vectors in physics.

  ▮   Physics defines the product of a real number k and a vector [arrow] **u**, say with reference point O, to be the vector (i) whose magnitude is equal to the magnitude of **u** multiplied by $|k|$; (ii) whose direction is that of **u** if $k \geq 0$,   but is opposite to **u** if $k < 0$. This is pictured in Fig. 1-3(*a*). Now let O be chosen as the origin of a coordinate system [of $\mathbf{R}^2$], and suppose (a, b) is the endpoint of **u**, as pictured in Fig. 1-3(*b*). Then, using geometry, one can easily show that the endpoint of $k\mathbf{u}$ is (ka, kb). On the other hand, our definition gives $k(a, b) = (ka, kb)$,   the same result.

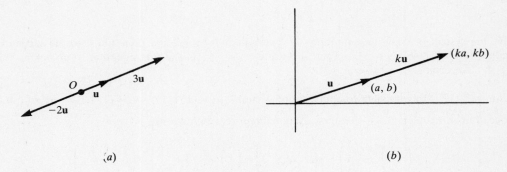

(*a*)   (*b*)   **Fig. 1-3**

1.16   Compute: (*a*) $-3(4, -5, -6)$,   (*b*) $-(6, 7, -8)$.

  ▮   (*a*) Multiply each component by the scalar: $-3(4, -5, -6) = (-12, 15, 18)$.   (*b*) Either multiply each component by -1 or take the negative of each component; either way we obtain $(6, -7, 8)$.

1.17   Compute: (*a*) $5\begin{pmatrix} -2 \\ 3 \\ 4 \end{pmatrix}$,   (*b*) $-2\begin{pmatrix} 7 \\ -5 \end{pmatrix}$.

  ▮   Multiply each component by the scalar:

$$(a) \quad 5\begin{pmatrix} -2 \\ 3 \\ 4 \end{pmatrix} = \begin{pmatrix} -10 \\ 15 \\ 20 \end{pmatrix} \qquad (b) \quad -2\begin{pmatrix} 7 \\ -5 \end{pmatrix} = \begin{pmatrix} -14 \\ 10 \end{pmatrix}.$$

1.18   Define the *difference*, $u - v$, of the vectors u and v in $\mathbf{R}^n$.

  ▮   The difference is obtained by adding the negative: $u - v \equiv u + (-v)$.

1.19 Compute: (a) $(3, -5, 6, 8) - (4, 1, -7, 9)$, (b) $\begin{pmatrix} 6 \\ -3 \end{pmatrix} - \begin{pmatrix} 2 \\ -5 \end{pmatrix}$.

❚ First find the negative of the second vector and then add.

(a) $(3, -5, 6, 8) - (4, 1, -7, 9) = (3, -5, 6, 8) + (-4, -1, 7, -9) = (-1, -6, 13, -1)$

(b) $\begin{pmatrix} 6 \\ -3 \end{pmatrix} - \begin{pmatrix} 2 \\ -5 \end{pmatrix} = \begin{pmatrix} 6 \\ -3 \end{pmatrix} + \begin{pmatrix} -2 \\ 5 \end{pmatrix} = \begin{pmatrix} 4 \\ 2 \end{pmatrix}$

1.20 Let $u = (2, -7, 1)$, $v = (-3, 0, 4)$, $w = (0, 5, -8)$. Find (a) $3u - 4v$, (b) $2u + 3v - 5w$.

❚ First perform the scalar multiplication and then the vector addition.

(a) $3u - 4v = 3(2, -7, 1) - 4(-3, 0, 4) = (6, -21, 3) + (12, 0, -16) = (18, -21, -13)$

(b) $2u + 3v - 5w = 2(2, -7, 1) + 3(-3, 0, 4) - 5(0, 5, -8)$
$\qquad = (4, -14, 2) + (-9, 0, 12) + (0, -25, 40)$
$\qquad = (4 - 9 + 0, -14 + 0 - 25, 2 + 12 + 40) = (-5, -39, 54)$

1.21 Let $u = (3, -2, 1, 4)$ and $v = (7, 1, -3, 6)$ in $\mathbf{R}^4$. Find (a) $u + v$, (b) $4u$, (c) $2u - 3v$.

❚ (a) $u + v = (3 + 7, -2 + 1, 1 - 3, 4 + 6) = (10, -1, -2, 10)$; (b) $4u = (4 \cdot 3, 4 \cdot (-2), 4 \cdot 1, 4 \cdot 4)$
$= (12, -8, 4, 16)$; (c) $2u - 3v = (6, -4, 2, 8) + (-21, -3, 9, -18) = (-15, -7, 11, -10)$.

1.22 Find x and y if $(4, y) = x(2, 3)$.

❚ (i) Multiply by the scalar x to obtain $(4, y) = x(2, 3) = (2x, 3x)$. (ii) Equate corresponding components: $4 = 2x$, $y = 3x$. (iii) Solve linear equations for x and y: $x = 2$ and $y = 6$.

1.23 Write $w = (1, 9)$ as a linear combination of the vectors $u = (1, 2)$ and $v = (3, -1)$.

❚ We want to find scalars x and y such that $w = xu + yv$; i.e.,

$$(1, 9) = x(1, 2) + y(3, -1) = (x, 2x) + (3y, -y) = (x + 3y, 2x - y)$$

Equality of corresponding components gives the two equations

$$x + 3y = 1 \qquad 2x - y = 9$$

To solve the system of equations, multiply the first equation by -2 and then add it to the second equation to obtain $-7y = 7$, or $y = -1$. Then substitute $y = -1$ in the first equation to obtain $x - 3 = 1$, or $x = 4$. Accordingly, $w = 4u - v$.

1.24 Write $v = (2, -3, 4)$ as a linear combination of vectors $u_1 = (1, 1, 1)$, $u_2 = (1, 1, 0)$, $u_3 = (1, 0, 0)$.

❚ Proceed as in Problem 1.23, this time using column representations.

$$\begin{pmatrix} 2 \\ -3 \\ 4 \end{pmatrix} = x \begin{pmatrix} 1 \\ 1 \\ 1 \end{pmatrix} + y \begin{pmatrix} 1 \\ 1 \\ 0 \end{pmatrix} + z \begin{pmatrix} 1 \\ 0 \\ 0 \end{pmatrix} = \begin{pmatrix} x \\ x \\ x \end{pmatrix} + \begin{pmatrix} y \\ y \\ 0 \end{pmatrix} + \begin{pmatrix} z \\ 0 \\ 0 \end{pmatrix} = \begin{pmatrix} x + y + z \\ x + y \\ x \end{pmatrix}$$

Now set corresponding components equal to each other:

$$x + y + z = 2 \qquad x + y = -3 \qquad x = 4$$

To solve the system of equations, substitute $x = 4$ into the second equation to obtain $4 + y = -3$, or $y = -7$. Then substitute into the first equation to find $z = 5$. Thus $v = 4u_1 - 7u_2 + 5u_3$.

1.25 Write $w = (1, 2, -5)$ as a linear combination of $u_1 = (1, -1, -1)$, $u_2 = (2, 1, 4)$ and $u_3 = (1, 1, 3)$.

❚ First multiply by scalars x, y, z and then add:

$$\begin{pmatrix} 1 \\ 2 \\ -5 \end{pmatrix} = x \begin{pmatrix} 1 \\ -1 \\ -1 \end{pmatrix} + y \begin{pmatrix} 2 \\ 1 \\ 4 \end{pmatrix} + z \begin{pmatrix} 1 \\ 1 \\ 3 \end{pmatrix} = \begin{pmatrix} x + 2y + z \\ -x + y + z \\ -x + 4y + 3z \end{pmatrix}$$

Then set corresponding components equal to each other to obtain the system:

$$x + 2y + z = 1 \qquad -x + y + z = 2 \qquad -x + 4y + 3z = -5$$

Add the first equation to the second equation to obtain (i) $3y + 2z = 3$.   Now add the first equation to the third equation to obtain $6y + 4z = -4$,   or (ii) $3y + 2z = -2$.   Equations (i) and (ii) are obviously inconsistent.   In other words, w is not a linear combination of u_1, u_2, u_3.

In Problems 1.26–1.36, u, v, w denote vectors in **R**^{n}; k, k' denote scalars in **R**; u_1, v_i and w_i denote the ith components of u, v and w, respectively.

1.26   Prove that $(u + v) + w = u + (v + w)$.

▌   By definition [Problem 1.12], $u_i + v_i$ is the ith component of $u + v$ and so $(u_i + v_i) + w_i$ is the ith component of $(u + v) + w$.   On the other hand, $v_i + w_i$ is the ith component of $v + w$ and so $u_i + (v_i + w_i)$ is the ith component of $u + (v + w)$.   But u_i, v_i, and w_i are real numbers for which the associative law holds; that is,

$$(u_i + v_i) + w_i = u_i + (v_i + w_i)   (i = 1, \ldots, n)$$

Accordingly, $(u + v) + w = u + (v + w)$,   since their corresponding components are equal.

1.27   Prove that $u + \mathbf{0} = u$.

▌ $u + \mathbf{0} = (u_1, u_2, \ldots, u_n) + (0, 0, \ldots, 0) = (u_1 + 0, u_2 + 0, \ldots, u_n) = (u_1, u_2, \ldots, u_n) = u$

1.28   Prove that $u + (-u) = \mathbf{0}$.

▌ $u + (-u) = (u_1, u_2, \ldots,) + (-u_1, -u_2, \ldots, -u_n)$
$\qquad\qquad = (u_1 - u_1, u_2 - u_2, \ldots, u_n - u_n) = (0, 0, \ldots, 0) = \mathbf{0}$

1.29   Prove that $u + v = v + u$.

▌   By definition [Problem 1.12], $u_i + v_i$ is the ith component of $u + v$,   and $v_i + u_i$ is the ith component of $v + u$.   But u_i and v_i are real numbers for which the commutative law holds, that is,

$$u_i + v_i = v_i + u_i   (i = 1, \ldots, n)$$

Hence $u + v = v + u$,   since their corresponding components are equal.

1.30   Prove that $k(u + v) = ku + kv$.

▌   Since $u_i + v_i$ is the ith component of $u + v$, $k(u_i + v_i)$ is the ith component of $k(u + v)$.   Since ku_i and kv_i are the ith components of ku and kv respectively, $ku_i + kv_i$ is the ith component of $ku + kv$. But k, u_i and v_i are real numbers; hence

$$k(u_i + v_i) = ku_i + kv_i   (i = 1, \ldots, n)$$

Thus $k(u + v) = ku + kv$,   as corresponding components are equal.

1.31   Prove that $(k + k')u = ku + k'u$.

▌   By definition [Problem 1.15], $(k + k')u_i$ is the ith component of the vector $(k + k')u$. Since ku_i and $k'u_i$ are the ith components of ku and $k'u$, respectively, $ku_i + k'u_i$ is the ith component of $ku + k'u$.   But k, k', and u_i are real numbers; hence

$$(k + k')u_i = ku_i + k'u_i   (i = 1, \ldots, n)$$

Thus $(k + k')u = ku + k'u$,   as corresponding components are equal.

1.32   Prove that $(kk')u = k(k'u)$.

▮ Since $k'u_i$ is the ith component of $k'u$, $k(k'u_i)$ is the ith component of $k(k'u)$.   But $(kk')u_i$ is the ith component of $(kk')u$ and, since k, k' and u_i are real numbers,

$$(kk')u_i = k(k'u_i)   (i = 1, \ldots, n)$$

Hence $(kk')u = k(k'u)$,   as corresponding components are equal.

1.33   Prove that $1 \cdot u = u$.

▮ $1 \cdot u = 1(u_1, u_2, \ldots, u_n) = (1u_1, 1u_2, \ldots, 1u_n) = (u_i, u_2, \ldots, u_n) = u$

1.34   Show that $0u = 0$ for any vector u.

▮   **Method 1.** $0u = 0(u_1, u_2, \ldots, u_n) = (0u_1, 0u_2, \ldots, 0u_n) = (0, 0, \ldots, 0) = \mathbf{0}$.

Method 2.   By Problem 1.31, $0u = (0 + 0)u = 0u + 0u$.   Adding $-(0u)$ to both sides gives

$$0v + (-(0v)) = (0v + 0v) + (-(0v))$$
$$\mathbf{0} = 0v + (0v + (-(0v)))   \text{[by Problems 1.28, 1.26]}$$
$$\mathbf{0} = 0v + \mathbf{0}   \text{[by Problem 1.28]}$$
$$\mathbf{0} = 0v   \text{[by Problem 1.27]}$$

[Note that Method 2, although lengthy, does not explicitly use coordinates.]

1.35   Show that $k0 = 0$ for any scalar k.

▮   **Method 1.** $k0 = k(0, 0, \ldots, 0) = (k \cdot 0, k \cdot 0, \ldots, k \cdot 0) = (0, 0, \ldots, 0) = \mathbf{0}$.

Method 2.   By Problem 1.34, $\mathbf{0} + \mathbf{0} = \mathbf{0}$.   Then, by Problem 1.31, $k0 = k(\mathbf{0} + \mathbf{0}) = k0 + k0$.   Adding $-(k0)$ to both sides leads to the required result, as in Problem 1.34.

1.36   Show that $(-1)u = -u$.

▮   From the previously proved properties,

$$u + (-u) = \mathbf{0} = 0u = (1 + (-1))u = 1u + (-1)u = u + (-1)u$$

and the result follows upon adding $-u$ to both sides.

1.3   SUMMATION SYMBOL

1.37   Let $f(k)$ be an algebraic expression involving an integer variable k.   Define the expression

$$S_n = \sum_{k=1}^{n} f(k)$$

where $n \geq 1$.   [Here 1 is called the *lower limit*, n is called the *upper limit*, and the Greek letter sigma functions as the *summation symbol*.]

▮ $S_n = f(1) + f(2) + \cdots + f(n-1) + f(n)$.   From this definition, it is obvious that, for $n \geq 2$, $S_n = S_{n-1} + f(n)$.

1.38   Suppose n_1 and n_2 are any integers such that $n_1 \leq n_2$.   Define $\displaystyle\sum_{k=n_1}^{n_2} f(k)$.

▮ $\displaystyle\sum_{k=n_1}^{n_2} f(k) = f(n_1) + f(n_1 + 1) + f(n_1 + 2) + \cdots + f(n_2)$.   For $n_2 < n_1$,   the summation is usually defined to be zero.

1.39   Compute $\displaystyle\sum_{k=1}^{4} k^3$.

■ $\sum_{k=1}^{4} k^3 = 1^3 + 2^3 + 3^3 + 4^3 = 1 + 8 + 27 + 64 = 100$

1.40   Compute $\sum_{j=2}^{5} j^2$.

■ $\sum_{j=2}^{5} j^2 = 2^2 + 3^2 + 4^2 + 5^2 = 4 + 9 + 16 + 25 = 54$

1.41   Find $\sum_{k=1}^{5} x_k$.

■ $\sum_{k=1}^{5} x_k = x_1 + x_2 + x_3 + x_4 + x_5$

1.42   Rewrite without the summation symbol: (a) $\sum_{i=1}^{n} a_i b_i$,   (b) $\sum_{i=0}^{n} a_i x^i$,   (c) $\sum_{k=1}^{p} a_{ik} b_{kj}$.

■   (a) $a_1 b_1 + a_2 b_2 + \cdots + a_n b_n$ (b) $a_0 + a_1 x + a_2 x^2 + \cdots + a_n x^n$

(c) $a_{i1} b_{1j} + a_{i2} b_{2j} + a_{i3} b_{3j} + \cdots + a_{ip} b_{pj}$

1.43   Prove: $\sum_{k=1}^{n} [f(k) + g(k)] = \sum_{k=1}^{n} f(k) + \sum_{k=1}^{n} g(k)$.

■   The proof is by induction on n.   For $n = 1$,

$$\sum_{k=1}^{1} [f(k) + g(k)] = f(1) + g(1) = \sum_{k=1}^{1} f(k) + \sum_{k=1}^{1} g(k)$$

Suppose $n > 1$,   and that the theorem holds for $n - 1$;   i.e.,

$$\sum_{k=1}^{n-1} [f(k) + g(k)] = \sum_{k=1}^{n-1} f(k) + \sum_{k=1}^{n-1} g(k)$$

Then [see Problem 1.37],

$$\sum_{k=1}^{n} [f(k) + g(k)] = \sum_{k=1}^{n-1} [f(k) + g(k)] + [f(n) + g(n)] = \sum_{k=1}^{n-1} f(k) + \sum_{k=1}^{n-1} g(k) + f(n) + g(n)$$

$$= \left[\sum_{k=1}^{n-1} f(k) + f(n) \right] + \left[\sum_{k=1}^{n-1} g(k) + g(n) \right] = \sum_{k=1}^{n} f(k) + \sum_{k=1}^{n} g(k)$$

Thus the theorem is proved.

1.44   Prove: $\sum_{k=1}^{n} c\, f(k) = c \sum_{k=1}^{n} f(k)$.

■   The proof is immediate from the distributive law, $a(b + c + \cdots) = ab + ac + \cdots$,   for real numbers.

1.4   DOT (INNER) PRODUCT

1.45   Suppose $u = (u_1, u_2, \ldots, u_n)$ and $v = (v_1, v_2, \ldots, v_n)$ are vectors in **R**^{n}. The *dot* [or *inner* or *scalar*] *product* of u and v, denoted by $u \cdot v$,   is defined to be the scalar obtained by multiplying corresponding components and adding the resulting products:

$$u \cdot v = \sum_{k=1}^{n} u_k v_k$$

Show how this definition conforms to that used in physics.

■   Suppose that **u** and **v** are vectors (arrows) in **R**^{3} emanating from the origin O, as pictured in Fig. 1-4. Moreover, suppose that $P(a_1, a_2, a_3)$ and $Q(b_1, b_2, b_3)$ are the endpoints of **u** and **v**, respectively, and let θ denote the angle between **u** and **v**. Physics defines the dot product as

$$\mathbf{u} \cdot \mathbf{v} = |\mathbf{u}|\, |\mathbf{v}| \cos \theta$$

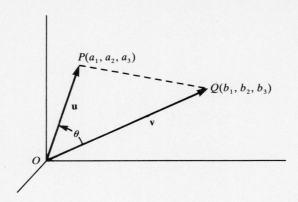

Fig. 1-4

Now $|\mathbf{u}|^2 = a_1^2 + a_2^2 + a_3^2$, $|\mathbf{v}|^2 = b_1^2 + b_2^2 + b_3^2$,   and

$$\overline{PQ}^2 = (a_1 - b_1)^2 + (a_2 - b_2)^2 + (a_3 - b_3)^2$$
$$= (a_1^2 + a_2^2 + a_3^2) + (b_1^2 + b_2^2 + b_3^2) - 2(a_1 b_1 + a_2 b_2 + a_3 b_3)$$
$$= |\mathbf{u}|^2 + |\mathbf{v}|^2 - 2 \sum_{k=1}^{3} u_k v_k$$

But, by the law of cosines, $\overline{PQ}^2 = |\mathbf{u}|^2 + |\mathbf{v}|^2 - 2|\mathbf{u}| \, |\mathbf{v}| \cos \theta$;   therefore, $|\mathbf{u}| \, |\mathbf{v}| \cos \theta = \sum_{k=1}^{3} u_k v_k$,   and the two definitions agree.

1.46   Compute $u \cdot v$,   where $u = (2, -3, 6)$ and $v = (8, 2, -3)$.

▮   Multiply corresponding components and add: $u \cdot v = (2)(8) + (-3)(2) + (6)(-3) = -8$.

1.47   Compute $u \cdot v$,   where $u = (1, -8, 0, 5)$ and $v = (3, 6, 4)$.

▮   The dot product is not defined between vectors with different numbers of components.

1.48   Compute $u \cdot v$,   where $u = (3, -5, 2, 1)$ and $v = (4, 1, -2, 5)$.

▮   Multiply corresponding components and add: $u \cdot v = (3)(4) + (-5)(1) + (2)(-2) + (1)(5) = 8$.

1.49   Compute $u \cdot v$,   where $u = (1, -2, 3, -4)$ and $v = (6, 7, 1, -2)$.

▮   Multiply corresponding components and add: $u \cdot v = (1)(6) + (-2)(7) + (3)(1) + (-4)(-2) = 3$.

1.50   Suppose $u = (3, 2, 1)$, $v = (5, -3, 4)$, $w = (1, 6, -7)$.   Find: (a) $(u + v) \cdot w$, (b) $u \cdot w + v \cdot w$.

▮   (a) First calculate $u + v$ by adding corresponding components: $u + v = (3 + 5, 2 - 3, 1 + 4) = (8, -1, 5)$.   Then compute the dot product $(u + v) \cdot w$ by multiplying corresponding components and adding: $(u + v) \cdot w = (8)(1) + (-1)(6) + (5)(-7) = -33$.
(b) First find $u \cdot w = 3 + 12 - 7 = 8$ and $v \cdot w = 5 - 18 - 28 = -41$.   Then $u \cdot w = 8 - 41 = -33$.   [For an explanation of the agreement between (a) and (b), see Problem 1.52.]

1.51   Let $u = (1, 2, 3, -4)$, $v = (5, -6, 7, 8)$;   and $k = 3$.   Find: (a) $k(u \cdot v)$, (b) $(ku) \cdot v$, (c) $u \cdot (kv)$.

▮   (a) First find $u \cdot v = 5 - 12 + 21 - 32 = -18$.   Then $k(u \cdot v) = 3(-18) = -54$.
(b) First find $ku = (3(1), 3(2), 3(3), 3(-4)) = (3, 6, 9, -12)$.   Then

$$(ku) \cdot v = (3)(5) + (6)(-6) + (9)(7) + (-12)(8) = 15 - 36 + 63 - 96 = -54$$

(c) First find $kv = (15, -18, 21, 24)$.   Then

$$u \cdot (kv) = (1)(15) + (2)(-18) + (3)(21) + (-4)(24) = 15 - 36 + 63 - 96 = -54$$

[See Problems 1.53 and 1.54.]

1.52   Prove that $(u + v) \cdot w = u \cdot w + v \cdot w$.

❚   Using Problem 1.43,

$$(u + v) \cdot w = \sum_{i=1}^{n} (u_i + v_i)w_i = \sum_{i=1}^{n} (u_i w_i + v_i w_i) = \sum_{i=1}^{n} u_i w_i + \sum_{i=1}^{n} v_i w_i = u \cdot w + v \cdot w$$

1.53   Prove that $(ku) \cdot v = k(u \cdot v)$.

❚   Using Problem 1.44,

$$(ku) \cdot v = \sum_{i=1}^{n} (ku_i)v_i = \sum_{i=1}^{n} k(u_i v_i) = k \sum_{i=1}^{n} u_i v_i = k(u \cdot v)$$

1.54   Prove that $u \cdot v = v \cdot u$.

❚ $u \cdot v = \sum_{i=1}^{n} u_i v_i = \sum_{i=1}^{n} v_i u_i = v \cdot u$

1.55   Prove: $u \cdot u \geq 0$,   and $u \cdot u = 0$ iff ["if and only if"] $u = \mathbf{0}$.

❚   Since u_i^2 is nonnegative for each i, $u \cdot u = u_1^2 + u_2^2 + \cdots + u_n^2 \geq 0$.   Furthermore, $u \cdot u = 0$ iff $u_i = 0$ for each i; that is, iff $u = \mathbf{0}$.

In Problems 1.56 and 1.57, u_i stands for the *i*th *vector* in a set of vectors (not the *i*th component of a vector u).   Similarly for v_j.

1.56   Let $u_1, u_2, \ldots, u_p$ and v be vectors in $\mathbf{R}^n$ and let $a_1, a_2, \ldots, a_p$ be scalars in $\mathbf{R}$.   Prove

$$(a) \quad \left(\sum_{k=1}^{p} a_k u_k \right) \cdot v = \sum_{k=1}^{p} a_k (u_k \cdot v) \qquad (b) \quad v \cdot \left(\sum_{k=1}^{k} a_k u_k \right) = \sum_{k=1}^{k} a_k (v \cdot u_k)$$

In words: Taking the dot product with a fixed vector v represents a *linear operation* on $\mathbf{R}^n$.

❚   (a) The proof is by induction on p.   The case $p = 1$ is true by Problem 1.53.   Suppose $p > 1$ and the theorem true for $p - 1$;   i.e.,

$$\left(\sum_{k=1}^{p-1} a_k u_k \right) \cdot v = \sum_{k=1}^{p-1} a_k (u_k \cdot v)$$

Then, using Problems 1.37, 1.52, 1.53, and the above inductive hypothesis, we have

$$\left(\sum_{k=1}^{p} a_k u_k \right) \cdot v = \left(\sum_{k=1}^{p-1} a_k u_k \right) \cdot v + (a_p u_p) \cdot v = \sum_{k=1}^{p-1} a_k (u_k \cdot v) + a_p (u_p \cdot v) = \sum_{k=1}^{p} a_k (u_k \cdot v)$$

(b) Using $u \cdot v = v \cdot u$ [Problem 1.54] and part (a), we have

$$v \cdot \left(\sum_{k=1}^{p} a_k u_k \right) = \left(\sum_{k=1}^{p} a_k u_k \right) \cdot v = \sum_{k=1}^{p} a_k (u_k \cdot v) = \sum_{k-1}^{p} a_k (v \cdot u_k)$$

1.57   Let $u_1, u_2, \ldots, u_p, v_1, v_2, \ldots, v_q$ be vectors $\mathbf{R}^n$ and let $a_1, a_2, \ldots, a_p, b_1, b_2, \ldots, b_q$ be any scalars in $\mathbf{R}$.   Prove that

$$\left(\sum_{j=1}^{p} a_j u_j \right) \cdot \left(\sum_{k=1}^{q} b_k v_k \right) = \sum_{j=1}^{p} \sum_{k=1}^{q} a_j b_k (u_j \cdot v_k)$$

(*bilinearity of the inner product*).

❚   Using Problem 1.56

$$\left(\sum_{j=1}^{p} a_j u_j \right) \cdot \left(\sum_{k=1}^{q} b_k v_k \right) = \sum_{j=1}^{p} a_j \left[u_j \cdot \left(\sum_{k=1}^{q} b_k v_k \right) \right] = \sum_{j=1}^{p} a_j \left[\sum_{k=1}^{q} b_k (u_j \cdot v_k) \right] = \sum_{j=1}^{p} \sum_{k=1}^{q} a_j b_k (u_j \cdot v_k)$$

1.5 NORM (LENGTH) IN $\mathbf{R}^n$

1.58 If $u = (u_1, u_2, \ldots, u_n)$ is a vector in $\mathbf{R}^n$, the *norm* or *length* of u, denoted $\|u\|$, is the nonnegative square root of $u \cdot u$:

$$\|u\| = \sqrt{u \cdot u} = \sqrt{u_1^2 + u_2^2 + \cdots + u_n^2}$$

Show that the above definition of the norm of a vector conforms to that of the length of a vector [arrow] in physics.

▮ Let **u** be a vector [arrow] in the plane $\mathbf{R}^2$ with endpoint $P(a, b)$, as pictured in Fig. 1-5. Then $|a|$ and $|b|$ are the lengths of the sides of the right triangle formed by **u** and the horizontal and vertical directions. By the Pythagorean theorem, the length of **u** is $|u| = \sqrt{a^2 + b^2}$. This value is the same as the norm of u.

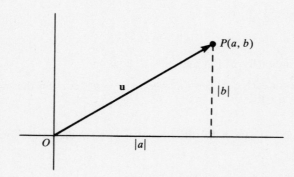

Fig. 1-5

1.59 $\mathbf{R}^n$, together with the definitions of vector addition, scalar multiplication, and inner product, is called *real Euclidean n-space*. Why?

▮ According to Problem 1.45, two vectors u and v in the inner-product space $\mathbf{R}^n$ may be termed *perpendicular* if $u \cdot v = 0$. For such vectors, we have from Problem 1.57

$$\|u + v\|^2 = (u + v) \cdot (u + v) = u \cdot u + 0 + 0 + v \cdot v = \|u\|^2 + \|v\|^2$$

which is the Pythagorean theorem. Since this theorem is a consequence of Euclidean geometry, we call $\mathbf{R}^n$ a "Euclidean" space.

1.60 Find $\|u\|$ if $u = (3, -12, -4)$.

▮ First find $\|u\|^2 = u \cdot u$ by squaring the components of u and adding: $\|u\|^2 = 3^2 + (-12)^2 + (-4)^2 = 9 + 144 + 16 = 169$. Then, $\|u\| = \sqrt{169} = 13$.

1.61 Find $\|v\|$ if $v = (2, -3, 8, -5)$.

▮ Square each component of v and then add to obtain $\|v\|^2 = v \cdot v$: $\|v\|^2 = 2^2 + (-3)^2 + 8^2 + (-5)^2 = 4 + 9 + 64 + 25 = 102$. Then $\|v\| = \sqrt{102}$.

1.62 Find $\|w\|$ if $w = (-3, 1, -2, 4, -5)$.

▮ $\|w\|^2 = (-3)^2 + 1^2 + (-2)^2 + 4^2 + (-5)^2 = 9 + 1 + 4 + 16 + 25 = 55$; hence $\|w\| = \sqrt{55}$.

1.63 Determine k such that $\|u\| = \sqrt{39}$ where $u = (1, k, -2, 5)$.

▮ $\|u\|^2 = 1^2 + k^2 + (-2)^2 + 5^2 = k^2 + 30$. Now solve $k^2 + 30 = 39$ and obtain $k = 3, -3$.

1.64 Give the definition of a *unit vector*.

▮ A vector u is a unit vector if $\|u\| = 1$ or, equivalently, if $u \cdot u = 1$.

1.65   Let v be an nonzero vector.   Show that

$$\hat{v} \equiv \frac{1}{\|v\|} v = \frac{v}{\|v\|}$$

is a unit vector in the same direction as v.   [The process of finding $\hat{v}$ is called *normalizing v*.]

▮ The vector $\hat{v}$ is a unit vector, since

$$\hat{v} \cdot \hat{v} = \left(\frac{v}{\|v\|}\right) \cdot \left(\frac{v}{\|v\|}\right) = \frac{1}{\|v\|^2}(v \cdot v) = \frac{1}{\|v\|^2}\|v\|^2 = 1$$

Moreover, $\hat{v}$ is in the same direction as v, since $\hat{v}$ is a positive scalar multiple of v.

1.66   Normalize $v = (12, -3, -4)$.

▮   First find $\|v\|^2 = v \cdot v = 12^2 + (-3)^2 + (-4)^2 = 144 + 9 + 16 = 169$.   Then divide each component of v by $\|v\| = \sqrt{169} = 13$ to obtain

$$\hat{v} = \frac{v}{\|v\|} = \left(\frac{12}{13}, \frac{-3}{13}, \frac{-4}{13}\right)$$

1.67   Normalize $w = (4, -2, -3, 8)$.

▮   First find $\|w\|^2 = w \cdot w = 4^2 + (-2)^2 + (-3)^2 + 8^2 = 16 + 4 + 9 + 64 = 93$.   Divide each component of w by $\|w\| = \sqrt{93}$ to obtain

$$\hat{w} = \frac{w}{|w|} = \left(\frac{4}{\sqrt{93}}, \frac{-2}{\sqrt{93}}, \frac{-3}{\sqrt{93}}, \frac{8}{\sqrt{93}}\right)$$

1.68   Normalize $v = \left(\frac{1}{2}, \frac{2}{3}, -\frac{1}{4}\right)$.

▮   Note that v and any positive multiple of v will have the same normalized form [see Problem 1.72].   Hence, first multiply v by 12 to "clear" fractions: $12v = (6, 8, -3)$.   Then $\|12v\|^2 = 36 + 64 + 9 = 109$.   Accordingly, the required unit vector is

$$\hat{v} = \widehat{12v} = \frac{12v}{\|12v\|} = \left(\frac{6}{\sqrt{109}}, \frac{8}{\sqrt{109}}, \frac{-3}{\sqrt{109}}\right)$$

1.69   Show that $\|u\| \geq 0$,   and $\|u\| = 0$ iff $u = 0$.

▮   Follows at once from Problem 1.55.

1.70   Prove the *Cauchy–Schwarz inequality*: $|u \cdot v| \leq \|u\|\,\|v\|$,   for arbitrary u and v in $\mathbf{R}^n$.

▮   We shall prove the following stronger statement: $|u \cdot v| \leq \sum\limits_{i=1}^{n} |u_i v_i| \leq \|u\|\,\|v\|$.   First, if $u = 0$ or $v = 0$,   then the inequality reduces to $0 \leq 0 \leq 0$ and is therefore true.   Hence we need only consider the case in which $u \neq 0$ and $v \neq 0$,   i.e., where $\|u\| \neq 0$ and $\|v\| \neq 0$.   Furthermore, because

$$|u \cdot v| = |\textstyle\sum u_i v_i| \leq \sum |u_i v_i|$$

we need only prove the second inequality.

Now, for any real numbers $x, y \in \mathbf{R}$, $0 \leq (x - y)^2 = x^2 - 2xy + y^2$ or, equivalently,

$$2xy \leq x^2 + y^2 \tag{1}$$

Set $x = |u_i|/\|u\|$ and $y = |v_i|/\|v\|$ in *(1)* to obtain, for any i,

$$2\,\frac{|u_i|}{\|u\|}\,\frac{|v_i|}{\|v\|} \leq \frac{|u_i|^2}{\|u\|^2} + \frac{|v_i|^2}{\|v\|^2} \tag{2}$$

But, by definition of the norm of a vector, $\|u\| = \sum u_i^2 = \sum |u_i|^2$ and $\|v\| = \sum v_i^2 = \sum |v_i|^2$. Thus summing (2) with respect to i and using $|u_i v_i| = |u_i|\,|v_i|$, we have

$$2\,\frac{\sum |u_i v_i|}{\|u\|\,\|v\|} \le \frac{\sum |u_i|^2}{\|u\|^2} + \frac{\sum |v_i|^2}{\|v\|^2} = \frac{\|u\|^2}{\|u\|^2} + \frac{\|v\|^2}{\|v\|^2} = 2$$

that is,

$$\frac{\sum |u_i v_i|}{\|u\|\,\|v\|} \le 1$$

Multiplying both sides by $\|u\|\,\|v\|$, we obtain the required inequality.

1.71 Prove *Minkowski's inequality*: $\|u + v\| \le \|u\| + \|v\|$, for arbitrary u and v in $\mathbf{R}^n$.

▌ By the Cauchy–Schwarz inequality (Problem 1.70) and the other properties of the inner product,

$$\|u + v\|^2 = (u + v) \cdot (u + v) = u \cdot u + 2(u \cdot v) + v \cdot v$$

$$\le \|u\|^2 + 2\,\|u\|\,\|v\| + \|v\|^2 = (\|u\| + \|v\|)^2$$

Taking the square roots of both sides yields the desired inequality.

1.72 Prove that the norm in $\mathbf{R}^n$ satisfies the following laws:

$[N_1]$: For any vector u, $\|u\| \ge 0$; and $\|u\| = 0$ iff $u = 0$.
$[N_2]$: For any vector u and any scalar k, $\|ku\| = |k|\,\|u\|$.
$[N_3]$: For any vectors u and v, $\|u + v\| \le \|u\| + \|v\|$.

▌ $[N_1]$ was proved in Problem 1.69 and $[N_3]$ in Problem 1.71. Hence we need only prove that $[N_2]$ holds. Suppose $u = (u_1, u_2, \ldots, u_n)$ and so $ku = (ku_1, ku_2, \ldots, ku_n)$. Then

$$\|ku\|^2 = (ku_1)^2 + (ku_2)^2 + \cdots + (ku_n)^2 = k^2(u_1^2 + u_2^2 + \cdots u_n^2) = k^2 \|u\|^2$$

Taking square roots gives $[N_2]$.

1.73 Show that $\|-u\| = \|u\|$, for any vector u in $\mathbf{R}^n$.

▌ Using property $[N_2]$ of Problem 1.72, we have $\|-u\| = \|(-1)u\| = |-1|\,\|u\| = \|u\|$.

1.74 Let $u = (1, 2, -2)$, $v = (3, -12, 4)$, and $k = -3$. (a) Find $\|u\|$, $\|v\|$, and $\|ku\|$. (b) Verify that $\|ku\| = |k|\,\|u\|$ and $\|u + v\| \le \|u\| + \|v\|$.

▌ (a) $\|u\| = \sqrt{1 + 4 + 4} = \sqrt{9} = 3$, $\|v\| = \sqrt{9 + 144 + 16} = \sqrt{169} = 13$, $ku = (-3, -6, 6)$, and $\|ku\| = \sqrt{9 + 36 + 36} = \sqrt{81} = 9$.
(b) Since $|k| = |-3| = 3$, we have $|k|\,\|u\| = 3 \cdot 3 = 9 = \|ku\|$. Also $u + v = (4, -10, 2)$. Thus

$$\|u + v\| = \sqrt{16 + 100 + 4} = \sqrt{120} \le 16 = 3 + 13 = \|u\| + \|v\|$$

1.6 DISTANCE, ANGLES, PROJECTIONS

1.75 Let u and v be any vectors in $\mathbf{R}^n$. The *distance* between u and v, denoted by $d(u, v)$, is defined as

$$d(u, v) \equiv \|u - v\|$$

Show that this definition corresponds to the usual notion of Euclidean distance in the plane $\mathbf{R}^2$.

▮ Let $u = (a, b)$ and $v = (c, d)$ in $\mathbf{R}^2$. As pictured in Fig. 1-6, the distance between the points $P(a, b)$ and $Q(c, d)$ is $d = \sqrt{(a - c)^2 + (b - d)^2}$. On the other hand, by the above definition,

$$d(u, v) = \|u - v\| = \|(a - c, b - d)\| = \sqrt{(a - c)^2 + (b - d)^2}$$

Both give the same value.

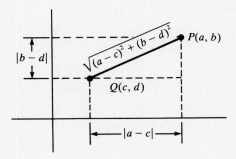

Fig. 1-6

1.76   Find $d(u, v)$, where (a) $u = (1, 7)$, $v = (6, -5)$; (b) $u = (3, -5, 4)$, $v = (6, 2, -1)$; (c) $u = (1, -2, 4, 1)$, $v = (3, 1, -5, 0)$.

▮ In each case use the formula $d(u, v) = \|u - v\| = \sqrt{(u_1 - v_1)^2 + \cdots + (u_n - v_n)^2}$.

(a) $d(u, v) = \sqrt{(1 - 6)^2 + (7 + 5)^2} = \sqrt{25 + 144} = \sqrt{169} = 13$

(b) $d(u, v) = \sqrt{(3 - 6)^2 + (-5 - 2)^2 + (4 + 1)^2} = \sqrt{9 + 49 + 25} = \sqrt{83}$

(c) $d(u, v) = \sqrt{(1 - 3)^2 + (-2 - 1)^2 + (4 + 5)^2 + (1 - 0)^2} = \sqrt{95}$

1.77   Find k such that $d(u, v) = 6$, if $u = (2, k, 1, -4)$ and $v = (3, -1, 6, -3)$.

▮ $(d(u, v))^2 = \|u - v\|^2 = (2 - 3)^2 + (k + 1)^2 + (1 - 6)^2 + (-4 + 3)^2 = k^2 + 2k + 28$. Now solve $k^2 + 2k + 28 = 6^2$ to obtain $k = 2, -4$.

1.78   From Problem 1.72, prove that the distance function $d(u, v)$ satisfies:

$[M_1]$ $d(u, v) \geq 0$, and $d(u, v) = 0$ iff $u = v$.
$[M_2]$ $d(u, v) = d(v, u)$.
$[M_3]$ $d(u, v) \leq d(u, w) + d(w, v)$ (triangle inequality).

▮ $[M_1]$ follows directly from $[N_1]$. By $[N_2]$,

$$d(u, v) = \|u - v\| = \|(-1)(v - u)\| = |-1| \, \|v - u\| = \|v - u\| = d(v, u)$$

which is $[M_2]$. By $[N_3]$,

$$d(u, v) = \|u - v\| = \|(u - w) + (w - v)\| \leq \|u - w\| + \|w - v\| = d(u, w) + d(w, v)$$

which is $[M_3]$.

1.79   Let u and v be vectors in $\mathbf{R}^n$. The *angle*, θ, between u and v if defined by

$$\cos \theta \equiv \frac{u \cdot v}{\|u\| \, \|v\|}$$

(a) Show that θ is a unique real number in $[0, \pi]$. (b) Show how this definition corresponds to that used in physics.

▮ (a) By the Cauchy–Schwarz inequality, $|u \cdot v| \le \|u\| \, \|v\|$. Hence, $-1 \le \cos \theta \le 1$, which uniquely defines a real angle $0 \le \theta \le \pi$. (b) Physics defines the dot product as $\mathbf{u} \cdot \mathbf{v} = |\mathbf{u}| \, |\mathbf{v}| \cos \theta$; dividing by $|\mathbf{u}| \, |\mathbf{v}|$ gives the above formula for $\cos \theta$.

1.80   Find $\cos \theta$, where θ is the angle between $u = (1, -2, 3)$ and $v = (3, -5, -7)$.

▮ First find

$$u \cdot v = 3 + 10 - 21 = -8 \qquad \|u\|^2 = 1 + 4 + 9 = 14 \qquad \|v\|^2 = 9 + 25 + 49 = 83$$

Then, using the formula in Problem 1.79,

$$\cos \theta = \frac{u \cdot v}{\|u\| \, \|v\|} = -\frac{8}{\sqrt{14} \, \sqrt{83}}$$

1.81   Find $\cos \theta$, where θ is the angle between $u = (4, -3, 1, 5)$ and $v = (2, 6, -1, 4)$.

▮ $u \cdot v = 8 - 18 - 1 + 20 = 9 \qquad \|u\|^2 = 16 + 9 + 1 + 25 = 51 \qquad \|v\|^2 = 4 + 36 + 1 + 16 = 57$

Then

$$\cos \theta = \frac{9}{\sqrt{51} \, \sqrt{57}}$$

1.82   Let u and $v \ne 0$ be vectors in $\mathbf{R}^n$. The [*vector*] *projection of u onto v* is the vector

$$\text{proj}\,(u, v) = \frac{u \cdot v}{\|v\|^2} \, v \qquad\qquad (1)$$

Show how this definition conforms to the notion of vector projection in physics.

▮ In the physical picture, Fig. 1-7, the [perpendicular] projection of $\mathbf{u}$ onto $\mathbf{v}$ is the vector $\mathbf{u}^*$, of magnitude

$$|\mathbf{u}^*| = |\mathbf{u}| \cos \theta = |\mathbf{u}| \, \frac{\mathbf{u} \cdot \mathbf{v}}{|\mathbf{u}| \, |\mathbf{v}|} = \frac{\mathbf{u} \cdot \mathbf{v}}{|\mathbf{v}|}$$

To obtain $\mathbf{u}^*$, we multiply its magnitude by the unit vector in the direction of $\mathbf{v}$:

$$\mathbf{u}^* = |\mathbf{u}^*| \, \frac{\mathbf{v}}{|\mathbf{v}|} = \frac{\mathbf{u} \cdot \mathbf{v}}{|\mathbf{v}|^2} \, \mathbf{v}$$

which corresponds to (1) above.

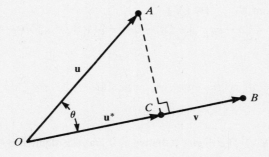

Fig. 1-7

1.83   Find $\text{proj}\,(u, v)$ where $u = (1, -2, 3)$ and $v = (2, 5, 4)$.

▮ First find $u \cdot v = 2 - 10 + 12 = 4$ and $\|v\|^2 = 4 + 25 + 16 = 45$. Then, by (1) of Problem 1.82,

$$\text{proj}\,(u, v) = \frac{u \cdot v}{\|v\|^2} \, v = \frac{4}{45} \, (2, 5, 4) = \left(\frac{8}{45}, \frac{20}{45}, \frac{16}{45} \right) = \left(\frac{8}{45}, \frac{4}{9}, \frac{16}{45} \right)$$

1.84   Find proj(u, v) where $u = (4, -3, 1, 5)$ and $v = (3, 6, -4, 1)$.

▮   First find $u \cdot v = 12 - 18 - 4 + 5 = -5$ and $\|v\|^2 = 9 + 36 + 16 + 1 = 62$.   Then,

$$\text{proj}(u, v) = \frac{u \cdot v}{\|v\|^2}\, v = -\frac{5}{62}(3, 6, -4, 1) = \left(\frac{-15}{62}, \frac{-30}{62}, \frac{20}{62}, \frac{-5}{62}\right) = \left(\frac{-15}{62}, \frac{-15}{31}, \frac{10}{31}, \frac{-5}{62}\right)$$

[Observe that when $u \cdot v < 0$,   proj(u, v) is in the opposite direction to v.]

1.7   ORTHOGONALITY

1.85   Let u and v be vectors in $\mathbf{R}^n$.   Then u is said to be *orthogonal* (*perpendicular*) to v if $u \cdot v = 0$.   Show that this definition conforms to the perpendicularity of vectors (arrows) in physics.

▮   Two arrows are perpendicular if and only if the angle between them is $\theta = 90°$.   But then, and only then, $u \cdot v = 0$,   by Problem 1.45.

1.86   Let $u = (5, 4, 1)$, $v = (3, -4, 1)$,   and $w = (1, -2, 3)$.   Which of the vectors, if any, are perpendicular?

▮   Find the dot product of each pair of vectors:

$$u \cdot v = 15 - 16 + 1 = 0 \qquad v \cdot w = 3 + 8 + 3 = 14 \qquad u \cdot w = 5 - 8 + 3 = 0$$

Hence u and v are orthogonal, u and w are orthogonal, but v and w are not orthogonal.

1.87   Determine k so that $u = (1, k, -3)$ and $v = (2, -5, 4)$ are orthogonal.

▮   Solve $u \cdot v = (1)(2) + (k)(-5) + (-3)(4) = 2 - 5k - 12 = 0$ for k; obtaining $k = -2$.

1.88   Determine k so that $u = (2, 3k, -4, 1, 5)$ and $v = (6, -1, 3, 7, 2k)$ are orthogonal.

▮ $u \cdot v = (2)6 + (3k)(-1) + (-4)(3) + (1)(7) + (5)(2k) = 12 - 3k - 12 + 7 + 10k = 7k + 7 = 0$.   Solving, $k = -1$.

1.89   Show that the zero vector $\mathbf{0}$ is orthogonal to every vector u in $\mathbf{R}^n$.

▮   Since $0u = \mathbf{0}$, $\mathbf{0} \cdot u = (0u) \cdot u = 0(u \cdot u) = 0$.

1.90   Show that the zero vector is the *only* vector orthogonal to every vector in $\mathbf{R}^n$.

▮   Suppose that u is orthogonal to every vector in $\mathbf{R}^n$.   Then u is orthogonal to itself; that is, $u \cdot u = 0$.   By Problem 1.55, $u = \mathbf{0}$.

1.91   Show that orthogonality of vectors is a symmetric relation but not a transitive relation.

▮   By Problem 1.54, $u \cdot v = v \cdot u$.   Therefore, u is orthogonal to v iff $u \cdot v = 0$ iff $v \cdot u = 0$ iff v is orthogonal to u.   Thus the relation is symmetric.   On the other hand, consider the vectors of Problem 1.86.   Here v is orthogonal to u, u is orthogonal to w, but v is not orthogonal to w.   Thus the relation is not transitive.

1.92   If u and v are orthogonal to w, show that $u + v$ is orthogonal to w.

▮ $(u + v) \cdot w = u \cdot w + v \cdot w = 0 + 0 = 0$

1.93   If u is orthogonal to w, show that any scalar multiple ku is orthogonal to w.

▮ $(ku) \cdot w = k(u \cdot w) = k(0) = 0$

1.94   Suppose u, u_2, u_3 are nonzero vectors that are pairwise orthogonal, and suppose w is the linear combination $w = xu_1 + yu_2 + zu_3$.   Show that

$$x = \frac{w \cdot u_1}{\|u_1\|^2} \qquad y = \frac{w \cdot u_2}{\|u_2\|^2} \qquad z = \frac{w \cdot u_3}{\|u_3\|^2}$$

❚ Take the dot product of w with u_1 to obtain:

$$w \cdot u_1 = (xu_1 + yu_2 + zu_3) \cdot u_1 = x(u_1 \cdot u_1) + y(u_2 \cdot u_1) + z(u_3 \cdot u_1)$$
$$= x(u_1 \cdot u_1) + y(0) + z(0) = x(u_1 \cdot u_1) \equiv x\|u_1\|^2$$

whence $x = (w \cdot u_1)/\|u_1\|^2$. Similarly, take the dot product of w with u_2 to obtain $y = (w \cdot u_2)/\|u_2\|^2$, and take the dot product of w with u_3 to obtain $z = (w \cdot u_3)/\|u_3\|^2$.

1.95 Show that $u_1 = (1, -2, 3)$, $u_2 = (1, 2, 1)$, $u_3 = (-8, 2, 4)$ are orthogonal to each other.

❚ Compute the dot product of each pair of vectors:

$$u_1 \cdot u_2 = 1 - 4 + 3 = 0 \qquad u_1 \cdot u_3 = -8 - 4 + 12 = 0 \qquad u_2 \cdot u_3 = -8 + 4 + 4 = 0$$

Thus the vectors are orthogonal to each other.

1.96 Write $w = (13, -4, 7)$ as a linear combination of the vectors u, u_2, u_3 in Problem 1.95; i.e., find x, y, z such that $w = xu_1 + yu_2 + zu_3$.

❚ **Method 1:** Multiply by the scalars x, y, z and then add:

$$\begin{pmatrix} 13 \\ -4 \\ 7 \end{pmatrix} = x \begin{pmatrix} 1 \\ -2 \\ 3 \end{pmatrix} + y \begin{pmatrix} 1 \\ 2 \\ 1 \end{pmatrix} + z \begin{pmatrix} -8 \\ 2 \\ 4 \end{pmatrix} = \begin{pmatrix} x + y - 8z \\ -2x + 2y + 2z \\ 3x + y + 4z \end{pmatrix}$$

Then set corresponding components equal to each other to obtain the system:

$$x + y - 8z = 13 \qquad -2x + 2y + 2z = -4 \qquad 3x + y + 4z = 7$$

Solve the system to obtain $x = 3$, $y = 2$, $z = -1$.

Method 2: Since u_1, u_2, u_3 are orthogonal to each other, use the formulas of Problem 1.94.

$$w \cdot u_1 = 13 + 8 + 21 = 42 \qquad \|u_1\|^2 = 1 + 4 + 9 = 14 \qquad x = 42/14 = 3$$
$$w \cdot u_2 = 13 - 8 + 7 = 12 \qquad \|u_2\|^2 = 1 + 4 + 1 = 6 \qquad y = 12/6 = 2$$
$$w \cdot u_3 = -104 - 8 + 28 = -84 \qquad \|u_3\|^2 = 64 + 4 + 16 = 84 \qquad z = -84/84 = -1$$

(Method 2, which uses orthogonality, is much simpler than Method 1.)

1.8 HYPERPLANES AND LINES IN $\mathbf{R}^n$

This section distinguishes between an n-tuple $P(a_1, a_2, \ldots, a_n) \equiv P(a_i)$ viewed as a point in $\mathbf{R}^n$ and an n-tuple $v = [c_1, c_2, \ldots, c_n]$ viewed as a vector (arrow) from the origin O to the point $C(c_1, c_2, \ldots, c_n)$.

1.97 Let $P(a_i)$ and $Q(b_i)$ be points in $\mathbf{R}^n$. The directed line segment from P to Q, written $\overrightarrow{PQ}$, is identified with the vector $v = [b_1 - a_1, b_2 - a_2, \ldots, b_n - a_n]$. Show geometrically that $\overrightarrow{PQ}$ and v have the same magnitude and direction.

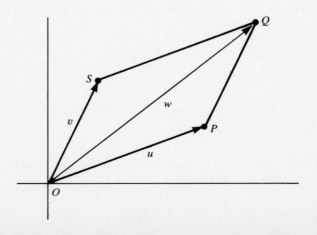

Fig. 1-8

▮   Consider points $P(a_1, a_2)$ and $Q(b_1, b_2)$ in a plane with origin O, as pictured in Fig. 1-8.   Let S be the point such that $OPQS$ forms a parallelogram.   Let u, w, and v be the vectors from O to the points P, Q, and S, respectively.   By the parallelogram law for the addition of vectors, $u + v = w$,   or

$$v = w - u = [b_1, b_2] - [a_1, a_2] = [b_1 - a_1, b_2 - a_2]$$

But, since $OPQS$ is a parallelogram, the vector v just obtained is identical in magnitude and direction to $\overrightarrow{PQ}$.

1.98   Find the vector v identified with $\overrightarrow{PQ}$ for the points $P(2, 5)$ and $Q(-3, 4)$.

▮ $v = [-3 - 2, 4 - 5] = [-5, -1]$

1.99   Find the vector v identified with $\overrightarrow{PQ}$ for the points $P(1, -2, 4)$ and $Q(6, 0, -3)$.

▮ $v = [6 - 1, 0 + 2, -3 - 4] = [5, 2, -7]$

1.100   Consider points $P(3, k, -2)$ and $Q(5, 3, 4)$ in $\mathbf{R}^3$.   Find k so that $\overrightarrow{PQ}$ is orthogonal to the vector $u = [4, -3, 2]$.

▮   First find $v = \overrightarrow{PQ} = [5 - 3, 3 - k, 4 + 2] = [2, 3 - k, 6]$.   Next compute $u \cdot v = (4)(2) - (3)(3 - k) + (2)(6) = 8 - 9 + 3k + 12 = 3k + 11$.   Lastly, set $u \cdot v = 0$ and solve for k, obtaining $k = -11/3$.

1.101   A *hyperplane* H in $\mathbf{R}^n$ is the set of points $(x_1, x_2, \ldots, x_n)$ that satisfy a linear equation

$$a_1 x_1 + a_2 x_2 + \cdots + a_n x_n = b \qquad\qquad (1)$$

where $\alpha = [a_1, \ldots, a_n] \neq \mathbf{0}$ is called a *normal* to H.   Justify this terminology by showing that any directed line segment $\overrightarrow{PQ}$, with $P, Q \in$ H,   is orthogonal to α.

▮   Let $u = \overrightarrow{OP}$, $w = \overrightarrow{OQ}$;   hence $v = w - u = \overrightarrow{PQ}$.   By *(1)*, $\alpha \cdot u = b$ and $\alpha \cdot w = b$.   But then $\alpha \cdot v = \alpha \cdot (w - u) = \alpha \cdot w - \alpha \cdot u = 0$.

1.102   Refer to <u>Problem 1.101.</u>   Prove that the distance from the origin O to the hyperplane *(1)* is given by $|b|/\sqrt{a_1^2 + a_2^2 + \cdots + a_n^2}$.

▮   Let $u = \overrightarrow{OP} = [x_1, x_2, \ldots, x_n]$ be the vector from O to the point $P(x_1, x_2, \ldots, x_n)$ of the hyperplane; we want to minimize $u \cdot u$ [which is the same thing as minimizing $\|u\|$] over H: $\alpha \cdot u = b$.   Using Problem 1.94, let us represent u as a linear combination of the vectors α and v, where v is orthogonal to α (and may therefore be considered as an arrow lying in the hyperplane):

$$u = \frac{u \cdot \alpha}{\|\alpha\|^2} \alpha + \frac{u \cdot v}{\|v\|^2} v = \frac{b}{\|\alpha\|^2} \alpha + v^*$$

where $\alpha \cdot v^* = 0$.   Then, using Problem 1.57

$$u \cdot u = \frac{b^2}{\|\alpha\|^4} \alpha \cdot \alpha + v^* \cdot v^* + 2 \frac{b}{\|\alpha\|^2} \alpha \cdot v^* = \left(\frac{|b|}{\|\alpha\|} \right)^2 + \|v^*\|^2$$

Obviously the minimum is assumed for $v^* = \mathbf{0}$,   and the desired distance is

$$\|u\|_{\min} = \frac{|b|}{\|\alpha\|} = \frac{|b|}{\sqrt{a_1^2 + \cdots + a_n^2}}$$

1.103   Find an equation of the plane H in $\mathbf{R}^3$ which passes through $P(2, -7, 1)$ and is normal to $\alpha = [3, 1, -11]$.

▮   An equation of H is of the form $3x + y - 11z = k$,   since α is normal to H.   Substitute $P(2, -7, 1)$ into this equation to obtain:

$$(3)(2) + (1)(-7) - (11)(1) = k \qquad \text{or} \qquad k = -12$$

Thus an equation of H is $3x + y - 11z = -12$.

1.104 Find an equation of the hyperplane H in $\mathbf{R}^4$ which passes through $P(3, -2, 1, -4)$ and is normal to $\alpha = [2, 5, -6, -2]$.

❚ An equation of H is the form $2x + 5y - 6z - 2w = k$. Substitute P into this equation to obtain $k = -2$.

1.105 Find an equation of the plane H in $\mathbf{R}^3$ which contains $P(1, -5, 2)$ and is parallel to the plane H′ determined by $3x - 7y + 4z = 5$.

❚ H and H′ are parallel if and only if their normals are parallel or antiparallel. Hence an equation of H is of the form $3x - 7y + 4z = k$. Substitute $P(1, -5, 2)$ into this equation to obtain $k = 46$.

1.106 Find the equation of the hyperplane H in $\mathbf{R}^n$ which intersects the x_i-axis at $a_i \neq 0$ $(i = 1, 2, \ldots, n)$.

❚ Since H does not pass through the origin, an equation of H has the form

$$k_1 x_1 + k_2 x_2 + \cdots + k_n x_n = 1$$

For $i = 1, 2, \ldots, n$, substitute $P_i(0, \ldots, a_i, \ldots, 0)$ into the equation to obtain $k_i = 1/a_i$. Hence the equation is $x_1/a_1 + x_2/a_2 + \cdots + x_n/a_n = 1$.

1.107 Find the normal to the plane H in $\mathbf{R}^3$ that intersects the coordinate axes at $x = 3$, $y = -4$, $z = 6$.

❚ By Problem 1.106, an equation of H is $x/3 - y/4 + z/6 = 1$, or $4x - 3y + 2z = 12$. Hence $\alpha = [4, -3, 2]$ is normal to H. [Note that any scalar multiple of a normal is also a normal.]

1.108 The *line* L in $\mathbf{R}^n$ passing through the point $P(a, a_2, \ldots, a_n)$ and in the direction of the nonzero vector $u = [u, u_2, \ldots, u_n]$ consists of the points $X(x_1, x_2, \ldots, x_n)$ which satisfy

$$X = P + tu \qquad \text{or} \qquad \begin{cases} x_1 = a_1 + u_1 t \\ x_2 = a_2 + u_2 t \\ \cdots\cdots\cdots \\ x_n = a_n + u_n t \end{cases} \qquad (1)$$

where the *parameter t* takes on all real numbers. [See Fig. 1-9.] Find a parametric representation *(1)* of the line in $\mathbf{R}^4$ passing through $P(4, -2, 3, 1)$ in the direction $u = [2, 5, -7, 11]$.

❚ $$\begin{cases} x = 4 + 2t \\ y = -2 + 5t \\ z = 3 - 7t \\ w = 1 + 11t \end{cases} \qquad \text{or} \qquad (4 + 2t, -2 + 5t, 3 - 7t, 1 + 11t)$$

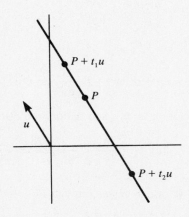

$P + t_1 u$

P

u

$P + t_2 u$

Fig. 1-9

1.109 Find the parametric representation of the line in $\mathbf{R}^2$ passing through the point $P(2, 5)$ and in the direction $u = [-3, 4]$.

❚ Use *(1)* of Problem 1.108 to obtain $x = 2 - 3t$, $y = 5 + 4t$.

1.110 Find a parametric equation of the line in $\mathbf{R}^3$ passing through the points $P(5, 4, -3)$ and $Q(1, -3, 2)$.

▌ First compute $u = \overrightarrow{PQ} = [1 - 5, -3 - 4, 2 - (-3)] = [-4, -7, 5]$. Then use *(1)* of Problem 1.108:
$x = 5 - 4t$, $y = 4 - 7t$, $z = -3 + 5t$.

1.111 Give *nonparametric* equations for the lines of (*a*) Problem 1.109, (*b*) Problem 1.110.

▌ Solve each coordinate equation for t and equate the results

(*a*) $$\frac{x - 2}{-3} = \frac{y - 5}{4} \quad \text{or} \quad 4x - 8 = -3y + 15 \quad \text{or} \quad 4x + 3y = 23$$

(*b*) $$\frac{x - 5}{-4} = \frac{y - 4}{-7} = \frac{z + 3}{5}$$

or the pair of linear equations $7x - 4y = 19$ and $5x + 4z = 13$.

1.112 Find a parametric equation of the line in $\mathbf{R}^3$ perpendicular to the plane $2x - 3y + 7z = 4$ and intersecting the plane at the point $P(6, 5, 1)$.

▌ Since the line is perpendicular to the plane, it must be in the direction of the normal vector, $\alpha = [2, -3, 7]$. Hence, $x = 6 + 2t$, $y = 5 - 3t$, $z = 1 + 7t$.

1.9 COMPLEX NUMBERS

In the following group of problems, $\mathbf{C}$ denotes the set of complex numbers; z and w denote complex numbers (elements of $\mathbf{C}$); a, b, x, y denote real numbers (elements of $\mathbf{R}$); and $i = \sqrt{-1}$ (meaning that $i^2 = -1$).

1.113 If $z = 2 + 3i$ and $w = 5 - 2i$, find (*a*) $z + w$, (*b*) $z - w$, (*c*) zw.

▌ Use the ordinary rules of algebra together with $i^2 = -1$ to obtain a result in the standard form $a + bi$. (*a*) $z + w = 2 - 3i + 4 + 5i = 6 + 2i$; (*b*) $z - w = (2 + 3i) - (5 - 2i) = 2 + 3i - 5 + 2i = -3 + 5i$; (*c*) $zw = (2 - 3i)(4 + 5i) = 8 - 12i + 10i - 15i^2 = 23 - 2i$.

1.114 Simplify: (*a*) i^0, i^3, i^4; (*b*) i^5, i^6, i^7, i^8.

▌ (*a*) $i^0 = 1$, $i^3 = i^2(i) = (-1)(i) = -i$, $i^4 = (i^2)(i^2) = (-1)(-1) = 1$
(*b*) $i^5 = (i^4)(i) = (1)(i) = i$, $i^6 = (i^4)(i^2) = (1)(i^2) = i^2 = -1$, $i^7 = i^3 = -i$, $i^8 = i^4 = 1$.

1.115 Simplify: (*a*) $(5 + 3i)(2 - 7i)$, (*b*) $(4 - 3i)^2$, (*c*) $(1 + 2i)^3$.

▌ (*a*) $(4 - 3i)^2 = 16 - 24i + 9i^2 = 7 - 24i$; (*b*) $(5 + 3i)(2 - 7i) = 10 + 6i - 35i - 21i^2 = 31 - 29i$;
(*c*) $(1 + 2i)^3 = 1 + 6i + 12i^2 + 8i^3 = 1 + 6i - 12 - 8i = -11 - 2i$.

1.116 Simplify: (*a*) i^{39}, (*b*) i^{174}, (*c*) i^{252}, (*d*) i^{317}

▌ (*a*) $i^{39} = i^{4 \cdot 9 + 3} = (i^4)^9 i^3 = 1^9 i^3 = i^3 = -i$; (*b*) $i^{174} = i^2 = -1$; (*c*) $i^{252} = i^0 = 1$; (*d*) $i^{317} = i^1 = i$.

1.117 Let $z = a + bi$. Then $a = \operatorname{Re} z$ and $b = \operatorname{Im} z$ are called, respectively, the real and imaginary parts of z. The *complex conjugate* of z is denoted and defined by

$$\bar{z} = a - bi \equiv \operatorname{Re} z - i \operatorname{Im} z \qquad \text{or} \qquad \operatorname{Re} \bar{z} = \operatorname{Re} z \qquad \operatorname{Im} \bar{z} = -\operatorname{Im} z$$

Find the complex conjugates of (*a*) $6 + 4i$, (*b*) $7 - 5i$, (*c*) $4 + i$, (*d*) $-3 - i$.

▌ (*a*) $6 - 4i$; (*b*) $7 + 5i$; (*c*) $4 - i$; (*d*) $-3 + i$.

1.118 Prove that z is a real number iff $z = \bar{z}$.

▌ $\operatorname{Im} z \equiv \dfrac{1}{2i}(z - \bar{z})$ vanishes iff $z = \bar{z}$.

1.119 Show that $z\bar{z}$ is real and nonnegative.

▌ $z\bar{z} = (\operatorname{Re} z + i \operatorname{Im} z)(\operatorname{Re} z - i \operatorname{Im} z) = (\operatorname{Re} z)^2 + (\operatorname{Im} z)^2 \geq 0$

1.120 The *absolute value* of a complex number z is defined by $|z| = \sqrt{z\bar{z}}$. Evaluate $|z|$ when (*a*) $z = 3 + 4i$, (*b*) $z = 5 - 2i$, (*c*) $z = -7 + i$, (*d*) $z = -1 - 4i$.

❙ Use the expression of Problem 1.119.
(*a*) $z\bar{z} = 3^2 + 4^2 = 25$, $|z| = \sqrt{25} = 5$
(*b*) $z\bar{z} = 5^2 + (-2)^2 = 29$, $|z| = \sqrt{26}$
(*c*) $z\bar{z} = (-7)^2 + 1^2 = 50$, $|z| = \sqrt{50} = 5\sqrt{2}$
(*d*) $z\bar{z} = (-1)^2 + (-4)^2 = 17$, $|z| = \sqrt{17}$

1.121 Prove that $|\bar{z}| = |z|$.

❙ From Problem 1.117 it is clear that $\bar{\bar{z}} = z$. Hence,

$$|\bar{z}| = \sqrt{\bar{z}\bar{\bar{z}}} = \sqrt{\bar{z}z} = \sqrt{z\bar{z}} = |z|$$

1.122 Express in the form $a + bi$:

$$(a)\ \frac{1}{3 - 4i} \qquad (b)\ \frac{2 - 7i}{5 + 3i}$$

❙ To simplify a fraction z/w of the complex numbers, multiply both numerator and denominator by $\bar{w}$, the conjugate of the denominator.

(*a*) $\dfrac{1}{3 - 4i} = \dfrac{(3 + 4i)}{(3 - 4i)(3 + 4i)} = \dfrac{3 + 4i}{25} = \dfrac{3}{25} + \dfrac{4}{25}i$

(*b*) $\dfrac{2 - 7i}{5 + 3i} = \dfrac{(2 - 7i)(5 - 3i)}{(5 + 3i)(5 - 3i)} = \dfrac{-11 - 41i}{34} = -\dfrac{11}{34} - \dfrac{41}{34}i$

1.123 Compute $\text{Im}\left(\dfrac{1}{2 - 3i}\right)^2$.

❙ $\left(\dfrac{1}{2 - 3i}\right)^2 = \dfrac{1}{-5 - 12i} = \dfrac{(-5 + 12i)}{(-5 - 12i)(-5 + 12i)} = \dfrac{-5 + 12i}{169} = -\dfrac{5}{169} + \dfrac{12}{169}i$

and so

$$\text{Im}\left(\frac{1}{2 - 3i}\right)^2 = \frac{12}{169}$$

1.124 Describe the geometrical representation of the complex number system **C**. [Such a representation is called the *complex plane*.]

❙ Each complex number $z = a + bi$ is identified with the point $P(a, b)$ in the cartesian plane, and vice versa. [See Fig. 1-10.] The x-axis [horizontal axis] is called the *real axis*, since its points correspond to those complex numbers $z = a + 0i = a$ which are real numbers; and the y-axis (vertical axis) is called the *imaginary axis*, since its points correspond to those complex numbers $z = 0 + bi = bi$ which are pure imaginary. Also, $|z| = (a^2 + b^2)^{1/2}$ is equal to the distance from the origin O to the point $z = P(a, b)$.

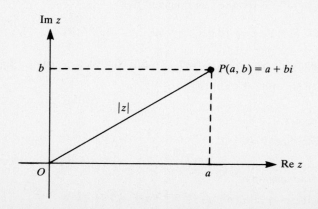

Fig. 1-10

1.125   Describe the geometrical relationship between the points z and $\bar{z}$ of the complex plane.

▮ They are mirror images of each other in the real axis.

1.126   Repeat Problem 1.125 for the points z and $i\bar{z}$.

▮ If $z = a + bi$, then $i\bar{z} = i(a - bi) = ia - bi^2 = b + ai$. Thus the two representative points, $P(a, b)$ and $Q(b, a)$, are mirror images in the 45° line $\operatorname{Re} z = \operatorname{Im} z$.

1.127   Suppose $z \neq 0$. Show that $\bar{z} = z^{-1}$ if and only if z lies on the unit circle in the complex plane **C**.

▮ $\bar{z} = z^{-1}$ iff $z\bar{z} = 1$ iff $|z| = 1$, which gives the result.

In Problems 1.128–1.130, $z = a + bi$ and $w = c + di$.

1.128   Prove: $\overline{z + w} = \bar{z} + \bar{w}$.

▮
$$\overline{z + w} = \overline{(a + bi) + (c + di)} = \overline{(a + c) + (b + d)i} = (a + c) - (b + d)i$$
$$= a + c - bi - di = (a - bi) + (c - di) = \bar{z} + \bar{w}$$

1.129   Prove: $\overline{zw} = \bar{z}\,\bar{w}$.

▮
$$\overline{zw} = \overline{(a + bi)(c + di)} = \overline{(ac - bd) + (ad + bc)i}$$
$$= (ac - bd) - (ad + bc)i = (a - bi)(c - di) = \bar{z}\,\bar{w}$$

1.130   Prove: $\bar{\bar{z}} = z$.

▮
$$\bar{\bar{z}} = \overline{\overline{a + bi}} = \overline{a - bi} = a - (-b)i = a + bi = z$$

1.131   Show that $|\bar{z}| = |z|$.

▮ $|\bar{z}|^2 = \bar{z}\,\bar{\bar{z}} = \bar{z}\,z = z\,\bar{z} = |z|^2$;   hence $|\bar{z}| = |z|$.

1.132   Prove: $|zw| = |z|\,|w|$.

▮ By Problem 1.129,

$$|zw|^2 = (zw)(\overline{zw}) = (zw)(\bar{z}\,\bar{w}) = (z\,\bar{z})(w\,\bar{w}) = |z|^2\,|w|^2$$

Now take square roots.

1.133   Prove: $zw = 0$ iff $z = 0$ or $w = 0$. [Thus **C** has no zero divisors.]

▮ Use Problem 1.132 [and the fact that $ab = 0$ iff $a = 0$ or $b = 0$ in **R**]:

$$zw = 0 \quad \text{iff} \quad |zw| = 0 \quad \text{iff} \quad |z|\,|w| = 0 \quad \text{iff} \quad |z| = 0 \quad \text{or} \quad |w| = 0 \quad \text{iff} \quad z = 0 \quad \text{or} \quad w = 0$$

1.134   Prove: For any complex numbers $z, w \in \mathbf{C}$, $|z + w| \le |z| + |w|$.

▮ Let $z = a + bi$ and $w = c + di$, where $a, b, c, d \in \mathbf{R}$. Consider the vectors $u = (a, b)$ and $v = (c, d)$ in $\mathbf{R}^2$. Note that

$$|z| = \sqrt{a^2 + b^2} = \|u\|, \quad |w| = |w| = \sqrt{c^2 + d^2} = \|v\|$$

and
$$|z + w| = |(a + c) + (b + d)i| = \sqrt{(a + c)^2 + (b + d)^2} = \|(\alpha + c, b + d)\| = \|u + v\|$$

By Minkowski's inequality (Problem 1.71), $\|u + v\| \le \|u\| + \|v\|$ and so

$$|z + w| = \|u + v\| \le \|u\| + \|v\| = |z| + |w|$$

1.10 VECTORS IN $\mathbf{C}^n$

In Problems 1.135–1.138, $u = (3 - 2i, 4i, 1 + 6i)$ and $v = (5 + i, 2 - 3i, 5)$ in $\mathbf{C}^3$.

1.135 Find $u + v$.

▌ Add corresponding components: $u + v = (8 - i, 2 + i, 6 + 6i)$.

1.136 Find $4iu$.

▌ Multiply each component of u by the scalar $4i$: $4iu = (8 + 12i, -16, -24 + 4i)$.

1.137 Find $(1 + i)v$.

▌ Multiply each component of v by the scalar $1 + i$:

$$(1 + i)v = (5 + 6i + i^2, 2 - i - 3i^2, 5 + 5i) = (4 + 6i, 5 - i, 5 + 5i)$$

1.138 Find $(1 - 2i)u + (3 + i)v$.

▌ First perform the scalar multiplications and then the vector addition: $(1 - 2i)u + (3 + i)v = (-1 - 8i, 8 + 4i, 13 + 4i) + (14 + 8i, 9 - 7i, 15 + 5i) = (13, 17 - 3i, 28 + 9i)$.

Problems 1.139–1.142 refer to the vectors $u = (7 - 2i, 2 + 5i)$ and $v = (1 + i, -3 - 6i)$ in $\mathbf{C}^2$.

1.139 Find $u + v$.

▌ $u + v = (7 - 2i + 1 + i, 2 + 5i - 3 - 6i) = (8 - i, -1 - i)$

1.140 Find $2iu$.

▌ $2iu = (14i - 4i^2, 4i + 10i^2) = (4 + 14i, -10 + 4i)$

1.141 Find $(3 - i)v$.

▌ $(3 - i)v = (3 + 3i - i - i^2, -9 - 18i + 3i + 6i^2) = (4 + 2i, -15 - 15i)$

1.142 Find $(1 + i)u + (2 - i)v$.

▌ $(1 + i)u + (2 - i)v = (9 + 5i, -3 + 6i) + (3 + i, -12 - 3i) = (12 + 6i, -15 + 3i)$

1.11 DOT (INNER) PRODUCT IN $\mathbf{C}^n$

1.143 Suppose $u = (z_1, z_2, \ldots, z_n)$ and $v = (w_1, w_2, \ldots, w_n)$ are vectors in $\mathbf{C}^n$. The *dot* or *inner product* of u and v is defined as

$$u \cdot v = \sum_{k=1}^{n} z_k \bar{w}_k$$

Show that this definition reduces to the one for $\mathbf{R}^n$ when all the components are real.

▌ By Problem 1.118, w_k is real if and only if $\bar{w}_k = w_k$. Accordingly, when all z_k and w_k are real,

$$u \cdot v = \sum_{k=1}^{n} z_k w_k$$

which is the same real number as is given by Problem 1.45.

1.144 Let $u = (z_1, z_2, \ldots, z_n)$ belong to $\mathbf{C}^n$. The *norm* or *length* of u, is defined by

$$\|u\| = \sqrt{u \cdot u} = \sqrt{z_1 \bar{z}_1 + z_2 \bar{z}_2 + \cdots + z_n \bar{z}_n}$$

Show that $\|u\|$ is real and nonnegative.

▌ By Problem 1.119, each $z_k \bar{z}_k$ is real and nonnegative. Hence the sum is real and nonnegative, and the square root is real and nonnegative.

1.145 Find $u \cdot v$ and $v \cdot u$, where $u = (1 - 2i, 3 + i)$ and $v = (4 + 2i, 5 - 6i)$ are vectors in $\mathbf{C}^2$.

▮ Recall that the conjugates of the components of the second vector appear in the dot product.

$$u \cdot v = (1 - 2i)(\overline{4 + 2i}) + (3 + i)(\overline{5 - 6i})$$
$$= (1 - 2i)(4 - 2i) + (3 + i)(5 + 6i) = -10i + 9 + 13i$$

$$v \cdot u = (4 + 2i)(\overline{1 - 2i}) + (5 - 6i)(\overline{3 + i})$$
$$= (4 + 2i)(1 + 2i) + (5 - 6i)(3 - i) = 10i + 9 - 23i = 9 - 13i$$

Note that $v \cdot u = \overline{u \cdot v}$. This is true in general, as proved in Problem 1.152.

1.146 Find $u \cdot v$ and $v \cdot u$, if $u = (3 - 2i, 4i, 1 + 6i)$ and $v = (5 + i, 2 - 3i, 7 + 2i)$ are vectors in $\mathbf{C}^3$.

▮ $$u \cdot v = (3 - 2i)(\overline{5 + i}) + (4i)(\overline{2 - 3i}) + (1 + 6i)(\overline{7 + 2i})$$
$$= (3 - 2i)(5 - i) + (4i)(2 + 3i) + (1 + 6i)(7 - 2i) = 20 + 35i$$

$$v \cdot u = (5 + i)(\overline{3 - 2i}) + (2 - 3i)(\overline{4i}) + (7 + 2i)(\overline{1 + 6i})$$
$$= (5 + i)(3 + 2i) + (2 - 3i)(-4i) + (7 + 2i)(1 - 6i) = 20 - 35i = \overline{u \cdot v}$$

1.147 Find $\|u\|$ for $u = (3 + 4i, 5 - 2i, 1 - 3i)$ in $\mathbf{C}^3$.

▮ $\|u\|^2 = u \cdot u = z_1\bar{z}_1 + z_2\bar{z}_2 + z_3\bar{z}_3 = [(3)^2 + (4)^2] + [(5)^2 + (-2)^2] + [(1)^2 + (-3)^2] = 64$; so $\|u\| = 8$.

1.148 Find $\|u\|$ for $u = (4 - i, 2i, 3 + 2i, 1 - 5i)$ in $\mathbf{C}^4$.

▮ $\|u\|^2 = [4^2 + (-1)^2] + [2^2] + [3^2 + 2^2] + [1^2 + (-5)^2] = 60$ or $\|u\| = \sqrt{60} = 2\sqrt{15}$

1.149 For $u = (7 - 2i, 2 + 5i)$ and $v = (1 + i, -3 - 6i)$ in $\mathbf{C}^2$, find (a) $u \cdot v$, (b) $\|u\|$, (c) $\|v\|$.

▮ (a) $u \cdot v = (7 - 2i)(\overline{1 + i}) + (2 + 5i)(\overline{-3 - 6i})$
$= (7 - 2i)(1 - i) + (2 + 5i)(-3 + 6i) = 5 - 9i - 36 - 3i = -31 - 12i$
(b) $\|u\| = \sqrt{7^2 + (-2)^2 + 2^2 + 5^2} = \sqrt{82}$
(c) $\|v\| = \sqrt{1^2 + 1^2 + (-3)^2 + (-6)^2} = \sqrt{47}$

1.150 Find $u \cdot v$ for $u = (2 + 3i, 4 - i, 2i)$ and $v = (3 - 2i, 5, 4\text{-}6i)$ in $\mathbf{C}^3$.

▮ $$u \cdot v = (2 + 3i)(\overline{3 - 2i}) + (4 - i)(\bar{5}) + (2i)(\overline{4 - 6i})$$
$$= (2 + 3i)(3 + 2i) + (4 - i)(5) + (2i)(4 + 6i) = 13i + 20 - 5i - 12 + 8i = 8 + 16i$$

1.151 Find $u \cdot u$ and $\|u\|$ for $u = (2 + 3i, 4 - i, 2i)$ in $\mathbf{C}^3$.

▮ $$u \cdot u = (2 + 3i)(\overline{2 + 3i}) + (4 - i)(\overline{4 - i}) + (2i)(\overline{2i})$$
$$= (2 + 3i)(2 - 3i) + (4 - i)(4 + i) + (2i)(-2i) = 13 + 17 + 4 = 34$$

so $\|u\| = \sqrt{u \cdot u} = \sqrt{34}$.

1.152 Prove that $u \cdot v = \overline{v \cdot u}$, for arbitrary u, v in $\mathbf{C}^n$.

▮ Apply Problems 1.128–1.130 to the definition of Problem 1.143:

$$\overline{u \cdot v} = \overline{\sum z_k \bar{w}_k} = \sum \bar{z}_k w_k = \sum w_k \bar{z}_k = v \cdot u$$

Now interchange u and v [or take conjugates of both sides].

1.153 Prove that $(zu) \cdot v = z(u \cdot v)$, for arbitrary u, v in $\mathbf{C}^n$ and arbitrary z in $\mathbf{C}$.

▮ Since $zu = (zz_1, zz_2, \ldots, zz_n)$, $(zu) \cdot v = zz_1\bar{w}_1 + zz_2\bar{w}_2 + \cdots + zz_n\bar{w}_n = z(z_1\bar{w}_1 + z_2\bar{w}_2 + \cdots + z_n\bar{w}_n) = z(u \cdot v)$ or in the notation of Problem 1.143,

$$(zu) \cdot v = \sum (zz_k)\bar{w}_k = z\left(\sum z_k \bar{w}_k\right) = z(u \cdot v)$$

1.154   Prove that $u \cdot (zv) = \bar{z}(u \cdot v)$.

▌ By Problems 1.152 and 1.153,

$$u \cdot (zv) = \overline{(zv) \cdot u} = \overline{z(v \cdot u)} = \bar{z}(\overline{v \cdot u}) = \bar{z}(u \cdot v)$$

1.155   Let $u_1 = (z_1, z_2, \ldots, z_n)$, $u_2 = (z_1', z_2', \ldots, z_n')$,   and $v = (w_1, w_2, \ldots, w_n)$.   Prove:

$$(u_1 + u_2) \cdot v = u_1 \cdot v + u_2 \cdot v$$

▌ Because $u_1 + u_2 = (z_1 + z_1', z_2 + z_2', \ldots, z_n + z_n')$,

$$(u_1 + u_2) \cdot v = \sum_{k=1}^{n} (z_k + z_k')\bar{w}_k = \sum_{k=1}^{n} (z_k \bar{w}_k + z_k'\bar{w}_k) = \sum_{k=1}^{n} z_k \bar{w}_k + \sum_{k=1}^{n} z_k'\bar{w}_k = u_1 \cdot v + u_2 \cdot v$$

1.156   For u, v_1, and v_2 in $\mathbf{C}^n$, prove: $u \cdot (v_1 + v_2) = u \cdot v_1 + u \cdot v_2$.

▌ $$u \cdot (v_1 + v_2) = \overline{(v_1 + v_2) \cdot u} = \overline{v_1 \cdot u + v_2 \cdot u} = \overline{v_1 \cdot u} + \overline{v_2 \cdot u} = u \cdot v_1 + u \cdot v_2$$

1.157   Suppose that $u_1, u_2, \ldots, u_p, v_1, v_2, \ldots, v_q$ belong to $\mathbf{C}^n$ and that $z_1, z_2, \ldots, z_p, w_1, w_2, \ldots, w_p$ belong to $\mathbf{C}$.   Prove

$$\left(\sum_{j=1}^{p} z_j u_j\right) \cdot \left(\sum_{k=1}^{q} w_k v_k\right) = \sum_{j=1}^{p} \sum_{k=1}^{q} z_j \bar{w}_k (u_j \cdot v_k)$$

(*bilinearity of the inner product*).

▌ In $\mathbf{R}^n$ we have $u \cdot kv = k(u \cdot v)$;   while in $\mathbf{C}^n$ $u \cdot kv = \bar{k}(u \cdot v)$.   The distributive laws are identical for both spaces.   It follows that the bilinearity formula of Problem 1.57 holds in $\mathbf{C}^n$ if the b_k in the double sum are replaced by their conjugates $\bar{b}_k$.

1.12   CROSS PRODUCT

The cross product is defined only for vectors in $\mathbf{R}^3$.

1.158   Evaluate the following determinants of order two:

$$(a) \quad \begin{vmatrix} 3 & 4 \\ 5 & 9 \end{vmatrix} \quad (b) \quad \begin{vmatrix} 2 & -1 \\ 4 & 3 \end{vmatrix} \quad (c) \quad \begin{vmatrix} 4 & 5 \\ 3 & -2 \end{vmatrix}$$

▌ (a) $\begin{vmatrix} 3 & 4 \\ 5 & 9 \end{vmatrix} = (3)(9) - (4)(5) = 7$ (b) $\begin{vmatrix} 2 & -1 \\ 4 & 3 \end{vmatrix} = 6 + 4 = 10$ (c) $\begin{vmatrix} 4 & 5 \\ 3 & -2 \end{vmatrix} = -8 - 15 = -23$

1.159   Evaluate the *negatives* of the following determinants of order two:

$$(a) \quad \begin{vmatrix} 3 & 6 \\ 4 & 2 \end{vmatrix} \quad (b) \quad \begin{vmatrix} 7 & -5 \\ 3 & 2 \end{vmatrix} \quad (c) \quad \begin{vmatrix} 4 & -1 \\ 8 & -3 \end{vmatrix} \quad (d) \quad \begin{vmatrix} -4 & -3 \\ 6 & -2 \end{vmatrix}$$

[*Hint*: $-\begin{vmatrix} a & b \\ c & d \end{vmatrix} = -(ad - bc) = bc - ad$, the product bc of the nondiagonal elements minus the product ad of the diagonal elements, called taking the determinant backward.]

▌ $\quad (a) \quad -\begin{vmatrix} 3 & 6 \\ 4 & 2 \end{vmatrix} = (6)(4) - (3)(2) = 18 \quad (b) \quad -\begin{vmatrix} 7 & -5 \\ 3 & 2 \end{vmatrix} = -15 - 14 = -29$

$\quad\quad (c) \quad -\begin{vmatrix} 4 & -1 \\ 8 & -3 \end{vmatrix} = -8 + 12 = 4 \quad (d) \quad -\begin{vmatrix} -4 & -3 \\ 2 & -2 \end{vmatrix} = -18 - 8 = -26$

1.160 Let $u = (a_1, a_2, a_3)$ and $v = (b_1, b_2, b_3)$ be vectors in **R**^{3}. Define the *cross product* of u and v, denoted $u \times v$.

❚ The cross product is the vector $u \times v \equiv (a_2b_3 - a_3b_2, a_3b_1 - a_1b_3, a_1b_2 - a_2b_1)$. The components of $u \times v$ may be expressed as determinants, as follows. Put the vector $v = (b_1, b_2, b_3)$ *under* the vector $u = (a_1, a_2, a_3)$ to form the array

$$\begin{pmatrix} a_1 & a_2 & a_3 \\ b_1 & b_2 & b_3 \end{pmatrix}$$

Then $\quad u \times v = (\begin{vmatrix} \boxed{a_1} & a_2 & a_3 \\ \boxed{b_1} & b_2 & b_3 \end{vmatrix}, \quad -\begin{vmatrix} a_1 & \boxed{a_2} & a_3 \\ b_1 & \boxed{b_2} & b_3 \end{vmatrix}, \quad \begin{vmatrix} a_1 & a_2 & \boxed{a_3} \\ b_1 & b_2 & \boxed{b_3} \end{vmatrix})$

That is, cover the first column of the array and take the determinant to obtain the first component of $u \times v$; cover the second column and take the determinant backward to obtain the second component; and cover the third column and take the determinant to obtain the third component.

1.161 Find $u \times v$, where $u = (1, 2, 3)$ and $v = (4, 5, 6)$.

❚ $u \times v = (\begin{vmatrix} \boxed{1} & 2 & 3 \\ \boxed{4} & 5 & 6 \end{vmatrix}, \quad -\begin{vmatrix} 1 & \boxed{2} & 3 \\ 4 & \boxed{5} & 6 \end{vmatrix}, \quad \begin{vmatrix} 1 & 2 & \boxed{3} \\ 4 & 5 & \boxed{6} \end{vmatrix}) = (12 - 15, 12 - 6, 5 - 8) = (-3, 6, -3)$

1.162 Find $u \times v$, where $u = (7, 3, 1)$ and $v = (1, 1, 1)$.

❚ $u \times v = (\begin{vmatrix} \boxed{7} & 3 & 1 \\ \boxed{1} & 1 & 1 \end{vmatrix}, \quad -\begin{vmatrix} 7 & \boxed{3} & 1 \\ 1 & \boxed{1} & 1 \end{vmatrix}, \quad \begin{vmatrix} 7 & 3 & \boxed{1} \\ 1 & 1 & \boxed{1} \end{vmatrix}) = (3 - 1, 1 - 7, 7 - 3) = (2, -6, 4)$

1.163 Find $u \times v$, where $u = (-4, 12, 2)$ and $v = (6, -18, -3)$.

❚ $u \times v = (\begin{vmatrix} \boxed{-4} & 12 & 2 \\ \boxed{6} & -18 & -3 \end{vmatrix}, \quad -\begin{vmatrix} -4 & \boxed{12} & 2 \\ 6 & \boxed{-18} & -3 \end{vmatrix}, \quad \begin{vmatrix} -4 & 12 & \boxed{2} \\ 6 & -18 & \boxed{-3} \end{vmatrix}) = (-36 - (-36), 12 - 12, 72 - 72)$

$\quad = (0, 0, 0) = \mathbf{0}$

Observe here that $v = -\frac{3}{2}u$; refer to Problem 1.171.

In Problems 1.164–1.169, $u = (a_1, a_2, a_3)$, $v = (b_1, b_2, b_3)$, and $w = (c_1, c_2, c_3)$ are vectors in **R**^{3}, and $k \in \mathbf{R}$.

1.164 Prove: $u \times v = -(v \times u)$.

❚ $u \times v = (a_2b_3 - a_3b_2, a_3b_1 - a_1b_3, a_1b_2 - a_2b_1)$. Then

$$-(v \times u) = -(b_2a_3 - b_3a_2, b_3a_1 - b_1a_3, b_1a_2 - b_2a_1)$$
$$= (b_3a_2 - b_2a_3, b_1a_3 - b_3a_1, b_2a_1 - b_1a_2) = u \times v$$

1.165 Prove: $(ku) \times v = k(u \times v) = u \times (kv)$.

❚ $ku = (ka_1, ka_2, ka_3)$; hence,

$$(ku) \times v = (ka_2b_3 - ka_3b_2, ka_3b_1 - ka_1b_3, ka_1b_2 - ka_2b_1)$$
$$= k(a_2b_3 - a_3b_2, a_3b_1 - a_1b_3, a_1b_2 - a_2b_1) = k(u \times v)$$

Similarly, $u \times (kv) = k(u \times v)$.

1.166 Prove: $u \times (v + w) = (u \times v) + (u \times w)$.

❚ The proof follows at once from the definition of vector addition and the fact that the components of a cross product are linear in the components of either vector.

1.167 Prove: $(v + w) \times u = (v \times u) + (w \times u)$.

❚ By Problems 1.164 and 1.166,

$$(v + w) \times u = -[u \times (v + w)] = -[(u \times v) - (u \times w)] = -(u \times v) - (u \times w) = (v \times u) + (w \times u)$$

1.168 Prove: $(u \times v) \times w = (u \cdot w)v - (v \cdot w)u$.

▍ Here $u \cdot w = a_1c_1 + a_2c_2 + a_3c_3$ and $v \cdot w = b_1c_1 + b_2c_2 + b_3c_3$. Let $(u \cdot w)v - (v \cdot w)u = (x_1, x_2, x_3)$; then

$$x_1 = (a_1c_1 + a_2c_2 + a_3c_3)b_1 - (b_1c_1 + b_2c_2 + b_3c_3)a_1 = (a_2c_2 + a_3c_3)b_1 - (b_2c_2 + b_3c_3)a_1$$

Similarly,

$$x_2 = (a_1c_1 + a_3c_3)b_2 - (b_1c_1 + b_3c_3)a_2 \qquad x_3 = (a_1c_1 + a_2c_2)b_3 - (b_1c_1 + b_2c_2)a_3$$

Write $(u \times v) \times w = (y_1, y_2, y_3)$; then

$$y_1 = (a_3b_1 - a_1b_3)c_3 - (a_1b_2 - a_2b_1)c_2 = a_3b_1c_3 - a_1b_3c_3 - a_1b_2c_2 + a_2b_1c_2$$
$$= b_1(a_3c_3 + a_2c_2) - a_1(b_3c_3 + b_2c_2)$$

Thus, $y_1 = x_1$. Similarly, $y_2 = x_2$ and $y_3 = x_3$.

1.169 Prove: $u \times v$ is orthogonal to both u and v.

▍ $\qquad u \cdot (u \times v) = a_1(a_2b_3 - a_3b_2) + a_2(a_3b_1 - a_1b_3) + a_3(a_1b_2 - a_2b_1)$
$$= a_1a_2b_3 - a_1a_3b_2 + a_2a_3b_1 - a_1a_2b_3 + a_1a_3b_2 - a_2a_3b_1 = 0$$

Thus, $u \times v$ is orthogonal to u. Similarly, $u \times v$ is orthogonal to v.

1.170 Show that $u \times u = \mathbf{0}$ for any vector u.

▍ By Problem 1.164, $u \times u = -(u \times u)$. Hence $u \times u = \mathbf{0}$.

1.171 Show that two vectors in $\mathbf{R}^3$ are *linearly dependent* if and only if their cross product is the zero vector.

▍ Linear dependence of two vectors u and v means that either $u = kv$, for some scalar k, or $v = lu$, for some scalar l. Suppose, then, that $u = kv$; by Problems 1.165 and 1.170, $u \times v = (kv) \times v = k(v \times v) = k\mathbf{0} = \mathbf{0}$, with the same result if $v = lu$. Conversely, suppose that $u \times v = \mathbf{0}$. If $u = \mathbf{0}$, then $u = kv$, for $k = 0$. If $u \neq \mathbf{0}$, set $w = u$ in Problem 1.168 to obtain

$$\mathbf{0} = (u \cdot u)v - (v \cdot u)u \qquad \text{or} \qquad v = \frac{v \cdot u}{u \cdot u}u \equiv lu$$

1.172 Find a unit vector u orthogonal to $v = (1, 3, 4)$ and $w = (2, -6, 5)$.

▍ In view of Problem 1.169, first find $v \times w$. The array

$$\begin{pmatrix} 1 & 3 & 4 \\ 2 & -6 & -5 \end{pmatrix} \qquad \text{gives} \qquad v \times w = (-15 + 24, 8 + 5, -6 - 6) = (9, 13, -12)$$

Now normalize $v \times w$ to get $u = (9/\sqrt{394}, 13/\sqrt{394}, -12/\sqrt{394})$.

Problems 1.173–1.176 refer to the points $P_1(1, 2, 3)$, $P_2(2, 5, -1)$, and $P_3(5, 3, 1)$ in $\mathbf{R}^3$.

1.173 Find the directed line segment (vector) u from P_1 to P_2.

▍ $u = P_2 - P_1 = (2, 5, -1) - (1, 2, 3) = (1, 3, -4)$

1.174 Find the directed line segment (vector) v from P_1 to P_3.

▍ $v = P_3 - P_1 = (5, 3, 1) - (1, 2, 3) = (4, 1, -2)$

1.175 Find a vector w normal to the plane H containing the points P_1, P_2, and P_3.

❙ H contains the vectors u and v determined above. Hence $u \times v$ is normal to H. The array

$$\begin{pmatrix} 1 & 3 & -4 \\ 4 & 1 & -2 \end{pmatrix} \quad \text{gives} \quad w = u \times v = (-6, +4, -16+2, 1-12) = (-2, -14, 11)$$

1.176 Give an equation for the plane H of Problem 1.175.

❙ Use the point $P_1(1, 2, 3)$ and the normal direction w to obtain

$$-2(x-1) - 14(y-2) - 11(z-3) = 0 \qquad \text{or} \qquad 2x + 14y + 11z = 63$$

1.177 Prove *Lagrange's identity*, $\|u \times v\|^2 = (u \cdot u)(v \cdot v) - (u \cdot v)^2$.

❙ If $u = (a_1, a_2, a_3)$ and $v = (b_1, b_2, b_3)$, then

$$\|u \times v\|^2 = (a_2 b_3 - a_3 b_2)^2 + (a_3 b_1 - a_1 b_3)^2 + (a_1 b_2 - a_2 b_1)^2 \tag{1}$$

$$(u \cdot u)(v \cdot v) - (u \cdot v)^2 = (a_1^2 + a_2^2 + a_3^2)(b_1^2 + b_2^2 + b_3^2) - (a_1 b_1 + a_2 b_2 + a_3 b_3)^2 \tag{2}$$

Expansion of the right-hand sides of *(1)* and *(2)* establishes the identity.

1.178 Show that $\|u \cdot v\| = \|u\| \, \|v\| \sin \theta$, where θ is the angle between u and v.

❙ By Problem 1.79, $u \cdot v = \|u\| \, \|v\| \cos \theta$. Then, by Problem 1.177,

$$\|u \times v\|^2 = \|u\|^2 \, \|v\|^2 - \|u\|^2 \, \|v\|^2 \cos^2 \theta = \|u\|^2 \, \|v\|^2 (1 - \cos^2 \theta) = \|u\|^2 \, \|v\|^2 \sin^2 \theta$$

Taking square roots gives us our result.

CHAPTER 2
Matrix Algebra

This chapter uses letters $A, B, C, \ldots$ to denote matrices and lowercase letters $a, b, c, x, y, \ldots$ to denote scalars. Unless otherwise stated or implied, scalars will be real numbers; in other words, the matrices will be over **R**.

2.1 MATRICES

2.1 Find the rows, columns and size of the matrix $A = \begin{pmatrix} 1 & 2 & 3 \\ 4 & 5 & 6 \end{pmatrix}$.

▌ The *rows* of A are the horizontal lines of numbers; there are two of them: $(1 \quad 2 \quad 3)$ and $(4 \quad 5 \quad 6)$. The *columns* of A are the vertical lines of numbers; there are three of them:

$$\begin{pmatrix} 1 \\ 4 \end{pmatrix} \qquad \begin{pmatrix} 2 \\ 5 \end{pmatrix} \qquad \begin{pmatrix} 3 \\ 6 \end{pmatrix}$$

The *size* of A is 2×3 [read: 2 by 3], the number of rows by the number of columns.

2.2 The notation $A = (a_{ij})_{m,n}$ [or simply $A = (a_{ij})$] is used to denote an $m \times n$ matrix. What is the significance of the first subscript, i, and the second subscript, j?

▌ The scalar a_{ij} is the element of A in row i and column j. Thus, the first subscript tells the row of the element and the second subscript tells the column.

2.3 Given the matrix $A = (a_{ij})$, find the location of (a) a_{35}, (b) $a_{4,12}$, (c) $a_{0,11}$, (d) $a_{13,-4}$.

▌ (a) a_{35} lies in the third row and fifth column. (b) $a_{4,12}$ appears in row 4 and column 12. [Note that here we need a comma to distinguish the subscripts.] (c) and (d) The scalars $a_{0,11}$ and $a_{13,-4}$ cannot be elements of A, since subscripts in a matrix are, by convention, positive integers.

2.4 Given matrices A and B, when is $A = B$?

▌ Two matrices are equal if and only if they have the same size and corresponding entries are equal.

2.5 Find x, y, z, w if $\begin{pmatrix} x+y & 2z+w \\ x-y & z-w \end{pmatrix} = \begin{pmatrix} 3 & 5 \\ 1 & 4 \end{pmatrix}$.

▌ Equate corresponding entries:

$$\begin{cases} x + y = 3 \\ x - y = 1 \\ 2z + w = 5 \\ z - w = 4 \end{cases}$$

The solution of the system of equations is $x = 2$, $y = 1$, $z = 3$, $w = -1$.

2.6 Which of the following matrices, if any, are equal?

$$A = \begin{pmatrix} 4 & 1 \\ 2 & 3 \end{pmatrix} \qquad B = \begin{pmatrix} 2 & 3 \\ 4 & 1 \end{pmatrix} \qquad C = \begin{pmatrix} 4 & 2 \\ 1 & 3 \end{pmatrix} \qquad D = \begin{pmatrix} 4 & 1 \\ 3 & 2 \end{pmatrix}$$

▌ Although all four matrices are 2×2 and contain the scalars 1, 2, 3, 4, no two of the matrices are equal *element by element*.

2.7 The $m \times n$ zero matrix, denoted by $\mathbf{0}_{m,n}$ or simply $\mathbf{0}$, is the matrix whose elements are all zero. Find x, y, z, t if

$$\begin{pmatrix} x+y & z+3 \\ y-4 & z+w \end{pmatrix} = \mathbf{0}$$

▌ Set all entries equal to zero to obtain the system

$$x + y = 0 \qquad z + 3 = 0 \qquad y - 4 = 0 \qquad z + w = 0$$

The solution of the system is $x = -4$, $y = 4$, $z = -3$, $w = 3$.

2.8 The *negative* of an $m \times n$ matrix $A = (a_{ij})$ is the $m \times n$ matrix $-A \equiv (-a_{ij})$. Find the negatives of

$$A = \begin{pmatrix} 1 & -3 & 4 & 7 \\ 2 & -5 & 0 & -8 \end{pmatrix} \qquad B = \begin{pmatrix} 2 & -3 \\ -6 & 1 \end{pmatrix} \qquad \mathbf{0} = \begin{pmatrix} 0 & 0 & 0 \\ 0 & 0 & 0 \end{pmatrix}$$

▮ Take the negative of each element:

$$-A = \begin{pmatrix} -1 & -(-3) & -4 & -7 \\ -2 & -(-5) & -0 & -(-8) \end{pmatrix} = \begin{pmatrix} -1 & 3 & -4 & -7 \\ -2 & 5 & 0 & 8 \end{pmatrix}$$

$$-B = \begin{pmatrix} -2 & 3 \\ 6 & -1 \end{pmatrix} \qquad -\mathbf{0} = \begin{pmatrix} -0 & -0 & -0 \\ -0 & -0 & -0 \end{pmatrix} = \begin{pmatrix} 0 & 0 & 0 \\ 0 & 0 & 0 \end{pmatrix} = \mathbf{0}$$

2.9 Show that, for any matrix A, we have $-(-A) = A$.

▮ $-(-A) = -(-a_{ij})_{m,n} = (-(-a_{ij}))_{m,n} = (a_{ij})_{m,n} = A$

2.10 A matrix A with only one row is called a *row matrix* or a *row vector* and is frequently denoted by $A = (a_1 \ a_2 \ \cdots \ a_n)$; we omit its first subscript since it must be one. Analogously, a matrix B with only one column is called a *column matrix* or a *column vector* and is frequently denoted by

$$B = \begin{pmatrix} b_1 \\ b_2 \\ \cdots \\ b_m \end{pmatrix}$$

Discuss the difference, if any, between the following objects:

$$u = (1 \ \ 2 \ \ 3) \qquad \text{and} \qquad v = \begin{pmatrix} 1 \\ 2 \\ 3 \end{pmatrix}$$

▮ Viewed as vectors in $\mathbf{R}^3$, u and v may be considered equal. However, as matrices, they cannot be equal, for they have different sizes.

2.2 MATRIX ADDITION AND SCALAR MULTIPLICATION

2.11 If $A = (a_{ij})_{m,n}$ and $B = (b_{ij})_{m,n}$ are matrices of the same size, their *sum* is defined as $A + B \equiv (a_{ij} + b_{ij})_{m,n}$. Find the sum of

$$A = \begin{pmatrix} 1 & -2 & 3 \\ 4 & 5 & -6 \end{pmatrix} \qquad \text{and} \qquad B = \begin{pmatrix} 3 & 0 & 2 \\ -7 & 1 & 8 \end{pmatrix}$$

▮ Add corresponding entries:

$$A + B = \begin{pmatrix} 1+3 & -2+0 & 3+2 \\ 4-7 & 5+1 & -6+8 \end{pmatrix} = \begin{pmatrix} 4 & -2 & 5 \\ -3 & 6 & 2 \end{pmatrix}$$

2.12 Find $A + B$ if $A = \begin{pmatrix} 1 & 2 & -3 \\ 0 & -4 & 1 \end{pmatrix}$ and $B = \begin{pmatrix} 3 & 5 \\ 1 & -2 \end{pmatrix}$.

▮ The sum is not defined, since the matrices have different sizes.

2.13 Find $A + B$ for $A = \begin{pmatrix} 1 & 2 & 3 \\ 4 & 5 & 6 \end{pmatrix}$ and $B = \begin{pmatrix} 1 & -1 & 2 \\ 0 & 3 & -5 \end{pmatrix}$.

▮ Add corresponding elements:

$$A + B = \begin{pmatrix} 1+1 & 2+(-1) & 3+2 \\ 4+0 & 5+3 & 6+(-5) \end{pmatrix} = \begin{pmatrix} 2 & 1 & 5 \\ 4 & 8 & 1 \end{pmatrix}$$

2.14 Add $C = \begin{pmatrix} 1 & 2 & -3 & 4 \\ 0 & -5 & 1 & -1 \end{pmatrix}$ and $D = \begin{pmatrix} 3 & -5 & 6 & -1 \\ 2 & 0 & -2 & -3 \end{pmatrix}$.

▮ $$C + D = \begin{pmatrix} 1+3 & 2+(-5) & (-3)+6 & 4+(-1) \\ 0+2 & (-5)+0 & 1+(-2) & (-1)+(-3) \end{pmatrix} = \begin{pmatrix} 4 & -3 & 3 & 3 \\ 2 & -5 & -1 & -4 \end{pmatrix}$$

2.15   Redefine the negative of a matrix [Problem 2.8] in terms of matrix addition.

▐   The negative of a given matrix A is the [unique] matrix whose sum with A is the zero matrix, that is, $A + (-A) = 0$.   [Note that this way of defining $-A$ avoids reference to the elements of A.]

2.16   If $A = (a_{ij})_{m,n}$ and k is a scalar, the matrix $kA \equiv (ka_{ij})_{m,n}$ is called the *product* of A by the scalar k.   Find $3A$ and $-5A$, where

$$A = \begin{pmatrix} 1 & -2 & 3 \\ 4 & 5 & -6 \end{pmatrix}$$

▐   Multiply each entry by the given scalar:

$$3A = \begin{pmatrix} 3\cdot1 & 3\cdot(-2) & 3\cdot3 \\ 3\cdot4 & 3\cdot5 & 3\cdot(-6) \end{pmatrix} = \begin{pmatrix} 3 & -6 & 9 \\ 12 & 15 & -18 \end{pmatrix}$$

$$-5A = \begin{pmatrix} -5\cdot1 & -5\cdot(-2) & -5\cdot3 \\ -5\cdot4 & -5\cdot5 & -5\cdot(-6) \end{pmatrix} = \begin{pmatrix} -5 & 10 & -15 \\ -20 & -25 & 30 \end{pmatrix}$$

2.17   Compute: (a) $3\begin{pmatrix} 2 & 4 \\ -3 & 1 \end{pmatrix}$, (b) $-2\begin{pmatrix} 1 & 7 \\ 2 & -3 \\ 0 & -1 \end{pmatrix}$.

▐   (a)

$$3\begin{pmatrix} 2 & 4 \\ -3 & 1 \end{pmatrix} = \begin{pmatrix} 3\cdot2 & 3\cdot4 \\ 3\cdot(-3) & 3\cdot1 \end{pmatrix} = \begin{pmatrix} 6 & 12 \\ -9 & 3 \end{pmatrix}$$

(b)

$$-2\begin{pmatrix} 1 & 7 \\ 2 & -3 \\ 0 & -1 \end{pmatrix} = \begin{pmatrix} (-2)\cdot1 & (-2)\cdot7 \\ (-2)\cdot2 & (-2)\cdot(-3) \\ (-2)\cdot0 & (-2)\cdot(-1) \end{pmatrix} = \begin{pmatrix} -2 & -14 \\ -4 & 6 \\ 0 & 2 \end{pmatrix}$$

2.18   The *difference*, $A - B$, of two matrices A and B of the same size is defined by $A - B \equiv A + (-B)$.   Find $A - B$ if

$$A = \begin{pmatrix} 4 & -5 & 6 \\ 2 & 3 & -1 \end{pmatrix} \quad \text{and} \quad B = \begin{pmatrix} 2 & -3 & 8 \\ 1 & -2 & -6 \end{pmatrix}$$

▐

$$A - B = A + (-B) = \begin{pmatrix} 4 & -5 & 6 \\ 2 & 3 & -1 \end{pmatrix} + \begin{pmatrix} -2 & 3 & -8 \\ -1 & 2 & 6 \end{pmatrix} = \begin{pmatrix} 2 & -2 & -2 \\ 1 & 5 & 5 \end{pmatrix}$$

2.19   Find $2A - 3B$,   where $A = \begin{pmatrix} 1 & -2 & 3 \\ 4 & 5 & -6 \end{pmatrix}$ and $B = \begin{pmatrix} 3 & 0 & 2 \\ -7 & 1 & 8 \end{pmatrix}$.

▐   First perform the scalar multiplications, and then a matrix addition:

$$2A - 3B = \begin{pmatrix} 2 & -4 & 6 \\ 8 & 10 & -12 \end{pmatrix} + \begin{pmatrix} -9 & 0 & -6 \\ 21 & -3 & -24 \end{pmatrix} = \begin{pmatrix} -7 & -4 & 0 \\ 29 & 7 & -36 \end{pmatrix}$$

[Note that we multiply B by -3 and then add, rather than multiplying B by 3 and subtracting.   This usually avoids errors.]

2.20   If $A = \begin{pmatrix} 2 & -5 & 1 \\ 3 & 0 & -4 \end{pmatrix}$, $B = \begin{pmatrix} 1 & -2 & -3 \\ 0 & -1 & 5 \end{pmatrix}$, $C = \begin{pmatrix} 0 & 1 & -2 \\ 1 & -1 & -1 \end{pmatrix}$,   find $3A + 4B - 2C$.

▐   First perform the scalar multiplications, and then the matrix additions:

$$3A + 4B - 2C = \begin{pmatrix} 6 & -15 & 3 \\ 9 & 0 & -12 \end{pmatrix} + \begin{pmatrix} 4 & -8 & -12 \\ 0 & -4 & 20 \end{pmatrix} + \begin{pmatrix} 0 & -2 & 4 \\ -2 & 2 & 2 \end{pmatrix} = \begin{pmatrix} 10 & -25 & -5 \\ 7 & -2 & 10 \end{pmatrix}$$

2.21   Find x, y, z, and w, if $3\begin{pmatrix} x & y \\ z & w \end{pmatrix} = \begin{pmatrix} x & 6 \\ -1 & 2w \end{pmatrix} + \begin{pmatrix} 4 & x+y \\ z+w & 3 \end{pmatrix}$.

❚ First write each side as a single matrix:

$$\begin{pmatrix} 3x & 3y \\ 3z & 3w \end{pmatrix} = \begin{pmatrix} x+4 & x+y+6 \\ z+w-1 & 2w+3 \end{pmatrix}$$

Set corresponding entries equal to each other to obtain the system of four equations,

$$\begin{array}{ll} 3x = x+4 & 2x = 4 \\ 3y = x+y+6 & \text{or} \quad 2y = 6+x \\ 3z = z+w-1 & 2z = w-1 \\ 3w = 2w+3 & w = 3 \end{array}$$

The solution is: $x=2$, $y=4$, $z=1$, $w=3$.

2.22 Let $B = \begin{pmatrix} 5 & -2 \\ 4 & 7 \end{pmatrix}$ and $C = \begin{pmatrix} 1 & 2 \\ 6 & -3 \end{pmatrix}$. Find $A = \begin{pmatrix} x & y \\ z & w \end{pmatrix}$ such that $2A = 3B - 2C$.

❚ **Method 1.** First compute $3B - 2C$:

$$3B - 2C = \begin{pmatrix} 15 & -6 \\ 12 & 21 \end{pmatrix} + \begin{pmatrix} -2 & -4 \\ -12 & 6 \end{pmatrix} = \begin{pmatrix} 13 & -10 \\ 0 & 27 \end{pmatrix}$$

Then set $2A = 3B - 2C$:

$$\begin{pmatrix} 2x & 2y \\ 2z & 2w \end{pmatrix} = \begin{pmatrix} 13 & -10 \\ 0 & 27 \end{pmatrix}$$

Equate corresponding entries: $2x = 13$, $2y = -10$, $2z = 0$, $2w = 27$. Hence $x = 13/2$, $y = -5$, $z = 0$, and $w = 27/2$; that is,

$$A = \begin{pmatrix} 13/2 & -5 \\ 0 & 27/2 \end{pmatrix}$$

Method 2. Apply Theorem 2.1 [proved in Problems 2.24–2.31] to obtain directly $A = (3/2)B - C$.

2.23 Find $2A + 5B$, where $A = \begin{pmatrix} 1 & 3 \\ 2 & -5 \end{pmatrix}$ and $B = \begin{pmatrix} 4 & -3 & -6 \\ 3 & 7 & -8 \end{pmatrix}$.

❚ Although $2A$ and $5B$ are defined, the sum $2A + 5B$ is not defined since $2A$ and $5B$ have different sizes.

Theorem 2.1: Let M be the collection of all $m \times n$ matrices over a field **K** of scalars. Then for any matrices $A = (a_{ij})$, $B = (b_{ij})$, and $C = (c_{ij})$ in M, and any scalars k_1, k_2 in **K**,

(i)	$(A+B)+C = A+(B+C)$	(v)	$k_1(A+B) = k_1 A + k_1 B$
(ii)	$A + 0 = A$	(vi)	$(k_1 + k_2)A = k_1 A + k_2 A$
(iii)	$A + (-A) = 0$	(vii)	$(k_1 k_2)A = k_1(k_2 A)$
(iv)	$A + B = B + A$	(viii)	$1A = A$

2.24 Prove (i) of Theorem 2.1

❚ The *ij*-entry of $A + B$ is $a_{ij} + b_{ij}$; hence, $(a_{ij} + b_{ij}) + c_{ij}$ is the *ij*-entry of $(A + B) + C$. The *ij*-entry of $B + C$ is $b_{ij} + c_{ij}$; hence, $a_{ij} + (b_{ij} + c_{ij})$ is the *ij*-entry of $A + (B + C)$. However, by the associative law of addition in **K**,

$$(a_{ij} + b_{ij}) + c_{ij} = a_{ij} + (b_{ij} + c_{ij})$$

Therefore, $(A + B) + C$ and $A + (B + C)$ have the same *ij*-entries, and hence $(A + B) + C = A + (B + C)$.

2.25 Prove (ii) of Theorem 2.1.

❚ The *ij*-entry of $A + 0$ is $a_{ij} + 0 = a_{ij}$. Therefore, $A + 0$ and A have the same *ij*-entries, and hence $A + 0 = A$.

2.26   Prove (iii) of Theorem 2.1.

▐   See Problem 2.15.

2.27   Prove (iv) of Theorem 2.1.

▐   The ij-entry of $A + B$ is $a_{ij} + b_{ij}$, and the ij-entry of $B + A$ is $b_{ij} + a_{ij}$. However, by the commutative law in **K**, $a_{ij} + b_{ij} = b_{ij} + a_{ij}$. Thus, $A + B$ and $B + A$ have the same ij-entries, and hence $A + B = B + A$.

2.28   Prove (v) of Theorem 2.1.

▐   The ij-entry of $A + B$ is $a_{ij} + b_{ij}$; hence $k_1(a_{ij} + b_{ij})$ is the ij-entry of $k_1(A + B)$. The ij-entry of $k_1 A$ is $k_1 a_{ij}$, and the ij-entry of $k_1 B$ is $k_1 b_{ij}$; hence $k_1 a_{ij} + k_1 b_{ij}$ is the ij-entry of $k_1 A + k_1 B$. However, by the distributive law in **K**, $k_1(a_{ij} + b_{ij}) = k_1 a_{ij} + k_1 b_{ij}$. Therefore, $k_1(A + B)$ and $k_1 A$ and $k_1 B$ have the same ij-entries; and hence $k_1(A + B) = k_1 A + k_1 B$.

2.29   Prove (vi) of Theorem 2.1.

▐   As in Problem 2.28, the proof is by the distributive law in **K**.

2.30   Prove (vii) of Theorem 2.1.

▐   The ij-entry of $(k_1 k_2)A$ is $(k_1 k_2)a_{ij}$. The ij-entry of $k_2 A$ is $k_2 a_{ij}$, and so $k_1(k_2 a_{ij})$ is the ij-entry of $k_1(k_2 A)$. However, by the associative law of multiplication in **K**, $(k_1 k_2)a_{ij} = k_1(k_2 a_{ij})$. Therefore, $(k_1 k_2)A$ and $k_1(k_2 A)$ have the same ij-entries, and hence $(k_1 k_2)A = k_1(k_2 A)$.

2.31   Prove (viii) of Theorem 2.1.

▐   The ij-entry of $1 \cdot A$ is $1 \cdot a_{ij} = a_{ij}$. Since $1 \cdot A$ and A have the same ij-entries, they are equal.

2.32   Comment on the difference, if any, between the $+$ signs in (vi) of Theorem 2.1.

▐   On the left, the $+$ sign refers to addition of scalars in **K**; on the right, to addition of matrices in M

2.33   Prove that $0A = \mathbf{0}$, for any matrix A.

▐   By (viii), (vi), and (ii) of Theorem 2.1,

$$A + 0A = 1A + 0A = (1 + 0)A = 1A = A$$

and the proof follows upon the addition of $-A$ to both sides.

2.34   Show that $(-1)A = -A$.

▐ $A + (-1)A = 1A + (-1)A = (1 + (-1))A = 0A = \mathbf{0}$, where the last step follows from Problem 2.33. Now add $-A$ to both sides.

2.35   Show that $A + A = 2A$ and $A + A + A = 3A$.

▐   Using (vi) and (viii) of Theorem 2.3, $2A = (1 + 1)A = 1A + 1A = A + A$. Similarly, $3A = (2 + 1)A = 2A + 1A = A + A + A$.

2.36   Prove that, for any positive integer n, $\displaystyle\sum_{k=1}^{n} A \equiv A + A + \cdots + A = nA$.

▐   The proof is by induction on n. The case $n = 1$ appears in Theorem 2.1(viii). Suppose $n > 1$, and the theorem holds for $n - 1$. Then

$$\sum_{k=1}^{n} A = \sum_{k=1}^{n-1} A \ + \ A = (n-1)A + 1A = [(n-1) + 1]A = nA$$

2.3 MATRIX MULTIPLICATION

2.37 The *product* of a row matrix and a column matrix with the same number of elements is their inner product as defined in Problem 1.45:

$$(a_1, a_2, \ldots, a_n) \begin{pmatrix} b_1 \\ b_2 \\ \cdots \\ b_n \end{pmatrix} = a_1 b_1 + a_2 b_2 + \cdots + a_n b_n \equiv \sum_{k=1}^{n} a_k b_k$$

Calculate:

$$(a) \quad (8, -4, 5) \begin{pmatrix} 3 \\ 2 \\ -1 \end{pmatrix} \qquad (b) \quad (6, -1, 7, 5) \begin{pmatrix} 4 \\ -9 \\ -3 \\ 2 \end{pmatrix} \qquad (c) \quad (3, 8, -2, 4) \begin{pmatrix} 5 \\ -1 \\ 6 \end{pmatrix}$$

and (d) $(1, 8, 3, 4)(6, 1, -3, 5)$.

▌ (a) Multiply corresponding entries and add:

$$(8, -4, 5) \begin{pmatrix} 3 \\ 2 \\ -1 \end{pmatrix} = (8)(3) + (-4)(2) + (5)(-1) = 24 - 8 - 5 = 11$$

(b) Multiply corresponding entries and add:

$$(6, -1, 7, 5) \begin{pmatrix} 4 \\ -9 \\ -3 \\ 2 \end{pmatrix} = 24 + 9 - 21 + 10 = 22$$

(c) The product is not defined when the row matrix and column matrix have different numbers of elements. (d) The product of a row matrix and a row matrix is not defined.

2.38 Let $(r \times s)$ denote a matrix with size $r \times s$. Find the size of each product, when the product is defined:

$$
\begin{array}{lll}
(a) \quad (2 \times 3)(3 \times 4) & (c) \quad (1 \times 2)(3 \times 1) & (e) \quad (3 \times 4)(3 \times 4) \\
(b) \quad (4 \times 1)(1 \times 2) & (d) \quad (5 \times 2)(2 \times 3) & (f) \quad (2 \times 2)(2 \times 4)
\end{array}
$$

▌ An $m \times p$ matrix is multipliable on the right by a $q \times n$ matrix only when $p = q$, and then the product is an $m \times n$ matrix. (a) 2×4; (b) 4×2; (c) not defined; (d) 5×3; (e) not defined; (f) 2×4.

2.39 Suppose that $A = (a_{ik})$ is an $m \times p$ matrix and $B = (b_{kj})$ is a $p \times n$ matrix. Then the *product* $AB \equiv (c_{ij})$ is the $m \times n$ matrix for which

$$c_{ij} = a_{i1} b_{1j} + a_{i2} b_{2j} + \cdots + a_{ip} b_{pj} = \sum_{k=1}^{p} a_{ik} b_{kj}$$

that is, the *ij*-entry of AB is the product of the *i*th row vector of A and the *j*th column vector of B. Find the product AB for

$$A = \begin{pmatrix} 1 & 3 \\ 2 & -1 \end{pmatrix} \qquad \text{and} \qquad B = \begin{pmatrix} 2 & 0 & -4 \\ 3 & -2 & 6 \end{pmatrix}$$

▌ Since A is 2×2 and B is 2×3, the product AB is defined as a 2×3 matrix. To obtain the

entries in the first row of AB, multiply the first row (1 3) of A by the columns $\begin{pmatrix} 2 \\ 3 \end{pmatrix}$, $\begin{pmatrix} 0 \\ -2 \end{pmatrix}$ and $\begin{pmatrix} -4 \\ 6 \end{pmatrix}$ of B, respectively:

$$\begin{pmatrix} \boxed{1 \quad 3} \\ 2 \quad -1 \end{pmatrix} \begin{pmatrix} \boxed{2} & \boxed{0} & \boxed{-4} \\ 3 & -2 & 6 \end{pmatrix} = \begin{pmatrix} (1)(2) + (3)(3) & (1)(0) + (3)(-2) & (1)(-4) + (3)(6) \end{pmatrix}$$

$$= \begin{pmatrix} 2 + 9 & 0 - 6 & -4 + 18 \end{pmatrix} = \begin{pmatrix} 11 & -6 & 14 \end{pmatrix}$$

To obtain the entries in the second row of AB, multiply the second row $(2, -1)$ of A by the columns of B, respectively:

$$\left(\begin{array}{cc} 1 & 3 \\ \boxed{2 & -1} \end{array}\right)\left(\boxed{\begin{array}{c} 2 \\ 3 \end{array}}\ \boxed{\begin{array}{c} 0 \\ -2 \end{array}}\ \boxed{\begin{array}{c} -4 \\ 6 \end{array}}\right) = \left(\begin{array}{ccc} 11 & -6 & 14 \\ (2)(2)+(-1)(3) & (2)(0)+(-1)(-2) & (2)(-4)+(-1)(6) \end{array}\right)$$

Thus
$$AB = \begin{pmatrix} 11 & -6 & 14 \\ 1 & 2 & -14 \end{pmatrix}$$

2.40 Find the product BA of the matrices A and B in Problem 2.39.

❚ Note that B is 2×3 and A is 2×2. Since the inner numbers 3 and 2 are not equal, the product BA is not defined.

2.41 Find the product AB, where $A = (2 \quad 1)$ and $B = \begin{pmatrix} 1 & -2 & 0 \\ 4 & 5 & -3 \end{pmatrix}$.

❚ Since A is 1×2 and B is 2×3, the product AB is defined as a 1×3 matrix, or row vector with 3 components. To obtain the components of AB, multiply the row of A by each column of B:

$$AB = (\boxed{2 \quad 1})\left(\boxed{\begin{array}{c} 1 \\ 4 \end{array}}\ \boxed{\begin{array}{c} -2 \\ 5 \end{array}}\ \boxed{\begin{array}{c} 0 \\ -3 \end{array}}\right) = ((2)(1)+(1)(4), (2)(-2)+(1)(5), (2)(0)+(1)(-3)) = (6, 1, -3)$$

2.42 Find the product AB, if

$$A = \begin{pmatrix} 2 & -1 \\ 1 & 0 \\ -3 & 4 \end{pmatrix} \qquad B = \begin{pmatrix} 1 & -2 & -5 \\ 3 & 4 & 0 \end{pmatrix}$$

❚ Since A is 3×2 and B is 2×3, the product AB is defined as a 3×3 matrix. To obtain the first row of AB, multiply the first row of A by each column of B, respectively:

$$\left(\begin{array}{cc} \boxed{2 & -1} \\ 1 & 0 \\ -3 & 4 \end{array}\right)\left(\boxed{\begin{array}{c} 1 \\ 3 \end{array}}\ \boxed{\begin{array}{c} -2 \\ 4 \end{array}}\ \boxed{\begin{array}{c} -5 \\ 0 \end{array}}\right) = \begin{pmatrix} 2-3 & -4-4 & -10+0 \\ & & \\ & & \end{pmatrix} = \begin{pmatrix} -1 & -8 & -10 \\ & & \\ & & \end{pmatrix}$$

To obtain the second row of AB, multiply the second row of A by each column of B, respectively:

$$\left(\begin{array}{cc} 2 & -1 \\ \boxed{1 & 0} \\ -3 & 4 \end{array}\right)\left(\boxed{\begin{array}{c} 1 \\ 3 \end{array}}\ \boxed{\begin{array}{c} -2 \\ 4 \end{array}}\ \boxed{\begin{array}{c} -5 \\ 0 \end{array}}\right) = \begin{pmatrix} -1 & -8 & -10 \\ 1+0 & -2+0 & -5+0 \end{pmatrix} = \begin{pmatrix} -1 & -8 & -10 \\ 1 & -2 & -5 \end{pmatrix}$$

To obtain the third row of AB, multiply the third row of A by each column of B, respectively

$$\left(\begin{array}{cc} 2 & -1 \\ 1 & 0 \\ \boxed{-3 & 4} \end{array}\right)\left(\boxed{\begin{array}{c} 1 \\ 3 \end{array}}\ \boxed{\begin{array}{c} -2 \\ 4 \end{array}}\ \boxed{\begin{array}{c} -5 \\ 0 \end{array}}\right) = \begin{pmatrix} -1 & -8 & -10 \\ 1 & -2 & -5 \\ -3+12 & 6+16 & 15+0 \end{pmatrix} = \begin{pmatrix} -1 & -8 & -10 \\ 1 & -2 & -5 \\ 9 & 22 & 15 \end{pmatrix}$$

Thus
$$AB = \begin{pmatrix} -1 & -8 & -10 \\ 1 & -2 & -5 \\ 9 & 22 & 15 \end{pmatrix}$$

2.43 Find the product BA, where A and B are the matrices of Problem 2.42.

❚ Since B is 2×3 and A is 3×2, the product BA is defined as a 2×2 matrix. To obtain the first row of BA, multiply the first row of B by each column of A, respectively:

$$\left(\boxed{1 \quad -2 \quad -5}\atop 3 \quad 4 \quad 0\right)\left(\boxed{\begin{array}{c} 2 \\ 1 \\ -3 \end{array}}\ \boxed{\begin{array}{c} -1 \\ 0 \\ 4 \end{array}}\right) = \begin{pmatrix} 2-2+15 & -1+0-20 \end{pmatrix} = \begin{pmatrix} 15 & -21 \end{pmatrix}$$

To obtain the second row of BA, multiply the second row of B by each column of A, respectively:

$$\left(\frac{1 \quad -2 \quad -5}{\boxed{3 \quad 4 \quad 0}}\right)\left(\boxed{\begin{matrix}2\\1\\-3\end{matrix}} \ \boxed{\begin{matrix}-1\\0\\4\end{matrix}}\right) = \left(\begin{matrix}15 & -21\\6+4+0 & -3+0+0\end{matrix}\right) = \left(\begin{matrix}15 & -21\\10 & -3\end{matrix}\right)$$

Thus
$$BA = \left(\begin{matrix}15 & -21\\10 & -3\end{matrix}\right)$$

2.44 Find the size of the product AB, where

$$A = \left(\begin{matrix}2 & -1 & 0\\1 & 0 & -3\end{matrix}\right) \qquad B = \left(\begin{matrix}1 & -4 & 0 & 1\\2 & -1 & 3 & -1\\4 & 0 & -2 & 0\end{matrix}\right)$$

▮ Since A is 2×3 and B is 3×4, the product AB is a 2×4 matrix.

2.45 Suppose $AB = (c_{ij})$ for the matrices A and B of Problem 2.44. Find: (a) c_{23}, (b) c_{14}, (c) c_{21}, (d) c_{32}.

▮ The element c_{ij}, the ij-entry of AB, is the product of row i of A by column j of B.

(a)
$$c_{23} = (1, 0, -3)\left(\begin{matrix}0\\3\\-2\end{matrix}\right) = (1)(0) + (0)(3) + (-3)(-2) = 0 + 0 + 6 = 6$$

(b)
$$c_{14} = (2, -1, 0)\left(\begin{matrix}1\\-1\\0\end{matrix}\right) = (2)(1) + (-1)(-1) + (0)(0) = 2 + 1 + 0 = 3$$

(c)
$$c_{21} = (1, 0, -3)\left(\begin{matrix}1\\2\\4\end{matrix}\right) = (1)(1) + (0)(2) + (-3)(4) = 1 + 0 - 12 = -11$$

(d) The element c_{32} does not exist, since A, and with it AB, has only two rows.

2.46 Find AB, where

$$A = \left(\begin{matrix}2 & 3 & -1\\4 & -2 & 5\end{matrix}\right) \qquad B = \left(\begin{matrix}2 & -1 & 0 & 6\\1 & 3 & -5 & 1\\4 & 1 & -2 & 2\end{matrix}\right)$$

▮ Since A is 2×3 and B is 3×4, the product is defined as a 2×4 matrix. Multiply the rows of A by the columns of B to obtain:

$$AB = \left(\begin{matrix}4+3-4 & -2+9-1 & 0-15+2 & 12+3-2\\8-2+20 & -4-6+5 & 0+10-10 & 24-2+10\end{matrix}\right) = \left(\begin{matrix}3 & 6 & -13 & 13\\26 & -5 & 0 & 32\end{matrix}\right)$$

2.47 Refer to Problem 2.46. Suppose that only the third column of the product AB were of interest. How could it be computed independently?

▮ By the rule for matrix multiplication, the jth column of a product is equal to the first factor times the jth column vector of the second. Thus,

$$\left(\begin{matrix}2 & 3 & -1\\4 & -2 & 5\end{matrix}\right)\left(\begin{matrix}0\\-5\\-2\end{matrix}\right) = \left(\begin{matrix}0-15+2\\0+10-10\end{matrix}\right) = \left(\begin{matrix}-13\\0\end{matrix}\right)$$

Similarly, the ith row of a product is equal to the ith row vector of the first factor times the second factor.

2.48 Find $\left(\begin{matrix}1 & 6\\-3 & 5\end{matrix}\right)\left(\begin{matrix}2\\-7\end{matrix}\right)$.

▮ The first factor is 2×2 and the second is 2×1, so the product is defined as a 2×1 matrix.

$$\left(\begin{matrix}1 & 6\\-3 & 5\end{matrix}\right)\left(\begin{matrix}2\\-7\end{matrix}\right) = \left(\begin{matrix}2-42\\-6-35\end{matrix}\right) = \left(\begin{matrix}-40\\-41\end{matrix}\right)$$

2.49 Find $\begin{pmatrix} 6 \\ 5 \end{pmatrix}\begin{pmatrix} 2 & 1 \\ -7 & -3 \end{pmatrix}$.

▌ The product is not defined since the first factor is 2×1 and the second factor is 2×2.

2.50 Find $(2, -7)\begin{pmatrix} 1 & 6 \\ -3 & 5 \end{pmatrix}$.

▌ The first factor is 1×2 and the second factor is 2×2, so the product is defined as a 1×2 (row) matrix.

$$(2, -7)\begin{pmatrix} 1 & 6 \\ -3 & 5 \end{pmatrix} = (2 + 21, 12 - 35) = (23, -23)$$

2.51 Find $\begin{pmatrix} 1 & 6 \\ -3 & 5 \end{pmatrix}(2, -7)$.

▌ The product is not defined, since the first factor is 2×2 and the second factor is 1×2.

2.52 Let A be an $m \times n$ matrix, with $m > 1$ and $n > 1$. Assuming u and v are vectors, discuss the conditions under which (a) Au, (b) vA is defined.

▌ (a) The product Au is defined only when u is a column vector with n components; i.e., an $n \times 1$ matrix. In such case, Au is a column vector with m components. (b) The product vA is defined only when v is a row vector with m components; i.e., a $1 \times m$ matrix. In such case, vA is a row vector with n components.

2.53 Compute

$$\begin{pmatrix} 2 \\ 3 \\ -1 \end{pmatrix}(6 \quad -4 \quad 5)$$

▌ The first factor is 3×1 and the second factor is 1×3, so the product is defined as a 3×3 matrix.

$$\begin{pmatrix} 2 \\ 3 \\ -1 \end{pmatrix}(6 \quad -4 \quad 5) = \begin{pmatrix} (2)(6) & (2)(-4) & (2)(5) \\ (3)(6) & (3)(-4) & (3)(5) \\ (-1)(6) & (-1)(-4) & (-1)(5) \end{pmatrix} = \begin{pmatrix} 12 & -8 & 10 \\ 18 & -12 & 15 \\ -6 & 4 & -5 \end{pmatrix}$$

2.54 Compute

$$(6, -4, 5)\begin{pmatrix} 2 \\ 3 \\ -1 \end{pmatrix}$$

▌ The first factor is 1×3 and the second factor is 3×1, so the product is defined as a 1×1 matrix, which we frequently write as a scalar.

$$(6, -4, 5)\begin{pmatrix} 2 \\ 3 \\ -1 \end{pmatrix} = (12 - 12 - 5) = (-5) = -5$$

Problems 2.55–2.58 establish the following theorem, where we assume that all products are defined.

Theorem 2.2: Suppose that A, B, C are matrices and k is a scalar. Then:

(i) $(AB)C = A(BC)$ *associative law*
(ii) $A(B + C) = AB + AC$ *left distributive law*
(iii) $(B + C)A = BA + CA$ *right distributive law*
(iv) $k(AB) = (kA)B = A(kB)$

2.55 Prove (i) of Theorem 2.2.

▌ Let $A = (a_{ij})$, $B = (b_{jk})$, and $C = (c_{kl})$. Furthermore, let $AB = S = (s_{ik})$ and $BC = T = (t_{jl})$.

Then

$$s_{ik} = \sum_{j=1}^{m} a_{ij}b_{jk} \qquad t_{jl} = \sum_{k=1}^{n} b_{jk}c_{kl}$$

Now multiplying S by C, i.e., (AB) by C, the element in the ith row and lth column of the matrix $(AB)C$ is

$$\sum_{k=1}^{n} s_{ik}c_{kl} = \sum_{k=1}^{n} \sum_{j=1}^{m} (a_{ij}b_{jk})c_{kl}$$

On the other hand, multiplying A by T, i.e., A by BC, the element in the ith row and lth column of the matrix $A(BC)$ is

$$\sum_{j=1}^{m} a_{ij}t_{jl} = \sum_{j=1}^{m} \sum_{k=1}^{n} a_{ij}(b_{jk}c_{kl})$$

The associative law in the field of scalars implies that the two double sums are equal, proving (i).

2.56   Prove (ii) of Theorem 2.20.

▌ Let $A = (a_{ik})$, $B = (b_{kj})$, and $C = (c_{kj})$. [Since AB and AC are defined, we can use the same index k for the columns of A and the rows of B and C.] Let $D = B + C = (d_{kj})$, $E = AB = (e_{ij})$, and $F = AC = (f_{ij})$. Then

$$d_{kj} = b_{kj} + c_{kj} \qquad e_{ij} = \sum_{k=1}^{p} a_{ik}b_{kj} \qquad f_{ij} = \sum_{k=1}^{p} a_{ik}c_{kj}$$

Hence the ij-entry of the matrix $AB + AC$ is

$$e_{ij} + f_{ij} = \sum_{k=1}^{p} a_{ik}b_{kj} + \sum_{k=1}^{p} a_{ik}c_{kj} = \sum_{k=1}^{p} (a_{ik}b_{kj} + a_{ik}c_{kj}) \qquad (1)$$

On the other hand, the ij-entry of the matrix $AD = A(B + C)$ is

$$\sum_{k=1}^{p} a_{ik}d_{kj} = \sum_{k=1}^{p} a_{ik}(b_{kk} + c_{kj}) \qquad (2)$$

The right sides of *(1)* and *(2)* are equal by virtue of the distributive law in the scalar field; this proves (ii).

2.57   Prove (iii) of Theorem 2.2.

▌ The proof is as in Problem 2.56. [There is no distinction between left and right multiplication in the field of scalars.]

2.58   Prove (iv) of Theorem 2.2.

▌
$$k\left(\sum_{r} a_{ir}b_{rj}\right) = \sum_{r} (ka_{ir})b_{rj} = \sum_{r} a_{ir}(kb_{rj})$$

2.59   Display two matrices A and B such that AB and BA are defined and have the same size, but $AB \neq BA$.

▌ Let $A = \begin{pmatrix} 1 & 6 \\ -3 & 5 \end{pmatrix}$ and $B = \begin{pmatrix} 4 & 0 \\ 2 & -1 \end{pmatrix}$. Then

$$AB = \begin{pmatrix} 1 & 6 \\ -3 & 5 \end{pmatrix}\begin{pmatrix} 4 & 0 \\ 2 & -1 \end{pmatrix} = \begin{pmatrix} 4+12 & 0-6 \\ -12+10 & 0-5 \end{pmatrix} = \begin{pmatrix} 16 & -6 \\ -2 & -5 \end{pmatrix}$$

$$BA = \begin{pmatrix} 4 & 0 \\ 2 & -1 \end{pmatrix}\begin{pmatrix} 1 & 6 \\ -3 & 5 \end{pmatrix} = \begin{pmatrix} 4+0 & 24+0 \\ 2+3 & 12-5 \end{pmatrix} = \begin{pmatrix} 4 & 24 \\ 5 & 7 \end{pmatrix}$$

Matrix multiplication does not obey the commutative law.

2.60   Show that $0A = 0$ [if A is not square, the two zero matrices will be of different sizes].

▌ Each entry of $0A$ is the inner product of a zero row of 0 and a column of A, and hence is the scalar 0. Thus, $0A = 0$.

2.61   Show that $A0 = 0$.

▌ Each entry of $A0$ is the inner product of a row of A and a zero column of 0, and hence is the scalar 0. Therefore, $A0 = 0$.

2.62   Show that we can have $AB = 0$,   with $A \neq 0$ and $B \neq 0$.

▌ Let $A = \begin{pmatrix} 1 & 2 \\ 2 & 4 \end{pmatrix}$ and $B = \begin{pmatrix} 6 & 2 \\ -3 & -1 \end{pmatrix}$.   Then

$$AB = \begin{pmatrix} 1 & 2 \\ 2 & 4 \end{pmatrix}\begin{pmatrix} 6 & 2 \\ -3 & -1 \end{pmatrix} = \begin{pmatrix} 6-6 & 2-2 \\ 12-12 & 4-4 \end{pmatrix} = \begin{pmatrix} 0 & 0 \\ 0 & 0 \end{pmatrix}$$

[In other words, matrix multiplication has zero divisors.]

2.63   Show that $(A + B)(C + D) = AC + AD + BC + BD$.

▌ **Method 1.**   Using the left and then the right distributive laws,

$$(A + B)(C + D) = (A + B)C + (A + B)D = AC + BC + AD + BD = AC + AD + BC + BD$$

Method 2.   Using the right and then the left distributive laws,

$$(A + B)(C + D) = A(C + D) + B(C + D) = AC + AD + BC + BD$$

2.4   TRANSPOSE OF A MATRIX

2.64   The *transpose* of a matrix A, denoted A^T, is the matrix obtained by writing the rows of A, in order, as columns. In other words, if $A = (a_{ij})$ is an $m \times n$ matrix, then $A^T = (a_{ij}^T)$ is the $n \times m$ matrix where $a_{ij}^T = a_{ji}$,   for all i and j.   Find A^T for

$$A = \begin{pmatrix} 1 & 2 & 3 \\ 4 & -5 & -6 \end{pmatrix}$$

▌ The first and second rows of A become the first and second columns of A^T:

$$A^T = \begin{pmatrix} 1 & 4 \\ 2 & -5 \\ 3 & -6 \end{pmatrix}$$

Equivalently, the first, second, and third columns of A become the first, second, and third rows of A^T.

2.65   Transpose:

$$(a) \quad \begin{pmatrix} a_1 & a_2 & a_3 & a_4 \\ a_4 & a_3 & a_2 & a_1 \end{pmatrix} \qquad (b) \quad \begin{pmatrix} 1 & 2 & 3 \\ -4 & -4 & -4 \\ 5 & 6 & 7 \end{pmatrix}$$

▌
$$(a) \quad \begin{pmatrix} a_1 & a_4 \\ a_2 & a_3 \\ a_3 & a_2 \\ a_4 & a_1 \end{pmatrix} \qquad (b) \quad \begin{pmatrix} 1 & -4 & 5 \\ 2 & -4 & 6 \\ 3 & -4 & 7 \end{pmatrix}$$

2.66   Find u^T, v^T, w^T for the row vectors $u = (2, 4)$, $v = (1, 3, 5)$, $w = (6, 6, 6)$.

▌ The transpose of a row vector will be a column vector:

$$u^T = \begin{pmatrix} 2 \\ 4 \end{pmatrix} \qquad v^T = \begin{pmatrix} 1 \\ 3 \\ 5 \end{pmatrix} \qquad w^T = \begin{pmatrix} 6 \\ 6 \\ 6 \end{pmatrix}$$

2.67 Find the transposes of the following column vectors:

$$u = \begin{pmatrix} 1 \\ 1 \end{pmatrix} \qquad v = \begin{pmatrix} 2 \\ 4 \\ 6 \end{pmatrix} \qquad w = \begin{pmatrix} -5 \\ -6 \\ 7 \end{pmatrix}$$

❚ The transpose of a column vector will be a row vector: $u^T = (1, 1)$, $v^T = (2, 4, 6)$, $w^T = (-5, -6, 7)$.

2.68 Given $A = \begin{pmatrix} 1 & 3 & 5 \\ 6 & -7 & -8 \end{pmatrix}$,   find A^T and $(A^T)^T$.

❚ Rewrite the rows of A as columns to obtain A^T, and then rewrite the rows of A^T as columns to obtain $(A^T)^T$:

$$A^T = \begin{pmatrix} 1 & 6 \\ 3 & -7 \\ 5 & -8 \end{pmatrix} \qquad (A^T)^T = \begin{pmatrix} 1 & 3 & 5 \\ 6 & -7 & -8 \end{pmatrix}$$

Observe that $(A^T)^T = A$; see Problem 2.76.

2.69 Show that the matrices AA^T and A^TA are defined for any matrix A.

❚ If A is an $m \times n$ matrix, then A^T is an $n \times m$ matrix. Hence AA^T is defined as an $m \times m$ matrix, and A^TA is defined as an $n \times n$ matrix.

2.70 Find AA^T, where $A = \begin{pmatrix} 1 & 2 & 0 \\ 3 & -1 & 4 \end{pmatrix}$.

❚ Obtain A^T by rewriting the rows of A as columns:

$$A^T = \begin{pmatrix} 1 & 3 \\ 2 & -1 \\ 0 & 4 \end{pmatrix} \quad \text{whence} \quad AA^T = \begin{pmatrix} 1 & 2 & 0 \\ 3 & -1 & 4 \end{pmatrix}\begin{pmatrix} 1 & 3 \\ 2 & -1 \\ 0 & 4 \end{pmatrix} = \begin{pmatrix} 5 & 1 \\ 1 & 26 \end{pmatrix}$$

2.71 Find A^TA, where A is the matrix of Problem 2.70.

❚
$$A^TA = \begin{pmatrix} 1 & 3 \\ 2 & -1 \\ 0 & 4 \end{pmatrix}\begin{pmatrix} 1 & 2 & 0 \\ 3 & -1 & 4 \end{pmatrix} = \begin{pmatrix} 1+9 & 2-3 & 0+12 \\ 2-3 & 4+1 & 0-4 \\ 0+12 & 0-4 & 0+16 \end{pmatrix} = \begin{pmatrix} 10 & -1 & 12 \\ -1 & 5 & -4 \\ 12 & -4 & 16 \end{pmatrix}$$

2.72 Find $(AB)^T$, if $A = \begin{pmatrix} 1 & 2 \\ 3 & -4 \end{pmatrix}$ and $B = \begin{pmatrix} 5 & 0 \\ -6 & 7 \end{pmatrix}$.

❚
$$AB = \begin{pmatrix} 5-12 & 0+14 \\ 15+24 & 0-28 \end{pmatrix} = \begin{pmatrix} -7 & 14 \\ 39 & -28 \end{pmatrix} \qquad \text{so} \qquad (AB)^T = \begin{pmatrix} -7 & 39 \\ 14 & -28 \end{pmatrix}$$

2.73 For the matrices of Problem 2.72, find A^TB^T.

❚ We have

$$A^T = \begin{pmatrix} 1 & 3 \\ 2 & -4 \end{pmatrix} \qquad B^T = \begin{pmatrix} 5 & -6 \\ 0 & 7 \end{pmatrix}$$

Then
$$A^TB^T = \begin{pmatrix} 5+0 & -6+21 \\ 10+0 & -12-28 \end{pmatrix} = \begin{pmatrix} 5 & 15 \\ 10 & -40 \end{pmatrix}$$

Note from Problem 2.72 that $(AB)^T \neq A^TB^T$.

2.74 For the matrices of Problem 2.72, find B^TA^T.

❚
$$B^TA^T = \begin{pmatrix} 5 & -6 \\ 0 & 7 \end{pmatrix}\begin{pmatrix} 1 & 3 \\ 2 & -4 \end{pmatrix} = \begin{pmatrix} 5-12 & 15+24 \\ 0+14 & 0-28 \end{pmatrix} = \begin{pmatrix} -7 & 39 \\ 14 & -28 \end{pmatrix}$$

From Problem 2.72, $(AB)^T = B^T A^T$; see Problem 2.78.

Theorem 2.3: The transpose operation on matrices satisfies

(i) $(A + B)^T = A^T + B^T$ (iii) $(kA)^T = kA^T$ (k a scalar)

(ii) $(A^T)^T = A$ (iv) $(AB)^T = B^T A^T$

2.75   Prove Theorem 2.3(i).

▌ If $A = (a_{ij})$ and $B = (b_{ij})$, then $a_{ij} + b_{ij}$ is the ij-entry of $A + B$; hence $a_{ij} + b_{ij}$ is the ji-entry (reverse order) of $(A + B)^T$. On the other hand, a_{ij} is the ji-entry of A^T and b_{ij} is the ji-entry of B^T; so $a_{ij} + b_{ij}$ is the ji-entry of $A^T + B^T$. Thus $(A + B)^T = A^T + B^T$, since corresponding entries are equal.

2.76   Prove Theorem 2.3(ii).

▌ Obviously, a double interchange of rows and columns is equivalent to no interchange.

2.77   Prove Theorem 2.3(iii).

▌ If $A = (a_{ij})$, then ka_{ij} is the ij-entry of kA, and so ka_{ij} is the ji-entry [reverse order] of $(kA)^T$. On the other hand, a_{ij} is the ji-entry of A^T, and hence ka_{ij} is the ji-entry of kA^T. Thus $(kA)^T = kA^T$, since corresponding entries are equal.

2.78   Prove Theorem 2.3(iv).

▌ If $A = (a_{ij})$ and $B = (b_{kj})$, the ij-entry of AB is

$$a_{i1}b_{1j} + a_{i2}b_{2j} + \cdots + a_{im}b_{mj} \tag{1}$$

Thus *(1)* is the ji-entry [reverse order] of $(AB)^T$.

On the other hand, column j of B becomes row j of B^T, and row i of A becomes column i of A^T. Consequently, the ji-entry of $B^T A^T$ is

$$\begin{pmatrix} b_{ij} & b_{2j} & \cdots & b_{mj} \end{pmatrix} \begin{pmatrix} a_{i1} \\ a_{i2} \\ \cdots \\ a_{im} \end{pmatrix} = b_{ij}a_{i1} + b_{2j}a_{i2} + \cdots + b_{mj}a_{im}$$

Thus, $(AB)^T = B^T A^T$, since corresponding entries are equal.

2.5 ELEMENTARY ROW OPERATIONS; PIVOTS

2.79   Show that each of the following *elementary row operations* has an inverse operation of the same type.
$[E_1]$: Interchange the ith row and the jth row: $R_i \leftrightarrow R_j$.
$[E_2]$: Multiply the ith row by a nonzero scalar k: $R_i \to kR_i$, $k \neq 0$.
$[E_3]$: Replace the ith row by k times the jth row plus the ith row: $R_i \to kR_j + R_i$.

▌ (*a*) Interchanging the same two rows twice, we obtain the original matrix; that is, this operation is its own inverse. (*b*) Multiplying the ith row by k and then by k^{-1}, or by k^{-1} and then by k, we obtain the original matrix. In other words, the operations $R_i \to kR_i$ and $R_i \to k^{-1}R_i$ are inverses. (*c*) Applying the operation $R_i \to kR_j + R_i$, and then the operation $R_i \to -kR_j + R_i$, or applying the operation $R_i \to -kR_j + R_i$ and then the operation $R_i \to kR_j + R_i$, we obtain the original matrix. In other words, the operations $R_i \to kR_j + R_i$ and $R_i \to -kR_j + R_i$ are inverses.

2.80   Express the following row operation in terms of the elementary row operations of Problem 2.79:
$[E]$: Replace the ith row by k' times the jth row plus k (nonzero) times the ith row: $R_i \to k'R_j + kR_i$, $k \neq 0$.

▌ E is equivalent to E_2 [with parameter k] followed by E_3 [with parameter k'].

2.81 Apply the operation $R_2 \leftrightarrow R_3$ to

$$A = \begin{pmatrix} 1 & 2 & 3 & 4 \\ 5 & 6 & 7 & 8 \\ 3 & -4 & 5 & -6 \end{pmatrix}$$

$\blacksquare$

$$\begin{pmatrix} 1 & 2 & 3 & 4 \\ 3 & -4 & 5 & -6 \\ 5 & 6 & 7 & 8 \end{pmatrix}$$

2.82 Apply the operation $R_1 \rightarrow 3R_1$ to the matrix of Problem 2.81.

$\blacksquare$

$$\begin{pmatrix} 3 & 6 & 9 & 12 \\ 5 & 6 & 7 & 8 \\ 3 & -4 & 5 & -6 \end{pmatrix}$$

2.83 Apply the operation $R_3 \rightarrow -3R_1 + R_3$ to the matrix of Problem 2.81.

$\blacksquare$

$$\begin{pmatrix} 1 & 2 & 3 & 4 \\ 5 & 6 & 7 & 8 \\ 0 & -10 & -4 & -18 \end{pmatrix}$$

2.84 Matrix A is *row equivalent* to matrix B, written $A \sim B$, if B can be obtained from A by a sequence of elementary row operations. Show that row equivalence is an equivalence relation. That is, show that (a) $A \sim A$; (b) if $A \sim B$, then $B \sim A$; (c) if $A \sim B$ and $B \sim C$, then $A \sim C$.

$\blacksquare$ (a) A can be obtained from A by applying E_2 with $k = 1$. (b) If B can be obtained from A by a sequence of elementary row operations, then applying the inverse operations to B in the reverse order will yield A. [By Problem 2.79, the inverse of an elementary row operation is an elementary row operation.] (c) If B can be obtained from A by a sequence of elementary row operations, and C can be obtained from B by a sequence of elementary row operations, then applying the sequences one after the other on A will result in C.

2.85 Suppose that a_{ij} is a nonzero element in a matrix A. Show that each of the following row operations, which change the kth row of A, yields a 0 in the kj-position of A:

$$(a) \quad R_k \rightarrow (-a_{kj}/a_{ij})R_i + R_k \qquad (b) \quad R_k \rightarrow -a_{kj}R_i + a_{ij}R_k$$

[The above element a_{ij} which is used to produce 0s above and/or below it is called the *pivot* of the operations.]

$\blacksquare$ The new scalar in the kj-position of A is

$$(a) \quad (-a_{kj}/a_{ij})a_{ij} + a_{kj} = 0 \qquad (b) \quad (-a_{kj})a_{ij} + (a_{ij})a_{kj} = 0$$

2.86 Refer to Problem 2.85. Discuss the advantages, if any, of using operation (b) instead of operation (a).

$\blacksquare$ Although (a) will involve fewer arithmetic operations (Problem 2.91), fractions may occur even if all a_{ij} are integers. Operation (b), which uses only multiplication and addition, will not produce fractions if all the entries a_{ij} are integers.

2.87 Discuss the advantage of using $a_{ij} = 1$ as a pivot.

$\blacksquare$ In this case, operations (a) and (b) are the same, and no fractions will be introduced if all the a_{ij} are integers.

2.88 Produce 0s above and below the (boxed) pivot:

$$\begin{pmatrix} 1 & 3 & -4 & 5 \\ 0 & \boxed{1} & 2 & -1 \\ 0 & -2 & 3 & 4 \end{pmatrix}$$

▮ Since the second row R_2 [which contains the pivot] will not change, first write down R_2:

$$\begin{pmatrix} 0 & 1 & 2 & -1 \end{pmatrix}$$

To obtain a 0 in R_1 above the pivot, multiply the second row R_2 by -3 and add it to R_1; that is, apply the operation $R_1 \rightarrow -3R_2 + R_1$:

$$\begin{pmatrix} 0+1 & -3+3 & -6-4 & 3+5 \\ 0 & 1 & 2 & -1 \end{pmatrix} = \begin{pmatrix} 1 & 0 & -10 & 8 \\ 0 & 1 & 2 & -1 \end{pmatrix}$$

To obtain a 0 in R_3 below the pivot, multiply R_2 by 2 and add it to R_3; that is, apply the operation $R_3 \rightarrow 2R_2 + R_3$:

$$\begin{pmatrix} 1 & 0 & -10 & 8 \\ 0 & 1 & 2 & -1 \\ 0+0 & 2-2 & 4+3 & -2+4 \end{pmatrix} = \begin{pmatrix} 1 & 0 & -10 & 8 \\ 0 & 1 & 2 & -1 \\ 0 & 0 & 7 & 2 \end{pmatrix}$$

This is the required matrix.

2.89 Produce 0s below the (boxed) pivot:

$$\begin{pmatrix} \boxed{2} & 1 & -3 & 4 \\ 3 & 4 & 1 & -2 \\ 5 & -2 & 3 & 0 \end{pmatrix}$$

▮ First write down R_1, since it will not change:

$$\begin{pmatrix} 2 & 1 & -3 & 4 \end{pmatrix}$$

To obtain a 0 in R_2 below the pivot, apply the operation $R_2 \rightarrow -3R_1 + 2R_2$: that is, calculate:

$$\begin{pmatrix} 2 & 1 & -3 & 4 \\ -6+6 & -3+8 & 9+2 & -12-4 \end{pmatrix} = \begin{pmatrix} 2 & 1 & -3 & 4 \\ 0 & 5 & 11 & -16 \end{pmatrix}$$

To obtain a 0 in R_3 below the pivot, apply the operation $R_3 \rightarrow -5R_1 + 2R_3$:

$$\begin{pmatrix} 2 & 1 & -3 & 4 \\ 0 & 5 & 11 & -16 \\ -10+10 & -5-4 & 15+6 & -20+0 \end{pmatrix} = \begin{pmatrix} 2 & 1 & -3 & 4 \\ 0 & 5 & 11 & -16 \\ 0 & -9 & 21 & -20 \end{pmatrix}$$

This is the required matrix.

2.90 Produce 0s below the (boxed) pivot:

$$\begin{pmatrix} \boxed{0} & 1 & -2 & 4 \\ 2 & 3 & 1 & -6 \\ 1 & -1 & 5 & 7 \end{pmatrix}$$

▮ Since the designated [boxed] element is zero, it cannot be used as a pivot.

2.91 Suppose A is an $m \times n$ matrix. Find the number of multiplications in (a) $R_k \rightarrow (-a_{kj}/a_{ij})R_i + R_k$, (b) $R_k \rightarrow -a_{kj}R_i + a_{ij}R_k$. [In computer applications, one counts only multiplications, not additions, in determining the complexity of a procedure.]

▮ (a) After calculating the multiplier $-a_{kj}/a_{ij}$, there will be only n multiplicants. [Each row contains n elements.] (b) Here there will be $2n$ multiplications.

2.6 ECHELON MATRICES, ROW REDUCTION, PIVOTING

2.92 Find the leading nonzero entries in the following matrices:

$$\begin{pmatrix} 0 & 1 & -3 & 4 & 6 \\ 4 & 0 & 2 & 5 & -3 \\ 0 & 0 & 7 & -2 & 8 \end{pmatrix} \quad \begin{pmatrix} 0 & 0 & 0 & 0 & 0 \\ 1 & 2 & 3 & 4 & 5 \\ 0 & 0 & 5 & -4 & 7 \end{pmatrix} \quad \begin{pmatrix} 0 & 2 & 2 & 2 & 2 \\ 0 & 3 & 1 & 0 & 0 \\ 0 & 0 & 0 & 0 & 0 \end{pmatrix}$$

$\blacksquare$ The leading nonzero entries are the first nonzero entries in the rows of the matrix:

$$\begin{pmatrix} 0 & \boxed{1} & -3 & 4 & 6 \\ \boxed{4} & 0 & 2 & 5 & -3 \\ 0 & 0 & \boxed{7} & -2 & 8 \end{pmatrix} \quad \begin{pmatrix} 0 & 0 & 0 & 0 & 0 \\ \boxed{1} & 2 & 3 & 4 & 5 \\ 0 & 0 & \boxed{5} & -4 & 7 \end{pmatrix} \quad \begin{pmatrix} 0 & \boxed{2} & 2 & 2 & 2 \\ 0 & \boxed{3} & 1 & 0 & 0 \\ 0 & 0 & 0 & 0 & 0 \end{pmatrix}$$

2.93 How many leading nonzero entries can an $m \times n$ matrix have?

$\blacksquare$ There are anywhere from 0 to m leading nonzero entries, one for each nonzero row.

2.94 A matrix A is called an *echelon matrix*, or is said to be in *echelon form*, if (i) any zero rows are on the bottom of the matrix; (ii) each leading nonzero entry is to the right of the leading nonzero entry in the preceding row.   Which, if any, of the matrices in Problem 2.92 are in echelon form?

$\blacksquare$ None of the three matrices is in echelon form.   [In the third matrix, the 3 is not *to the right of* the 2.]

2.95 Give an algorithm that row reduces an arbitrary $A = (a_{ij})$ to echelon form.   [The term "row reduce" or simply "reduce" shall mean to transform a matrix by row operations.]

$\blacksquare$ **Step 1.** Find the first column with a nonzero entry; call it the j_1-column.
Step 2. Interchange the rows so that a nonzero entry appears in the first row of the j_1-column; that is, so that $a_{1j_1} \neq 0$.
Step 3. Use a_{1j_1} as a pivot to obtain 0s below a_{1j_1}; that is, for each $i > 1$,   apply the row operation $R_i \to -a_{ij_1} R_1 + a_{1j_1} R_i$ or $R_i \to (-a_{ij_1}/a_{1j_1}) R_1 + R_i$.
Step 4. Repeat Steps 1, 2, and 3 with the submatrix formed by all the rows except the first.
Step 5. Continue the above process until the matrix is in echelon form.

2.96 Row reduce

$$A = \begin{pmatrix} 1 & 2 & -3 & 0 \\ 2 & 4 & -2 & 2 \\ 3 & 6 & -4 & 3 \end{pmatrix}$$

to echelon form.

$\blacksquare$ Use $a_{11} = 1$ as a pivot to obtain 0s below a_{11}; that is, apply the row operations $R_2 \to -2R_1 + R_2$ and $R_3 \to -3R_1 + R_3$ to obtain the matrix

$$\begin{pmatrix} 1 & 2 & -3 & 0 \\ 0 & 0 & 4 & 2 \\ 0 & 0 & 5 & 3 \end{pmatrix}$$

Now use $a_{23} = 4$ as a pivot to obtain a 0 below a_{23}; that is, apply the row operation $R_3 \to -5R_2 + 4R_3$ to obtain the matrix

$$\begin{pmatrix} 1 & 2 & -3 & 0 \\ 0 & 0 & 4 & 2 \\ 0 & 0 & 0 & 2 \end{pmatrix}$$

which is in echelon form.

2.97 Row reduce

$$A = \begin{pmatrix} 1 & -2 & 3 & -1 \\ 2 & -1 & 2 & 2 \\ 3 & 1 & 2 & 3 \end{pmatrix}$$

to echelon form.

▮ Apply the operations $R_2 \rightarrow -2R_1 + R_2$ and $R_3 \rightarrow -3R_1 + R_3$, and then the operation $R_3 \rightarrow -7R_2 + 3R_3$:

$$A \sim \begin{pmatrix} 1 & -2 & 3 & -1 \\ 0 & 3 & -4 & 4 \\ 0 & 7 & -7 & 6 \end{pmatrix} \sim \begin{pmatrix} 1 & -2 & 3 & -1 \\ 0 & 3 & -4 & 4 \\ 0 & 0 & 7 & -10 \end{pmatrix}$$

The matrix is now in echelon form.

2.98 Row reduce

$$A = \begin{pmatrix} 0 & 1 & 3 & -2 \\ 2 & 1 & -4 & 3 \\ 2 & 3 & 2 & -1 \end{pmatrix}$$

to echelon form.

▮ First interchange R_1 and R_2 to obtain a nonzero pivot in the first row; then apply $R_3 \rightarrow -R_1 + R_3$; and finally apply $R_3 \rightarrow -2R_2 + R_3$:

$$A \sim \begin{pmatrix} 2 & 1 & -4 & 3 \\ 0 & 1 & 3 & -2 \\ 2 & 3 & 2 & -1 \end{pmatrix} \sim \begin{pmatrix} 2 & 1 & -4 & 3 \\ 0 & 1 & 3 & -2 \\ 0 & 2 & 6 & -4 \end{pmatrix} \sim \begin{pmatrix} 2 & 1 & -4 & 3 \\ 0 & 1 & 3 & -2 \\ 0 & 0 & 0 & 0 \end{pmatrix}$$

The matrix is now in echelon form.

2.99 Row reduce

$$A = \begin{pmatrix} -4 & 1 & -6 \\ 1 & 2 & -5 \\ 6 & 3 & -4 \end{pmatrix}$$

to echelon form.

▮ Hand calculations are usually simpler if the pivot element equals 1. Therefore, first interchange R_1 and R_2; then apply $R_2 \rightarrow 4R_1 + R_2$ and $R_3 \rightarrow -6R_1 + R_3$; and then apply $R_3 \rightarrow R_2 + R_3$:

$$A \sim \begin{pmatrix} 1 & 2 & -5 \\ -4 & 1 & -6 \\ 6 & 3 & -4 \end{pmatrix} \sim \begin{pmatrix} 1 & 2 & -5 \\ 0 & 9 & -26 \\ 0 & -9 & 26 \end{pmatrix} \sim \begin{pmatrix} 1 & 2 & -5 \\ 0 & 9 & -26 \\ 0 & 0 & 0 \end{pmatrix}$$

The matrix is now in echelon form.

2.100 The algorithm of Problem 2.95 becomes the *pivoting algorithm* if, in step 2, the entry in column j, *of greatest absolute value* is chosen as the pivot a_{1j_1} and if, in step 3, the row operation $R_i \rightarrow (-a_{ij_1}/a_{1j_1})R_1 + R_i$ is specified. Use the pivoting algorithm to reduce the following matrix A to echelon form:

$$A = \begin{pmatrix} 2 & -2 & 2 & 1 \\ -3 & 6 & 0 & -1 \\ 1 & -7 & 10 & 2 \end{pmatrix}$$

▮ First interchange R_1 and R_2 so that -3 can be used as the pivot, and then apply $R_2 \rightarrow (2/3)R_1 + R_2$ and $R_3 \rightarrow (1/3)R_1 + R_3$:

$$A \sim \begin{pmatrix} -3 & 6 & 0 & -1 \\ 2 & -2 & 2 & 1 \\ 1 & -7 & 10 & 2 \end{pmatrix} \sim \begin{pmatrix} -3 & 6 & 0 & -1 \\ 0 & 2 & 2 & 1/3 \\ 0 & -5 & 10 & 5/3 \end{pmatrix}$$

Now interchange R_2 and R_3 so that -5 may be used as the pivot, and apply $R_3 \rightarrow (2/5)R_2 + R_3$:

$$A \sim \begin{pmatrix} -3 & 6 & 0 & -1 \\ 0 & -5 & 10 & 5/3 \\ 0 & 2 & 2 & 1/3 \end{pmatrix} \sim \begin{pmatrix} -3 & 6 & 0 & -1 \\ 0 & -5 & 10 & 5/3 \\ 0 & 0 & 6 & 1 \end{pmatrix}$$

The matrix has been brought to echelon form.

2.101   Describe the advantages, if any, of using the pivoting algorithm.

   ▌   The row operation $R_i \rightarrow (-a_{ij_1}/a_{1j_1})R_1 + R_i$ involves division by the [current] pivot a_{1j_1}. On the computer, roundoff errors may be substantially reduced when one divides by a number as large in absolute value as possible.

2.102   If A and B are echelon matrices with the same size, show that the sum $A + B$ need not be an echelon matrix.

   ▌

$$\begin{pmatrix} 3 & 4 & 5 \\ 0 & 2 & 1 \end{pmatrix} + \begin{pmatrix} -3 & -4 & 1 \\ 0 & 1 & -3 \end{pmatrix} = \begin{pmatrix} 0 & 0 & 6 \\ 0 & 3 & -2 \end{pmatrix}$$

2.103   Show that if A is an echelon matrix, then kA, for any scalar k, is also an echelon matrix.

   ▌   If $k = 0$, then kA is the zero matrix, which is in echelon form. If $k \neq 0$, then multiplying the entries of A by k does not change the positions of the zero rows, and does not change the positions of the leading nonzero entries.

2.7   ROW CANONICAL FORM, GAUSS ELIMINATION

2.104   A matrix A is said to be in *row canonical form* if (i) A is an echelon matrix; (ii) each leading nonzero entry is 1; (iii) each leading nonzero entry is the only nonzero entry in its column. Which of the following echelon matrices, whose leading nonzero entries have been boxed, are in row canonical form?

$$(a) \begin{pmatrix} \boxed{2} & 3 & 2 & 0 & 4 & 5 & -6 \\ 0 & 0 & \boxed{7} & 1 & -3 & 2 & 0 \\ 0 & 0 & 0 & 0 & 0 & \boxed{6} & 2 \\ 0 & 0 & 0 & 0 & 0 & 0 & 0 \end{pmatrix} \qquad (b) \begin{pmatrix} \boxed{1} & 2 & 3 \\ 0 & 0 & \boxed{1} \\ 0 & 0 & 0 \\ 0 & 0 & 0 \end{pmatrix}$$

$$(c) \begin{pmatrix} 0 & \boxed{1} & 3 & 0 & 0 & 4 & 0 \\ 0 & 0 & 0 & \boxed{1} & 0 & -3 & 0 \\ 0 & 0 & 0 & 0 & \boxed{1} & 2 & 0 \\ 0 & 0 & 0 & 0 & 0 & 0 & \boxed{1} \end{pmatrix}$$

   ▌   (*a*) The matrix is not in row canonical form since leading nonzero entries are not 1.   (*b*) The leading nonzero entry in the second row is not the only nonzero entry in its column; thus the matrix is not in row canonical form.   (*c*) This matrix is in row canonical form.

2.105   Which of the matrices in Problem 2.92 are in row canonical form?

   ▌   These matrices are not echelon matrices and hence automatically cannot be in row canonical form.

2.106   Which of the matrices are in row canonical form?

$$\begin{pmatrix} 1 & 2 & -3 & 0 & 1 \\ 0 & 0 & 5 & 2 & -4 \\ 0 & 0 & 0 & 7 & 3 \end{pmatrix} \qquad \begin{pmatrix} 0 & 1 & 7 & -5 & 0 \\ 0 & 0 & 0 & 0 & 1 \\ 0 & 0 & 0 & 0 & 0 \end{pmatrix} \qquad \begin{pmatrix} 1 & 0 & 5 & 0 & 2 \\ 0 & 1 & 2 & 0 & 4 \\ 0 & 0 & 0 & 1 & 7 \end{pmatrix}$$

   ▌   The second and third.

2.107   Give the *Gauss elimination algorithm* for the reduction of an arbitrary matrix A to row canonical form.

   ▌   The algorithm consists of two main steps:

Step 1.   Reduce the matrix A to an echelon form [Problem 2.95]; denote the leading nonzero entries $a_{1j_1}, a_{2j_2}, \ldots, a_{rj_r}$.

Step 2.   If $a_{rj_r} \neq 1$, multiply the last nonzero row, R_r, by $1/a_{rj_r}$. Then use $a_{rj_r} = 1$ as pivot to obtain 0s above the pivot. Repeat the process with $R_{r-1}, R_{r-2}, \ldots, R_2$. Finally, if necessary, multiply R_1 by $1/a_{1j_1}$ to make $a_{1j_1} = 1$.

The matrix is now in row canonical form. Step 2 is sometimes called *back-substitution*, since the leading nonzero entries are used as pivots in the reverse order, from the bottom up.

2.108 Put the following echelon matrix in row canonical form:

$$A = \begin{pmatrix} 2 & 3 & 4 & 5 & 6 \\ 0 & 0 & 3 & 2 & 5 \\ 0 & 0 & 0 & 0 & 4 \end{pmatrix}$$

▮ Multiply R_3 by 1/4 so that the leading nonzero entry, a_{35}, equals 1. Produce 0s above a_{35} by applying the operations $R_2 \to -5R_3 + R_2$ and $R_1 \to -6R_3 + R_1$:

$$A \sim \begin{pmatrix} 2 & 3 & 4 & 5 & 6 \\ 0 & 0 & 3 & 2 & 5 \\ 0 & 0 & 0 & 0 & 1 \end{pmatrix} \sim \begin{pmatrix} 2 & 3 & 4 & 5 & 0 \\ 0 & 0 & 3 & 2 & 0 \\ 0 & 0 & 0 & 0 & 1 \end{pmatrix}$$

Multiply R_2 by 1/3 so that the leading nonzero entry, a_{23}, equals 1. Produce a 0 above a_{23} with the operation $R_1 \to -4R_2 + R_1$:

$$A \sim \begin{pmatrix} 2 & 3 & 4 & 5 & 0 \\ 0 & 0 & 1 & 2/3 & 0 \\ 0 & 0 & 0 & 0 & 1 \end{pmatrix} \sim \begin{pmatrix} 2 & 3 & 0 & 7/3 & 0 \\ 0 & 0 & 1 & 2/3 & 0 \\ 0 & 0 & 0 & 0 & 1 \end{pmatrix}$$

Finally, multiply R_1 by 1/2 to obtain the row canonical form

$$A \sim \begin{pmatrix} 1 & 3/2 & 0 & 7/6 & 0 \\ 0 & 0 & 1 & 2/3 & 0 \\ 0 & 0 & 0 & 0 & 1 \end{pmatrix}$$

2.109 Reduce the matrix

$$B = \begin{pmatrix} 2 & 2 & -1 & 6 & 4 \\ 4 & 4 & 1 & 10 & 13 \\ 6 & 6 & 0 & 20 & 19 \end{pmatrix}$$

to row canonical form.

▮ First, reduce B to an echelon form by applying $R_2 \to -2R_1 + R_2$ and $R_3 \to -3R_1 + R_3$, and then $R_3 \to -R_2 + R_3$:

$$B \sim \begin{pmatrix} 2 & 2 & -1 & 6 & 4 \\ 0 & 0 & 3 & -2 & 5 \\ 0 & 0 & 3 & 2 & 7 \end{pmatrix} \sim \begin{pmatrix} 2 & 2 & -1 & 6 & 4 \\ 0 & 0 & 3 & -2 & 5 \\ 0 & 0 & 0 & 4 & 2 \end{pmatrix}$$

Now apply step 2 of the Gauss algorithm. Multiply R_3 by 1/4, so the pivot $b_{34} = 1$, and then apply $R_2 \to 2R_3 + R_2$ and $R_1 \to -6R_3 + R_1$:

$$B \sim \begin{pmatrix} 2 & 2 & -1 & 6 & 4 \\ 0 & 0 & 3 & -2 & 5 \\ 0 & 0 & 0 & 1 & 1/2 \end{pmatrix} \sim \begin{pmatrix} 2 & 2 & -1 & 0 & 1 \\ 0 & 0 & 3 & 0 & 6 \\ 0 & 0 & 0 & 1 & 1/2 \end{pmatrix}$$

Now multiply R_2 by 1/3, making the pivot $b_{23} = 1$, and apply $R_1 \to R_2 + R_1$:

$$B \sim \begin{pmatrix} 2 & 2 & -1 & 0 & 1 \\ 0 & 0 & 1 & 0 & 2 \\ 0 & 0 & 0 & 1 & 1/2 \end{pmatrix} \sim \begin{pmatrix} 2 & 2 & 0 & 0 & 3 \\ 0 & 0 & 1 & 0 & 2 \\ 0 & 0 & 0 & 1 & 1/2 \end{pmatrix}$$

Finally, multiply R_1 by 1/2 to obtain the row canonical form

$$B \sim \begin{pmatrix} 1 & 1 & 0 & 0 & 3/2 \\ 0 & 0 & 1 & 0 & 2 \\ 0 & 0 & 0 & 1 & 1/2 \end{pmatrix}$$

2.110 Reduce to row canonical form

$$A = \begin{pmatrix} 1 & -2 & 3 & 1 & 2 \\ 1 & 1 & 4 & -1 & 3 \\ 2 & 5 & 9 & -2 & 8 \end{pmatrix}$$

▌ First reduce A to echelon form by applying $R_2 \to -R_1 + R_2$ and $R_3 \to -2R_1 + R_3$, and then applying $R_3 \to -3R_2 + R_3$:

$$A \sim \begin{pmatrix} 1 & -2 & 3 & 1 & 2 \\ 0 & 3 & 1 & -2 & 1 \\ 0 & 9 & 3 & -4 & 4 \end{pmatrix} \sim \begin{pmatrix} 1 & -2 & 3 & 1 & 2 \\ 0 & 3 & 1 & -2 & 1 \\ 0 & 0 & 0 & 2 & 1 \end{pmatrix}$$

Now use back-substitution. Multiply R_3 by $1/2$ to obtain the pivot $a_{34} = 1$, and then apply $R_2 \to 2R_3 + R_2$ and $R_1 \to -R_3 + R_1$:

$$A \sim \begin{pmatrix} 1 & -2 & 3 & 1 & 2 \\ 0 & 3 & 1 & -2 & 1 \\ 0 & 0 & 0 & 1 & 1/2 \end{pmatrix} \sim \begin{pmatrix} 1 & -2 & 3 & 0 & 3/2 \\ 0 & 3 & 1 & 0 & 2 \\ 0 & 0 & 0 & 1 & 1/2 \end{pmatrix}$$

Now multiply R_2 by $1/3$ to obtain the pivot $a_{22} = 1$, and then apply $R_1 \to 2R_2 + R_1$:

$$A \sim \begin{pmatrix} 1 & -2 & 3 & 0 & 3/2 \\ 0 & 1 & 1/3 & 0 & 2/3 \\ 0 & 0 & 0 & 1 & 1/2 \end{pmatrix} \sim \begin{pmatrix} 1 & 0 & 11/3 & 0 & 17/6 \\ 0 & 1 & 1/3 & 0 & 2/3 \\ 0 & 0 & 0 & 1 & 1/2 \end{pmatrix}$$

Since $a_{11} = 1$, the last matrix is the desired row canonical form.

2.111 Describe the *Gauss–Jordan elimination algorithm* which reduces an arbitrary matrix A to its row canonical form.

▌ The algorithm is similar to the algorithm in Problem 2.95 except that here the algorithm first normalizes a row to obtain a unit pivot and then uses the pivot to place 0s both below and above the pivot before obtaining the next pivot.

2.112 One speaks of "an" echelon form of a matrix A, "the" row canonical form of A. Why?

▌ An arbitrary matrix A may be row equivalent to many echelon matrices. On the other hand, regardless of the algorithm that is used, a matrix A is row equivalent to a unique matrix in row canonical form. [The term "canonical" usually connotes uniqueness.]

2.113 Use Gauss–Jordan elimination to obtain *the* row canonical form of the matrix of Problem 2.110

▌ Use the leading nonzero entry $a_{11} = 1$ as pivot to put 0s below it, applying $R_2 \to -R_1 + R_2$ and $R_3 \to -2R_1 + R_3$; this yields

$$A \sim \begin{pmatrix} 1 & -2 & 3 & 1 & 2 \\ 0 & 3 & 1 & -2 & 1 \\ 0 & 9 & 3 & -4 & 4 \end{pmatrix}$$

Multiply R_2 by $1/3$ to get the pivot $a_{22} = 1$ and produce 0s below and above a_{22} by applying $R_3 \to -9R_2 + R_3$ and $R_1 \to 2R_2 + R_1$:

$$A \sim \begin{pmatrix} 1 & -2 & 3 & 1 & 2 \\ 0 & 1 & 1/3 & -2/3 & 1/3 \\ 0 & 9 & 3 & -4 & 4 \end{pmatrix} \sim \begin{pmatrix} 1 & 0 & 11/3 & -1/3 & 8/3 \\ 0 & 1 & 1/3 & -2/3 & 1/3 \\ 0 & 0 & 0 & 2 & 1 \end{pmatrix}$$

Lastly, multiply R_3 by $1/2$ to get the pivot $a_{34} = 1$ and produce 0s above a_{34} by applying $R_2 \to (2/3)R_3 + R_2$ and $R_1 \to (1/3)R_3 + R_1$:

$$A \sim \begin{pmatrix} 1 & 0 & 11/3 & -1/3 & 8/3 \\ 0 & 1 & 1/3 & -2/3 & 1/3 \\ 0 & 0 & 0 & 1 & 1/2 \end{pmatrix} \sim \begin{pmatrix} 1 & 0 & 11/3 & 0 & 17/6 \\ 0 & 1 & 1/3 & 0 & 2/3 \\ 0 & 0 & 0 & 1 & 1/2 \end{pmatrix}$$

2.114 Exhibit all the row canonical forms for 2×2 matrices.

▌

$$\begin{pmatrix} 1 & 0 \\ 0 & 1 \end{pmatrix} \qquad \begin{pmatrix} 1 & k \\ 0 & 0 \end{pmatrix} \qquad \begin{pmatrix} 0 & 0 \\ 0 & 0 \end{pmatrix}$$

where k is an arbitrary scalar.

2.115 Reduce the echelon matrix

$$C = \begin{pmatrix} 5 & -9 & 6 \\ 0 & 2 & 3 \\ 0 & 0 & 7 \end{pmatrix}$$

to row canonical form.

▌ Use back-substitution to obtain:

$$C \sim \begin{pmatrix} 5 & -9 & 6 \\ 0 & 2 & 3 \\ 0 & 0 & 1 \end{pmatrix} \sim \begin{pmatrix} 5 & -9 & 0 \\ 0 & 2 & 0 \\ 0 & 0 & 1 \end{pmatrix} \sim \begin{pmatrix} 5 & -9 & 0 \\ 0 & 1 & 0 \\ 0 & 0 & 1 \end{pmatrix} \sim \begin{pmatrix} 5 & 0 & 0 \\ 0 & 1 & 0 \\ 0 & 0 & 1 \end{pmatrix} \sim \begin{pmatrix} 1 & 0 & 0 \\ 0 & 1 & 0 \\ 0 & 0 & 1 \end{pmatrix}$$

2.116 Given an $n \times n$ echelon matrix in *triangular form*,

$$A = \begin{pmatrix} a_{11} & a_{12} & a_{13} & \cdots & a_{1,n-1} & a_{1n} \\ 0 & a_{22} & a_{23} & \cdots & a_{2,n-1} & a_{2n} \\ 0 & 0 & a_{33} & \cdots & a_{3,n-1} & a_{3n} \\ \cdots\cdots\cdots\cdots\cdots\cdots\cdots\cdots \\ 0 & 0 & 0 & \cdots & 0 & a_{nn} \end{pmatrix}$$

with all $a_{ii} \neq 0$. Find the row canonical form of A (generalization of Problem 2.115).

▌ Multiplying R_n by $1/a_{nn}$ and using the new $a_{nn} = 1$ as pivot, we obtain the matrix

$$\begin{pmatrix} a_{11} & a_{12} & a_{13} & \cdots & a_{1,n-1} & 0 \\ 0 & a_{22} & a_{23} & \cdots & a_{2,n-1} & 0 \\ 0 & 0 & a_{33} & \cdots & a_{3,n-1} & 0 \\ \cdots\cdots\cdots\cdots\cdots\cdots\cdots\cdots \\ 0 & 0 & 0 & \cdots & 0 & 1 \end{pmatrix}$$

Observe that the last column of A has been converted into a unit vector. Each succeeding back-substitution yields a new unit column vector, and the end result is

$$A \sim \begin{pmatrix} 1 & 0 & \cdots & 0 \\ 0 & 1 & \cdots & 0 \\ \cdots\cdots\cdots\cdots \\ 0 & 0 & \cdots & 1 \end{pmatrix}$$

i.e., A has the $n \times n$ *identity matrix* I as its row canonical form.

2.8 BLOCK MATRICES

2.117 A matrix A may be partitioned into a system of smaller matrices, called *blocks*, by a set of horizontal and vertical lines. The matrix A is then called a *block matrix*. Give the size of each block matrix:

(a) $\begin{pmatrix} 1 & -2 & 0 & 1 & 3 \\ 2 & 3 & 5 & 7 & -2 \\ 3 & 1 & 4 & 5 & 9 \end{pmatrix}$ (b) $\begin{pmatrix} 1 & -2 & 0 & 1 & 3 \\ 2 & 3 & 5 & 7 & -2 \\ 3 & 1 & 4 & 5 & 9 \end{pmatrix}$

(which are partitionings of the same matrix).

▌ (a) The block matrix has two rows, $\begin{pmatrix} 1 & -2 & 0 & 1 & 3 \\ 2 & 3 & 5 & 7 & -2 \end{pmatrix}$ and $(3 \quad 1 \mid 4 \quad 5 \mid 9)$, and three columns,

$$\begin{pmatrix} 1 & -2 \\ 2 & 3 \\ 3 & 1 \end{pmatrix} \qquad \begin{pmatrix} 0 & 1 \\ 5 & 7 \\ 4 & 5 \end{pmatrix} \qquad \begin{pmatrix} 3 \\ -2 \\ 9 \end{pmatrix}$$

Therefore, its size is 2×3. [There are four *block sizes*: 2×2, 2×1, 1×2, and 1×1.] (b) 3×2.

2.118 Suppose matrices A and B are partitioned into block matrices, say $A = (A_{ij})$ and $B = (B_{ij})$, where corresponding blocks A_{ij} and B_{ij} have the same size. Find the sum $A + B$.

▌ The sum $A + B$ may be obtained by adding corresponding blocks:

$$A + B = \begin{pmatrix} A_{11} + B_{11} & A_{12} + B_{12} & \cdots & A_{1n} + B_{1n} \\ A_{21} + B_{21} & A_{22} + B_{22} & \cdots & A_{2n} + B_{2n} \\ \hdotsfor{4} \\ A_{m1} + B_{m1} & A_{m2} + B_{m2} & \cdots & A_{mn} + B_{mn} \end{pmatrix}$$

The justification is that adding the corresponding blocks adds the corresponding elements of A and B.

2.119 Let a matrix A be partitioned into blocks; say, $A = (A_{ij})$. Find the scalar multiple kA.

▌
$$kA = \begin{pmatrix} kA_{11} & kA_{12} & \cdots & kA_{1n} \\ kA_{21} & kA_{22} & \cdots & kA_{2n} \\ \hdotsfor{4} \\ kA_{m1} & kA_{m2} & \cdots & kA_{mn} \end{pmatrix}$$

because multiplying each block by k effects the multiplication of each element of A by k.

2.120 Suppose matrices U and V are partitioned into blocks as follows:

$$U = \begin{pmatrix} U_{11} & U_{12} & \cdots & U_{1p} \\ U_{21} & U_{22} & \cdots & U_{2p} \\ \hdotsfor{4} \\ U_{m1} & U_{m2} & \cdots & U_{mp} \end{pmatrix} \qquad V = \begin{pmatrix} V_{11} & V_{12} & \cdots & V_{1n} \\ V_{21} & V_{22} & \cdots & V_{2n} \\ \hdotsfor{4} \\ V_{p1} & V_{22} & \cdots & V_{pn} \end{pmatrix}$$

where the number of columns of each block U_{ik} is equal to the number of rows in each block V_{kj}. Find the product UV.

▌ The product UV may be obtained by multiplying the corresponding block matrices; that is,

$$UV = \begin{pmatrix} W_{11} & W_{12} & \cdots & W_{1n} \\ W_{21} & W_{22} & \cdots & W_{2n} \\ \hdotsfor{4} \\ W_{m1} & W_{m2} & \cdots & W_{mn} \end{pmatrix} \qquad \text{where} \qquad W_{ij} = U_{i1}V_{1j} + U_{i2}V_{2j} + \cdots + U_{ip}V_{pj}$$

To convince yourself of the validity of block multiplication, consider the following computation of the $(1, 1)$-element of UV:

$$w_{11} = \overbrace{u_{11}v_{11} + u_{12}v_{21} + u_{13}v_{31} + \cdots}^{(1,1)\text{-element of } U_{11}V_{11}} + \overbrace{u_{1r}v_{r1} + v_{1,r+1}v_{r+1,1} + u_{1,r+1}v_{r+1,1} + \cdots}^{(1,1)\text{-element of } U_{12}V_{21}} + \overbrace{\cdots}$$

Thus, the partitioning of U and V merely partitions the sums defining the elements of UV.

2.121 Compute AB using block multiplication, where

$$A = \left(\begin{array}{cc|c} 1 & 2 & 1 \\ 3 & 4 & 0 \\ \hline 0 & 0 & 2 \end{array}\right) \qquad \text{and} \qquad B = \left(\begin{array}{ccc|c} 1 & 2 & 3 & 1 \\ 4 & 5 & 6 & 1 \\ \hline 0 & 0 & 0 & 1 \end{array}\right)$$

▌ Here $A = \begin{pmatrix} E & F \\ \mathbf{0}_{1 \times 2} & G \end{pmatrix}$ and $B = \begin{pmatrix} R & S \\ \mathbf{0}_{1 \times 3} & T \end{pmatrix}$ where $E, F, G, R, S,$ and T are the given blocks. Hence

$$AB = \begin{pmatrix} ER & ES + FT \\ \mathbf{0}_{1 \times 3} & GT \end{pmatrix} = \begin{pmatrix} \begin{pmatrix} 9 & 12 & 15 \\ 19 & 26 & 33 \end{pmatrix} & \begin{pmatrix} 3 \\ 7 \end{pmatrix} + \begin{pmatrix} 1 \\ 0 \end{pmatrix} \\ (0 \quad 0 \quad 0) & (2) \end{pmatrix} = \begin{pmatrix} 9 & 12 & 15 & 4 \\ 19 & 26 & 33 & 7 \\ 0 & 0 & 0 & 2 \end{pmatrix}$$

2.122 Compute CD using block multiplication, where

$$C = \left(\begin{array}{ccc|c} 1 & 0 & 0 & 1 \\ 0 & 1 & 0 & 2 \\ 0 & 0 & 1 & 3 \end{array}\right) \qquad \text{and} \qquad D = \left(\begin{array}{ccc} 1 & 0 & 0 \\ 0 & 1 & 0 \\ 0 & 0 & 0 \\ \hline 4 & 5 & 6 \end{array}\right)$$

❚ $CD = \begin{pmatrix} 1 & 0 & 0 \\ 0 & 1 & 0 \\ 0 & 0 & 1 \end{pmatrix}\begin{pmatrix} 1 & 0 & 0 \\ 0 & 1 & 0 \\ 0 & 0 & 1 \end{pmatrix} + \begin{pmatrix} 1 \\ 2 \\ 3 \end{pmatrix}(4,5,6) = \begin{pmatrix} 1 & 0 & 0 \\ 0 & 1 & 0 \\ 0 & 0 & 1 \end{pmatrix} + \begin{pmatrix} 4 & 5 & 6 \\ 8 & 10 & 12 \\ 12 & 15 & 18 \end{pmatrix} = \begin{pmatrix} 5 & 5 & 6 \\ 8 & 1 & 12 \\ 12 & 15 & 19 \end{pmatrix}$

2.123 Compute EF by block multiplication, where

$$E = \begin{pmatrix} 1 & 2 & 0 & 0 & 0 \\ 3 & 4 & 0 & 0 & 0 \\ 0 & 0 & 5 & 1 & 2 \\ 0 & 0 & 3 & 4 & 1 \end{pmatrix} \quad \text{and} \quad F = \begin{pmatrix} 3 & -2 & 0 & 0 \\ 2 & 4 & 0 & 0 \\ 0 & 0 & 1 & 2 \\ 0 & 0 & 2 & -3 \\ 0 & 0 & -4 & 1 \end{pmatrix}$$

❚ $EF = \begin{pmatrix} \begin{pmatrix} 1 & 2 \\ 3 & 4 \end{pmatrix}\begin{pmatrix} 3 & -2 \\ 2 & 4 \end{pmatrix} & \mathbf{0}_{2\times 2} \\ \mathbf{0}_{2\times 2} & \begin{pmatrix} 5 & 1 & 2 \\ 3 & 4 & 1 \end{pmatrix}\begin{pmatrix} 1 & 2 \\ 2 & -3 \\ -4 & 1 \end{pmatrix} \end{pmatrix} = \begin{pmatrix} \begin{pmatrix} 3+4 & -2+8 \\ 9+8 & -6+16 \end{pmatrix} & \mathbf{0}_{2\times 2} \\ \mathbf{0}_{2\times 2} & \begin{pmatrix} 5+2-8 & 10-3+2 \\ 3+8-4 & 6-12+1 \end{pmatrix} \end{pmatrix}$

$= \begin{pmatrix} 7 & 6 & 0 & 0 \\ 17 & 10 & 0 & 0 \\ 0 & 0 & -1 & 9 \\ 0 & 0 & 7 & -5 \end{pmatrix}$

2.124 Multiply

$$A = \begin{pmatrix} 3 & 1 & 0 & 0 \\ -2 & 1 & 0 & 0 \\ 0 & 0 & 6 & -4 \\ 0 & 0 & 2 & -3 \end{pmatrix} \quad \text{and} \quad B = \begin{pmatrix} 0 & 0 & 4 & 1 \\ 0 & 0 & -3 & -1 \\ 1 & 2 & 0 & 0 \\ -5 & 3 & 0 & 0 \end{pmatrix}$$

❚ $AB = \begin{pmatrix} \mathbf{0}_{2\times 2} & \begin{pmatrix} 3 & 1 \\ -2 & 1 \end{pmatrix}\begin{pmatrix} 4 & 1 \\ -3 & -1 \end{pmatrix} \\ \begin{pmatrix} 6 & -4 \\ 2 & -3 \end{pmatrix}\begin{pmatrix} 1 & 2 \\ -5 & 3 \end{pmatrix} & \mathbf{0}_{2\times 2} \end{pmatrix} = \begin{pmatrix} 0 & 0 & 9 & 2 \\ 0 & 0 & -11 & -3 \\ 26 & 0 & 0 & 0 \\ -17 & -5 & 0 & 0 \end{pmatrix}$

2.125 In how many ways can a 5×8 matrix be partitioned into a 3×4 block matrix?

❚ Required will be $3 - 1 = 2$ horizontal dividing lines and $4 - 1 = 3$ vertical dividing lines. Now, there are $5 - 1 = 4$ places to put the horizontal lines, and $8 - 1 = 7$ places for the vertical lines. The 2 horizontal lines can be put in 4 places in $\binom{4}{2} = 6$ ways, and the 3 vertical lines can be put in 7 places in $\binom{7}{3} = 35$ ways. Accordingly, there are $6 \cdot 35 = 210$ ways to accomplish the partitioning.

2.126 In how many ways can an $m \times n$ matrix be partitioned into an $r \times s$ block matrix?

❚ $\binom{m-1}{r-1}\binom{n-1}{s-1}$ ways (generalization of Problem 2.125).

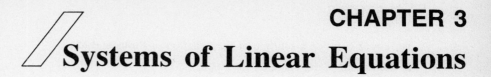

CHAPTER 3
Systems of Linear Equations

3.1 LINEARITY. SOLUTIONS

3.1 When is an equation in n unknowns, $x_1, x_2, \ldots, x_n$, said to be *linear*?

▌ When it has the following *standard form*:

$$a_1 x_1 + a_2 x_2 + \cdots + a_n x_n = b \qquad (1)$$

In *(1)*, the constant a_k is called the *coefficient of* x_k and b is called the *constant* of the equation. In the problems of this chapter, all constants are assumed to belong to **R**.

3.2 Determine whether the equation $5x + 7y - 8yz = 16$ is linear.

▌ No, since the product yz of two unknowns is of second degree.

3.3 Is the equation $x + \pi y + ez = \log 5$ linear?

▌ Yes, since π, e, and $\log 5$ are constants.

3.4 Determine whether $\dfrac{y+8}{x-2} = x + 6$ is equivalent to a linear equation.

▌ Assuming $x \neq 2$, clear the fraction:

$$y + 8 = (x-2)(x+6) = x^2 + 4x - 12 \qquad \text{or} \qquad x^2 + 4x - y = 20$$

No, since x^2 is of second degree.

3.5 Determine whether the following equation is linear: $3x + ky - 8z = 16$.

▌ As it stands, there are four unknowns: x, y, z, k. Because of the term ky it is not a linear equation. However, assuming k is a constant, the equation is linear in the unknowns x, y, z.

3.6 Is $x = 3$, $y = 2$, $z = 1$ a *solution* of the [linear] equation $x + 2y - 3z = 4$?

▌ In general, the n-tuple [vector in **R**^{n}] $u = (u_1, u_2, \ldots, u_n)$ is a solution of (or *solves* or *satisfies*) the equation $F(x_1, x_2, \ldots, x_n) = 0$ if $F(u_1, u_2, \ldots, u_n) = 0$. So we substitute in the given equation to obtain:

$$3 + 2(2) - 3(1) \overset{?}{=} 4 \qquad \text{or} \qquad 3 + 4 - 3 \overset{?}{=} 4 \qquad \text{or} \qquad 4 \overset{?}{=} 4$$

Yes, it is a solution.

3.7 Is $x = 1$, $y = 2$, $z = 3$ a solution of the equation of Problem 3.6?

▌ Substitute in the equation to obtain:

$$1 + 2(2) - 3(3) \overset{?}{=} 4 \qquad \text{or} \qquad 1 + 4 - 9 \overset{?}{=} 4 \qquad \text{or} \qquad -4 \overset{?}{=} 4$$

No, it is not a solution.

3.8 Is $u = (8, 1, 2)$ a solution of the equation $x + 2y - 3z = 4$?

▌ Since x, y, z is the ordering of the unknowns, $u = (8, 1, 2)$ is short for $x = 8$, $y = 1$, $z = 2$. Substitute in the equation to obtain:

$$8 + 2(1) - 3(2) \overset{?}{=} 4 \qquad \text{or} \qquad 8 + 2 - 6 \overset{?}{=} 4 \qquad \text{or} \qquad 4 \overset{?}{=} 4$$

Yes, it is a solution.

3.9 Is $v = (2, -1, 5)$ a solution of the equation of Problem 3.8?

▮ Substitute in the equation to obtain:

$$2 + 2(-1) - 3(5) \overset{?}{=} 4 \quad \text{or} \quad 2 - 2 - 15 \overset{?}{=} 4 \quad \text{or} \quad -15 \overset{?}{=} 4$$

No, it is not a solution.

3.10 Is $w = (3, -1, 2, 5)$ a solution of the equation of Problem 3.8?

▮ No, only a vector with 3 components can be a solution, since the equation has only three unknowns.

3.11 Is $u = (3, 2, 1, 0)$ a solution of the equation $x_1 + 2x_2 - 4x_3 + x_4 = 3$?

▮ Substitute to obtain $3 + 2(2) - 4(1) + 0 \overset{?}{=} 3$, or $3 \overset{?}{=} 3$; yes, it is a solution.

3.12 Is $v = (1, 2, 4, 5)$ a solution of the equation of Problem 3.11?

▮ Substitute to obtain $1 + 2(2) - 4(4) + 5 \overset{?}{=} 3$, or $-6 \overset{?}{=} 3$; not a solution.

3.13 Is $u = (3, -5, 2)$ a solution of the equation $y + 2x = z - 1$?

▮ We will assume that x, y, z is the ordering of the unknowns in u, regardless of their order in the equation. Thus $u = (3, -5, 2)$ is short for $x = 3$, $y = -5$, $z = 2$. Substitute in the equation to obtain:

$$-5 + 2(3) \overset{?}{=} 2 - 1 \quad \text{or} \quad -5 + 6 \overset{?}{=} 2 - 1 \quad \text{or} \quad 1 \overset{?}{=} 1$$

Yes, it is a solution.

3.14 Is $u = (6, 4, -2)$ a solution of the equation $3x_2 + x_3 - x_1 = 4$?

▮ By convention, the components of u are ordered according to the subscripts on the unknowns. That is, $u = (6, 4, -2)$ is short for $x_1 = 6$, $x_2 = 4$, $x_3 = -2$. Substitute in the equation to obtain $3(4) - 2 - 6 \overset{?}{=} 4$, or $4 \overset{?}{=} 4$. Yes, it is a solution.

Theorem 3.1: Consider the *degenerate linear equation* $0x_1 + 0x_2 + \cdots + 0x_n = b$. (i) If the constant $b \neq 0$, the equation has no solution. (ii) If the constant $b = 0$, every vector in $\mathbf{R}^n$ is a solution.

3.15 Prove (i) of Theorem 3.1.

▮ Consider any vector $u = (k_1, k_2, \ldots, k_n)$. Because $0k_i = 0$, for any i, we obtain upon substitution:

$$0k_1 + 0k_2 + \cdots + 0k_n = b \quad \text{or} \quad 0 + 0 + \cdots + 0 = b \quad \text{or} \quad 0 = b$$

This is not a true statement, since $b \neq 0$. Hence there is no vector u that is a solution.

3.16 Prove (ii) of Theorem 3.1.

▮ For any $u = (k_1, k_2, \ldots, k_n)$ in $\mathbf{R}^n$, $0k_1 + 0k_2 + \cdots + 0k = 0$; thus u is a solution.

3.17 Describe the solutions of the equation $x + 3y + x - 3 = 2y + 2x + y$.

▮ Rewrite in standard form by collecting terms and transposing:

$$2x + 3y - 3 = 2x + 3y \quad \text{or} \quad 2x + 3y - 2x - 3y = 3 \quad \text{or} \quad 0x + 0y = 3$$

The equation is degenerate with a nonzero constant; thus the equation has no solution.

3.18 Describe the solutions of the equation $4y - x - 3y + 3 = 2 + x - 2x + y + 1$.

▮ Rewrite in standard form by collecting terms and transposing:

$$y - x + 3 = y - x + 3 \quad \text{or} \quad y - x - y + x = 3 - 3 \quad \text{or} \quad 0x + 0y = 0$$

The equation is degenerate with a zero constant; thus every vector $u = (a, b)$ in $\mathbf{R}^2$ is a solution.

3.2 LINEAR EQUATIONS IN ONE UNKNOWN

Theorem 3.2: Consider the linear equation $ax = b$. (i) If $a \neq 0$, then $x = b/a$ is the unique solution. (ii) If $a = 0$ but $b \neq 0$, there is no solution. (iii) if $a = 0$ and $b = 0$, every scalar k is a solution.

3.19 Prove (i) of Theorem 3.2.

❚ Since $a \neq 0$, the scalar b/a exists. Substituting b/a in $ax = b$ yields $a(b/a) = b$, or $b = b$; hence b/a is a solution. On the other hand, suppose x_0 is a solution to $ax = b$, so that $ax_0 = b$. Multiplying both sides by $1/a$ yields $x_0 = b/a$. Hence b/a is the unique solution of $ax = b$.

3.20 Prove (ii) of Theorem 3.2.

❚ Proved by Theorem 3.1(i).

3.21 Prove (iii) of Theorem 3.2.

❚ Proved by Theorem 3.1(ii).

3.22 Solve $4x = -12$.

❚ Multiply by $1/4$ to obtain the unique solution $x = -12/4 = -3$.

3.23 Solve $5x = 0$.

❚ Multiply by $1/5$ to obtain the unique solution $x = 0/5 = 0$.

3.24 Solve $kx = \pi$.

❚ Assuming that k is a constant and $k \neq 0$, then π/k is the unique solution. If $k = 0$, there is no solution.

3.25 Solve $4x - 1 = x + 6$.

❚ Transpose to rewrite the equation in standard form: $4x - x = 6 + 1$, or $3x = 7$. Multiply by $1/3$ to obtain the unique solution $x = 7/3$.

3.26 Solve $2x - 5 - x = x + 3$.

❚ Rewrite the equation in standard form $x - 5 = x + 3$, or $x - x = 3 + 8$, or $0x = 8$. The equation has no solution [Theorem 3.2(ii)].

3.27 Solve $4 + x - 3 = 2x + 1 - x$.

❚ Rewrite the equation in standard form: $x + 1 = x + 1$, or $x - x = 1 - 1$, or $0x = 0$. Every scalar k is a solution [Theorem 3.2(iii)].

3.3 LINEAR EQUATIONS IN TWO UNKNOWNS

3.28 Determine three distinct solutions of $2x + y = 4$.

❚ Choose any value for either unknown, say $x = -2$. Substitute $x = -2$ into the equation to obtain $2(-2) + y = 4$, or $-4 + y = 4$, or $y = 8$. Thus $x = -2$, $y = 8$ or, in other words, the point $(-2, 8)$ in $\mathbf{R}^2$ is a solution. Now substitute $x = 3$ in the equation to obtain $2(3) + y = 4$, or $6 + y = 4$, or $y = -2$; hence $(3, -2)$ is a solution. Lastly, substitute $y = 0$ in the equation to obtain $2x + 0 = 4$, or $2x = 4$, or $x = 2$; Thus $(2, 0)$ is a solution.

3.29 Plot the graph of the equation $2x + y = 4$ in Problem 3.28.

❚ Plot the three solutions, $(-2, 8)$, $(3, -2)$, and $(2, 0)$, in the cartesian plane $\mathbf{R}^2$, as pictured in Fig. 3-1. Draw the straight line $\mathscr{L}$ determined by two of the solutions—say, $(-2, 8)$ and $(2, 0)$—and note that the third solution also lies on $\mathscr{L}$. Indeed, $\mathscr{L}$ is the set of *all* solutions; that is to say, $\mathscr{L}$ is the graph of the given equation.

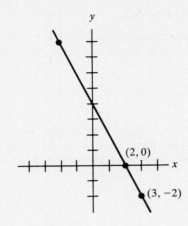

Fig. 3-1

3.30　Find three distinct solutions of　$2x - 3y = 14$.

❚　Choose any value for either unknown.　[Normally, one finds two solutions by choosing　$x = 0$　to get the x-intercept, and then　$y = 0$　to get the y-intercept.]　Substitute　$x = 0$　into the equation to obtain

$$2(0) - 3y = 14 \quad \text{or} \quad -3y = 14 \quad \text{or} \quad y = -14/3$$

Thus　$x = 0$,　$y = -14/3$　or, in other words, the pair $(0, -14/3)$ is a solution.
　　Substitute　$y = 0$　into the equation to obtain

$$2x - 3(0) = 14 \quad \text{or} \quad 2x = 14 \quad \text{or} \quad x = 7$$

Hence $(7, 0)$ is another solution.　Substitute　$x = -2$　into the equation to obtain

$$2(-2) - 3y = 14 \quad \text{or} \quad -4 - 3y = 14 \quad \text{or} \quad -3y = 18 \quad \text{or} \quad y = -6$$

Thus　$x = -2$　and　$y = -6$　or, in other words, the pair $(-2, -6)$ is a solution.

3.31　Plot the graph of the equation　$2x - 3y = 14$　in Problem 3.30.

❚　Plot the three solutions on the cartesian plane $\mathbf{R}^2$, as pictured in Fig. 3-2.　The line [cf. Problem 3.29] passing through these three points is the graph of the equation.

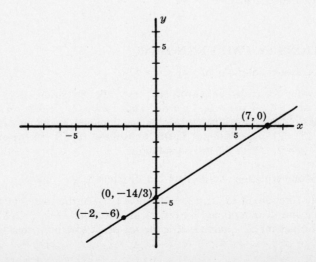

Fig. 3-2

3.32 Find three distinct solutions in $\mathbf{R}^2$ of $x = 3$, and graph the equation.

▮ As an $\mathbf{R}^2$-equation, $x = 3$ is short for $x + 0y = 3$. Here, any value of y yields $x = 3$; for example, $(3, 2)$, $(3, -4)$, $(3, 0)$ are solutions. The graph of $x = 3$ is a vertical line which intersects the x-axis at $x = 3$, as pictured in Fig. 3-3.

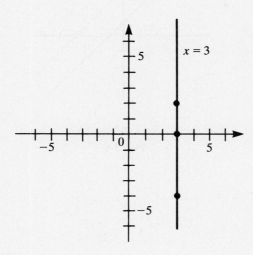

Fig. 3-3

3.33 Find three distinct solutions in $\mathbf{R}^2$ of $y = -2$, and graph the equation.

▮ As an $\mathbf{R}^2$-equation, $y = -2$ is short for $0x + y = -2$. Thus any value of x yields $y = -2$; for example, $(3, -2)$, $(-4, -2)$, $(0, -2)$ are solutions. The graph of $y = -2$ is a horizontal line which intersects the y-axis at $y = -2$, as pictured in Fig. 3-4.

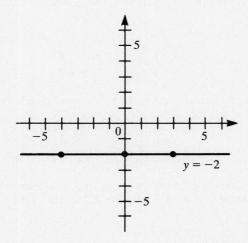

Fig. 3-4

3.34 Consider a system of two linear equations, call them $\mathbf{L}_1$ and $\mathbf{L}_2$, in unknowns x and y. A *solution* of the system is a pair $u = (k_1, k_2)$ which satisfies both equations.
(a) Describe geometrically the case where the system has a unique solution, and give an example.
(b) Describe geometrically the case where the system has no solution, and give an example.
(c) Describe geometrically the case where the system has an infinite number of solutions, and give an example.

▮ (a) The lines corresponding to the linear equations $\mathbf{L}_1$ and $\mathbf{L}_2$ intersect in a single point, as in Fig. 3-5.
(b) The lines corresponding to the linear equations $\mathbf{L}_1$ and $\mathbf{L}_2$ are parallel, as in Fig. 3-6.
(c) The lines corresponding to the linear equations $\mathbf{L}_1$ and $\mathbf{L}_2$ coincide, as in Fig. 3-7.

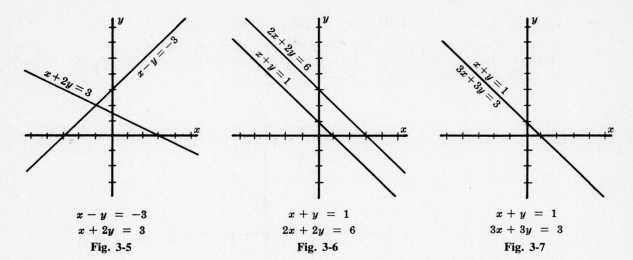

$$x - y = -3$$
$$x + 2y = 3$$
Fig. 3-5

$$x + y = 1$$
$$2x + 2y = 6$$
Fig. 3-6

$$x + y = 1$$
$$3x + 3y = 3$$
Fig. 3-7

3.35 Solve the system

$$\mathbf{L_1}: \quad 3x - 2y = 7$$
$$\mathbf{L_2}: \quad x + 2y = 1$$

▮ As the coefficients of y are the negatives of each other, add the equations:

$$3x - 2y = 7$$
$$\underline{x + 2y = 1}$$
Addition: $\quad 4x \qquad = 8 \quad$ or $\quad x = 2$

Substitute $x = 2$ into the second equation to obtain $2 + 2y = 1$, or $y = -\frac{1}{2}$. Thus $x = 2$, $y = -\frac{1}{2}$ or, in other words, the pair $(2, -\frac{1}{2})$, is the solution of the system.

3.36 Solve

$$\mathbf{L_1}: \quad 2x + 5y = \quad 8$$
$$\mathbf{L_2}: \quad 3x - 2y = -7$$

▮ To eliminate x, multiply $\mathbf{L_1}$ by 3 and multiply $\mathbf{L_2}$ by -2, and then add the resulting equations:

$$3\mathbf{L_1}: \quad 6x + 15y = 24$$
$$\underline{-2\mathbf{L_2}: \quad -6x + 4y = 14}$$
Addition: $\qquad 19y = 38 \quad$ or $\quad y = 2$

Substitute $y = 2$ into one of the original equations, say $\mathbf{L_1}$, to obtain $2x + 5(2) = 8$, or $2x + 10 = 8$, or $2x = -2$, or $x = -1$. Hence $x = -1$ and $y = 2$, or the pair $(-1, 2)$, is the unique solution to the system.

3.37 Solve Problem 3.36 by first eliminating y.

▮ Multiply $\mathbf{L_1}$ by 2 to get $4x + 10y = 16$; multiply $\mathbf{L_2}$ by 5 to get $15x - 10y = -35$; then add to get $19x = -19$, or $x = -1$. Substitute $x = -1$ in $\mathbf{L_1}$ to obtain $2(-1) + 5y = 8$, or $-2 + 5y = 8$, or $5y = 10$, or $y = 2$. Again we get $(-1, 2)$ as the solution.

3.38 Solve

$$\mathbf{L_1}: \quad 5x - 2y = \quad 8$$
$$\mathbf{L_2}: \quad 3x + 4y = 10$$

▮ To eliminate y, multiply $\mathbf{L_1}$ by 2 to get $10x - 4y = 16$; then add it to $\mathbf{L_2}$ to obtain $13x = 26$, or

$x = 2$. Substitute $x = 2$ in $\mathbf{L}_2$ to obtain $3(2) + 4y = 10$, or $6 + 4y = 10$, or $4y = 4$, or $y = 1$. Thus the pair $(2, 1)$ is the unique solution to the system.

3.39 Solve

$$\begin{aligned} \mathbf{L}_1: && x - 2y &= 5 \\ \mathbf{L}_2: && -3x + 6y &= -10 \end{aligned}$$

❚ To eliminate x, multiply $\mathbf{L}_1$ by 3 to get $3x - 6y = 15$; then add it to $\mathbf{L}_2$ to get $0x + 0y = 5$. This is a degenerate equation which has a nonzero constant; therefore, the system has no solution. [Geometrically speaking, the lines are parallel.]

3.40 Solve

$$\begin{aligned} \mathbf{L}_1: && x - 2y &= 5 \\ \mathbf{L}_2: && -3x + 6y &= -15 \end{aligned}$$

❚ To eliminate x, multiply $\mathbf{L}_1$ by 3 to get $3x - 6y = 15$; then add it to $\mathbf{L}_2$ to obtain $0x + 0y = 0$. This is a degenerate equation where the constant term is also zero. Hence the system has an infinite number of solutions, which correspond to the solutions of either equation. [Geometrically speaking, the lines coincide.] To find the *general solution*, which encompasses this infinity of *particular solutions*, let $y = a$ and substitute in $\mathbf{L}_1$ to obtain $x - 2y = 5$, or $x = 5 + 2a$. Accordingly, the general solution to the system is $(5 + 2a, a)$, where a is any real number.

3.41 Suppose $ad - bc \neq 0$ for the system

$$\begin{aligned} \mathbf{L}_1: && ax + by &= e \\ \mathbf{L}_2: && cx + dy &= f \end{aligned}$$

Show that the system has the unique solution $x = (de - bf)/(ad - bc)$, $y = (af - ce)/(ad - bc)$.

❚ Multiply $\mathbf{L}_1$ by d, multiply $\mathbf{L}_2$ by $-b$, and then add to obtain $(ab - bc)x = de - bf$. Since $ad - bc \neq 0$, we obtain the unique value $x = (de - bf)/(ad - bc)$. Now multiply $\mathbf{L}_1$ by $-c$, multiply $\mathbf{L}_2$ by a, and then add to obtain $(ad - bc)y = af - ce$. Since $ad - bc \neq 0$, we obtain the unique value $y = (af - ce)/(ad - bc)$.

3.42 Give a geometrical interpretation of the result of Problem 3.41. Assume, for simplicity, that $bd \neq 0$.

❚ The condition $ad - bc \neq 0$ can be rewritten as

$$ad \neq bc \quad \text{or} \quad \frac{a}{b} \neq \frac{c}{d} \quad \text{or} \quad -\frac{a}{b} \neq -\frac{c}{d}$$

But $-(a/b)$ is the slope of $\mathscr{L}_1$: $y = -(a/b)x + (e/b)$ and $-(c/d)$ is the slope of $\mathscr{L}_2$: $y = -(c/d)x + (f/d)$. When the slopes are distinct, the two lines must intersect in a single point.

Conversely, we see that when $ad - bc = 0$, the two lines are parallel or coincident, so that the system of Problem 3.41 has no solution.

3.4 ONE EQUATION IN MANY UNKNOWNS

This section deals further with the linear equation *(1)* of Problem 3.1.

3.43 Find the *leading unknown* and its position p in the equation

$$0x_1 + 0x_2 + 8x_3 - 4x_4 + 0x_5 - 7x_6 = 2$$

❚ By the *leading unknown* in a linear equation we mean the first unknown with a nonzero coefficient. Its position p is therefore the smallest integral value of j for which $a_j \neq 0$. Here, x_3 is the leading unknown, since $a_1 = 0$, $a_2 = 0$, but $a_3 \neq 0$. Thus $p = 3$.

3.44 Find the leading unknown and its position p in the equation $0x - 7y + 2z = 4$.

▮ The unknown y is the leading unknown and $p = 2$.

3.45 Find the leading unknown and its position p in the equation $4y - 7z = 6$.

▮ If the unknowns are x, y, z, then y is the leading unknown and $p = 2$. However, if only y and z are the unknowns, then $p = 1$.

3.46 Find the leading unknown and its position p in the equation $0x + 0y + 0z = 6$.

▮ The equation is degenerate; hence it does not have a leading unknown.

Theorem 3.3: Consider a nondegenerate linear equation $a_1 x_1 + a_2 x_2 + \cdots + a_n x_n = b$, where $n > 1$; let the leading unknown be x_p.
  (i) Any set of values for the unknowns x_j with $j \neq p$ will yield a unique solution of the equation. [The unknowns x_j are called *free variables*, since one can assign any values to them.]
  (ii) Every solution of the equation is obtained in (i). [The set of all solutions is called the *general solution* of the equation.]

3.47 Prove (i) of Theorem 3.3.

▮ Set $x_j = k_j$ for $j \neq p$. Because $a_j = 0$ for $j < p$, substitution in the equation yields

$$a_p x_p + a_{p+1} k_{p+1} + \cdots + a_n k_n = b \quad \text{or} \quad a_p x_p = b - a_{p+1} k_{p+1} - \cdots - a_n k_n$$

with $a_p \neq 0$. By Theorem 3.2(i), x_p is uniquely determined as

$$x_p = \frac{1}{a_p} (b - a_{p+1} k_{p+1} - \cdots - a_n b_n)$$

3.48 Prove (ii) of Theorem 3.3.

▮ Suppose $u = (k_1, k_2, \ldots, k_n)$ is a solution. Then

$$a_p k_p + a_{p+1} k_{p+1} + \cdots + a_n k_n = b \quad \text{or} \quad k_p = \frac{1}{a_p} (b - a_{p+1} k_{p+1} - \cdots - a_n k_n)$$

This, however, is precisely the solution

$$u = \left(k_1, \ldots, k_{p-1}, \frac{b - a_{p+1} k_{p+1} - \cdots - a_n b_n}{a_p}, k_{p+1}, \ldots, k_n \right)$$

obtained in Problem 3.47.

3.49 Find three particular solutions to the equation $2x - 4y + z = 8$.

▮ Here x is the leading unknown. Accordingly, assign any values to the free variables y and z, and then solve for x to obtain a solution. For example: (1) Set $y = 1$ and $z = 1$. Substitution in the equation yields $2x - 4(1) + 1 = 8$, or $2x - 4 + 1 = 8$, or $2x = 11$, or $x = 11/2$; thus $u_1 = (11/2, 1, 1)$ is a solution. (2) Set $y = 1$, $z = 0$. Substitution yields $x = 6$; hence $u_2 = (6, 1, 0)$ is a solution. (3) Set $y = 0$, $z = 1$. Substitution yields $x = 7/2$; thus $u_3 = (7/2, 0, 1)$ is a solution.

3.50 Find the general solution of the equation of Problem 3.49

▮ To find the general solution, assign arbitrary values to the free variables, say, $y = a$ and $z = b$. [We call a and b *parameters* of the solution.] Then substitute in the equation to obtain $2x - 4a + b = 8$, or $2x = 8 + 4a - b$, or $x = 4 + 2a - \frac{1}{2} b$. Thus $u = (4 + 2a - \frac{1}{2} b, a, b)$ is the general solution.

3.51 Find three particular solutions of the equation $0x + 3y - 4z = 5$ or, simply, $3y - 4z = 5$.

$\blacksquare$ Here y is the leading unknown, and x and z are free variables. (1) Set $x = 1$ and $y = 1$. Substitution yields $y = 1/3$; hence $u_1 = (1, 1/3, 1)$ is a solution. (2) Set $x = 0$ and $y = 1$. Substitution yields $y = 1/3$; hence $u_2 = (0, 1/3, 1)$ is a solution. (3) Set $x = 1$ and $y = 0$. Substitution yields $y = 5/3$; hence $u_3 = (1, 5/3, 0)$ is a solution.

3.52   Find the general solution of the equation of Problem 3.51.

$\blacksquare$ Set $x = a$ and $z = b$ [where a and b are parameters]. Substitute to obtain $3y - 4b = 5$, or $3y = 5 + 4b$, or $y = (5 + 4b)/3$. Thus $u = (a, (5 + 4b)/3, b)$ is the general solution.

3.5 m EQUATIONS IN n UNKNOWNS

We consider a system with standard form

$$\begin{aligned}
\mathbf{L}_1: \quad & a_{11}x_1 + a_{12}x_2 + \cdots + a_{1n}x_n = b_1 \\
\mathbf{L}_2: \quad & a_{21}x_1 + a_{22}x_2 + \cdots + a_{2n}x_n = b_b \\
& \cdots\cdots\cdots\cdots\cdots\cdots\cdots\cdots\cdots\cdots\cdots \\
\mathbf{L}_m: \quad & a_{m1}x_i + a_{m2}x_2 + \cdots + a_{mn}x_n = b_m
\end{aligned}$$

(3.1)

3.53   Find the number of unknowns in the system

$$\begin{aligned}
x + 2z &= 7 \\
3x - 5y &= 4
\end{aligned}$$

$\blacksquare$ Although each equation shows only 2 unknowns, the system has 3 unknowns, x, y and z. [We assume that there is no unknown with only zero coefficients.]

3.54   Determine whether $x_1 = -8$, $x_2 = 4$, $x_3 = 1$, $x_4 = 2$ is a solution of the system

$$\begin{aligned}
x_1 + 2x_2 - 5x_3 + 4x_4 &= 3 \\
2x_2 + 3x_2 + x_3 - 2x_4 &= 1
\end{aligned}$$

$\blacksquare$ Substitute in each equation to obtain:

(1) $-8 + 2(4) - 5(1) + 4(2) \stackrel{?}{=} 3$ or $-8 + 8 - 5 + 8 \stackrel{?}{=} 3$ or $3 \stackrel{?}{=} 3$
(2) $2(-8) + 3(4) + 1 - 2(2) \stackrel{?}{=} 1$ or $-16 + 14 + 1 - 4 \stackrel{?}{=} 1$ or $-5 \stackrel{?}{=} 3$

No, the second equation is not satisfied.

3.55   Is $u = (-8, 6, 1, 1)$ a solution of the system of Problem 3.54?

$\blacksquare$ Substitute u in each equation to obtain:

(1) $-8 + 2(6) - 5(1) + 4(1) \stackrel{?}{=} 3$ or $-8 + 12 - 5 + 4 \stackrel{?}{=} 3$ or $3 \stackrel{?}{=} 3$
(2) $2(-8) + 3(6) + 1 - 2(1) \stackrel{?}{=} 1$ or $-16 + 18 + 1 - 2 \stackrel{?}{=} 1$ or $1 \stackrel{?}{=} 1$

Yes, since it is a solution of both equations.

3.56   Is $v = (1, 2, 3, 4, 5)$ a solution of the system of Problem 3.54?

$\blacksquare$ No, a vector with 5 components cannot solve a system with only 4 unknowns.

3.57   Rewrite the following system in standard form:

$$\begin{aligned}
2x + 2y - z &= 7 \\
z + 3x - y &= 4
\end{aligned}$$

$\blacksquare$ As always, we order the unknowns as (x, y, z). Hence the standard form of the system is

$$\begin{aligned}
2x + 2y - z &= 7 \\
3x - y + z &= 4
\end{aligned}$$

3.58   Is $u = (1, 6, 7)$ a solution of the system of Problem 3.57?

❙   Yes, by substitution it is a solution.

     Problems 3.59–3.63 concern the following *elementary operations* on system *(3.1)*:

$[E_1]$ Interchange the ith equation and the jth equation: $L_i \leftrightarrow L_j$.

$[E_2]$ Multiply the ith equation by a nonzero scalar: $L_i \rightarrow kL_i$ $(k \neq 0)$.

$[E_3]$ Replace the ith equation by k times the jth equation plus the ith equation: $L_i \rightarrow kL_j + L_i$.

$[E]$ Replace the ith equation by k' times the jth equation plus k (nonzero) times the ith equation: $L_i \rightarrow k'L_j + kL_i$ $(k \neq 0)$.

3.59   Show that $[E_1]$ has an inverse operation of the same type.

❙   Interchanging the same two equations twice, we obtain the original system; that is, $L_i \rightarrow L_j$ is its own inverse.

3.60   Show that $[E_2]$ has an inverse operation of the same type.

❙   Multiplying the ith equation by $k \neq 0$ and then by k^{-1}, or by $k \neq 0$ and then k, we obtain the original system.   In other words, the operations $L_i \rightarrow kL_i$ and $L_i \rightarrow k^{-1}L_i$ are inverses.

3.61   Show that $[E_3]$ has an inverse operation of the same type.

❙   Applying the operation $L_i \rightarrow kL_j + L_i$ and then the operation $L_i \rightarrow -kL_j + L_i$,   or visa versa, we obtain the original system.   In other words, the operations $L_i \rightarrow kL_j + L_i$ and $L_i \rightarrow k^{-1}L_i$ are inverses.

3.62   Show that the effect of applying $[E]$ can be obtained by applying $[E_2]$ and then $[E_3]$.

❙   Applying $L_i \rightarrow kL_i$ and then applying $L_i \rightarrow k'L_j + L_i$ has the same result as applying the operation $L_i \rightarrow k'L_j + kL_i$.

3.63   Show that $[E]$ has an inverse operation of the same type.

❙   By Problems 3.60–3.62, applying the inverse of $[E]$ is equivalent to applying the inverse of $[E_3]$, $L_i \rightarrow -k'L_j + L_i$,   and then applying the inverse of $[E_2]$, $L_i \rightarrow k^{-1}L_i$.   Thus the desired operation is

$$L_i \rightarrow k^{-1}(-k'L_j + L_i) = (-k^{-1}k')L_j + k^{-1}L_i \qquad (k^{-1} \neq 0)$$

which has the form of $[E]$.

3.64   Apply the operation $L_2 \leftrightarrow L_3$ to

$$
\begin{aligned}
L_1: &\quad x - 2y + 3z = 5 \\
L_2: &\quad 2x + y - 4z = 1 \\
L_3: &\quad 3x + 2y - 7z = 3
\end{aligned}
$$

❙

$$
\begin{aligned}
L_1: &\quad x - 2y + 3z = 5 \\
L_3: &\quad 3x + 2y - 7z = 3 \\
L_2: &\quad 2x + y - 4z = 1
\end{aligned}
$$

3.65   Apply the operation $L_2 \rightarrow 3L_2$ to the original system in Problem 3.64.

❙

$$
\begin{aligned}
L_1: &\quad x - 2y + 3z = 5 \\
3L_2: &\quad 6x + 3y - 12z = 3 \\
L_3: &\quad 3x + 2y - 7z = 3
\end{aligned}
$$

3.66   Apply the operation $L_3 \rightarrow -3L_1 + L_3$ to the original system in Problem 3.64.

❙   Multiply L_1 by -3 and add it to L_3:

$$
\begin{array}{rl}
-3L_1: & -3x + 6y - 9z = -15 \\
L_3: & 3x + 2y - 7z = 3 \\
\hline
\text{Addition:} & 8y - 16z = -12
\end{array}
$$

This last equation replaces the third equation in the original system to yield

$$\begin{aligned}
\mathbf{L_1}: & \quad x - 2y + 3z = \quad 5 \\
\mathbf{L_2}: & \quad 2x + y - 4z = \quad 1 \\
-3\mathbf{L_1} + \mathbf{L_3}: & \qquad\qquad 8y - 16z = -12
\end{aligned}$$

[Observe that the unknown x has been eliminated from the third equation.]

3.67 Suppose that each equation $\mathbf{L}_i$ in the system *(3.1)* is multiplied by a constant c_i, and that the resulting equations are added to yield

$$(c_1 a_{11} + \cdots + c_m a_{m1})x_1 + \cdots + (c_1 a_{1n} + \cdots + c_m a_{mn})x_n = c_1 b_1 + \cdots + c_m b_m \qquad (1)$$

Such an equation is termed a *linear combination* of the equations $\mathbf{L}_i$. Show that any solution of the system *(3.1)* is also a solution of the linear combination *(1)*.

❚ Suppose $u = (k_1, k_2, \ldots, k_n)$ is a solution of *(3.1)*:

$$a_{i1}k_1 + a_{i2}k_2 + \cdots + a_{in}k_n = b_i \qquad (i = 1, \ldots, m) \qquad (2)$$

To show that u is a solution of *(1)*, we must verify the equation

$$(c_1 a_{11} + \cdots + c_m a_{m1})k_1 + \cdots + (c_1 a_{1n} + \cdots + c_m a_{mn})k_n = c_1 b_1 + \cdots + c_m b_m$$

But this can be rearranged into

$$c_1(a_{11}k_1 + \cdots + a_{1n}k_n) + \cdots + c_m(a_{m1} + \cdots + a_{mn}k_n) = c_1 b_1 + \cdots + c_m b_m$$

or, by *(2)*, $\qquad\qquad\qquad\qquad c_1 b_1 + \cdots + c_m b_m = c_1 b_1 + \cdots + c_m b_m$

which is clearly a true statement.

3.68 Suppose that a system (#) of linear equations is obtained from a system (*) of linear equations by applying a single elementary operation—$[E_1]$, $[E_2]$, or $[E_3]$. Show that (#) and (*) have all solutions in common [the two systems are *equivalent*].

❚ Each equation in (#) is a linear combination of the equations in (*). Therefore, by Problem 3.67, any solution of (*) will be a solution of all the equations in (#). In other words, the solution set of (*) is contained in the solution set of (#). On the other hand, since the operations $[E_1]$, $[E_2]$ and $[E_3]$ have inverse elementary operations, the system (*) can be obtained from (#) by a single elementary operation. Accordingly, the solution set of (#) is contained in the solution set of (*). Thus (#) and (*) have the same solutions.

3.69 Show that if a system (#) of linear equations is obtained from a system (*) of linear equations by a finite sequence of elementary operations, then (#) and (*) are equivalent systems.

❚ By Problem 3.68, each step preserves the solution set. Hence the original system (*) and the final system (#) [and any system in between] are equivalent. [This result forms the basis of the solution techniques of Sections 3.6 and 3.7.]

3.70 If a system of linear equations contains the degenerate equation

$$\mathbf{L}: \quad 0x_1 + 0x_2 + \cdots + 0x_n = b \qquad (b \neq 0)$$

what can one say about the solution set of the system?

❚ $\mathbf{L}$ does not have any solution, and hence the system does not have any solution; the solution set is empty.

3.71 If a system of linear equations contains the degenerate equation

$$\mathbf{L}: \quad 0x_1 + 0x_2 + \cdots + 0x_n = 0$$

what can one say about the solution set of the system?

❚ Every vector in $\mathbf{R}^n$ satisfies $\mathbf{L}$. Hence we may delete $\mathbf{L}$ from the system without changing its solution set.

3.6 SYSTEMS IN TRIANGULAR AND ECHELON FORM

3.72 What is meant by *triangular form*?

▮ A system of linear equations is in triangular form if the number of equations is equal to the number of unknowns and if x_k is the leading unknown of the kth equation. The paradigm is:

$$\begin{aligned}
a_{11}x_1 + a_{12}x_2 + \cdots\cdots\cdots + a_{1,n-1}x_{n-1} + a_{1n}x_n &= b_1 \\
a_{22}x_2 + \cdots\cdots\cdots + a_{2,n-1}x_{n-1} + a_{2n}x_n &= b_2 \\
\cdots\cdots\cdots\cdots\cdots\cdots\cdots\cdots\cdots & \\
a_{n-1,n-1}x_{n-1} + a_{n-1,n}x_n &= b_{n-1} \\
a_{nn}x_n &= b_n
\end{aligned}$$

(3.2)

where all $a_{kk} \neq 0$.

3.73 Describe the *back-substitution* algorithm for the unique solution of a triangular system of linear equations.

▮ The technique of back substitution is as follows. First, we solve the last equation of *(3.2)* for the last unknown, x_n:

$$x_n = \frac{b_n}{a_{nn}}$$

Second, we substitute this value for x_n in the next-to-last equation and solve it for the next-to-last unknown, x_{n-1}:

$$x_{n-1} = \frac{b_{n-1} - a_{n-1,n}(b_n/a_{nn})}{a_{n-1,n-1}}$$

Third, we substitute these values for x_n and x_{n-1} in the third-from-last equation and solve it for the third-from-last unknown, x_{n-2}:

$$x_{n-2} = \frac{b_{n-2} - (a_{n-2,n-1}/a_{n-1,n-1})[b_{n-1} - a_{n-1,n}(b_n/a_{nn})] - (a_{n-2,n}/a_{nn})b_n}{a_{a-2,n-2}}$$

In general, we determine x_k by substituting the previously obtained values of $x_n, x_{n-1}, \ldots, x_{k+1}$ in the kth equation:

$$x_k = \frac{b_k - \displaystyle\sum_{m=k+1}^{n} a_{km}x_m}{a_{kk}}$$

The process ceases when we have determined the first unknown, x_1. The solution is unique because, at each step of the algorithm, the value of x_k is uniquely determined.

3.74 Find the solution of the system

$$\begin{aligned}
2x + 4y - z &= 11 \\
5y + z &= 2 \\
3z &= -9
\end{aligned}$$

▮ The system is in triangular form, so solve by back substitution. (i) The last equation yields $z = -3$. (ii) Substitute in the second equation to obtain $5y - 3 = 2$, or $5y = 5$, or $y = 1$. (iii) Substitute $z = -3$ and $y = 1$ in the first equation to obtain $2x + 4(1) - (-3) = 11$, or $2x + 4 + 3 = 11$, or $2x = 4$, or $x = 2$. Thus the vector $u = (2, 1, -3)$ is the unique solution of the system.

3.75 Solve the system

$$\begin{aligned}
5x - 3y + 2z &= 1 \\
2y - 5z &= 2 \\
4z &= 8
\end{aligned}$$

▮ The system is in triangular form; hence solve by back substitution. (i) The last equation gives $z = 2$. (ii) Substitute in the second equation to obtain $2y - 5(2) = 2$, or $2y - 10 = 2$, or $2y = 12$, or $y = 6$. (iii) Substitute $z = 2$ and $y = 6$ in the first equation to obtain $5x - 3(6) + 2(2) = 1$, or $5x - 18 + 4 = 1$, or $5x = 15$, or $x = 3$. Thus the vector $(3, 6, 2)$ is the unique solution of the system.

3.76   Solve

$$2x - 3y + 5z - 2t = 9$$
$$5y - z + 3t = 1$$
$$7z - t = 3$$
$$2t = 8$$

▌ The system is in triangular form; hence we solve by back substitution.   (i) The last equation gives $t = 4$.   (ii) Substituting in the third equation gives $7z - 4 = 3$, or $7z = 7$, or $z = 1$.   (iii) Substituting $z = 1$ and $t = 4$ in the second equation gives $5y - 1 + 3(4) = 1$, or $5y - 1 + 12 = 1$, or $5y = -10$, or $y = -2$.   (iv) Substituting $y = -2$, $z = 1$, $t = 4$ in the first equation gives $2x - 3(-2) + 5(1) - 2(4) = 9$, or $2x + 6 + 5 - 8 = 9$, or $2x = 6$, or $x = 3$.   Thus, $x = 3$, $y = -2$, $z = 1$, $t = 4$ is the unique solution of the system.

3.77   What is meant by *echelon form*?

▌ A system of linear equations is in echelon form if no equation is degenerate and if the leading unknown in each equation is to the right of the leading unknown of the preceding equation.   The paradigm is:

$$a_{11}x_1 + a_{12}x_2 + a_{13}x_3 + \cdots\cdots\cdots\cdots + a_{1n}x_n = b_1$$
$$a_{2j_2}x_{j_2} + a_{2,j_2+1}x_{j_2+1} + \cdots\cdots\cdots\cdots + a_{2n}x_n = b_2$$
$$\cdots\cdots\cdots\cdots\cdots\cdots\cdots\cdots\cdots\cdots\cdots\cdots\cdots\cdots$$
$$a_{rj_r}x_{j_r} + a_{r,j_r+1}x_{j_r+1} + \cdots + a_{rn}x_n = b_r$$

$$(3.3)$$

where $1 < j_2 < \cdots < j_r$,   and where $a_{11} \neq 0$, $a_{2j_2} \neq 0, \ldots, a_{rj_r} \neq 0$.   Note that $r \leq n$.

3.78   Determine the *free variables* in the system

$$3x + 2y - 5z - 6s + 2t = 4$$
$$z + 8s - 3t = 6$$
$$s - 5t = 5$$

▌ In the echelon form, any unknown that is not a leading unknown is termed a free variable.   Here, y and t are the free variables.

3.79   Determine the free variables in the system

$$5x - 3y + 7z = 1$$
$$4y + 5z = 6$$
$$4z = 9$$

▌ The leading unknowns are x, y, z.   Hence there are no free variables (in any triangular system).

3.80   Determine the free variables in the system

$$x + 2y - 3z = 2$$
$$2x - 3y + z = 1$$
$$5x - 4y - z = 4$$

▌ The notion of free variable applies only to a system in echelon form.

Theorem 3.4:   The system *(3.3)* of linear equations in echelon form has a unique solution, if $r = n$;   and has one solution for each specification of the $n - r$ free variables, if $r < n$.

3.81   Prove Theorem 3.4.

▌ The proof is by induction on the number r of equations in the system.   If $r = 1$,   then we have a single, nondegenerate, linear equation, to which Theorem 3.3 applies when $n > r = 1$ and Theorem 3.2(i) applies when $n = r = 1$.   Thus the theorem holds for $r = 1$.

Now assume that $r > 1$ and that the theorem is true for a system of $r - 1$ equations. We view the $r - 1$ equations

$$a_{2j_2}x_{j_2} + a_{2,j_2+1}x_{j_2+1} + \cdots\cdots\cdots\cdots + a_{2n}x_n = b_2$$
$$\cdots\cdots\cdots\cdots\cdots\cdots\cdots\cdots\cdots\cdots\cdots\cdots\cdots\cdots$$
$$a_{rj_r}x_{j_r} + a_{r,j_r+1}x_{j_r+1} + \cdots + a_{rn}x_n = b_r$$

as a system in the unknowns $x_{j_2}, \ldots, x_n$. Note that the system is in echelon form. By the induction hypothesis, we can arbitrarily assign values to the $(n - j_2 + 1) - (r - 1)$ free variables in the reduced system to obtain a solution [say, $x_{j_2} = k_{j_2}, \ldots, x_n = k_n$]. As in case $r = 1$, these values and arbitrary values for the additional $j_2 - 2$ free variables [say, $x_2 = k_2, \ldots, x_{j_2-1} = k_{j_2-1}$], yield a solution of the first equation with

$$x_1 = \frac{1}{a_{11}}(b_1 - a_{12}k_2 - \cdots - a_{1n}k_n)$$

[Note that there are $(n - j_2 + 1) - (r - 1) + (j_2 - 2) = n - r$ free variables.] Furthermore, these values for $x_1, \ldots, x_n$ also satisfy the other equations since, in these equations, the coefficients of $x_1, \ldots, x_{j_2-1}$ are zero.

Now if $r = n$, then $j_2 = 2$. Thus by induction we obtain a unique solution of the subsystem and then a unique solution of the entire system. Accordingly, the theorem is proven.

3.82 Show how to obtain the *parametric form* of the general solution of the echelon system *(3.3)* when $n > r$.

▮ Replace the $n - r$ free variable by parameters $t_1, t_2, \ldots, t_{n-r}$. Then use back substitution to obtain values of the leading unknowns in terms of the parameters. The solution will be in the form

$$x_1 = c_1 + c_{11}t_1 + c_{12}t_2 + \cdots + c_{1,n-r}t_{n-r}$$
$$x_2 = c_2 + c_{21}t_1 + c_{22}t_{22} + \cdots + c_{2,n-r}t_{n-r}$$
$$\cdots\cdots\cdots\cdots\cdots\cdots\cdots\cdots\cdots\cdots\cdots\cdots\cdots$$
$$x_n = c_n + c_{n1}t_1 + c_{n2}t_2 + \cdots + c_{n,n-r}t_{n-r}$$

3.83 Show how to obtain the *free-variable form* of the general solution of the echelon system *(3.3)* when $n > r$.

▮ Suppose $x_{k_1}, x_{k_2}, \ldots, x_{k_{n-r}}$ are the free variables. Use back substitution to solve for the nonfree variables $x_1, x_{j_2}, \ldots, x_{j_r}$ in terms of the free variables. The solution will be in the form

$$x_1 = d_1 + d_{11}x_{k_1} + d_{12}x_{k_2} + \cdots + d_{1,n-r}x_{k_{n-r}}$$
$$x_{j_2} = d_2 + d_{21}x_{k_1} + d_{22}x_{k_2} + \cdots + d_{2,n-r}x_{k_{n-r}}$$
$$\cdots\cdots\cdots\cdots\cdots\cdots\cdots\cdots\cdots\cdots\cdots\cdots\cdots$$
$$x_{j_r} = d_r + d_{r1}x_{k_1} + d_{r2}x_{k_2} + \cdots + d_{r,n-r}x_{k_{n-r}}$$

3.84 Find three particular solutions of the system

$$x + 4y - 3z + 2t = 5$$
$$z - 4t = 2$$

▮ The system is in echelon form. The leading unknowns are x and z; hence the free variables are y and t. Accordingly, assign any values to y and t, and then solve by back substitution for x and z to obtain a solution. For example:

(1) Let $y = 1$ and $t = 1$. Substitute $t = 1$ in the last equation to obtain $z - 4 = 2$, or $z = 6$. Substitute $y = 1$, $z = 6$, $t = 1$ in the first equation to obtain $x + 4(1) - 3(6) + 2(1) = 5$, or $x + 4 - 18 + 2 = 5$, or $x = 22$. Thus $u_1 = (22, 1, 6, 1)$ is a particular solution.

(2) Let $y = 1$, $t = 0$. Substitute $t = 0$ in the last equation to get $z = 2$. Substitute $y = 1$, $z = 2$, $t = 0$ in the first equation to get $x = 6$. Thus $u_2 = (6, 1, 2, 0)$ is a particular solution.

(3) Let $y = 0$, $t = 1$. Substitute $t = 1$ in the last equation to get $z = 6$. Substitute $y = 0$, $z = 6$, $t = 1$ in the first equation to get $x = 21$. Thus $u_3 = (21, 0, 6, 1)$ is a particular solution.

3.85   Express the general solution to the system of Problem 3.84 (*a*) in the free-variable form, (*b*) in parametric form.

▎ (*a*) Use back substitution to solve for the leading unknowns *x* and *z* in terms of the free variables *y* and *t*. The last equation gives $z = 2 + 4t$.   Substitute in the first equation to obtain $x + 4y - 3(2 + 4t) + 2t = 5$, or $x + 4y - 6 - 12t + 2t = 5$,   or $x = 11 - 4y + 10t$.   Accordingly,

$$x = 11 - 4y + 10t$$
$$z = \ \ 2 + 4t$$

is the free-variable form of the general solution. (*b*) In (*a*), set $y = a$ and $t = b$ to obtain the parametric form

$$x = 11 - 4a + 10b$$
$$y = a$$
$$z = 2 + 4b$$
$$t = b$$

i.e., the solution vector $u = (11 - 4a + 10b, a, 2 + 4b, b)$.

3.86   Repeat Problem 3.85 for the echelon system

$$2x - 3y + 6z + 2s - 5t = 3$$
$$y - 4z + \ \ s \qquad = 1$$
$$s - 3t = 2$$

▎ (*a*) Use back substitution to solve for the leading unknowns *x*, *y*, and *s* in terms of the free variables *z* and *t*. The third equation gives $s = 2 + 3t$.   Substitute in the second equation to obtain $y - 4z + (2 + 3t) = 1$,   or $y = -1 + 4z - 3t$.   Substitute for *y* and *s* in the first equation to obtain $2x - 3(-1 + 4z - 3t) + 6z + 2(2 + 3t) = 3$,   or $x = -2 + 3z - 5t$.   Thus the free-variable form of the general solution is

$$x = -2 + 3z - 5t \qquad y = -1 + 4z - 3t \qquad s = 2 + 3t$$

(*b*) Set $z = a$ and $t = b$ in (*a*) to obtain the solution vector

$$u = (-2 + 3a - 5b, -1 + 4a - 3b, a, 2 + 3b, b)$$

3.87   Find three particular solutions of the system in Problem 3.86.

▎ Specialize *a* and *b* in the parametric solution.   (1) Let $a = 2$ and $b = 1$,   obtaining $u_1 = (-1, 4, 2, 5, 1)$. (2) Let $a = 1$ and $b = 0$,   obtaining $u_2 = (1, 3, 1, 2, 0)$. (3) Let $a = 0$ and $b = 1$, obtaining $u_3 = (-7, -4, 0, 5, 0)$.

3.7   GAUSSIAN ELIMINATION

3.88   Consider the system *(3.1)* of *m* equations in *n* unknowns [Section 3.5].   Describe the *Gaussian elimination algorithm* which reduces the system to echelon [possibly triangular] form, or which determines that the system has no solution.

▎ The algorithm follows.

Step 1.   Interchange equations so that the first unknown, x_1, appears with a nonzero coefficient in the first equation; i.e., arrange that $a_{11} \neq 0$.

Step 2.   Use a_{11} as a *pivot* to eliminate x_1 from all the equations except the first equation.   That is, for each $i > 1$,   apply the elementary operation [Section 3.5]

$$[E_3]: \ \ \mathbf{L}_i \to -(a_{i1}/a_{11})\mathbf{L}_1 + \mathbf{L}_i \qquad \text{or} \qquad [E]: \ \ \mathbf{L}_i \to -a_{i1}\mathbf{L}_1 + a_{11}\mathbf{L}_i$$

Step 3. Examine each new equation **L** to see if it is degenerate:

    (a) If **L** has the form $0x_1 + 0x_2 + \cdots + 0x_n = 0$, then delete **L** from the system. [See Problem 3.71]

    (b) If **L** has the form $0x_1 + 0x_2 + \cdots + 0x_n = b \neq 0$, Then EXIT from the algorithm, as the system has no solution. [See Problem 3.70.]

Step 4. Repeat Steps 1, 2, and 3 with the subsystem formed by all the equations, excluding the first equation.

Step 5. Continue the above process until the system is in echelon form or a degenerate equation is obtained in Step 3(b).

3.89 Show that Step 3(a) in the Gaussian algorithm may be replaced by:

Step 3(a'). If **L** has the form $0x_1 + 0x_2 + \cdots + 0x_n = 0$ or if **L** is a multiple of another equation, then delete **L** from the system.

❚ If $\mathbf{L} = k\mathbf{L}'$ for some other equation $\mathbf{L}'$ in the system, the operation $\mathbf{L} \rightarrow -k\mathbf{L}' + \mathbf{L}$ replaces **L** by $0x_1 + 0x_2 + \cdots + 0x_n = 0$, which will be deleted under Step 3(a). In other words, in both cases **L** is deleted from the system.

3.90 Solve the system

$$\begin{array}{rcr} 2x + y - 2z &=& 10 \\ 3x + 2y + 2z &=& 1 \\ 5x + 4y + 3z &=& 4 \end{array}$$

❚ Reduce to echelon form. To eliminate x from the second and third equations, apply the operations $\mathbf{L}_2 \rightarrow -3\mathbf{L}_1 + 2\mathbf{L}_2$ and $\mathbf{L}_3 \rightarrow -5\mathbf{L}_1 + 2\mathbf{L}_3$:

$$\begin{array}{lr} -3\mathbf{L}_1: & -6x - 3y + 6z = -30 \\ 2\mathbf{L}_2: & 6x + 4y + 4z = 2 \\ \hline -3\mathbf{L}_1 + 2\mathbf{L}_2: & y + 10z = -28 \end{array} \qquad \begin{array}{lr} -5\mathbf{L}_1: & -10x - 5y + 10z = -50 \\ 2\mathbf{L}_3: & 10x + 8y + 6z = 8 \\ \hline -5\mathbf{L}_1 + 2\mathbf{L}_3: & 3y + 16z = -42 \end{array}$$

This yields the following system, from which y is eliminated from the third equation by the operation $\mathbf{L}_3 \rightarrow -3\mathbf{L}_2 + \mathbf{L}_3$:

$$\left. \begin{array}{rcr} 2x + y - 2z &=& 10 \\ y + 10z &=& -28 \\ 3y + 16z &=& -42 \end{array} \right\} \rightarrow \left\{ \begin{array}{rcr} 2x + y - 2z &=& 10 \\ y + 10z &=& -28 \\ -14z &=& 42 \end{array} \right.$$

The system is now in triangular form, and hence has the unique solution [back substitution] $u = (1, 2, -3)$.

3.91 Solve the system

$$\begin{array}{rcr} x - 2y + z &=& 7 \\ 2x - y + 4z &=& 17 \\ 3x - 2y + 2z &=& 14 \end{array}$$

❚ Reduce to echelon form. Apply $\mathbf{L}_2 \rightarrow -2\mathbf{L}_1 + \mathbf{L}_2$ and $\mathbf{L}_3 \rightarrow -3\mathbf{L}_1 + \mathbf{L}_3$ to eliminate x from the second and third equations, and then apply $\mathbf{L}_3 \rightarrow -4\mathbf{L}_2 + 3\mathbf{L}_3$ to eliminate y from the third equation. These operations yield:

$$\left. \begin{array}{rcr} x - 2y + z &=& 7 \\ 3y + 2z &=& 3 \\ 4y - z &=& -7 \end{array} \right\} \rightarrow \left\{ \begin{array}{rcr} x - 2y + z &=& 7 \\ 3y + 2z &=& 3 \\ -11z &=& -33 \end{array} \right.$$

The system is in triangular form, and hence has the unique solution [back substitution] $u = (2, -1, 3)$.

3.92 Solve the system

$$
\begin{aligned}
x + 2y - z &= 3 \\
2x + 5y - 4z &= 5 \\
3x + 4y + 2z &= 12
\end{aligned}
$$

▌ Reduce to echelon form. Apply $L_2 \to -2L_1 + L_2$ and $L_3 \to -3L_1 + L_3$, and then $L_3 \to 2L_2 + L_3$ to obtain:

$$
\left.\begin{aligned}
x + 2y - z &= 3 \\
y - 2z &= -1 \\
-2y + 5z &= 3
\end{aligned}\right\} \rightarrow
\left\{\begin{aligned}
x + 2y - z &= 3 \\
y - 2z &= -1 \\
z &= 1
\end{aligned}\right.
$$

The system is in triangular form, and hence has the unique solution [back substitution] $u = (2, 1, 1)$.

3.93 Solve the system

$$
\begin{aligned}
2x + y - 3z &= 1 \\
5x + 2y - 6z &= 5 \\
3x - y - 4z &= 7
\end{aligned}
$$

▌ Reduce to echelon form. Apply $L_2 \to -5L_1 + 2L_2$ and $L_3 \to -3L_1 + 2L_3$, and then $L_3 \to -5L_2 + L_3$ to obtain:

$$
\left.\begin{aligned}
2x + y - 3z &= 1 \\
- y + 3z &= 5 \\
-5y + z &= 11
\end{aligned}\right\} \rightarrow
\left\{\begin{aligned}
2x + y - 3z &= 1 \\
- y + 3z &= 5 \\
- 14z &= -14
\end{aligned}\right.
$$

The system is in triangular form, and hence has the unique solution [back substitution] $u = (3, -2, 1)$.

3.94 Solve the system

$$
\begin{aligned}
2x + y - 2z &= 8 \\
3x + 2y - 4z &= 15 \\
5x + 4y - z &= 1
\end{aligned}
$$

▌ Reduce to echelon form. Apply $L_2 \to -3L_1 + 2L_2$ and $L_3 \to -5L_1 + 2L_3$, and then $L_3 \to -3L_2 + L_3$ to obtain:

$$
\left.\begin{aligned}
2x + y - 2z &= 8 \\
y - 2z &= 6 \\
3y + 8z &= -38
\end{aligned}\right\} \rightarrow
\left\{\begin{aligned}
2x + y - 2z &= 8 \\
y - 2z &= 6 \\
14z &= -56
\end{aligned}\right.
$$

The system is in triangular form, and hence has the unique solution [back substitution] $u = (1, -2, -4)$.

3.95 Solve the system

$$
\begin{aligned}
x + 2y - 3z &= 1 \\
2z + 5y - 8z &= 4 \\
3x + 8y - 13z &= 7
\end{aligned}
$$

▮ Reduce to echelon form. To eliminate x from the second and third equations, apply the operations $L_2 \to -2L_1 + L_2$ and $L_3 \to -3L_1 + L_3$ to obtain [using Problem 3.89]:

$$\left. \begin{array}{r} x + 2y - 3z = 1 \\ y - 2z = 2 \\ 2y - 4z = 4 \end{array} \right\} \quad \to \quad \left\{ \begin{array}{r} x + 2y - 3z = 1 \\ y - 2z = 2 \end{array} \right.$$

The system is now in echelon form, with free variable z.

To obtain the general solution in parametric form, let $z = a$ and solve by back substitution: $x = -3 - a$, $y = 2 + 2a$, $z = a$; or $u = (-3 - a, 2 + 2a, a)$.

3.96   Solve the system

$$\begin{array}{r} x + 2y - 2z = -1 \\ 3x - y + 2z = 7 \\ 5x + 3y - 4z = 2 \end{array}$$

▮ Reduce to echelon form. To eliminate x from the second and third equations, apply the operations $L_2 \to -3L_1 + L_2$ and $L_3 \to -5L_1 + L_3$ to obtain the equivalent system

$$\begin{array}{r} x + 2y - 3z = -1 \\ -7y + 11z = 10 \\ -7y + 11z = 7 \end{array}$$

The operation $L_3 \to -L_2 + L_3$ yields the degenerate equation $0 = -3$. Thus the system has no solution.

3.97   Solve the system

$$\begin{array}{r} x + 2y - 3z - 4t = 2 \\ 2x + 4y - 5z - 7t = 7 \\ -3x - 6y + 11z + 14t = 0 \end{array}$$

▮ Reduce the system to echelon form. To eliminate x from L_2 and L_3, apply the operations $L_2 \to -2L_1 + L_2$ and $L_3 \to 3L_1 + L_3$. Problem 3.89 applies, and we obtain

$$\left. \begin{array}{r} x + 2y - 3z - 4t = 2 \\ z + t = 3 \\ 2z + 2t = 6 \end{array} \right\} \quad \to \quad \left\{ \begin{array}{r} x + 2y - 3z - 4t = 2 \\ z + t = 3 \end{array} \right.$$

The system is now in echelon form, with free variables y and t. Solving for x and z, we obtain the free-variable form of the general solution: $x = 11 - 2y + t$, $z = 3 + t$.

3.98   Solve the system

$$\begin{array}{r} 2x - 5y + 3z - 4s + 2t = 4 \\ 3x - 7y + 2z - 5s + 4t = 9 \\ 5x - 10y - 5z - 4s + 7t = 22 \end{array}$$

▮ Reduce the system to echelon form. Apply the operations $L_2 \to -3L_1 + 2L_2$ and $L_3 \to -5L_1 + 2L_3$, and then $L_3 \to -5L_2 + L_3$ to obtain:

$$\left. \begin{array}{r} 2x - 5y + 3z - 4s + 2t = 4 \\ y - 5z + 2s + 2t = 6 \\ 5y - 25z + 12s + 4t = 24 \end{array} \right\} \quad \to \quad \left\{ \begin{array}{r} 2x - 5y + 3z - 4s + 2t = 4 \\ y - 5z + 2s + 2t = 6 \\ 2s - 6t = -6 \end{array} \right.$$

The system is now in echelon form. Solving for the leading unknowns, x, y, and s, in terms of the free variables, z and t, we obtain the free-variable form of the general solution:

$$x = 26 + 11z - 15t \qquad y = 12 + 5z - 8t \qquad s = -3 + 3t$$

From this follows at once the parametric form

$$x = 26 + 11a - 15b \qquad y = 12 + 5a - 8b \qquad z = a \qquad s = -3 + 3b \qquad t = b$$

3.99   Solve the system

$$x - 3y + 2z - s + 2t = 2$$
$$3x - 9y + 7z - s + 3t = 7$$
$$2x - 6y + 7z + 4s - 5t = 7$$

$\blacksquare$ Reduce the system to echelon form. Apply the operations $L_2 \rightarrow -3L_1 + L_2$ and $L_3 \rightarrow -2L_1 + L_3$, and then $L_3 \rightarrow -3L_2 + L_3$ to obtain [use Problem 3.89]

$$\left. \begin{array}{l} x - 3y + 2z - s + 2t = 2 \\ z + 2s - 3t = 1 \\ 3z + 6s - 9t = 3 \end{array} \right\} \rightarrow \left\{ \begin{array}{l} x - 3y + 2z - s + 2t = 2 \\ z + 2s - 3t = 1 \end{array} \right.$$

The system is now in echelon form. Solve for the leading unknowns, x and z, in terms of the free variables, y, s, and t, to obtain the general solution as $x = 3y + 5s - 8t$, $z = 1 - 2s + 3t$.

3.100   Solve the system

$$x + 2y - 3z + 4t = 2$$
$$2x + 5y - 2z + t = 1$$
$$5x + 12y - 7z + 6t = 7$$

$\blacksquare$ Reduce the system to echelon form. Eliminate x from the second and third equations by the operations $L_2 \rightarrow -2L_1 + L_2$ and $L_3 \rightarrow -5L_1 + L_3$; this yields the system

$$x + 2y - 3z + 4t = 2$$
$$ y + 4z - 7t = -3$$
$$ 2y + 8z - 14t = 3$$

The operation $L_3 \rightarrow -2L_2 + L_3$ yields the degenerate equation $0 = 9$. Thus the system has no solution [even though the system has more unknowns than equations].

Theorem 3.5:   Any system of linear equations has either: (i) a unique soluton, (ii) no solution, or (iii) an infinite number of solutions.

3.101   Prove Theorem 3.5.

$\blacksquare$ Applying the Gaussian elimination algorithm to the system, we can either reduce it to echelon form or determine that it has no solution. If the echelon form has no free variables, then the system has a unique solution. If the echelon form has free variables, then the system has an infinite number of solutions.

3.102   Determine the values of k so that the following system in unknowns x, y, z has: (i) a unique solution, (ii) no solution, (iii) an infinite number of solutions:

$$x - 2y = 1$$
$$x - y + kz = -2$$
$$ ky + 4z = 6$$

$\blacksquare$ Reduce the system to echelon form. Eliminate x from the second equation by the operation $L_2 \rightarrow -L_1 + L_2$, and then eliminate y from the third equation by $L_3 \rightarrow -kL_2 + L_3$; this yields

$$\left. \begin{array}{l} x - 2y = 1 \\ y + kz = -3 \\ ky + 4z = 6 \end{array} \right\} \rightarrow \left\{ \begin{array}{l} x - 2y = 1 \\ y + kz = -3 \\ (4 - k^2)z = 6 + 3k \end{array} \right.$$

The system has a unique solution if the coefficient of z in the third equation is not zero; that is, if $4 - k^2 \neq 0$. However, $4 - k^2 = 0$ if and only if $k = 2$ or $k = -2$. Hence the system has a unique solution if $k \neq 2$ and $k \neq -2$. If $k = 2$, then the third equation reduces to $0 = 12$; in which case the system has no solution. If $k = -2$, then the third equation becomes $0 = 0$, which can be deleted; the reduced system has a free variable z and hence an infinite number of solutions. Summarizing: (i) $k \neq 2$ and $k \neq -2$, (ii) $k = 2$, (iii) $k = -2$.

3.103   Determine the values of k so that the following system in unknowns x, y, z has: (i) a unique solution, (ii) no solution, (iii) an infinite number of solutions:

$$\begin{aligned} x + y - z &= 1 \\ 2x + 3y + kz &= 3 \\ x + ky + 3z &= 2 \end{aligned}$$

▐   Reduce the system to echelon form.   Eliminate x from the second and third equations by the operations $L_2 \to -2L_1 + L_2$ and $L_3 \to -L_1 + L_3$ to obtain:

$$\begin{aligned} x + y - z &= 1 \\ y + (k+2)z &= 1 \\ (k-1)y + 4z &= 1 \end{aligned}$$

To eliminate y from the third equation, apply the operation $L_3 \to -(k-1)L_2 + L_3$ to obtain:

$$\begin{aligned} x + y - z &= 1 \\ y + (k+2)z &= 1 \\ (3+k)(2-k)z &= 2-k \end{aligned}$$

The system has a unique solution if the coefficient of z in the third equation is not zero; that is, if $k \neq 2$ and $k \neq -3$.   In case $k = 2$, the third equation reduces to $0 = 0$ and the system has an infinite number of solutions [one for each value of z].   In case $k = -3$, the third equation reduces to $0 = 5$ and the system has no solution.   Summarizing: (i) $k \neq 2$ and $k \neq 3$, (ii) $k = -3$, (iii) $k = 2$.

3.104   Determine the values of k so that the following system in unknowns x, y, z has: (i) a unique solution, (ii) no solution, (iii) an infinite number of solutions:

$$\begin{aligned} kx + y + z &= 1 \\ x + ky + z &= 1 \\ x + y + kz &= 1 \end{aligned}$$

▐   First interchange L_1 and L_2 to ensure a nonzero pivot in the first equation.   Then eliminate y from L_2 and L_3 by applying $L_2 \to -L_1 + L_2$ and $L_3 \to -kL_1 + L_3$; this yields

$$\left. \begin{aligned} x + y + kz &= 1 \\ x + ky + z &= 1 \\ kx + y + z &= 1 \end{aligned} \right\} \to \left\{ \begin{aligned} x + y + kz &= 1 \\ (k-1)y + (1-k)z &= 0 \\ (1+k)y + (1-k^2)z &= 1-k \end{aligned} \right.$$

To eliminate y from the third equation, apply $L_3 \to L_2 + L_3$ to obtain

$$\begin{aligned} x + y + kz &= 1 \\ (k-1)y + (1-k)z &= 0 \\ (2-k-k^2)z &= 1-k \end{aligned}$$

The system has a unique solution if $2 - k - k^2 = (2+k)(1-k)$, the coefficient of z in L_3, is not zero; that is, if $k \neq -2$ and $k \neq 1$.   In case $k = 1$, the third and second equations reduce to $0 = 0$ and the system has an infinite number of solutions.   In case $k = -2$, the third equation reduces to $0 = 3$ and the system has no solution.   Summarizing: (i) $k \neq -2$ and $k \neq 1$, (ii) $k = -2$, (iii) $k = 1$.

3.105   What condition must be placed on $a, b,$ and c so that the following system in unknowns $x, y,$ and z has a solution?

$$\begin{aligned} x + 2y - 3z &= a \\ 2x + 6y - 11z &= b \\ x - 2y + 7z &= c \end{aligned}$$

▐   Reduce to echelon form.   Eliminating x from the second and third equation by the operations $L_2 \to -2L_1 + L_2$ and $L_3 \to -L_1 + L_3$,   we obtain the equivalent system

$$\begin{aligned} x + 2y - 3z &= a \\ 2y - 5z &= b - 2a \\ -4y + 10z &= c - a \end{aligned}$$

Eliminating y from the third equation by the operation $L_3 \rightarrow 2L_2 + L_3$, we finally obtain the equivalent system

$$x + 2y - 3z = a$$
$$2y - 5z = b - 2a$$
$$0 = c + 2b - 5a$$

The system will have no solution if $c + 2b - 5a \neq 0$. Thus the system will have at least one solution if $c + 2b - 5a = 0$, or $5a = 2b + c$. Note, in this case, that the system will have infinitely many solutions. In other words, the system cannot have a unique solution.

3.106 Prove that the following three statements about a system of linear equations are equivalent: (i) The system is *consistent* (has a solution). (ii) No linear combination of the equations is the equation

$$0x_1 + 0x_2 + \cdots + 0x_n = b \neq 0 \tag{*}$$

(iii) The system is reducible to echelon form.

❚ Suppose the system is reducible to echelon form. The echelon form has a solution, and hence the original system has a solution. Thus (iii) implies (i).

Suppose the system has a solution. By Problem 3.67, any linear combination of the equations also has a solution. But (*) has no solution; hence (*) is not a linear combination of the equations. Thus (i) implies (ii).

Finally, suppose the system is not reducible to echelon form. Then, in the gaussian algorithm, it must yield an equation of the form (*). Hence (*) is a linear combination of the equations. Thus not-(iii) implies not-(ii), or, equivalently, (ii) implies (iii).

3.107 Suppose $\mathcal{S}$ is a system of linear equations with more unknowns than equations. Show that $\mathcal{S}$ cannot have a unique solution.

❚ Reducing $\mathcal{S}$ to echelon form, we can never obtain a triangular system because $\mathcal{S}$ has more unknowns than equations. In other words, we either get an inconsistent degenerate equation, in which case $\mathcal{S}$ has no solution; or an echelon form with free variables, in which case $\mathcal{S}$ has an infinite number of solutions.

3.8 SYSTEMS OF LINEAR EQUATIONS IN MATRIX FORM

3.108 Use a matrix product to represent system *(3.1)* of Section 3.5.

❚ $AX = B$, where $A = (a_{ij})$ is the $m \times n$ *coefficient matrix*,

$$X = \begin{pmatrix} x_1 \\ x_2 \\ \cdots \\ x_n \end{pmatrix} \qquad \text{and} \qquad B = \begin{pmatrix} b_1 \\ b_2 \\ \cdots \\ b_m \end{pmatrix}$$

The block matrix

$$(A, B) = \begin{pmatrix} a_{11} & a_{12} & \cdots & a_{1n} & b_1 \\ a_{21} & a_{22} & \cdots & a_{2n} & b_2 \\ \cdots\cdots\cdots\cdots\cdots\cdots\cdots\cdots \\ a_{m1} & a_{m2} & \cdots & a_{mn} & b_m \end{pmatrix}$$

is called the *augmented matrix* of the system.

3.109 Rewrite the following system as a matrix equation: $\begin{array}{l} 2x + 3y - 4z = 7 \\ x - 2y - 5z = 3 \end{array}$.

❚
$$\begin{pmatrix} 2 & 3 & -4 \\ 1 & -2 & -5 \end{pmatrix} \begin{pmatrix} x \\ y \\ z \end{pmatrix} = \begin{pmatrix} 7 \\ 3 \end{pmatrix}$$

[Note that the size of the column of unknowns is not equal to the size of the column of constants.]

3.110   Find the augmented matrix of the system in Problem 3.109.

❚
$$\begin{pmatrix} 2 & 3 & -4 & 7 \\ 1 & -2 & -5 & 3 \end{pmatrix}$$

Observe that the elementary row operations on the augmented matrix [Problem 2.79] of a system of linear equations correspond precisely to the elementary operations on the linear equations [Section 3.5]. Accordingly, one may solve a system of linear equations by applying the Gaussian elimination algorithm to the augmented matrix rather than to the system itself.

3.111   Describe the relationship between the solvability of a system of linear equations and the echelon form of its augmented matrix.

❚   By the observation made in Problem 3.110, the system has a solution if and only if the echelon form does not have a row vector $(0, 0, \ldots, 0, b)$, where $b \neq 0$.

3.112   Describe the relationship between the solution of a system of linear equations and the row canonical form [Problem 2.104] of its augmented matrix.

❚   The row canonical form of the augmented matrix (excluding zero rows) gives the free-variable form of the solution of the system (when the system is consistent); one simply transposes the free-variable terms to the side of the constants. This comes from the fact that, in row canonical form, the coefficients of the leading unknowns are the leading nonzero entries in the matrix, which are equal to one and are the only nonzero entries in their respective columns.

3.113   Solve the system $\begin{array}{l} x - 2y - 3z = 4 \\ 2x - 3y + z = 5 \end{array}$ by use of the augmented matrix.

❚   Reduce the augmented matrix to echelon form and then to row canonical form:
$$\begin{pmatrix} 1 & -2 & -3 & 4 \\ 2 & -3 & 1 & 5 \end{pmatrix} \sim \begin{pmatrix} 1 & -2 & -3 & 4 \\ 0 & 1 & 7 & -3 \end{pmatrix} \sim \begin{pmatrix} 1 & 0 & 11 & -2 \\ 0 & 1 & 7 & -3 \end{pmatrix}$$

Thus the free-variable form of the general solution is

$$\begin{array}{ll} x \quad + 11z = -2 & \quad x = -2 - 11z \\ y + 7z = -3 & \text{or} \quad y = -3 - 7z \end{array}$$

(Note that z is the free variable.)

3.114   Solve, using the augmented matrix:

$$\begin{array}{l} x + y - 2z + 4t = 5 \\ 2x + 2y - 3z + t = 3 \\ 3x + 3y - 4z - 2t = 1 \end{array}$$

❚   Reduce the augmented matrix to echelon form and then to row canonical form:

$$\begin{pmatrix} 1 & 1 & -2 & 4 & 5 \\ 2 & 2 & -3 & 1 & 3 \\ 3 & 3 & -4 & -2 & 1 \end{pmatrix} \sim \begin{pmatrix} 1 & 1 & -2 & 4 & 5 \\ 0 & 0 & 1 & -7 & -7 \\ \cdots\cdots\cdots\cdots\cdots\cdots \end{pmatrix} \sim \begin{pmatrix} 1 & 1 & 0 & -10 & -9 \\ 0 & 0 & 1 & -7 & -7 \end{pmatrix}$$

[The third row is deleted from the second matrix since it is a multiple of the second row and will result in a zero row.]   Thus the free-variable form of the general solution of the system is as follows:

$$\begin{array}{ll} x + y \quad - 10t = -9 & \quad x = -9 - y + 10t \\ z - 7t = -7 & \text{or} \quad z = -7 + 7t \end{array}$$

Here the free variables are y and t.

3.115   Solve, using the augmented matrix:

$$\begin{array}{l} x + 2y + z = 3 \\ 2x + 5y - z = -4 \\ 3x - 2y - z = 5 \end{array}$$

❚ Reduce the augmented matrix to echelon form and then to row canonical form:

$$\begin{pmatrix} 1 & 2 & 1 & 3 \\ 2 & 5 & -1 & -4 \\ 3 & -2 & -1 & 5 \end{pmatrix} \sim \begin{pmatrix} 1 & 2 & 1 & 3 \\ 0 & 1 & -3 & -10 \\ 0 & -8 & -4 & -4 \end{pmatrix} \sim \begin{pmatrix} 1 & 2 & 1 & 3 \\ 0 & 1 & -3 & -10 \\ 0 & 0 & -28 & -84 \end{pmatrix}$$

$$\sim \begin{pmatrix} 1 & 2 & 1 & 3 \\ 0 & 1 & -3 & -10 \\ 0 & 0 & 1 & 3 \end{pmatrix} \sim \begin{pmatrix} 1 & 2 & 0 & 0 \\ 0 & 1 & 0 & -1 \\ 0 & 0 & 1 & 3 \end{pmatrix} \sim \begin{pmatrix} 1 & 0 & 0 & 2 \\ 0 & 1 & 0 & -1 \\ 0 & 0 & 1 & 3 \end{pmatrix}$$

Because the row canonical form turns out to be triangular, the solution is unique: $x = 2$, $y = -1$, $z = 3$.

3.116 Solve, using the augmented matrix:

$$\begin{aligned} x_1 + x_2 - 2x_3 + 3x_4 &= 4 \\ 2x_1 + 3x_2 + 3x_3 - x_4 &= 3 \\ 5x_1 + 7x_2 + 4x_3 + x_4 &= 5 \end{aligned}$$

❚ Reduce the augmented matrix to echelon form:

$$\begin{pmatrix} 1 & 1 & -2 & 3 & 4 \\ 2 & 3 & 3 & -1 & 3 \\ 5 & 7 & 4 & 1 & 5 \end{pmatrix} \sim \begin{pmatrix} 1 & 1 & -2 & 3 & 4 \\ 0 & 1 & 7 & -7 & -5 \\ 0 & 2 & 14 & -14 & -15 \end{pmatrix} \sim \begin{pmatrix} 1 & 1 & -2 & 3 & 4 \\ 0 & 1 & 7 & -7 & -5 \\ 0 & 0 & 0 & 0 & -5 \end{pmatrix}$$

The third row of the echelon matrix corresponds to the degenerate equation $0 = 5$; hence, the system has no solution.

3.117 Solve, using the augmented matrix:

$$\begin{aligned} x + 2y - 3z - 2s + 4t &= 1 \\ 2x + 5y - 8z - s + 6t &= 4 \\ x + 4y - 7z + 5s + 2t &= 8 \end{aligned}$$

❚ Reduce the augmented matrix to echelon form and then to row canonical form:

$$\begin{pmatrix} 1 & 2 & -3 & -2 & 4 & 1 \\ 2 & 5 & -8 & -1 & 6 & 4 \\ 1 & 4 & -7 & 5 & 2 & 8 \end{pmatrix} \sim \begin{pmatrix} 1 & 2 & -3 & -2 & 4 & 1 \\ 0 & 1 & -2 & 3 & -2 & 2 \\ 0 & 2 & -4 & 7 & -2 & 7 \end{pmatrix} \sim \begin{pmatrix} 1 & 2 & -3 & -2 & 4 & 1 \\ 0 & 1 & -2 & 3 & -2 & 2 \\ 0 & 0 & 0 & 1 & 2 & 3 \end{pmatrix}$$

$$\sim \begin{pmatrix} 1 & 2 & -3 & 0 & 8 & 7 \\ 0 & 1 & -2 & 0 & -8 & -7 \\ 0 & 0 & 0 & 1 & 2 & 3 \end{pmatrix} \sim \begin{pmatrix} 1 & 0 & 1 & 0 & 24 & 21 \\ 0 & 1 & -2 & 0 & -8 & -7 \\ 0 & 0 & 0 & 1 & 2 & 3 \end{pmatrix}$$

Thus the free-variable form of the solution is

$$\begin{aligned} x + z + 24t &= 21 \\ y - 2z - 8t &= -7 \qquad \text{or} \\ s + 2t &= 3 \end{aligned} \qquad \begin{aligned} x &= 21 - z + 24t \\ y &= -7 + 2z + 8t \\ s &= 3 \end{aligned}$$

3.118 The *rank* of a matrix A, written rank (A), is the number of row vectors in a maximal, linearly independent set of row vectors [see Chapter 8]. How is rank (A) related to the size of an echelon form of A?

❚ It can be shown that rank (A) equals the number of (nonzero) rows in any echelon form of A.

Theorem 3.6: A system of linear equations, $AX = B$, has a solution if and only if the rank of the coefficient matrix is equal to the rank of its augmented matrix.

3.119 Prove Theorem 3.6.

❚ The only case where rank $(A) \neq$ rank (A, B) is when the process of reducing (A, B) to echelon form produces a row $(0, 0, \ldots, 0, b)$, with $b \neq 0$. This, however, is the condition for the system to be inconsistent [see Problem 3.106].

3.9 HOMOGENEOUS SYSTEMS

3.120 Define a *homogeneous* system of linear equations.

❚ A system of linear equations is homogeneous if all constant terms are equal to zero:

$$\begin{array}{l} a_{11}x_1 + a_{12}x_2 + \cdots + a_{1n}x_n = 0 \\ a_{21}x_1 + a_{22}x_2 + \cdots + a_{2n}x_n = 0 \\ \cdots\cdots\cdots\cdots\cdots\cdots\cdots\cdots\cdots \\ a_{m1}x_1 + a_{m2}x_2 + \cdots + a_{mn}x_n = 0 \end{array} \qquad (3.4)$$

or, in matrix form, $AX = 0$.

3.121 Prove: The zero vector $\mathbf{0} = (0, 0, \ldots, 0)$ is a solution (the *zero solution*) of any homogeneous system $AX = 0$.

❚ $A\mathbf{0} = \mathbf{0}$

3.122 Prove: If $u_1, u_2, \ldots, u_q$ are solutions of a homogeneous system $AX = 0$, then any linear combination of the vectors, say $k_1u_1 + k_2u_2 + \cdots + k_qu_q$, is also a solution of $AX = 0$.

❚ Using Theorem 2.2, we have:

$$A(k_1u_1 + k_2u_2 + \cdots + k_qu_q) = A(k_1u_1) + A(k_2u_2) + \cdots + A(k_qu_q) = k_1(Au_1) + k_2(Au_2) + \cdots + k_q(Au_q)$$
$$= k_1\mathbf{0} + k_2\mathbf{0} + \cdots + k_q\mathbf{0} = \mathbf{0} + \mathbf{0} + \cdots + \mathbf{0} = \mathbf{0}$$

Problems 3.123–3.128 use the following theorem [see Problem 8.105].

Theorem 3.7: Suppose the echelon form of a homogeneous system $AX = 0$ has s free variables. Let $u_1, u_2, \ldots, u_s$ be the solutions obtained by setting one of the free variables equal to one and the remaining free variables equal to zero. Then $u_1, u_2, \ldots, u_s$ form a *basis* for the solution space $\mathcal{W}$ of $AX = 0$. [This means that any solution of the system can be expressed as a *unique* linear combination of $u_1, u_2, \ldots, u_s$; furthermore, the *dimension* of $\mathcal{W}$ is $\dim(\mathcal{W}) = s$.]

3.123 Let $\mathcal{W}$ be the solution space of the following homogeneous system:

$$\begin{array}{r} x + 3y - 2z + 5s - 3t = 0 \\ 2x + 7y - 3z + 7s - 5t = 0 \\ 3x + 11y - 4z + 10s - 9t = 0 \end{array}$$

Find the dimension and a basis for $\mathcal{W}$.

❚ Reduce the system to echelon form. Apply the operations $L_2 \to -2L_1 + L_2$ and $L_3 \to -3L_1 + L_3$, and then $L_3 \to -2L_2 + L_3$ to obtain:

$$\left.\begin{array}{r} x + 3y - 2z + 5s - 3t = 0 \\ y + z - 3s + t = 0 \\ 2y + 2z - 5s = 0 \end{array}\right\} \to \left\{\begin{array}{r} x + 3y - 2z + 5s - 3t = 0 \\ y + z - 3s + t = 0 \\ s - 2t = 0 \end{array}\right.$$

In echelon form, the system has two free variables, z and t; hence $\dim(\mathcal{W}) = 2$. A basis $\{u_1, u_2\}$ for $\mathcal{W}$ may be obtained as follows: (1) Set $z = 1$, $t = 0$. Back substitution yields $s = 0$, then $y = -1$, and then $x = 5$. Thus, $u_1 = (5, -1, 1, 0, 0)$. (2) Set $z = 0$, $t = 1$. Back substitution yields $s = 2$, then $y = 5$, and then $x = -2$. Thus, $u_2 = (-2, 5, 0, 2, 1)$.

3.124 Find the general solution of the homogeneous system of Problem 3.123.

❚ By Theorem 3.7, the general solution is the vector

$$au_1 + bu_2 = a(5, -1, 1, 0, 0) + b(-2, 5, 0, 2, 1) = (5a - 2b, -a + 5b, a, 2b, b)$$

where a and b are arbitrary constants. Observe that this is nothing other than the parametric form of the general solution under the choice of parameters $z = a$ [we set $z = 1$ to get u_1] and $t = b$ [we set $t = 1$ to get u_2].

3.125 Let $\mathcal{W}$ be the solution space of the homogeneous system

$$
\begin{aligned}
x + 2y - 3z + 2s - 4t &= 0 \\
2x + 4y - 5z + s - 6t &= 0 \\
5x + 10y - 13z + 4s - 16t &= 0
\end{aligned}
$$

Find the dimension and a basis for $\mathcal{W}$.

▮ Reduce to echelon form. Apply the operations $L_2 \to -2L_1 + L_2$ and $L_3 \to -5L_2 + L_3$, and then $L_3 \to -2L_2 + L_3$ to obtain:

$$
\left.\begin{aligned}
x + 2y - 3z + 2s - 4t &= 0 \\
z - 3s + 2t &= 0 \\
2z - 6s + 4t &= 0
\end{aligned}\right\} \to \left\{\begin{aligned}
x + 2y - 3z + 2s - 4t &= 0 \\
z - 3s + 2t &= 0
\end{aligned}\right.
$$

In echelon form, the system has three free variables, y, s and t; hence $\dim(\mathcal{W}) = 3$. A basis $\{u_1, u_2, u_3\}$ for $\mathcal{W}$ is obtained as follows: (1) Set $y = 1$, $s = 0$, $t = 0$. Back substitution yields the solution $u_1 = (-2, 2, 0, 0, 0)$. (2) Set $y = 0$, $s = 1$, $t = 0$. Back substitution yields the solution $u_2 = (7, 0, 3, 1, 0)$. (3) Set $y = 0$, $s = 0$, $t = 1$. Back substitution yields the solution $u_3 = (-2, 0, -2, 0, 1)$.

3.126 Let $\mathcal{W}$ be the solution space of the system

$$
\begin{aligned}
x + 2y - 3z &= 0 \\
2x + 5y + 2z &= 0 \\
3x - y - 4z &= 0
\end{aligned}
$$

Find the dimension and a basis for $\mathcal{W}$.

▮ Reduce the system to echelon form. Apply $L_2 \to -2L_1 + L_2$ and $L_3 \to -3L_1 + L_3$, and then $L_3 \to 7L_2 + L_3$ to obtain:

$$
\left.\begin{aligned}
x + 2y - 3z &= 0 \\
y + 8z &= 0 \\
-7y + 5z &= 0
\end{aligned}\right\} \to \left\{\begin{aligned}
x + 2y - 3z &= 0 \\
y + 8z &= 0 \\
61z &= 0
\end{aligned}\right.
$$

There are no free variables (the system is in triangular form.) Hence $\dim(\mathcal{W}) = 0$ and $\mathcal{W}$ has no basis. Specifically, $\mathcal{W}$ consists only of the zero solution, $\mathcal{W} = \{0\}$.

3.127 Let $\mathcal{W}$ be the solution space of the system

$$
\begin{aligned}
2x + 4y - 5z + 3t &= 0 \\
3x + 6y - 7z + 4t &= 0 \\
5x + 10y - 11z + 6t &= 0
\end{aligned}
$$

Find the dimension and a basis for $\mathcal{W}$.

▮ Reduce the system to echelon form. Apply $L_2 \to -3L_1 + 2L_2$ and $L_3 \to -5L_1 + 2L_3$, and then $L_3 \to -3L_2 + L_3$ to obtain:

$$
\left.\begin{aligned}
2x + 4y - 5z + 3t &= 0 \\
z - t &= 0 \\
3z - 3t &= 0
\end{aligned}\right\} \to \left\{\begin{aligned}
2x + 4y - 5z + 3t &= 0 \\
z - t &= 0
\end{aligned}\right.
$$

In echelon form, the system has two free variables, y and t; hence $\dim(\mathcal{W}) = 2$. A basis $\{u_1, u_2\}$ for $\mathcal{W}$ is obtained as follows: (1) Set $y = 1$, $t = 0$. Back substitution yields the solution $u_1 = (-2, 1, 0, 0)$. (2) Set $y = 0$, $t = 1$. Back substitution yields the solution $u_2 = (1, 0, 1, 1)$.

3.128 Let $\mathcal{W}$ be the solution space of the system

$$
\begin{aligned}
x + 2y - z &= 0 \\
2x + 5y + 2z &= 0 \\
x + 4y + 7z &= 0 \\
x + 3y + 3z &= 0
\end{aligned}
$$

Find the dimension and a basis for $\mathcal{W}$.

▌ Reduce the system to echelon form, obtaining:

$$x + 2y - z = 0 \\ y + 4z = 0 \\ 2y + 8z = 0 \\ y + 4z = 0 \quad \rightarrow \quad \begin{cases} x + 2y - z = 0 \\ y + 4z = 0 \end{cases}$$

In echelon form, there is only one free variable, z. Hence $\dim(\mathcal{W}) = 1$. To obtain a basis $\{u_1\}$ for $\mathcal{W}$, set $z = 1$. Back substitution yields $y = -4$ and then $x = 9$. Thus, $u_1 = (9, -4, 1)$.

Theorem 3.8: A homogeneous system of linear equations with more unknowns than equations has a nonzero solution.

3.129 Prove Theorem 3.8.

▌ Since 0 is a solution, the system is consistent and can be brought into echelon form. Also, in echelon form, the system will have free variables and hence a nonzero solution.

3.130 Determine whether the following homogeneous system has a nonzero solution:

$$x_1 - 2x_2 + 3x_3 - 2x_4 = 0 \\ 3x_1 - 7x_2 - 2x_3 + 4x_4 = 0 \\ 4x_1 + 3x_2 + 5x_3 + 2x_4 = 0$$

▌ Yes, by Theorem 3.8.

3.10 NONHOMOGENEOUS AND ASSOCIATED HOMOGENEOUS SYSTEMS

3.131 Define the homogeneous system associated with a given nonhomogeneous system $AX = B$.

▌ $AX = 0$

3.132 Find the homogeneous system associated with the nonhomogeneous system

$$x + 3y - 5z + 7t = 3 \\ 2x - 5y + 2z - 8t = 2 \\ 4x - 2y - 6z + 9t = 8$$

▌ Replace the constants by zeros to obtain:

$$x + 3y - 5z + 7t = 0 \\ 2x - 5y + 2z - 8t = 0 \\ 4x - 2y - 6z + 9t = 0$$

3.133 Prove: If u and v are solutions of a nonhomogeneous system $AX = B$, then the difference $w = v - u$ is a solution of the associated homogeneous system $AX = 0$.

▌ $Aw = A(v - u) = Av - Au = B - B = 0$

Theorem 3.9: The general solution of a nonhomogeneous system $AX = B$ may be obtained by adding the general solution of the homogeneous system $AX = 0$ to a particular solution v_0 of $AX = B$.

3.134 Prove Theorem 3.9.

▌ Let w be any solution of $AX = 0$; then $A(v_0 + w) = Av_0 + Aw = B + 0 = B$. That is, the sum $v_0 + w$ is a solution of $AX = B$. On the other hand, suppose v is a solution of $AX = B$. Then the identity $v = v_0 + (v - v_0)$ and Problem 3.133 show that *any* solution of $AX = B$ can be obtained by adding a solution of $AX = 0$ to the particular solution v_0 of $AX = B$.

As we shall see [Problem 3.135], the general solution given by Theorem 3.9 essentially coincides with the free-variable or parametric forms [Problem 3.85].

3.135   Consider the system

$$x - 3y - 2z + 4t = 5$$
$$3x - 8y - 3z + 8t = 18$$
$$2x - 3y + 5z - 4t = 19$$

(a) Find the parametric form of the general solution of the system.   (b) Show that the result of (a) may be rewritten in the form given by Theorem 3.9.

▌ (a) Reduce the system to echelon form.   Apply $L_2 \to -3L_1 + L_2$ and $L_3 \to -2L_1 + L_3$,   and then $L_3 \to -3L_2 + L_3$ to obtain:

$$\left. \begin{array}{r} x - 3y - 2z + 4t = 5 \\ y + 3z - 4t = 3 \\ 3y + 9z - 12t = 9 \end{array} \right\} \quad \to \quad \left\{ \begin{array}{l} x - 3y - 2z + 4t = 5 \\ \quad\quad y + 3z - 4t = 3 \end{array} \right.$$

In echelon form, the free variables are z and t.   Set $z = a$ and $t = b$,   where a and b are parameters. Back substitution yields $y = 3 - 3a + 4b$,   and then $x = 14 - 7a + 8b$.   Thus the parametric form of the solution is

$$x = 14 - 7a + 8b \qquad y = 3 - 3a + 4b \qquad z = a \qquad t = b \tag{*}$$

(b) Let $v_0 = (14, 3, 0, 0)$ be the vector of constant terms in (*), let $u_1 = (-7, 3, 1, 0)$ be the vector of coefficients of a in (*), and let $u_2 = (8, 4, 0, 1)$ be the vector of coefficients of b in (*).   Then the general solution (*) may be rewritten in vector form as

$$(x, y, z, t) = v_0 + au_1 + bu_2 \tag{**}$$

We next show that (**) is the general solution per Theorem 3.9.   First note that v_0 is the solution of the inhomogeneous system obtained by setting $a = 0$ and $b = 0$.   Consider the associated homogeneous system, in echelon form:

$$x - 3y - 2z + 4t = 0$$
$$y + 3z - 4t = 0$$

The free variables are z and t.   Set $z = 1$ and $t = 0$ to obtain the solution $u_1 = (-7, -3, 1, 0)$.   Set $z = 0$ and $t = 1$ to obtain the solution $u_2 = (8, 4, 0, 1)$.   By Theorem 3.7 $\{u_1, u_2\}$ is a basis for the solution space of the associated homogeneous system.   Thus (**) has the desired form.

3.11   SYSTEMS OF LINEAR EQUATIONS AS VECTOR EQUATIONS

3.136   Replace the standard system *(3.1)* by a single vector equation.

▌
$$x_1 \begin{pmatrix} a_{11} \\ a_{21} \\ \vdots \\ a_{m1} \end{pmatrix} + x_2 \begin{pmatrix} a_{12} \\ a_{22} \\ \vdots \\ a_{m2} \end{pmatrix} + \cdots + x_n \begin{pmatrix} a_{1n} \\ a_{2n} \\ \vdots \\ a_{mn} \end{pmatrix} = \begin{pmatrix} b_1 \\ b_2 \\ \vdots \\ b_m \end{pmatrix}$$

or, if $u_1, u_2, \ldots, u_n$ and v denote the (column) vectors,

$$x_1 u_1 + x_2 u_2 + \cdots + x_n u_n = v$$

Thus, v is a linear combination of $u_1, u_2, \ldots, u_n$ if and only if the system has a solution.

3.137   Convert the following vector equation to an equivalent system of linear equations and solve:

$$\begin{pmatrix} 1 \\ -6 \\ 5 \end{pmatrix} = x \begin{pmatrix} 1 \\ 2 \\ 3 \end{pmatrix} + y \begin{pmatrix} 2 \\ 5 \\ 8 \end{pmatrix} + z \begin{pmatrix} 3 \\ 2 \\ 3 \end{pmatrix}$$

▌
$$\begin{pmatrix} 1 \\ -6 \\ 5 \end{pmatrix} = \begin{pmatrix} x \\ 2x \\ 3x \end{pmatrix} + \begin{pmatrix} 2y \\ 5y \\ 8y \end{pmatrix} + \begin{pmatrix} 3z \\ 2z \\ 3z \end{pmatrix} = \begin{pmatrix} x + 2y + 3z \\ 2x + 5y + 2z \\ 3x + 8y + 3z \end{pmatrix}$$

Set corresponding components of the vectors equal to each other, and reduce the system to echelon form:

$$\left.\begin{array}{r} x + 2y + 3z = 1 \\ 2x + 5y + 2z = -6 \\ 3x + 8y + 3z = 5 \end{array}\right\} \rightarrow \left\{\begin{array}{r} x + 2y + 3z = 1 \\ y - 4z = -8 \\ 2y - 6z = 2 \end{array}\right\} \rightarrow \left\{\begin{array}{r} x + 2y + 3z = 2 \\ y - 4z = -8 \\ 2z = 18 \end{array}\right.$$

The system is triangular, and back substitution yields the unique solution $x = -81$, $y = 28$, $z = 9$.

3.138 Write the vector $v = (1, -2, 5)$ as a linear combination of the vectors $u_1 = (1, 1, 1)$, $u_2 = (1, 2, 3)$, and $u_3 = (2, -1, 1)$.

▮ Find the equivalent system of linear equations and solve. Writing

$$v = xu_1 + yu_2 + zu_3 = (x + y + 2z, x + 2y - z, x + 3y + z)$$

we obtain the system

$$\left.\begin{array}{r} x + y + 2z = 1 \\ x + 2y - z = -2 \\ x + 3y + z = 5 \end{array}\right\} \rightarrow \left\{\begin{array}{r} x + y + 2z = 1 \\ y - 3z = -3 \\ 2y - z = 4 \end{array}\right\} \rightarrow \left\{\begin{array}{r} x + y + 2z = 1 \\ y - 3z = -3 \\ 5z = 10 \end{array}\right.$$

The unique solution of the triangular form is $x = -6$, $y = 3$, $z = 2$; thus $v = -6u_1 + 3u_2 + 2u_3$.

3.139 Write $v = (2, 3, -5)$ as a linear combination of $u_1 = (1, 2, -3)$, $u_2 = (2, -1, -4)$, and $u_3 = (1, 7, -5)$.

▮
$$(2, 3, -5) = xu_1 + yu_2 + zu_3 = (x + 2y + z, 2x - y + 7z, -3x - 4y - 5z)$$

or

$$\left.\begin{array}{r} x + 2y + z = 2 \\ 2x - y + 7z = 3 \\ -3x - 4y - 5z = -5 \end{array}\right\} \rightarrow \left\{\begin{array}{r} x + 2y + z = 2 \\ -5y + 5z = -1 \\ 2y - 2z = 1 \end{array}\right\} \rightarrow \left\{\begin{array}{r} x + 2y + z = 2 \\ -5y + 5z = -1 \\ 0 = 3 \end{array}\right.$$

The system is inconsistent and so has no solution. Accordingly, v cannot be written as a linear combination of the vectors u_1, u_2 and u_3.

3.140 Consider the following vector equation where $x_1, x_2, \ldots, x_n$ are unknown scalars:

$$x_1u_1 + x_2u_2 + \cdots + x_nu_n = 0 \qquad (1)$$

The vectors $u_1, u_2, \ldots, u_n$ are *linearly dependent* or *linearly independent* according as the equation *(1)* has a nonzero solution or only the zero solution. Determine whether the vectors $(1, 1, 1)$, $(2, -1, 3)$, and $(1, -5, 3)$ are linearly dependent or linearly independent.

▮ First set a linear combination of the vectors equal to the zero vector:

$$\begin{pmatrix} 0 \\ 0 \\ 0 \end{pmatrix} = x\begin{pmatrix} 1 \\ 1 \\ 1 \end{pmatrix} + y\begin{pmatrix} 2 \\ -1 \\ 3 \end{pmatrix} + z\begin{pmatrix} 1 \\ -5 \\ 3 \end{pmatrix} = \begin{pmatrix} x + 2y + z \\ x - y - 5z \\ x + 3y + 3z \end{pmatrix}$$

Set corresponding components equal to each other, and reduce the system to echelon form:

$$\left.\begin{array}{r} x + 2y + z = 0 \\ x - y - 5z = 0 \\ x + 3y + 3z = 0 \end{array}\right\} \rightarrow \left\{\begin{array}{r} x + 2y + z = 0 \\ -3y - 6z = 0 \\ y + 2z = 0 \end{array}\right\} \rightarrow \left\{\begin{array}{r} x + 2y + z = 0 \\ y + 2z = 0 \end{array}\right.$$

The system in echelon form has a free variable; hence the system has a nonzero solution. Accordingly, the original vectors are linearly dependent.

3.141 Determine whether or not the vectors $(1, -2, -3)$, $(2, 3, -1)$, and $(3, 2, 1)$ are linearly dependent.

▮ Set a linear combination (with coefficients x, y, z) of the vectors equal to the zero vector;

$$(0, 0, 0) = (x + 2y + 3z, -2x + 3y + 2z, -3x - y + z)$$

or

$$\left.\begin{array}{r} x + 2y + 3z = 0 \\ -2x + 3y + 2z = 0 \\ -3x - y + x = 0 \end{array}\right\} \rightarrow \left\{\begin{array}{r} x + 2y + 3z = 0 \\ 7y + 8z = 0 \\ 5y + 10z = 0 \end{array}\right. \begin{array}{c} x + 2y + 3z = 0 \\ y + 2z = 0 \\ 7y + 8z = 0 \end{array} \rightarrow \left\{\begin{array}{r} x + 2y + 3z = 0 \\ y + 2z = 0 \\ -6z = 0 \end{array}\right.$$

The homogeneous system is in triangular form, with no free variables; hence it has only the zero solution. Thus the original vectors are linearly independent.

3.142 Determine whether the vectors $(1, 1, -1)$, $(2, -3, 1)$, and $(8, -7, 1)$ are linearly dependent or linearly independent.

▌ Set a linear combination (with coefficients x, y, z) of the vectors equal to the zero vector:

$$(0, 0, 0) = (x + 2y + 8z, x - 3y - 7z, -x + y + z)$$

or

$$\left.\begin{array}{r} x + 2y + 8z = 0 \\ x - 3y - 7z = 0 \\ -x + y + z = 0 \end{array}\right\} \rightarrow \left\{\begin{array}{r} x + 2y + 8z = 0 \\ -5y - 15z = 0 \\ 3y + 9z = 0 \end{array}\right\} \rightarrow \left\{\begin{array}{r} x + 2y + 8z = 0 \\ y + 3z = 0 \\ y + 3z = 0 \end{array}\right\} \rightarrow \left\{\begin{array}{r} x + 2y + 8z = 0 \\ y + 3z = 0 \end{array}\right.$$

In echelon form, the system has a free variable, and hence the system has a nonzero solution. Thus the original vectors are linearly dependent.

Theorem 3.10: Any $n + 1$ or more vectors in $\mathbf{R}^n$ are linearly dependent.

3.143 Prove Theorem 3.10.

▌ Suppose $u_1, u_2, \ldots, u_q$ are vectors in $\mathbf{R}^n$ and $q > n$. The vector equation

$$x_1 u_1 + x_2 u_2 + \cdots + x_q u_q = \mathbf{0}$$

is equivalent to a homogeneous system of n equations in $q > n$ unknowns. By Theorem 3.8, this system has a nonzero solution. Therefore $u_1, u_2, \ldots, u_q$ are linearly dependent.

3.144 Show that the vectors $(1, -2, 3, -4)$, $(1, 2, 1, 5)$, $(2, 0, -6, -5)$ $(3, -7, 0, 2)$, and $(-8, 1, -7, 4)$ are linearly dependent.

▌ These are five vectors in $\mathbf{R}^4$; by Theorem 3.10, the vectors are linearly dependent.

3.145 Show that any set of q vectors that includes the zero vector is linearly dependent.

▌ Denoting the vectors as $\mathbf{0}, u_2, u_3, \ldots, u_q$, we have $1\mathbf{0} + 0u_2 + 0u_3 + \cdots + 0u_q = \mathbf{0}$.

CHAPTER 4
Square Matrices

4.1 DIAGONAL, TRACE

4.1   Define the *diagonal* [or *main diagonal*] of an *n*-square matrix $A = (a_{ij})$.

▮ The diagonal of A consists of the elements $a_{11}, a_{22}, \ldots, a_{nn}$.

4.2   Find the diagonal of the matrix

$$A = \begin{pmatrix} 1 & 2 & 3 \\ 4 & 5 & 6 \\ 7 & 8 & 9 \end{pmatrix}$$

▮ The diagonal consists of the elements from the upper left corner to the lower right corner of the matrix; here, the scalars 1, 5, and 9.

4.3   Find the diagonal of the matrix $B = \begin{pmatrix} t-2 & 3 \\ -4 & t+5 \end{pmatrix}$.

▮ The pair $[t-2, t+5]$.

4.4   Find the diagonal of the matrix $C = \begin{pmatrix} 1 & 2 & -3 \\ 4 & -5 & 6 \end{pmatrix}$.

▮ The diagonal is defined only for square matrices.

4.5   Define the *trace* of an *n*-square matrix $A = (a_{ij})$.

▮ The trace of A is the sum of its diagonal elements; that is,

$$\operatorname{tr}(A) = a_{11} + a_{22} + \cdots + a_{nn} \equiv \sum_{i=1}^{n} a_{ii}$$

4.6   Find the trace of the A in Problem 4.2.

▮ The trace is the sum of the diagonal elements: $\operatorname{tr}(A) = 1 + 5 + 9 = 15$.

4.7   Find the trace of the matrix B in Problem 4.3.

▮ Add the diagonal elements: $\operatorname{tr}(B) = (t-2) + (t+5) = 2t+3$.

Theorem 4.1:   Suppose $A = (a_{ij})$ and $B = (b_{ij})$ are *n*-square matrices and k is a scalar. Then: (i) $\operatorname{tr}(A+B) = \operatorname{tr}(A) + \operatorname{tr}(B)$, (ii) $\operatorname{tr}(kA) = k \cdot \operatorname{tr}(A)$, (iii) $\operatorname{tr}(AB) = \operatorname{tr}(BA)$.

4.8   Prove (i) of Theorem 4.1.

▮ Let $A + B = (c_{ij})$. Then $c_{ij} a_{ij} + b_{ij}$, so that

$$\operatorname{tr}(A+B) = \sum_{k=1}^{n} c_{kk} = \sum_{k=1}^{n} (a_{kk} + b_{kk}) = \sum_{k=1}^{n} a_{kk} + \sum_{k=1}^{n} b_{kk} = \operatorname{tr}(A) + \operatorname{tr}(B)$$

4.9   Prove (ii) of Theorem 4.1.

▮ Let $kA = (c_{ij})$. Then $c_{ij} = ka_{ij}$, and

$$\operatorname{tr}(kA) = \sum_{j=1}^{n} ka_{jj} = k \sum_{j=1}^{n} a_{jj} = k \cdot \operatorname{tr}(A)$$

4.10   Prove (iii) of Theorem 4.1.

    ▌ Let $AB = (c_{ij})$ and $BA = (d_{ij})$. Then

$$c_{ij} = \sum_{k=1}^{n} a_{ik}b_{kj} \qquad \text{and} \qquad d_{ij} = \sum_{k=1}^{n} b_{ik}a_{kj}$$

whence

$$\operatorname{tr}(AB) = \sum_{i=1}^{n} c_{ii} = \sum_{i=1}^{n}\sum_{k=1}^{n} a_{ik}b_{ki} = \sum_{k=1}^{n}\sum_{i=1}^{n} b_{ki}a_{ik} = \sum_{k=1}^{n} d_{kk} = \operatorname{tr}(BA)$$

4.11   Establish that, in general, $\operatorname{tr}(AB) \neq \operatorname{tr}(A)\operatorname{tr}(B)$.

    ▌ Use the matrices of Problem 2.62.

4.12   Let $A = (a_{ij})$ be a square matrix of order n, with entries in **R**, having the property $a_{ij} = a_{ji}$ for all i, j [see Section 4.10]. Show that $\operatorname{tr}(A^2) \geq 0$.

    ▌ Let $A^2 = (c_{ij})$. Then $c_{ij} = \sum_{k=1} a_{ik}a_{kj}$ and so

$$\operatorname{tr}(A^2) = \sum_{i=1}^{n} c_{ii} = \sum_{i=1}^{n}\sum_{k=1}^{n} a_{ik}a_{ki} = \sum_{i=1}^{n}\sum_{k=1}^{n} (a_{ik})^2 \geq 0$$

with equality iff $A = \mathbf{0}$.

4.2   IDENTITY, SCALAR, AND DIAGONAL MATRICES

4.13   Define the n-square *identity* [or *unit*] matrix, denoted I_n or simply I.

    ▌ I_n is the n-square matrix with 1s on the diagonal and 0s elsewhere.

4.14   Exhibit the identity matrices of order 2, 3, and 4.

    ▌
$$I_2 = \begin{pmatrix} 1 & 0 \\ 0 & 1 \end{pmatrix} \qquad I_3 = \begin{pmatrix} 1 & 0 & 0 \\ 0 & 1 & 0 \\ 0 & 0 & 1 \end{pmatrix} \qquad I_4 = \begin{pmatrix} 1 & 0 & 0 & 0 \\ 0 & 1 & 0 & 0 \\ 0 & 0 & 1 & 0 \\ 0 & 0 & 0 & 1 \end{pmatrix}$$

4.15   Indicate the identity matrix using *Kronecker delta notation*.

    ▌ The Kronecker delta is defined as

$$\delta_{ij} = \begin{cases} 0 & \text{if } i \neq j \\ 1 & \text{if } i = j \end{cases}$$

Accordingly, $I = (\delta_{ij})$.

4.16   Find the trace of I_n.

    ▌ I_n has n 1s on the diagonal; hence $\operatorname{tr}(I_n) = n$.

4.17   If A is an $m \times n$ matrix, show that $I_m A = A$.

    ▌ Note first that $I_m A$ is also an $m \times n$ matrix, say $I_m A = (f_{ij})$. But

$$f_{ij} = \sum_{k=1}^{m} \delta_{ik}a_{kj} = \delta_{ii}a_{ij} = a_{ij}$$

Thus $I_m A = A$, since corresponding entries are equal.

4.18   If A is an $m \times n$ matrix, show that $AI_n = A$.

    ▌ Note first that AI_n is also an $m \times n$ matrix, say $AI_n = (g_{ij})$. But

$$g_{ij} = \sum_{k=1}^{n} a_{ik}\delta_{kj} = a_{ij}\delta_{jj} = a_{ij}$$

Thus $AI_n = A$, since corresponding entries are equal.

4.19   Define the *scalar matrix* D_k belonging to a scalar k.

▮ $D_k \equiv kI$.

4.20   Find the scalar matrices of orders 2, 3, and 4 corresponding to the scalar $k = 5$.

▮   In each case, put 5s on the diagonal and 0s elsewhere:

$$\begin{pmatrix} 5 & 0 \\ 0 & 5 \end{pmatrix} \qquad \begin{pmatrix} 5 & 0 & 0 \\ 0 & 5 & 0 \\ 0 & 0 & 5 \end{pmatrix} \qquad \begin{pmatrix} 5 & & & \\ & 5 & & \\ & & 5 & \\ & & & 5 \end{pmatrix}$$

[It is common practice to omit blocks or patterns of 0s as in the third matrix.]

4.21   Show that $D_k A = kA$, for a scalar matrix D_k of proper order.

▮ $D_k A = (kI)A = k(IA) = kA$.

4.22   Show that $BD_k = kB$, for a scalar matrix D_k of proper order.

▮ $BD_k = B(kI) = k(BI) = kB$.   [The upshot of Problems 4.21 and 4.22 is that multiplication by a scalar can be replaced with a special matrix multiplication.]

4.23   Establish the following algebraic properties of scalar matrices of the same order: (i) $D_k + D_l = D_{k+l}$; (ii) $D_k D_l = D_{kl}$.

▮   (i) $D_k + D_l = kI + lI = (k + l)I = D_{k+l}$;   (ii) $D_k D_l = (kI)(lI) = k(I)(lI) = kl(I)(I) = klI = D_{kl}$.

4.24   Define a *diagonal matrix*.

▮   A square matrix $D = (d_{ij})$ is diagonal if its nondiagonal entries are all zero. Such a matrix is frequently notated as $D = \text{diag}(d_{11}, d_{22}, \ldots, d_{nn})$, where some or all of the d_{ii} may be zero.

4.25   Write out $\text{diag}(3, -7, 2)$, $\text{diag}(4, -5)$, and $\text{diag}(6, -3, -9, 1)$

▮   Put the given scalars on the diagonal, with 0s elsewhere:

$$\begin{pmatrix} 3 & 0 & 0 \\ 0 & -7 & 0 \\ 0 & 0 & 2 \end{pmatrix} \qquad \begin{pmatrix} 4 & 0 \\ 0 & -5 \end{pmatrix} \qquad \begin{pmatrix} 6 & & & \\ & -3 & & \\ & & -9 & \\ & & & 1 \end{pmatrix}$$

4.26   Find AB, where $A = \text{diag}(2, -3, 5)$ and $B = \text{diag}(7, 4, 6)$

▮   The product is a diagonal matrix obtained by multiplying corresponding diagonal entries: $AB = \text{diag}(2 \cdot 7, -3 \cdot 4, 5 \cdot 6) = \text{diag}(14, -12, 30)$.

4.27   Let $D = (d_{ij})$ be an m-square diagonal matrix, and let $A = (a_{ij})$ be an $m \times n$ matrix. Show that DA may be obtained by multiplying each row R_i of A by d_{ii}.

▮   The ith row of DA is obtained by premultiplying A by the ith row $(0, 0, \ldots, d_{ii}, \ldots, 0)$ of D:

$$(0, 0, \ldots, \underset{\underset{i\text{th entry}}{\uparrow}}{d_{ii}}, \ldots, 0)\begin{pmatrix} a_{11} & a_{12} & \cdots & a_{1n} \\ \cdots\cdots\cdots\cdots\cdots\cdots \\ a_{i1} & a_{i2} & \cdots & a_{in} \\ \cdots\cdots\cdots\cdots\cdots\cdots \\ a_{m1} & a_{m2} & \cdots & a_{mn} \end{pmatrix} = (d_{ii}a_{i1}, d_{ii}a_{i2}, \ldots, d_{ii}a_{in})$$

$$= d_{ii}(a_{i1}, a_{i2}, \ldots, a_{in}) = d_{ii}R_i$$

4.28   Let $D = (d_{ij})$ be an n-square diagonal matrix, and let $B = (b_{ij})$ be an $m \times n$ matrix. Show that BD may be obtained by multiplying each column C_j of B by d_{jj}.

▮   Proceed as in Problem 4.27, but this time postmultiply by the jth column vector of D.

4.29   Show that $D^T = D$,   for any diagonal $D = (d_{ij})$.

▌ Let $D^T = (a_{ij})$.   If $i \neq j$,   then $a_{ij} = d_{ji} = 0 = d_{ij}$;   if $i = j$,   then $a_{ii} = d_{ii}$.   Thus $D^T = D$.

4.30   Show that $I^T = I$.

▌ Since I is diagonal, $I^T = I$.

4.31   Does $0^T = 0$?

▌ If $\mathbf{0}$ is a square matrix, then $\mathbf{0}^T = \mathbf{0}$.   Otherwise, $\mathbf{0}^T$ and $\mathbf{0}$ have different sizes, and hence cannot be equal.

4.3   ALGEBRA OF SQUARE MATRICES. COMMUTING MATRICES

4.32   What is an *algebra* of matrices?

▌ A nonempty collection $\mathscr{C}$ of matrices is called an *algebra* [of matrices] if $\mathscr{C}$ is closed under the operations of matrix addition, scalar multiplication of a matrix, and matrix multiplication.

4.33   Show that the collection $\mathscr{A}_n$ of all n-square matrices forms an algebra of matrices.

▌ Clearly, the collection $\mathscr{A}_n$ is nonempty.   The sum of any two n-square matrices is an n-square matrix. Any scalar multiple of an n-square matrix is an n-square matrix.   Lastly, the product of two n-square matrices is an n-square matrix.   Thus $\mathscr{A}_n$ is an algebra of matrices.

4.34   Does the set $\mathscr{D}_n$ of all n-square diagonal matrices form an algebra of matrices?

▌ Yes, $\mathscr{D}_n$ is nonempty and the sum, scalar product, and product of diagonal matrices are diagonal matrices.

4.35   Is the set of all n-square scalar matrices an algebra?

▌ Yes, by Problem 4.23 together with $\alpha D_k = D_{\alpha k}$.

4.36   Does the set of all 2×3 matrices form an algebra of matrices?

▌ No, the product of two 2×3 matrices is not defined.

4.37   Show that an algebra $\mathscr{A}$ of matrices contains a zero matrix.

▌ Since $\mathscr{A}$ is nonempty, it contains some matrix A.   Then, by scalar multiplication, $0A = \mathbf{0}$ belongs to $\mathscr{A}$.

4.38   Show that the collection $\mathscr{B}$ of all 2×2 matrices of the form $\begin{pmatrix} s & t \\ t & s \end{pmatrix}$ is an algebra of matrices.

▌ Clearly, $\mathscr{B}$ is nonempty.   If $A = \begin{pmatrix} a & b \\ b & a \end{pmatrix}$ and $B = \begin{pmatrix} c & d \\ d & c \end{pmatrix}$ belong to $\mathscr{B}$, then

$$A + B = \begin{pmatrix} a + c & b + d \\ b + d & a + c \end{pmatrix} \qquad kA = \begin{pmatrix} ka & kb \\ kb & ka \end{pmatrix} \qquad AB = \begin{pmatrix} ac + bd & ad + bc \\ bc + ad & bd + ac \end{pmatrix}$$

also belong to $\mathscr{B}$.   Thus $\mathscr{B}$ is an algebra of matrices.

4.39   When do matrices A and B *commute*?

▌ Matrices A and B commute if $AB = BA$,   a condition that applies only for square matrices of the same order.

4.40   Show that $A = \begin{pmatrix} 1 & 2 \\ 3 & 4 \end{pmatrix}$ and $B = \begin{pmatrix} 5 & 4 \\ 6 & 11 \end{pmatrix}$ commute.

▌ $$AB = \begin{pmatrix} 5 + 12 & 4 + 22 \\ 15 + 24 & 12 + 44 \end{pmatrix} = \begin{pmatrix} 17 & 26 \\ 39 & 56 \end{pmatrix} \quad \text{and} \quad BA = \begin{pmatrix} 5 + 12 & 10 + 16 \\ 6 + 33 & 12 + 44 \end{pmatrix} = \begin{pmatrix} 17 & 26 \\ 39 & 56 \end{pmatrix}$$

Since $AB = BA$,   the matrices commute.

4.41   Find all matrices $M = \begin{pmatrix} x & y \\ z & t \end{pmatrix}$ that commute with $A = \begin{pmatrix} 1 & 1 \\ 0 & 1 \end{pmatrix}$.

▎ $AM = \begin{pmatrix} x+z & y+t \\ z & t \end{pmatrix}$ and $MA = \begin{pmatrix} x & x+y \\ z & z+t \end{pmatrix}$

Set $AM = MA$,   to obtain the four equations

$$x + z = x \qquad y + t = x + y \qquad z = z \qquad t = z + t$$

From the first or last equation, $z = 0$;   from the second equation, $x = t$.   Thus M is any matrix of the form

$$\begin{pmatrix} x & y \\ 0 & x \end{pmatrix}$$

4.42   Show that the scalar matrix kI_n commutes with any n-square matrix A.

▎   We have: $(kI)A = k(IA) = kA$ and $A(kI) = k(AI) = kA = (kI)A$.

4.43   Show that $\mathcal{B}$ [Problem 4.38] is a *commutative algebra*.

▎   Using the notation of Problem 4.38, make the calculation

$$BA = \begin{pmatrix} ca + db & cb + da \\ da + cb & db + ca \end{pmatrix}$$

Thus $BA = AB$.

4.4 POWERS OF MATRICES

4.44   The nonnegative integral powers of a square matrix M may be defined recursively by

$$M^0 = I \qquad M^1 = M \qquad \text{and} \qquad M^{r+1} = MM^r \quad (r = 1, 2, 3, \ldots)$$

Prove the following *theorem*:   (a) $A^p A^q = A^{p+q}$.   (b) If A and B commute, so do A^p and B^q.

▎   (a) The proof is by induction on p.   The case $p = 0$ is true since $A^0 = I$,   and the case $p = 1$ is true by definition.   Suppose $p > 1$ and the result holds for $p - 1$.   Then

$$A^p A^q = a(A^{p-1})A^q = AA^{p+q-1} = A^{p+q}$$

(b) First we show that A commutes with B^q by induction on q.   The case $q = 0$ is true since $B^0 = I$, and the case $q = 1$ is true by hypothesis.   Suppose $q > 1$ and A commutes with B^{q-1}.   Then

$$B^q A = BB^{q-1}A = BAB^{q-1} = ABB^{q-1} = AB^q$$

Thus A commutes with B^q.   Similarly, by induction on p, B^q commutes with A^p.

Problems 4.45–4.48 refer to the matrix $A = \begin{pmatrix} 1 & 2 \\ 4 & -3 \end{pmatrix}$.

4.45   Calculate A^2.

▎ $A^2 = AA = \begin{pmatrix} 1 & 2 \\ 4 & -3 \end{pmatrix}\begin{pmatrix} 1 & 2 \\ 4 & -3 \end{pmatrix} = \begin{pmatrix} 1+8 & 2-6 \\ 4-12 & 8+9 \end{pmatrix} = \begin{pmatrix} 9 & -4 \\ -8 & 17 \end{pmatrix}$

4.46 Calculate A^3.

❙
$$A^3 = AA^2 = \begin{pmatrix} 1 & 2 \\ 4 & -3 \end{pmatrix} \begin{pmatrix} 9 & -4 \\ -8 & 17 \end{pmatrix} = \begin{pmatrix} 9-16 & -4+34 \\ 36+24 & -16-51 \end{pmatrix} = \begin{pmatrix} -7 & 30 \\ 60 & -67 \end{pmatrix}$$

[The theorem in Problem 4.44 guarantees the same result from the computation A^2A.]

4.47 Evaluate $f(A)$ for the polynomial $f(x) = 2x^2 - 4x + 5$.

❙
$$f(A) = 2A^3 - 4A + 5I = 2\begin{pmatrix} -7 & 30 \\ 60 & -67 \end{pmatrix} - 4\begin{pmatrix} 1 & 2 \\ 4 & -3 \end{pmatrix} + 5\begin{pmatrix} 1 & 0 \\ 0 & 1 \end{pmatrix}$$
$$= \begin{pmatrix} -14 & 60 \\ 120 & -134 \end{pmatrix} + \begin{pmatrix} -4 & -8 \\ -16 & 12 \end{pmatrix} + \begin{pmatrix} 5 & 0 \\ 0 & 5 \end{pmatrix}$$
$$= \begin{pmatrix} -14-4+5 & 60-8+0 \\ 120-16+0 & -134+12+5 \end{pmatrix} = \begin{pmatrix} -13 & 52 \\ 104 & -117 \end{pmatrix}$$

4.48 Show that A is a zero of the polynomial $g(x) = x^2 + 2x - 11$.

❙
$$g(A) = A^2 + 2A - 11I = \begin{pmatrix} 9 & -4 \\ -8 & 17 \end{pmatrix} + 2\begin{pmatrix} 1 & 2 \\ 4 & -3 \end{pmatrix} - 11\begin{pmatrix} 1 & 0 \\ 0 & 1 \end{pmatrix}$$
$$= \begin{pmatrix} 9 & -4 \\ -8 & 17 \end{pmatrix} + \begin{pmatrix} 2 & 4 \\ 8 & -6 \end{pmatrix} + \begin{pmatrix} -11 & 0 \\ 0 & -11 \end{pmatrix}$$
$$= \begin{pmatrix} 9+2-11 & -4+4+0 \\ -8+8+0 & 17-6-11 \end{pmatrix} = \begin{pmatrix} 0 & 0 \\ 0 & 0 \end{pmatrix}$$

[To the Interested Reader: Look up the *Cayley–Hamilton theorem*.]

Problems 4.49–4.52 refer to the matrix $A = \begin{pmatrix} 2 & 2 \\ 3 & -1 \end{pmatrix}$.

4.49 Calculate A^2.

❙
$$A^2 = \begin{pmatrix} 2 & 2 \\ 3 & -1 \end{pmatrix} \begin{pmatrix} 2 & 2 \\ 3 & -1 \end{pmatrix} = \begin{pmatrix} 4+6 & 4-2 \\ 6-3 & 6+1 \end{pmatrix} = \begin{pmatrix} 10 & 2 \\ 3 & 7 \end{pmatrix}$$

4.50 Calculate A^3.

❙
$$A^3 = AA^2 = \begin{pmatrix} 2 & 2 \\ 3 & -1 \end{pmatrix} \begin{pmatrix} 10 & 2 \\ 3 & 7 \end{pmatrix} = \begin{pmatrix} 20+6 & 4+14 \\ 30-3 & 6-7 \end{pmatrix} = \begin{pmatrix} 26 & 18 \\ 27 & -1 \end{pmatrix}$$

4.51 Find $f(A)$, where $f(x) = x^3 - 3x^2 - 2x + 4$.

❙
$$f(A) = A^3 - 3A^2 - 2A + 4I = \begin{pmatrix} 26 & 18 \\ 27 & -1 \end{pmatrix} - 3\begin{pmatrix} 10 & 2 \\ 3 & 7 \end{pmatrix} - 2\begin{pmatrix} 2 & 2 \\ 3 & -1 \end{pmatrix} + 4\begin{pmatrix} 1 & 0 \\ 0 & 1 \end{pmatrix}$$
$$= \begin{pmatrix} 26 & 18 \\ 27 & -1 \end{pmatrix} + \begin{pmatrix} -30 & -6 \\ -9 & -21 \end{pmatrix} + \begin{pmatrix} -4 & -4 \\ -6 & 2 \end{pmatrix} + \begin{pmatrix} 4 & 0 \\ 0 & 4 \end{pmatrix}$$
$$= \begin{pmatrix} -4 & 8 \\ 12 & -16 \end{pmatrix}$$

4.52 Find $g(A)$, where $g(x) = x^2 - x - 8$.

❙
$$g(A) = A^2 - A - 8I = \begin{pmatrix} 10 & 2 \\ 3 & 7 \end{pmatrix} - \begin{pmatrix} 2 & 2 \\ 3 & -1 \end{pmatrix} - 8\begin{pmatrix} 1 & 0 \\ 0 & 1 \end{pmatrix}$$
$$= \begin{pmatrix} 10 & 2 \\ 3 & 7 \end{pmatrix} + \begin{pmatrix} -2 & -2 \\ -3 & 1 \end{pmatrix} + \begin{pmatrix} -8 & 0 \\ 0 & -8 \end{pmatrix} = \begin{pmatrix} 0 & 0 \\ 0 & 0 \end{pmatrix}$$

Thus A is a zero of $g(x)$.

Problems 4.53–4.55 refer to the matrix $B = \begin{pmatrix} 1 & 3 \\ 5 & 3 \end{pmatrix}$.

4.53 Calculate B^2.

▌ $$B^2 = BB = \begin{pmatrix} 1 & 3 \\ 5 & 3 \end{pmatrix}\begin{pmatrix} 1 & 3 \\ 5 & 3 \end{pmatrix} = \begin{pmatrix} 1+15 & 3+9 \\ 5+15 & 15+9 \end{pmatrix} = \begin{pmatrix} 16 & 12 \\ 20 & 24 \end{pmatrix}$$

4.54 Find $f(B)$, where $f(x) = 2x^2 - 4x + 3$.

▌ $$f(B) = 2B^2 - 4B + 3I = 2\begin{pmatrix} 16 & 12 \\ 20 & 24 \end{pmatrix} - 4\begin{pmatrix} 1 & 3 \\ 5 & 3 \end{pmatrix} + 3\begin{pmatrix} 1 & 0 \\ 0 & 1 \end{pmatrix}$$

$$= \begin{pmatrix} 32 & 24 \\ 40 & 48 \end{pmatrix} + \begin{pmatrix} -4 & -12 \\ -20 & -12 \end{pmatrix} + \begin{pmatrix} 3 & 0 \\ 0 & 3 \end{pmatrix} = \begin{pmatrix} 31 & 12 \\ 20 & 39 \end{pmatrix}$$

4.55 Find $g(B)$, where $g(x) = x^2 - 4x - 12$.

▌ $$g(B) = B^2 - 4B - 12I = \begin{pmatrix} 16 & 12 \\ 20 & 24 \end{pmatrix} - 4\begin{pmatrix} 1 & 3 \\ 5 & 3 \end{pmatrix} - 12\begin{pmatrix} 1 & 0 \\ 0 & 1 \end{pmatrix}$$

$$= \begin{pmatrix} 16 & 12 \\ 20 & 24 \end{pmatrix} + \begin{pmatrix} -4 & -12 \\ -20 & -12 \end{pmatrix} + \begin{pmatrix} -12 & 0 \\ 0 & -12 \end{pmatrix} = \begin{pmatrix} 0 & 0 \\ 0 & 0 \end{pmatrix}$$

i.e., B is a zero of $g(x)$.

Problems 4.56–4.59 refer to the matrix $A = \begin{pmatrix} 1 & 2 \\ 0 & 1 \end{pmatrix}$.

4.56 Calculate A^2.

▌ $$A^2 = AA = \begin{pmatrix} 1 & 2 \\ 0 & 1 \end{pmatrix}\begin{pmatrix} 1 & 2 \\ 0 & 1 \end{pmatrix} = \begin{pmatrix} 1+0 & 2+2 \\ 0+0 & 0+1 \end{pmatrix} = \begin{pmatrix} 1 & 4 \\ 0 & 1 \end{pmatrix}$$

4.57 Calculate A^3.

▌ $$A^3 = AA^2 = \begin{pmatrix} 1 & 2 \\ 0 & 1 \end{pmatrix}\begin{pmatrix} 1 & 4 \\ 0 & 1 \end{pmatrix} = \begin{pmatrix} 1+0 & 4+2 \\ 0+0 & 0+1 \end{pmatrix} = \begin{pmatrix} 1 & 6 \\ 0 & 1 \end{pmatrix}$$

4.58 Let $S_k = \begin{pmatrix} 1 & k \\ 0 & 1 \end{pmatrix}$. Show that $AS_k = S_kA = S_{k+2}$.

▌ $$AS_k = \begin{pmatrix} 1 & 2 \\ 0 & 1 \end{pmatrix}\begin{pmatrix} 1 & k \\ 0 & 1 \end{pmatrix} = \begin{pmatrix} 1+0 & k+2 \\ 0+0 & 0+1 \end{pmatrix} = S_{k+2}$$

$$S_kA = \begin{pmatrix} 1 & k \\ 0 & 1 \end{pmatrix}\begin{pmatrix} 1 & 2 \\ 0 & 1 \end{pmatrix} = \begin{pmatrix} 1+0 & 2+k \\ 0+0 & 0+1 \end{pmatrix} = S_{k+2}$$

4.59 Calculate A^n.

▌ By Problem 4.58, multiplying A^m by A adds 2 to the upper right entry; hence

$$A^n = \begin{pmatrix} 1 & 2n \\ 0 & 1 \end{pmatrix}$$

4.60 Define an *idempotent* matrix.

▌ A matrix E is idempotent if $E^2 = E$.

4.61 Show that the identity matrix I is idempotent.

▌ $I^2 = II = I$

4.62 Show that any square zero matrix $\mathbf{0}$ is idempotent.

▌ $\mathbf{0}^2 = \mathbf{00} = \mathbf{0}$

4.63   Show that

$$E = \begin{pmatrix} 2 & -2 & -4 \\ -1 & 3 & 4 \\ 1 & -2 & -3 \end{pmatrix}$$

is idempotent.

▌ $E^2 = \begin{pmatrix} 2 & -2 & -4 \\ -1 & 3 & 4 \\ 1 & -2 & -3 \end{pmatrix}\begin{pmatrix} 2 & -2 & -4 \\ -1 & 3 & 4 \\ 1 & -2 & -3 \end{pmatrix} = \begin{pmatrix} 4+2-4 & -4-6+8 & -8-8+12 \\ -2-3+4 & 2+9-8 & 4+12-12 \\ 2+2-3 & -2-6+6 & -4-8+9 \end{pmatrix} = \begin{pmatrix} 2 & -2 & -4 \\ -1 & 3 & 4 \\ 1 & -2 & -3 \end{pmatrix} = E$

4.64   Show that if $AB = A$ and $BA = B$, then A and B are idempotent.

▌ $\qquad A = AB = A(BA) = (AB)A = AA = A^2$

$\qquad B = BA = B(AB) = (BA)B = BB = B^2$

4.65   Show that the product of *commutative* idempotent matrices is idempotent.

▌ $\qquad (AB)(AB) = A(BA)B = A(AB)B = (AA)(BB) = AB$

4.66   Define a *nilpotent matrix of class p* for a positive integer p.

▌ A is nilpotent of class p if $A^p = 0$ but $A^{p-1} \neq 0$.

4.67   Show that if A is nilpotent of class p, then $A^q = 0$ for $q > p$.

▌ $A^q = A^p A^{q-p} = 0 A^{q-p} = 0$

4.68   Show that

$$A = \begin{pmatrix} 1 & 1 & 3 \\ 5 & 2 & 6 \\ -2 & -1 & -3 \end{pmatrix}$$

is nilpotent of class 3.

▌ $\qquad A^2 = \begin{pmatrix} 1 & 1 & 3 \\ 5 & 2 & 6 \\ -2 & -1 & -3 \end{pmatrix}\begin{pmatrix} 1 & 1 & 3 \\ 5 & 2 & 6 \\ -2 & -1 & -3 \end{pmatrix} = \begin{pmatrix} 0 & 0 & 0 \\ 3 & 3 & 9 \\ -1 & -1 & -3 \end{pmatrix}$

and $\qquad A^2 = A^2 A = \begin{pmatrix} 0 & 0 & 0 \\ 3 & 3 & 9 \\ -1 & -1 & -3 \end{pmatrix}\begin{pmatrix} 1 & 1 & 3 \\ 5 & 2 & 6 \\ -2 & -1 & -3 \end{pmatrix} = 0$

4.69   Define an *involutory matrix*.

▌ A matrix A is involutory if $A^2 = I$, the identity matrix.

4.70   Show that

$$A = \begin{pmatrix} 4 & 3 & 3 \\ -1 & 0 & -1 \\ -4 & -4 & -3 \end{pmatrix}$$

is involutory.

▌ $\qquad A^2 = \begin{pmatrix} 16-3-12 & 12+0-12 & 12-3-9 \\ -4+0+4 & -3+0+4 & -3+0+3 \\ -16+4+12 & -12+0+12 & -12+4+9 \end{pmatrix} = \begin{pmatrix} 1 & 0 & 0 \\ 0 & 1 & 0 \\ 0 & 0 & 1 \end{pmatrix} = I$

4.71   Establish a connection between the involutory matrices and the idempotent matrices.

▌ Consider the decomposition $A = \frac{1}{2}(I + A) - \frac{1}{2}(I - A) \equiv A^+ - A^-$ of an arbitrary involutory matrix A.

We have:

$$A^+A^+ = \tfrac{1}{2}(I+A)\tfrac{1}{2}(I+A) = \tfrac{1}{4}(I^2 + AI + IA + A^2)$$
$$= \tfrac{1}{4}(2I + 2A) = \tfrac{1}{2}(I+A) = A^+$$

likewise, $A^-A^- = A^-$. Thus, *any involutory matrix is expressible as a difference of idempotent matrices.*

4.5 SQUARE MATRICES AS FUNCTIONS

4.72 Show that an n-square matrix A defines a function from $\mathbf{R}^n$ into $\mathbf{R}^n$ in two different ways.

▮ Let u be a vector in $\mathbf{R}^n$. With u as a column vector, A defines a function $A : \mathbf{R}^n \to \mathbf{R}^n$ by $A(u) = Au$. On the other hand, with u as a row vector, A defines a function $A : \mathbf{R}^n \to \mathbf{R}^n$ by $A(u) = uA$.

Unless otherwise stated or implied, in subsequent problems vectors in $\mathbf{R}^n$ will be defined as column vectors, and the function defined by the matrix A will be $A(u) = Au$. For typographical reasons, column vectors will often be indicated as transposed row vectors. For Problems 4.73–4.76,

$$A = \begin{pmatrix} 1 & -2 & 3 \\ 4 & 5 & -6 \\ 2 & 0 & -1 \end{pmatrix}$$

4.73 Find $A(u)$, where $u = (1, -3, 7)^T$.

▮ $A(u) = Au = \begin{pmatrix} 1 & -2 & 3 \\ 4 & 5 & -6 \\ 2 & 0 & -1 \end{pmatrix}\begin{pmatrix} 1 \\ -3 \\ 7 \end{pmatrix} = \begin{pmatrix} 1+6+21 \\ 4-15-42 \\ 2+0-7 \end{pmatrix} = \begin{pmatrix} 28 \\ -53 \\ -5 \end{pmatrix}$

4.74 Find $A(v)$, where $v = (2, -5, 6, -4)^T$.

▮ $A(v)$ is not defined since v does not belong to $\mathbf{R}^3$.

4.75 Find $A(w)$, where $w = (2, -1, 4)^T$.

▮ $\qquad A(w) = Aw = \begin{pmatrix} 1 & -2 & 3 \\ 4 & 5 & -6 \\ 2 & 0 & -1 \end{pmatrix}\begin{pmatrix} 2 \\ -1 \\ 4 \end{pmatrix} = \begin{pmatrix} 2+2+12 \\ 8-5-24 \\ 4+0-4 \end{pmatrix} = \begin{pmatrix} 16 \\ -21 \\ 0 \end{pmatrix}$

4.76 Find $A(u)$, where $u = (3, -7, 8)$.

▮ By our convention, $A(u)$ is not defined for a row vector u.

4.77 Given $A = \begin{pmatrix} 1 & 3 \\ 4 & -3 \end{pmatrix}$. Find a *nonzero* column vector $u = \begin{pmatrix} x \\ y \end{pmatrix}$ such that $A(u) = 3u$.

▮ First set up the matrix equation $Au = 3u$:

$$\begin{pmatrix} 1 & 3 \\ 4 & -3 \end{pmatrix}\begin{pmatrix} x \\ y \end{pmatrix} = 3\begin{pmatrix} x \\ y \end{pmatrix}$$

Write each side as a single matrix (column vector):

$$\begin{pmatrix} x+3y \\ 4x-3y \end{pmatrix} = \begin{pmatrix} 3x \\ 3y \end{pmatrix}$$

Set corresponding elements equal to each other to obtain the system of equations, and reduce it to echelon form:

$$\begin{rcases} x+3y=3x \\ 4x-3y=3y \end{rcases} \to \begin{cases} 2x-3y=0 \\ 4x-6y=0 \end{cases} \to \begin{cases} 2x-3y=0 \\ 0=0 \end{cases} \to 2x-3y=0$$

The system reduces to one homogeneous equation in two unknowns, and so has an infinite number of solutions. To obtain a nonzero solution let, say, $y = 2$; then $x = 3$. That is, $u = (3, 2)^T$ is the desired vector.

4.78 Given $B = \begin{pmatrix} 1 & 3 \\ 5 & 3 \end{pmatrix}$. Find a *nonzero* vector $u = \begin{pmatrix} x \\ y \end{pmatrix}$ such that $B(u) = 6u$.

▌ Proceed as in Problem 4.77:

Then:
$$\begin{pmatrix} 1 & 3 \\ 5 & 3 \end{pmatrix}\begin{pmatrix} x \\ y \end{pmatrix} = 6\begin{pmatrix} x \\ y \end{pmatrix} \qquad \text{or} \qquad \begin{pmatrix} x + 3y \\ 5x + 3y \end{pmatrix} = \begin{pmatrix} 6x \\ 6y \end{pmatrix}$$

$$\left.\begin{array}{r} x + 3y = 6x \\ 5x + 3y = 6y \end{array}\right\} \rightarrow \left\{\begin{array}{r} -5x + 3y = 0 \\ 5x - 3y = 0 \end{array}\right\} \rightarrow 5x - 3y = 0$$

There are an infinite number of solutions. To obtain a nonzero solution, set $y = 5$; hence $x = 3$. Thus, $u = (3, 5)^T$.

4.79 Given

$$A = \begin{pmatrix} 1 & 2 & -3 \\ 2 & 5 & -1 \\ 5 & 12 & -5 \end{pmatrix}$$

Find all vectors $u = (x, y, z)^T$ such that $A(u) = \mathbf{0}$.

▌ Set up the equation $Au = \mathbf{0}$ and then write each side as a single matrix:

$$\begin{pmatrix} 1 & 2 & -3 \\ 2 & 5 & -1 \\ 5 & 12 & -5 \end{pmatrix}\begin{pmatrix} x \\ y \\ z \end{pmatrix} = \begin{pmatrix} 0 \\ 0 \\ 0 \end{pmatrix} \qquad \text{or} \qquad \begin{pmatrix} x + 2y - 3z \\ 2x + 5y - z \\ 5x + 12y - 5z \end{pmatrix} = \begin{pmatrix} 0 \\ 0 \\ 0 \end{pmatrix}$$

Set corresponding elements equal to each other to obtain a homogeneous system, and reduce the system to echelon form:

$$\left.\begin{array}{r} x + 2y - 3z = 0 \\ 2x + 5y - z = 0 \\ 5x + 12y - 5z = 0 \end{array}\right\} \rightarrow \left\{\begin{array}{r} x + 2y - 3z = 0 \\ y + 5z = 0 \\ 2y + 10z = 0 \end{array}\right\} \rightarrow \left\{\begin{array}{r} x + 2y - 3z = 0 \\ y + 5z = 0 \end{array}\right.$$

In the echelon form, z is the free variable. To obtain the general solution, set $z = a$, where a is a parameter. Back substitution yields $y = -5a$, and then $x = 13a$. Thus, $u = (13a, -5a, a)^T$ represents all vectors such that $Au = \mathbf{0}$.

4.6 INVERTIBLE MATRICES, INVERSES

4.80 Define an *invertible* matrix.

▌ A square matrix A is invertible if there exists a [square] matrix B such that $AB = BA = I$, where I is the identity matrix.

4.81 Show that the matrix B in Problem 4.80 is unique.

▌ If $AB_1 = B_1A = I$ and $AB_2 = B_2A = I$, then $B_1 = B_1I = B_1(AB_2) = (B_1A)B_2 = IB_2 = B_2$.

4.82 Define the *inverse* of an invertible matrix.

▌ If A is invertible, then the unique matrix B such that $AB = BA = I$ is called the *inverse* of A and is denoted by A^{-1}.

4.83 Show that the inverse relation is symmetric; i.e., $(A^{-1})^{-1} = A$.

▌ If $AB = BA = I$, then $BA = AB = I$; so, if B is the inverse of A, then B is invertible and A is the inverse of B. In other words, $(A^{-1})^{-1} = A$.

4.84 Show that $A = \begin{pmatrix} 2 & 5 \\ 1 & 3 \end{pmatrix}$ and $B = \begin{pmatrix} 3 & -5 \\ -1 & 2 \end{pmatrix}$ are inverses.

$$AB = \begin{pmatrix} 2 & 5 \\ 1 & 3 \end{pmatrix}\begin{pmatrix} 3 & -5 \\ -1 & 2 \end{pmatrix} = \begin{pmatrix} 6-5 & -10+10 \\ 3-3 & -5+6 \end{pmatrix} = \begin{pmatrix} 1 & 0 \\ 0 & 1 \end{pmatrix} = I$$

$$BA = \begin{pmatrix} 3 & -5 \\ -1 & 2 \end{pmatrix}\begin{pmatrix} 2 & 5 \\ 1 & 3 \end{pmatrix} = \begin{pmatrix} 6-5 & 15-15 \\ -2+2 & -5+6 \end{pmatrix} = \begin{pmatrix} 1 & 0 \\ 0 & 1 \end{pmatrix} = I$$

4.85 Show that $A = \begin{pmatrix} 1 & 0 & 2 \\ 2 & -1 & 3 \\ 4 & 1 & 8 \end{pmatrix}$ and $B = \begin{pmatrix} -11 & 2 & 2 \\ -4 & 0 & 1 \\ 6 & -1 & -1 \end{pmatrix}$ are inverses.

$$ AB = \begin{pmatrix} -11+0+12 & 2+0-2 & 2+0-2 \\ -22+4+18 & 4+0-3 & 4-1-3 \\ -44-4+48 & 8+0-8 & 8+1-8 \end{pmatrix} = \begin{pmatrix} 1 & 0 & 0 \\ 0 & 1 & 0 \\ 0 & 0 & 1 \end{pmatrix} = I $$

By Problem 4.121, $AB = I$ if and only if $BA = I$; hence we do not need to test if $BA = I$. Thus A and B are inverses of each other.

4.86 Prove the following restricted version of Problem 4.121: If A *is symmetric* and there exists a matrix B such that $AB = I$, then A is invertible, with inverse B.

▮ If $AB = I$, then $B^T A = I$. But $B^T = B^T (AB) = (B^T A)B = B$.

4.87 When is the general 2×2 matrix $A = \begin{pmatrix} a & b \\ c & d \end{pmatrix}$ invertible? What then is its inverse?

▮ We seek scalars x, y, z, t such that

$$ \begin{pmatrix} a & b \\ c & d \end{pmatrix}\begin{pmatrix} x & y \\ z & t \end{pmatrix} = \begin{pmatrix} 1 & 0 \\ 0 & 1 \end{pmatrix} \quad \text{or} \quad \begin{pmatrix} ax + bz & ay + bt \\ cx + dz & cy + dt \end{pmatrix} = \begin{pmatrix} 1 & 0 \\ 0 & 1 \end{pmatrix} $$

which reduces to solving the following two systems

$$ \begin{cases} ax + bz = 1 \\ cx + dz = 0 \end{cases} \qquad \begin{cases} ay + bt = 0 \\ cy + dt = 1 \end{cases} $$

both of which have coefficient matrix A. Set $|A| = ad - bc$ [the *determinant* of A]. By Problems 3.41 and 3.42, the two systems are solvable—and A is invertible—when and only when $|A| \neq 0$. In that case, the first system has the unique solution $x = d/|A|$, $z = -c/|A|$, and the second system has the unique solution $y = -b/|A|$, $t = a/|A|$. Accordingly,

$$ A^{-1} = \begin{pmatrix} d/|A| & -b/|A| \\ -c/|A| & a/|A| \end{pmatrix} = \frac{1}{|A|}\begin{pmatrix} d & -b \\ -c & a \end{pmatrix} $$

In words: When $|A| \neq 0$, the inverse of a 2×2 matrix A is obtained by (i) interchanging the elements on the main diagonal, (ii) taking the negatives of the other elements, and (iii) multiplying the matrix by $1/|A|$.

4.88 Find the inverse of $A = \begin{pmatrix} 3 & 5 \\ 2 & 3 \end{pmatrix}$.

▮ Use the explicit formula of Problem 4.87. Thus, first find $|A| = (3)(3) - (5)(2) = -1 \neq 0$. Next interchange the diagonal elements, take the negatives of the other elements, and multiply by $1/|A|$:

$$ A^{-1} = -1\begin{pmatrix} 3 & -5 \\ -2 & 3 \end{pmatrix} = \begin{pmatrix} -3 & 5 \\ 2 & -3 \end{pmatrix} $$

4.89 Find the inverse of $A = \begin{pmatrix} 5 & 3 \\ 4 & 2 \end{pmatrix}$.

▮ First find $|A| = (5)(2) - (3)(4) = -2$. Next interchange the diagonal elements, take the negatives of the nondiagonal elements, and multiply by $1/|A|$:

$$ A^{-1} = -\frac{1}{2}\begin{pmatrix} 2 & -3 \\ -4 & 5 \end{pmatrix} = \begin{pmatrix} -1 & 3/2 \\ 2 & -5/2 \end{pmatrix} $$

4.90 Find the inverse of $B = \begin{pmatrix} 2 & -3 \\ 1 & 3 \end{pmatrix}$.

▮ First find $|B| = (2)(3) - (-3)(1) = 9$. Next interchange the diagonal elements, take the negatives of the nondiagonal elements, and multiply by $1/|B|$:

$$ B^{-1} = \frac{1}{9}\begin{pmatrix} 3 & 3 \\ -1 & 2 \end{pmatrix} = \begin{pmatrix} 1/3 & 1/3 \\ -1/9 & 2/9 \end{pmatrix} $$

4.91   Try to find the inverse of $A = \begin{pmatrix} -2 & 6 \\ 3 & -9 \end{pmatrix}$.

▮   First find $|A| = (-2)(-9) - (6)(3) = 0$.   Since $|A| = 0$, A has no inverse.

4.92   Give the Gaussian elimination algorithm which either finds the inverse of an n-square matrix A or determines that A is not invertible.

▮   **Step 1.**   Form the $n \times 2n$ [block] matrix $M = (A : I)$;   that is, A is in the left half of M and I is in the right half of M.
   Step 2.   Row reduce M to echelon form.   If the process generates a zero now in the A-half of M, STOP (A is not invertible).   Otherwise, the A-half will assume triangular form.
   Step 3.   Further row reduce M to the row canonical form $(I : B)$, where I has replaced A in the left half of the matrix.
   Step 4.   Set $A^{-1} = B$.
   The justification of this algorithm is found in Problem 4.122.

4.93   Find the inverse of

$$A = \begin{pmatrix} 1 & 0 & 2 \\ 2 & -1 & 3 \\ 4 & 1 & 8 \end{pmatrix}$$

▮   Form the block matrix $M = (A : I)$ and reduce M to echelon form:

$$M = \left(\begin{array}{ccc:ccc} 1 & 0 & 2 & 1 & 0 & 0 \\ 2 & -1 & 3 & 0 & 1 & 0 \\ 4 & 1 & 8 & 0 & 0 & 1 \end{array}\right) \sim \left(\begin{array}{ccc:ccc} 1 & 0 & 2 & 1 & 0 & 0 \\ 0 & -1 & -1 & -2 & 1 & 0 \\ 0 & 1 & 0 & -4 & 0 & 1 \end{array}\right) \sim \left(\begin{array}{ccc:ccc} 1 & 0 & 2 & 1 & 0 & 0 \\ 0 & -1 & -1 & -2 & 1 & 0 \\ 0 & 0 & -1 & -6 & 1 & 1 \end{array}\right)$$

In echelon form, the left half of M is in triangular form; hence A is invertible.   Further row reduce M to row canonical form:

$$M \sim \left(\begin{array}{ccc:ccc} 1 & 0 & 0 & -11 & 2 & 2 \\ 0 & -1 & 0 & 4 & 0 & -1 \\ 0 & 0 & 1 & 6 & -1 & -1 \end{array}\right) \sim \left(\begin{array}{ccc:ccc} 1 & 0 & 0 & -11 & 2 & 2 \\ 0 & 1 & 0 & -4 & 0 & 1 \\ 0 & 0 & 1 & 6 & -1 & -1 \end{array}\right)$$

The final block matrix is in the form $(I : A^{-1})$.

4.94   Find the inverse of

$$B = \begin{pmatrix} 1 & -2 & 2 \\ 2 & -3 & 6 \\ 1 & 1 & 7 \end{pmatrix}$$

▮   Form the block matrix $M = (B : I)$ and reduce M to echelon form:

$$M = \left(\begin{array}{ccc:ccc} 1 & -2 & 2 & 1 & 0 & 0 \\ 2 & -3 & 6 & 0 & 1 & 0 \\ 1 & 1 & 7 & 0 & 0 & 1 \end{array}\right) \sim \left(\begin{array}{ccc:ccc} 1 & -2 & 2 & 1 & 0 & 0 \\ 0 & 1 & 2 & -2 & 1 & 0 \\ 0 & 3 & 5 & -1 & 0 & 1 \end{array}\right) \sim \left(\begin{array}{ccc:ccc} 1 & -2 & 2 & 1 & 0 & 0 \\ 0 & 1 & 2 & -2 & 1 & 0 \\ 0 & 0 & -1 & 5 & -3 & 1 \end{array}\right)$$

In echelon form, the left half of M is in triangular form; hence B has an inverse.   Further row reduce M to row canonical form:

$$M \sim \left(\begin{array}{ccc:ccc} 1 & -2 & 0 & 11 & -6 & 2 \\ 0 & 1 & 0 & 8 & -5 & 2 \\ 0 & 0 & 1 & -5 & 3 & -1 \end{array}\right) \sim \left(\begin{array}{ccc:ccc} 1 & 0 & 0 & 27 & -16 & 6 \\ 0 & 1 & 0 & 8 & -5 & 2 \\ 0 & 0 & 1 & -5 & 3 & -1 \end{array}\right)$$

The final matrix has the form $(I : B^{-1})$.

4.95   Find the inverse of

$$A = \begin{pmatrix} 1 & 2 & -4 \\ -1 & -1 & 5 \\ 2 & 7 & -3 \end{pmatrix}$$

▮ Form the block matrix $M = (A : I)$ and row reduce M to echelon form:

$$M = \begin{pmatrix} 1 & 2 & -4 & 1 & 0 & 0 \\ -1 & -1 & 5 & 0 & 1 & 0 \\ 2 & 7 & -3 & 0 & 0 & 1 \end{pmatrix} \sim \begin{pmatrix} 1 & 2 & -4 & 1 & 0 & 0 \\ 0 & 1 & 1 & 1 & 1 & 0 \\ 0 & 3 & 5 & -2 & 0 & 1 \end{pmatrix} \sim \begin{pmatrix} 1 & 2 & -4 & 1 & 0 & 0 \\ 0 & 1 & 1 & 1 & 1 & 0 \\ 0 & 0 & 2 & -5 & -3 & 1 \end{pmatrix}$$

The left half of M is now in triangular form; hence A has an inverse. Further row reduce M to row canonical form:

$$M \sim \begin{pmatrix} 1 & 2 & 0 & -9 & -6 & 2 \\ 0 & 1 & 0 & 7/2 & 5/2 & -1/2 \\ 0 & 0 & 1 & -5/2 & -3/2 & 1/2 \end{pmatrix} \sim \begin{pmatrix} 1 & 0 & 0 & -16 & -11 & 3 \\ 0 & 1 & 0 & 7/2 & 5/2 & -1/2 \\ 0 & 0 & 1 & -5/2 & -3/2 & 1/2 \end{pmatrix} = (I : A^{-1})$$

4.96 Apply the Gaussian algorithm to

$$B = \begin{pmatrix} 1 & 3 & -4 \\ 1 & 5 & -1 \\ 3 & 13 & -6 \end{pmatrix}$$

▮ Form the block matrix $M = (B : I)$ and row reduce to echelon form:

$$\begin{pmatrix} 1 & 3 & -4 & 1 & 0 & 0 \\ 1 & 5 & -1 & 0 & 1 & 0 \\ 3 & 13 & -6 & 0 & 0 & 1 \end{pmatrix} \sim \begin{pmatrix} 1 & 3 & -4 & 1 & 0 & 0 \\ 0 & 2 & 3 & -1 & 1 & 0 \\ 0 & 4 & 6 & -3 & 0 & 1 \end{pmatrix} \sim \begin{pmatrix} 1 & 3 & -4 & 1 & 0 & 0 \\ 0 & 2 & 3 & -1 & 1 & 0 \\ 0 & 0 & 0 & -1 & -2 & 1 \end{pmatrix}$$

In echelon form, M has a zero row in its left half; that is, B is not row reducible to triangular form. Accordingly, B is not invertible.

4.97 Let A and B be invertible matrices of the same order. Show that the product AB is also invertible and $(AB)^{-1} = B^{-1}A^{-1}$.

▮
$$(AB)(B^{-1}A^{-1}) = A(BB^{-1})A^{-1} = AIA^{-1} = AA^{-1} = I$$
$$(B^{-1}A^{-1})(AB) = B^{-1}(A^{-1}A)B = B^{-1}IB = B^{-1}B = I$$

4.98 Let $A_1, A_2, \ldots, A_n$ be p-square invertible matrices. Show that $(A_1 A_2 \cdots A_n)^{-1} = A_n^{-1} \cdots A_2^{-1} A_1^{-1}$.

▮ The proof is by induction on n. For $n = 1$, we have $A_1^{-1} = A_1^{-1}$. Suppose $n > 1$ and the theorem holds for n. We prove it is true for $n + 1$. Using Problem 4.97, we have

$$(A_1 A_2 \cdots A_n A_{n+1})^{-1} = [(A_1 A_2 \cdots A_n) A_{n+1}]^{-1} = A_{n+1}^{-1} (A_1 A_2 \cdots A_n)^{-1} = A_{n+1}^{-1} A_n^{-1} \cdots A_2^{-1} A_1^{-1}$$

Thus the theorem holds for $n + 1$. Accordingly, the theorem holds for every positive n.

4.99 Show that if A has a zero row, then AB has a zero row.

▮ If row r of A is zero, so is row r of AB (see Problem 2.47).

4.100 Show that if A has a zero row, then A is not invertible.

▮ If A were invertible, then $AA^{-1} = I$ would imply a zero row in I.

4.101 Show that if B has a zero column, then AB has a zero column.

▮ If column c of B is zero, so is column c of AB (see Problem 2.47).

4.102 Show that if B has a zero column, then B is not invertible.

▮ If B were invertible, then $B^{-1}B = I$ would imply a zero column in I.

4.103 If A is invertible, show that kA is invertible when $k \neq 0$, with inverse $k^{-1}A^{-1}$.

▮ Since $k \neq 0$, $k^{-1} = 1/k$ exists. Then $(kA)(k^{-1}A^{-1}) = (kk^{-1})(AA^{-1}) = 1 \cdot I = I$. Hence $k^{-1}A^{-1}$ is the inverse of kA.

4.104 Suppose A and B are invertible. Show, by example, that $A + B$ need not be invertible.

▮ Choose $B = (-1)A$. Then $A + B = 0$ is not invertible.

4.105 Show that a diagonal matrix $D = \text{diag}\,(a_1, a_2, \ldots, a_n)$ is invertible if and only if no $a_i = 0$.

▌ If any $a_i = 0$, then D has a zero row and hence [Problem 4.100] D is not invertible. If no $a_i = 0$, so that each a_i^{-1} exists,

$$\text{diag}\,(a_1, a_2, \ldots, a_n) \cdot \text{diag}\,(a_1^{-1}, a_2^{-1}, \ldots, a_n^{-1}) = \text{diag}\,(1, 1, \ldots, 1) = I$$

Hence $D^{-1} = \text{diag}\,(a_1^{-1}, a_2^{-1}, \ldots, a_n^{-1})$.

4.106 Show that A is invertible if and only if A^T is invertible.

▌ If A is invertible, then there exists a matrix B such that $AB = BA = I$. Then $(AB)^T = (BA)^T = I^T$ and so $B^T A^T = A^T B^T = I$. Hence A^T is invertible, with inverse B^T. The converse follows from the fact that $(A^T)^T = A$.

4.107 Show that the operations of inversion and transposition commute; that is, $(A^T)^{-1} = (A^{-1})^T$.

▌ In Problem 4.106, B^T is the inverse of A^T; that is $B^T = (A^T)^{-1}$. But $B = A^{-1}$; hence $(A^{-1})^T = (A^T)^{-1}$.

4.7 ELEMENTARY MATRICES

4.108 Define *elementary matrix*.

▌ Let E be the matrix obtained by applying an elementary row operation e [Problem 2.79] to the identity matrix I; that is, let $E = e(i)$. Then E is called the elementary matrix corresponding to the row operation e.

4.109 Find the 3-square elementary matrix E_1 corresponding to the operation $R_1 \leftrightarrow R_2$. Apply the operation $R_1 \leftrightarrow R_2$ to I_3; that is, interchange the first and second rows of I_3 to obtain

$$E_1 = \begin{pmatrix} 0 & 1 & 0 \\ 1 & 0 & 0 \\ 0 & 0 & 1 \end{pmatrix}$$

4.110 Find the 3-square elementary matrix E_2 corresponding to the operation $R_3 \rightarrow -7R_3$.

▌ Apply the operation $R_3 \rightarrow -7R_3$ to I_3; that is, multiply the third row of I_3 by -7 to obtain

$$E_2 = \begin{pmatrix} 1 & 0 & 0 \\ 0 & 1 & 0 \\ 0 & 0 & -7 \end{pmatrix}$$

4.111 Find the 3-square elementary matrix E_3 corresponding to the operation $R_2 \rightarrow -3R_1 + R_2$.

▌ Apply the operation $R_2 \rightarrow -3R_1 + R_2$ to I_3; that is, replace the second row of I_3 by $-3R_1 + R_2$ to obtain

$$E_3 = \begin{pmatrix} 1 & 0 & 0 \\ -3 & 1 & 0 \\ 0 & 0 & 1 \end{pmatrix}$$

4.112 Let $e_i = (0, \ldots, 1, \ldots, 0)$ be the row vector with 1 in the ith position and 0 elsewhere. Show that $e_i A = R_i$, the ith row of A.

▌ Observe that e_i is the ith row of I, the identity matrix. By Problem 2.47, the ith row of $IA = A$ is $e_i A$.

> *Theorem 4.2*: Let e be an elementary row operation and E corresponding m-square elementary matrix; i.e., $E = e(I_m)$. Then for any $m \times n$ matrix A, $e(A) = EA$. That is, the result $e(A)$ of applying the operation e on the matrix A can be obtained by premultiplying A by the corresponding elementary matrix E.

4.113 Prove Theorem 4.2 if e is the elementary row operation $R_i \leftrightarrow R_j$.

▮ Let us use the sign $\frown$ to mark the ith component of a row vector; e.g., the ith row of I will be indicated as $e_i = (0, \ldots, \widehat{1}, \ldots, 0)$. Similarly, the sign will mark the jth component. Then

$$E = e(I) = (e_1, \ldots, \widehat{e_j}, \ldots, \widehat{\widehat{e_i}}, \ldots, e_m)^T \quad \text{and} \quad e(A) = (R_1, \ldots, \widehat{R_j}, \ldots, \widehat{\widehat{R_i}}, \ldots, R_m)^T$$

But, by Problem 2.47, the kth row of EA is the kth row of E, times A; hence, recalling Problem 4.112,

$$EA = (e_1 A, \ldots, \widehat{e_j A}, \ldots, \widehat{\widehat{e_i A}}, \ldots, e_m A)^T = (R_1, \ldots, \widehat{R_j}, \ldots, \widehat{\widehat{R_i}}, \ldots, R_m)^T = e(A)$$

4.114 Prove Theorem 4.2 if e is the elementary row operation $R_i \rightarrow kR_i$ $(k \neq 0)$.

▮ Using the notation of Problem 4.113,

$$E = e(I) = (e_1, \ldots, \widehat{ke_i}, \ldots, e_m)^T \quad \text{and} \quad e(A) = (R_1, \ldots, \widehat{kR_i}, \ldots, R_m)^T$$

Thus $$EA = (e_1 A, \ldots, \widehat{ke_i A}, \ldots, e_m A)^T = (R_1, \ldots, \widehat{kR_i}, \ldots, R_m)^T = e(A)$$

4.115 Prove Theorem 4.2 if e is the elementary row operation $R_i \rightarrow kR_j + R_i$.

▮ $$E = e(I) = (e_1, \ldots, \widehat{ke_j + e_i}, \ldots, e_m)^T \quad \text{and} \quad e(A) = (R_1, \ldots, \widehat{kR_j + R_i}, \ldots, R_m)^T$$

Using $(ke_j + e_i)A = k(e_j A) + e_i A = kR_j + R_i$, we have

$$EA = (e_1 A, \ldots, \widehat{(ke_j + e_i)A}, \ldots, e_m A)^T = (R_1, \ldots, \widehat{kR_j + R_i}, \ldots, R_m)^T = e(A)$$

4.116 Show that A is row equivalent to B if and only if there exist elementary matrices $E_1, \ldots, E_s$ such that $E_s \cdots E_2 E_1 A = B$.

▮ By definition, A is row equivalent to B if there exist elementary row operations $e_1, \ldots, e_s$ for which $e_s(\ldots(e_2(e_1(A)))\ldots) = B$. But, by Theorem 4.2, the above holds if and only if $E_s \cdots E_2 E_1 A = B$ where E_i is the elementary matrix corresponding to e_i.

4.117 Show that the elementary matrices are invertible and that their inverses are also elementary matrices.

▮ Let E be the elementary matrix corresponding to the elementary row operation e: $e(I) = E$. Let e' be the inverse operation of e and let E' be its corresponding elementary matrix. Then $I = e'(e(I)) = e'(E) = E'E$ and $I = e(e'(I)) = e(E') = EE$. Therefore E' is the inverse of E.

Theorem 4.3: The following are equivalent: (a) A is invertible; (b) A is row equivalent to the identity matrix I; (c) A is a product of elementary matrices.

4.118 Prove that (a) implies (b) in Theorem 4.3.

▮ Suppose A is invertible and suppose A is row equivalent to a matrix B in row canonical form. Then there exist elementary matrices $E_1, E_2, \ldots, E_s$ such that $E_s \cdots E_2 E_1 A = B$. Since A is invertible and each elementary matrix E_i is invertible, B is invertible [Problem 4.97]. But if $B \neq I$, then B has a zero row; hence B is not invertible [Problem 4.100]. Thus $B = I$, and (a) implies (b).

4.119 Prove that (b) implies (c) in Theorem 4.3.

▮ If (b) holds, then there exist elementary matrices $E_1, E_2, \ldots, E_s$ such that $E_s \cdots E_2 E_1 A = I$, and so $A = (E_s \cdots E_2 E_1)^{-1} = E_1^{-1} E_2^{-1} \cdots E_s^{-1}$. But the E_i^{-1} are also elementary matrices. Thus (b) implies (c).

4.120 Prove that (c) implies (a) in Theorem 4.3.

▮ If (c) holds, then $A = E_1 E_2 \cdots E_s$. The E_i are invertible matrices; hence their product, A, is also invertible. Thus (c) implies (a). Accordingly, the theorem is proved.

4.121 Let A and B be square matrices of the same order. Show that if $AB = I$, then $B = A^{-1}$. Accordingly, $AB = I$ if and only if $BA = I$.

   ▌ Suppose A is not invertible. Then A is not row equivalent to the identity matrix I, and so A is row equivalent to a matrix with a zero row. In other words, there exist elementary matrices $E_1, \ldots, E_s$ such that $E_s \cdots E_2 E_1 A$ has a zero row. Hence $E_s \cdots E_2 E_1 A B = E_s \cdots E_2 E_1$, an invertible matrix, also has a zero row. But this contradicts Problem 4.100. Thus, A is invertible and

$$B = IB = (A^{-1}A)B = A^{-1}(AB) = A^{-1}I = A^{-1}$$

4.122 Suppose A is the invertible and, say, it is row reducible to the identity matrix I by the sequence of elementary operations $e_1, \ldots, e_n$. Show that this sequence of elementary row operations applied to I yields A^{-1}.

   ▌ Let E_i be the elementary matrix corresponding to the operation e_i. Then, by hypothesis $E_n \cdots E_2 E_1 A = I$. Thus $(E_n \cdots E_2 E_1 I)A = I$ and hence $A^{-1} = E_n \cdots E_2 E_1 I$. In other words, A^{-1} can be obtained from I by applying the elementary row operations $e_1, \ldots, e_n$.

4.123 Show that B is row equivalent to A if and only if there exists an invertible matrix P such that $B = PA$.

   ▌ If $B \sim A$, then $B = e_s(\ldots(e_2(e_1(A)))\ldots) = E_s \cdots E_2 E_1 A = PA$, where $P = E_s \cdots E_2 E_1$ is invertible. The converse follows from the fact that each step is reversible.

4.124 Show that if AB is invertible, A is invertible. [Thus, if A has no inverse, AB has no inverse.]

   ▌ If AB is invertible, then there exists a matrix C such that $(AB)C = I$. Hence $A(BC) = I$ and BC is the inverse of A [by Problem 4.121].

4.8 COLUMN OPERATIONS. MATRIX EQUIVALENCE

4.125 List the three *elementary column operations*.

   ▌ $[F_1]$ Interchange the ith column and the jth column: $C_i \leftrightarrow C_j$.
      $[F_2]$ Multiply the ith column by a nonzero scalar: $k C_i \to kC_i$ $(k \neq 0)$.
      $[F_3]$ Replace the ith column by k times the jth column plus the ith column: $C_c i \to kC_j + C_i$.

4.126 Find the inverse of $[F_1]$ in Problem 4.125.

   ▌ Interchanging the same two columns twice yields the original matrix; hence $C_i \leftrightarrow C_j$ is its own inverse.

4.127 Find the inverse of $[F_2]$ in Problem 4.125.

   ▌ Since $k \neq 0$, the scalar k^{-1} exists. Then $C_i \to kC_i$ and $C_i \to k^{-1}C_i$ are inverses.

4.128 Find the inverse of $[F_3]$ in Problem 4.125.

   ▌ Applying $C_i \to kC_j + C_i$ and then $C_i \to -kC_j + C_i$, or vice versa, yields the original matrix. Hence they are inverses.

4.129 Find the 3-square elementary matrix F_1 corresponding to the column operation $C_1 \leftrightarrow C_2$.

   ▌ Apply $C_1 \to C_2$ to I_3:

$$\begin{pmatrix} 1 & 0 & 0 \\ 0 & 1 & 0 \\ 0 & 0 & 1 \end{pmatrix} \to \begin{pmatrix} 0 & 1 & 0 \\ 1 & 0 & 0 \\ 0 & 0 & 1 \end{pmatrix} = F_1$$

4.130 Find the 3-square elementary matrix F_2 corresponding to the column operation $C_2 \to -5C_2$.

   ▌ Apply $C_2 \to -5C_2$ to I_3 to obtain

$$F_2 = \begin{pmatrix} 1 & 0 & 0 \\ 0 & -5 & 0 \\ 0 & 0 & 1 \end{pmatrix}$$

4.131 Find the 3-square elementary matrix F_3 corresponding to the column operation $C_2 \rightarrow -4C_1 + C_2$.

▮ Apply the operation to I_3 to obtain

$$F_3 = \begin{pmatrix} 1 & -4 & 0 \\ 0 & 1 & 0 \\ 0 & 0 & 1 \end{pmatrix}$$

Notation: Let e and f denote, respectively, corresponding elementary row and column operations; let E and F denote the respective elementary matrices.

4.132 Show that $f(A) = [e(A^T)]^T$; that is, applying the column operation f to a matrix A gives the same result as applying the corresponding row operation e to A^T and then taking the transpose.

▮ This follows directly from the fact that the columns of A are the rows of A^T, and vice versa.

4.133 Show that F is the transpose of E.

▮
$$F = f(I) = [e(I^T)]^T = [e(I)]^T = E^T$$

Theorem 4.4: $f(A) = AF$.

4.134 Prove Theorem 4.4.

▮ By Problem 4.132 and Theorem 4.2, $f(A) = [e(A^T)]^T = [EA^T]^T = (A^T)^T E^T = AF$.

4.135 Under what conditions is B *column equivalent* to A?

▮ B is column equivalent to A if B can be obtained from A by applying a sequence of elementary column operations.

4.136 Show that B is column equivalent to A if and only if there exists an invertible matrix Q such that $B = AQ$.

▮ If B is column equivalent to A, then $B = f_s(\ldots(f_2(f_1(A)))\ldots) = AF_1 F_2 \cdots F_s \equiv AQ$, where $Q = F_1 F_2 \cdots F_s$ is invertible. The converse follows from the fact that each step is reversible.

4.137 When is B equivalent to A?

▮ B is equivalent to A if B can be obtained from A by a sequence of elementary row and/or column operations.

4.138 Show that B is equivalent to A if and only if there exist invertible matrices P and Q such that $B = PAQ$.

▮ If B is equivalent to A, then $B = E_s \cdots E_2 E_1 A F_1 F_2 \cdots F_t \equiv PAQ$ where $P = E_s \cdots E_2 E_1$ and $Q = F_1 F_2 \cdots F_t$ are invertible. The converse follows from the fact that each step is reversible.

Theorem 4.5: An $m \times n$ matrix A is equivalent to a block matrix $\left(\begin{array}{c|c} I_r & 0 \\ \hline 0 & 0 \end{array}\right)$. [The nonnegative integer r is called the *rank* of A.]

4.139 Prove Theorem 4.5.

▮ The proof is constructive, in the form of an algorithm.

Step 1. Row reduce A to row canonical form, with leading nonzero entries $a_{11}, a_{2j_2}, \ldots, a_{rj_r}$.

Step 2. Interchange C_2 and C_{j_2}, interchange C_3 and $C_{j_3}, \ldots$, and interchange C_r and C_{j_r}. This gives a matrix in the form $\left(\begin{array}{c|c} I_r & B \\ \hline 0 & 0 \end{array}\right)$, with leading nonzero entries $a_{11}, a_{22}, \ldots, a_{rr}$.

Step 3. Use column operations, with the a_{ii} as pivots, to replace each entry in B with a zero; i.e., for $i = 1, 2, \ldots, r$ and $j = r+1, r+2, \ldots, n$, apply the operation $C_j \rightarrow -b_{ij} C_i + C_j$.

4.9 UPPER TRIANGULAR AND OTHER SPECIAL MATRICES

4.140 Define an *upper triangular* matrix.

▌ A square matrix $A = (a_{ij})$ is upper triangular if all entries below the main diagonal are equal to zero; that is, if $a_{ij} = 0$ for $i > j$.

4.141 Display the generic upper triangular matrices of orders 2, 3, and 4.

▌
$$\begin{pmatrix} a_{11} & a_{12} \\ 0 & a_{22} \end{pmatrix} \qquad \begin{pmatrix} b_{11} & b_{12} & b_{13} \\ & b_{22} & b_{23} \\ & & b_{33} \end{pmatrix} \qquad \begin{pmatrix} c_{11} & c_{12} & c_{13} & c_{14} \\ & c_{22} & c_{23} & c_{24} \\ & & c_{33} & c_{34} \\ & & & c_{44} \end{pmatrix}$$

[As in diagonal matrices, it is common practice to omit patterns of 0s.]

Problems 4.142–4.147 refer to n-square upper triangular matrices $A = (a_{ij})$ and $B = (b_{ij})$.

4.142 Show that $A + B$ is upper triangular, with diagonal $[a_{11} + b_{11}, a_{22} + b_{22}, \ldots, a_{nn} + b_{nn}]$.

▌ Let $A + B = (c_{ij})$. If $i > j$, then $c_{ij} = a_{ij} + b_{ij} = 0 + 0 = 0$. Hence $A + B$ is upper triangular. Also, the $c_{ii} = a_{ii} + b_{ii}$ are the diagonal elements.

4.143 Show that kA is upper triangular, with diagonal $[ka_{11}, ka_{22}, \ldots, ka_{nn}]$.

▌ Let $kA = (c_{ij})$. If $i > j$, then $c_i = ka_{ij} = k \cdot 0 = 0$. Hence kA is upper triangular. Also, the $c_{ii} = ka_{ii}$ are the diagonal elements.

4.144 Show that the product AB is upper triangular.

▌ Let $AB = (c_{ij})$; then

$$C_{ij} = \sum_{k=1}^{n} a_{ik} b_{kj}$$

If $i > j$, then, for any k, either $i > k$ or $k > j$, so that either $a_{ik} = 0$ or $b_{kj} = 0$. Thus, $c_{ik} = 0$, and AB is upper triangular.

4.145 Show that the diagonal entries of AB are $a_{11}b_{11}, a_{22}b_{22}, \ldots, a_{nn}b_{nn}$.

▌ In the notation of Problem 4.144,

$$c_{ii} = \sum_{k=1}^{n} a_{ik} b_{ki}$$

But, for $k < i$, $a_{ik} = 0$; and, for $k > i$, $b_{ki} = 0$. Hence $c_{ii} = a_{ii} b_{ii}$, as claimed.

4.146 Show that the diagonal entries of A^p and $a_{11}^p, a_{22}^p, \ldots, a_{nn}^p$.

▌ This is an immediate consequence of Problem 4.145.

4.147 Show that, for any polynomial $f(x)$, the diagonal entries of $f(A)$ are $f(a_{11}), f(a_{22}), \ldots, f(a_{nn})$.

▌ This follows, by induction on the degree of $f(x)$, Problems 4.142, 4.143, and 4.146.

4.148 Show that the collection $\mathcal{T}_n$ of n-square upper triangular matrices form an algebra of matrices.

▌ This follows from the fact that $\mathcal{T}_n$ is nonempty and from Problems 4.142–4.144.

4.149 Show by example that the algebra $\mathcal{T}_2$ of 2-square upper triangular matrices is not commutative.

▌
$$\begin{pmatrix} 1 & 2 \\ 0 & 3 \end{pmatrix}\begin{pmatrix} 4 & 5 \\ 0 & 6 \end{pmatrix} = \begin{pmatrix} 4 & 17 \\ 0 & 18 \end{pmatrix} \quad \text{and} \quad \begin{pmatrix} 4 & 5 \\ 0 & 6 \end{pmatrix}\begin{pmatrix} 1 & 2 \\ 0 & 3 \end{pmatrix} = \begin{pmatrix} 4 & 23 \\ 0 & 18 \end{pmatrix}$$

4.150 Prove: If A is an n-square upper triangular matrix with a zero on its diagonal, then A is not invertible.

❚ Let $A = (a_{ij})$ and let k be the smallest integer such that $a_{kk} = 0$. Then A may be partitioned as

$$A = \left(\begin{array}{c|c} B & C \\ \hline \mathbf{0} & D \end{array}\right)$$

where B is of size $(k-1) \times k$, and where the size of D is $(n-k+1) \times (n-k)$. Thus D has more rows than columns, so row reducing D into echelon form will result in a zero row. Therefore row reducing A into echelon form will result in a zero row. Thus A is not invertible.

4.151 Suppose that A is triangular; that is, A is an upper triangular matrix with no diagonal entry equal to zero. Show that A is invertible and that its inverse is also triangular, with diagonal entries $a_{11}^{-1}, a_{22}^{-1}, \ldots, a_{nn}^{-1}$.

❚ Applying the Gaussian algorithm of Problem 4.92 to the block matrix $M = (A \quad I)$, we normalize the leading nonzero entries to unity by multiplying the ith row of M by a_{ii}^{-1} $(i = 1, 2, \ldots, n)$; this replaces I with $\mathrm{diag}\,[a_{11}^{-1}, a_{22}^{-1}, \ldots, a_{nn}^{-1}]$. Now we complete the conversion of A to I by adding suitable multiples of lower rows of M to upper rows; this induces the transformation of $\mathrm{diag}\,[a_{11}^{-1}, a_{22}^{-1}, \ldots, a_{nn}^{-1}]$ into a triangular matrix, A^{-1}, with the same diagonal entries.

4.152 Using only the elements 0 and 1, find (a) all 2×2 diagonal matrices, (b) all 2×2 upper triangular matrices.

❚ (a) Diagonal matrices must have 0s off the diagonal:

$$\begin{pmatrix} 1 & 0 \\ 0 & 1 \end{pmatrix} \quad \begin{pmatrix} 1 & 0 \\ 0 & 0 \end{pmatrix} \quad \begin{pmatrix} 0 & 0 \\ 0 & 1 \end{pmatrix} \quad \begin{pmatrix} 0 & 0 \\ 0 & 0 \end{pmatrix}$$

(b) Upper triangular matrices must have 0s below the diagonal. This gives the four matrices of (a), plus the four matrices obtained from them by changing the $(1, 2)$-element to 1.

Problems 4.153–4.155 are true/false questions. If false, give a counter example.

4.153 All square echelon matrices are upper triangular matrices.

❚ True.

4.154 All upper triangular matrices are in echelon form.

❚ False: $A = \begin{pmatrix} 0 & 1 \\ 0 & 1 \end{pmatrix}$ is upper triangular but not in echelon form.

4.155 If A^2 is an upper triangular matrix, then A is also upper triangular.

❚ False: Consider the matrix of Problem 4.70.

4.156 Find an upper triangular matrix A such that $A^3 = \begin{pmatrix} 8 & -57 \\ 0 & 27 \end{pmatrix}$.

❚ Set $A = \begin{pmatrix} x & y \\ 0 & z \end{pmatrix}$. Then [Problem 4.146], $x^3 = 8$, so $x = 2$; $y^3 = 27$, so $y = 3$. Next calculate A^3 using $x = 2$ and $z = 3$:

$$A^2 = \begin{pmatrix} 2 & y \\ 0 & 3 \end{pmatrix}\begin{pmatrix} 2 & y \\ 0 & 3 \end{pmatrix} = \begin{pmatrix} 4 & 5y \\ 0 & 9 \end{pmatrix} \quad \text{and} \quad A^3 = \begin{pmatrix} 2 & y \\ 0 & 3 \end{pmatrix}\begin{pmatrix} 4 & 5y \\ 0 & 9 \end{pmatrix} = \begin{pmatrix} 8 & 19y \\ 0 & 27 \end{pmatrix}$$

Thus $19y = -57$, or $y = -3$. Accordingly, $A = \begin{pmatrix} 2 & -3 \\ 0 & 3 \end{pmatrix}$.

4.157 Define a *lower triangular* matrix.

❚ A square matrix $A = (a_{ij})$ is lower triangular if all entries above the main diagonal are equal to zero; that is, if $a_{ij} = 0$ for $i < j$.

4.158 Exhibit the generic lower triangular matrices of orders 2, 3, and 4.

❚ In each case put 0s above the diagonal:

$$\begin{pmatrix} a_{11} & 0 \\ a_{21} & a_{22} \end{pmatrix} \quad \begin{pmatrix} b_{11} & & \\ b_{21} & b_{22} & \\ b_{31} & b_{32} & b_{33} \end{pmatrix} \quad \begin{pmatrix} c_{11} & & & \\ c_{21} & c_{22} & & \\ c_{31} & c_{32} & c_{33} & \\ c_{41} & c_{42} & c_{43} & c_{44} \end{pmatrix}$$

4.159 A is lower triangular if and only if A^T is upper triangular: True or false?

 ▮ True. [Because of this, the theory of lower triangular matrices is essentially the same as that of upper triangular matrices.]

4.160 With reference to Problem 4.159, verify that the product of lower diagonal matrices is lower diagonal.

 ▮ If A and B are lower diagonal, then B^T, A^T, and with them $B^T A^T = (AB)^T$, are upper diagonal; hence, $((AB)^T)^T = AB$ is lower diagonal.

4.161 What kinds of matrices are both upper triangular and lower triangular?

 ▮ If A is both upper and lower triangular, then every entry off the main diagonal must be zero. Hence A is diagonal.

4.162 Define a *tridiagonal* matrix.

 ▮ A square matrix is tridiagonal if the nonzero entries occur only on the diagonal, directly above the diagonal [on the *superdiagonal*], or directly below the diagonal [on the *subdiagonal*].

4.163 Display the generic tridiagonal matrices of orders 4 and 5.

 ▮ In each case put 0s outside the diagonal, superdiagonal, or subdiagonal:

$$\begin{pmatrix} a_{11} & a_{21} & & \\ a_{21} & a_{22} & a_{31} & \\ & a_{32} & a_{33} & a_{34} \\ & & a_{43} & a_{44} \end{pmatrix} \qquad \begin{pmatrix} b_{11} & b_{21} & & & \\ b_{21} & b_{22} & b_{23} & & \\ & b_{32} & b_{33} & b_{34} & \\ & & b_{43} & b_{44} & b_{45} \\ & & & b_{54} & b_{55} \end{pmatrix}$$

4.164 Show that the product of tridiagonal matrices need not be tridiagonal.

 ▮

$$\begin{pmatrix} 1 & 1 & 0 \\ 1 & 1 & 1 \\ 0 & 1 & 1 \end{pmatrix}\begin{pmatrix} 1 & 1 & 0 \\ 1 & 1 & 1 \\ 0 & 1 & 1 \end{pmatrix} = \begin{pmatrix} 2 & 2 & 1 \\ 2 & 3 & 2 \\ 1 & 2 & 2 \end{pmatrix}$$

4.10 SYMMETRIC MATRICES

4.165 Define a *symmetric* matrix.

 ▮ A real matrix A is symmetric if $A^T = A$. Equivalently, $A = (a_{ij})$ is symmetric if each $a_{ij} = a_{ji}$. [Note that A must be square in order for $A^T = A$.]

4.166 Define a *skew-symmetric* (or *antisymmetric*) matrix.

 ▮ A real matrix A is skew-symmetric if $A^T = -A$. Equivalently, $A = (a_{ij})$ is skew-symmetric if each $a_{ij} = -a_{ji}$.

4.167 Show that the diagonal elements of a skew-symmetric matrix must be zero.

 ▮ If $A = (a_{ij})$ is skew-symmetric, then $a_{ii} = -a_{ii}$. Hence each $a_{ii} = 0$.

   Problems 4.168–4.174 involve the following matrices:

$$A = \begin{pmatrix} 0 & 5 & -2 \\ -5 & 0 & 3 \\ 2 & -3 & 0 \end{pmatrix} \qquad B = \begin{pmatrix} 4 & -7 & 1 \\ -7 & 3 & 2 \\ 1 & 2 & -5 \end{pmatrix} \qquad C = \begin{pmatrix} 1 & 1 & 1 \\ 1 & 1 & 1 \end{pmatrix}$$

$$D = \begin{pmatrix} 1 & 1 \\ 1 & 0 \end{pmatrix} \qquad E = \begin{pmatrix} 1 & 1 \\ -1 & 0 \end{pmatrix} \qquad F = \begin{pmatrix} 0 & 1 \\ -1 & 0 \end{pmatrix} \qquad G = \begin{pmatrix} 0 & 0 \\ 0 & 0 \end{pmatrix}$$

4.168 Is A symmetric or skew-symmetric?

 ▮ By inspection, $A^T = -A$; thus A is skew-symmetric.

4.169 Is B symmetric or skew-symmetric?

⫿ By inspection, $B^T = B$; thus B is symmetric.

4.170 Is C symmetric or skew-symmetric?

⫿ Since C is not square, C is neither symmetric nor skew-symmetric.

4.171 Is D symmetric or skew-symmetric?

⫿ By inspection $D^T = D$; hence D is symmetric.

4.172 Is E symmetric or skew-symmetric?

⫿ We see that $E^T \neq \pm E$. Accordingly, E is neither symmetric nor skew-symmetric.

4.173 Is F symmetric or skew-symmetric?

⫿ By inspection, $F^T = -F$; hence F is skew-symmetric.

4.174 Is G symmetric or skew-symmetric?

⫿ Both, since $\mathbf{0}^T = \mathbf{0} = -\mathbf{0}$ when $\mathbf{0}$ is square.

4.175 Is the identity matrix I symmetric?

⫿ Since $I^T = I$, the identity matrix is symmetric.

4.176 Is every $\mathbf{0}$ matrix symmetric?

⫿ If $\mathbf{0}$ is not square, then $\mathbf{0}^T$ and $\mathbf{0}$ cannot be equal, being of different sizes, and so $\mathbf{0}$ is not symmetric.

4.177 Find x and A, if $A = \begin{pmatrix} 4 & x+2 \\ 2x-3 & x+1 \end{pmatrix}$ is symmetric.

⫿ Set the symmetric elements [mirror images in the diagonal] $x+2$ and $2x-3$ equal to each other, to obtain $x = 5$; hence $A = \begin{pmatrix} 4 & 7 \\ 7 & 6 \end{pmatrix}$.

4.178 Find x, y, z, t, if

$$B = \begin{pmatrix} 5 & 2 & x \\ y & z & -3 \\ 4 & t & -7 \end{pmatrix}$$

is symmetric.

⫿ Equate symmetric elements to obtain $x = 4$, $y = 2$, $t = -3$. The unknown z, on the diagonal, is indeterminate.

4.179 Suppose $A = (a_{ij})$ and $B = (b_{ij})$ are symmetric. Show that $A + B$ is symmetric.

⫿ If $A + B = (c_{ij})$, then $c_{ij} = a_{ij} + b_{ij} = a_{ji} + b_{ji} = c_{ji}$.

4.180 Suppose $A = (a_{ij})$ is symmetric. Show that kA is symmetric

⫿ If $kA = (c_{ij})$, then $c_{ij} = ka_{ij} = ka_{ji} = c_{ji}$.

4.181 Show that AB need not be symmetric, even though A and B are symmetric.

⫿ Let $A = \begin{pmatrix} 1 & 2 \\ 2 & 3 \end{pmatrix}$ and $B = \begin{pmatrix} 4 & 5 \\ 5 & 6 \end{pmatrix}$. Then $AB = \begin{pmatrix} 14 & 17 \\ 23 & 28 \end{pmatrix}$ is not symmetric. [See Problem 4.182.]

4.182 Let A and B be symmetric matrices. Show that AB is symmetric if and only if A and B commute.

⫿ If $AB = BA$, then $(AB)^T = B^T A^T = BA = AB$, and hence AB is symmetric. Conversely, $(AB)^T = AB$, then $AB = (AB)^T = B^T A^T = BA$, and so A and B commute.

4.183 Suppose $A = (a_{ij})$ is skew-symmetric. Show that kA is skew-symmetric.

▌ If $kA = (c_{ij})$, then $c_{ij} = ka_{ij} = k(-a_{ji}) - = (ka_{ji}) = -c_{ji}$.

Theorem 4.6: If A is a square matrix, then (i) $A + A^T$ is symmetric; (ii) $A - A^T$ is skew-symmetric; (iii) $A = B + C$, for some symmetric matrix B and some skew-symmetric matrix C.

4.184 Prove (i) of Theorem 4.6.

▌ $(A + A^T)^T = A^T + (A^T)^T = A^T + A = A + A^T$

4.185 Prove (ii) of Theorem 4.6.

▌ $(A - A^T)^T = A^T - (A^T)^T = A^T - A = -(A - A^T)$

4.186 Prove (iii) of Theorem 4.6.

▌ Set $B = \frac{1}{2}(A + A^T)$ and $C = \frac{1}{2}(A - A^T)$. Then $A = B + C$, where A is symmetric by Problems 4.184 and 4.180, and C is skew-symmetric by Problems 4.185 and 4.183.

4.187 Prove that the decomposition of Theorem 4.6 (iii) is unique.

▌ $A = B + C$ and $A = B' + C'$ imply, by subtraction,

$$0 = B - B' + C - C' \tag{1}$$

and, taking the transpose of *(1)*,

$$0 = B - B' - C + C' \tag{2}$$

Addition of *(1)* and *(2)* gives $2(B - B') = 0$, or $B = B'$; then also $C = C'$.

4.188 Write $A = \begin{pmatrix} 2 & 3 \\ 7 & 8 \end{pmatrix}$ as the sum of a symmetric matrix B and a skew-symmetric matrix C.

▌ Calculate

$$A^T = \begin{pmatrix} 2 & 7 \\ 3 & 8 \end{pmatrix} \qquad A + A^T = \begin{pmatrix} 4 & 10 \\ 10 & 16 \end{pmatrix} \qquad A - A^T = \begin{pmatrix} 0 & -4 \\ 4 & 0 \end{pmatrix}$$

Then, by Problem 4.187,

$$B = \tfrac{1}{2}(A + A^T) = \begin{pmatrix} 2 & 5 \\ 5 & 8 \end{pmatrix} \qquad C = \tfrac{1}{2}(A - A^T) = \begin{pmatrix} 0 & -2 \\ 2 & 0 \end{pmatrix}$$

4.189 Show that A is symmetric if and only if A^T is symmetric.

▌ Follows from the fact that $(A^T)^T = A$.

4.190 Suppose A is symmetric. Show that A^2 and, in general, A^n is symmetric.

▌ $(A^2)^T = (AA)^T = A^T A^T = AA = A^2$. Also, by induction, $(A^n)^T = (AA^{n-1})^T = (A^{n-1})^T A^T = A^{n-1}A = A^n$.

4.191 Show that if A is symmetric, then $f(A)$ is symmetric for any polynomial $f(x)$.

▌ Follows from Problems 4.190, 4.179, and 4.180.

4.192 Let A be an n-square symmetric matrix and P any $n \times m$ matrix. Show that $P^T A P$ is also symmetric.

▌ $(P^T A P)^T = P^T A^T (P^T)^T = P^T A P$

4.11 COMPLEX MATRICES

This section assumes that the scalars and, in particular, the entries in the matrices are complex numbers. Recall [Problem 1.117] that if $z = a + bi$ is a complex number, then $\bar{z} = a - bi$ is its conjugate.

4.193 Define the (*complex*) *conjugate* of a matrix A, denoted $\bar{A}$.

▎ If $A = (a_{ij})$, then $\bar{A} \equiv (\bar{a}_{ij})$.

4.194 Find the conjugate of $\begin{pmatrix} 2+i & 3-5i & 4+8i \\ 6-i & 2-9i & 5+6i \end{pmatrix}$.

▎
$$\overline{\begin{pmatrix} 2+i & 3-5i & 4+8i \\ 6-i & 2-9i & 5+6i \end{pmatrix}} = \begin{pmatrix} \overline{2+i} & \overline{3-5i} & \overline{4+8i} \\ \overline{6-i} & \overline{2-9i} & \overline{5+6i} \end{pmatrix} = \begin{pmatrix} 2-i & 3+5i & 4-8i \\ 6+i & 2+9i & 5-6i \end{pmatrix}$$

4.195 Find the conjugate of $\begin{pmatrix} 6+7i & 5i \\ 9 & 4-i \end{pmatrix}$.

▎
$$\overline{\begin{pmatrix} 6+7i & 5i \\ 9 & 4-i \end{pmatrix}} = \begin{pmatrix} \overline{6+7i} & \overline{5i} \\ \overline{9} & \overline{4-i} \end{pmatrix} = \begin{pmatrix} 6-7i & -5i \\ 9 & 4+i \end{pmatrix}$$

4.196 When does $\bar{A} = A$?

▎ When, and only when, $\overline{a_{ij}} = a_{ij}$ for all i, j; that is, when and only when A is a real matrix.

4.197 Find $\bar{A}$, if $A = \begin{pmatrix} 3 & 7 & -5 \\ 6 & -2 & 8 \end{pmatrix}$.

▎ Since A is real, $\bar{A} = A$.

Theorem 4.7: (i) $\overline{A+B} = \bar{A} + \bar{B}$, (ii) $\overline{kA} = \bar{k}\,\bar{A}$, (iii) $\bar{\bar{A}} = A$, (iv) $\overline{AB} = \bar{A}\,\bar{B}$, (v) $(\bar{A})^T = \overline{A^T}$.

4.198 Prove (i) of Theorem 4.7.

▎ As usual, write $A = (a_{ij})$, $B = (b_{ij})$, and $\overline{A+B} = (c_{ij})$. Then $c_{ij} = \overline{a_{ij} + b_{ij}} = \overline{a_{ij}} + \overline{b_{ij}}$. Hence $\overline{A+B} = \bar{A} + \bar{B}$.

4.199 Prove (ii) of Theorem 4.7.

▎ Since $kA = (ka_{ij})$, $\overline{kA} = (\overline{ka_{ij}}) = (\bar{k}\,\overline{a_{ij}}) = \bar{k}(\overline{a_{ij}}) = \bar{k}\,\bar{A}$.

4.200 Prove (iii) of Theorem 4.7.

▎ Write $\bar{\bar{A}} = (c_{ij})$. Then $c_{ij} = \overline{\overline{a_{ij}}} = a_{ij}$. Thus $\bar{\bar{A}} = A$.

4.201 Prove (iv) of Theorem 4.7

▎ Let $\overline{AB} = (c_{ij})$. Then $c_{ij} = \overline{\Sigma_k a_{ik} b_{kj}} = \Sigma_k \overline{a_{ik} b_{kj}} = \Sigma_k \overline{a_{ik}}\,\overline{b_{kj}}$. Thus $\overline{AB} = \bar{A}\,\bar{B}$.

4.202 Prove (v) of Theorem 4.7.

▎ Let $(\bar{A})^T = (b_{ij})$ and $\overline{A^T} = (c_{ij})$. Using the notation $q_{ij}^{\;T} = q_{ji}$, we have: $b_{ij} = \overline{a_{ij}}^{\;T} = \overline{a_{ji}}$ and $c_{ij} = a_{ij}^{\;T} = \overline{a_{ji}}$. Thus $b_{ij} = c_{ij}$, whence $(\bar{A})^T = \overline{A^T}$. [In words: the operations of transposition and conjugation commute.]

4.203 The conjugate transpose [= transposed conjugate] of a matrix A is indicated as A^H [after the mathematician Hermite]. Find A^H if $A = \begin{pmatrix} 2+8i & 5-3i & 4-7i \\ 6i & 1-4i & 3+2i \end{pmatrix}$.

▎
$$A^H = \begin{pmatrix} \overline{2+8i} & \overline{6i} \\ \overline{5-3i} & \overline{1-4i} \\ \overline{4-7i} & \overline{3+2i} \end{pmatrix} = \begin{pmatrix} 2-8i & -6i \\ 5+3i & 1+4i \\ 4+7i & 3-2i \end{pmatrix}$$

4.204 When will $A^H = A^T$?

▎ If A is real, then $\bar{A} = A$, and so $A^H = A^T$; and conversely.

In Problems 4.205–4.208 it is shown that Theorem 2.3 remains valid when the transpose is replaced by the conjugate transpose. As before, matrices A and B need not be square, but merely compatible for addition or for matrix multiplication.

4.205 Prove (i): $(A + B)^H = A^H + B^H$.

▍ $(A + B)^H = (\overline{A + B})^T = (\bar{A} + \bar{B})^T = \bar{A}^T + \bar{B}^T = A^H + B^H$

4.206 Prove (ii): $(A^H)^H = A$.

▍ $(A^H)^H = (\overline{\bar{A}^T})^T = ((\bar{\bar{A}})^T)^T = (A^T)^T = A$

4.207 Prove (iii): $(kA)^H = \bar{k} A^H$. [Note that in Theorem 2.3(iii), $\bar{k} = k$.]

▍ $(kA)^H = (\overline{kA})^T = (\bar{k} \bar{A})^T = \bar{k} \bar{A}^T = \bar{k} A^H$

4.208 Prove (iv): $(AB)^H = B^H A^H$.

▍ $(A^B)^H = (\overline{AB})^T = (\bar{A} \bar{B})^T = \bar{B}^T \bar{A}^T = B^H A^H$

4.12 HERMITIAN MATRICES

4.209 Define: (a) Hermitian matrix, (b) skew-Hermitian matrix.

▍ (a) A complex matrix A is Hermitian if $A^H = A$; equivalently, $A = (a_{ij})$ is Hermitian if each $a_{ij} = \overline{a_{ji}}$. Note that only square matrices can be Hermitian. (b) A complex matrix A is skew-Hermitian if $A^H = -A$; equivalently $A = (a_{ij})$ is skew-Hermitian if each $a_{ij} = -\overline{a_{ji}}$.

4.210 Show that the diagonal elements of a Hermitian [skew-Hermitian] matrix are real [pure imaginary].

▍ $\text{Im } a_{rr} = \frac{1}{2i}(a_{rr} - \overline{a_{rr}}) = \frac{1}{2i}(a_{rr} - a_{rr}) = 0$ [see Problem 1.118] when $A = (a_{ij})$ is Hermitian, and $\text{Re } a_{rr} = \frac{1}{2}(a_{rr} + \overline{a_{rr}}) = \frac{1}{2}(a_{rr} - a_{rr}) = 0$ when A is skew-Hermitian.

4.211 Show that if A is Hermitian and skew-Hermitian, then $A = 0$.

▍ If $A^H = A = -A$, then $2A = 0$, or $A = 0$.

Problems 4.212–4.217 involve the following matrices:

$$A = \begin{pmatrix} 4 & 3 - 5i \\ 3 + 5i & -7 \end{pmatrix} \qquad B = \begin{pmatrix} 4i & 3 + 2i \\ -3 + 2i & -7i \end{pmatrix}, \qquad C = \begin{pmatrix} 0 & 5 - 7i \\ 4 + 3i & 0 \end{pmatrix}$$

$$D = \begin{pmatrix} 3 & 1 - 2i & 4 + 7i \\ 1 + 2i & -4 & -2i \\ 4 - 7i & 2i & 2 \end{pmatrix} \qquad E = \begin{pmatrix} 5i & -4 + i & 3 - 2i \\ 4 + i & 2i & -2 + i \\ -3 - 2i & 2 + i & -6i \end{pmatrix} \qquad F = \begin{pmatrix} 6 & 7 & -2 \\ 7 & -1 & 8 \\ -2 & 8 & -3 \end{pmatrix}$$

4.212 Is A Hermitian or skew-Hermitian?

▍ The diagonal elements, 4 and −7, are real, and $3 - 5i$ and $3 + 5i$ are conjugates; hence A is Hermitian.

4.213 Is B Hermitian or skew-Hermitian?

▍ The diagonal elements, $4i$ and $-7i$, are pure imaginary, and $3 + 2i$ and $-3 + 2i$ are negative conjugates [the real parts are negatives and the imaginary parts are equal]. Hence B is skew-Hermitian.

4.214 Is C Hermitian or skew-Hermitian?

▍ The symmetric elements $5 - 7i$ and $4 + 3i$ are neither conjugates nor negative conjugates. Hence C is neither Hermitian nor skew-Hermitian.

4.215   Is D Hermitian or skew-Hermitian?

I   The diagonal elements, 3, -4 and 2, are real, and the symmetric elements, $1-2i$ and $1+2i$, $4+7i$ and $4-7i$, and $-2i$ and $2i$, are conjugates.   Hence D is Hermitian.

4.216   Is E Hermitian or skew-Hermitian?

I   The diagonal elements, $5i$, $2i$, and $-6i$, are pure imaginary, and the symmetric pairs of elements, $-4+i$ and $4+i$, $3-2i$ and $-3-2i$, and $-2+i$ and $2+i$, are negative conjugates.   Hence E is skew-Hermitian.

4.217   If F Hermitian or skew-Hermitian?

I F is real and symmetric; viewed as a complex matrix, F is Hermitian.

4.218   Is $G = \operatorname{diag}[5, 2i, -3, 4+i]$ Hermitian or skew-Hermitian.

I   The diagonal elements of G are neither all real nor all imaginary.   Hence G is neither Hermitian nor skew-Hermitian.

4.219   If A and B are Hermitian, show that $A+B$ is Hermitian.

I   By Problem 4.205, $(A+B)^H = A^H + B^H = A + B$.

4.220   Suppose A Hermitian and k real.   Show that kA is Hermitian.

I   By Problem 4.207, $(kA)^H = \bar{k}A^H = kA$.

4.221   Suppose A skew-Hermitian and k real.   Show that kA is skew-Hermitian.

I   By Problem 4.207, $(kA)^H = \bar{k}A^H = k(-A) = -(kA)$.

4.222   If A is an arbitrary complex matrix, show that the matrices AA^H and A^HA are Hermitian.

I   First of all, A and A^H are compatible for multiplication in either order, yielding a square product.   Then, by Problems 4.208 and 4.206,

$$(AA^H)^H = (A^H)^H A^H = AA^H \quad \text{and} \quad (A^HA)^H = A^H(A^H)^H = A^HA$$

4.223   When do two Hermitian matrices have a Hermitian product?

I   We have $(AB)^H = B^HA^H = BA$.   Hence AB is Hermitian if and only if $AB = BA$.   [Compare Problem 4.182.]

4.224   Show that if A is Hermitian, so is $A^p (p = 2, 3, \ldots)$.

I   Since A and A^{k-1} commute an induction immediately gives the result.

Following is the complex analogue to Theorem 4.6.

Theorem 4.8:   If A is a square matrix, then (i) $A + A^H$ is Hermitian; (ii) $A - A^H$ is skew-Hermitian; (iii) $A = B + C$, where B is Hermitian and C is skew-Hermitian.

4.225   Prove (i) of Theorem 4.8.

I $(A + A^H)^H = A^H + (A^H)^H = A^H + A = A + A^H$

4.226   Prove (ii) of Theorem 4.8.

I $(A - A^H)^H = A^H - (A^H)^H = A^H - A = -(A - A^H)$

4.227   Prove (iii) of Theorem 4.8.

I   Set $B + \frac{1}{2}(A + A^H)$ and $C = \frac{1}{2}(A - A^H)$.   Then $A = B + C$, where B is Hermitian by Problems 4.225 and 4.220, and C is skew-Hermitian by Problems 4.226 and 4.221.   [Uniqueness of B and C is proved as in Problem 4.187.]

4.228 Write $A = \begin{pmatrix} 2+6i & 5+3i \\ 9-i & 4-2i \end{pmatrix}$ in the form $A = B + C$, where B is Hermitian and C is skew-Hermitian.

▮ $A^H = \begin{pmatrix} 2-6i & 9+i \\ 5-3i & 4+2i \end{pmatrix}$ $A + A^H = \begin{pmatrix} 4 & 14+4i \\ 14-4i & 8 \end{pmatrix}$ $A - A^H = \begin{pmatrix} 12i & -4+2i \\ 4+2i & -4i \end{pmatrix}$

and the desired matrices are

$$B = \frac{1}{2}(A + A^H) = \begin{pmatrix} 2 & 7+2i \\ 7-2i & 4 \end{pmatrix} \quad \text{and} \quad C = \frac{1}{2}(A - A^H) = \begin{pmatrix} 6i & -2+i \\ 2+i & -2i \end{pmatrix}$$

4.229 Suppose A is an n-square Hermitian matrix. Show that $P^H A P$ is Hermitian for any $n \times m$ matrix P.

▮ The product is defined as an m-square matrix; we have $(P^H A P)^H = P^H A^H (P^H)^H = P^H A P$.

4.13 ORTHOGONAL MATRICES

4.230 Define *orthogonal* matrix.

▮ A real matrix A is said to be orthogonal if $AA^T = A^T A = I$. Observe that an orthogonal matrix A is necessarily square and invertible, with inverse $A^{-1} = A^T$.

4.231 Show that

$$A = \begin{pmatrix} 1/9 & 8/9 & -4/9 \\ 4/9 & -4/9 & -7/9 \\ 8/9 & 1/9 & 4/9 \end{pmatrix}$$

is orthogonal.

▮ By Problem 4.121, we need only show that $AA^T = I$:

$$AA^T = \begin{pmatrix} 1/9 & 8/9 & -4/9 \\ 4/9 & -4/9 & -7/9 \\ 8/9 & 1/9 & 4/9 \end{pmatrix} \begin{pmatrix} 1/9 & 4/9 & 8/9 \\ 8/9 & -4/9 & 1/9 \\ -4/9 & -7/9 & 4/9 \end{pmatrix} = \frac{1}{81} \begin{pmatrix} 1+64+16 & 4-32+28 & 8+8-16 \\ 4-32+28 & 16+16+49 & 32-4-28 \\ 8+8-16 & 32-4-28 & 64+1+16 \end{pmatrix}$$

$$= \frac{1}{81} \begin{pmatrix} 81 & 0 & 0 \\ 0 & 81 & 0 \\ 0 & 0 & 81 \end{pmatrix} = \begin{pmatrix} 1 & 0 & 0 \\ 0 & 1 & 0 \\ 0 & 0 & 1 \end{pmatrix} = I$$

4.232 Define an *orthogonal set* of vectors in $\mathbf{R}^n$.

▮ The vectors $u_1, u_2, \ldots, u_r$ form an orthonormal set if the vectors are pairwise orthogonal $[u_i \cdot u_j = 0$ for $i \neq j]$ and if the vectors have unit lengths $[u_i \cdot u_i = 1$ for $i = 1, 2, \ldots, r]$. In terms of the Kronecker delta, the condition for orthogonality is $u_i \cdot u_j = \delta_{ij}$.

4.233 Show that

$$A = \begin{pmatrix} a_1 & a_2 & a_3 \\ b_1 & b_2 & b_3 \\ c_1 & c_2 & c_3 \end{pmatrix}$$

is orthogonal if and only if its rows $u_1 = (a_1, a_2, a_3)$, $u_2 = (b_1, b_2, b_3)$, $u_3 = (c_1, c_2, c_3)$ form an orthonormal set.

▮ If A is orthogonal, then

$$AA^T = \begin{pmatrix} a_1 & a_2 & a_3 \\ b_1 & b_2 & b_3 \\ c_1 & c_2 & c_3 \end{pmatrix} \begin{pmatrix} a_1 & b_1 & c_1 \\ a_2 & b_2 & c_2 \\ a_3 & b_3 & c_3 \end{pmatrix} = \begin{pmatrix} 1 & 0 & 0 \\ 0 & 1 & 0 \\ 0 & 0 & 1 \end{pmatrix} = I$$

This yields

$$a_1^2 + a_2^2 + a_3^2 = u_1 \cdot u_1 = 1 \qquad a_1 b_1 + a_2 b_2 + a_3 b_3 = u_1 \cdot u_2 = 0 \quad a_1 c_1 + a_2 c_2 + a_3 c_3 = u_1 \cdot u_3 = 0$$

$$b_1 a_1 + b_2 a_2 + b_3 a_3 = u_2 \cdot u_1 = 0 \qquad b_1^2 + b_2^2 + b_3^2 = u_2 \cdot u_2 = 1 \qquad b_1 c_1 + b_2 c_2 + b_3 c_3 = u_2 \cdot u_3 = 0$$

$$c_1 a_1 + c_2 a_2 + c_3 a_3 = u_3 \cdot u_1 = 0 \qquad c_1 b_1 + c_2 b_2 + c_3 b_3 = u_3 \cdot u_2 = 0 \quad c_1^2 + c_2^2 + c_3^2 = u_3 \cdot u_3 = 1$$

that is, $u_i \cdot u_j = \delta_{ij}$. Accordingly, u_1, u_2, u_3 form an orthonormal set. The converse follows from the fact that each step is reversible.

4.234   Show that A is orthogonal iff A^T is orthogonal.

▎ $AA^T = A^TA = I$ iff $A^T(A^T)^T = (A^T)^TA^T = I$.

Problems 4.235 and 4.236 establish

> **Theorem 4.9:**   Let A be a real matrix.   Then the following are equivalent: (a) A is orthogonal; (b) The rows of A form an orthonormal set; (c) the columns of A form an orthonormal set.

4.235   Prove that (a) and (b) are equivalent.

▎ Let $R_1, R_2, \ldots, R_n$ denote the rows of A; then $R_1^T, R_2^T, \ldots, R_n^T$ are the columns of A^T.   Now let $AA^T = (c_{ij})$.   By matrix multiplication, $c_{ij} = R_i R_j^T = R_i \cdot R_j$.   Thus $AA^T = I$ iff $R_i \cdot R_j = \delta_{ij}$ iff the rows $R_1, R_2, \ldots, R_n$ form an orthonormal set.

4.236   Prove that (a) and (c) are equivalent.

▎ By Problems 4.234 and 4.235, A is orthogonal iff A^T is orthogonal iff the rows of A^T form an orthonormal set iff the columns of A form an orthonormal set.

4.237   If $A = \begin{pmatrix} 1/\sqrt{5} & 2/\sqrt{5} \\ x & y \end{pmatrix}$ is orthogonal, find x and y.

▎ Let R_1, R_2 and C_1, C_2 denote, respectively, the rows and columns of A.   Since $R_1 \cdot R_2 = 0$,   we get $x/\sqrt{5} + 2y/\sqrt{5} = 0$ or $x + 2y = 0$.   Since C_1 is a unit vector, we get $x^2 + 1/5 = 1$ or $x = \pm 2/\sqrt{5}$.
Case (i): $x = 2/\sqrt{5}$.   Then $x + 2y = 0$ yields $y = -1/\sqrt{5}$.
Case (ii): $x = -2/\sqrt{5}$.   Then $x + 2y = 0$ yields $y = 1/\sqrt{5}$.
In other words, there are exactly two possibilities:

$$A = \begin{pmatrix} 1/\sqrt{5} & 2/\sqrt{5} \\ -2/\sqrt{5} & 1/\sqrt{5} \end{pmatrix} \quad \text{and} \quad A = \begin{pmatrix} 1/\sqrt{5} & 2/\sqrt{5} \\ 2/\sqrt{5} & -1/\sqrt{5} \end{pmatrix}$$

4.238   If

$$A = \begin{pmatrix} x & 2/3 & 2/3 \\ 2/3 & 1/3 & y \\ z & s & t \end{pmatrix}$$

is orthogonal, find x, y, z, s, t.

▎ Let R_1, R_2, R_3 denote the rows of A, and let C_1, C_2, C_3 denote the columns of A.   Since R_1 is a unit vector, $x^2 + 4/9 + 4/9 = 1$,   or $x = \pm 1/3$.   Since R_2 is a unit vector, $4/9 + 1/9 + y^2 = 1$,   or $y = \pm 2/3$. Since $R_1 \cdot R_2 = 0$,   we get $2x/3 + 2/9 + 2y/3 = 0$,   or $3x + 3y = -1$.   The only possibility is that $x = 1/3$ and $y = -2/3$.   Thus,

$$A = \begin{pmatrix} 1/3 & 2/3 & 2/3 \\ 2/3 & 1/3 & -2/3 \\ z & s & t \end{pmatrix}$$

Since the columns are unit vectors,

$$\frac{1}{9} + \frac{4}{9} + z^2 = 1 \qquad \frac{4}{9} + \frac{1}{9} + s^2 = 1 \qquad \frac{4}{9} + \frac{4}{9} + t^2 = 1$$

Thus $z = \pm 2/3$, $s = \pm 2/3$ and $t = \pm 1/3$.
Case (i): $z = 2/3$.   Since C_1 and C_2 are orthogonal, $s = -2/3$; since C_1 and C_3 are orthogonal, $t = 1/3$.
Case (ii): $z = -2/3$.   Since C_1 and C_2 are orthogonal, $s = 2/3$; since C_1 and C_3 are orthogonal, $t = -1/3$.
Hence there are exactly two possible solutions:

$$A = \begin{pmatrix} 1/3 & 2/3 & 2/3 \\ 2/3 & 1/3 & -2/3 \\ 2/3 & -2/3 & 1/3 \end{pmatrix} \quad \text{and} \quad A = \begin{pmatrix} 1/3 & 2/3 & 2/3 \\ 2/3 & 1/3 & -2/3 \\ -2/3 & 2/3 & -1/3 \end{pmatrix}$$

4.239 If A is orthogonal, show that A^{-1} is orthogonal.

▌ Since A is orthogonal, $A^T = A^{-1}$. By Problem 4.234, A^T is orthogonal. Hence A^{-1} is orthogonal.

4.240 Suppose A and B are n-square orthogonal matrices. Show that AB is an [n-square] orthogonal matrix.

▌ It suffices to show that $(AB)(AB)^T = I$:

$$(AB)(AB)^T = (AB)(B^TA^T) = A(BB^T)A^T = AIA^T = AA^T = I$$

4.241 Exhibit the most general 2×2 orthogonal matrix.

▌ Let $A = \begin{pmatrix} a & b \\ c & d \end{pmatrix}$. Then a, b, c, d are real numbers, and the rows of A form an orthonormal set. Hence,

$$a^2 + b^2 = 1 \qquad c^2 + d^2 = 1 \qquad ac + bd = 0$$

Similarly, the columns form an orthonormal set, so

$$a^2 + c^2 = 1 \qquad b^2 + d^2 = 1 \qquad ab + cd = 0$$

Therefore, $c^2 = 1 - a^2 = b^2$, whence $c = \pm b$.

Case (i): $c = +b$. Then $b(a + d) = 0$, or $d = -a$; the corresponding matrix is $\begin{pmatrix} a & b \\ b & -a \end{pmatrix}$.

Case (ii): $c = -b$. Then $b(d - a) = 0$, or $d = a$; the corresponding matrix is $\begin{pmatrix} a & b \\ -b & a \end{pmatrix}$.

4.242 Show that every 2×2 orthogonal matrix has the form

$$\begin{pmatrix} \cos\theta & \sin\theta \\ -\sin\theta & \cos\theta \end{pmatrix} \qquad \text{or} \qquad \begin{pmatrix} \cos\theta & \sin\theta \\ \sin\theta & -\cos\theta \end{pmatrix}$$

for some real number θ.

▌ Let a and b be any real numbers such that $a^2 + b^2 = 1$. Then there exists a real number θ such that $a = \cos\theta$ and $b = \sin\theta$. The result now follows from Problem 4.241.

4.14 UNITARY MATRICES

4.243 Define *unitary* matrix.

▌ A complex matrix A is said to be unitary if $AA^H = A^HA = I$. Observe that a unitary matrix is necessarily square and invertible.

4.244 Show that the following matrix is unitary:

$$A = \frac{1}{2}\begin{pmatrix} 1 & -i & -1+i \\ i & 1 & 1+i \\ 1+i & -1+i & 0 \end{pmatrix}$$

▌ We need only show that $AA^H = I$:

$$AA^H = \frac{1}{4}\begin{pmatrix} 1 & -i & -1+i \\ i & 1 & 1+i \\ 1+i & -1+i & 0 \end{pmatrix}\begin{pmatrix} 1 & -i & 1-i \\ i & 1 & -1-i \\ -1-i & 1-i & 0 \end{pmatrix}$$

$$= \frac{1}{4}\begin{pmatrix} 1+1+2 & -i-i+2i & 1-i+i-1+0 \\ i+i-2i & 1+1+2 & i+1-1-i \\ 1+i-i-1+0 & -i+1-1+i+0 & 2+2+0 \end{pmatrix} = \begin{pmatrix} 1 & 0 & 0 \\ 0 & 1 & 0 \\ 0 & 0 & 1 \end{pmatrix}$$

4.245 Show that A is unitary if and only if A^H is unitary.

▌ $AA^H = A^HA = I$ iff $A^H(A^H)^H = (A^H)^HA^H = I$.

4.246 Define an *orthonormal set* of vectors in $\mathbf{C}^n$.

▮ The vectors $u_1, u_2, \ldots, u_r$ in $\mathbf{C}^n$ form an orthonormal set if $u_i \cdot u_j = \delta_{ij}$. Recall that the dot product in $\mathbf{C}^n$ is defined by

$$(a_1, a_2, \ldots, a_n) \cdot (b_1, b_2, \ldots, b_n) = a_1 \overline{b_1} + a_2 \overline{b_2} + \cdots + a_n \overline{b_n}$$

4.247 Show that $\bar{u} \cdot \bar{v} = \overline{u \cdot v}$ in $\mathbf{C}^n$.

▮ Let $u = (a_1, a_2, \ldots, a_n)$ and $v = (b_1, b_2, \ldots, b_n)$. Then

$$\bar{u} \cdot \bar{v} = (\overline{a_1}, \ldots, \overline{a_n}) \cdot (\overline{b_1}, \ldots, \overline{b_n}) = \bar{a}_1 \bar{\bar{b}}_1 + \cdots + \bar{a} \bar{\bar{b}}_n$$
$$= \bar{a}_1 b_1 + \cdots + \bar{a}_n b_n = \overline{a_1 \overline{b_1}} + \cdots + \overline{a_n \overline{b_n}} = \overline{u \cdot v}$$

4.248 Show that $u_1, u_2, \ldots, u_r$ is an orthonormal set of vectors in $\mathbf{C}^n$ if and only if $\bar{u}_1, \bar{u}_2, \ldots, \bar{u}_r$ is an orthonormal set.

▮ By Problem 4.247, we have $u_i \cdot u_j = 0$ if and only if $\bar{u}_i \cdot \bar{u}_j = \bar{0} = 0$, and $u_i \cdot u_i = 1$ if and only if $\bar{u}_i \cdot \bar{u}_i = \bar{1} = 1$.

Following is the complex analogue of Theorem 4.9.

Theorem 4.10: Let A be a complex matrix. Then the following are equivalent: (a) A is unitary; (b) the rows of A form an orthonormal set; (c) the columns of A form an orthonormal set.

4.249 Prove that (a) and (b) are equivalent.

▮ Let $R_1, \ldots, R_n$ denote the rows of A; then $\bar{R}_1^T, \ldots, \bar{R}_n^T$ are the columns of A^H. Let $AA^H = (c_{ij})$. By matrix multiplication, $c_{ij} = R_i \bar{R}_j^T = R_i \cdot R_j$. Then $AA^H = I$ iff $R_i \cdot R_j = \delta_{ij}$ iff $R_1, R_2, \ldots, R_n$ form an orthonormal set.

4.250 Prove that (a) and (c) are equivalent.

▮ By Problems 4.245, 4.249, and 4.248, A is unitary iff A^H is unitary iff the rows of A^H are orthonormal iff the conjugates of the columns of A are orthonormal iff the columns of A are orthonormal.

4.251 Show that

$$A = \begin{pmatrix} \dfrac{1}{3} - \dfrac{2}{3}i & \dfrac{2}{3}i \\ -\dfrac{2}{3}i & -\dfrac{1}{3} - \dfrac{2}{3}i \end{pmatrix}$$

is unitary.

▮ The rows form an orthonormal set:

$$\left(\frac{1}{3} - \frac{2}{3}i, \frac{2}{3}i\right) \cdot \left(\frac{1}{3} - \frac{2}{3}i, \frac{2}{3}i\right) = \left(\frac{1}{9} + \frac{4}{9}\right) + \frac{4}{9} = 1$$

$$\left(\frac{1}{3} - \frac{2}{3}i, \frac{2}{3}i\right) \cdot \left(-\frac{2}{3}i, -\frac{1}{3} - \frac{2}{3}i\right) = \left(\frac{2}{9}i + \frac{4}{9}\right) + \left(-\frac{2}{9}i - \frac{4}{9}\right) = 0$$

$$\left(-\frac{2}{3}i, -\frac{1}{3} - \frac{2}{3}i\right) \cdot \left(-\frac{2}{3}i, -\frac{1}{3} - \frac{2}{3}i\right) = \frac{4}{9} + \left(\frac{1}{9} + \frac{4}{9}\right) = 1$$

4.252 If A is unitary, show that A^{-1} is unitary.

▮ Since A is unitary, $A^H = A^{-1}$. But [Problem 4.245] A^H is unitary.

4.253 Show that the product AB of unitary matrices A and B is also unitary

▮ $$(AB)(AB)^H = (AB)(B^H A^H) = A(BB^H)A^H = AIA^H = AA^H = I$$

4.15 NORMAL MATRICES

4.254 Define *normal* matrix.

▌ Matrix A is normal if A is real and $AA^T = A^TA$, or if A is complex and $AA^H = A^HA$. Thus, real matrices that are symmetric or orthogonal, and complex matrices that are Hermitian or unitary, are all special cases of normal matrices.

Problems 4.255–4.258 refer to the following matrices:

$$A = \begin{pmatrix} 6 & -3 \\ 3 & 6 \end{pmatrix} \quad B = \begin{pmatrix} 2 & 5 \\ 3 & -1 \end{pmatrix} \quad C = \begin{pmatrix} 2+3i & 1 \\ i & 1+2i \end{pmatrix} \quad D = \begin{pmatrix} 1 & i \\ 1+i & 0 \end{pmatrix}$$

4.255 Is A normal?

▌ $$AA^T = \begin{pmatrix} 6 & -3 \\ 3 & 6 \end{pmatrix}\begin{pmatrix} 6 & 3 \\ -3 & 6 \end{pmatrix} = \begin{pmatrix} 45 & 0 \\ 0 & 45 \end{pmatrix} \quad \text{and} \quad A^TA = \begin{pmatrix} 6 & 3 \\ -3 & 6 \end{pmatrix}\begin{pmatrix} 6 & -3 \\ 3 & 6 \end{pmatrix} = \begin{pmatrix} 45 & 0 \\ 0 & 45 \end{pmatrix}$$

Since $AA^T = A^TA$, the matrix A is normal.

4.256 Is B normal?

▌ $$BB^T = \begin{pmatrix} 2 & 5 \\ 3 & -1 \end{pmatrix}\begin{pmatrix} 2 & 3 \\ 5 & -1 \end{pmatrix} = \begin{pmatrix} 29 & 1 \\ 1 & 10 \end{pmatrix} \quad \text{and} \quad B^TB = \begin{pmatrix} 2 & 3 \\ 5 & -1 \end{pmatrix}\begin{pmatrix} 2 & 5 \\ 3 & -1 \end{pmatrix} = \begin{pmatrix} 13 & 7 \\ 7 & 26 \end{pmatrix}$$

Since $BB^T \neq B^TB$, the matrix B is not normal.

4.257 Is C normal?

▌ $$CC^H = \begin{pmatrix} 2+3i & 1 \\ i & 1+2i \end{pmatrix}\begin{pmatrix} 2-3i & -i \\ 1 & 1-2i \end{pmatrix} = \begin{pmatrix} 14 & 4-4i \\ 4+4i & 6 \end{pmatrix}$$
$$C^HC = \begin{pmatrix} 2-3i & -i \\ 1 & 1-2i \end{pmatrix}\begin{pmatrix} 2+3i & 1 \\ i & 1+2i \end{pmatrix} = \begin{pmatrix} 14 & 4-4i \\ 4+4i & 6 \end{pmatrix}$$

Since $CC^H = C^HC$, the complex matrix C is normal.

4.258 Is D normal?

▌ $$DD^H = \begin{pmatrix} 1 & i \\ 1+i & 0 \end{pmatrix}\begin{pmatrix} 1 & 1-i \\ -i & 0 \end{pmatrix} = \begin{pmatrix} 2 & 1-i \\ 1+i & 2 \end{pmatrix}$$
$$D^HD = \begin{pmatrix} 1 & 1-i \\ -i & 0 \end{pmatrix}\begin{pmatrix} 1 & i \\ 1+i & 0 \end{pmatrix} = \begin{pmatrix} 3 & i \\ -i & 1 \end{pmatrix}$$

Since $DD^H \neq D^HD$, the complex matrix D is not normal.

4.259 Show that a [real] skew-symmetric matrix A is normal.

▌ $$AA^T = A(-A) = (-A)A = A^TA$$

4.260 Show that a [complex] skew-Hermitian matrix A is normal.

▌ $AA^H = A(-A) = (-A)A = A^HA$

4.261 Exhibit a real matrix that is normal but is not symmetric, skew-symmetric, or orthogonal.

▌ One such is the matrix A of Problem 4.255.

4.262 Show that the sum of a real scalar matrix [Problem 4.19] and a skew-symmetric matrix is normal; that is, show that $B = kI + A$ is normal whenever A is skew-symmetric.

▌ $B^T = (kI + A)^T = (kI)^T + A^T = kI - A$. Then: $BB^T = (kI + A)(kI - A) = k^2I - A^2$ and $B^TB = (kI - A)(kI + A) = k^2I - A^2$. Since $BB^T = B^TB$, the matrix B is normal.

Theorem 4.11: A real 2×2 normal matrix is either symmetric or the sum of a scalar matrix and a skew-symmetric matrix.

4.263 Prove Theorem 4.11.

▌ If $A = \begin{pmatrix} a & b \\ c & d \end{pmatrix}$, then

$$AA^T = \begin{pmatrix} a & b \\ c & d \end{pmatrix}\begin{pmatrix} a & c \\ b & d \end{pmatrix} = \begin{pmatrix} a^2 + b^2 & ac + bd \\ ac + bd & c^2 + d^2 \end{pmatrix}$$

$$A^TA = \begin{pmatrix} a & c \\ b & d \end{pmatrix}\begin{pmatrix} a & b \\ c & d \end{pmatrix} = \begin{pmatrix} a^2 + c^2 & ab + cd \\ ab + cd & b^2 + d^2 \end{pmatrix}$$

Since $AA^T = A^TA$, we get the equations

$$a^2 + b^2 = a^2 + c^2 \qquad c^2 + d^2 = b^2 + d^2 \qquad ac + bd = ab + cd$$

The first equation yields $b^2 = c^2$; hence $b = c$ or $b = -c$.

Case (i): $b = c$ [which includes the case $b = -c = 0$]. Then we obtain the symmetric matrix

$$A = \begin{pmatrix} a & b \\ b & d \end{pmatrix}.$$

Case (ii): $b = -c \neq 0$. Then $ac + bd = b(d - a)$ and $ab + cd = b(a - d)$. Thus $b(d - a) = b(a - d)$, and so $2b(d - a) = 0$. Since $b \neq 0$, we get $a = d$. Thus A has the form

$$A = \begin{pmatrix} a & b \\ -b & a \end{pmatrix} = \begin{pmatrix} a & 0 \\ 0 & a \end{pmatrix} + \begin{pmatrix} 0 & b \\ -b & 0 \end{pmatrix}$$

which is the sum of a scalar matrix and a skew-symmetric matrix.

4.264 Exhibit a normal complex matrix that is neither Hermitian, skew-Hermitian, nor unitary.

▌ See Problem 4.257.

4.16 SQUARE BLOCK MATRICES

4.265 Define *square block matrix*.

▌ A block matrix A is called a square block matrix if (i) A is a square matrix, (ii) the blocks form a square matrix [i.e., the numbers of horizontal and vertical partition lines are equal], and (iii) the diagonal blocks are square matrices.

Problems 4.266–4.268 refer to the following block matrices:

$$A = \begin{pmatrix} 1 & 2 & 3 & 4 & 5 \\ 1 & 1 & 1 & 1 & 1 \\ 9 & 8 & 7 & 6 & 5 \\ 3 & 3 & 3 & 3 & 3 \\ 1 & 3 & 5 & 7 & 9 \end{pmatrix} \qquad \begin{pmatrix} 1 & 2 & 3 & 4 & 5 \\ 1 & 1 & 1 & 1 & 1 \\ 9 & 8 & 7 & 6 & 5 \\ 3 & 3 & 3 & 3 & 3 \\ 1 & 3 & 5 & 7 & 9 \end{pmatrix} \qquad C = \begin{pmatrix} 1 & 2 & 3 & 4 & 5 \\ 1 & 1 & 1 & 1 & 1 \\ 9 & 8 & 7 & 6 & 5 \\ 3 & 3 & 3 & 3 & 3 \\ 1 & 3 & 5 & 7 & 9 \end{pmatrix}$$

4.266 Is A a square block matrix?

▌ No: Although A is a 5×5 square matrix and is a 3×3 block matrix, the second and third diagonal blocks are not square matrices.

4.267 Is B a square block matrix?

▌ Yes.

4.268 Complete the partitioning of C into a square block matrix.

▌ One horizontal line is between the second and third rows; hence add a vertical line between the second and third columns. The other horizontal line is between the fourth and fifth rows; hence add a vertical line between the fourth and fifth columns. [The horizontal lines and the vertical lines must be symmetrically placed to obtain a square block matrix.] This yields the square block matrix

$$C = \begin{pmatrix} 1 & 2 & 3 & 4 & 5 \\ 1 & 1 & 1 & 1 & 1 \\ 9 & 8 & 7 & 6 & 5 \\ 3 & 3 & 3 & 3 & 3 \\ 1 & 3 & 5 & 7 & 9 \end{pmatrix}$$

4.269   Define *block diagonal matrix*.

▮ A square block matrix M is a block diagonal matrix if all blocks off the diagonal are zero matrices; that is, if M has the form

$$M = \begin{pmatrix} A_1 & 0 & \cdots & 0 \\ 0 & A_2 & \cdots & 0 \\ \multicolumn{4}{c}{\dotfill} \\ 0 & 0 & \cdots & A_r \end{pmatrix} \qquad \text{(with } A_i \text{ square)}$$

Such a block diagonal matrix M is frequently written $M = \text{diag}\,[A_1, A_2, \ldots, A_r]$.

Problems 4.270–4.272 refer to the following matrices:

$$A = \begin{pmatrix} 1 & 0 & 0 \\ 0 & 0 & 2 \\ 0 & 0 & 3 \end{pmatrix} \qquad B = \begin{pmatrix} 1 & 2 & 0 & 0 & 0 \\ 3 & 0 & 0 & 0 & 0 \\ 0 & 0 & 4 & 0 & 0 \\ 0 & 0 & 5 & 0 & 0 \\ 0 & 0 & 0 & 0 & 6 \end{pmatrix} \qquad C = \begin{pmatrix} 0 & 1 & 0 \\ 0 & 0 & 0 \\ 0 & 2 & 0 \end{pmatrix}$$

4.270   Partition A so that it becomes a block diagonal matrix with as many diagonal blocks as possible.

▮

$$A = \begin{pmatrix} 1 & 0 & 0 \\ 0 & 0 & 2 \\ 0 & 0 & 3 \end{pmatrix}$$

4.271   Partition B so that it becomes a block diagonal matrix with as many diagonal blocks as possible.

▮

$$B = \begin{pmatrix} 1 & 2 & 0 & 0 & 0 \\ 3 & 0 & 0 & 0 & 0 \\ 0 & 0 & 4 & 0 & 0 \\ 0 & 0 & 5 & 0 & 0 \\ 0 & 0 & 0 & 0 & 6 \end{pmatrix}$$

4.272   Partition C so it becomes a block diagonal matrix with as many diagonal blocks as possible.

▮ Considered as a single 3×3 block, C [or any other 3×3 matrix] is a block diagonal matrix; no further partitioning of C is possible.

Problems 4.272–4.276 refer to the following block diagonal matrices of which corresponding diagonal blocks have the same size: $M = \text{diag}\,[A_1, A_2, \ldots, A_r]$ and $N = \text{diag}\,[B_1, B_2, \ldots, B_r]$.

4.273   Find $M + N$.

▮ Simply add the diagonal blocks: $M + N = \text{diag}\,[A_1 + B_1, A_2 + B_2, \ldots, A_r + B_r]$.

4.274   Find kM.

▮ Simply multiply the diagonal blocks by k: $kM = \text{diag}\,[kA_1, kA_2, \ldots, kA_r]$.

4.275   Find MN.

▮ Simply multiply corresponding diagonal blocks: $MN = \text{diag}\,[A_1 B_1, A_2 B_2, \ldots, A_r B_r]$.

4.276   Find $f(M)$ for a given polynomial $f(x)$.

▮ Find $f(A_i)$ for each diagonal block A_i.   Then, by Problems 4.273–4.275, $f(M) = \text{diag}[\,f(A_1), f(A_2), \ldots, f(A_r)]$.

4.277 Define *block upper triangular matrix*.

▮ A square block matrix is a block upper triangular matrix if all blocks below the diagonal are zero matrices.

4.278 Define *block lower triangular matrix*.

▮ A square block matrix is a block lower triangular matrix if all blocks above the diagonal are zero blocks. [Alternatively, a block lower triangular matrix is the transpose of a block upper triangular matrix.]

Problems 4.279–4.281 refer to the following block matrices:

$$A = \left(\begin{array}{cc|c} 1 & 2 & 0 \\ 3 & 4 & 5 \\ \hline 0 & 0 & 6 \end{array}\right) \qquad B = \left(\begin{array}{c|ccc} 1 & 0 & 0 & 0 \\ \hline 2 & 3 & 4 & 0 \\ 5 & 0 & 6 & 0 \\ \hline 0 & 7 & 8 & 9 \end{array}\right) \qquad C = \left(\begin{array}{cc|c} 1 & 2 & 0 \\ 3 & 4 & 5 \\ \hline 0 & 6 & 7 \end{array}\right)$$

4.279 Is A upper triangular? lower triangular?

▮ A is upper triangular, but not lower triangular.

4.280 Is B upper triangular? lower triangular?

▮ B is lower triangular, but not upper triangular.

4.281 Is C upper triangular? lower triangular?

▮ C is neither upper triangular nor lower triangular. Furthermore, no other partitioning of C will make it into either a block upper triangular matrix or a block lower triangular matrix.

4.282 Decompose an arbitrary square block matrix into the sum of a block upper triangular matrix and a block lower triangular matrix. Is the decomposition unique?

▮
$$\begin{pmatrix} A_{11} & A_{22} & \cdots & A_{1r} \\ A_{21} & A_{22} & \cdots & A_{2r} \\ \cdots\cdots\cdots\cdots\cdots \\ A_{r1} & A_{r2} & \cdots & A_{rr} \end{pmatrix} = \begin{pmatrix} \frac{1}{2}A_{11} & A_{12} & \cdots & A_{1r} \\ & \frac{1}{2}A_{22} & \cdots & A_{2r} \\ & & \cdots\cdots\cdots \\ & & & \frac{1}{2}A_{rr} \end{pmatrix} + \begin{pmatrix} \frac{1}{2}A_{11} & & & \\ A_{21} & \frac{1}{2}A_{22} & & \\ \cdots\cdots\cdots\cdots \\ A_{r1} & A_{r2} & \cdots & \frac{1}{2}A_{rr} \end{pmatrix}$$

The decomposition is not unique, for the diagonal blocks can be split in an infinity of ways.

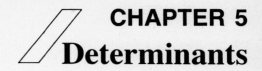

CHAPTER 5
Determinants

5.1 DETERMINANT FUNCTION. DETERMINANTS OF ORDER ONE

5.1   What notation is used for the *determinant function*?

▌ The *determinant* of an *n*-square matrix $A = (a_{ij})$ is denoted by det A or det (a_{ij}) or $|A|$ or

$$\begin{vmatrix} a_{11} & a_{12} & \cdots & a_{1n} \\ a_{21} & a_{22} & \cdots & a_{2n} \\ \cdots\cdots\cdots\cdots\cdots\cdots \\ a_{n1} & a_{n2} & \cdots & a_{nn} \end{vmatrix}$$

The *order* of the determinant is the integer *n*.

5.2   Give the domain and range of the determinant function.

▌ The determinant function assigns a scalar, det A, to each square matrix A. Accordingly, the domain of the determinant function consists of all square matrices and the range of the determinant function consists of all square matrices and the range of the determinant functions consists of all the scalars.

5.3   The determinant function is said to be *multiplicative*. What does this statement mean?

▌ For any *n*-square matrices A and B, det $AB = (\det A)(\det B)$. This basic result will be proved as Theorem 5.3.

5.4   Define the *determinant of order one*.

▌ The determinant of a 1×1 matrix $A = (a_{11})$ is the scalar a_{11} itself. [Note that this definition is consistent with Problem 5.3.]

5.5   Find det (24), det (-6), and det $(t + 2)$.

▌ The determinant is the scalar itself; hence, det $24 = 24$, det $(-6) = -6$, and det $(t + 2) = t + 2$.

5.6   Show that the equation $ax = b$ has a unique solution if and only if det $(a) \neq 0$.

▌ By Theorem 3.2, $ax = b$ has a unique solution if and only if det $(a) \equiv a \neq 0$.

5.2 DETERMINANTS OF ORDER TWO

5.7   Define the *determinant of order two*.

▌ The determinant of a 2×2 matrix $A = (a_{ij})$ is det $A \equiv \begin{vmatrix} a_{11} & a_{12} \\ a_{21} & a_{22} \end{vmatrix} = a_{11}a_{22} - a_{12}a_{21}$.

5.8   Give a mnemonic for the evaluation of the determinant of order two.

▌

The determinant is equal to the product of the elements along the plus-labeled arrow minus the product of the elements along the minus-labeled arrow. [There is an analogous scheme for determinants of order three, but not for higher-order determinants.]

Problems 5.9–5.13 concern the following matrices:

$$A = \begin{pmatrix} 5 & 4 \\ 2 & 3 \end{pmatrix} \quad B = \begin{pmatrix} 2 & 1 \\ -4 & 6 \end{pmatrix} \quad C = \begin{pmatrix} 3 & -2 \\ 4 & 5 \end{pmatrix} \quad D = \begin{pmatrix} 4 & -5 \\ -1 & -2 \end{pmatrix} \quad E = \begin{pmatrix} a & b \\ c & d \end{pmatrix}$$

113

5.9   Find det A.

▌ $\begin{vmatrix} 5 & 4 \\ 2 & 3 \end{vmatrix} = (5)(3) - (4)(2) = 15 - 8 = 7$

5.10   Find det B.

▌ $\begin{vmatrix} 2 & 1 \\ -4 & 6 \end{vmatrix} = (2)(6) - (1)(-4) = 12 + 4 = 16$

5.11   Find det C.

▌ $\begin{vmatrix} 3 & -2 \\ 4 & 5 \end{vmatrix} = 15 + 8 = 23$

5.12   Find det D.

▌ $\begin{vmatrix} 4 & -5 \\ -1 & -2 \end{vmatrix} = -8 - 5 = -13$

5.13   Find det E.

▌ $\begin{vmatrix} a & b \\ c & d \end{vmatrix} = ad - bc$

5.14   Find det A, where $A = \begin{pmatrix} a - b & a \\ a & a + b \end{pmatrix}$.

▌ $$\begin{vmatrix} a - b & a \\ a & a + b \end{vmatrix} = (a - b)(a + b) - (a)(a) = -b^2$$

5.15   Find det B, where $B = \begin{pmatrix} t - 5 & 7 \\ -1 & t + 3 \end{pmatrix}$.

▌ $$\begin{vmatrix} t - 5 & 7 \\ -1 & t + 3 \end{vmatrix} = (t - 5)(t + 3) + 7 = t^2 - 2t - 15 + 7 = t^2 - 2t - 8$$

5.16   Determine those values of k for which $\begin{vmatrix} k & k \\ 4 & 2k \end{vmatrix} = 0$.

▌ $\begin{vmatrix} k & k \\ 4 & 2k \end{vmatrix} = 2k^2 - 4k = 0$,   or $2k(k - 2) = 0$.   Hence, $k = 0$ or $k = 2$.

5.17   Determine those values of t for which $\begin{vmatrix} t - 2 & 3 \\ 4 & t - 1 \end{vmatrix} = 0$.

▌ $$\begin{vmatrix} t - 2 & 3 \\ 4 & t - 1 \end{vmatrix} = t^2 - 3t + 2 - 12 = t^2 - 3t - 10 = 0, \quad \text{or} \quad (t - 5)(t + 2) = 0.$$

Hence, $t = 5$ or $t = -2$.

5.18   Consider two linear equations in two unknowns:

$$a_1 x + b_1 y = c_1$$
$$a_2 x + b_2 y = c_2$$

By Problem 3.14, the system has a unique solution if and only if $D \equiv a_1 b_2 - a_2 b_1 \neq 0;$ that solution is

$$x = \frac{b_2 c_1 - b_1 c_2}{a_1 b_2 - a_2 b_1} \qquad y = \frac{a_1 c_1 - a_2 c_1}{a_1 b_2 - a_2 b_1}$$

Express the solution completely in terms of determinants.

❙
$$x = \frac{N_x}{D} = \frac{b_2 c_1 - b_1 c_2}{a_1 b_2 - a_2 b_1} = \frac{\begin{vmatrix} c_1 & b_1 \\ c_2 & b_2 \end{vmatrix}}{\begin{vmatrix} a_1 & b_1 \\ a_2 & b_2 \end{vmatrix}} \qquad y = \frac{N_y}{D} = \frac{a_1 c_2 - a_2 c_1}{a_1 b_2 - a_2 b_1} = \frac{\begin{vmatrix} a_1 & c_1 \\ a_2 & c_2 \end{vmatrix}}{\begin{vmatrix} a_1 & b_1 \\ a_2 & b_2 \end{vmatrix}}$$

Here D, the determinant of the matrix of coefficients, appears in the denominator of both quotients. The numerators N_x and N_y of the quotients for x and y, respectively, can be obtained by substituting the column of constant terms in place of the column of coefficients of the given unknown in the matrix of coefficients.

5.19 Solve by determinants: $\begin{cases} 2x - 3y = 7 \\ 3x + 5y = 1 \end{cases}$.

❙ The determinant D of the matrix of coefficients is

$$D = \begin{vmatrix} 2 & -3 \\ 3 & 5 \end{vmatrix} = (2)(5) - (3)(-3) = 10 + 9 = 19$$

Since $D \neq 0$, the system has a unique solution. To obtain the numerator N_x replace, in the matrix of coefficients, the coefficients of x by the constant terms:

$$N_x = \begin{vmatrix} 7 & -3 \\ 1 & 5 \end{vmatrix} = (7)(5) - (1)(-3) = 35 + 3 = 38$$

To obtain the numerator N_y replace, in the matrix of coefficients, the coefficients of y by the constant terms:

$$N_y = \begin{vmatrix} 2 & 7 \\ 3 & 1 \end{vmatrix} = (2)(1) - (3)(7) = 2 - 21 = -19$$

Thus the unique solution of the system is

$$x = \frac{N_x}{D} = \frac{38}{19} = 2 \qquad \text{and} \qquad y = \frac{N_y}{D} = \frac{-19}{19} = -1$$

5.20 Solve by determinants: $\begin{cases} 2x = 5 + y \\ 3 + 2y + 3x = 0 \end{cases}$.

❙ First arrange the system in standard form:

$$\begin{aligned} 2x - y &= 5 \\ 3x + 2y &= -3 \end{aligned}$$

The determinant D of the matrix of coefficients is

$$D = \begin{vmatrix} 2 & -1 \\ 3 & 2 \end{vmatrix} = (2)(2) - (3)(-1) = 4 + 3 = 7$$

Since $D \neq 0$, the system has a unique solution. Now,

$$N_x = \begin{vmatrix} 5 & -1 \\ -3 & 2 \end{vmatrix} = (5)(2) - (-3)(-1) = 10 - 3 = 7$$

and

$$N_y = \begin{vmatrix} 2 & 5 \\ 3 & -3 \end{vmatrix} = (2)(-3) - (3)(5) = -6 - 15 = -21$$

Thus the unique solution of the system is

$$x = \frac{N_x}{D} = \frac{7}{7} = 1 \qquad \text{and} \qquad y = \frac{N_y}{D} = \frac{-21}{7} = -3$$

5.21 Solve by determinants: $\begin{cases} 2x - 4y = 7 \\ 3x - 6y = 5 \end{cases}$.

❚ The determinant D of the matrix of coefficients is

$$D = \begin{vmatrix} 2 & -4 \\ 3 & -6 \end{vmatrix} = (2)(-6) - (3)(-4) = -12 + 12 = 0$$

Since $D = 0$, the system does not have a unique solution, and we cannot solve the system by determinants.

5.22 If $ab \neq 0$, solve using determinants: $\begin{cases} ax - 2by = c \\ 3ax - 5by = 2c \end{cases}$.

❚ First find $D = \begin{vmatrix} a & -2b \\ 3a & -5b \end{vmatrix} = -5ab + 6ab = ab$. Since $D = ab \neq 0$, the system has a unique solution. Next find

$$N_x = \begin{vmatrix} c & -2b \\ 2c & -5b \end{vmatrix} = -5bc + 4bc = -bc \quad \text{and} \quad N_y = \begin{vmatrix} a & c \\ 3a & 2c \end{vmatrix} = 2ac - 3ac = -ac$$

Then $x = N_x/D = -bc/ab = -c/a$ and $y = N_y/D = -ac/ab = -c/b$.

5.23 Verify the multiplicative property [Problem 5.3] for determinants of order two.

❚ $$AB = \begin{pmatrix} a_{11} & a_{12} \\ a_{21} & a_{22} \end{pmatrix}\begin{pmatrix} b_{11} & b_{12} \\ b_{21} & b_{22} \end{pmatrix} = \begin{pmatrix} a_{11}b_{11} + a_{12}b_{21} & a_{11}b_{12} + a_{12}b_{22} \\ a_{21}b_{11} + a_{22}b_{21} & a_{21}b_{12} + a_{22}b_{22} \end{pmatrix}$$

$$\det AB = (a_{11}b_{11} + a_{12}b_{21})(a_{21}b_{12} + a_{22}b_{22}) - (a_{11}b_{12} + a_{12}b_{22})(a_{21}b_{11} + a_{22}b_{21})$$

$$= a_{11}b_{11}a_{21}b_{12} \overset{(1)}{} + a_{11}b_{11}a_{22}b_{22} + a_{12}b_{21}a_{21}b_{12} \overset{(4)}{} + a_{12}b_{21}a_{22}b_{22}$$

$$- a_{11}b_{12}a_{21}b_{11} - a_{11}b_{12}a_{22}b_{21} \overset{(2)}{} - a_{12}b_{22}a_{21}b_{11} \overset{(3)}{} - a_{12}b_{22}a_{22}b_{21}$$

On the other hand,

$$(\det A)(\det B) = (a_{11}a_{22} - a_{12}a_{21})(b_{11}b_{22} - b_{12}b_{21})$$

$$= a_{11}a_{22}b_{11}b_{22} \overset{(1)}{} - a_{11}a_{22}b_{12}b_{21} \overset{(2)}{} - a_{12}a_{21}b_{11}b_{22} \overset{(3)}{} + a_{12}a_{21}b_{12}b_{21} \overset{(4)}{}$$

5.3 DETERMINANTS OF ORDER THREE

5.24 Define the *determinant of order three*.

❚ The determinant of a 3-square matrix $A = (a_{ij})$ is

$$\det A \equiv \begin{vmatrix} a_{11} & a_{12} & a_{13} \\ a_{21} & a_{22} & a_{23} \\ a_{31} & a_{32} & a_{33} \end{vmatrix} = a_{11}a_{22}a_{33} + a_{12}a_{23}a_{31} + a_{13}a_{21}a_{32} - a_{13}a_{22}a_{31} - a_{12}a_{21}a_{33} - a_{11}a_{23}a_{32}$$

5.25 Use Fig. 5-1 to obtain the determinant of

$$A = \begin{pmatrix} a_1 & b_1 & c_1 \\ a_2 & b_2 & c_2 \\ a_3 & b_3 & c_3 \end{pmatrix}$$

❚ Form the product of each of the three numbers joined by an arrow in the diagram on the left, and precede each product by a plus sign as follows:

$$+ a_1b_2c_3 + b_1c_2a_3 + c_1a_2b_3$$

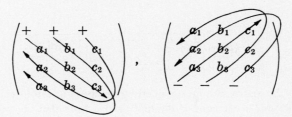

Fig. 5-1

Now form the product of each of the three numbers joined by an arrow in the diagram on the right, and precede each product by a minus sign as follows:

$$- a_3 b_2 c_1 - b_3 c_2 a_1 - c_3 a_2 b_1$$

Then the determinant of A is precisely the sum of the above two expressions:

$$|A| = \begin{vmatrix} a_1 & b_1 & c_1 \\ a_2 & b_2 & c_2 \\ a_3 & b_3 & c_3 \end{vmatrix} = a_1 b_2 c_3 + b_1 c_2 a_3 + c_1 a_2 b_3 - a_3 b_2 c_1 - b_3 c_2 a_1 - c_3 a_2 b_1$$

[The above method of computing $|A|$ does not hold for determinants of order greater than 3.]

Problems 5.26–5.29 concern the following matrices:

$$A = \begin{pmatrix} 2 & 1 & 1 \\ 0 & 5 & -2 \\ 1 & -3 & 4 \end{pmatrix} \qquad B = \begin{pmatrix} 3 & -2 & -4 \\ 2 & 5 & -1 \\ 0 & 6 & 1 \end{pmatrix} \qquad C = \begin{pmatrix} -2 & -1 & 4 \\ 6 & -3 & -2 \\ 4 & 1 & 2 \end{pmatrix} \qquad D = \begin{pmatrix} 7 & 6 & 5 \\ 1 & 2 & 1 \\ 3 & -2 & 1 \end{pmatrix}$$

5.26 Find det A.

▌ Use Fig. 5-1.

$$\begin{vmatrix} 2 & 1 & 1 \\ 0 & 5 & -2 \\ 1 & -3 & 4 \end{vmatrix} = (2)(5)(4) + (1)(-2)(1) + (1)(-3)(0) - (1)(5)(1) - (-3)(-2)(2) - (4)(1)(0)$$

$$= 40 - 2 + 0 - 5 - 12 - 0 = 21$$

5.27 Find det B.

▌
$$\begin{vmatrix} 3 & -2 & -4 \\ 2 & 5 & -1 \\ 0 & 6 & 1 \end{vmatrix} = (3)(5)(1) + (-2)(-1)(0) + (-4)(6)(2) - (0)(5)(-4) - (6)(-1)(3) - (1)(-2)(2)$$

$$= 15 + 0 - 48 - 0 + 18 + 4 = -11$$

5.28 Find det C.

▌
$$\begin{vmatrix} -2 & -1 & 4 \\ 6 & -3 & -2 \\ 4 & 1 & 2 \end{vmatrix} = 12 + 8 + 24 + 48 - 4 + 12 = 100$$

5.29 Find det D.

▌
$$\begin{vmatrix} 7 & 6 & 5 \\ 1 & 2 & 1 \\ 3 & -2 & 1 \end{vmatrix} = 14 + 18 - 10 - 30 + 14 - 6 = 0$$

5.30 Show that

$$\begin{vmatrix} a_{11} & a_{12} & a_{13} \\ a_{21} & a_{22} & a_{23} \\ a_{31} & a_{32} & a_{33} \end{vmatrix} = a_{11}\begin{vmatrix} \boxed{a_{11}} & \boxed{a_{12}} & \boxed{a_{13}} \\ \boxed{a_{21}} & a_{22} & a_{23} \\ \boxed{a_{31}} & a_{32} & a_{33} \end{vmatrix} - a_{12}\begin{vmatrix} \boxed{a_{11}} & \boxed{a_{12}} & \boxed{a_{13}} \\ a_{21} & \boxed{a_{22}} & a_{23} \\ a_{31} & \boxed{a_{32}} & a_{33} \end{vmatrix} + a_{13}\begin{vmatrix} \boxed{a_{11}} & \boxed{a_{12}} & \boxed{a_{13}} \\ a_{21} & a_{22} & \boxed{a_{23}} \\ a_{31} & a_{32} & \boxed{a_{33}} \end{vmatrix}$$

Note that each 2×2 matrix can be obtained by deleting, in the original matrix, the row and column containing its coefficient [an element of the first row]. Note that the coefficients are taken with alternating signs.

❚ Expand the determinants of order two to obtain

$$a_{11}(a_{22}a_{33} - a_{23}a_{32}) - a_{12}(a_{21}a_{33} - a_{23}a_{31}) + a_{13}(a_{21}a_{32} - a_{22}a_{31})$$
$$= a_{11}a_{22}a_{33} - a_{12}a_{23}a_{32} - a_{12}a_{21}a_{33} + a_{12}a_{23}a_{31} + a_{13}a_{21}a_{32} - a_{13}a_{22}a_{31}$$

Except for order of terms, this is the expansion given in Problem 5.24.

Problems 5.31–5.33 refer to the matrices

$$A = \begin{pmatrix} 1 & 2 & 3 \\ 4 & -2 & 3 \\ 0 & 5 & -1 \end{pmatrix} \qquad B = \begin{pmatrix} 2 & 3 & 4 \\ 5 & 6 & 7 \\ 8 & 9 & 1 \end{pmatrix} \qquad C = \begin{pmatrix} 2 & 3 & -4 \\ 0 & -4 & 2 \\ 1 & -1 & 5 \end{pmatrix}$$

5.31 Find det A.

❚ Expand by the first row, as in Problem 5.30.

$$\begin{vmatrix} 1 & 2 & 3 \\ 4 & -2 & 3 \\ 0 & 5 & -1 \end{vmatrix} = \begin{vmatrix} \boxed{1} & 2 & 3 \\ \boxed{4} & -2 & 3 \\ \boxed{0} & 5 & -1 \end{vmatrix} - 2\begin{vmatrix} 1 & \boxed{2} & 3 \\ 4 & -2 & 3 \\ 0 & \boxed{5} & -1 \end{vmatrix} + 3\begin{vmatrix} 1 & 2 & \boxed{3} \\ 4 & -2 & 3 \\ 0 & 5 & \boxed{-1} \end{vmatrix}$$
$$= 1\begin{vmatrix} -2 & 3 \\ 5 & -1 \end{vmatrix} - 2\begin{vmatrix} 4 & 3 \\ 0 & -1 \end{vmatrix} + 3\begin{vmatrix} 4 & -2 \\ 0 & 5 \end{vmatrix}$$
$$= 1(2 - 15) - 2(-4 + 0) + 3(20 + 0) = -13 + 8 + 60 = 55$$

5.32 Find det B.

❚
$$\begin{vmatrix} 2 & 3 & 4 \\ 5 & 6 & 7 \\ 8 & 9 & 1 \end{vmatrix} = 2\begin{vmatrix} 6 & 7 \\ 9 & 1 \end{vmatrix} - 3\begin{vmatrix} 5 & 7 \\ 8 & 1 \end{vmatrix} + 4\begin{vmatrix} 5 & 6 \\ 8 & 9 \end{vmatrix}$$
$$= 2(6 - 63) - 3(5 - 56) + 4(45 - 48) = 27$$

5.33 Find det C.

❚
$$\begin{vmatrix} 2 & 3 & -4 \\ 0 & -4 & 2 \\ 1 & -1 & 5 \end{vmatrix} = 2\begin{vmatrix} -4 & 2 \\ -1 & 5 \end{vmatrix} - 3\begin{vmatrix} 0 & 2 \\ 1 & 5 \end{vmatrix} + (-4)\begin{vmatrix} 0 & -4 \\ 1 & -1 \end{vmatrix}$$
$$= 2(-20 + 2) - 3(0 - 2) - 4(0 + 4) = -46$$

Problems 5.34–5.36 refer to the following matrices:

$$A = \begin{pmatrix} 2 & 0 & 1 \\ 4 & 2 & -3 \\ 5 & 3 & 1 \end{pmatrix} \qquad B = \begin{pmatrix} 2 & 0 & 1 \\ 3 & 2 & -3 \\ -1 & -3 & 5 \end{pmatrix} \qquad C = \begin{pmatrix} 1 & 0 & 0 \\ 3 & 2 & -4 \\ 4 & 1 & 3 \end{pmatrix}$$

5.34 Find det A.

❚
$$\begin{vmatrix} 2 & 0 & 1 \\ 4 & 2 & -3 \\ 5 & 3 & 1 \end{vmatrix} = 2\begin{vmatrix} 2 & -3 \\ 3 & 1 \end{vmatrix} - 0\begin{vmatrix} 4 & -3 \\ 5 & 1 \end{vmatrix} + 1\begin{vmatrix} 4 & 2 \\ 5 & 3 \end{vmatrix} = 2(2 + 9) + (12 - 10) = 24$$

5.35   Find det b.

$$\begin{vmatrix} 2 & 0 & 1 \\ 3 & 2 & -3 \\ -1 & -3 & 5 \end{vmatrix} = 2(10-9) + 1(-9+2) = -5$$

5.36   Find det C.

$$\begin{vmatrix} 1 & 0 & 0 \\ 3 & 2 & -4 \\ 4 & 1 & 3 \end{vmatrix} = 1(6+4) = 10$$

[The calculations become more and more simple as there are more and more 0s in the expanding row.]

Problems 5.37–5.39 involve the matrix $A = \begin{pmatrix} a_1 & b_1 & c_1 \\ a_2 & b_2 & c_2 \\ a_2 & b_3 & c_3 \end{pmatrix}$.

5.37   Express det A as a linear combination of determinants of order two with coefficients from the second row.

▮   From Problem 5.24,

$$\det A = a_1 b_2 c_3 + a_2 b_3 c_1 + a_3 b_1 c_2 - a_1 b_3 c_2 - a_2 b_1 c_3 - a_3 b_2 c_1$$

$$= -a_2(b_1 c_3 - b_3 c_1) + b_2(a_1 c_3 - a_3 c_1) - c_2(a_1 b_3 - a_3 b_1)$$

$$= -a_2 \begin{vmatrix} a_1 & b_1 & c_1 \\ a_2 & b_2 & c_2 \\ a_3 & b_3 & c_3 \end{vmatrix} + b_2 \begin{vmatrix} a_1 & b_1 & c_1 \\ a_2 & b_2 & c_2 \\ a_3 & b_3 & c_3 \end{vmatrix} - c_2 \begin{vmatrix} a_1 & b_1 & c_1 \\ a_2 & b_2 & c_2 \\ a_3 & b_3 & c_3 \end{vmatrix}$$

5.38   Express det A as a linear combination of determinants of order two with coefficients from the third row.

▮

$$\det A = a_1 b_2 c_3 + b_1 c_2 a_3 + c_1 a_2 b_3 - a_3 b_2 c_1 - b_3 c_2 a_1 - c_3 a_2 b_1$$

$$= a_3(b_1 c_2 - b_2 c_1) - b_3(a_1 c_2 - a_2 c_1) + c_3(a_1 b_2 - a_2 b_1)$$

$$= a_3 \begin{vmatrix} a_1 & b_1 & c_1 \\ a_2 & b_2 & c_2 \\ a_3 & b_3 & c_3 \end{vmatrix} - b_3 \begin{vmatrix} a_1 & b_1 & c_1 \\ a_2 & b_2 & c_2 \\ a_3 & b_3 & c_3 \end{vmatrix} + c_3 \begin{vmatrix} a_1 & b_1 & c_1 \\ a_2 & b_2 & c_2 \\ a_3 & b_3 & c_3 \end{vmatrix}$$

5.39   As shown by Problems 5.30, 5.37, and 5.38, the signs of the coefficients in a row-expansion of a third-order determinant form a checkerboard pattern in the original matrix:

$$\begin{pmatrix} + & - & + \\ - & + & - \\ + & - & + \end{pmatrix}$$

One can also expand the determinant so that the coefficients come from a column rather than from a row. The same checkerboard pattern works for the columns.   Exhibit the three column-expansions of det A.

▮ $\det A = a_1 \begin{vmatrix} b_2 & c_2 \\ b_3 & c_3 \end{vmatrix} - a_2 \begin{vmatrix} b_1 & c_1 \\ b_3 & c_3 \end{vmatrix} + a_3 \begin{vmatrix} b_1 & c_1 \\ b_2 & c_2 \end{vmatrix}$ *first column*

$= -b_1 \begin{vmatrix} a_2 & c_2 \\ a_3 & c_3 \end{vmatrix} + b_2 \begin{vmatrix} a_1 & c_1 \\ a_3 & c_3 \end{vmatrix} - b_3 \begin{vmatrix} a_1 & c_1 \\ a_2 & c_2 \end{vmatrix}$ *second column*

$= c_1 \begin{vmatrix} a_2 & b_2 \\ a_3 & b_3 \end{vmatrix} - c_2 \begin{vmatrix} a_1 & b_1 \\ a_3 & b_3 \end{vmatrix} + c_3 \begin{vmatrix} a_1 & b_1 \\ a_2 & b_2 \end{vmatrix}$ *third column*

5.40   Give a criterion in terms of determinants which tells when the system

$$a_1 x + b_1 y + c_1 z = d_1$$
$$a_2 x + b_2 y + c_2 z = d_2$$
$$a_3 x + b_3 y + c_3 z = d_3$$

has a unique solution, and express such a unique solution in terms of determinants.

▌ The system has a unique solution if and only if the determinant of the matrix of coefficients is not zero:

$$D \equiv \begin{vmatrix} a_1 & b_1 & c_1 \\ a_2 & b_2 & c_2 \\ a_3 & b_3 & c_3 \end{vmatrix} \neq 0$$

In this case, the unique solution of the system can be expressed as quotients of determinants,

$$x = \frac{N_x}{D} \qquad y = \frac{N_y}{D} \qquad z = \frac{N_z}{D}$$

where the numerators N_x, N_y, and N_z are obtained by replacing the column of coefficients of the unknown in the matrix of coefficients by the column of constant terms:

$$N_x = \begin{vmatrix} d_1 & b_1 & c_1 \\ d_2 & b_2 & c_2 \\ d_3 & b_3 & c_3 \end{vmatrix} \qquad N_y = \begin{vmatrix} a_1 & d_1 & c_1 \\ a_2 & d_2 & c_2 \\ a_3 & d_3 & c_3 \end{vmatrix} \qquad N_z = \begin{vmatrix} a_1 & b_1 & d_1 \\ a_2 & b_2 & d_2 \\ a_3 & b_3 & d_3 \end{vmatrix}$$

5.41 Solve by determinants:

$$\begin{aligned} 2x + y - z &= 3 \\ x + y + z &= 1 \\ x - 2y - 3z &= 4 \end{aligned}$$

▌ First compute the determinant D of the matrix of coefficients:

$$D = \begin{vmatrix} 2 & 1 & -1 \\ 1 & 1 & 1 \\ 1 & -2 & -3 \end{vmatrix} = -6 + 1 + 2 + 1 + 4 + 3 = 5$$

Since $D \neq 0$, the system has a unique solution. Next evaluate N_x, N_y, and N_z, the numerators for x, y and z, respectively:

$$N_x = \begin{vmatrix} 3 & 1 & -1 \\ 1 & 1 & 1 \\ 4 & -2 & -3 \end{vmatrix} = -9 + 4 + 2 + 4 + 6 + 3 = 10$$

$$N_y = \begin{vmatrix} 2 & 3 & -1 \\ 1 & 1 & 1 \\ 1 & 4 & -3 \end{vmatrix} = -6 + 3 - 4 + 1 - 8 + 9 = -5$$

$$N_z = \begin{vmatrix} 2 & 1 & 3 \\ 1 & 1 & 1 \\ 1 & -2 & 4 \end{vmatrix} = 8 + 1 - 6 - 3 + 4 - 4 = 0$$

Thus the unique solution is $x = N_x/D = 2$, $y = N_y/D = -1$, $z = N_z/D = 0$.

5.42 Solve, using determinants:

$$\begin{aligned} 3y + 2x &= z + 1 \\ 3x + 2z &= 8 - 5y \\ 3z - 1 &= x - 2y \end{aligned}$$

▌ First put the system in standard form with the unknowns appearing in columns:

$$\begin{aligned} 2x + 3y - z &= 1 \\ 3x + 5y + 2z &= 8 \\ x - 2y - 3z &= -1 \end{aligned}$$

Compute the determinant D of the matrix of coefficients:

$$D = \begin{vmatrix} 2 & 3 & -1 \\ 3 & 5 & 2 \\ 1 & -2 & -3 \end{vmatrix} = -30 + 6 + 6 + 5 + 8 + 27 = 22$$

Since $D \neq 0$, the system has a unique solution. To compute N_x, N_y, and N_z, replace the coefficients of x, y, and z in the matrix of coefficients by the constant terms:

$$N_x = \begin{vmatrix} 1 & 3 & -1 \\ 8 & 5 & 2 \\ -1 & -2 & -3 \end{vmatrix} = -15 - 6 + 16 - 5 + 4 + 72 = 66$$

$$N_y = \begin{vmatrix} 2 & 1 & -1 \\ 3 & 8 & 2 \\ 1 & -1 & -3 \end{vmatrix} = -48 + 2 + 3 + 8 + 4 + 9 = -22$$

$$N_z = \begin{vmatrix} 2 & 3 & 1 \\ 3 & 5 & 8 \\ 1 & -2 & -1 \end{vmatrix} = -10 + 24 - 6 - 5 + 32 + 9 = 44$$

Hence $x = N_x/D = 3$, $y = N_y/D = -1$, $z = N_z/D = 2$.

5.4 PERMUTATIONS

5.43 Define a *permutation*.

▮ A permutation σ of a finite set $\mathcal{F}$ is a one-to-one mapping of $\mathcal{F}$ into itself. In the usual case, we have $\mathcal{F} = \{1, 2, \ldots, n\}$ and we employ the expanded notation

$$\sigma = \begin{pmatrix} 1 & 2 & \cdots & n \\ j_1 & j_2 & \cdots & j_n \end{pmatrix} \quad \text{or} \quad \sigma = j_1 j_2 \cdots j_n$$

where $j_i = \sigma(i)$. There are $n!$ such permutations σ, and they compose a group [Section 6.5], denoted S_n.

5.44 List the permutations in S_2.

▮ There are $2! = 2 \cdot 1 = 2$ permutations in S_2: 12 and 21.

5.45 List the permutations in S_3.

▮ There are $3! = 3 \cdot 2 \cdot 1 = 6$ permutations in S_3: 123, 132, 213, 231, 312, 321.

5.46 Define the *parity* of a permutation.

▮ A permutation σ is said to be *even* or *odd* according as there is an even or odd number of *inversions* in σ. By an inversion in σ we mean a pair of integers (i, k) such that $i > k$ but i precedes k in σ. The *sign* of σ, written sgn σ, is defined by

$$\text{sgn } \sigma = \begin{cases} 1 & \text{if } \sigma \text{ is even} \\ -1 & \text{if } \sigma \text{ is odd} \end{cases}$$

5.47 Find the parity of $\sigma = 35142$ (in S_5).

▮ For each element, count the number of elements smaller than it and to the right of it. Thus: 3 produces the inversions $(3, 1)$ and $(3, 2)$; 5 produces the inversions $(5, 1)$, $(5, 4)$, $(5, 2)$; 1 produces no inversion; 4 produces the inversion $(4, 2)$; 2 produces no inversion. Since there are, in all, six inversions, σ is even and sgn $\sigma = 1$.

5.48 Find the parity of the identity permutation ε.

▮ The identity permutation $\varepsilon = 123 \ldots n$ is even because there are no inversions in ε.

5.49 Find the parities of the permutations in S_2.

▮ The permutation 12 is even and 21 is odd.

5.50 Find the parities of the permutations in S_3.

▮ The permutations 123, 231, and 312 are even, and the permutations 132, 213 and 321 are odd.

5.51 Prove that S_n $(n \geq 2)$ consists of equal numbers $(n!/2)$ of even and odd permutations.

▮ Use induction. By Problem 5.49, the theorem holds for $n = 2$. With $n > 2$, consider the arbitrary element

$$\sigma = j_1 \ldots j_r j_{r+1} \ldots j_{n-1} \qquad (1)$$

of S_{n-1}. There are n ways of inserting the symbol "n" into (1)—just before j_1, just before $j_2, \ldots$, just before j_{n-1}, and just after j_{n-1}. Each way generates a distinct element σ' of S_n, and every element of S_n is thereby accounted for.

Now, because in (1) all $j_i < n$, the number of inversions in σ' equals the number of inversions in σ plus the *number of* the j_i to the right of the inserted "n." The latter number being independent of the particular σ, we conclude that *each element σ in S_{n-1} gives rise to x elements σ' in S_n of like parity to σ, and to $n - x$ elements of unlike parity.* By the inductive hypothesis, S_{n-1} consists of α even permutations and β odd permutations, where $\alpha = \beta$. Then S_n consists of $\alpha x + \beta(n - x)$ even permutations and $\beta x + \alpha(n - x) = \alpha x + \beta(n - x)$ odd permutations; and the proof is complete.

5.52 Define a *transposition* and determine its parity.

▮ A transposition is a permutation τ which interchanges two numbers, i and $j > i$, and leaves the other numbers fixed:

$$\tau = 12 \ldots (i-1)j(i+1) \ldots (j-1)i(j+1) \ldots n$$

There are $2(j - i - 1) + 1$ inversions in τ; namely, $(j, i), (j, x), (x, i)$, where $x = i + 1, \ldots, j - 1$. Thus any transposition τ is odd.

5.53 Using transpositions, determine the parity of $\sigma = 542163$ (in S_6).

▮ Transform σ to the identity permutation using transpositions; such as,

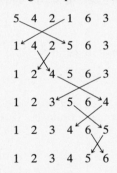

Since an odd number, 5, of transpositions was used [and since odd × odd = odd], σ is an odd permutation.

5.54 Let $\sigma = 24513$ and $\tau = 41352$ be permutations in S_5. Find the composition $\tau \circ \sigma$.

▮ Recall that $\sigma = 24513$ and $\tau = 41352$ are short ways of writing

$$\sigma = \begin{pmatrix} 1 & 2 & 3 & 4 & 5 \\ 2 & 4 & 5 & 1 & 3 \end{pmatrix} \quad \text{and} \quad \tau = \begin{pmatrix} 1 & 2 & 3 & 4 & 5 \\ 4 & 1 & 3 & 5 & 2 \end{pmatrix}$$

Accordingly, the effect of σ and then τ on $1, 2, \ldots, 5$ is as follows:

$$\begin{array}{ccccccc}
& 1 & 2 & 3 & 4 & 5 \\
\sigma & \downarrow & \downarrow & \downarrow & \downarrow & \downarrow \\
& 2 & 4 & 5 & 1 & 3 \\
\tau & \downarrow & \downarrow & \downarrow & \downarrow & \downarrow \\
& 1 & 5 & 2 & 4 & 3
\end{array}$$

Thus $\tau \circ \sigma = \begin{pmatrix} 1 & 2 & 3 & 4 & 5 \\ 1 & 5 & 2 & 4 & 3 \end{pmatrix}$, or $\tau \circ \sigma = 15243$.

5.55 Let σ and τ be the permutation in Problem 5.54. Find the composition $\sigma \circ \tau$.

❚ The effect of τ and then σ on $1, 2, \ldots, 5$ is as follows:

$$
\begin{array}{cccccc}
 & 1 & 2 & 3 & 4 & 5 \\
\tau & \downarrow & \downarrow & \downarrow & \downarrow & \downarrow \\
 & 4 & 1 & 3 & 5 & 2 \\
\sigma & \downarrow & \downarrow & \downarrow & \downarrow & \downarrow \\
 & 1 & 2 & 5 & 3 & 4
\end{array}
$$

Thus $\sigma \circ \tau = 12534$.

5.56 Find the inverse σ^{-1} of the permutation σ in Problem 4.10.

❚ By definition, $\sigma^{-1}(j) = k$ if and only if $\sigma(k) = j$; hence,

$$
\sigma^{-1} = \begin{pmatrix} 2 & 4 & 5 & 1 & 3 \\ 1 & 2 & 3 & 4 & 5 \end{pmatrix} = \begin{pmatrix} 1 & 2 & 3 & 4 & 5 \\ 4 & 1 & 5 & 2 & 3 \end{pmatrix} \quad \text{or} \quad \sigma^{-1} = 41523
$$

5.57 Consider any permutation $\sigma = j_1 j_2 \ldots j_n$. Show that for each inversion (i, k) in σ, there is a pair (i^*, k^*) such that

$$
i^* < k^* \quad \text{and} \quad \sigma(i^*) > \sigma(k^*) \tag{1}
$$

and vice versa. Thus σ is even or odd according as to whether there is an even or odd number of pairs satisfying *(1)*.

❚ Choose i^* and k^* so that $\sigma(i^*) = i$ and $\sigma(k^*) = k$. Then $i > k$ if and only if $\sigma(i^*) > \sigma(k^*)$, and i precedes k in σ if and only if $i^* < k^*$.

5.58 Consider the polynomial $g(x_1, \ldots, x_n) = \prod_{i < j}(x_i - x_j)$. Write out explicitly $g(x_1, x_2, x_3, x_4)$.

❚ $\prod_{i < j}(x_i - x_j)$ means the product of all terms $(x_i - x_j)$ for which $i < j$. Hence $g(x_1, \ldots, x_4) =$

$(x_1 - x_2)(x_1 - x_3)(x_1 - x_4)(x_2 - x_3)(x_2 - x_4)(x_3 - x_4)$.

5.59 Let σ be an arbitrary permutation. For the polynomial g of Problem 5.58, define $\sigma(g) = \prod_{i < j}(x_{\sigma(i)} - x_{\sigma(j)})$. Show that $\sigma(g) = (\text{sgn } \sigma)g$.

❚ In expanded form,

$$
\sigma(g) = (x_{\sigma(1)} - x_{\sigma(2)})(x_{\sigma(1)} - x_{\sigma(3)}) \cdots (x_{\sigma(1)} - X_{\sigma(n)})
$$

$$
\cdot (x_{\sigma(2)} - x_{\sigma(3)})(x_{\sigma(2)} - x_{\sigma(4)}) \cdots (x_{\sigma(2)} - x_{\sigma(n)})
$$

$$
\cdot \cdots
$$

$$
\cdot (x_{\sigma(n-1)} - x_{\sigma(n)})
$$

Because σ is one-one, each factor on the right has the form $x_i - x_j$, with $i \neq j$; however, if the pair $(i, j) = (\lambda, \nu)$ is represented in the product, the pair $(i, j) = (\nu, \lambda)$ is *not* represented. Consequently, $\sigma(g) = g$ or $\sigma(g) = -g$ according as there are an even or odd number of factors for which

$$
i < j \quad \text{and} \quad \sigma(i) > \sigma(j) \tag{1}
$$

But then, by Problem 5.57, $\sigma(g) = g$ if σ is an even permutation $(\text{sgn } \sigma = 1)$ and $\alpha(g) = -g$ if σ is an odd permutation $(\text{sgn } \sigma = -1)$.

5.60 Let $\sigma, \tau \in S_n$. Show that $\text{sgn }(\tau \circ \sigma) = (\text{sgn } \tau)(\text{sgn } \sigma)$. Thus the product [composition] of two even or two odd permutations is even, and the product of an odd and an even permutation is odd.

❚ By Problem 5.59, $\text{sgn }(\tau \circ \sigma)g = (\tau \circ \sigma)(g) = \tau(\sigma(g)) = \tau((\text{sgn } \sigma)g) = (\text{sgn } \tau)(\text{sgn } \sigma)g$. Accordingly, $\text{sgn }(\tau \circ \sigma) = (\text{sgn } \tau)(\text{sgn } \sigma)$.

5.61   Given the permutation $\sigma = j_1 j_2 \cdots j_n$, write $\sigma^{-1} = k_1 k_2 \cdots k_n$. Show that (*a*) sgn σ^{-1} = sgn σ;   (*b*) for any scalars a_{ij}, $a_{j_1 1} a_{j_2 2} \cdots a_{j_n n} = a_{1 k_1} a_{2 k_2} \cdots a_{n k_n}$.

▌   (*a*) By Problem 5.60, (sgn σ^{-1})(sgn σ) = sgn ε = 1.   (*b*) Since $\sigma = j_1 j_2 \cdots j_n$ is a permutation, $a_{j_1 1} a_{j_2 2} \cdots a_{j_n n} = a_{1 k_1} a_{2 k_2} \cdots a_{n k_n}$, for some permutation $k_1 k_2 \cdots k_n \equiv \tau$. But, by the very notation,

$$j_{k_1} = \sigma(k_1) = 1 \qquad j_{k_2} = \sigma(k_2) = 2 \qquad \cdots \qquad j_{k_n} = \sigma(k_n) = n$$

which means that $\sigma \circ \tau = \varepsilon$, or $\tau = \sigma^{-1}$.

5.5 DETERMINANTS OF ARBITRARY ORDER

5.62   Define the determinant of a general *n*-square matrix $A = (a_{ij})$.

▌   Consider a product of *n* elements of A such that no two elements are in the same row or the same column. Such a product can be written in the form

$$a_{1 j_1} a_{2 j_2} \cdots a_{n j_n}$$

where we have chosen to order the factors by now, making the sequence of second subscripts a permutation $\sigma = j_1 j_2 \cdots j_n$ in S_n. [The equivalent ordering by column is given by Problem 5.61, when the notation there is changed appropriately.] The *determinant* of A, denoted det A or $|A|$, is the sum of all such products, with each product multiplied by sgn σ. That is,

$$|A| = \sum_{\sigma \in S_n} (\text{sgn } \sigma) a_{1 j_1} a_{2 j_2} \cdots a_{n j_n} \equiv \sum_{\sigma \in S_n} (\text{sgn } \sigma) a_{1 \sigma(1)} a_{2 \sigma(2)} \cdots a_{n \sigma(n)} \qquad (1)$$

Such a determinant is said to be of *order n*. By Problem 5.51, half of the summands in *(1)* carry a +, and half a −.

5.63   Find det A, where $A = (a_{11})$ is a 1×1 matrix.

▌   Since S_1 consists of the even identity permutation, det $A = a_{11}$, in agreement with Problem 5.4.

5.64   Find det A, where $A = (a_{ij})$ is a 2×2 matrix.

▌   In S_2, the permutation 12 is even and the permutation 21 is odd. Hence, det $A = a_{11} a_{22} - a_{12} a_{21}$, in agreement with Problem 5.7.

5.65   Find det A, where $A = (a_{ij})$ is a 3×3 matrix.

▌   In S_3, the permutations 123, 231, and 312 are even, and the permutations 321, 213, and 132 are odd. Accordingly,

$$\det A = a_{11} a_{22} a_{33} + a_{12} a_{23} a_{31} + a_{13} a_{21} a_{32} - a_{13} a_{22} a_{31} - a_{12} a_{21} a_{33} - a_{11} a_{23} a_{32}$$

This agrees with Problem 5.24.

5.66   Prove: The determinant of a matrix A and its transpose A^T are equal: $|A| = |A^T|$.

▌   If $A = (a_{ij})$, then $A^T = (b_{ij})$, with $b_{ij} = a_{ji}$. Hence,

$$|A^T| = \sum_{\sigma \in S_n} (\text{sgn } \sigma) b_{1 \sigma(2)} b_{2 \sigma(2)} \cdots b_{n \sigma(n)} = \sum_{\sigma \in S_n} (\text{sgn } \sigma) a_{\sigma(1), 1} a_{\sigma(2), 2} \cdots a_{\sigma(n), n}$$

Let $\tau = \sigma^{-1}$. By Problem 5.61, sgn τ = sgn σ, and $a_{\sigma(1), 1} a_{\sigma(2), 2} \cdots a_{\sigma(n), n} = a_{1 \tau(1)} a_{2 \tau(2)} \cdots a_{n \tau(n)}$. Hence,

$$|A^T| = \sum_{\sigma \in S_n} (\text{sgn } \tau) a_{1 \tau(1)} a_{2 \tau(2)} \cdots a_{n \tau(n)}$$

However, as σ runs through all the elements of S_n, $\tau = \sigma^{-1}$ also runs through all the elements of S_n. Thus $|A^T| = |A|$. [In consequence, for any theorem about the determinant of a matrix A which concerns the rows of A, there will be an analogous theorem concerning the columns of A. This applies in particular to Theorem 5.1.]

5.67 Prove: If B is obtained from a square matrix A by interchanging two rows (columns) of A, then $|B| = -|A|$.

▌ We prove the theorem for the case that two columns are interchanged. Let τ be the transposition which interchanges the two numbers corresponding to the two columns of A that are interchanged. If $A = (a_{ij})$ and $B = (b_{ij})$, then $b_{ij} = a_{i\tau(j)}$. Hence, for any permutation σ,

$$b_{1\sigma(1)}b_{2\sigma(2)} \cdots b_{n\sigma(n)} = a_{1\tau\sigma(1)}a_{2\tau\sigma(2)} \cdots a_{n\tau\sigma(n)} \qquad \text{(with} \quad \tau\sigma \equiv \tau \circ \sigma\text{)}$$

Thus
$$|B| = \sum_{\sigma \in S_n} (\text{sgn } \sigma)a_{1\tau\sigma(1)}a_{2\tau\sigma(2)} \cdots a_{n\tau\sigma(n)}$$

Since the transposition τ is an odd permutation [Problem 5.52], $\text{sgn } \tau\sigma = (\text{sgn } \tau)(\text{sgn } \sigma) = -\text{sgn } \sigma$, and so

$$|B| = - \sum_{\sigma \in S_n} (\text{sgn } \tau\sigma)a_{1\tau\sigma(1)}a_{2\tau\sigma(2)} \cdots a_{n\tau\sigma(n)}$$

But as σ runs through all the elements of S_n, $\tau\sigma$ also runs through all the elements of S_n; hence $|B| = -|A|$.

5.68 Prove: If B is obtained from a square matrix A by multiplying a row (column) of A by a scalar k, then $|B| = k|A|$.

▌ If the ith row of A is multiplied by k, then every term in *(1)* of Problem 5.62 is multiplied by k, and so $|B| = k|A|$.

5.69 Show that if A has a row [column] of zeros, then $|A| = 0$.

▌ Each summand in *(1)* of Problem 5.62 contains a factor from every row, and so from the row of zeros.

5.70 Show that if A has two identical rows [columns], then $|A| = 0$.

▌ If we interchange the two identical rows of A, we obtain the matrix A. Hence, by Problem 5.67, $|A| = -|A|$ and so $|A| = 0$.

5.71 Let σ be any permutation in S_n other than the identity ε. Prove that $\sigma(i) < i$, for some i.

▌ Suppose, on the contrary, that $\sigma(i) \geq i$ for all i. Then, $\sigma(n) = n$ is forced. But now $\sigma(n-1) = n-1$ is forced [because $\sigma(n-1) = n$ is ruled out by one-oneness]. Continuing the argument, we establish that $\sigma(i) = i$, for all i. Thus, $\sigma = \varepsilon$—a contradiction.

5.72 For $A = (a_{ij})$ upper or lower triangular, show that $|A| = a_{11}a_{22} \cdots a_{nn}$, the product of the diagonal elements of A. [Recall that a triangular matrix is a special upper triangular matrix.]

▌ Suppose A upper triangular; then $a_{ij} = 0$ whenever $i > j$. Let σ be any permutation other than the identity permutation ε. By Problem 5.71, there exists an i such that $\sigma(i) < i$, so that

$$a_{1\sigma(1)}a_{2\sigma(2)} \cdots a_{n\sigma(n)} = 0$$

for every permutation except ε. Hence, $|A| = a_{11}a_{22} \cdots a_{nn}$.

Problems 5.73–5.75 refer to the following matrices:

$$A = \begin{pmatrix} 5 & 6 & 7 & 8 \\ 0 & 0 & 0 & 0 \\ 1 & -3 & 5 & -7 \\ 8 & 4 & 2 & 6 \end{pmatrix} \qquad B = \begin{pmatrix} 5 & 6 & 7 & 6 \\ 1 & -3 & 5 & -3 \\ 4 & 9 & -3 & 9 \\ 2 & 7 & 8 & 7 \end{pmatrix} \qquad C = \begin{pmatrix} 2 & 3 & 4 & 5 \\ 0 & -3 & 7 & -8 \\ 0 & 0 & 5 & 6 \\ 0 & 0 & 0 & 4 \end{pmatrix}$$

5.73 Find $\det A$.

▌ Since A has a row of zeros, $\det A = 0$, by Problem 5.69.

5.74   Find det B.

▮   Since the second and fourth columns of B are equal, det $B = 0$, by Problem 5.70.

5.75   Find det C.

▮   Since C is triangular, det $C = -120$, the product of the diagonal entries.

> **Theorem 5.1:**   Let B be obtained from a square matrix A by an elementary row operation. (i) If two rows of A were interchanged, then $|B| = -|A|$. (ii) If a row of A was multiplied by a scalar k, then $|B| = k|A|$. (iii) If a multiple of a row was added to another row, then $|B| = |A|$.

5.76   Prove Theorem 5.1.

▮   Parts (i) and (ii) were proved as Problems 5.67 and 5.68. We now prove (iii). Suppose c times the kth row is added to the jth row of A. Using the symbol $\frown$ to denote the jth position in a determinant term, we have

$$|B| = \sum_{\sigma} (\operatorname{sgn} \sigma) a_{1i_1} a_{2i_2} \cdots \overbrace{(ca_{ki_k} + a_{ji_j})} \cdots a_{ni_n}$$

$$= c \sum_{\sigma} (\operatorname{sgn} \sigma) a_{1i_1} a_{2i_2} \cdots \overbrace{a_{ki_k}} \cdots a_{ni_n} + \sum_{\sigma} (\operatorname{sgn} \sigma) a_{1i_1} a_{2i_2} \cdots \overbrace{a_{ji_j}} \cdots a_{ni_n}$$

The first sum is the determinant of a matrix whose kth and jth rows are identical; hence, by Problem 5.70, the sum is zero. The second sum is just the determinant of A.

5.77   Let B be row equivalent to a square matrix A. Show that $|B| = 0$ if and only if $|A| = 0$.

▮   By Theorem 5.1, the effect of an elementary row operation is to change the sign of the determinant or to multiply the determinant by a nonzero scalar. Therefore $|B| = 0$ if and only if $|A| = 0$.

5.78   Define a *singular* and a *nonsingular* matrix.

▮   A square matrix A is singular if det $A = 0$ and is nonsingular if det $A \neq 0$.

> **Theorem 5.2:**   For a square matrix A, the following are equivalent: (i) A is invertible; (ii) A is nonsingular; (iii) $AX = \mathbf{0}$ has only the zero solution.

5.79   Prove Theorem 5.2.

▮   The proof is by the Gaussian algorithm. If A is invertible, it is row equivalent to I [det $I = 1 \neq 0$]; hence, det $A \neq 0$, and A is nonsingular. If A is not invertible, it is row equivalent to a matrix with a zero row; hence, det $A = 0$, and A is singular. Thus, (i) and (ii) are equivalent.

Similarly, if $AX = \mathbf{0}$ has a unique solution $(X = \mathbf{0})$, A must be row equivalent to a triangular matrix and therefore [Problem 5.72] nonsingular, and therefore [by the above] invertible. Conversely, if A is invertible, with inverse A^{-1}, then

$$AX = \mathbf{0} \Rightarrow A^{-1}AX = \mathbf{0} \Rightarrow X = \mathbf{0}$$

Thus (iii) and (i) are equivalent.

Problems 5.80–5.82 establish

> **Lemma 1:**   For any elementary matrix E [Problem 4.108], $|EA| = |E||A|$.

5.80   Prove that $|E_1 A| = |E_1||A|$, where E_1 is the elementary matrix corresponding to the elementary row operation e_1 which interchanges two rows.

■ By Theorem 5.1(i), $|e_1(A)| = -|A|$. We have $|E_1| = |e_1(I)| = -|I| = -1$. Hence $|E_1 A| = |e_1(A)| = -|A| = |E_1| |A|$.

5.81 Prove that $|E_2 A| = |E_2| |A|$, where E_2 corresponds to the elementary row operation e_2 which multiplies a row by a nonzero scalar k.

■ By Theorem 5.1(ii), $|e_2(A)| = k|A|$. We have $|E_2| = |e_2(I)| = k|I| = k$. Hence $|E_2 A| = |e_2(A)| = k|A| = |E_2| |A|$.

5.82 Prove that $|E_3 A| = |E_3| |A|$, where E_3 corresponds to the elementary row operation e_3 which adds a multiple of one row to another row.

■ By Theorem 5.1(iii), $|e_3(A)| = |A|$. We have $|E_3| = |e_3(I)| = |I| = 1$. Hence $|E_3 A| = |e_3(a)| = |A| = |E_3| |A|$.

Theorem 5.3: Let A and B be n-square matrices. Then $\det AB = (\det A)(\det B)$.

5.83 Prove Theorem 5.3.

■ If A is singular, then AB is also singular [by Problem 4.124 and Theorem 5.2], and so $|AB| = 0 = |A| |B|$. On the other hand if A is nonsingular, then [Theorem 4.3] $A = E_p \cdots E_2 E_1$, a product of elementary matrices. Thus, by Lemma 1,

$$|A| = |E_p \cdots E_2 E_1 I| = |E_p| \cdots |E_2| |E_1| |I| = |E_p| \cdots |E_2| |E_1|$$

and so
$$|AB| = |E_p \cdots E_2 E_1 B| = |E_p| \cdots |E_2| |E_1| |E_1| |B| = |A| |B|.$$

5.84 Verify Theorem 5.3 for the matrices of Problems 5.34 and 5.35.

■ As calculated, $\det A = 24$ and $\det B = -5$. We have

$$AB = \begin{pmatrix} 2 & 0 & 1 \\ 4 & 2 & -3 \\ 5 & 3 & 1 \end{pmatrix}\begin{pmatrix} 2 & 0 & 1 \\ 3 & 2 & -3 \\ -1 & -3 & 5 \end{pmatrix} = \begin{pmatrix} 3 & -3 & 7 \\ 17 & 13 & -17 \\ 18 & 3 & 1 \end{pmatrix}$$

Making use of Theorem 5.1 (iii), we find

$$\det AB = \begin{vmatrix} 3 & -3 & 7 \\ 17 & 13 & -17 \\ 18 & 3 & 1 \end{vmatrix} = \begin{vmatrix} 3 & 0 & 10 \\ 17 & 30 & 0 \\ 18 & 21 & 19 \end{vmatrix} = \begin{vmatrix} 3 & 0 & 10 \\ 17 & 30 & 0 \\ 0 & 21 & -41 \end{vmatrix}$$

$$= 3(30)(-41) + 10(17)(21) = -3690 + 3570 = -120$$

and $-120 = (24)(-5)$.

5.85 If P is nonsingular, show that $|P^{-1}| = |P|^{-1}$.

■ $P^{-1}P = I$. Hence $1 = |I| = |P^{-1}P| = |P^{-1}| |P|$, and so $|P^{-1}| = |P|^{-1}$.

5.86 Suppose that B is *similar* to A; that is, suppose that there is a nonsingular matrix P such that $B = P^{-1}AP$. Show that $|B| = |A|$.

■ Using Theorem 5.3 and Problem 5.85, $|B| = |P^{-1}AP| = |P^{-1}| |A| |P| = |A| |P^{-1}| |P| = |A|$. [Although the matrices P^{-1} and A need not commute, their determinants, as scalars, do commute.]

5.87 If A is orthogonal, show that $|A| = -\pm 1$.

■ By the definition $AA^T = I$ and Problem 5.56, $|AA^T| = |A| |A^T| = |A|^2 = |I| = 1$; hence $|A| = \pm 1$.

5.88 Let $D_k = kI$ be an n-square scalar matrix. Show that $|D_k| = k^n$.

■ By Problem 5.72, $|D_k| = k \cdot k \cdot k \cdots \cdot k = k^n$.

5.89   If A is an n-square matrix, show that $|kA| = k^n |A|$.

▮   By Problem 5.88, $|kA| = |kIA| = |kI||A| = k^n |A|$.

5.6   EVALUATION OF DETERMINANTS; THE LAPLACE EXPANSION

5.90   Define the *minors* of a square matrix.

▮   Let $A = (a_{ij})$ be an n-square matrix.   Let M_{ij} be the $(n-1)$-square submatrix of A obtained by deleting the ith row and the jth column of A.   Then the determinant $|M_{ij}|$ is called the *ij-minor of A*.

5.91   Define the *cofactors* of a square matrix.

▮   Let $A = (a_{ij})$ be a square matrix.   The *ij-cofactor of A* (or: *cofactor of a_{ij}*), denoted here by A_{ij}, is the "signed" minor

$$A_{ij} = (-1)^{i+j} |M_{ij}|$$

Observe that the sign of the term $(-1)^{i+j}$ alternates in a checkerboard pattern, with +'s on the main diagonal:

$$\begin{pmatrix} + & - & + & - & \cdots \\ - & + & - & + & \cdots \\ + & - & + & - & \cdots \\ - & + & - & + & \cdots \\ \cdots & \cdots & \cdots & \cdots & \cdots \end{pmatrix}$$

Problems 5.92–5.99 refer to the matrix

$$A = \begin{pmatrix} 2 & 3 & 4 \\ 5 & 6 & 7 \\ 8 & 9 & 1 \end{pmatrix} \qquad B = \begin{pmatrix} 2 & 1 & -3 & 4 \\ 5 & -4 & 7 & -2 \\ 4 & 0 & 6 & -3 \\ 3 & -2 & 5 & 2 \end{pmatrix}$$

5.92   Find M_{21} and the minor $|M_{21}|$ of A.

▮   Delete the second row and the first column of A:

$$M_{21} = \begin{pmatrix} \boxed{\begin{matrix} 2 \\ 5 \\ 8 \end{matrix}} \begin{matrix} 3 & 4 \\ 6 & 7 \\ 9 & 1 \end{matrix} \end{pmatrix} = \begin{pmatrix} 3 & 4 \\ 9 & 1 \end{pmatrix} \qquad \text{and so} \quad |M_{21}| = 3 - 36 = -33$$

5.93   Find the cofactor A_{21}.

▮   Multiply the minor $|M_{21}|$ by the sign $(-1)^{2+1} = -1$; that is, $A_{21} = (-1)(-33) = 33$.

5.94   Find the minor $|M_{22}|$ of A.

▮   Delete the second row and second column from A and then find the determinant:

$$|M_{22}| = \begin{vmatrix} 2 & 3 & 4 \\ 5 & 6 & 7 \\ 8 & 9 & 1 \end{vmatrix} = \begin{vmatrix} 2 & 4 \\ 8 & 1 \end{vmatrix} = 2 - 32 = -30$$

5.95   Find the cofactor A_{22}.

▮   Multiply the minor $|M_{22}|$ by the appropriate sign: $A_{22} = (-1)^{2+2} |M_{22}| = (+1)(-30) = -30$.

5.96   Find the minor $|M_{23}|$ of A.

▮

$$|M_{23}| = \begin{vmatrix} 2 & 3 & 4 \\ 5 & 6 & 7 \\ 8 & 9 & 1 \end{vmatrix} = \begin{vmatrix} 2 & 3 \\ 8 & 9 \end{vmatrix} = 18 - 24 = -6$$

5.97 Find the cofactor A_{23}.

▮ $$A_{23} = (-1)^{2+3}|M_{23}| = (-1)(-6) = 6$$

5.98 Find the cofactor of the 7 in B; that is, find B_{23}.

▮ $$B_{23} = (-1)^{2+3} \begin{vmatrix} 2 & 1 & \boxed{-3} & 4 \\ \boxed{5 & -4 & 7 & -2} \\ 4 & 0 & \boxed{6} & -3 \\ 3 & -2 & \boxed{5} & 2 \end{vmatrix} = - \begin{vmatrix} 2 & 1 & 4 \\ 4 & 0 & -3 \\ 3 & -2 & 2 \end{vmatrix}$$

$$= -(0 - 9 - 32 - 0 - 12 - 8) = -(-61) = 61$$

5.99 Find the cofactor of the 2 in the last column of B; that is, find B_{44}.

▮ $$B_{44} = (-1)^{4+4} \begin{vmatrix} 2 & 1 & -3 & \boxed{4} \\ 5 & -4 & 7 & \boxed{-2} \\ 4 & 0 & 6 & \boxed{-3} \\ \boxed{3 & -2 & 5 & 2} \end{vmatrix} = \begin{vmatrix} 2 & 1 & -3 \\ 5 & -4 & 7 \\ 4 & 0 & 6 \end{vmatrix}$$

$$= -48 + 28 + 0 - 48 - 0 - 30 = -98$$

Theorem 5.4: (*Laplace Expansion Theorem*): *The determinant of the n-square matrix $A = (a_{ij})$ is equal to the sum of the products obtained by multiplying the elements of any row (column) by their respective cofactors:*

$$|A| = \sum_{j=1}^{n} a_{ij}A_{ij} \qquad \text{and} \qquad |A| = \sum_{i=1}^{n} a_{ij}A_{ij}$$

5.100 Prove Theorem 5.4.

▮ Each term in $|A|$ [see *(1)* of Problem 5.62] contains one and only one entry of the ith row $(a_{i1}, a_{i2}, \ldots, a_{in})$ of A. Hence we can write $|A|$ in the form

$$|A| = a_{i1}A_{i1}^* + a_{i2}A_{i2}^* + \cdots + a_{in}A_{in}^*$$

where A_{ij}^* is the sum of terms involving no entry of the ith row of A. Thus the theorem is proved if we can show that

$$A_{ij}^* = A_{ij} \equiv (-1)^{i+j}|M_{ij}|$$

First we consider the case that $i = n$, $j = n$. Then the sum of terms in $|A|$ containing a_{nn} is

$$a_{nn}A_{nn}^* = a_{nn} \sum_{\sigma} (\text{sgn } \sigma)a_{1\sigma(1)}a_{2\sigma(2)} \cdots a_{n-1,\sigma(n-1)}$$

where we sum over all permutations $\sigma \in S_n$ for which $\sigma(n) = n$. However, this is equivalent to summing over all permutations of $\{1, \ldots, n-1\}$. [By Problem 5.51, sgn $\sigma(1) \ldots \sigma(n-1)n = $ sgn $\sigma(1) \ldots \sigma(n-1)$.] Thus $A_{nn}^* = |M_{nn}| = (-1)^{n+n}|M_{nn}|$.

Now we consider any i and j. We interchange the ith row with each succeeding row until it is last, and we interchange the jth column with each succeeding column until it is last. Note that the determinant $|M_{ij}|$ is not affected since the relative positions of the other rows and columns are not affected by these interchanges. However, the sign of $|A|$ and of A_{ij}^* is changed $n-i$ and then $n-j$ times. Accordingly,

$$A_{ij}^* = (-1)^{n-i+n-j}|M_{ij}| = (-1)^{i+j}|M_{ij}|$$

5.101 Let $A = (a_{ij})$ be a nonzero n-square matrix with $n > 1$. Give an algorithm which reduces the determinant of A to a determinant of order $n-1$.

Step 1. Choose an element $a_{ij} = 1$ or, if lacking, $a_{ij} \neq 0$.

Step 2. Using a_{ij} as a pivot, apply elementary row [column] operations to put 0s in all the other positions in column j [row i].

Step 3. Expand the determinant cofactors of column j [row i].

If Step 2 involves multiplication of a row [column] by a scalar, the final answer must be adjusted according to Theorem 5.1(ii).

5.102 Compute the determinant of

$$A = \begin{pmatrix} 5 & 4 & 2 & 1 \\ 2 & 3 & 1 & -2 \\ -5 & -7 & -3 & 9 \\ 1 & -2 & -1 & 4 \end{pmatrix}$$

▮ Use $a_{23} = 1$ as a pivot to put 0s in the other entries of the third column; that is, apply the row operations $R_1 \to -2R_2 + R_1$, $R_3 \to 3R_2 + R_3$, and $R_4 \to R_2 + R_4$. By Theorem 5.1(iii), the value of the determinant does not change by these operations.

$$|A| = \begin{vmatrix} 5 & 4 & 2 & 1 \\ 2 & 3 & 1 & -2 \\ -5 & -7 & -3 & 9 \\ 1 & -2 & -1 & 4 \end{vmatrix} = \begin{vmatrix} 1 & -2 & 0 & 5 \\ 2 & 3 & 1 & -2 \\ 1 & 2 & 0 & 3 \\ 3 & 1 & 0 & 2 \end{vmatrix}$$

Now if we expand by the third column, we may neglect all terms which contain 0. Thus

$$|A| = (-1)^{2+3} \begin{vmatrix} 1 & -2 & \boxed{0} & 5 \\ \boxed{2} & \boxed{3} & \boxed{1} & \boxed{-2} \\ 1 & 2 & \boxed{0} & 3 \\ 3 & 1 & \boxed{0} & 2 \end{vmatrix} = - \begin{vmatrix} 1 & -2 & 5 \\ 1 & 2 & 3 \\ 3 & 1 & 2 \end{vmatrix}$$

$$= -(4 - 18 + 5 - 30 - 3 + 4) = -(-38) = 38$$

5.103 Evaluate the determinant of

$$A = \begin{pmatrix} 2 & 5 & -3 & -2 \\ -2 & -3 & 2 & -5 \\ 1 & 3 & -2 & 2 \\ -1 & -6 & 4 & 3 \end{pmatrix}$$

▮ Use $a_{31} = 1$ as a pivot and apply the row operations $R_1 \to -2R_3 + R_1$, $R_2 \to 2R_3 + R_2$, and $R_4 \to R_3 + R_4$:

$$|A| = \begin{vmatrix} 2 & 5 & -3 & -2 \\ -2 & -3 & 2 & -5 \\ 1 & 3 & -2 & 2 \\ -1 & -6 & 4 & 3 \end{vmatrix} = \begin{vmatrix} 0 & -1 & 1 & -6 \\ 0 & 3 & -2 & -1 \\ 1 & 3 & -2 & 2 \\ 0 & -3 & 2 & 5 \end{vmatrix} = + \begin{vmatrix} -1 & 1 & -6 \\ 3 & -2 & -1 \\ -3 & 2 & 5 \end{vmatrix} = 10 + 3 - 36 + 36 - 2 - 15 = -4$$

5.104 Evaluate the determinant of

$$B = \begin{pmatrix} 3 & -2 & -5 & 4 \\ 1 & -2 & -2 & 3 \\ -2 & 4 & 7 & -3 \\ 2 & -3 & -5 & 8 \end{pmatrix}$$

▮ Use $b_{21} = 1$ as a pivot and put 0s in the other entries in the second row by the column operations $C_2 \to 2C_1 + C_2$, $C_3 \to 2C_1 + C_3$, and $C_4 \to -3C_1 + C_4$. Then

$$|B| = \begin{vmatrix} 3 & -2 & -5 & 4 \\ 1 & -2 & -2 & 3 \\ -2 & 4 & 7 & -3 \\ 2 & -3 & -5 & 8 \end{vmatrix} = \begin{vmatrix} 3 & -2+2(3) & -5+2(3) & 4-3(3) \\ 1 & -2+2(1) & -2+2(1) & 3-3(1) \\ -2 & 4+2(-2) & 7+2(-2) & -3-3(-2) \\ 2 & -3+2(2) & -5+2(2) & 8-3(2) \end{vmatrix}$$

$$= \begin{vmatrix} 3 & 4 & 1 & -5 \\ 1 & 0 & 0 & 0 \\ -2 & 0 & 3 & 3 \\ 2 & 1 & -1 & 2 \end{vmatrix} = - \begin{vmatrix} 4 & 1 & -5 \\ 0 & 3 & 3 \\ 1 & -1 & 2 \end{vmatrix} = -(24+3+0+15+12-0) = -54$$

Problems 5.105–5.107 refer to the following matrices:

$$A = \begin{pmatrix} 1 & 2 & 2 & 3 \\ 1 & 0 & -2 & 0 \\ 3 & -1 & 1 & -2 \\ 4 & -3 & 0 & 2 \end{pmatrix} \qquad B = \begin{pmatrix} 2 & 1 & 3 & 2 \\ 3 & 0 & 1 & -2 \\ 1 & -1 & 4 & 3 \\ 2 & 2 & -1 & 1 \end{pmatrix} \qquad C = \begin{pmatrix} 6 & 2 & 1 & 0 & 5 \\ 2 & 1 & 1 & -2 & 1 \\ 1 & 1 & 2 & -2 & 3 \\ 3 & 0 & 2 & 3 & -1 \\ -1 & -1 & -3 & 4 & 2 \end{pmatrix}$$

5.105 Find det A.

▌ Use $a_{21} = 1$ as a pivot, and apply $C_3 \to 2C_1 + C_3$:

$$|A| = \begin{vmatrix} 1 & 2 & 4 & 3 \\ 1 & 0 & 0 & 0 \\ 3 & -1 & 7 & -2 \\ 4 & -3 & 8 & 2 \end{vmatrix} = - \begin{vmatrix} 2 & 4 & 3 \\ -1 & 7 & -2 \\ -3 & 8 & 2 \end{vmatrix} = -(28+24-24+63+32+8) = -131$$

5.106 Find det B.

▌ Use $b_{12} = 1$ as a pivot, and apply $R_3 \to R_1 + R_3$ and $R_4 \to -2R_1 + R_4$:

$$|B| = \begin{vmatrix} 2 & 1 & 3 & 2 \\ 3 & 0 & 1 & -2 \\ 3 & 0 & 7 & 5 \\ -2 & 0 & -7 & -3 \end{vmatrix} = - \begin{vmatrix} 3 & 1 & -2 \\ 3 & 7 & 5 \\ -2 & -7 & -3 \end{vmatrix} = \begin{vmatrix} 3 & 1 & -2 \\ 3 & 7 & 5 \\ 2 & 7 & 3 \end{vmatrix} = 63+10-42+28-105-9 = -55$$

[Observe that we factored out -1 from the third row, so that the minus sign in front of the determinant became a plus sign.]

5.107 Find det C.

▌ First reduce $|C|$ to a determinant of order four, and then to a determinant of order three. Use $c_{22} = 1$ as a pivot and apply $R_1 \to -2R_2 + R_1$, $R_3 \to -R_2 + R_3$, and $R_5 \to R_2 + R_5$:

$$|C| = \begin{vmatrix} 2 & 0 & -1 & 4 & 3 \\ 2 & 1 & 1 & -2 & 1 \\ -1 & 0 & 1 & 0 & 2 \\ 3 & 0 & 2 & 3 & -1 \\ 1 & 0 & -2 & 2 & 3 \end{vmatrix} = \begin{vmatrix} 2 & -1 & 4 & 3 \\ -1 & 1 & 0 & 2 \\ 3 & 2 & 3 & -1 \\ 1 & -2 & 2 & 3 \end{vmatrix} = \begin{vmatrix} 1 & -1 & 4 & -5 \\ 0 & 1 & 0 & 0 \\ 5 & 2 & 3 & -5 \\ -1 & -2 & 2 & 7 \end{vmatrix}$$

$$= \begin{vmatrix} 1 & 4 & -5 \\ 5 & 3 & -5 \\ -1 & 2 & 7 \end{vmatrix} = 21+20+50+15+10-140 = -24$$

5.108 Evaluate the determinant of $A = \begin{pmatrix} \frac{1}{2} & -1 & -\frac{1}{3} \\ \frac{3}{4} & \frac{1}{2} & -1 \\ 1 & -4 & 1 \end{pmatrix}$.

▌ First multiply the first row by 6 and the second row by 4. Then

$$6 \cdot 4 |A| = 24|A| = \begin{vmatrix} 3 & -6 & -2 \\ 3 & 2 & -4 \\ 1 & -4 & 1 \end{vmatrix} = 6+24+24+4-48+18 = 28$$

Hence $|A| = 28/24 = 7/6$. [Observe that the original multiplications eliminated the fractions, so the arithmetic is simpler.]

5.109 Evaluate the determinant of

$$A = \begin{pmatrix} t+3 & -1 & 1 \\ 5 & t-3 & 1 \\ 6 & -6 & t+4 \end{pmatrix}$$

▮ Add the second column to the first column, and then add the third column to the second column to produce 0s:

$$|A| = \begin{vmatrix} t+2 & 0 & 1 \\ t+2 & t-2 & 1 \\ 0 & t-2 & t+4 \end{vmatrix}$$

Now factor $t+2$ from the first column and $t-2$ from the second column to get

$$|A| = (t+2)(t-2) \begin{vmatrix} 1 & 0 & 1 \\ 1 & 1 & 1 \\ 0 & 1 & t+4 \end{vmatrix}$$

Finally subtract the first column from the third column to obtain

$$|A| = (t+2)(t-2) \begin{vmatrix} 1 & 0 & 0 \\ 1 & 1 & 0 \\ 0 & 1 & t+4 \end{vmatrix} = (t+2)(t-2)(t+4)$$

5.110 Describe the Gaussian elimination algorithm for calculating the determinant of an n-square matrix $A = (a_{ij})$.

▮ The algorithm uses Gaussian elimination to transform A into an upper triangular matrix [whose determinant is the product of its diagonal entries; Problem 5.72]. Since the algorithm involves exchanging rows, which changes the sign of the determinant, one must keep track of such changes using some variable, say SIGN. The algorithm will also use "pivoting"; that is, the element with the greatest absolute value will be used as the pivot. The algorithm follows.

Step 1. Set SIGN$=0$. [This initializes the variable SIGN.]
Step 2. Find the entry a_{i1} in the first column with greatest absolute value. (*a*) If $a_{i1} = 0$, then set det $A = 0$ and EXIT. (*b*) If $i \neq 1$, then interchange the first and ith rows and set SIGN$=$SIGN$+1$.
Step 3. Use a_{11} as a pivot and elementary row operations of the form $R_p \rightarrow kR_q + R_p$ to put 0s below a_{11}.
Step 4. Repeat Steps 2 and 3 with the submatrix obtained by deleting the first row and the first column.
Step 5. Continue the above process until A is an upper triangular matrix.
Step 6. Set det $A = (-1)^{\text{SIGN}} a_{11} a_{22} \cdots a_{nn}$, and EXIT.

Note that the operation $R_p \rightarrow kR_p$, which is permitted in the Gaussian algorithm for a system of linear equations, is barred here, as it changes the value of the determinant.

5.111 Find the determinant of

$$A = \begin{pmatrix} 3 & 8 & 6 \\ -2 & -3 & 1 \\ 5 & 10 & 15 \end{pmatrix}$$

using the Gaussian elimination algorithm of Problem 5.110

▮
$$A \sim \begin{pmatrix} 5 & 10 & 15 \\ -2 & -3 & 1 \\ 3 & 8 & 6 \end{pmatrix} \sim \begin{pmatrix} 5 & 10 & 15 \\ 0 & 1 & 7 \\ 0 & 2 & -3 \end{pmatrix} \sim \begin{pmatrix} 5 & 10 & 15 \\ 0 & 2 & -3 \\ 0 & 1 & 7 \end{pmatrix} \sim \begin{pmatrix} 5 & 10 & 15 \\ 0 & 2 & -3 \\ 0 & 0 & 17/2 \end{pmatrix}$$

A is now in (upper) triangular form and SIGN$=2$, since there were two interchanges of rows. Hence, $|A| = (-1)^{\text{SIGN}}(5)(2)(17/2) = 85$.

5.112 Let $A = (a_{ij})$ be an $n \times n$ matrix and let B be the matrix obtained from A by replacing the ith row of A by the row vector $(b_{i1} \quad \ldots \quad b_{in})$. Show that

$$|B| = \sum_{j=1}^{n} b_{ij} A_{ij}$$

▌ Let $B = (b_{ij})$. By Theorem 5.4,

$$|B| = \sum_{j=1}^{n} b_{ij} B_{ij}$$

But B_{ij} does not depend upon the ith row of B, $B_{ij} = A_{ij}$ for $j = 1, \ldots, n$.

5.113 Use Problem 5.112 to establish the *generalized Laplace expansion theorem*:

$$\sum_{j=1}^{n} a_{ij} A_{kj} = \delta_{ik} |A| \qquad (1)$$

▌ If $i = k$, (1) is just Theorem 5.4. If $i \neq k$, choose $(b_{i1}, \ldots, b_{in}) = (a_{k1}, \ldots, a_{km})$ in Problem 5.112, so that

$$|B| = \sum_{j=1}^{n} a_{kj} A_{ij}$$

But $|B| = 0$, since B has two identical rows; thus (1) again holds [with the free indices i and k interchanged].

Applying (1) to A^T gives the analogous "column" result

$$\sum_{i=1}^{n} a_{ij} A_{ik} = \delta_{jk} |A| \qquad (2)$$

5.7 CLASSICAL ADJOINT

5.114 Define the *classical adjoint* of a square matrix A.

▌ The classical adjoint [traditionally, just "adjoint"] of A, denoted adj A, is the transpose of the matrix of cofactors of A.

5.115 Find adj A for the matrix

$$A = \begin{pmatrix} 2 & 3 & 4 \\ 5 & 6 & 7 \\ 8 & 9 & 1 \end{pmatrix}$$

▌ First find the nine cofactors A_{ij} of A:

$$A_{11} = + \begin{vmatrix} 6 & 7 \\ 9 & 1 \end{vmatrix} = -57 \qquad A_{12} = - \begin{vmatrix} 5 & 7 \\ 8 & 1 \end{vmatrix} = 51 \qquad A_{13} = + \begin{vmatrix} 5 & 6 \\ 8 & 9 \end{vmatrix} = -3$$

$$A_{21} = - \begin{vmatrix} 3 & 4 \\ 9 & 1 \end{vmatrix} = 33 \qquad A_{22} = + \begin{vmatrix} 2 & 4 \\ 8 & 1 \end{vmatrix} = -30 \qquad A_{23} = - \begin{vmatrix} 2 & 3 \\ 8 & 9 \end{vmatrix} = 6$$

$$A_{31} = + \begin{vmatrix} 3 & 4 \\ 6 & 7 \end{vmatrix} = -3 \qquad A_{32} = - \begin{vmatrix} 2 & 4 \\ 5 & 7 \end{vmatrix} = 6 \qquad A_{33} = + \begin{vmatrix} 2 & 3 \\ 5 & 6 \end{vmatrix} = -3$$

Take the transpose of the above matrix of cofactors:

$$\text{adj } A = \begin{pmatrix} -57 & 33 & -3 \\ 51 & -30 & 6 \\ -3 & 6 & -3 \end{pmatrix}$$

5.116 Find adj A for the arbitary 2-square matrix $A = \begin{pmatrix} a & b \\ c & d \end{pmatrix}$.

▌ $$\text{adj } A = \begin{pmatrix} +|d| & -|c| \\ -|b| & +|a| \end{pmatrix}^T = \begin{pmatrix} d & -c \\ -b & a \end{pmatrix}^T = \begin{pmatrix} d & -b \\ -c & a \end{pmatrix}$$

5.117 Show that adj (adj A) = A for the matrix in Problem 5.116.

▌ $$\text{adj(adj } A) = \text{adj} \begin{pmatrix} d & -b \\ -c & a \end{pmatrix} = \begin{pmatrix} +|a| & -|-c| \\ -|-b| & +|d| \end{pmatrix}^T = \begin{pmatrix} a & c \\ b & d \end{pmatrix}^T = \begin{pmatrix} a & b \\ c & d \end{pmatrix} = A$$

Theorem 5.5: For any square matrix A,

$$A\,(\text{adj}\,A) = (\text{adj}\,A)\,A = \text{diag}\,(|A|, |A|, \ldots, |A|) \equiv |A|\,I$$

5.118 Prove Theorem 5.5.

▋ Simply rewrite *(1)* and *(2)* of Problem 5.113 in matrix form.

5.119 Assuming that $|A| \neq 0$, show that $A^{-1} = (1/|A|)(\text{adj}\,A)$.

▋ By Theorem 5.5, when $|A| \neq 0$,

$$A\,\frac{1}{|A|}\,(\text{adj}\,A) = \frac{1}{|A|}\,(\text{adj}\,A)\,A = I$$

whence, by definition, $A^{-1} = (1/|A|)(\text{adj}\,A)$.

In Problems 5.120–5.123, $A = \begin{pmatrix} 1 & 2 & 3 \\ 2 & 3 & 4 \\ 1 & 5 & 7 \end{pmatrix}$.

5.120 Find det A.

▋ $|A| = 21 + 8 + 30 - 9 - 20 - 28 = 2$

5.121 Find adj A.

$$\text{adj}\,A = \begin{pmatrix} +\begin{vmatrix} 3 & 4 \\ 5 & 7 \end{vmatrix} & -\begin{vmatrix} 2 & 4 \\ 1 & 7 \end{vmatrix} & +\begin{vmatrix} 2 & 3 \\ 1 & 5 \end{vmatrix} \\ -\begin{vmatrix} 2 & 3 \\ 5 & 7 \end{vmatrix} & +\begin{vmatrix} 1 & 3 \\ 1 & 7 \end{vmatrix} & -\begin{vmatrix} 1 & 2 \\ 1 & 5 \end{vmatrix} \\ +\begin{vmatrix} 2 & 3 \\ 3 & 4 \end{vmatrix} & -\begin{vmatrix} 1 & 3 \\ 2 & 4 \end{vmatrix} & +\begin{vmatrix} 1 & 2 \\ 2 & 3 \end{vmatrix} \end{pmatrix}^T = \begin{pmatrix} 1 & -10 & 7 \\ 1 & 4 & -3 \\ -1 & 2 & -1 \end{pmatrix}^T = \begin{pmatrix} 1 & 1 & -1 \\ -10 & 4 & 2 \\ 7 & -3 & -1 \end{pmatrix}$$

5.122 Verify that $A\,(\text{adj}\,A) = |A|\,I$.

$$A\,(\text{adj}\,A) = \begin{pmatrix} 1 & 2 & 3 \\ 2 & 3 & 4 \\ 1 & 5 & 7 \end{pmatrix}\begin{pmatrix} 1 & 1 & -1 \\ -10 & 4 & 2 \\ 7 & -3 & -1 \end{pmatrix} = \begin{pmatrix} 2 & 0 & 0 \\ 0 & 2 & 0 \\ 0 & 0 & 2 \end{pmatrix} = 2\begin{pmatrix} 1 & 0 & 0 \\ 0 & 1 & 0 \\ 0 & 0 & 1 \end{pmatrix} = |A|\,I$$

5.123 Use adj A to find A^{-1}.

▋ Since $|A| \neq 0$,

$$A^{-1} = \frac{1}{|A|}\,(\text{adj}\,A) = \frac{1}{2}\begin{pmatrix} 1 & 1 & -1 \\ -10 & 4 & 2 \\ 7 & -3 & -1 \end{pmatrix} = \begin{pmatrix} \frac{1}{2} & \frac{1}{2} & -\frac{1}{2} \\ -5 & 2 & 1 \\ \frac{7}{2} & -\frac{3}{2} & -\frac{1}{2} \end{pmatrix}$$

In Problems 5.124–5.127, $A = \begin{pmatrix} 2 & 3 & -4 \\ 0 & -4 & 2 \\ 1 & -1 & 5 \end{pmatrix}$.

5.124 Find det A.

▋ $|A| = -40 + 6 + 0 - 16 + 4 + 0 = -46$

5.125 Find adj A.

▋ First find the nine cofactors A_{ij} of A:

$$A_{11} = +\begin{vmatrix} -4 & 2 \\ -1 & 5 \end{vmatrix} = -18 \qquad A_{12} = -\begin{vmatrix} 0 & 2 \\ 1 & 5 \end{vmatrix} = 2 \qquad A_{13} = +\begin{vmatrix} 0 & -4 \\ 1 & -1 \end{vmatrix} = 4$$

$$A_{21} = - \begin{vmatrix} 3 & -4 \\ -1 & 5 \end{vmatrix} = -11 \qquad A_{22} = + \begin{vmatrix} 2 & -4 \\ 1 & 5 \end{vmatrix} = 14 \qquad A_{23} = - \begin{vmatrix} 2 & 3 \\ 1 & -1 \end{vmatrix} = 5$$

$$A_{31} = + \begin{vmatrix} 3 & -4 \\ -4 & 2 \end{vmatrix} = -10 \qquad A_{32} = - \begin{vmatrix} 2 & -4 \\ 0 & 2 \end{vmatrix} = -4 \qquad A_{33} = + \begin{vmatrix} 2 & 3 \\ 0 & -4 \end{vmatrix} = -8$$

Then adj A is the transpose of the matrix of cofactors:

$$\text{adj } A = \begin{pmatrix} -18 & -11 & -10 \\ 2 & 14 & -4 \\ 4 & 5 & -8 \end{pmatrix}$$

5.126 Verify that $A\,(\text{adj } A) = |A|\,I$.

❚ $A(\text{adj } A) = \begin{pmatrix} 2 & 3 & -4 \\ 0 & -4 & 2 \\ 1 & -1 & 5 \end{pmatrix} \begin{pmatrix} -18 & -11 & -10 \\ 2 & 14 & -4 \\ 4 & 5 & -8 \end{pmatrix} = \begin{pmatrix} -46 & 0 & 0 \\ 0 & -46 & 0 \\ 0 & 0 & -46 \end{pmatrix} = -46 \begin{pmatrix} 1 & 0 & 0 \\ 0 & 1 & 0 \\ 0 & 0 & 1 \end{pmatrix} = -46I = |A|\,I$

5.127 Use adj A to find A^{-1}.

❚ Since $|A| \neq 0$,

$$A^{-1} = \frac{1}{|A|}\,(\text{adj } A) = \begin{pmatrix} -18/-46 & -111/-46 & -10/-46 \\ 2/-46 & 14/-46 & -4/-46 \\ 4/-46 & 5/-46 & -8/-46 \end{pmatrix} = \begin{pmatrix} 9/23 & 11/46 & 5/23 \\ -1/23 & -7/23 & 2/23 \\ -2/23 & -5/46 & 4/23 \end{pmatrix}$$

In Problems 5.128–5.130, $B = \begin{pmatrix} 1 & 1 & 1 \\ 2 & 3 & 4 \\ 5 & 8 & 9 \end{pmatrix}$.

5.128 Find $|B|$.

❚ $|B| = 27 + 20 + 16 - 15 - 32 - 18 = -2$

5.129 Find adj B.

$$\text{adj } B = \begin{pmatrix} \begin{vmatrix} 3 & 4 \\ 8 & 9 \end{vmatrix} & -\begin{vmatrix} 2 & 4 \\ 5 & 9 \end{vmatrix} & \begin{vmatrix} 2 & 3 \\ 5 & 8 \end{vmatrix} \\ -\begin{vmatrix} 1 & 1 \\ 8 & 9 \end{vmatrix} & \begin{vmatrix} 1 & 1 \\ 5 & 9 \end{vmatrix} & -\begin{vmatrix} 1 & 1 \\ 5 & 8 \end{vmatrix} \\ \begin{vmatrix} 1 & 1 \\ 3 & 4 \end{vmatrix} & -\begin{vmatrix} 1 & 1 \\ 2 & 4 \end{vmatrix} & \begin{vmatrix} 1 & 1 \\ 2 & 3 \end{vmatrix} \end{pmatrix}^T = \begin{pmatrix} -5 & 2 & 1 \\ -1 & 4 & -3 \\ 1 & -2 & 1 \end{pmatrix}^T = \begin{pmatrix} -5 & -1 & 1 \\ 2 & 4 & -2 \\ 1 & -3 & 1 \end{pmatrix}$$

5.130 Find B^{-1}, using adj B.

❚ Because $|B| \neq 0$,

$$B^{-1} = \frac{1}{|B|}\,(\text{adj } B) = \frac{1}{-2} \begin{pmatrix} -5 & -1 & 1 \\ 2 & 4 & -2 \\ 1 & -3 & 1 \end{pmatrix} = \begin{pmatrix} 5/2 & 1/2 & -1/2 \\ -1 & -2 & 1 \\ -1/2 & 3/2 & -1/2 \end{pmatrix}$$

5.131 Prove that a square matrix is singular if and only if its classical adjoint is singular.

❚ If matrix A is nonsingular, so that $|A| \neq 0$, Theorem 5.5 shows that adj A has the inverse $|A|^{-1} A$; hence adj A is also nonsingular.

On the other hand, if A is singular but nonzero, Theorem 5.5 gives A adj $A = \mathbf{0}$, which implies that adj A is not invertible and is therefore singular. (If an inverse B existed, then $\mathbf{0} = \mathbf{0}B = A[(\text{adj } A)B] = A$.) Finally, if $A = \mathbf{0}$ [singular], then adj $A = \mathbf{0}$ [singular].

5.132 Let A be n-square $(n \geq 2)$. Show that

$$|\text{adj } A| = |A|^{n-1} \tag{1}$$

▮ By Theorem 5.5, $A \text{ adj } A = |A| I$, whence $|A| |\text{adj } A| = |A|^n$. If $|A| \neq 0$, cancellation gives (1). If $|A| = 0 = |\text{adj } A|$ [Problem 5.131], then (1) holds trivially.

5.133 Let A be n-square $(n \geq 2)$ and nonsingular. Show that

$$\text{adj (adj } A) = |A|^{n-2} A \tag{1}$$

▮ By Theorem 5.5 and Problem 5.132, $(\text{adj } A)[\text{adj (adj } A)] = |A|^{n-1} I$, and premultiplication by A yields

$$|A| [\text{adj (adj } A)] = |A|^{n-1} A$$

Now multiply through by $|A|^{-1}$ to obtain (1).

5.134 Suppose that A is upper [lower] triangular. Show that adj A is upper [lower] triangular.

▮ Let $A = (a_{ij})$ and adj $A = (b_{ij})$. If A is upper triangular, then, for $i < j$, the submatrix M_{ij} of A is also upper triangular and M_{ij} has one or more 0s on its diagonal. Hence $|M_{ij}| = 0$. Therefore, for $i < j$, $b_{ji} = A_{ij} = |M_{ij}| = 0$, and so adj A is upper triangular.

5.135 If A is diagonal, show that adj A is diagonal.

▮ If A is diagonal, then A is both upper and lower triangular. Hence adj A is both upper and lower triangular, and so adj A is diagonal.

5.8 VOLUME AS A DETERMINANT

5.136 How are determinants related to area and volume?

▮ Let $u_1, u_2, \ldots, u_n$ be vectors (arrows) in $\mathbf{R}^n$. Let $\mathscr{S}$ be the parallelopiped formed by the vectors and let A be the matrix with rows $u_1, u_2, \ldots, u_n$. Then the volume of $\mathscr{S}$ (or: area of $\mathscr{S}$, when $n = 2$), denoted $V(\mathscr{S})$, is equal to the absolute value of det A.

5.137 Let $u_1 = (2, 3)$ and $u_2 = (5, 1)$ be vectors in $\mathbf{R}^2$. Draw the parallelogram $\mathscr{S}$ determined by the vectors (arrows).

▮ See Fig. 5-2.

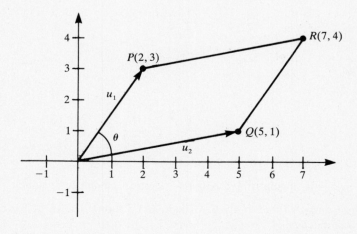

Fig. 5-2

5.138 Verify the result of Problem 5.136 for the parallelogram of Problem 5.137.

▮ With $A = \begin{pmatrix} 2 & 3 \\ 5 & 1 \end{pmatrix} \equiv \begin{pmatrix} u_{11} & u_{12} \\ u_{21} & u_{22} \end{pmatrix}$, we have

$$AA^T = \begin{pmatrix} u_1 \cdot u_1 & u_1 \cdot u_2 \\ u_2 \cdot u_1 & u_2 \cdot u_2 \end{pmatrix} \quad \text{whence} \quad (\det A)^2 = (u_1 \cdot u_1)(u_2 \cdot u_2) - (u_1 \cdot u_2)^2$$

But, by Problem 1.177 [valid for the subspace $\mathbf{R}^2$ of $\mathbf{R}^3$],

$$(u_1 \cdot u_1)(u_2 \cdot u_2) - (u_1 \cdot u_2)^2 = \|u_1 \times u_2\|^2 = (\|u_1\| \, \|u_2\| \sin \vartheta)^2 = [V(\mathscr{S})]^2$$

Consequently, $V(\mathscr{S}) = \sqrt{(\det A)^2} = |\det A|$.

5.139 Find $V(\mathscr{S})$ for the parallelopiped $\mathscr{S}$ in $\mathbf{R}^3$ determined by the vectors $u_1 = (2, 5, 2)$, $u_2 = (4, 2, 3)$, and $u_3 = (1, 1, 4)$ in $\mathbf{R}^3$.

▮ Evaluate the determinant of the matrix

$$A = \begin{pmatrix} 2 & 5 & 2 \\ 4 & 2 & 3 \\ 1 & 1 & 4 \end{pmatrix}$$

[whose rows are the vectors u_1, u_2, u_3]:

$$|A| = 16 + 15 + 8 - 4 - 6 - 80 = -51 \quad \text{whence} \quad V(\mathscr{S}) = 51$$

5.140 Let u_1 and u_2 be vectors [arrows] in $\mathbf{R}^2$ and let A be the matrix with rows u_1 and u_2. Give a geometric condition that determines whether det A is positive, zero or negative.

▮ If the smallest rotation that brings u_1 into u_2 is counterclockwise [clockwise], det A is positive [negative]. If u_1 is in the same or opposite direction as u_2, then det A is zero.

5.141 Let u_1, u_2, and u_3 be vectors (arrows) in $\mathbf{R}^3$. Let A be the matrix with rows u_1, u_2, u_3. Give a geometric condition that determines whether det A is positive, zero, or negative.

▮ det A is positive or negative according as u_1, u_2, u_3 form a right- or left-handed coordinate system. If the three vectors lie in a plane, then det A is zero.

5.142 Show that the vectors $u_1 = (1, 2, 4)$, $u_2 = (2, 1, -3)$, and $u_3 = (5, 7, 9)$ lie in a plane.

▮
$$\begin{vmatrix} 1 & 2 & 4 \\ 2 & 1 & -3 \\ 5 & 7 & 9 \end{vmatrix} = 9 - 30 + 56 - 20 + 21 - 36 = 0$$

[or else note that $u_3 = 3u_1 + u_2$].

5.143 Find the volume $V(\mathscr{S})$ of the parallelopiped $\mathscr{S}$ in $\mathbf{R}^4$ determined by the vectors $u_1 = (2, -1, 4, -3)$, $u_2 = (-1, 1, 0, 2)$, $u_3 = (3, 2, 3, -1)$, $u_4 = (1, -2, 2, 3)$.

▮ Evaluate the following determinant:

$$\begin{vmatrix} 2 & -1 & 4 & -3 \\ -1 & 1 & 0 & 2 \\ 3 & 2 & 3 & -1 \\ 1 & -2 & 2 & 3 \end{vmatrix} = \begin{vmatrix} 1 & -1 & 4 & -1 \\ 0 & 1 & 0 & 0 \\ 5 & 2 & 3 & -5 \\ -1 & -2 & 2 & 7 \end{vmatrix} = \begin{vmatrix} 1 & 4 & -1 \\ 5 & 3 & -5 \\ -1 & 2 & 7 \end{vmatrix}$$

$$= 21 + 20 - 10 - 3 + 10 - 140 = -102$$

Hence $V(\mathscr{S}) = 102$.

5.9 CRAMER'S RULE. BLOCK MATRICES

Theorem 5.6 (*Cramer's Rule*): Let $AX = B$ be an $n \times n$ system of linear equations with nonsingular coefficient matrix $A = (a_{ij})$. Let A_i be the matrix obtained from A by replacing the ith column of A by the column vector B. Let $D \equiv |A|$, and let $N_i \equiv |A_i|$ for $i = 1, 2, \ldots, n$. Then the system has the unique solution $x_i = N_i/D$ $(i = 1, 2, \ldots, n)$.

5.144 Prove Theorem 5.6.

❚ By Problem 5.112 [column version], $N_i = \sum_{k=1}^{n} b_k A_{ki}$. Hence,

$$\sum_{j=1}^{n} a_{ij} x_j = \sum_{j=1}^{n} a_{ij} \left(\frac{1}{D} \sum_{k=1}^{n} b_k A_{kj} \right) = \frac{1}{D} \sum_{k=1}^{n} \left(\sum_{j=1}^{n} a_{ij} A_{kj} \right) b_k$$

By *(1)* of Problem 5.113, the sum over j has the value $\delta_{ij} D$; we then have

$$\sum_{j=1}^{n} a_{ij} x_j = \frac{1}{D} \sum_{k=1}^{n} \delta_{ik} D b_k = \frac{1}{D} (D b_i) = b_i$$

This solution is unique because A is nonsingular.

5.145 Use Cramer's Rule to solve the system:

$$\begin{array}{l} x_1 + x_2 + x_3 + x_4 = 2 \\ x_1 + 2x_2 + 3x_3 + 4x_4 = 2 \\ 2x_1 + 3x_2 + 5x_3 + 9x_4 = 2 \\ x_1 + x_2 + 2x_3 + 7x_4 = 2 \end{array}$$

❚ Compute

$$D = |A| = \begin{vmatrix} 1 & 1 & 1 & 1 \\ 1 & 2 & 3 & 4 \\ 2 & 3 & 5 & 9 \\ 1 & 1 & 2 & 7 \end{vmatrix} = 2 \qquad N_1 = |A_1| = \begin{vmatrix} 2 & 1 & 1 & 1 \\ 2 & 2 & 3 & 4 \\ 2 & 3 & 5 & 9 \\ 2 & 1 & 2 & 7 \end{vmatrix} = -4$$

$$N_2 = \begin{vmatrix} 1 & 2 & 1 & 1 \\ 1 & 2 & 3 & 4 \\ 2 & 2 & 5 & 9 \\ 1 & 2 & 2 & 7 \end{vmatrix} = 18 \qquad N_3 = \begin{vmatrix} 1 & 1 & 2 & 1 \\ 1 & 2 & 2 & 4 \\ 2 & 3 & 2 & 9 \\ 1 & 1 & 2 & 7 \end{vmatrix} = -12 \qquad N_4 = \begin{vmatrix} 1 & 1 & 1 & 2 \\ 1 & 2 & 3 & 2 \\ 2 & 3 & 5 & 2 \\ 1 & 1 & 2 & 2 \end{vmatrix} = 2$$

Then, $x_1 = N_1/D = -2$, $x_2 = N_2/D = 9$, $x_3 = N_3/D = -6$, $x_4 = N_4/D = 1$.

5.146 Use Cramer's Rule to solve the system

$$\begin{array}{l} 2x_1 + x_2 + 5x_3 + x_4 = 5 \\ x_1 + x_2 - 3x_3 - 4x_4 = -1 \\ 3x_1 + 6x_2 - 2x_3 + x_4 = 8 \\ 2x_1 + 2x_2 + 2x_3 - 3x_4 = 2 \end{array}$$

❚ Compute

$$D = \begin{vmatrix} 2 & 1 & 5 & 1 \\ 1 & 1 & -3 & -4 \\ 3 & 6 & -2 & 1 \\ 2 & 2 & 2 & -3 \end{vmatrix} = -120 \qquad N_1 = \begin{vmatrix} 5 & 1 & 5 & 1 \\ -1 & 1 & -3 & -4 \\ 8 & 6 & -2 & 1 \\ 2 & 2 & 2 & -3 \end{vmatrix} = -240$$

$$N_2 = \begin{vmatrix} 2 & 5 & 5 & 1 \\ 1 & -1 & -3 & -4 \\ 3 & 8 & -2 & 1 \\ 2 & 2 & 2 & -3 \end{vmatrix} = -24 \qquad N_3 = \begin{vmatrix} 2 & 1 & 5 & 1 \\ 1 & 1 & -1 & -4 \\ 3 & 6 & 8 & 1 \\ 2 & 2 & 2 & -3 \end{vmatrix} = 0 \qquad N_4 = \begin{vmatrix} 2 & 1 & 5 & 5 \\ 1 & 1 & -3 & -1 \\ 3 & 6 & -2 & 8 \\ 2 & 2 & 2 & 2 \end{vmatrix} = -96$$

Then, $x_1 = N_1/D = 2$, $x_2 = N_2/D = 1/5$, $x_3 = N_3/D = 0$, $x_4 = N_4/D = 4/5$.

5.147 Investigate the system

$$\begin{array}{l} x_1 + x_2 + x_3 + x_4 = 5 \\ x_1 + 2x_2 + 3x_3 + 4x_4 = 3 \\ 4x_1 + x_2 + 2x_3 + 3x_4 = 7 \\ 3x_1 + 2x_2 + 3x_3 + 4x_4 = 2 \end{array}$$

❚ Since

$$D = \begin{vmatrix} 1 & 1 & 1 & 1 \\ 1 & 2 & 3 & 4 \\ 4 & 1 & 2 & 3 \\ 3 & 2 & 3 & 4 \end{vmatrix} = 0$$

Cramer's Rule cannot be used to solve the system. In fact, the system is inconsistent and has no solution. To see this, subtract twice the fourth equation from the sum of the first three, obtaining $0 = 13$.

5.148 Suppose $M = \begin{pmatrix} A & C \\ 0 & B \end{pmatrix}$ is a square block matrix. Show that $\det M = (\det A)(\det B)$.

❚ Suppose $A = (a_{ij})$ is r-square, $B = (b_{ij})$ is s-square, and $M = (m_{ij})$ is n-square, where $n = r + s$. By definition,

$$\det M = \sum_{\sigma \in S_n} (\operatorname{sgn} \sigma) m_{1\sigma(1)} m_{2\sigma(2)} \cdots m_{n\sigma(n)}$$

If $i > r$ and $j \le r$, then $m_{ij} = 0$. Thus we need only consider those permutations σ such that $\sigma\{r+1, r+2, \ldots, r+s\} = \{r+1, r+2, \ldots, r+s\}$ and, therefore $\sigma\{1, 2, \ldots, r\} = \{1, 2, \ldots, r\}$. Let $\sigma_1(k) = \sigma(k)$ for $k \le r$, and let $\sigma_2(k) = \sigma(r+k) - r$ for $k \le s$. Then

$$(\operatorname{sgn} \sigma) m_{1\sigma(1)} m_{2\sigma(2)} \cdots m_{n\sigma(n)} = (\operatorname{sgn} \sigma_1) a_{1\sigma_1(1)} a_{2\sigma_1(2)} \cdots a_{r\sigma_2(r)} (\operatorname{sgn} \sigma_2) b_{1\sigma_2(1)} b_{2\sigma_2(2)} \cdots b_{s\sigma_2(s)}$$

which implies $\det M = (\det A)(\det B)$.

5.149 Suppose M is an upper [lower] triangular block matrix with [square] diagonal blocks $A_1, A_2, \ldots, A_n$. Show that $\det M = (\det A_1)(\det A_2) \cdots (\det A_n)$.

❚ The proof is by induction on n, using Problem 5.148 for the case $n = 2$. Write

$$M = \begin{pmatrix} B & C \\ 0 & A_n \end{pmatrix}$$

By the induction hypothesis, $|B| = |A_1| |A_2| \cdots |A_{n-1}|$. Hence $|M| = |B| |A_n| = |A_1| |A_2| \cdots |A_{n-1}| |A_n|$.

5.150 Find $|M|$ if

$$M = \begin{pmatrix} 2 & 3 & 4 & 7 & 8 \\ -1 & 5 & 3 & 2 & 1 \\ 0 & 0 & 2 & 1 & 5 \\ 0 & 0 & 3 & -1 & 4 \\ 0 & 0 & 5 & 2 & 6 \end{pmatrix}$$

❚ Note that M is an upper triangular block matrix. Evaluate the determinant of each diagonal block:

$$\begin{vmatrix} 2 & 3 \\ -1 & 5 \end{vmatrix} = 10 + 3 = 13 \qquad \begin{vmatrix} 2 & 1 & 5 \\ 3 & -1 & 4 \\ 5 & 2 & 6 \end{vmatrix} = -12 + 20 + 30 + 25 - 16 - 18 = 29$$

Then $|M| = (13)(29) = 377$.

5.151 Find $\det M$, where

$$M = \begin{pmatrix} 3 & 4 & 0 & 0 & 0 \\ 2 & 5 & 0 & 0 & 0 \\ 0 & 9 & 2 & 0 & 0 \\ 0 & 5 & 0 & 6 & 7 \\ 0 & 0 & 4 & 3 & 4 \end{pmatrix}$$

▮ Partition M into a lower triangular block matrix, as follows:

$$M = \begin{pmatrix} 3 & 4 & 0 & 0 & 0 \\ 2 & 5 & 0 & 0 & 0 \\ 0 & 9 & 2 & 0 & 0 \\ 0 & 5 & 0 & 6 & 7 \\ 0 & 0 & 4 & 3 & 4 \end{pmatrix}$$

Evaluate the determinant of each diagonal block:

$$\begin{vmatrix} 3 & 4 \\ 2 & 5 \end{vmatrix} = 15 - 8 = 7 \qquad |2| = 2 \qquad \begin{vmatrix} 6 & 7 \\ 3 & 4 \end{vmatrix} = 24 - 21 = 3$$

Hence $|M| = (7)(2)(3) = 42$.

5.10 SUBMATRICES, GENERAL MINORS, PRINCIPAL MINORS

5.152 Let $A = (a_{ij})$ be an n-square matrix. Let $i_1, i_2, \ldots, i_r$ $(r < n)$ be an ordered set of row indices, and let $j_1, j_2, \ldots, j_r$ be an ordered set of column indices. Define the *submatrix* of A corresponding to these index sets.

▮

$$A_{i_1, i_2, \ldots, i_r}^{j_1, j_2, \ldots, j_r} \equiv \begin{pmatrix} a_{i_1, j_2} & a_{i_1, j_2} & \cdots & a_{i_1, j_r} \\ a_{i_2, j_1} & a_{i_2, j_2} & \cdots & a_{i_2, j_2} \\ \cdots\cdots\cdots\cdots\cdots\cdots\cdots \\ a_{i_r, j_1} & a_{i_r, j_2} & \cdots & a_{i_r, j_r} \end{pmatrix}$$

The submatrix is of *order r*.

5.153 Define a *minor of order r*, and the associated *signed minor*, of an n-square matrix A.

▮ The determinant $\left| A_{i_1, i_2, \ldots, i_r}^{j_1, j_2, \ldots, j_r} \right|$ of a submatrix of order r is termed a minor of order r, and

$$(-1)^{i_1 + i_2 + \cdots + i_r + j_1 + j_2 + \cdots + j_r} \left| A_{i_1, i_2, \ldots, i_r}^{j_1, j_2, \ldots, j_r} \right|$$

is the corresponding signed minor. Note that a minor of order $n - 1$ is a *minor* in the sense of Problem 5.90.

5.154 Refer to Problem 5.153. Show that a signed minor of order $n - 1$ is just a *cofactor*, as defined in Problem 5.91.

▮ Let $\left| A_{i_1, i_2, \ldots, i_{n-1}}^{j_1, j_2, \ldots, j_{n-1}} \right|$ omit the ith row and jth column of A. Then, with $s \equiv 1 + 2 + \cdots + n$, we have

$$(i + j) + (i_1 + i_2 + \cdots + i_{n-1} + j_1 + j_2 + \cdots + j_{n-1}) = (i + i_1 + \cdots + i_{n-1}) + (j + j_1 + \cdots + j_{n-1}) = 2s$$

which implies that

$$(-1)^{i_1 + i_2 + \cdots + i_{n-1} + j_1 + j_2 + \cdots + j_{n-1}} = (-1)^{i+j}$$

5.155 Compute the minor $\left| A_{3,5}^{1,4} \right|$ and its signed minor if $A = (a_{ij})$ is a 5-square matrix.

▮ The subscripts 3 and 5 refer to the rows of A, and the superscripts 1 and 4 refer to the columns of A. Hence

$$\left| A_{3,5}^{1,4} \right| = \begin{vmatrix} a_{31} & a_{34} \\ a_{51} & a_{54} \end{vmatrix} = a_{31}a_{54} - a_{3451} \qquad \text{and} \qquad (-1)^{3+5+1+4} \left| A_{3,5}^{1,4} \right| = - \left| A_{3,5}^{1,4} \right|$$

5.156 Find the *complementary minor* of $\left| A_{3,5}^{1,4} \right|$ in Problem 5.155.

▮ Find the complement in A of the submatrix $A_{3,5}^{1,4}$, and take its determinant:

$$\left| A_{1,2,4}^{2,3,5} \right| = \begin{vmatrix} a_{12} & a_{13} & a_{15} \\ a_{22} & a_{23} & a_{35} \\ a_{42} & a_{43} & a_{45} \end{vmatrix}$$

5.157 When is a minor a *principal minor*?

$\blacksquare$ When the row and column indices of the submatrix are the same; i.e., when the diagonal elements of the minor come from the diagonal of the matrix.

Problems 5.158–5.162 refer to the following minors of a 5-square matrix $A = (a_{ij})$:

$$M_1 = \begin{vmatrix} a_{22} & a_{24} & a_{25} \\ a_{42} & a_{44} & a_{45} \\ a_{52} & a_{54} & a_{55} \end{vmatrix} \qquad M_2 = \begin{vmatrix} a_{11} & a_{13} & a_{15} \\ a_{21} & a_{23} & a_{25} \\ a_{51} & a_{53} & a_{55} \end{vmatrix} \qquad M_3 = \begin{vmatrix} a_{22} & a_{25} \\ a_{52} & a_{55} \end{vmatrix}$$

5.158 Is M_1 a principal minor?

$\blacksquare$ Yes, since its diagonal elements belong to the diagonal of A.

5.159 Is M_2 a principal minor?

$\blacksquare$ No, since a_{23} belongs to the diagonal of M_2 but not to that of A.

5.160 Is M_3 a principal minor?

$\blacksquare$ Yes; its diagonal elements belong to the diagonal of A.

5.161 Find the complement of M_1.

$\blacksquare$ The missing row indices are 1 and 3, and the missing column indices are 1 and 3. Hence

$$\begin{vmatrix} a_{11} & a_{13} \\ a_{31} & a_{33} \end{vmatrix}$$

is the complement of M_1. [In general, a minor is a principal minor iff its complement is a principal minor.]

5.162 Find the complement of M_2.

$\blacksquare$ $\qquad \begin{vmatrix} a_{32} & a_{34} \\ a_{42} & a_{44} \end{vmatrix} \qquad$ (not principal)

5.11 MISCELLANEOUS PROBLEMS

5.163 Let $\mathscr{A}$ be an algebra of n-square matrices whose elements are drawn from a field $\mathbf{K}$. Show that the defterminant function $D: \mathscr{A} \to \mathbf{K}$ is multilinear.

$\blacksquare$ Suppose that the ith row of $A \in \mathscr{A}$ has the form, $(\alpha_{i1} + \beta_{i1}, \alpha_{i2} + \beta_{i2}, \ldots, \alpha_{in} + \beta_{in})$; then

$$\det A = \sum_{S_n} (\text{sgn } \sigma) a_{1\sigma(1)} \cdots a_{i-1,\sigma(i-1)}(\alpha_{i\sigma(i)} + \beta_{i\sigma(i)}) \cdots a_{n\sigma(n)}$$

$$= \sum_{S_n} (\text{sgn } \sigma) a_{i\sigma(1)} \cdots \alpha_{i\sigma(i)} \cdots a_{n\sigma(n)} + \sum_{S_n} (\text{sgn } \sigma) a_{1\sigma(1)} \cdots \beta_{i\sigma(i)} \cdots a_{n\sigma(n)}$$

$$\equiv \det A_\alpha + \det A_\beta$$

Thus $D(\ )$ is additive with respect to any row. Further, by Problem 5.68, $D(\ )$ is homogeneous of order 1 in any row. Thus $D(\ )$ is multilinear.

5.164 Evaluate

$$|A| = \begin{vmatrix} 0 & 1+i & 1+2i \\ 1-i & 0 & 2-3i \\ 1-2i & 2+3i & 0 \end{vmatrix}$$

$\blacksquare$ Multiply the second row by $1+i$ and the third row by $1+2i$; then

$$(1+i)(1+2i)|A| = (-1+3i)|A| = \begin{vmatrix} 0 & 1+i & 1+2i \\ 2 & 0 & 5-i \\ 5 & -4+7i & 0 \end{vmatrix} = \begin{vmatrix} 0 & 1+i & 1+2i \\ 2 & 0 & 5-i \\ 1 & -4+7i & -10+2i \end{vmatrix} = \begin{vmatrix} 0 & 1+i & 1+2i \\ 0 & 8-14i & 25-5i \\ 1 & -4+7i & -10+2i \end{vmatrix}$$

$$= \begin{vmatrix} 1+i & 1+2i \\ 8-14i & 25-5i \end{vmatrix} = -6+18i$$

and $|A| = 6$.

5.165 Evaluate

$$|B| = \begin{vmatrix} 0.921 & 0.185 & 0.476 & 0.614 \\ 0.782 & 0.157 & 0.527 & 0.138 \\ 0.872 & 0.484 & 0.637 & 0.799 \\ 0.312 & 0.555 & 0.841 & 0.448 \end{vmatrix}$$

$$|B| = 0.921 \begin{vmatrix} 1 & 0.201 & 0.517 & 0.667 \\ 0.782 & 0.157 & 0.527 & 0.138 \\ 0.872 & 0.484 & 0.637 & 0.799 \\ 0.312 & 0.555 & 0.841 & 0.448 \end{vmatrix} = 0.921 \begin{vmatrix} 1 & 0.201 & 0.517 & 0.667 \\ 0 & 0 & 0.123 & -0.384 \\ 0 & 0.309 & 0.196 & 0.217 \\ 0 & 0.492 & 0.680 & 0.240 \end{vmatrix}$$

$$= 0.921 \begin{vmatrix} 0 & 0.123 & -0.384 \\ 0.309 & 0.196 & 0.217 \\ 0.492 & 0.680 & 0.240 \end{vmatrix} = 0.921(-0.384) \begin{vmatrix} 0 & -0.320 & 1 \\ 0.309 & 0.196 & 0.217 \\ 0.492 & 0.680 & 0.240 \end{vmatrix}$$

$$= 0.921(-0.384) \begin{vmatrix} 0 & 0 & 1 \\ 0.309 & 0.265 & 0.217 \\ 0.492 & 0.757 & 0.240 \end{vmatrix} = 0.921(-0.384) \begin{vmatrix} 0.309 & 0.265 \\ 0.492 & 0.757 \end{vmatrix}$$

$$= 0.921(-0.384)(0.104) = -0.037$$

5.166 Without expanding the determinant, show that

$$\begin{vmatrix} 1 & a & b+c \\ 1 & b & c+a \\ 1 & c & a+b \end{vmatrix} = 0$$

▮ Add the second column to the third column, and remove the common factor from the third column; this yields:

$$\begin{vmatrix} 1 & a & b+c \\ 1 & b & c+a \\ 1 & c & a+b \end{vmatrix} = \begin{vmatrix} 1 & a & a+b+c \\ 1 & b & a+b+c \\ 1 & c & a+b+c \end{vmatrix} = (a+b+c) \begin{vmatrix} 1 & a & 1 \\ 1 & b & 1 \\ 1 & c & 1 \end{vmatrix} = (a+b+c)(0) = 0$$

5.167 Show that the difference product $g(x_1, \ldots, x_n)$ of Problem 5.58 can be represented as a determinant.

▮ Consider the *Vandermonde determinant* of $x_1, x_2, \ldots, x_{n-1}, x$:

$$V_{n-1}(x) \equiv \begin{vmatrix} 1 & 1 & \cdots & 1 & 1 \\ x_1 & x_2 & \cdots & x_{n-1} & x \\ x_1^2 & x_2^2 & \cdots & X_{n-1}^2 & x^2 \\ \cdots\cdots\cdots\cdots\cdots\cdots\cdots \\ x_1^{n-1} & x_2^{n-1} & \cdots & x_{n-1}^{n-1} & x^{n-1} \end{vmatrix}$$

This is a polynomial in x of degree $n-1$, of which the roots are $x_1, x_2, \ldots, x_{n-1}$; moreover, the leading coefficient [the cofactor of x^{n-1}] is equal to $V_{n-2}(x_{n-1})$. Thus, from algebra,

$$V_{n-1}(x) = (x-x_1)(x-x_2)\cdots(x-x_{n-1})V_{n-2}(x_{n-1})$$

so that, by recursion,

$$V_{n-1}(x) = [(x-x_1)\cdots(x-x_{n-1})][(x_{n-1}-x_1)\cdots(x_{n-1}-x_{n-2})]V_{n-3}(x_{n-2})$$

$$= \cdots$$

$$= [(x-x_1)\cdots(x-x_{n-1})][(x_{n-1}-x_1)\cdots(x_{n-1}-x_{n-2})]\cdots[(x_2-x_1)]$$

It follows that

$$V_{n-1}(x_n) = \prod_{n \geq i > j \geq 1} (x_i - x_j) = (-1)^{n(n-1)/2} \prod_{1 \leq i < j \leq n} (x_i - x_j) \quad \text{or} \quad g(x_1, \ldots, x_n) = (-1)^{n(n-1)/2} V_{n-1}(x_n)$$

CHAPTER 6

Algebraic Structures

6.1 SETS, MATHEMATICAL INDUCTION, PRODUCTS SETS

In the following problems, $\mathbf{N}$, $\mathbf{Z}$, $\mathbf{Q}$, $\mathbf{R}$, $\mathbf{C}$ denote, respectively, the sets of positive integers, integers, rational numbers, real numbers, and complex numbers.

6.1 Which among these sets are equal: $\{1, 3, 4\}$, $\{4, 3, 1, 4\}$, $\{3, 4, 3, 1\}$, $\{4, 1, 4, 3\}$?

▌ They are all equal. Order and repetition do not change a set.

6.2 List the elements of $A = \{x: x \in \mathbf{N}, 3 < x < 12\}$.

▌ A consists of the positive integers between 3 and 12; hence $A = \{4, 5, 6, 7, 8, 9\,10, 11\}$.

6.3 List the elements of $B = \{x: x \in \mathbf{N}, x \text{ is even}, x < 15\}$

▌ B consists of even positive integers less than 15; hence $B = \{2, 4, 6, 8, 10, 12, 14\}$.

6.4 List the elements of $C = \{x: x \in \mathbf{N}, 4 + x = 3\}$.

▌ There is no positive integer that satisfies the condition $4 + x = 3$; hence C contains no element. In other words, $C = \emptyset$, the empty set.

6.5 Prove that $A = \{2, 3, 4, 5\}$ is not a subset of $B = \{x: x \in \mathbf{N}, \ x \text{ is even}\}$.

▌ It is only necessary to show that one element of A does not belong to B. Now $3 \in A$ and, since B consists of even numbers, $3 \notin B$.

Problems 6.6–6.12 concern the sets $A = \{2, 3, 4, 5\}$, $B = \{3, 5, 7\}$, $C = \{1, 4\}$, $D = \{3, 4, 6\}$.

6.6 Find $A \cup B$.

▌ $A \cup B$ consists of elements in either A or B (or both); hence, $A \cup B = \{2, 3, 4, 5, 7\}$.

6.7 Find $A \cap B$.

▌ $A \cap B$ consists of the elements in both A and B; hence, $A \cap B = \{3, 5\}$.

6.8 Find $3 + A$.

▌ Add 3 to each element of A to obtain $3 + A = \{5, 6, 7, 8\}$.

6.9 Find $4 \cdot B$.

▌ Multiply each element of B by 4 to obtain $4 \cdot B = \{12, 20, 28\}$.

6.10 Find $C + D$.

▌ Add each element of C to each element of D [neglecting repetitions], to obtain $C + D = \{1 + 3, 1 + 4, 1 + 6, 4 + 3, 4 + 4, 4 + 6\} = \{4, 5, 7, 8, 10\}$.

6.11 Find $C + C$.

▌ Add each element of C to each element of C, to obtain $C + C = \{1 + 1, 1 + 4, 4 + 1, 4 + 4\} = \{2, 5, 5, 8\} = \{2, 5, 8\}$.

6.12 Find $D + D$.

▌ Add each element of D to each element of D, to obtain $D + D = \{3 + 3, 3 + 4, 3 + 6, 4 + 3, 4 + 4, 4 + 6, 6 + 3, 6 + 4, 6 + 6\} = \{6, 7, 9, 8, 10, 12\}$.

6.13   Find an infinite set A such that $A + A$ and A are disjoint; that is, $(A + A) \cap A = \emptyset$. [Here $\emptyset$ denotes the empty set.]

    ▮   Let $A = \{1, 3, 5, \ldots\} = \{\text{positive odd integers}\}$. Then $A + A$ consists only of even integers.

6.14   Find an infinite set B such that $B + B = B$.

    ▮ $B = \{0, 1, 2, \ldots\} = \{\text{nonnegative integers}\}$.

6.15   Find a finite set C such that $C + C = C$.

    ▮ $C = \{0\}$.

6.16   Determine the *power set*, $\mathscr{P}(A)$, of $A = \{a, b, c, d\}$.

    ▮   The elements of $\mathscr{P}(A)$ are the subsets of A: $\mathscr{P}(A) = \{A, \{a, b, c\}, \{a, b, d\}, \{a, c, d\}, \{b, c, d\},$ $\{a, b\}, \{a, c\}, \{a, d\}, \{b, c\}, \{b, d\}, \{c, d\}, \{a\}, \{b\}, \{c\}, \{d\}, \emptyset\}$. Note that $\mathscr{P}(A)$ has $2^4 = 16$ elements.

6.17   State the *principle of mathematical induction* in two equivalent forms.

    ▮   *Form I*:   Let P be a proposition defined on the positive integers $\mathbf{N}$; i.e., $P(n)$ is either true or false for each n in $\mathbf{N}$. Suppose P has the following two properties:
        (i) $P(1)$ is true.
        (ii) $P(n + 1)$ is true whenever $P(n)$ is true.
        Then P is true for every positive integer.

    Form II ("*Complete Induction*"):   Let P be a proposition defined on the positive integers $\mathbf{N}$, such that:
        (i) $P(1)$ is true.
        (ii) $P(n)$ is true whenever $P(k)$ is true for all $1 \le k < n$.
        Then P is true for every positive integer.

6.18   Show that the principle of mathematical induction [complete form] is equivalent to the assertion that every nonempty set of positive integers has a smallest member [the *well-ordering principle* for $\mathbf{N}$].

    ▮   Suppose that $\mathbf{N}$ is well-ordered, and that we are given a proposition $P(n)$ satisfying the hypotheses (i) and (ii) of the induction principle. Let $\mathbf{F}$ denote the subset of $\mathbf{N}$ on which P is false. If $\mathbf{F}$ is nonempty, it has a smallest member, q; by (i), $q \ge 2$. Then $P(1), \ldots, P(q - 1)$ are all true; hence, by (ii), $P(q)$ is true. This contradiction shows that $\mathbf{F}$ must be empty. Thus, P is true for every positive integer, and the induction principle is valid.
    Conversely, suppose that the induction principle holds and that there exists a subset, $\mathbf{S}$, of $\mathbf{N}$ that has no smallest member. Let $\mathbf{S}^*$ be the complement of $\mathbf{S}$, and define the proposition $P(n)$: n belongs to $\mathbf{S}^*$. $P(n)$ satisfies (i) and (ii) of ladder induction [if it did not, $\mathbf{S}$ would have a smallest member]; consequently, $\mathbf{S}^* = \mathbf{N}$, which means that $\mathbf{S}$ is empty. Thus, $\mathbf{N}$ is well-ordered.

6.19   Prove that the sum of the first n odd integers is n^2; that is, prove $P(n)$: $1 + 3 + 5 + \cdots + (2n - 1) = n^2$.

    ▮   Since $1 = 1^2$, $P(1)$ is true. Assuming $P(n)$ is true, we add $2n + 1$ to both sides of $P(n)$, obtaining:

$$1 + 3 + 5 + \cdots + (2n - 1) + (2n + 1) = n^2 + (2n + 1) = (n + 1)^2$$

which is $P(n + 1)$. That is, $P(n + 1)$ is true whenever $P(n)$ is true. By the principle of mathematical induction, P is true of all n.

6.20   Define the *product set* of sets A and B.

    ▮   The product set of A and B, denoted by $A \times B$, consists of all ordered pairs (a, b) where $a \in A$ and $b \in B$: $A \times B = \{(a, b): a \in A, b \in B\}$. The product of a set with itself, say $A \times A$, is denoted by A^2.

    Problems 6.21–6.23 refer to sets $A = \{1, 2, 3\}$ and $B = \{a, b\}$.

6.21 Find $A \times B$.

▌ $A \times B$ consists of all ordered pairs (x, y) where $x \in A$ and $y \in B$. Hence

$$A \times B = \{(1, a), (1, b), (2, a), (2, b), (3, a), (3, b)\}$$

6.22 Find $B \times A$.

▌ $B \times A$ consists of all ordered pairs (y, x) where $y \in B$ and $x \in A$. Hence

$$B \times A = \{(a, 1), (a, 2), (a, 3), (b, 1), (b, 2), (b, 3)\} \neq A \times B$$

6.23 Find B^2.

▌ $B^2 = B \times B$ consists of all ordered pairs (x, y) where $x, y \in B$. Hence

$$B \times B = \{(a, a), (a, b), (b, a), (b, b)\}$$

6.24 Given $A = \{1, 2\}$, $B = \{a, b, c\}$, and $C = \{c, d\}$. Find $(A \times B) \cap (A \times C)$ and $A \times (B \cap C)$.

▌ $A \times B = \{(1, a), (1, b), (1, c), (2, a), (2, b), (2, c)\}$ and $A \times C = \{(1, c), (1, d), (2, c), (2, d)\}$

Hence $(A \times B) \cap (A \times C) = \{(1, c), (2, c)\}$. Since $B \cap C = \{c\}$, $A \times (B \cap C) = \{(1, c,), (2, c)\}$. Observe that $(A \times B) \cap (A \times C) = A \times (B \cap C)$. [This is true for any sets A, B and C.]

6.25 Given $A = \{1, 2\}$, $B = \{x, y, z\}$, and $C = \{3, 4\}$. Find $A \times B \times C$.

▌ $A \times B \times C$ consists of all ordered triplets (a, b, c) where $A \in A$, $b \in B$, $c \in C$. Since A, B, and C are finite sets, $A \times B \times C$ can be listed by use of the *tree diagram* of Fig. 6-1. That is, the elements of $A \times B \times C$ are precisely the 12 ordered triplets to the right of the tree diagram.

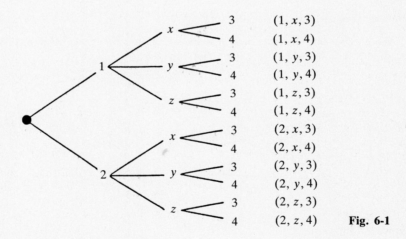

Fig. 6-1

6.26 Describe the geometrical representation of the product set $\mathbf{R} \times \mathbf{R}$.

▌ $\mathbf{R} \times \mathbf{R}$ is identified with the points in the plane, as in Fig. 6-2. Here each point P represents an ordered pair (a, b) of real numbers and vice versa; the vertical line through P meets the x axis at a, and the horizontal line through P meets the y axis at b. $\mathbf{R}^2$ is frequently called the *cartesian plane*.

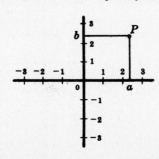

Fig. 6-2

6.27 Which among these ordered n-tuples are equal: $(1, 3, 4)$, $(4, 3, 1, 4)$, $(3, 4, 3, 1)$, $(4, 1, 4, 3)$?   [Compare with Problem 6.1.]

▮ None of the sets are equal.   Order and repetition do count with ordered n-tuples.

6.2 RELATIONS

6.28 Define [*binary*] *relation*.

▮ A binary relation, or simply *relation*, from a set A to a set B is a subset $R \subset A \times B$.   Given $a \in A$, $b \in B$,   we write $a R b$ [real "*a* is related to *b*"] iff $(a, b) \in R$.

6.29 Define a relation R on a set A.

▮ R is a relation on A if R is a relation from A to A; i.e., if $R \subset A \times A$.

6.30 Define the *inverse* of a relation.

▮ Let R be a relation from A to B.   The inverse of R, denoted R^{-1}, is the relation from B to A which consists of those ordered pairs which when reversed belong to R:

$$R^{-1} = \{(b, a): (a, b) \in R\}$$

In other words, $b R^{-1} a$ if and only if $a R b$.   Note that every relation has an inverse, not merely those relations that correspond to one-one mappings.

Problems 6.31–6.34 involve the sets $A = \{1, 2, 3, 4\}$ and $B = \{x, y, z\}$,   and the relation $R = \{(1, y), (1, z), (3, y), (4, x), (4, z)\}$ from A to B.

6.31 Draw the "arrow diagram" of the relation R.

▮ Write down the elements of A and the elements of B in two disjoint disks, and then draw an arrow from $a \in B$ to $b \in B$ whenever $a R b$.   See Fig. 6-3.

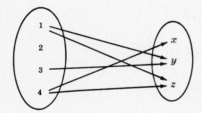

$$\begin{array}{c} \\ 1 \\ 2 \\ 3 \\ 4 \end{array} \begin{array}{ccc} x & y & x \\ \begin{pmatrix} 0 & 1 & 1 \\ 0 & 0 & 0 \\ 0 & 1 & 0 \\ 1 & 0 & 1 \end{pmatrix} \end{array}$$

Fig. 6-3   **Fig. 6-4**

6.32 Represent R by a matrix.

▮ The matrix M_R of the relation R appears in Fig. 6-4.   Observe that the rows of the matrix are labeled by the elements of A and the columns by the elements of B.   Also observe that the entry in the matrix corresponding to $a \in A$ and $b \in B$ is 1 if $a R b$ and 0 otherwise.

6.33 Determine the *domain* and the *range* of R.

▮ The domain of R is the subset of A consisting of the first elements of the ordered pairs of R, and the range of R is the subset of B consisting of the second elements:

$$\text{domain of } R = \{1, 3, 4\} \quad \text{and} \quad \text{range of } R = \{x, y, z\}$$

6.34 Find the inverse relation R^{-1} of R.

▮ Reverse the ordered pairs of R to obtain R^{-1}:

$$R^{-1} = \{(y, 1), (z, 1), (y, 3), (x, 4), (z, 4)\}$$

[Reversing the arrows in Fig. 6-3 yields the arrow diagram of R^{-1}, and taking the transpose of the matrix in Fig. 6-4 gives the matrix of R^{-1}.]

6.35 Let $A = \{1, 2, 3, 4\}$ and let $R = \{(1, 2), (2, 2), (2, 4), (3, 2), (3, 4), (4, 1), (4, 3)\}$ be a relation on A. Draw the *directed graph* of R.

▌ Write down the elements of A, and then draw an arrow from an element x to an element y whenever xRy. See Fig. 6-5.

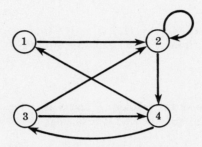

$$R = \{(1, 2), (2, 2), (2, 4), (3, 2), (3, 4), (4, 1), (4, 3)\} \qquad \textbf{Fig. 6-5}$$

Problems 6.36–6.39 concern the set $A = \{1, 2, 3, 4, 6\}$ and the relation R on A defined by "x divides y," written $x \mid y$. [Note $x \mid y$ iff there exists an integer z such that $xz = y$.]

6.36 Write R as a set of ordered pairs.

▌ $R = \{(1, 1), (1, 2), (1, 3), (1, 4), (1, 6), (2, 2), (2, 4), (2, 6), (3, 3), (3, 6), (4, 4), (6, 6)\}$

6.37 Find the inverse relation R^{-1} of R, and describe in words?

▌ Reverse the ordered pairs of R to obtain R^{-1}:

$$R^{-1} = \{(1, 1), (2, 1), (3, 1), (4, 1), (6, 1), (2, 2), (4, 2), (6, 2), (3, 3), (6, 3), (4, 4), (6, 6)\}$$

R^{-1} can be described by the statement "x is a multiple of y".

6.38 Find the matrix representation of R.

▌
$$M_R = \begin{pmatrix} 1 & 1 & 1 & 1 & 1 \\ 0 & 1 & 0 & 1 & 1 \\ 0 & 0 & 1 & 0 & 1 \\ 0 & 0 & 0 & 1 & 0 \\ 0 & 0 & 0 & 0 & 1 \end{pmatrix}$$

[We assume the rows and columns of M_R are labeled by the elements 1, 2, 3, 4, 6, respectively. Clearly, a different ordering of the elements of A yields a different matrix.]

6.39 Find the directed graph of R.

▌ See Fig. 6-6.

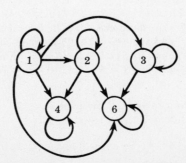

Fig. 6-6

6.40   Define the *composition* of relations.

    ▌ Let A, B, and C be sets, and let R be a relation from A and B and let S be a relation from B to C; that is, R is a subset of $A \times B$ and S is a subset of $B \times C$. Then R and S give rise to a relation, $R \circ S$, from A to C; namely, the subset of $A \times C$ defined by

$$(a, c) \in R \circ S \quad \text{iff} \quad (a, b) \in R \quad \text{and} \quad (b, c) \in S \quad \text{for some} \quad b \in B$$

The relation $R \circ S$ is called the *composition* of R and S; it is sometimes denoted simply by RS.

    Problems 6.41–6.43 concern sets $A = \{1, 2, 3\}$, $B = \{a, b, c\}$, and $C = \{x, y, z\}$ and two relations: $R = \{(1, b), (2, a), (2, c)\}$ (from A to B) and $S = \{(a, y), (b, x), (c, y), (c, z)\}$ (from B to C).

6.41   Find the composition $R \circ S$.

    ▌ Draw the arrow diagrams of the relations R and S as in Fig. 6-7. Observe that 1 in A is "connected" to x in C by the path $1 \to b \to x$; hence $(1, x)$ belongs to $R \circ S$. Similarly, $(2, y)$ and $(2, z)$ belong to $R \circ S$. Thus $R \circ S = \{(1, x), (2, y), (2, z)\}$.

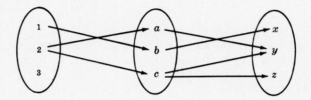

Fig. 6-7

6.42   Find the matrices M_R, M_S, and $M_{R \circ S}$ of the respective relations R, S, and $R \circ S$.

    ▌

$$M_R = \begin{matrix} & \begin{matrix} a & b & c \end{matrix} \\ \begin{matrix} 1 \\ 2 \\ 3 \end{matrix} & \begin{pmatrix} 0 & 1 & 0 \\ 1 & 0 & 1 \\ 0 & 0 & 0 \end{pmatrix} \end{matrix} \qquad M_S = \begin{matrix} & \begin{matrix} x & y & z \end{matrix} \\ \begin{matrix} a \\ b \\ c \end{matrix} & \begin{pmatrix} 0 & 1 & 0 \\ 1 & 0 & 0 \\ 0 & 1 & 1 \end{pmatrix} \end{matrix} \qquad M_{R \circ S} = \begin{matrix} & \begin{matrix} x & y & z \end{matrix} \\ \begin{matrix} 1 \\ 2 \\ 3 \end{matrix} & \begin{pmatrix} 1 & 0 & 0 \\ 0 & 1 & 1 \\ 0 & 0 & 0 \end{pmatrix} \end{matrix}$$

6.43   Compare the matrix product $M_R M_S$ to the matrix $M_{R \circ S}$.

    ▌

$$M_R M_S = \begin{pmatrix} 1 & 0 & 0 \\ 0 & 2 & 1 \\ 0 & 0 & 0 \end{pmatrix}$$

Observe that $M_{R \circ S}$ and $M_R M_S$ have nonzero elements in corresponding positions. This result holds for any ordering of A, B, and C.

    Theorem 6.1 (*Associative Law*): Let A, B, C, and D be sets. Suppose R is a relation from A to B, S is a relation from B to C, and T is a relation from C to D. Then $(R \circ S) \circ T = R \circ (S \circ T)$.

6.44   Prove Theorem 6.1.

    ▌ We need to show that each ordered pair in $(R \circ S) \circ T$ belongs to $R \circ (S \circ T)$, and vice versa. Suppose then, that (a, d) belongs to $(R \circ S) \circ T$. Then there exists a c in C such that $(a, c) \in R \circ S$ and $(c, d) \in T$. Since $(a, c) \in R \circ S$, there exists a b in B such that $(a, b) \in R$ and $(b, c) \in S$. Since $(b, c) \in S$ and $(c, d) \in T$, we have $(b, d) \in S \circ T$; and since $(a, b) \in R$ and $(b, d) \in S \circ T$, we have $(a, d) \in R \circ (S \circ T)$. Thus $(R \circ S) \circ T \subset R \circ (S \circ T)$. Similarly $R \circ (S \circ T) \subset (R \circ S) \circ T$. Both inclusion relations prove $(R \circ S) \circ T = T = R \circ (S \circ T)$.

    Problems 6.45–6.49 refer to a relation R on a set A.

6.45   When is R *reflexive*?

    ▌ R is reflexive if $a R b$ for every a in A.

6.46   When is *R symmetric*?

▐   *R* is symmetric if aRb implies bRa.

6.47   When is *R antisymmetric*?

▐   *R* is antisymmetric if aRb and bRa implies $a = b$.

6.48   When is *R transitive*?

▐   *R* is transitive if aRb and bRc implies aRc.

6.49   Criticize the following argument: Let *R* be symmetric and transitive.   Then aRb implies bRa, and together these imply aRa.   Therefore, *R* is reflexive.

▐   Reflexivity means aRa for *every* a.   The above argument establishes aRa merely for those *a* that are related to some *b*.

Problems 6.50–6.53 refer to the following five relations on the set $A = \{1, 2, 3\}$:

$$R = \{(1, 1), (1, 2), (1, 3), (3, 3)\}$$
$$S = \{(1, 1), (1, 2), (2, 1), (2, 2), (3, 3)\}$$
$$T = \{(1, 1), (1, 2), (2, 2), (2, 3)\}$$

$\varnothing$ = empty relation
$A \times A$ = universal relation

6.50   Which of the five relations are reflexive?

▐   *R* is not reflexive since $2 \in A$ but $(2, 2) \notin R$.   *T* is not reflexive since $(3, 3) \notin T$ and, similarly, $\varnothing$ is not reflexive.   *S* and $A \times A$ are reflexive.

6.51   Which of the five relations are symmetric?

▐   *R* is not symmetric since $(1, 2) \in R$ but $(2, 1) \notin R$,   and similarly *T* is not symmetric.   *S*, $\varnothing$, and $A \times A$ are symmetric.

6.52   Which of the five relations are transitive?

▐   *T* is not transitive since $(1, 2)$ and $(2, 3)$ belong to *T*, but $(1, 3)$ does not belong to *T*.   The other four relations are transitive.

6.53   Which of the five relations are antisymmetric?

▐   *S* is not antisymmetric since $(1, 2)$ and $(2, 1)$ both belong to *S*, yet $1 \neq 2$.   Similarly, $A \times A$ is not antisymmetric.   The other three relations are antisymmetric.

6.54   Given $R = \{(1, 1), (2, 2), (2, 3), (3, 2), (4, 2), (4, 4)\}$ on $A = \{1, 2, 3, 4\}$.   Draw the directed graph of *R*.

▐   See Fig. 6-8.

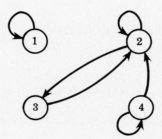

Fig. 6-8

6.55   Is *R* of Problem 6.54 reflexive?

▐   *R* is not reflexive, because $3 \in A$ but $(3, 3) \notin R$.

6.56   Is R of Problem 6.54 symmetric?

▮ R is not symmetric, because $(4, 2) \in R$ but $(2, 4) \notin R$.

6.57   Is R of Problem 6.54 transitive?

▮ R is not transitive, because $(4, 2) \in R$ and $(2, 3) \in R$ but $(4, 3) \notin R$.

6.58   Is R of Problem 6.54 antisymmetric?

▮ R is not antisymmetric, because both $(2, 3) \in R$ and $(3, 2) \in R$.

6.59   Suppose R and S are transitive relations on a set A. Show that $R \cap S$ is transitive.

▮   Suppose (a, b) and (b, c) are in $R \cap S$. Then (a, b) and (b, c) are in both R and S. Since both relations are transitive, $(a, c) \in R$ and $(a, c) \in S$. Thus $(a, c) \in R \cap S$, and so $R \cap S$ is transitive.

6.60   Suppose R and S are antisymmetric relations on a set A. Show that $R \cap S$ is antisymmetric.

▮   Suppose (a, b) and (b, a) are both in $R \cap S$. Then, in particular, (a, b) and (b, a) are both in R. Since R is antisymmetric, $a = b$. Hence $R \cap S$ is antisymmetric.

6.61   Give a relation R on $A = \{1, 2, 3\}$ with the property that (a) R is both symmetric and antisymmetric; (b) R is neither symmetric nor antisymmetric; (c) R is transitive but $R \cup R^{-1}$ is not transitive.

▮   (a) $R = \{(1, 1), (2, 2)\}$;   (b) $R = \{(1, 2), (2, 1), (2, 3)\}$;   (c) $R = \{(1, 2)\}$.

6.62   Let $\perp$ denote the relation of orthogonality in $\mathbf{R}^3$. Is $\perp$ reflexive?

▮   No: If $u \neq \mathbf{0}$ then $u \not\perp u$; that is, $u \cdot u \neq 0$.

6.63   Is $\perp$ symmetric?

▮   Yes: If $u \cdot v = 0$, then $v \cdot u = 0$.

6.64   Is $\perp$ transitive?

▮   No: $u = (1, 1, 1)$ is orthogonal to $v = (1, 1, -2)$, and v is orthogonal to $w = (4, 0, 2)$; but $u \cdot w = 6 \neq 0$.

6.65   Prove that a relation R is symmetric if and only if $R = R^{-1}$.

▮   If R is symmetric,

$$(a, b) \in R \Leftrightarrow (b, a) \in R \Leftrightarrow (a, b) \in R^{-1}$$

so that $R = R^{-1}$. Conversely, if $R = R^{-1}$,

$$(a, b) \in R \Rightarrow (a, b) \in R^{-1} \Rightarrow (b, a) \in R$$

so that R is symmetric.

6.3 PARTITIONS AND EQUIVALENCE RELATIONS

6.66   Define a *partition* of a set.

▮   Let S be any nonempty set. A partition of S is a collection of nonempty subsets of S, called *cells*, such that each a in S belongs to one and only one cell.

Problems 6.67–6.69 refer to the following collections of subsets of $X = \{1, 2, 3, \ldots, 8, 9\}$:

$$P_1 = [\{1, 3, 6\}, \{2, 8\}, \{5, 7, 9\}] \qquad P_2 = [\{1, 5, 7\}, \{2, 4, 8, 9\}, \{3, 5, 6\}] \qquad P_3 = [\{2, 4, 5, 8\}, \{1, 9\}, \{3, 6, 7\}]$$

6.67   Is P_1 a partition of X?

▮   No; because $4 \in X$ does not belong to any cell.

6.68   Is P_2 a partition of X?

❚ No; because $5 \in X$ belongs to two distinct cells.

6.69   Is P_3 a partition of X?

❚ Yes; because each element of X belongs to exactly one cell.   Equivalently: the cells are disjoint and their union is X.

6.70   Find all the partitions of $X = \{a, b, c, d\}$.

❚ Note first that each partition of X contains either 1, 2, 3, or 4 cells.   The partitions are as follows:

(1) $[\{a, b, c, d\}]$
(2) $[\{a\}, \{b, c, d\}], [\{b\}, \{a, c, d\}], [\{c\}, \{a, b, d\}], [\{d\}, \{a, b, c\}],$
      $[\{a, b\}, \{c, d\}], [\{a, c\}, \{b, d\}], [\{a, d\}, \{b, c\}]$
(3) $[\{a\}, \{b\}, \{c, d\}], [\{a\}, \{c\}, \{b, d\}], [\{a\}, \{d\}, \{b, c\}],$
      $[\{b\}, \{c\}, \{a, d\}], [\{b\}, \{d\}, \{a, c\}], [\{c\}, \{d\}, \{a, b\}]$
(4) $[\{a\}, \{b\}, \{c\}, \{d\}]$
There are fifteen different partitions of X.

6.71   Let $f(n, k)$ represent the number of partitions of a set S of n elements into k cells $(k = 1, 2, \ldots, n)$.   Find a recursion formula for $f(n, k)$ and use it to check the results of Problem 6.70.

❚ Let b be some distinguished element of S.   If b itself constitutes a cell, then $S - b$ can be partitioned into $k - 1$ cells in $f(n - 1, k - 1)$ ways.   On the other hand, each partition of $S - b$ into k cells allows b to be admitted into a cell in k ways.   We have thus shown that

$$f(n, k) = f(n - 1, k - 1) + kf(n - 1, k) \qquad (1)$$

which is the desired recursion formula.
  Solution of (1) in the form of Pascal's triangle,

$$
\begin{array}{c}
k \to \\
\begin{array}{cccc}
1 & & & \\
1 & 1 & & \\
1 & 3 & 1 & \\
1 & 7 & 6 & 1
\end{array}
\end{array}
$$

$m \downarrow$

$\ldots\ldots\ldots\ldots$

confirms Problem 6.70.

6.72   What is an *equivalence relation*?

❚ A relation R on a set A is called an equivalence relation if it is reflexive, symmetric, and transitive.   [Ordinary equality is obviously the model for equivalence relations.]

6.73   Let L be the set of lines in the euclidean plane.   Show that R : "is parallel to ($\parallel$) or coincident with ($=$)" is an equivalence relation on L.

❚ Since $a = a$, for any line A in L, R is reflexive.   If $a \parallel b$, then $b \parallel a$; so R is symmetric.   If $a \parallel b$ and $b \parallel c$, then $a \parallel c$ or $a = c$; hence R is transitive.   Thus R is an equivalence relation.

6.74   On the set **L** of Problem 6.73, is the relation S : "has a point in common with" an equivalence relation?

❚ No.   For example, if a and c are distinct horizontal lines and b is a vertical line, then aSb and bSc, but $a\cancel{S}c$.

6.75   Let T be the set of triangles in the euclidian plane.   Show that the relation R of similarity is an equivalence relation on T.

❚ Every triangle is similar to itself, so R is reflexive.   If triangle a is similar to triangle b, then b is similar to a; hence R is symmetric.   If a is similar to b, and b is similar to c, then a is similar to c.   Hence R is an equivalence relation.

6.76   Show that the relation $\subseteq$ of set inclusion is not an equivalence relation.

▌ The relation $\subseteq$ is reflexive and transitive, but $\subseteq$ is not symmetric; that is, $A \subseteq B$ does not imply that $B \subseteq A$.

6.77   Consider the set **Z** of integers and an integer $m > 1$. We say that *x is congruent to y, modulo m*, written

$$x \equiv y (\text{mod } m)$$

if $x - y$ is divisible by m. Show that this defines an equivalence relation on **Z**.

▌ For any x in **Z**, we have $x \equiv x(\text{mod } m)$ because $x - x = 0$ is divisible by m. Hence the relation is reflexive.

Suppose $x \equiv y(\text{mod } m)$, so $x - y$ is divisible by m. Then $-(x - y) = y - x$ is also divisible by m, so $y \equiv x(\text{mod } m)$. Thus the relation is symmetric.

Now suppose $x \equiv y(\text{mod } m)$ and $y \equiv z(\text{mod } m)$ so $x - y$ and $y - z$ are each divisible by m. Then the sum

$$(x - y) + (y - z) = x - z$$

is also divisible by m; hence $x \equiv z(\text{mod } m)$. Thus the relation is transitive.

Theorem 6.2: Similarity of matrices is an equivalence relation.

6.78   Prove the reflexivity part of Theorem 6.2. [Recall that A is similar to B if there exists an invertible matrix P such that $A = P^{-1}BP$.]

▌ The identity matrix I is invertible and $I = I^{-1}$. Since $A = I^{-1}AI$, A is similar to A.

6.79   Prove the symmetry part of Theorem 6.2.

▌ If $A = P^{-1}BP$, then $B = PAP^{-1} = (P^{-1})^{-1}AP^{-1}$.

6.80   Prove the transitivity part of Theorem 6.2.

▌ If $A = P^{-1}BP$ and $B = Q^{-1}CQ$, then $A = P^{-1}(Q^{-1}CQ)P = (P^{-1}Q^{-1})C(QP) = (QP)^{-1}C(QP)$.

Theorem 6.3: Congruence of matrices is an equivalence relation.

6.81   Prove Theorem 6.3. [A is *congruent* to B if $A = P^TBP$, for some invertible P.]

▌ Because of the properties $(XY)^T = Y^TX^T$ and $(X^T)^{-1} = (X^{-1})^T$, the proof echoes Problems 6.78–6.80.

6.82   Let R be an equivalence relation on a set A. Define the *equivalence class* of an element $a \in A$, denoted $[a]$.

▌ The equivalence class $[a]$ is the set of elements of A to which a is related; that is, $[a] = \{x : (a, x) \in R\}$.

6.83   Let R be an equivalence relation on a set A. Define the *quotient of A by R*, denoted A/R.

▌ A/R is the collection of equivalence classes; that is, $A/R = \{[a] : a \in A\}$.

Theorem 6.4: Let R be an equivalence relation on a set A. Then the quotient set A/R is a partition of A.

6.84   Prove Theorem 6.4.

▌ Let a denote an arbitrary element of A. Since R is reflexive, $a \in [a]$. Suppose that also $a \in [b]$; we shall show that $[b] = [a]$. In fact,

$$\left.\begin{array}{l} x \in [b] \;\Rightarrow\; bRx \\ a \in [b] \;\Rightarrow\; bRa \;\Rightarrow\; aRb \end{array}\right\} \Rightarrow aRx \Rightarrow x \in [a]$$

and conversely. Therefore, each element of A belongs to one and only one equivalence class, making A/R a partition of A.

6.85   Let R be the following equivalence relation of the set $A = \{1, 2, 3, 4, 5, 6\}$:

$$R = \{(1, 1), (1, 5), (2, 2), (2, 3), (2, 6), (3, 2), (3, 3), (3, 6), (4, 4), (5, 1), (5, 5), (6, 2), (6, 3), (6, 6)\}$$

Find the partition of A *induced by* R; i.e., find the equivalence classes of R.

▌   Those elements related to 1 are 1 and 5 hence $[1] = \{1, 5\}$. We pick an element which does not belong to $[1]$, say 2. Those elements related to 2 are 2, 3, and 6; hence $[2] = \{2, 3, 6\}$. The only element which does not belong to $[1]$ or $[2]$ is 4, and the only element related to 4 is 4. Thus $[4] = \{4\}$.
     Accordingly, $\{\{1, 5\}, \{2, 3, 6\}, \{4\}\}$ is the partition of A induced by R.

6.86   The relation $R = \{(1, 1), (1, 2), (2, 1), (3, 3)\}$ is an equivalence relation of the set $S = \{1, 2, 3\}$. Find the quotient set S/R.

▌   Under the relation R, $[1] = \{1, 2\}$, $[2] = \{1, 2\}$,   and $[3] = \{3\}$. Noting that $[1] = [2]$,   we have $S/R = \{[1], [3]\}$.

6.87   Let R_5 be the relation on $\mathbf{Z}$, the set of integers, defined by $x \equiv y \pmod 5$. By Problem 6.77, R_5 is an equivalence relation on $\mathbf{Z}$. Find the induced equivalence classes.

▌   There are exactly five distinct equivalence classes in $\mathbf{Z}/R_5$:

$$\begin{array}{ll} A_0 = \{\ldots, -10, -5, 0, 5, 10, \ldots\} & A_3 = \{\ldots, -7, -2, 3, 8, 13, \ldots\} \\ A_1 = \{\ldots, -9, -4, 1, 6, 11, \ldots\} & A_4 = \{\ldots, -6, -1, 4, 9, 14, \ldots\} \\ A_2 = \{\ldots, -8, -3, 2, 7, 12, \ldots\} & \end{array}$$

Any integer x is uniquely expressible in the form $x = 5q + r$ where $0 \le r \le 4$; $x \in A_r$.

6.4   OPERATIONS AND SEMIGROUPS

6.88   Define a (*binary*) *operation*.

▌   A binary operation [or *operation*] on a nonempty set S is a function $*$ from $S \times S$ into S.

If $*$ is a binary operation on a set S, then we write $a * b$ or simply ab instead of $*(a, b)$. If S is a finite set, then the operation can be given by its operation table where the entry in the row labeled a and the column labeled b is $a * b$. If A is a subset of S, then A is said to be *closed under* $*$ if $a * b$ belongs to A for any elements a and b in A.

Problems 6.89–6.100 concern the following subsets of the positive integers $\mathbf{N}$:

$$\begin{array}{ll} A = \{0, 1\} & D = \{2, 4, 6, \ldots\} = \{x : x \text{ is even}\} \\ B = \{1, 2\} & E = \{1, 3, 5, \ldots\} = \{x : x \text{ is odd}\} \\ C = \{x : x \text{ is prime}\} & F = \{2, 4, 8, \ldots\} = \{x : x = 2^n, n \in \mathbf{N}\} \end{array}$$

6.89   Is A closed under multiplication?

▌   Compute: $0 \cdot 0 = 0$, $0 \cdot 1 = 0$, $1 \cdot 0 = 0$,   and $1 \cdot 1 = 1$. Yes, A is closed under multiplication.

6.90   Is A closed under addition?

▌   No, since $1 + 1 = 2$ does not belong to A.

6.91   Is B closed under multiplication?

▌   Since $2 \cdot 2 = 4$,   which does not belong to B, the set B is not closed under multiplication.

6.92 Is B closed under addition?

▮ No, since $1 + 2 = 3$ does not belong to B.

6.93 Is C closed under multiplication?

▮ Note that 2 and 3 are prime but $2 \cdot 3 = 6$ is not prime; hence C is not closed under multiplication.

6.94 Is C closed under addition?

▮ No, since $3 + 5 = 8$ does not belong to C.

6.95 Is D closed under multiplication?

▮ The product of even numbers is even; hence D is closed under multiplication.

6.96 Is D closed under addition?

▮ Yes, since the sum of even integers is even.

6.97 Is E closed under multiplication?

▮ The product of odd numbers is odd; hence E is closed under multiplication.

6.98 Is E closed under addition?

▮ No, since $3 + 5 = 8$ does not belong to E.

6.99 If F closed under multiplication?

▮ Since $2^r \cdot 2^s = 2^{r+s}$, F is closed under multiplication.

6.100 Is F closed under addition?

▮ No, since $2 + 4 = 6$ does not belong to F.

6.101 Define an *associative* operation.

▮ An operation $*$ on a set S is associative if, for any a, b, c in S, we have $(a * b) * c = a * (b * c)$.

6.102 Is addition in $\mathbf{Z}$ (the set of all integers) associative?

▮ Yes.

6.103 Is subtraction $\mathbf{Z}$ associative?

▮ No. For example, $(12 - 6) - 2 = 6 - 2 = 4$, but $12 - (6 - 2) = 12 - 4 = 8$.

6.104 Is multiplication in $\mathbf{Z}$ associative?

▮ Yes.

6.105 Is the operation $p * q = \max(p, q)$, defined on $\mathbf{Z}$, associative?

▮ Yes: Given $p, q, r \in \mathbf{Z}$, let $a \le b \le c$ denote their rearrangement in natural order. Then

$$(p * q) * r = (c * q) * r \text{ or } (p * c) * r \text{ or } (p * q) * c$$
$$= c * r \text{ or } c * r \text{ or } c = c \text{ or } c \text{ or } c = c$$

and, similarly, $p * (q * r) = c$.

6.106 Is exponentiation in $\mathbf{Z}$ associative?

▮ No. For example, if we let $a * b = a^b$, then

$$(2 * 2) * 3 = (2^2)^3 = 4^3 = 64 \qquad \text{but} \qquad 2 * (2 * 3) = 2^{2^3} = 2^8 = 256$$

6.107 Suppose an operation [written as a product] on a set S is not associative. How many ways can the product $abcd$ of the four elements be formed?

▌ There are five ways to insert parentheses: $((ab)c)d$, $(ab)(cd)$, $(a(bc))d$, $a((bc)d)$, and $a(b(cd))$.

Theorem 6.5: Suppose $*$ is an associative operation on a set S. Then all possible "products" of n ordered elements of S are equal.

6.108 Prove Theorem 6.5.

▌ The proof is by induction on n. The cases $n = 1$ and $n = 2$ are trivially true, and the case $n = 3$ is true since $*$ is associative. Suppose $n > 3$ and use the notations

$$(a_1 a_2 \cdots a_n) \equiv (\cdots ((a_1 a_2) a_3) \cdots) a_n \qquad \text{and} \qquad [a_1 a_2 \cdots a_n] \equiv \text{any product}$$

We shall show that $[a_1 a_2 \cdots a_n] = (a_1 a_2 \cdots a_n)$. In fact, since $[a_1 a_2 \cdots a_n]$ denotes some product, there exists an $r < n$ such that $[a_1 a_2 \cdots a_n] = [a_1 a_2 \cdots a_r][a_{r+1} \cdots a_n]$. Therefore, by induction [Problem 6.17],

$$[a_1 a_2 \cdots a_n] = [a_1 a_2 \cdots a_r][a_{r+1} \cdots a_n] = [a_1 a_2 \cdots a_r](a_{r+1} \cdots a_n)$$
$$= [a_1 \cdots a_r]((a_{r+1} \cdots a_{n-1}) a_n) = ([a_1 \cdots a_r](a_{r-1} \cdots a_{n-1})) a_n$$
$$= [a_1 \cdots a_{n-1}] a_n = (a_1 \cdots a_{n-1}) a_n = (a_1 a_2 \cdots a_n).$$

Thus the theorem is proved.

In consequence, when dealing with an associative operation, we can dispense with parentheses and simply write $a_1 * a_2 * \cdots * a_m$.

6.109 Define *semigroup*.

▌ A set S together with an associative operation $*$ on S constitute a semigroup. We denote the semigroup by $(S, *)$ or simply by S when the operation is understood.

6.110 Define an *identity element* for an operation $*$ on a set S.

▌ An element e in S is an identity element for $*$ if $a * e = e * a = a$, for every element a in S. More generally, e a *right identity* if $a * e = a$ for every a in S, and a *left identity* if $e * a = a$ for every a in S. [Be aware that an operation need not possess either a right or a left identity.]

6.111 Suppose e is a left identity and f is a right identity for an operation $*$. Show that $e = f$.

▌ Since e is a left identity, $e * f = f$; but since f is a right identity, $e * f = e$. Hence $e = f$. This result tells us, in particular, that an identity element is unique, and that if an operation has more than one left [right] identity then it has no right [left] identity.

6.112 Does the operation of Problem 6.105 have an identity element when defined on **Z**? on **N**?

▌ No identity over **Z**; $e = 1$ for **N**.

6.113 Define the *left* and *right cancellation laws* for an operation $*$ on a set S.

▌ The operation $*$ on S satisfies the left cancellation law if

$$a * b = a * c \qquad \text{implies} \qquad b = c$$

and the right cancellation law if

$$b * a = c * a \qquad \text{implies} \qquad b = c$$

6.114 Define a *commutative* operation.

▌ An operation $*$ on a set S is said to be commutative [or, to satisfy the *commutative law*] if $a * b = b * a$ for all a, b in S.

6.115 Let $*$ on S have a (unique) identity element e. What is meant by an *inverse* of an element a of S?

▌ An element b is an inverse of element a if $a*b = b*a = e$.

6.116 Suppose S has an associative operation with identity element e. Show that $a \in S$ has at most one inverse.

▌ Let b and b' be inverses of a. Then

$$b*(a*b') = b*e = b \qquad \text{and} \qquad (b*a)*b' = e*b' = b'$$

Since S is associative, $(b*a)*b' = b*(a*b')$; hence $b = b'$.

Problems 6.117–6.120 concern the operation of taking the least common multiple: $p*q =$ l.c.m. (p, q) $(p, q \in \mathbf{N})$.

6.117 Find $4*6$, $3*5$, $9*18$, and $1*6$.

▌ Since $x*y$ means the least common multiple of x and y, we have $4*6 = 12$, $3*5 = 15$, $9*18 = 18$, $1*6 = 6$.

6.118 Is $(\mathbf{N}, *)$ a semigroup? Is it commutative?

▌ One proves in number theorem that $(a*b)*c = a*(b*c)$, i.e., that the operation of l.c.m. is associative, and that $a*b = b*a$, i.e., that the operation of l.c.m. is commutative. Hence $(\mathbf{N}, *)$ is a commutative semigroup.

6.119 Find the identity element of $*$.

▌ The integer 1 is the identity element since the l.c.m. of 1 and any positive integer a is a.

6.120 Which elements in $\mathbf{N}$, if any, have inverses and what are they?

▌ Since l.c.m. $(a, b) = 1$ if and only if $a = 1$ and $b = 1$, the only number which has an inverse is 1, and it is its own inverse.

Problems 6.121–6.125 refer to the set $\mathbf{Q}$ of rational numbers and the operation $*$ on $\mathbf{Q}$ defined by $a*b = a + b - ab$.

6.121 Find $3*4, 2*(-5)$, and $7*\frac{1}{2}$.

▌
$$3*4 = 3 + 4 - (3)(4) = 3 + 4 - 12 = -5$$
$$2*(-5) = 2(-5) - (2)(-5) = 2 - 5 + 10 = 7$$
$$7*\tfrac{1}{2} = 7 + \tfrac{1}{2} - 7(\tfrac{1}{2}) = 4$$

6.122 Is $(\mathbf{Q}, *)$ a semigroup?

▌ Determine whether or not $*$ is associative:

$$(a*b)*c = (a + b - ab)*c = (a + b - ab) + c - (a + b - ab)c$$
$$= a + b - ab + c - ac - bc + abc = a + b + c - ab - ac - bc + abc$$

and

$$a*(b*c) = a*(b + c - bc) = a + (b + c - bc) - a(b + c - bc)$$
$$= a + b + c - bc - ab - ac + abc$$

Hence $*$ is associative and $(\mathbf{Q}, *)$ is a semigroup.

6.123 Is $*$ commutative?

▌ $a*b = a + b - ab = b + a - ba = b*a$; hence $*$ is commutative.

6.124   Find the identity element for $*$.

    ▮ $a * 0 = 0 * a = a$.   Thus 0 is the identity element.

6.125   Do any of the elements in **Q** have an inverse?   What is it?

    ▮   In order for a to have an inverse x, we must have $a * x = 0$,   since 0 is the identity element. Compute as follows:

$$a * x = 0, \quad a + x - ax = 0, \quad a = ax - x, \quad a = x(a - 1), \quad x = a/(a - 1)$$

Thus if $a \neq 1$,   then a has the unique inverse $a/(a - 1)$.

     Problems 6.126–6.128 refer to a nonempty set S with the operation $a * b = a$.

6.126   Is the operation associative?

    ▮   Yes, in fact, $(a * b) * c = a * c = a$ and $a * (b * c) = a * b = a$.

6.127   Is the operation commutative?

    ▮   If S has more than one element, then $*$ is not commutative: Specifically, for $a \neq b$, $a * b = a$ but $b * a = b$.

6.128   Show that the right cancellation law holds.   Does the left cancellation law hold?

    ▮   Suppose $a * c = b * c$.   We have $a * c = a$ and $b * c = b$;   hence $a = b$.   The left cancellation law does not hold.   For example, when $b \neq c$,   it is still the case that $a * b = a * c (= a)$.

6.129   Let S be a set of symbols.   Define the *free semigroup* on S.

    ▮   A *word* on S is a finite sequence of its elements.   For example, $U = ababb$ and $V = accba$ are words on $S = \{a, b, c\}$.   When discussing words on S, we frequently call S the *alphabet* and its elements *letters*.   For convenience, the empty sequence, denoted by ε or 1, is also considered a word on S.   We shall also abbreviate our notation by writing a^2 for aa, a^3 for aaa, and so on.   The set of all words on S is usually denoted by S^*.

     Now consider two words U and V on S.   We can form the word UV obtained by writing the letters of V after the letters of U.   For example, if U and V are the words above, then

$$UV = ababbaccba = abab^2ac^2ba$$

This operation is called *concatenation*.   Clearly the operation is associative.   Thus the set of words on S is a semigroup under the concatenation operation.   This semigroup is called the *free semigroup* on S [or, generated by S].   Clearly the empty word ε is an identity element for the semigroup, and the semigroup satisfies both right and left cancellation laws.

6.5   GROUPS AND SUBGROUPS

6.130   Define a *group*.

    ▮   Let G be a nonempty set with a binary operation [denoted by juxtaposition].   Then G is called a *group* if the following axioms hold:

[G_1]   *Associative law*, i.e., for any a, b, c in G, we have $(ab)c = a(bc)$.
[G_2]   *Identity element*, i.e., there exists an element e in G such that $ae = ea = a$ for any element a in G.
[G_3]   *Inverses*, i.e., for each a in G, there exists an element a^{-1} [the *inverse* of a] in G such that $aa^{-1} = a^{-1}a = e$.

([G_2] and [G_3] make a semigroup into a group.)

6.131   Define an *abelian* group.

    ▮   A group G is said to be abelian [or *commutative*] if the commutative law holds; i.e., if $ab = ba$ for every $a, b \in G$.

When the binary operation is denoted by juxtaposition as above, the group G is said to be written *multiplicatively*. When G is abelian, the binary operation is frequently denoted by $+$ and G is said to be written *additively*. In such a case the identity element is denoted by 0 and is called the *zero* element, and the inverse is denoted by $-a$ and is called the *negative* of a. If A and B are subsets of G then we write

$$AB = \{ab : a \in A, b \in B\} \quad \text{or} \quad A + B = \{a + b : \alpha \in A, b \in B\}$$

The number of elements in a group G, denoted $|G|$, is called the order of G. G is a *finite group* if its order is finite.

6.132 Which of the following are groups under addition: **N**, **Z**, **Q**, **R**, **C**?

⫾ The integers **Z**, the rationals **Q**, the reals **R**, and the complex numbers **C** are each a (abelian) group under addition. The positive integers **N** do not form a group under addition, e.g., $0 \notin \mathbf{N}$.

6.133 The nonzero rational numbers **Q**\\{0} form an abelian group under multiplication. What is the identity element, and what are the inverses?

⫾ The rational number 1 is the identity element and q/p is the multiplicative inverse of the rational number p/q.

6.134 Let S be the set of $n \times n$ matrices with rational entries, under the operation of matrix multiplication. Is S a group?

⫾ No. Although matrix multiplication is associative and matrix multiplication has an identity element I [with rational entries], S is not a group since inverses do not always exist.

6.135 The set G of nonsingular $n \times n$ matrices does form a group under matrix multiplication. What is the identity element, and what are inverses?

⫾ The identity element is the identity matrix I, and the inverse of A is its inverse matrix A^{-1}. This is an example of a nonabelian group, since matrix multiplication is noncommutative.

6.136 What is the *symmetric group of degree n*?

⫾ This is another name for S_n of permutations of $\{1, 2, \ldots, n\}$ under the operation of composition [Problem 5.54].

6.137 Find the elements and the multiplication table of the symmetric group S_3.

⫾ S_3 has $3! = 6$ elements, as follows:

$$\varepsilon = \begin{pmatrix} 1 & 2 & 3 \\ 1 & 2 & 3 \end{pmatrix} \qquad \sigma_2 = \begin{pmatrix} 1 & 2 & 3 \\ 3 & 2 & 1 \end{pmatrix} \qquad \phi_1 = \begin{pmatrix} 1 & 2 & 3 \\ 2 & 3 & 1 \end{pmatrix}$$

$$\sigma_1 = \begin{pmatrix} 1 & 2 & 3 \\ 1 & 3 & 2 \end{pmatrix} \qquad \sigma_3 = \begin{pmatrix} 1 & 2 & 3 \\ 2 & 1 & 3 \end{pmatrix} \qquad \phi_2 = \begin{pmatrix} 1 & 2 & 3 \\ 3 & 1 & 2 \end{pmatrix}$$

The multiplication table of S_3 appears in Fig. 6-9.

	ϵ	σ_1	σ_2	σ_3	ϕ_1	ϕ_2
ϵ	ϵ	σ_1	σ_2	σ_3	ϕ_1	ϕ_2
σ_1	σ_1	ϵ	ϕ_1	ϕ_2	σ_2	σ_3
σ_2	σ_2	ϕ_2	ϵ	ϕ_1	σ_3	σ_1
σ_3	σ_3	ϕ_1	ϕ_2	ϵ	σ_1	σ_2
ϕ_1	ϕ_1	σ_3	σ_1	σ_2	ϕ_2	ϵ
ϕ_2	ϕ_2	σ_2	σ_3	σ_1	ϵ	ϕ_1

Fig. 6-9

Problem 6.138–6.142 refer to a group G with identity element e.

6.138   Show that the identity element e is unique.

▮ Follows from Problem 6.111.

6.139   Show that the inverse a^{-1} of any element a of G is unique.

▮ Follows from Problem 6.116.

6.140   Show that the left and right cancellation laws hold in G.

▮ If $ab = ac$, then $b = eb = (a^{-1}a)b = a^{-1}(ab) = a^{-1}(ac) = (a^{-1}a)c = ec = c$. Similarly, if $ba = ca$, then $b = c$.

6.141   Show that $(a^{-1})^{-1} = a$ for any element a in G.

▮ Since a^{-1} is the inverse of a, we have $aa^{-1} = a^{-1}a = e$. Hence a is the inverse of a^{-1}; that is, $a = (a^{-1})^{-1}$.

6.142   Show that $(ab)^{-1} = b^{-1}a^{-1}$.

▮ $(b^{-1}a^{-1})(ab) = b^{-1}(a^{-1}a)b = b^{-1}eb = b^{-1}b = e$. Similarly, $(ab)(b^{-1}a^{-1}) = e$. Therefore, $b^{-1}a^{-1}$ is the inverse of ab; that is, $b^{-1}a^{-1} = (ab)^{-1}$.

6.143   Define a *subgroup* of a group.

▮ A subset H of a group G is a subgroup of G if H itself forms a group under the operation of G.

6.144   Suppose H is a subset of a group G. Show that H is a subgroup of G if (i) the identity element e belongs to H; (ii) H is closed under the operation of G; (iii) H is closed under inverses [i.e., if $a \in H$ then $a^{-1} \in H$].

▮ H is nonempty and has an identity element, by (i). The operation is well-defined on H, by (ii). Inverses exist in H, by (iii). Lastly, the associative law holds in H, since it holds in G. Thus H is a subgroup of G.

6.145   Consider the group $\mathbf{Z}$ of integers under addition. Let H be the subset of $\mathbf{Z}$ consisting of all multiples of an integer $m > 1$: $H = \{\ldots, -3m, -2m, -m, 0, m, 2m, 3m, \ldots\}$. Show that H is a subgroup of $\mathbf{Z}$.

▮ (i) H contains the identity element 0 of $\mathbf{Z}$. (ii) If rm and sm are any elements of H, then $rm + sm = (r + s)m$ is also an element of H. (iii) If rm is any element of H, then its negative $-rm$ also belongs to H.

6.146   Let G be any group, and let a be any element of G. Define the *cyclic subgroup generated by a*, denoted $gp(a)$.

▮ As usual, we define $a^0 \equiv e$, $(a)^{-1} \equiv a^{-1}$, and $a^{n+1} = a^n a$. Clearly, $a^m a^n = a^{m+n}$ and $(a^m)^n = a^{mn}$, for any integers m and n. Let $gp(a)$ denote the set of all powers of a:

$$gp(a) = \{\ldots, a^{-2}, a^{-1}, e, a, a^2, a^3, \ldots\}$$

Then $gp(a)$ contains e, is closed under the group operation, and contains inverses. Thus $gp(a)$ is a subgroup of G.

6.147   Let a be any element in a group G. Describe the cyclic subgroup $gp(a)$ when $gp(a)$ is finite, and define the *order* of a.

▮ If $gp(a)$ is finite, then some of the powers of a are not distinct; say, $a^r = a^s$, with $r > s$. Then $a^{r-s} = e$, where $r - s > 0$. The smallest positive integer m such that $a^m = e$ is called the *order* of a and will be denoted $|a|$.

   If $|a| = m$, then its cyclic subgroup $gp(a)$ has m elements: $gp(a) = \{e, a, a^2, a^3, \ldots, a^{m-1}\}$. [If $gp(a)$ is not finite, then we define $|a| = 0$.]

Problems 6.148–6.151 refer to the group $G = \{1, 2, 3, 4, 5, 6\}$ under multiplication modulo 7.

6.148   Find the multiplication table of G.

▮ To find $a*b$ in G, find the remainder when the product ab is divided by 7. For example, $5 \cdot 6 = 30$ which yields a remainder of 2 when divided by 7; hence $5*6=2$ in G. The multiplication table of G appears in Fig. 6-10.

$*$	1	2	3	4	5	6
1	1	2	3	4	5	6
2	2	4	6	1	3	5
3	3	6	2	5	1	4
4	4	1	5	2	6	3
5	5	3	1	6	4	2
6	6	5	4	3	2	1

Fig. 6-10

6.149   Find 2^{-1}, 3^{-1}, 6^{-1}.

▮ Figure 6-10 shows that 1 is the identity element of G. Recall that a^{-1} is that element of G such that $aa^{-1} = 1$. Hence $2^{-1} = 4$, $3^{-1} = 5$, and $6^{-1} = 6$.

6.150   Find the orders of, and subgroups generated by, 2 and 3.

▮ We have $2^1 = 2$, $2^2 = 4$, but $2^3 = 1$. Hence $|2| = 3$ and $gp(2) = \{1, 2, 4\}$. We have $3^1 = 3$, $3^2 = 2$, $3^3 = 6$, $3^4 = 4$, $3^5 = 5$, $3^6 = 1$. Hence $|3| = 6$ and $gp(3) = G$.

6.151   Is G cyclic?

▮ G is cyclic since $G = gp(3)$.

6.152   Let H be a subgroup of a group G. Define a *right* (*left*) *coset* of H.

▮ Let a be any element of G. Then the set $Ha = \{ha: h \in H\}$ is called a right coset of H. Analogously, aH is called a left coset of H.

Theorem 6.6: Let H be a subgroup of a group G. Then the right cosets Ha form a partition of G.

6.153   Prove Theorem 6.6.

▮ Define a relation R on G by $aRb \Leftrightarrow b \in Ha$. We show that R is an equivalence relation.
(1) $e \in H \Rightarrow a \in Ha \Rightarrow aRa$ [R is reflexive]
(2) $aRb \Rightarrow b = ha \Rightarrow a = h^{-1}b \Rightarrow a \in Hb \Rightarrow bRa$ [R is symmetric]
(3) $\left. \begin{array}{l} aRb \\ bRc \end{array} \right\} \Rightarrow \left. \begin{array}{l} b = h_1 a \\ c = h_2 b \end{array} \right\} \Rightarrow c = (h_2 h_1)a \Rightarrow c \in Ha \Rightarrow aRc$ [R is transitive]
Under R, $[a] = Ha$, so that Theorem 6.6 follows immediately from Theorem 6.4.

6.154   Let H be a finite subgroup of G. Show that H and any coset Ha have the same number of elements.

▮ Let $H = \{h_1, h_2, \ldots, h_k\}$, where H has k elements. Then $Ha = \{h_1 a, h_2 a, \ldots, h_k a\}$. However, $h_i a = h_j a$ implies $h_i = h_j$; hence the k elements listed in Ha are distinct.

Theorem 6.7 (*Lagrange*): Let H be a subgroup of a finite group G. Then the order of H divides the order of G.

6.155   Prove Theorem 6.7.

▮ Suppose H has r elements and there are s distinct right cosets. By Theorem 6.6, the cosets partition G, and by Problem 6.154, each coset has r elements. Therefore, G has rs elements, and so the order of H divides the order of G.

6.156   Let H be a subgroup of a group G. Define the *index of H in G*, denoted $[G:H]$.

▮ The index of H in G is equal to the number of distinct right (left) cosets of H in G. If G is finite, then $[G:H] = |G|/|H|$.

6.157   Let H be a subgroup of a group G.   Define a *coset representative system* for H in G.

▮   A subset C of G is a coset representative system of H if C contains exactly one element from each coset.   Such an element is called a representative of the coset.

6.158   Let H be a subgroup of a finite group G.   How many coset representative systems exist for the cosets of H?

▮   These are $|H|$ ways of choosing an element from any coset [see Problem 6.154], and there are $[G : H]$ distinct cosets.   Hence the desired number is $|H|^{[G \, : \, H]}$.

In Problems 6.159–6.161, $\mathbf{Z}$ denotes the group of integers under addition, and $H = \{\ldots, -10, -5, 0, 5, 5, 10, \ldots\}$ is the subgroup of $\mathbf{Z}$ consisting of the multiples of 5.

6.159   Find the cosets of H in $\mathbf{Z}$.

▮   There are five distinct [left] cosets of H in $\mathbf{Z}$, as follows:

$$0 + H = H = \{\ldots, -10, -5, 0, 5, 10, \ldots\}$$
$$1 + H = \{\ldots, -9, -4, 1, 6, 11, \ldots\}$$
$$2 + H = \{\ldots, -8, -3, 2, 7, 12, \ldots\}$$
$$3 + H = \{\ldots, -7, -2, 3, 8, 13, , \ldots\}$$
$$4 + H = \{\ldots, -6, -1, 4, 9, 14, \ldots\}$$

Any other coset $n + H$ coincides with one of these.

6.160   Find the index of H in $\mathbf{Z}$.

▮   Although $\mathbf{Z}$ and H are both infinite, the index of H in $\mathbf{Z}$ is finite.   Specifically, $[\mathbf{Z} : H] = 5$,   the number of cosets.

6.161   Find coset representatives of H in $\mathbf{Z}$.

▮   Choose exactly one element from each coset; e.g., $\{0, 1, 2, 3, 4\}$ or $\{-1, 0, 1, 2, 3\}$.

Problems 6.162–6.166 refer to the symmetric group S_3, whose multiplication table appears in Fig. 6-9.

6.162   Find the order of, and the subgroup generated by, each element of S_3.

▮ $\varepsilon^1 = \varepsilon$,   so $|\varepsilon| = 1$ and $gp(\varepsilon) = \{\varepsilon\}$. $\sigma_1^1 = \sigma_1$, $\sigma_1^2 = \varepsilon$;   so $|\sigma_1| = 2$ and $gp(\sigma_1) = \{\sigma_1, \varepsilon\}$. Similarly, $|\sigma_2| = 2$, $gp(\sigma_2) = \{\sigma_2, \varepsilon\}$;   and $|\sigma_3| = 2$, $gp(\sigma_3) = \{\sigma_3, \varepsilon\}$.   We have

$$\phi_1^1 = \phi_1, \quad \phi_1^2 = \phi_2, \quad \phi_1^3 = \phi_2 \cdot \phi_1 = \varepsilon$$

Hence $|\phi_1| = 3$ and $gp(\phi_1) = \{\varepsilon, \phi_1, \phi_2\}$.   Also, $\phi_2^1 = \phi_2$, $\phi_2^2 = \phi_1$, $\phi_2^3 = \phi_1 \cdot \phi_2 = \varepsilon$;   hence $|\phi_2| = 3$ and $gp(\phi_2) = \{\varepsilon, \phi_2, \phi_1\}$.

6.163   Can you find a subgroup H of order four?

▮   The order of S_3 is six.   By Lagrange's theorem, the order of H must divide the order of S_3.   Hence there is no subgroup of order four.

6.164   Let $A = \{\sigma_1, \sigma_2\}$ and $B = \{\phi_1, \phi_2\}$.   Find (a) AB, (b) $\sigma_3 A$, and (c) $A\sigma_3$.

▮   (a) Multiply each element of A by each element of B: $\sigma_1\phi_1 = \sigma_2$, $\sigma_1\phi_2 = \sigma_3$, $\sigma_2\phi_1 = 3$, $\sigma_2\phi_2 = \sigma_1$. Hence $AB = \{\sigma_1, \sigma_2, \sigma_3\}$.   ($b$) Multiply σ_3 by each element of A: $\sigma_3\sigma_1 = \phi_1$, $\sigma_3\sigma_2 = \phi_2$.   Hence $c_3 A = \{\phi_1, \phi_2\}$.   ($c$) Multiply each element of A by σ_3: $\sigma_1\sigma_3 = \phi_2$, $\sigma_2\sigma_3 = \phi_1$.   Hence $A\sigma_3 = \{\phi_1, \phi_2\}$.

6.165   Let $H = gp(\sigma_1)$ and $K = gp(\sigma_2)$.   Is HK a subgroup of S_3?

▮ $H = \{\varepsilon, \sigma_1\}$, $K = \{\varepsilon, \sigma_2\}$ and then $HK = \{\varepsilon, \sigma_1, \sigma_2, \phi_1\}$,   which is not a subgroup of S_3 since HK has four elements.   [Compare with Problem 6.163.]

6.166   Is S_3 cyclic?

$\blacksquare$ S_3 is not cyclic, since S_3 is not generated by any of its elements.

6.167   If H is a subgroup of G, show that $HH = H$.

$\blacksquare$ Since H is closed under the operation of G, we have $HH \subseteq H$.   On the other hand, suppose $h \in H$. Since H is a subgroup, the identity element e belongs to H.   Hence $eh = h \in HH$,   and so $H \subseteq HH$. The two inclusions give $HH = H$.

6.168   Show that $Ha = Hb$ if and only if $ab^{-1} \in H$.

$\blacksquare$ If $Ha = Hb$,   then $a \in Ha = Hb$.   Hence there exists $h \in H$ such that $a = hb$,   and $ab^{-1} = h$ belongs to H.   On the other hand, suppose $h \equiv ab^{-1} \in H$.   Then $a = hb \in Hb$.   But $a \in Ha$.   Thus $Ha = Hb$,   since the cosets form a partition of G.

6.169   Let G be a finite group of order n.   Show that $a^n = e$ for any $a \in G$.

$\blacksquare$ If $|gp(a)| = m$,   then $a^m = e$.   By Lagrange's theorem, m divides n; say, $n = mr$.   Then $a^n = a^{mr} = (a^m)^r = e^r = e$.

6.6   NORMAL SUBGROUPS, FACTOR GROUPS, GROUP HOMOMORPHISMS

6.170   Define a *normal* subgroup of a group G.

$\blacksquare$ A subgroup H of G is normal subgroup if $a^{-1}Ha \subset H$ for every $a \in G$.   Equivalently, H is normal if $aH = Ha$ for every $a \in G$.

6.171   Let G be the group of nonsingular 2×2 matrices under matrix multiplication.   Let H be the subset of G consisting of the lower triangular matrices; i.e., matrices of the form $\begin{pmatrix} a & 0 \\ c & d \end{pmatrix}$.   Show that H is a subgroup of G, but not a normal subgroup.

$\blacksquare$ H is closed under matrix multiplication and inverses, and the identity matrix I belongs to H.   Hence H is a subgroup of G.   However, H is not a normal subgroup since, for example,

$$\begin{pmatrix} 1 & 2 \\ 1 & 3 \end{pmatrix}^{-1} \begin{pmatrix} 1 & 0 \\ 1 & 1 \end{pmatrix} \begin{pmatrix} 1 & 2 \\ 1 & 3 \end{pmatrix} = \begin{pmatrix} 3 & -2 \\ -1 & 1 \end{pmatrix} \begin{pmatrix} 1 & 0 \\ 1 & 1 \end{pmatrix} \begin{pmatrix} 1 & 2 \\ 1 & 3 \end{pmatrix} = \begin{pmatrix} -1 & -4 \\ 1 & 3 \end{pmatrix}$$

does not belong to H.

6.172   Let G be a group of matrices in Problem 6.171.   Let K be the subset of G consisting of matrices with determinant 1.   Show that K is a normal subgroup of G.

$\blacksquare$ Since $\det I = 1$, I belongs to K.   If A and B belong to K, then $\det (AB) = (\det A)(\det B) = (1)(1) = 1$ and so AB belongs to K.   Also, $\det A^{-1} = 1/\det A = 1$ and so A^{-1} belongs to K.   Thus K is a subgroup. Moreover, for any matrix X in G and any matrix A in K, $\det (X^{-1}AX) = 1$.   Hence $X^{-1}AX$ belongs to K, so K is a normal subgroup of G.

6.173   Consider the permutation group S_3, whose multiplication table appears in Fig. 6-9.   Is the subgroup $H = \{\varepsilon, \sigma_1\}$ normal?

$\blacksquare$ The right and left cosets of H are as follows:

Right Cosets	Left Cosets
$H\quad = \{\varepsilon, \sigma_1\}$	$H = \{\varepsilon, \sigma_1\}$
$H\phi_1 = \{\phi_1, \sigma_2\}$	$\phi_1 H = \{\phi_1, \sigma_3\}$
$H\phi_2 = \{\phi_2, \sigma_3\}$	$\phi_2 H = \{\phi_2, \sigma_2\}$

Since $H\phi_1 \neq \phi_1 H$, H is not a normal subgroup of S_3.

6.174   Show that any subgroup H of an abelian group G is normal.

$\blacksquare$ Let h be any element of H and let g be any element of G.   Then $g^{-1}hg = hg^{-1}g = h$ belongs to H. Hence H is a normal subgroup.

6.175 Let H be a subgroup, and K a normal subgroup, of a group G. Prove that HK is a subgroup of G [cf. Problem 6.165].

▮ We must show that $e \in HK$ and that HK is closed under multiplication and inverses. Since H and K are subgroups, $e \in H$ and $e \in K$. Hence $e = ee$ belongs to HK. Suppose $x, y \in HK$. Then $x = h_1 k_1$ and $y = h_2 k_2$ where $h_1, h_2 \in H$ and $k_1, k_2 \in K$. Then

$$xy = h_1 k_1 h_2 k_2 = h_1 h_2 (h_2^{-1} k_1 h_2) k_2$$

Since K is normal, $h_2^{-1} k_1 h_2 \in K$; and since H and K are subgroups, $h_1 h_2 \in H$ and $(h_2^{-1} k_2 h_2) k_2 \in K$. Thus $xy \in HK$, and so HK is closed under multiplication. We also have that

$$x^{-1} = (h_1 k_1)^{-1} = k_1^{-1} h_1^{-1} = h_1^{-1} (h_1 k_1^{-1} h_1^{-1})$$

Since K is a normal subgroup $h_1 k_1^{-1} h_1^{-1}$ belongs to K. Also h_1^{-1} belongs to H. Therefore $x^{-1} \in HK$, and hence HK, is closed under inverses. Consequently, HK is a subgroup.

The following theorem defines the *quotient group*, G/H, corresponding to a normal subgroup H of G.

Theorem 6.8: Let H be a normal subgroup of a group G. Then the cosets of H in G form a group under *coset multiplication*, as defined by $(aH)(bH) = abH$.

6.176 Prove Theorem 6.8.

▮ Coset multiplication is well-defined, since

$$(aH)(bH) = a(Hb)H = a(bH)H = ab(HH) = abH$$

[Here we used the fact that H is normal, so $Hb = bH$, and, from Problem 6.167, that $HH = H$.] Associativity of coset multiplication follows from the fact that associativity holds in G. H is the identity element of G/H, since

$$(aH)H = a(HH) = aH \qquad \text{and} \qquad H(aH) = (Ha)H = (aH)H = aH$$

Lastly, $a^{-1}H$ is the inverse of aH since

$$(a^{-1}H)(aH) = a^{-1}aH = eH = H \qquad \text{and} \qquad (aH)(a^{-1}H) = aa^{-1}H = eH = H$$

Thus G/H is a group under coset multiplication.

6.177 Let $\mathbf{Z}$ be the group of integers under addition, and let H be the subgroup of $\mathbf{Z}$ consisting of the multiples of 5. Show that H is a normal subgroup of $\mathbf{Z}$, and find the quotient group $\mathbf{Z}/H$.

▮ Since $\mathbf{Z}$ is abelian, H is automatically a normal subgroup. Let $\bar{0}, \bar{1}, \bar{2}, \bar{3}$, and $\bar{4}$ denote, respectively, the five cosets listed in Problem 6.159. The addition table for the quotient group $\mathbf{Z}/H = \{\bar{0}, \bar{1}, \bar{2}, 3, \bar{4}\}$ appears in Fig. 6-11. [This group is usually called *the integers modulo 5* and is frequently denoted $\mathbf{Z}_5$.]

+	$\bar{0}$	$\bar{1}$	$\bar{2}$	$\bar{3}$	$\bar{4}$
$\bar{0}$	$\bar{0}$	$\bar{1}$	$\bar{2}$	$\bar{3}$	$\bar{4}$
$\bar{1}$	$\bar{1}$	$\bar{2}$	$\bar{3}$	$\bar{4}$	$\bar{0}$
$\bar{2}$	$\bar{2}$	$\bar{3}$	$\bar{4}$	$\bar{0}$	$\bar{1}$
$\bar{3}$	$\bar{3}$	$\bar{4}$	$\bar{0}$	$\bar{1}$	$\bar{2}$
$\bar{4}$	$\bar{4}$	$\bar{0}$	$\bar{1}$	$\bar{2}$	$\bar{3}$

Fig. 6-11

6.178 Define a group *homomorphism*. Also, define a group *isomorphism*.

▮ A mapping f from a group G [with operation $*$] into a group G' [with operation $*'$] is a homomorphism if

$$f(a * b) = f(a) *' f(b)$$

for every a, b in G.   In addition, if f is one-to-one and onto, then f is an isomorphism and G and G' are said to be *isomorphic*, written $G \simeq G'$.

6.179   Let G be the group of real numbers under addition, and let G' be the group of positive real numbers under multiplication.   Show that the mapping $f: G \to G'$ defined by $f(a) = 2^a$ is a homomorphism.   Is it an isomorphism?

▮   The mapping f is a homomorphism since $f(a + b) = 2^{a+b} = 2^a 2^b = f(a)f(b)$.   Moreover, since f is one-to-one and onto, f is an isomorphism.

6.180   Let G be the group of real n-square matrices under addition.   Show that the trace function is a homomorphism of G into the group R of real numbers under addition.

▮   Let A and B be matrices in G.   Then $\operatorname{tr}(A + B) = \operatorname{tr}(A) + \operatorname{tr}(B)$.   Thus the trace function is a homomorphism.

6.181   Let G be the group of real, nonsingular, n-square matrices under multiplication.   Show that the determinant function is a homomorphism of G into the group G' of nonzero real numbers under multiplication.

▮   Let A and B be matrices in G.   Then $\det(AB) = (\det A)(\det B)$.   Hence the determinant function is a homomorphism.

6.182   Given a homomorphism $f: G \to G'$,   show that $f(e) = e'$ where e and e' are the identity elements of G and G', respectively.

▮   Since $e = e * e$ and f is a homomorphism, $f(e) = f(e * e) = f(e) *' f(e)$.   Hence, $e' = f(e)^{-1} *' f(e) = [f(e)^{-1} *' f(e)] *' f(e) = e' *' f(e) = f(e)$.

6.183   Given a homomorphism $f: G \to G'$, show that $f(a^{-1}) = f(a)^{-1}$,   for any element a in G.

▮   By Problem 6.182,

$$f(a) *' f(a^{-1}) = f(a * a^{-1}) = f(e) = e' = f(e) = f(a^{-1} * a) = f(a^{-1}) *' f(a)$$

Thus, $f(a)^{-1} = f(a^{-1})$.

6.184   Define the *kernel* and the *image* of a group homomorhism $f: G \to G'$.

▮   The kernel of f, written Ker f, is the set of elements of G whose image is the identity element e' of G': Ker $f = \{a \in G : f(a) = e'\}$.

The image of f, written $f(G)$ or Im f, consists of the images of elements of G under f:

$$\operatorname{Im} f = \{b \in G' : \quad b = f(a) \text{ for some } \quad a \in G\}$$

[The term *range* is also used for image.]

Theorem 6.9:   Let $f: G \to G'$ be a homomorphism with kernel K.   Then (i) K is a normal subgroup of G, and (ii) the quotient group G/K is isomorphic to the image of f.

6.185   Prove (i) of Theorem 6.9.

▮   By Problem 6.182, $f(e) = e'$,   so $e \in K$.   Now suppose $a, b \in K$ and $g \in G$.   Then $f(a) = e'$ and $f(b) = e'$.   Hence [using juxtaposition to indicate either group operation],

$$
\begin{aligned}
f(ab) \quad &= f(a)f(b) = e'e' = e' \\
f(a^{-1}) \quad &= f(a)^{-1} = e'^{-1} = e' \\
f(gag^{-1}) &= f(g)f(a)f(g^{-1}) = f(g)e'f(g)^{-1} = e'
\end{aligned}
$$

Hence ab, a^{-1}, and gag^{-1} belong to K, so K is a normal subgroup.

6.186   Prove (ii) of Theorem 6.9.

▌ Let $H \subseteq G'$ be the image of f, and define a mapping $\phi: G/K \to H$ by $\phi(Ka) = f(a)$. We show that ϕ is well-defined; i.e., if $Ka = Kb$ then $\phi(Ka) = \phi(Kb)$. Suppose $Ka = Kb$. Then $ab^{-1} \in K$ [Problem 6.168]. Then $f(ab^{-1}) = e'$, and so

$$f(a)f(b)^{-1} = f(a)f(b^{-1}) = f(ab^{-1}) = e'$$

Hence $f(a) = f(b)$, and so $\phi(Ka) = \phi(Kb)$. Thus ϕ is well-defined. We next show that ϕ is a homomorphism:

$$\phi(KaKb) = \phi(Kab) = f(ab) = f(a)f(b) = \phi(Ka)\phi(Kb)$$

Thus ϕ is a homomorphism. We next show that ϕ is one-to-one. Suppose $\phi(Ka) = \phi(Kb)$. Then

$$f(a) = f(b) \quad \text{or} \quad f(a)f(b)^{-1} = e' \quad \text{or} \quad f(a)f(b^{-1}) = e' \quad \text{or} \quad f(ab^{-1}) = e'$$

Thus $ab^{-1} \in K$, and again by Problem 6.168, $Ka = Kb$. Thus ϕ is one-to-one. Finally, we show that ϕ is onto. Let $h \in H$. Since H is the image of f, there exists $a \in G$ such that $f(a) = h$. Thus $\phi(Ka) = f(a) = h$, and so ϕ is onto. Consequently $G/K \simeq H$ and the theorem is proved.

Problems 6.187–6.189 refer to the following groups and mapping:
G = group of nonzero complex numbers under multiplication,
G' = group of nonzero real numbers under multiplication, $f: G \to G'$ defined by $f(z) = |z|$.

6.187 Show that f is a group homomorphism.

▌ $f(z_1 z_2) = |z_1 z_2| = |z_1||z_2| = f(z_1)f(z_2)$

6.188 Describe geometrically the kernel K of the homomorphism f.

▌ K consists of those complex numbers z such that $|z| = 1$; i.e., K is the unit circle.

6.189 Describe the quotient group G/K.

▌ G/K is isomorphic to the image of f, which is the group of positive real numbers under multiplication.

6.190 Show that any cyclic group is isomorphic either to the integers $\mathbf{Z}$ under addition, or to $\mathbf{Z}_m$, the integers under addition modulo m.

▌ Let a be any element in a group G. The function $f: \mathbf{Z} \to G$ defined by $f(n) = a^n$ is a homomorphism since $f(m + n) = a^{m+n} = a^m \cdot a^n = f(m)f(n)$. The image of f is $gp(a)$, the cyclic subgroup generated by a. Thus, $gp(a) \simeq \mathbf{Z}/K$, where K is the kernel of f. If $K = \{0\}$, then $gp(a) \simeq \mathbf{Z}$. On the other hand, if m is the order of a, then $K = \{\text{multiples of } m\}$, and so $gp(a) \simeq \mathbf{Z}_m$.

6.7 RINGS AND IDEALS

6.191 Define a *ring*.

▌ Let R be a nonempty set with two binary operations, an operation of addition (denoted by $+$) and an operation of multiplication (denoted by juxtaposition). Then R is called a *ring* if the following axioms are satisfied:
[R_1] For any $a, b, c \in R$, we have $(a + b) + c = a + (b + c)$.
[R_2] There exists an element $0 \in R$, called the *zero* element, such that $a + 0 = 0 + a = a$ for every $a \in R$.
[R_3] For each $a \in R$ there exists an element $-a \in R$, called the *negative* of a, such that $a + (-a) = (-a) + a = 0$.
[R_4] For any $a, b \in R$, we have $a + b = b + a$.
[R_5] For any $a, b, c \in R$, we have $(ab)c = a(bc)$.
[R_6] For any $a, b, c \in R$, we have:
　(i) $a(b + c) = ab + ac$, and (ii) $(b + c)a = ba + ca$.
Axioms [R_1] through [R_4] make R an abelian group under addition.

6.192 How is subtraction defined in a ring R?

▌ $a - b \equiv a + (-b)$

6.193   Define a *commutative* ring.

▐ A ring R is commutative if $ab = ba$ for every $a, b \in R$.

6.194   Define a *unity element* in a ring R.

▐ A nonzero element $1 \in R$ is called a *unity element* if $a \cdot 1 = 1 \cdot a = a$ for every element $a \in R$.

6.195   Let R be a ring with an identity element 1. Define a *unit* in R.

▐ An element $a \in R$ is a unit if a has a multiplicative inverse, $a^{-1} \in R$, such that $aa^{-1} = a^{-1}a = 1$.

6.196   Consider the ring $\mathbf{Z}$ of integers. (*a*) Is $\mathbf{Z}$ commutative? (*b*) Does $\mathbf{Z}$ have a unity element? (*c*) What are the units in $\mathbf{Z}$?

▐ (*a*) $\mathbf{Z}$ is a commutative ring since $ab = ba$ for any integers $a, b \in \mathbf{Z}$. (b) The number 1 is a unity element in $\mathbf{Z}$. (*c*) The only units in $\mathbf{Z}$ are 1 and -1.

6.197   Find the units of $\mathbf{Z}_m$, the ring of integers modulo m.

▐ If a is a unit in $\mathbf{Z}_m$, then $a^{-1}a \equiv 1 \pmod{m}$, or, in $\mathbf{Z}$,

$$a^{-1}a = 1 + rm \qquad \text{or} \qquad a^{-1}a - rm = 1$$

This shows that any common division of a and m must divide 1; i.e., that a and m are relatively prime. Conversely, if a and m are relatively prime in $\mathbf{Z}$, then

$$1 = gcd\,(a, m) = pa + qm \qquad \text{or} \qquad pa \equiv 1 \pmod{m}$$

which shows that a is a unit of $\mathbf{Z}_m$ [with inverse p]. Thus the units of $\mathbf{Z}_m$ are precisely those integers which are relatively prime to m.

6.198   In $\mathbf{Z}_{10}$ find -3, -8, and 3^{-1}.

▐ By $-a$ in a ring R we mean that element such that $a + (-a) = (-a) + a = 0$. Hence $-3 = 7$ since $3 + 7 = 7 + 3 = 0$ in $\mathbf{Z}_{10}$. Similarly $-8 = 2$. By a^{-1} in a ring R we mean that element such that $a \cdot a^{-1} = a^{-1} \cdot a = 1$. Hence $3^{-1} = 7$ since $3 \cdot 7 = 7 \cdot 3 = 1$ in $\mathbf{Z}_{10}$.

6.199   Let $f(x) = 2x^2 + 4x + 4$. Find the roots of $f(x)$ over $\mathbf{Z}_{10}$.

▐ Substitute each of the ten elements of $\mathbf{Z}_{10}$ into $f(x)$ to see which elements yield 0. We have:

$$f(0) = 4, \quad f(2) = 0, \quad f(4) = 2, \quad f(6) = 0, \quad f(8) = 4$$
$$f(1) = 0, \quad f(3) = 4, \quad f(5) = 4, \quad f(7) = 0, \quad f(9) = 2$$

Thus the roots are $1, 2, 6$, and 7. [This example shows that a polynomial of degree n can have more than n roots over an arbitrary ring. This cannot happen if the ring is a field.]

Problems 6.200–6.202 refer to the ring R of real n-square matrices.

6.200   Is R commutative?

▐ No; matrix multiplication is not commutative.

6.201   Does R have a unity element?

▐ Yes; the identity matrix I.

6.202   Find the units in R.

▐ The nonsingular or invertible matrices are the units in R.

6.203   Prove that $a \cdot 0 = 0 \cdot a = 0$ in a ring R.

▮   Since $0 = 0 + 0$,   we have $a \cdot 0 = a(0 + 0) = a \cdot 0 + a \cdot 0$.   Adding $-(a \cdot 0)$ to both sides yields $0 = a \cdot 0$. Similarly, $0 \cdot a = 0$.

6.204   Show that negatives are unique in any ring.

▮   Given an element a, suppose element x has the property that $a + x = 0$ [which automatically makes $x + a = 0$].   We have:

$$-a = -a + 0 = -a + (a + x) = (-a + a) + x = 0 + x = x$$

6.205   Show that $a(-b) = (-a)b = -ab$ in a ring R.

▮ $ab + a(-b) = a(b + (-b)) = a \cdot 0 = 0$;   hence [Problem 6.204], $a(-b) = -ab$.   Similarly, $(-a)b = -ab$.

6.206   Show that $(-1)a = -a$ in a ring R with a unity element 1.

▮ $a + (-1)a = 1 \cdot a + (-1)a = (1 + (-1))a = 0 \cdot a = 0$;   hence [Problem 6.204], $(-1)a = -a$.

6.207   Let R be a ring with a unity element 1.   Show that the set R^* of units in R is a group under multiplication.

▮   If a and b are units in R, then ab is a unit, since $b^{-1}a^{-1}$ is the inverse of ab.   Thus R^* is closed under multiplication.   Also, R^* is nonempty, since $1 \in R^*$;   and R^* is associative, since R is associative. Finally, if a is a unit in R, so is a^{-1} [since it has the inverse a]; consequently, R^* is closed under inverses. Thus R^* is a group under multiplication.

6.208   Define a *subring* of a ring R.

▮   A nonempty subset S of R is a subring of R if S itself forms a ring under the operations of R.   It is clear that S is a subring of R if and only if $a, b \in S$ implies $a - b \in S$ and $ab \in S$.   [Closure under subtraction implies inclusion of 0, inclusion of negatives, and hence closure under addition.]

6.209   Define an *ideal* in a ring R.

▮   A subset J of R is an ideal in R if
(i) $0 \in J$ (or: J is nonempty).
(ii) J is closed under subtraction; i.e., $a - b \in J$ for any $a, b \in J$.
(iii) J is closed under multiples from R; i.e., $ra, ar \in J$ for $a \in J, r \in R$.
With respect to (iii), J is called a *left ideal* if only $ra \in J$,   and a *right ideal* if only $ar \in J$.   Thus the term *ideal* shall mean two-sided ideal, as above.   In a commutative ring, any left or right ideal is an ideal.

6.210   Show that $\{0\}$ is an ideal in any ring R.

▮   Follows from the fact that $0 - 0 = 0$ belongs to $\{0\}$, and, for any $r \in R$,   we have $r \cdot 0 = 0 \cdot r = 0$ belongs to $\{0\}$.

6.211   Let **Z** be the ring of integers and let J_m consist of the multiples of $m \geq 2$.   Show that J_m is an ideal in **Z**.

▮   Clearly $0 \in J_m$.   Suppose ma and mb are arbitrary elements in J_m.   Then $ma - mb = m(a - b)$ also belongs to J_m.   Also, for any $r \in \mathbf{Z}$,   we have $r(ma) = (ma)r = m(ar)$ as an element of J_m.   Thus J_m is an ideal in **Z**.

6.212   Let M be the ring of real 2×2 matrices.   Give an example of a left ideal, J, which is not a right ideal, and an example of a right ideal, K, which is not a left ideal.

▮
$$J = \left\{ \begin{pmatrix} 0 & a \\ 0 & b \end{pmatrix} \right\} \qquad K = \left\{ \begin{pmatrix} a & b \\ 0 & 0 \end{pmatrix} \right\}$$

6.213   Suppose J and K are ideals in a ring R.   Prove that $J \cap K$ is an ideal in R.

▮   Since J and K are ideal, $0 \in J$ and $0 \in K$.   Hence $0 \in J \cap K$.   Now let $a, b \in J \cap K$ and let $r \in R$.   Then $a, b \in J$ and $a, b \in K$.   Since J and K are ideals,

$$a - b, ra, ar \in J \quad \text{and} \quad a - b, ra, ar \in K$$

Hence $a - b, ra, ar \in J \cap K$. Therefore $J \cap K$ is an ideal.

6.214 Let J be an ideal in a ring R with an identity element 1. Prove: (a) If $1 \in J$ then $J = R$. (b) If any unit $u \in J$ then $J = R$.

▮ (a) If $1 \in J$, then for any $r \in R$ we have $r \cdot 1 \in J$, or $r \in J$. Hence $J = R$. (b) If $u \in J$, then $u^{-1} \cdot u \in J$, or $1 \in J$. Hence $J = R$, by (a).

The following theorem uses the fact that an ideal J in a ring R is a subgroup [necessarily normal] of the additive group of R. Thus the collection of cosets $\{a + J: a \in R\}$ forms a partition of R.

Theorem 6.10: Let J be an ideal in a ring R. Then the cosets $\{a + J: a \in R\}$ form a ring under the coset operations

$$(a + J) + (b + J) = (a + b) + J \quad \text{and} \quad (a + J)(b + J) = ab + J$$

6.215 Prove Theorem 6.10. [The ring of cosets is denoted by R/J and is called the quotient ring.]

▮ The analogous Theorem 6.8 for groups shows that R/J is a commutative group under addition, with J as the zero element. Coset multiplication is well-defined, since

$$(a + J)(b + J) = ab + aJ + Jb + JJ \subseteq ab + J + J + J \subseteq ab + J$$

Associativity and the distributive laws hold in R/J, since they hold in R. Thus R/J is a ring.

6.216 Suppose J is an ideal in a commutative ring R. Show that R/J is commutative.

▮ $$(a + J)(b + J) = ab + J = ba + J = (b + J)(a + J)$$

6.217 Suppose that J is an ideal in a ring R with unity element 1, and suppose that $1 \notin J$. Show that $1 + J$ is a unity element for R/J.

▮ For any coset $a + J$, we have $(a + J)(1 + J) = a \cdot 1 + J = a + J$ and $(1 + J)(a + J) = 1 \cdot a + J = a + J$. Thus $1 + J$ is a unity element in R/J.

6.218 Define ring *homomorphism*, ring *isomorphism*.

▮ A mapping f from a ring R into a ring R' is called a *homomorphism* if $f(a + b) = f(a) + f(b)$ and $f(ab) = f(a)f(b)$, for every $a, b \in R$. [Through here notated as if they are the same, the ring operations in R' will generally differ from those in R.] In addition, if f is one-to-one and onto, then f is called an *isomorphism*.

6.219 Discuss the relation between ring homomorphisms and group homomorphisms [Section 6.6], and state the ring analog to Theorem 6.9.

▮ A ring homomorphism $f: R \to R'$ is automatically a group homomorphism on the additive structures of R and R'. Thus $f(0) = 0'$. If R and R' have unity elements 1 and $1'$, respectively, then we also require that $f(1) = 1'$ in order for f to be a ring homomorphism. We also define the kernel of f by $\text{Ker } f = \{a \in R : f(a) = 0'\}$.
The fundamental theorem on ring homomorphisms follows.

Theorem 6.11: Let $f: R \to R'$ be a ring homomorphism with kernel J. Then J is an ideal in R, and R/J is isomorphic to the image of f.

6.220 Consider the rings $R = 2\mathbf{Z}$ and $R' = 3\mathbf{Z}$ [that is, R consists of all multiples of 2, and R' consists of all multiples of 3]. Show that R is not isomorphic to R'.

▮ If $f: R \to R'$ is a ring homomorphism, then $f(2) = 3k$ for some integer k. Since f is a homomorphism, $f(4) = f(2 + 2) = f(2) + f(2) = 3k + 3k = 6k$; further, $f(4) = f(2 \cdot 2) = f(2) \cdot f(2) =$

$(3k) \cdot (3k) = 9k^2$. But $9k^2 = 6k$ and k integral imply $k = 0$. Hence $f(2) = 0$. But $f(0) = 0$. Thus f is not an isomorphism.

Problems 6.221–6.223 concern an ideal J in a ring R, and the (canonical) map $f: R \to R/J$ (recall Theorem 6.10) defined by $f(a) = a + J$.

6.221 Show that f is a ring homomorphism.

$$f(a + b) = (a + b) + J = (a + J) + (b + J) = f(a) + f(b)$$

$$f(ab) = ab + J = (a + J)(b + J) = f(a)(b)$$

6.222 Show that f is an onto mapping.

▌ Any coset $a + J$ in R/J is the image of $a \in R$.

6.223 Find the kernel K of f.

▌ The zero element of R/J is J. Thus K consists of those $a \in R$ such that $f(a) = J$, or $a + J = J$. But $a + J = J$ if and only if $a \in J$. Thus J is the kernel of f.

6.8 INTEGRAL DOMAINS, PID, UFD

All rings R in this section, and in Section 6.9, are assumed to be commutative and to have a unity element 1, unless otherwise specified.

6.224 Define a *zero divisor* in a ring R.

▌ A nonzero element $a \in R$ is a zero divisor if there exists a nonzero element b such that $ab = 0$.

6.225 Define *integral domain*.

▌ A commutative ring D with a unity element 1 is an integral domain if D has no zero divisors.

6.226 Show that the ring $\mathbf{Z}_{105}$ of the integers modulo 105 is not an integral domain.

▌ Any $\mathbf{Z}_m$ with m composite has divisors of zero; for $m = ab$ $(1 < a, b < m)$ implies $ab = 0$ in $\mathbf{Z}_m$.

6.227 Show that the ring $\mathbf{Z}_{29}$ of integers modulo 29 is an integral domain.

▌ Conversely to Problem 6.226, if m is prime then $\mathbf{Z}_m$ has no zero divisors. In fact, for $1 < a, b < m$,

$$ab = 0 + km \;\Rightarrow\; m \mid a \text{ or } m \mid b \;\Rightarrow\; a = 0 \text{ or } b = 0$$

6.228 Suppose D is an integral domain. Show that if $ab = ac$, with $a \neq 0$, then $b = c$.

▌ If $ab = ac$, then $ab - ac = 0$ and so $a(b - c) = 0$. Since $a \neq 0$ and D has no zero divisors, we must have $b - c = 0$, or $b = c$, as claimed. Thus, multiplication in D obeys the *cancellation law*.

6.229 Define a *principal ideal* in a commutative ring R with an identity element 1.

▌ Let a be any element in R. Then the set $(a) = \{ra : r \in R\}$ is an ideal; it is called the *principal ideal generated by a*.

6.230 What is a PID?

▌ PID abbreviates a *Principal Ideal Domain*. A ring is a PID if R is an integral domain and if every ideal in R is principal.

6.231 Show that $\mathbf{Z}$ is a PID.

▌ $\mathbf{Z}$ is an integral domain, since $\mathbf{Z}$ has no zero divisors. Suppose J is an ideal in $\mathbf{Z}$. If $J = \{0\}$, then $J = (0)$, the ideal generated by 0. Suppose that $J \neq \{0\}$ and that $x \neq 0$ belongs to J. Then

$-x = (-1)x$ belongs to J; hence J contains at least one positive integer. Let a be the smallest positive integer in J [recall Problem 6.18]. We claim that $J = (a)$; i.e., that J consists of all the multiples of a. Suppose $x \in J$. By the division algorithm, $x = qa + r$ where $0 \le r \le a$. Since J is an ideal and $a, x \in J$, we have $r = x - qa$ belongs to J. Since a is the smallest positive integer in J and $r < a$, we must have $r = 0$, making x a multiple of a. Thus $J = (a)$ and hence $\mathbf{Z}$ is a PID.

6.232 Define *associates* in a ring R.

▌ An element $b \in R$ is called an associate of $a \in R$ if $b = ua$ for some unit $u \in R$.

6.233 Find the associates of 4 in $\mathbf{Z}_{10}$ (the integers modulo 10).

▌ The units in $\mathbf{Z}_{10}$ are 1, 3, 7, and 9 [see Problem 6.197]. Multiply 4 by each of the units to obtain $1 \cdot 4 = 4$, $3 \cdot 4 = 2$, $7 \cdot 4 = 8$, and $9 \cdot 4 = 6$. [The multiplication is done modulo 10.]. Thus 2, 4, 6, and 8 are the associates of 4 in $\mathbf{Z}_{10}$.

6.234 Find the associates of 5 in $\mathbf{Z}_{10}$.

▌ Multiply 5 by each of the units to obtain $1 \cdot 5 = 5$, $3 \cdot 5 = 5$, $7 \cdot 5 = 5$, and $9 \cdot 5 = 5$. Thus only 5 is an associate of 5 in $\mathbf{Z}_{10}$.

6.235 Show that the relation of being associates is an equivalence relation in a ring R.

▌ Any element a is an associate of itself, since $a = 1 \cdot a$ [*reflexive law*]. Suppose b is an associate of a. Then $b = ua$, where u is a unit. Then $a = u^{-1}b$, where u^{-1} is a unit; hence a is an associate of b [*symmetric law*]. Lastly, suppose a is an associate of b and b is an associate of c: $a = u_1 b$ and $b = u_2 c$, where u_1 and u_2 are units. Then $a = u_1(u_2 c) = (u_1 u_2)c$, where the product $u_1 u_2$ is also a unit. Hence a is an associate of c [*transitive law*].

6.236 Define an *irreducible* element in an integral domain D.

▌ A nonunit $p \in D$ is irreducible if $p = ab$ implies a or b is a unit. [This is obviously an extension of the notion of "prime" in $\mathbf{Z}$.]

6.237 Define a *unique factorization domain (UFD)*.

▌ An integral domain D is a UFD if every nonunit $a \in D$ can be written uniquely [up to associates and order] as a product of irreducible elements.

6.238 Find the associates of $n \in \mathbf{Z}$

▌ The only units in $\mathbf{Z}$ are 1 and -1 [Problem 6.196]. Hence n and $-n$ are the only associates of n.

6.239 What are the irreducible elements in $\mathbf{Z}$?

▌ The prime numbers [and their negatives] are the irreducible elements in $\mathbf{Z}$.

6.240 Express 12 in $\mathbf{Z}$ as a product of irreducible elements.

▌ There are twelve such products:

$$12 = 2 \cdot 2 \cdot 3 = (-2) \cdot (-2) \cdot 3 = (-2) \cdot 2 \cdot (-3) = 2 \cdot (-2) \cdot (-3)$$
$$= 2 \cdot 3 \cdot 2 = (-2) \cdot (-3) \cdot 2 = (-2) \cdot 3 \cdot (-2) = 2 \cdot (-3) \cdot (-2)$$
$$= 3 \cdot 2 \cdot 2 = (-3) \cdot (-2) \cdot 2 = (-3) \cdot 2 \cdot (-2) = 3 \cdot (-2) \cdot (-2)$$

6.241 Is $\mathbf{Z}$ a UFD?

▌ Yes. [Although 12, etc., can be written in many ways as a product of irreducible elements, all such products differ only with respect to order or associates.]

6.242 The set $D = \{a + b\sqrt{13} : a, b \text{ integers}\}$ is an integral domain. The units of D are ± 1, $18 \pm 5\sqrt{13}$, and $-18 \pm 5\sqrt{13}$. The elements 2, $3 - \sqrt{13}$, and $-3 - \sqrt{13}$ are irreducible in D. Show that D is not a UFD.

▌ $4 = 2 \cdot 2$ and $4 = (3 - \sqrt{13})(-3 - \sqrt{13})$.

6.9 FIELDS

6.243 Define a *field*.

❚ A commutative ring F with an identity element 1 is called a *field* if every nonzero $a \in F$ is a unit. Alternatively, F is a field if its nonzero elements form a group under multiplication.

6.244 Show that a field F is an integral domain; i.e., has no zero divisors.

❚ If $ab = 0$ and $a \neq 0$, then $b = 1 \cdot b = a^{-1}ab = a^{-1} \cdot 0 = 0$.

6.245 Which of the following are fields with respect to the usual operations of addition and multiplication: the integers **Z**, the rational numbers **Q**, the real numbers **R**, the complex numbers **C**?

❚ **Z** is the classical example of an integral domain which is not a field [only 1 and -1 are units]. **Q, R**, and **C** are fields.

6.246 Let S be the set of real numbers of the form $a + \sqrt{3}$, where a and b are rational numbers. Show that S is a field.

❚ A set S of real or complex numbers is a field if S contains 0 and 1 and S is closed under addition, subtraction, multiplication, and division [except by zero]. Since $0 = 0 + 0\sqrt{3}$ and $1 = 1 + 0\sqrt{3}$, both 0 and 1 belong to S. Also:

$$(a + b\sqrt{3}) + (c + d\sqrt{3}) = (a + c) + (b + d)\sqrt{3}$$
$$(a + b\sqrt{3}) - (c + d\sqrt{3}) = (a - c) + (b - d)\sqrt{3}$$
$$(a + b\sqrt{3})(c + d\sqrt{3}) = (ac + 3bd) + (ad + bc)\sqrt{3}$$

Hence S is closed under addition, subtraction, and multiplication. We show that S is closed under division [making every nonzero element a unit] as follows:

$$\frac{(a + b\sqrt{3})}{(c + d\sqrt{3})} = \frac{(a + b\sqrt{3})(c - d\sqrt{3})}{(c + d\sqrt{3})(c - d\sqrt{3})} = \frac{ac - 3bd}{c^2 - 3d^2} + \frac{bc - ad}{c^2 - 3d^2}\sqrt{3}$$

Thus S is a field.

6.247 Let D be the ring of real 2×2 matrices of the form $\begin{pmatrix} a & -b \\ b & a \end{pmatrix}$. Show that D is isomorphic to the complex numbers **C**, whence D is a field.

❚ Let $f : \mathbf{C} \to D$ be defined by $f(a + bi) = \begin{pmatrix} a & -b \\ b & a \end{pmatrix}$. Clearly f is one-to-one and onto. Suppose $z_1 = a + bi$ and $z_2 = c + di$; then,

$$z_1 + z_2 = (a + c) + (b + d)i \qquad \text{and} \qquad z_1 z_2 = (ac - bd) + (ad + bc)i$$

Therefore,

$$f(z_1) + f(z_2) = \begin{pmatrix} a & -b \\ b & a \end{pmatrix} + \begin{pmatrix} c & -d \\ d & c \end{pmatrix} = \begin{pmatrix} a + c & -(b + d) \\ b + d & a + c \end{pmatrix} = f(z_1 + z_2)$$

$$f(z_1)f(z_2) = \begin{pmatrix} a & -b \\ b & a \end{pmatrix}\begin{pmatrix} c & -d \\ d & c \end{pmatrix} = \begin{pmatrix} ac - bd & -(ad + bc) \\ ad + bc & ac - bd \end{pmatrix} = f(z_1 z_2)$$

Lastly, $f(1) = f(1 + 0i) = I$, the identity matrix. Thus f is an isomorphism.

Theorem 6.12: A finite integral domain D is a field.

6.248 Prove Theorem 6.12.

❚ Suppose D has n elements, say $D = \{a_1, a_2, \ldots, a_n\}$. Let a be any nonzero element of D, and consider the n elements $aa_1, aa_2, \ldots, aa_n$. Since $a \neq 0$, we have $aa_i = aa_j$ implies $a_i = a_j$ [Problem 6.244]. Thus the n elements above are distinct, and so they must be a rearrangement of the elements D. One of them, say aa_k, must equal the identity element 1 of D; that is, $aa_k = 1$. Thus a_k is the inverse of a. Since a was any nonzero element of D, we have that D is a field.

6.249 Show that $\mathbf{Z}_p$ is a field where p is a prime number.

❚ $\mathbf{Z}_p$ is an integral domain [Problem 6.243] and finite.

6.250 Show that the only ideal J in a field F is $\{0\}$ or F itself.

❚ If $J \neq \{0\}$, then J contains a nonzero element a. Since F is a field, a is a unit. By Problem 6.214(b), $J = F$.

6.251 Suppose $f: K \to K'$ is a homomorphism from a field K to a field K'. Show that f is an *embedding*; i.e., f is one-to-one.

❚ $J = \operatorname{Ker} f$, which is an ideal in K, by Theorem 6.11. If $J = K$, then $f(1) = 0'$. But, since f is a homomorphism, we require $f(1) = 1'$. Hence $J \neq K$, and so $J = \{0\}$ [by Problem 6.250]. Suppose $f(a) = f(b)$. Then $f(a - b) = f(a) - f(b) = 0$. Hence $a - b$ belongs to J, and so $a - b = 0$, or $a = b$. Accordingly, f is one-to-one.

6.252 Let D be an integral domain. Define the *field of quotients of D*.

❚ Let S consist of all ordered pairs [quotients] a/b, where $a, b \in D$ and $b \neq 0$. Define $a/b = c/d$ if $ad = bc$. [This is an equivalence relation.] Let $F(D)$ be the set of equivalence classes $[a/b]$, with the operations of addition and multiplication defined by

$$[a/b] + [c/d] = [(ad + bc)/(bd)] \qquad \text{and} \qquad [a/b] \cdot [c/d] = [(ac)/(bd)]$$

Then $F(D)$ is the desired field.

6.253 What is the field of quotients of the integral domain $\mathbf{Z}$ of integers?

❚ $F(\mathbf{Z}) = \mathbf{Q}$, the field of rational numbers.

6.254 Let $K = D[x]$, the integral domain of polynomials in x with real coefficients. What is the field of quotients of K?

❚ $F(K)$ is the field of rational functions of the form $f(x)/g(x)$, where $f(x)$ and $g(x) \neq 0$ are polynomials.

6.255 Let D be an integral domain. Show how D is embedded in its field of quotients $F(D)$.

❚ Let $f: D \to F(D)$ be defined by $f(a) = [a/1]$. Then f is an embedding; i.e., f is a homomorphism and f is one-to-one. [For example, we identify an integer n in $\mathbf{Z}$ with the fraction $n/1$ in $\mathbf{Q}$.]

6.256 Define a *maximal* ideal K in a ring R.

❚ K is a maximal ideal in R if $K \neq R$ and if no ideal J lies strictly between K and R; i.e., if $K \subseteq J \subseteq R$ implies $K = J$ or $J = R$.

6.257 Suppose K is a maximal ideal in a commutative ring R with identity element 1. Prove that the quotient ring R/K is a field.

❚ Since $K \neq R$, we have $1 \notin K$ [Problem 6.214(a)]. By Problem 6.217, the coset $1 + K$ is an identity element for R/K. By Problem 6.216, R/K is commutative. It remains to show that any coset other than K [the zero element of R/K] has a multiplicative inverse in R/K. Suppose $a + K \neq K$. Then $a \notin K$. Let $J = \{ra + sk : r, s \in R, k \in K\}$. Then J is an ideal containing both a and K. Since $a \notin K$, we have $K \neq J$. Therefore, since K is maximal, $J = R$. Thus $1 \in J$, and so there exist $r_0, s_0 \in R$ and $k_0 \in K$ such that $1 = r_0 a + s_0 k_0$. It follows that

$$1 + k = r_0 a + s_0 k_0 + K = r_0 a + K = (r_0 + K)(a + K)$$

i.e., $r_0 + K$ is the multiplicative inverse of $a + K$. Thus R/K is a field.

CHAPTER 7
Vector Spaces and Subspaces

The definition of a vector space involves an arbitrary field (see Section 6.9) whose elements are called *scalars*. We adopt the following notation (unless otherwise stated or implied):

$$K \quad \text{the field of scalars}$$
$$a, b, c, \text{ or } k \quad \text{the elements of } K$$
$$V \quad \text{the given vector space}$$
$$u, v, w \quad \text{the elements of } V$$

Nothing essential is lost if the reader assumes that K is the real field **R** or the complex field **C**.

7.1 VECTOR SPACES

7.1 Define a vector space.

▌ Let K be a given field and let V be a nonempty set with rules of addition and scalar multiplication which assigns to any $u, v \in V$ a *sum* $U + v \in V$ and to any $u \in V$, $k \in K$ a *product* $ku \in V$. Then V is called a *vector space over K* (and the elements of V are called *vectors*) if the following axioms hold:

$[A_1]$: For any vectors $u, v, w \in V$, $(u + v) + w = u + (v + w)$.

$[A_2]$: There is a vector in V, denoted by 0 and called the *zero vector*, for which $u + 0 = u$ for any vector $u \in V$.

$[A_3]$ For each vector $u \in V$ there is a vector in V, denoted by $-u$, for which $u + (-u) = 0$.

$[A_4]$: For any vectors $u, v \in V$, $u + v = v + u$.

$[M_1]$: For any scalar $k \in K$ and any vectors $u, v \in V$, $k(u + v) = ku + kv$.

$[M_2]$: For any scalars $a, b \in K$ and any vector $u \in V$, $(a + b)u = au + bu$.

$[M_3]$: For any scalars $a, b \in K$ and any vector $u \in V$, $(ab)u = a(bu)$.

$[M_4]$: For the unit scalar $1 \in K$, $1u = u$ for any vector $u \in V$.

7.2 Show that in a vector space V. (a) Any sum of vectors of the form $v_1 + v_2 + \cdots + v_m$ requires no parentheses and does not depend upon the order of the summands. (b) The zero vector 0 is unique. (c) The negative $-u$ of a vector u is unique. (d) The cancellation law holds, that is, for any vectors $u, v, w \in V$, $u + w = v + w$ implies $u = v$.

▌ The first four axioms of a vector space V indicate that V is a commutative group under addition. The above properties follow from this fact.

7.3 How is subtraction defined in a vector space V?

▌ Subtraction is defined by $u - v \equiv u + (-v)$.

7.4 In the statement of axiom $[M_2]$, $(a + b)u = au + bu$, which operation does each plus sign represent?

▌ The $+$ in $(a + b)u$ denotes the addition of the two scalars a and b; hence it represents the addition operation in the field K. On the other hand, the $+$ in $au + bu$ denotes the addition of the two vectors au and bu; hence it represents the operation of vector addition. Thus each $+$ represents a different operation.

7.5 In the statement of axiom $[M_3]$, $(ab)u = a(bu)$, which operation does each product represent?

▌ In $(ab)u$ the product ab of the scalars a and b denotes multiplication in the field K, whereas the product of the scalar ab and the vector u denotes scalar multiplication.

In $a(bu)$ the product bu of the scalar b and the vector u denotes scalar multiplication; also, the product of the scalar a and the vector bu denotes scalar multiplication.

7.6   Let $V = K^n$, where K is an arbitrary field. Show how V is made into a vector space over K.

▌ Vector addition and scalar multiplication is defined by $(a_1, a_2, \ldots, a_n) + (b_1, b_2, \ldots, b_n) = (a_1 + b_1,$ $a_2 + b_2, \ldots, a_n + b_n)$ and $k(a_1, a_2, \ldots, a_n) = (ka_1, ka_2, \ldots, ka_n)$ where $a_i, b_i, k \in K$. The zero vector in V is the n-tuple of zeros, $0 = (0, 0, \ldots, 0)$. The proof that $V = K^n$ is a vector space is identical to the proofs in Section 1.3 for $\mathbf{R}^n$.

7.7   Let $K = \mathbf{Z}_3$, the integers modulo 3. How many elements are there in the vector space $V = K^4$?

▌ There are three choices, 0, 1 or 2, for each of the four components of a vector in V. Hence V has $3 \cdot 3 \cdot 3 \cdot 3 = 3^4 = 81$ elements.

7.8   Let V be the set of all $m \times n$ matrices with entries from an arbitrary field K. Show how V is made into a vector space.

▌ V is a vector space over K with respect to the operations of matrix addition and scalar multiplication. The proof of this fact is identical to the proof of Theorem 2.3 on $m \times n$ matrices over $\mathbf{R}$.

7.9   Let V be the set of all polynomials $a_0 + a_1 t + a_2 t^2 + \cdots + a_n t^n$ with coefficients a_i from a field K. Show how V is made into a vector space.

▌ V is a vector space over k with respect to the usual operations of addition of polynomials and multiplication by a constant.

7.10   Show that $V = \mathbf{R}^2$ is not a vector space over $\mathbf{R}$ with respect to the following operations of vector addition and scalar multiplication: $(a, b) + (c, d) = (a + c, b + d)$ and $k(a, b) = (ka, b)$. Show that one of the axioms of a vector space does not hold.

▌ Let $r = 1$, $s = 2$, $v = (3, 4)$. Then

$$(r + s)v = 3(3, 4) = (9, 4)$$
$$rv + sv = 1(3, 4) + 2(3, 4) = (3, 4) + (6, 4) = (9, 8)$$

Since $(r + s)v \neq rv + sv$, axiom $[M_2]$ does not hold.

7.11   Show that $V = \mathbf{R}^2$ is not a vector space over $\mathbf{R}$ with respect to the operations: $(a, b) + (c, d) = (a, b)$ and $k(a, b) = (ka, kb)$. Show that one of the axioms of a vector space does not hold.

▌ Let $v = (1, 2)$, $w = (3, 4)$. Then

$$v + w = (1, 2) + (3, 4) = (1, 2)$$
$$w + v = (3, 4) + (1, 2) = (3, 4)$$

Since $v + w \neq w + v$, axiom $[A_4]$ does not hold.

7.12   Show that $V = \mathbf{R}^2$ is not a vector space over $\mathbf{R}$ with respect to the operations: $(a, b) + (c, d) = (a + c, b + d)$ and $k(a, b) = (k^2 a, k^2 b)$. Show that one of the axioms of a vector space does not hold.

▌ Let $r = 1$, $s = 2$, $v = (3, 4)$. Then

$$(r + s)v = 3(3, 4) = (27, 36)$$
$$rv + sv = 1(3, 4) + 2(3, 4) = (3, 4) + (12, 16) = (15, 20)$$

Thus $(r + s)v \neq rv + sv$, and so axiom $[M_2]$ does not hold.

7.13   Suppose E is a field which contains a subfield K. Show how E may be viewed as a vector space over K.

▌ Let the usual addition in E be the vector addition and let the scalar product kv of $k \in K$ and $v \in E$ be the product of k and v as elements of the field E. Then E is a vector space over K.

7.14   Is the real field $\mathbf{R}$ a vector space: (a) Over $\mathbf{Q}$? (b) Over $\mathbf{Z}$? (c) Over $\mathbf{C}$?

▌ (a) Yes, since $\mathbf{Q}$ is a subfield of $\mathbf{R}$. (b) No, since $\mathbf{Z}$ is not a field. (c) No, since $\mathbf{C}$ is not a subfield of $\mathbf{R}$.

7.15   Is the complex field **C** a vector space: (a) Over **R**?   (b) Over **Q**?   (c) Over **Z**?   (d) Over **C**?

▌ (a) Yes, since **R** is a subfield of **C**.   (b) Yes, since **Q** is a subfield of **C**.   (c) No, since **Z** is not a field. (d) Yes, every field is a vector space over itself.

7.16   Is $\mathbf{Z}_7$ a vector space over $\mathbf{Z}_5$?

▌ No. $\mathbf{Z}_5 = \{0, 1, 2, 3, 4\}$ is not a subfield of $\mathbf{Z}_7 = \{0, 1, 2, \ldots, 6\}$ since the operations are different, e.g., $2 + 3 = 0$ in $\mathbf{Z}_5$ but $2 + 3 \neq 0$ in $\mathbf{Z}_7$.   Hence $\mathbf{Z}_7$ is not a vector space over $\mathbf{Z}_5$.

> **Theorem 7.1:**   Let V be a vector space over a field K.
>   (i)   For any scalar $k \in K$ and $0 \in V$, $k0 = 0$.
>   (ii)   For $0 \in K$ and any vector $u \in V$, $0u = 0$.
>   (iii) If $ku = 0$, where $k \in K$ and $u \in V$, then $k = 0$ or $u = 0$.
>   (iv) For any $k \in K$ and any $u \in V$, $(-k)u = k(-u) = -ku$.

7.17   Prove (i) of Theorem 7.1: $k0 = 0$.

▌ By axiom $[A_2]$ with $u = 0$,   we have $0 + 0 = 0$.   Hence by axiom $[M_1]$, $k0 = k(0 + 0) = k0 + k0$. Adding $-k0$ to both sides gives the desired result.

7.18   Prove (ii) of Theorem 7.1: $0u = 0$.

▌ By a property of K, $0 + 0 = 0$.   Hence by axiom $[M_2]$, $0u = (0 + 0)u = 0u + 0u$.   Adding $-0u$ to both sides yields the required result.

7.19   Prove (iii) of Theorem 7.1:   If $ku = 0$,   then $k = 0$ or $u = 0$.

▌ Suppose $ku = 0$ and $k \neq 0$.   Then there exists a scalar k^{-1} such that $k^{-1}k = 1$;   hence $u = 1u = (k^{-1}k)u = k^{-1}(ku) = k^{-1}0 = 0$.

7.20   Prove (iv) of Theorem 7.1: $(-k)u = k(-u) = -ku$.

▌ Using $u + (-u) = 0$,   we obtain $0 = k0 = k(u + (-u)) = ku + k(-u)$.   Adding $-ku$ to both sides gives $-ku = k(-u)$.
    Using $k + (-k) = 0$,   we obtain $0 = 0u = (k + (-k))u = ku + (-k)u$.   Adding $-ku$ to both sides yields $-ku = (-k)u$.   Thus $(-k)u = k(-u) = -ku$.

7.21   Show that for any scalar k and any vectors u and v, $k(u - v) = ku - kv$.

▌ Use the definition of subtraction, $u - v \equiv u + (-v)$,   and the result $k(-v) = -kv$ to obtain $k(u - v) = k(u + (-v)) = ku + k(-v) = ku + (-kv) = ku - kv$.

> **Theorem 7.2:**   Let K be an arbitrary field and let X be any nonempty set.   Let V be the set of all functions from X into K.   The sum of any two functions $f, g \in V$ is the function $f + g \in V$ defined by
>
> $$(f + g)(x) = f(x) + g(x) \qquad \forall x \in X$$
>
> [The symbol ∀ means "for every."]   Then V is a vector space over K, that is, V satisfies the eight axioms of a vector space.   [V is nonempty since X is nonempty.]

7.22   Prove V in Theorem 7.2 satisfies axiom $[A_1]$.

▌ Let $f, g, h \in V$.   To show that $(f + g) + h = f + (g + h)$,   it is necessary to show that the function $(f + g) + h$ and the function $f + (g + h)$ both assign the same value to each $x \in X$.   Now,

$$((f + g) + h)(x) = (f + g)(x) + h(x) = (f(x) + g(x)) + h(x) \qquad \forall x \in X$$
$$(f + (g + h))(x) = f(x) + (g + h)(x) = f(x) + (g(x) + h(x)) \qquad \forall x \in X$$

But $f(x)$, $g(x)$, and $h(x)$ are scalars in the field K where addition of scalars is associative; hence $(f(x) + g(x)) + h(x) = f(x) + (g(x) + h(x))$.   Accordingly, $(f + g) + h = f + (g + h)$.

7.23   Prove V in Theorem 7.2 satisfies axiom $[A_2]$.

▌   Let 0 denote the zero function: $0(x) = 0$, $\forall x \in X$.   Then for any function $f \in V$,

$$(f + 0)(x) = f(x) + 0(x) = f(x) + 0 = f(x)   \qquad \forall x \in X$$

Thus $f + 0 = f$,   and 0 is the zero vector in V.

7.24   Prove V in Theorem 7.2 satisfies axiom $[A_3]$.

▌   For any function $f \in V$,   let $-f$ be the function defined by $(-f)(x) = f(x)$.   Then,

$$(f + (f))(x) = f(x) + (-f)(x) = f(x) - f(x) = 0 = 0(x)   \qquad \forall x \in X$$

Hence $f + (-f) = 0$.

7.25   Prove V in Theorem 7.2 satisfies axiom $[A_4]$.

▌   Let $f, g \in V$.   Then

$$(f + g)(x) = f(x) + g(x) = g(x) + f(x) = (g + f)(x)   \qquad \forall x \in X$$

Hence $f + g = g + f$.   [Note that $f(x) + g(x) = g(x) + f(x)$ follows from the fact that $f(x)$ and $g(x)$ are scalars in the field K where addition is commutative.]

7.26   Prove V in Theorem 7.2 satisfies axiom $[M_1]$.

▌   Let $f, g \in V$ and $k \in K$.   Then

$$(k(f + g))(x) = k((f + g)(x)) = k(f(x) + g(x)) = kf(x) + kg(x)$$
$$= (kf)(x) + (kg)(x) = (kf + kg)(x)   \qquad \forall x \in X$$

Hence $k(f + g) = kf + kg$.   [Note that $k(f(x) + g(x)) = kf(x) + kg(x)$ follows from the fact that k, $f(x)$, and $g(x)$ are scalars in the field K where multiplication is distributive over addition.]

7.27   Prove V in Theorem 7.2 satisfies axiom $[M_2]$.

▌   Let $f \in V$ and $a, b \in K$.   Then

$$((a + b)f)(x) = (a + b)f(x) = af(x) + bf(x) = (af)(x) + bf(x)$$
$$= (af + bf)(x)   \qquad \forall x \in X$$

Hence $(a + b)f = af + bf$.

7.28   Prove V in Theorem 7.2 satisfies axiom $[M_3]$.

▌   Let $f \in V$ and $a, b \in K$.   Then,

$$((ab)f)(x) = (ab)f(x) = a(bf(x)) = a(bf)(x) = (a(bf))(x)   \qquad \forall x \in X$$

Hence $(ab)f = a(bf)$.

7.29   Prove V in Theorem 7.2 satisfies axiom $[M_4]$.

▌   Let $f \in V$.   Then, for the unit $1 \in K$, $(1f)(x) = 1f(x) = f(x)$, $\forall x \in X$.   Hence $1f = f$.

7.30   Let V be the set of infinite sequences $(a_1, a_2, \ldots)$ with entries from a field K.   Show how V is made into a vector space.

▌   Vector addition in V and scalar multiplication on V is defined by

$$(a_1, a_2, \ldots) + (b_1, b_2, \ldots) = (a_1 + b_1, a_2 + b_2, \ldots)$$
$$k(a_1, a_2, \ldots) = (ka_1, ka_2, \ldots)$$

where $a_1, b_j, k \in K$.   The proof that V is a vector space is similar to the proofs in Section 1.3 for $\mathbf{R}^n$.

7.31 What is the zero vector 0 and the negative of a vector $u = (a_1, a_2, \ldots)$ in the vector space V of Problem 7.30?

▮ $0 = (0, 0, \ldots)$, the sequences of 0s, and $-u = (-a_1, a_2, \ldots)$, the sequence of negatives of the entries in u.

7.32 Let V be the set of ordered pairs (a, b) of real numbers with addition in V and scalar multiplication on V defined by $(a, b) + (c, d) = (a + c, b + d)$ and $k(a, b) = (ka, 0)$. Which of the eight axioms of a vector space are satisfied by V?

▮ V satisfies all of the axioms of a vector space except $[M_4]$: $1u = u$.

7.33 Show that axiom $[M_4]$ is not a consequence of the other axioms of a vector space.

▮ Since the algebraic structure V in Problem 7.32 satisfies all the axioms except $[M_4]$, one cannot derive $[M_4]$ from the other axioms.

7.34 Suppose E is a field containing a subfield K. Show how the set $V = E^n$ may be viewed as a vector space over K.

▮ Define vector addition and scalar multiplication in V as follows:

$$(a_1, a_2, \ldots, a_n) + (b_1, b_2, \ldots, b_n) = (a_1 + b_1, a_2 + b_2, \ldots, a_n + b_n)$$
$$k(a_1, a_2, \ldots, a_n) = (ka_1, ka_2, \ldots, ka_n)$$

where $a_i, b_j \in E$ and $k \in K$. Then V is a vector space over K. [This vector space is different than the vector space E^n viewed as a vector space over E.]

7.35 Can $\mathbf{C}^2$ (pairs of complex numbers) be defined as a vector space: (a) Over $\mathbf{R}$? (b) Over $\mathbf{Q}$? (c) Over $\mathbf{C}$? (d) Over $\mathbf{Z}$?

▮ By Problem 7.34: (a) yes, (b) yes, (c) yes. (d) Since $\mathbf{Z}$ is not a field, no.

7.36 Can $\mathbf{R}^2$ be defined as a vector space: (a) Over $\mathbf{Q}$? (b) Over $\mathbf{R}$? (c) Over $\mathbf{C}$?

▮ By Problem 7.34: (a) yes, (b) yes. (c) Since $\mathbf{C}$ is not a subfield of $\mathbf{R}$, no.

7.37 How are "dot product," length, and orthogonality defined in an abstract vector space V?

▮ The dot product, and related notions of length and orthogonality, are not considered as part of the fundamental vector space structure, but as an additional structure which may or may not be introduced. Such spaces shall be investigated in Chapters 14 and 20.

7.2 SUBSPACES OF VECTOR SPACES

7.38 Define a subspace of a vector space.

▮ Let W be a subset of a vector space over a field K. W is called a *subspace* of V if W is itself a vector space over K with respect to the operations of vector addition and scalar multiplication V.

> ***Theorem 7.3:*** W is a subspace of V if and only if
> (i) W is nonempty (or: $0 \in W$).
> (ii) W is closed under vector addition: $v, w \in W$ implies $v + w \in W$.
> (iii) W is closed under scalar multiplication: $v \in W$ implies $kv \in W$ for every $k \in K$.

7.39 Prove Theorem 7.3.

▮ Suppose W satisfies (i), (ii), and (iii). By (i), W is nonempty; and by (ii) and (iii), the operations of vector addition and scalar multiplication are well-defined for W. Moreover, the axioms $[A_1]$, $[A_4]$, $[M_1]$, $[M_2]$, $[M_3]$, and $[M_4]$ hold in W since the vectors in W belong to V. Hence we need only show that $[A_2]$ and $[A_3]$ also hold in W. By (i), W is nonempty, say $u \in W$. Then by (iii), $0u = 0 \in W$ and $v + 0 = v$ for every $v \in W$. Hence W satisfies $[A_2]$. Lastly, if $v \in W$ then $(-1)v = -v \in W$ and $v + (-v) = 0$; hence W satisfies $[A_3]$. Thus W is a subspace of V.

Conversely, if W is a subspace of V then clearly (i), (ii), and (iii) hold.

Corollary 7.4: W is a subspace of V if and only if (i) $0 \in W$ (or $W \neq \emptyset$)   and (ii) $v, w \in W$ implies $av + bw \in W$ for every $a, b \in K$.

7.40   Prove Corollary 7.4.

⬛   Suppose W satisfies (i) and (ii).   Then, by (i), W is nonempty.   Furthermore, if $v, w \in W$ then, by (ii), $v + w = 1v + 1w \in W$;   and if $v \in W$ and $k \in K$ then, by (ii), $kv = kv + 0v \in W$.   Thus by Theorem 7.3, W is a subspace of V.
Conversely, if W is a subspace of V, then clearly (i) and (ii) hold in W.

7.41   Let V be any vector space.   Describe the "smallest" and "largest" subspaces of V.

⬛   The set $\{0\}$ consisting of the zero vector alone is a subspace of V contained in every other subspace of V, and the entire space V is a subspace of V which contains every other subspace of V.

Problems 7.42–7.46 refer to the vector space $V = \mathbf{R}^3$.

7.42   Show that W is a subspace of $V = \mathbf{R}^3$ where W is the xy plane which consists of those vectors whose third component is 0, i.e., $W = \{(a, b, 0): a, b \in \mathbf{R}\}$.

⬛ $0 = (0, 0, 0) \in W$ since the third component of 0 is 0.   For any vectors $v = (a, b, 0)$, $w = (c, d, 0)$ in W, and any scalars (real numbers) k and k', $kv + k'w = k(a, b, 0) + k'(c, d, 0) = (ka, kb, 0) + (k'c, k'd, 0) = (ka + k'c, kb + k'd, 0)$.   Thus $kv + k'w \in W$,   and so W is a subspace of V.

7.43   Show that W is a subspace of $V = \mathbf{R}^3$ where W consists of those vectors each whose sum of components is zero, i.e., $W = \{(a, b, c): a + b + c = 0\}$.

⬛ $0 = (0, 0, 0) \in W$ since $0 + 0 + 0 = 0$.   Suppose $v = (a, b, c)$, $w = (a', b', c')$ belong to W, i.e., $a + b + c = 0$ and $a' + b' + c' = 0$.   Then for any scalars k and k', $kv + k'w = k(a, b, c) + k'(a', b', c') = (ka, kb, kc) + (k'a', k'b', k'c') = (ka + k'a', kb + k'b', kc + k'c')$ and, furthermore,

$$(ka + k'a') + (kb + k'b') + (kc + k'c') = k(a + b + c) + k'(a' + b' + c') = k0 + k'0 = 0.$$

Thus $kv + k'w \in W$,   and so W is a subspace of V.

7.44   Show that W is not a subspace of $V = \mathbf{R}^3$ where W consists of those vectors whose first component is nonnegative, i.e., $W = \{(a, b, c): a \geq 0\}$.

⬛   Show that one of the properties of, say, Theorem 7.3 does not hold. $v = (1, 2, 3) \in W$ and $k = -5 \in \mathbf{R}$.   But $kv = -5(1, 2, 3) = (-5, -10, -15)$ does not belong to W since -5 is negative.   Hence W is not a subspace of V.

7.45   Show that W is not a subspace of $V = \mathbf{R}^3$ where W consists of those vectors whose length does not exceed 1,   i.e., $W = \{(a, b, c): a^2 + b^2 + c^2 \leq 1\}$.

⬛ $v = (1, 0, 0) \in W$ and $w = (0, 1, 0) \in W$.   But $v + w = (1, 0, 0) + (0, 1, 0) = (1, 1, 0)$ does not belong to W since $1^2 + 1^2 + 0^2 = 2 > 1$.   Hence W is not a subspace of V.

7.46   Show that W is not a subspace of $V = \mathbf{R}^3$ where W consists of those vectors whose components are rational numbers, i.e., $W = \{(a, b, c): a, b, c \in \mathbf{Q}\}$.

⬛ $v = (1, 2, 3) \in W$ and $k = \sqrt{2} \in \mathbf{R}$.   But $kv = \sqrt{2}(1, 2, 3) = (\sqrt{2}, 2\sqrt{2}, 3\sqrt{2})$ does not belong to W since its components are not rational numbers.   Hence W is not a subspace of V.

Problems 7.47–7.48 refer to the vector space V of all n-square matrices over a field K.

7.47   Show that W is a subspace of V where W consists of the symmetric matrices, i.e., all matrices $A = (a_{ij})$ for which $a_{ij} = a_{ji}$.

▮ $0 \in W$ since all entries of 0 are 0 and hence equal. Now suppose $A = (a_{ij})$ and $B = (b_{ij})$ belong to W, i.e., $a_{ji} = a_{ij}$ and $b_{ji} = b_{ij}$. For any scalars $a, b \in K$, $aA + bB$ is the matrix whose ij-entry is $aa_{ij} + bb_{ij}$. But $aa_{ji} + bb_{ji} = aa_{ij} + bb_{ij}$. Thus $aA + bB$ is also symmetric, and so W is a subspace of V.

7.48   Show that W is a subspace of V where W consists of all matrices which commute with a given matrix T; that is, $W = \{A \in V : AT = TA\}$.

▮ $0 \in W$ since $0T = 0 = T0$. Now suppose $A, B \in W$; that is, $AT = TA$ and $BT = TB$. For any scalars $a, b \in K$, $(aA + bB)T = (aA)T + (bB)T = a(AT) + b(BT) = a(TA) + b(TB) = T(aA) + T(bB) = T(aA + bB)$. Thus $aA + bB$ commutes with T, i.e., belongs to W; hence W is a subspace of V.

Problems 7.48–7.50 refer to the vector space V of all 2×2 matrices over the real field **R**.

7.49   Show that W is not a subspace of V where W consists of all matrices with zero determinant.

▮ [Recall that $\det\begin{pmatrix} a & b \\ c & d \end{pmatrix} = ad - bc$.] The matrices $A = \begin{pmatrix} 1 & 0 \\ 0 & 0 \end{pmatrix}$ and $B = \begin{pmatrix} 0 & 0 \\ 0 & 1 \end{pmatrix}$ belong to W since $\det(A) = 0$ and $\det(B) = 0$. But $A + B = \begin{pmatrix} 1 & 0 \\ 0 & 1 \end{pmatrix}$ does not belong to W since $\det(A + B) = 1$. Hence W is not a subspace of V.

7.50   Show that W is not a subspace of V where W consists of all matrices A for which $A^2 = A$.

▮ The unit matrix $I = \begin{pmatrix} 1 & 0 \\ 0 & 1 \end{pmatrix}$ belongs to W since

$$I^2 = \begin{pmatrix} 1 & 0 \\ 0 & 1 \end{pmatrix}\begin{pmatrix} 1 & 0 \\ 0 & 1 \end{pmatrix} = \begin{pmatrix} 1 & 0 \\ 0 & 1 \end{pmatrix} = I$$

But $2I = \begin{pmatrix} 2 & 0 \\ 0 & 2 \end{pmatrix}$ does not belong to W since

$$(2I)^2 = \begin{pmatrix} 2 & 0 \\ 0 & 2 \end{pmatrix}\begin{pmatrix} 2 & 0 \\ 0 & 2 \end{pmatrix} = \begin{pmatrix} 4 & 0 \\ 0 & 4 \end{pmatrix} \neq 2I$$

Hence W is not a subspace of V.

Problems 7.51–7.57 refer to the vector space V of all functions from the real field **R** into **R**. Here **0** denotes the zero function: $\mathbf{0}(x) = 0$, for every $x \in \mathbf{R}$.

7.51   Show that W is a subspace of V where $W = \{f : f(3) = 0\}$, i.e., W consists of those functions which map 3 into 0.

▮ $0 \in W$ since $\mathbf{0}(3) = 0$. Suppose $f, g \in W$, i.e., $f(3) = 0$ and $g(3) = 0$. Then for any real numbers a and b, $(af + bg)(3) = af(3) + bg(3) = a0 + b0 = 0$. Hence $af + bg \in W$, and so W is a subspace of V.

7.52   Show that W is a subspace of V where $W = \{f : f(7) = f(1)\}$, i.e., W consists of those functions which assign the same value to 7 and 1.

▮ $0 \in W$ since $\mathbf{0}(7) = 0 = \mathbf{0}(1)$. Suppose $f, g \in W$, i.e., $f(7) = f(1)$ and $g(7) = g(1)$. Then, for any real numbers a and b, $(af + bg)(7) = af(7) + bg(7) = af(1) + bg(1) = (af + bg)(1)$. Hence $af + bg \in W$, and so W is a subspace of V.

7.53   Show that W is a subspace of V where W consists of the odd functions, i.e., those functions f for which $f(-x) = -f(x)$.

▮ $0 \in W$ since $\mathbf{0}(-x) = 0 = -0 = -\mathbf{0}(x)$. Suppose $f, g \in W$, i.e., $f(-x) = -f(x)$ and $g(-x) = -g(x)$. Then for any real numbers a and b, $(af + bg)(-x) = af(-x) + bg(-x) = -af(x) - bg(x) = -(af(x) + bg(x)) = -(af + bg)(x)$. Hence $af + bg \in W$, and so W is a subspace of V.

7.54   Show that W is not a subspace of V where $W = \{f : f(7) = 2 + f(1)\}$.

▌ Suppose $f, g \in W$, i.e., $f(7) = 2 + f(1)$ and $g(7) = 2 + g(1)$. Then $(f + g)(7) = f(7) + g(7) = 2 + f(1) + 2 + g(1) = 4 + f(1) + g(1) = 4 + (f + g)(1) \neq 2 + (f + g)(1)$. Hence $f + g \notin W$, and so W is not a subspace of V.

7.55 Show that W is not a subspace where W consists of all nonnegative functions, i.e., all function f for which $f(x) \geq 0$, $\forall x \in \mathbf{R}$.

▌ Let $k = -2$ and let $f \in V$ be defined by $f(x) = x^2$. Then $f \in W$ since $f(x) = x^2 \geq 0$, $\forall x \in \mathbf{R}$. But $(kf)(5) = kf(5) = (-2)(5^2) = -50 < 0$. Hence $kf \notin W$, and so W is not a subspace of V.

7.56 Show that W is a subspace of V where W consists of the bounded functions. [A function $f \in V$ is *bounded* if there exists $M \in \mathbf{R}$ such that $|f(x)| \leq M$ for every $x \in X$.]

▌ Clearly $\mathbf{0}$ is bounded since $\mathbf{0}(x) = 0$ for every $x \subseteq \mathbf{R}$. Now let $f, g \in W$ with M_f and M_g bounds for f and g, respectively. Then for any scalars a, b and $\forall x \in \mathbf{R}$, $|(af + bg)(x)| = |af(x) + bg(x)| \leq |af(x)| + |bg(x)| = |a| \, |f(x)| + |b| \, |g(x)| \leq |a| M_f + |b| M_g$. That is, $|a| M_f + |b| M_g$ is a bound for the function $af + bg$. Thus W is a subspace of V.

7.57 Is W a subspace of V where (a) W consists of the continuous functions? (b) W consists of the differentiable functions?

▌ One proves in calculus that the constant function $\mathbf{0}$ is continuous and differentiable. Also, one proves in calculus that if f and g are continuous (differentiable) functions then, for any real numbers a and b, the function $af + bg$ is continuous (differentiable). Thus (a) yes, (b) yes.

7.58 Let V be the vector space of polynomials $a_0 + a_1 t + a_2 t^2 + \cdots + a_n t^n$ with real coefficients, i.e., $a_i \in \mathbf{R}$. Determine whether or not W is a subspace of V where
(a) W consists of all polynomials with integral coefficients.
(b) W consists of all polynomials with degree ≤ 3.
(c) W consists of all polynomials $b_0 + b_1 t^2 + b_2 t^4 + \cdots + b_n t^{2n}$, i.e., polynomials with only even powers of t.

▌ (a) No, since scalar multiples of polynomials in W do not always belong to W. For example, $v = 3 + 5t + 7t^2 \in W$ but $\frac{1}{2}v = \frac{3}{2} + \frac{5}{2}t + \frac{7}{2}t^2 \notin W$. [Observe that W is "closed" under vector addition, i.e., sums of elements in W belong to W.] (b) and (c). Yes. For, in each case, W is nonempty, the sum of elements in W belong to W, and the scalar multiples of any element in W belong to W.

Theorem 7.5: The intersection of any number of subspaces of a vector space V is a subspace of V.

7.59 Prove Theorem 7.5.

▌ Let $\{W_i : i \in \mathbf{I}\}$ be a collection of subspaces of V and let $W = \cap (W_i : i \in \mathbf{I})$. Since each W_i is a subspace, $0 \in W_i$ for every $i \in \mathbf{I}$. Hence $0 \in W$. Suppose $u, v \in W$. Then $u, v \in W_i$ for every $i \in \mathbf{I}$. Since each W_i is a subspace, $au + bv \in W_i$ for each $i \in \mathbf{I}$. Hence $au + bv \in W$. Thus W is a subspace of V.

7.60 Show that the union $W_1 \cup W_2$ of subspaces of a vector space V need not be a subspace of V.

▌ Let $V = \mathbf{R}^2$ and let $W_1 = \{(a, 0): a \in \mathbf{R}\}$ and $W_2 = \{(0, b): b \in \mathbf{R}\}$. That is, W_1 is the x axis and W_2 is the y axis in $\mathbf{R}^2$. Then W_1 and W_2 are subspaces of V. Let $u = (1, 0)$ and $v = (0, 1)$. Then u and v both belong to the union $W_1 \cup W_2$, but $u + v = (1, 1)$ does not belong to $W_1 \cup W_2$. Hence $W_1 \cup W_2$ is not a subspace of V.

Theorem 7.6: Consider a homogeneous system of linear equations in n unknowns $x_1, x_2, \ldots, x_n$ over a field K:

$$a_{11}x_1 + a_{12}x_2 + \cdots + a_{1n}x_n = 0$$
$$a_{21}x_1 + a_{22}x_2 + \cdots + a_{2n}x_n = 0$$
$$\cdots \cdots \cdots \cdots \cdots \cdots \cdots \cdots \cdots \cdots \cdots$$
$$a_{m1}x_1 + a_{m2}x_2 + \cdots + a_{mn}x_n = 0$$

Then the solution set W is a subspace of the vector space K^n.

7.61   Prove Theorem 7.6.

  ▮ The system is equivalent to the matrix equation $AX = 0$. Since $A0 = 0$, the zero vector $0 \in W$. Suppose u and v are vectors in W, i.e., u and v are solutions of the system. Then $Au = 0$ and $Av = 0$. Thus, for any scalars a and b in K, we have $A(au + bv) = aAu + bAv = a0 + b0 = 0 + 0 = 0$. Hence $au + bv$ is a solution of the system, i.e., $au + bv \in W$. Thus W is a subspace of K^n.

7.62   Let $AX = B$ be a nonhomogeneous system of linear equations in n unknowns over a field K. Show that the solution of the system is not a subspace of K^n

  ▮ If $B \neq 0$, then $A0 \neq B$; hence 0 is not a solution of $AX = B$. Thus the solution set is not a subspace.

7.63   Discuss whether or not $\mathbf{R}^2$ is a subspace of $\mathbf{R}^3$.

  ▮ No. Although one may "identify" the vector $(a, b) \in \mathbf{R}^2$ with, say, $(a, b, 0)$ in the xy plane in $\mathbf{R}^3$, they are distinct elements belonging to distinct, disjoint sets.

  Problems 7.64–7.65 refer to the vector space V consisting of infinite sequences $(a_1, a_2, \ldots)$ in a field K. (See Problem 7.30.)

7.64   Show that W is a subspace of V where W consists of all sequences with 0 as the first entry.

  ▮ Clearly $0 = (0, 0, \ldots)$ belongs to W. Suppose $u, v \in W$. Then the first entries of u and v equal 0. Then the first entry of $u + v$ is $0 + 0 = 0$, and, for any scalar $k \in K$, the first entry of ku is $k \cdot 0 = 0$. Thus $u + v$ and ku belong to W, and hence W is a subspace of V.

7.65   Show that W is a subspace of V where W consists of all sequences with only a finite number of nonzero entries.

  ▮ $0 = (0, 0, \ldots)$ has no nonzero entries, hence $0 \in W$. Suppose $u, v \in W$. Then u and v have only a finite number of nonzero entries. Hence $u + v$ and, for any scalar $k \in K$, ku have only a finite number of nonzero entries. Thus $u + v$ and ku belong to W, and so W is a subspace of V.

7.3 LINEAR COMBINATIONS, LINEAR SPANS

7.66   Define linear combinations in a vector space.

  ▮ Let V be a vector space over a field K and let $v_1, \ldots, v_m \in V$. Any vector in V of the form $a_1v_1 + a_2v_2 + \cdots + a_mv_m$ where the $a_1 \in K$, is called a *linear combination* of $v_1, \ldots, v_m$.

7.67   Let S be a subset of a vector space V. Define the *linear span* of S, denoted by span(S) or $L(S)$.

  ▮ If $S = \emptyset$, then span(S) = $\{0\}$. Otherwise, span(S) consists of all the linear combinations of vectors in S.

7.68   Describe geometrically span(u) where u is a nonzero vector in $\mathbf{R}^3$.

  ▮ The set span(u) consists of all scalar multiples of u; geometrically, span(u) is the line in $\mathbf{R}^3$ through the origin 0 and the point u as pictured in Fig. 7-1.

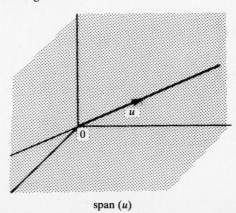

span (u)   **Fig. 7-1**

7.69   Describe geometrically span(u, v) where u and v are nonzero vectors in $\mathbf{R}^3$ which are not multiples of each other.

▌ The set span(u, v) consists of all vectors of the form $au + bv$ where $a, b \in \mathbf{R}$; geometrically, span(u, v) is the plane in $\mathbf{R}^3$ through the origin 0 and the points u and v as pictured in Fig. 7-2.

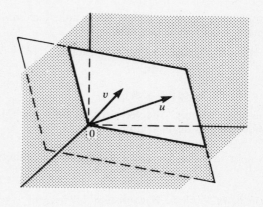

span (u, v) **Fig. 7-2**

Theorem 7.7:   Let S be a subset of a vector space V.
(i)   The set span(S) is a subspace of V which contains S.
(ii)   If W is any subspace of V containing S, then span$(S) \subseteq W$.

7.70   Prove (i) of Theorem 7.7: The set span(S) is a subspace of V containing S.

▌ If $S = \emptyset$, then span$(S) = \{0\}$, which is a subspace of V containing the empty set $\emptyset$. Now suppose $S \neq \emptyset$. If $v \in S$, then $1v = v \in \text{span}(S)$; hence S is a subset of span(S). Also, span$(S) \neq \emptyset$ since $S \neq \emptyset$. Now suppose $v, w \in \text{span}(S)$; say $v = a_1 v_1 + \cdots + a_m v_m$ and $w = b_1 w_1 + \cdots + b_n w_n$ where $v_i, w_j \in S$ and a_i, b_j are scalars. Then $v + w = a_1 v_1 + \cdots + a_m v_m + b_1 w_1 + \cdots + b_n w_n$ and, for any scalar k, $kv = k(a_1 v_1 + \cdots + a_m v_m) = ka_1 v_1 + \cdots + ka_m v_m$ belong to span(S) since each is a linear combination of vectors in S. Thus span(S) is a subspace of V.

7.71   Prove (ii) of Theorem 7.7:   If W is a subspace of V containing S, then span$(S) \subseteq W$.

▌ If $S = \emptyset$, then any subspace W contains S, and span$(S) = \{0\}$ is contained in W. Now suppose $S \neq \emptyset$ and suppose $v_1, \ldots, v_m \in S \subset W$. Then all multiples $a_1 v_1, \ldots, a_m v_m \in W$, where $a_i \in K$, and hence the sum $a_1 v_1 + \cdots + a_m v_m \in W$. That is, W contains all linear combinations of elements of S. Consequently, span$(S) \subseteq W$, as claimed.

7.72   Suppose u is a linear combination of the vectors $v_1, \ldots, v_m$ and suppose each v_i is a linear combination of the vectors $w_1, \ldots, w_n$: $u = a_1 v_1 + a_2 v_2 + \cdots + a_m v_m$ and $v_i = b_{i1} w_1 + b_{i2} w_2 + \cdots + b_{in} w_n$. Show that u is also a linear combination of the w_j. Thus if $S \subseteq \text{span}(T)$, then span$(S) \subseteq \text{span}(T)$.

▌ We have

$$u = a_1 v_1 + a_2 v_2 + \cdots + a_m v_m$$
$$= a_1(b_{11} w_1 + \cdots + b_{1n} w_n) + a_2(b_{21} w_1 + \cdots + b_{2n} w_n) + \cdots + a_m(b_{m1} w_1 + \cdots + b_{mn} w_n)$$
$$= (a_1 b_{11} + a_2 b_{21} + \cdots + a_m b_{m1}) w_1 + \cdots + (a_1 b_{1n} + a_2 b_{2n} + \cdots + a_m b_{mn}) w_n$$

or simply

$$u = \sum_{i=1}^{m} a_i v_i = \sum_{i=1}^{m} a_i \left(\sum_{j=1}^{n} b_{ij} w_j \right) = \sum_{j=1}^{n} \left(\sum_{i=1}^{m} a_i b_{ij} \right) w_j$$

7.73   Write the vector $v = (1, -2, 5)$ as a linear combination of the vectors $e_1 = (1, 1, 1)$, $e_2 = (1, 2, 3)$, and $e_3 = (2, -1, 1)$.

▮ We wish to express v as $v = xe_1 + ye_2 + ze_3$, with x, y, and z as yet unknown scalars. Thus we require

$$(1, -2, 5) = x(1, 1, 1) + y(1, 2, 3) + z(2, -1, 1)$$
$$= (x, x, x) + (y, 2y, 3y) + (2z, -z, z)$$
$$= (x + y + 2z, x + 2y - z, x + 3y + z)$$

Form the equivalent system of equations by setting corresponding components equal to each other, and then reduce to echelon form:

$$\begin{array}{lll} x + y + 2z = 1 & x + y + 2z = 1 & x + y + 2z = 1 \\ x + 2y - z = -2 \quad \text{or} & y - 3z = -3 \quad \text{or} & y - 3z = -3 \\ x + 3y + z = 5 & 2y - z = 4 & 5z = 10 \end{array}$$

Note that the above system is consistent and so has a solution. Solve for the unknowns to obtain $x = -6$, $y = 3$, $z = 2$. Hence $v = -6e_1 + 3e_2 + 2e_3$.

7.74 Write the vector $v = (2, -5, 3)$ in $\mathbf{R}^3$ as a linear combination of the vectors $e_1 = (1, -3, 2)$, $e_2 = (2, -4, -1)$, and $e_3 = (1, -5, 7)$.

▮ Set v as a linear combination of the e_i using the unknowns x, y, and z: $v = xe_1 + ye_2 + ze_3$.

$$(2, -5, 3) = x(1, -3, 2) + y(2, -4, -1) + z(1, -5, 7)$$
$$= (x + 2y + z, -3x - 4y - 5z, 2x - y + 7z)$$

Form the equivalent system of equations and reduce to echelon form:

$$\begin{array}{lll} x + 2y + z = 2 & x + 2y + z = 2 & x + 2y + z = 2 \\ -3x - 4y - 5z = -5 \quad \text{or} & 2y - 2z = 1 \quad \text{or} & 2y - 2z = 1 \\ 2x - y + 7z = 3 & -5y + 5z = -1 & 0 = 3 \end{array}$$

The system is inconsistent and so has no solution. Accordingly, v cannot be written as a linear combination of the vectors e_1, e_2, and e_3.

7.75 For which value of k will the vector $u = (1, -2, k)$ in $\mathbf{R}^3$ be a linear combination of the vectors $v = (3, 0, -2)$ and $w = (2, -1, -5)$?

▮ Set $u = xv + yw$: $(1, -2, k) = x(3, 0, -2) + y(2, -1, -5) = (3x + 2y, -y, -2x - 5y)$. Form the equivalent system of equations:

$$3x + 2y = 1 \qquad -y = -2 \qquad -2x - 5y = k$$

By the first two equations, $x = -1$, $y = 2$. Substitute into the last equation to obtain $k = -8$.

7.76 Write the polynomial $v = t^2 + 4t - 3$ over $\mathbf{R}$ as a linear combination of the polynomials $e_1 = t^2 - 2t + 5$, $e_2 = 2t^2 - 3t$, and $e_3 = t + 3$.

▮ Set v as a linear combination of the e_i using the unknowns x, y, and z: $v = xe_1 + ye_2 + ze_3$.

$$t^2 + 4t - 3 = x(t^2 - 2t + 5) + y(2t^2 - 3t) + z(t + 3)$$
$$= xt^2 - 2xt + 5x + 2yt^2 - 3yt + zt + 3z$$
$$= (x + 2y)t^2 + (-2x - 3y + z)t + (5x + 3z)$$

Set coefficients of the same powers of t equal to each other, and reduce the system to echelon form:

$$\begin{array}{lll} x + 2y = 1 & x + 2y = 1 & x + 2y = 1 \\ -2x - 3y + z = 4 \quad \text{or} & y + z = 6 \quad \text{or} & y + z = 6 \\ 5x + 3z = -3 & -10y + 3z = -8 & 13z = 52 \end{array}$$

Note that the system is consistent and so has a solution. Solve for the unknowns to obtain $x = -3$, $y = 2$, $z = 4$. Thus $v = -3e_1 + 2e_2 + 4e_3$.

7.77 Write the matrix $E = \begin{pmatrix} 3 & 1 \\ 1 & -1 \end{pmatrix}$ as a linear combination of the matrices $A = \begin{pmatrix} 1 & 1 \\ 1 & 0 \end{pmatrix}$, $B = \begin{pmatrix} 0 & 0 \\ 1 & 1 \end{pmatrix}$, and $C = \begin{pmatrix} 0 & 2 \\ 0 & -1 \end{pmatrix}$.

▮ Set E as a linear combination of A, B, C using the unknowns x, y, z: $E = xA + yB + zC$.

$$\begin{pmatrix} 3 & 1 \\ 1 & -1 \end{pmatrix} = x\begin{pmatrix} 1 & 1 \\ 1 & 0 \end{pmatrix} + y\begin{pmatrix} 0 & 0 \\ 1 & 1 \end{pmatrix} + z\begin{pmatrix} 0 & 2 \\ 0 & -1 \end{pmatrix}$$
$$= \begin{pmatrix} x & x \\ x & 0 \end{pmatrix} + \begin{pmatrix} 0 & 0 \\ y & y \end{pmatrix} + \begin{pmatrix} 0 & 2z \\ 0 & -z \end{pmatrix} = \begin{pmatrix} x & x+2z \\ x+y & y-z \end{pmatrix}$$

Form the equivalent system of equations by setting corresponding entries equal to each other: $x = 3$, $x + y = 1$, $x + 2z = 1$, $y - z = -1$. Substitute $x = 3$ in the second and third equations to obtain $y = -2$ and $z = -1$. Since these values also satisfy the last equation, they form a solution of the system. Hence $E = 3A - 2B - C$.

7.78 Determine whether or not $v = (3, 9, -4, -2)$ in $\mathbf{R}^4$ is a linear combination of $u_1 = (1, -2, 0, 3)$, $u_2 = (2, 3, 0, -1)$, and $u_3 = (2, -1, 2, 1)$, that is, whether or not $v \in \text{span}(u_1, u_2, u_3)$.

▮ Set v as a linear combination of the u_i using unknowns $x, y,$ and z; that is, set $v = xu_1 + yu_2 + zu_3$:

$$(3, 9, -4, -2) = x(1, -2, 0, 3) + y(2, 3, 0 -1) + z(2, -1, 2, 1)$$
$$= (x + 2y + 2z, -2x + 3y - z, 2z, 3x - y + z)$$

Form the equivalent system of equations by setting corresponding components equal to each other, and then reduce to echelon form:

$$\begin{array}{l} x + 2y + 2z = 3 \\ -2x + 3y - z = 9 \\ \qquad\quad 2z = -4 \\ 3x - y + z = -2 \end{array} \quad\text{or}\quad \begin{array}{l} x + 2y + 2z = 3 \\ \quad 7y + 3z = 15 \\ \qquad\quad 2z = -4 \\ -7y - 5z = -11 \end{array} \quad\text{or}\quad \begin{array}{l} x + 2y + 2z = 3 \\ \quad 7y + 3z = 15 \\ \qquad\quad 2z = -4 \\ \qquad -2z = 4 \end{array} \quad\text{or}\quad \begin{array}{l} x + 2y + 2z = 3 \\ \quad 7y + 3z = 15 \\ \qquad\quad 2z = -4 \end{array}$$

Note that the above system is consistent and so has a solution; hence v is a linear combination of the u_i. Solving for the unknowns we obtain $x = 1$, $y = 3$, $z = -2$. Thus $v = u_1 + 3u_2 - 2u_3$.

Note that if the system of linear equations were not consistent, i.e., had no solution, then the vector v would not be a linear combination of the u_i.

Problems 7.79–7.82 refer to the vectors $u = (1, -3, 2)$ and $v = (2, -1, 1)$ in $\mathbf{R}^3$.

7.79 Write $w = (1, 7, -4)$ as a linear combination of u and v.

▮ Set $w = xu + yv$ using unknowns x and y: $(1, 7, -4) = x(1, -3, 2) + y(2, -1, 1) = (x + 2y, -3x - y, 2x + y)$. Form the equivalent system of equations: $x + 2y = 1$, $-3x - y = 7$, $2x + y = -4$. Solve the system to obtain $x = -3$ and $y = 2$. Hence $w = -3u + 2v$.

7.80 Write $w = (2, -5, 4)$ as a linear combination of u and v.

▮ Set $w = xu + yv$ using unknowns x and y: $(2, -5, 4) = x(1, -3, 2) + y(2, -1, 1) = (x + 2y, -3x - y, 2x + y)$. Form the equivalent system and reduce to echelon form:

$$\begin{array}{l} x + 2y = 2 \\ -3x - y = -5 \\ 2x + y = 4 \end{array} \quad\text{or}\quad \begin{array}{l} x + 2y = 2 \\ \quad 5y = 1 \\ -3y = 0 \end{array} \quad\text{or}\quad \begin{array}{l} x + 2y = 2 \\ \quad 5y = 1 \\ \quad 0 = \frac{3}{5} \end{array}$$

The last equation shows the system is inconsistent. Hence w is not a linear combination of u and v.

7.81 Find k so that $w = (1, k, 5)$ is a linear combination of u and v.

▮ Set $w = xu + yv$: $(1, k, 5) = x(1, -3, 2) + y(2, -1, 1) = (x + 2y, -3x - y, 2x + y)$. Form the equivalent system of equations: $x + 2y = 1$, $-3x - y = k$, $2x + y = 5$. By the first and third equations, $x = 2$, $y = -1$. Substitute in the second equation to obtain $k = -8$.

7.82 Find a condition on a, b, c so that $w = (a, b, c)$ is a linear combination of u and v, i.e., so that $w \in \text{span}(u, v)$.

▌ Set $w = xu + yv$ using unknowns x and y: $(a, b, c) = x(1, -3, 2) + y(2, -1, 1) = (x + 2y, -3x - y, 2x + y)$. Form the equivalent system and reduce to echelon form:

$$\begin{aligned} x + 2y &= a \\ -3x - y &= b \quad \text{or} \\ 2x + y &= c \end{aligned} \qquad \begin{aligned} x + 2y &= a \\ 5y &= 3a + b \quad \text{or} \\ -3y &= -2a + c \end{aligned} \qquad \begin{aligned} x + 2y &= a \\ 5y &= 3a + b \\ 0 &= -a + 3b + 5c \end{aligned}$$

The system is consistent if and only if $a - 3b - 5c = 0$ and hence w is a linear combination of u and v iff $a - 3b - 5c = 0$.

7.83 Find conditions on $a, b,$ and c so that $(a, b, c) \in \mathbf{R}^3$ belongs to the space spanned by $u = (2, 1, 0)$, $v = (1, -1, 2)$, and $w = (0, 3, -4)$.

▌ Set (a, b, c) as a linear combination of $u, v,$ and w using unknowns x, y and z: $(a, b, c) = x(2, 1, 0) + y(1, -1, 2) + z(0, 3, -4) = (2x + y, x - y + 3z, 2y - 4z)$. Form the equivalent system of linear equations and reduce it to echelon form:

$$\begin{aligned} 2x + y &= a \\ x - y + 3z &= b \quad \text{or} \\ 2y - 4z &= c \end{aligned} \qquad \begin{aligned} 2x + y &= a \\ 3y - 6z &= a - 2b \quad \text{or} \\ 2y - 4z &= c \end{aligned} \qquad \begin{aligned} 2x + y &= a \\ 3y - 6z &= a - 2b \\ 0 &= 2a - 4b - 3c \end{aligned}$$

The vector (a, b, c) belongs to the space generated by $u, v,$ and w if and only if the above system is consistent, and it is consistent if and only if $2a - 4b - 3c = 0$.

7.84 Suppose W is a subspace of V. Show that $\text{span}(W) = W$.

▌ Since W is a subspace of V, W is closed under linear combinations. Hence $\text{span}(W) \subseteq W$. But $W \subseteq \text{span}(W)$. Both inclusions yield $\text{span}(W) = W$.

7.85 Show $\text{span}(\text{span}(S)) = \text{span}(S)$.

▌ Since $\text{span}(S)$ is a subspace of V, Problem 7.84 implies that $\text{span}(\text{span}(S)) = \text{span}(S)$.

7.86 Suppose S and T are subsets of a vector space V such that $S \subseteq T$. Show that $\text{span}(S) \subseteq \text{span}(T)$.

▌ Suppose $v \in \text{span}(S)$. Then $v = a_1 u_1 + \cdots + a_r u_r$ where $a_i \in K$ and $u_i \in S$. But $S \subseteq T$; hence every $u_i \in T$. Thus $v \in \text{span}(T)$. Accordingly, $\text{span}(S) \subseteq \text{span}(T)$.

7.87 Show that $\text{span}(S)$ is the intersection of all the subspaces of V containing S.

▌ Let $\{W_i\}$ be the collection of all subspaces of V containing S, and let $W = \cap W_i$. Since each W_i is a subspace of V, the set W is a subspace of V. Also, since each W_i contains S, the intersection W contains S. Hence $\text{span}(S) \subseteq W$. On the other hand, $\text{span}(S)$ is a subspace of V containing S; so $\text{span}(S) = W_k$ for some k. Then $W \subseteq W_k = \text{span}(S)$. Both inclusions give $\text{span}(S) = W$.

7.88 Show that $\text{span}(S) = \text{span}(S \cup \{0\})$. Thus one may delete the zero vector from any spanning set.

▌ By Problem 7.86, $\text{span}(S) \subseteq \text{span}(S \cup \{0\})$. Suppose $v \in \text{span}(S \cup \{0\})$, say, $v = a_1 u_1 + \cdots + a_n u_n + b \cdot 0$ where $a_i, b \in K$ and $u_i \in S$. Then $v = a_1 u_1 + \cdots + a_n u_n$, and so $v \in \text{span}(S)$. Thus $\text{span}(S \cup \{0\}) \subseteq \text{span}(S)$. Both inclusions give $\text{span}(S) = \text{span}(S \cup \{0\})$.

7.4 SPANNING SETS, GENERATORS

7.89 Define a spanning set or generators of a vector space V.

▌ The vectors $u_1, u_2, \ldots, u_r$ are said to span or generate V or are said to form a spanning set of V if $V = \text{span}(u_1, \ldots, u_r)$. Alternatively, $u_1, u_2, \ldots, u_r$ span V if, for every vector $v \in V$, there exist scalars $a_1, a_2, \ldots, a_r \in K$ such that $v = a_1 u_1 + a_2 u_2 + \cdots + a_r u_r$, i.e., v is a linear combination of $u_1, u_2, \ldots, u_r$.

7.90 Show that the vectors $e_1 = (1, 0, 0)$, $e_2 = (0, 1, 0)$, and $e_3 = (0, 0, 1)$ span the vector space $\mathbf{R}^3$.

l Let $v = (a, b, c)$ be an arbitrary vector in $\mathbf{R}^3$. Set $v = xe_1 + ye_2 + ze_3$ where x, y, z are unknown scalars: $(a, b, c) = x(1, 0, 0) + y(0, 1, 0) + z(0, 0, 1) = (x, y, z)$. Thus $x = a$, $y = b$, $z = c$. Hence v is a linear combination of e_1, e_2, e_3; specifically, $v = ae_1 + be_2 + ce_3$. Thus e_1, e_2, e_3 span $\mathbf{R}^3$.

7.91 Show that the vectors $u = (1, 2, 3)$, $v = (0, 1, 2)$, and $w = (0, 0, 1)$ span $\mathbf{R}^3$.

l We need to show that an arbitrary vector $(a, b, c) \in \mathbf{R}^3$ is a linear combination of u, v, and w. Set $(a, b, c) = xu + yv + zw$: $(a, b, c) = x(1, 2, 3) + y(0, 1, 2) + z(0, 0, 1) = (x, 2x + y, 3x + 2y + z)$. Then form the system of equations

$$\begin{array}{rl} x & = a \\ 2x + y & = b \\ 3x + 2y + z & = c \end{array} \quad \text{or} \quad \begin{array}{r} z + 2y + 3x = a \\ y + 2x = b \\ x = a \end{array}$$

The above system is in echelon form and is consistent; in fact $x = a$, $y = b - 2a$, $z = c - 2b + a$ is a solution. Thus u, v, and w span $\mathbf{R}^3$.

7.92 Show that $u_1 = (1, 2, 5)$, $u_2 = (1, 3, 7)$, and $u_3 = (1, -1, -1)$ do not span $\mathbf{R}^3$.

l Set $w = (a, b, c)$ as a linear combination of u_1, u_2, and u_3: $(a, b, c) = x(1, 2, 5) + y(1, 3, 7) + z(1, -1, -1) = (x + y + z, 2x + 3y - z, 5x + 7y - z)$. Form the equivalent system of linear equations and reduce it to echelon form:

$$\begin{array}{l} x + y + z = a \\ 2x + 3y - z = b \\ 5x + 7y - z = c \end{array} \quad \text{or} \quad \begin{array}{rl} x + y + z = & a \\ y - 3z = & -2a + b \\ 2y - 6z = & -5a + c \end{array} \quad \text{or} \quad \begin{array}{rl} x + y + z = & a \\ y - 3z = & -2a + b \\ 0 = & -a - 2b + c \end{array}$$

The last equation shows that w belongs to $L(u_1, u_2, u_3)$ only if $a + 2b - c = 0$. Therefore, there are vectors in $\mathbf{R}^3$ which do not belong to span(u_1, u_2, u_3). Accordingly, u_1, u_2, u_3 do not span $\mathbf{R}^3$.

Problems 7.93–7.95 refer to the xy plane $W = \{(a, b, 0)\}$ in $\mathbf{R}^3$.

7.93 Show that $u = (1, 2, 0)$ and $v = (0, 1, 0)$ span W. Show that an arbitrary vector $(a, b, 0) \in W$ is a linear combination of u and v.

l Set $(a, b, 0) = xu + yv$: $(a, b, 0) = x(1, 2, 0) + y(0, 1, 0) = (x, 2x + y, 0)$. Then form the system of equations

$$\begin{array}{rl} x & = a \\ 2x + y & = b \\ 0 & = 0 \end{array} \quad \text{or} \quad \begin{array}{r} y + 2x = b \\ x = a \end{array}$$

The system is consistent; in fact $x = a$, $y = b - 2a$ is a solution. Hence u and v span W.

7.94 Show that $u = (2, -1, 0)$ and $v = (1, 3, 0)$ span W.

l Set $(a, b, 0) = xu + yv$: $(a, b, 0) = x(2, -1, 0) + y(1, 3, 0) = (2x + y, -x + 3y, 0)$. Form the following system and reduce it to echelon form:

$$\begin{array}{rl} 2x + y & = a \\ -x + 3y & = b \\ 0 & = 0 \end{array} \quad \text{or} \quad \begin{array}{r} 2x + y = a \\ 7y = a + 2b \end{array}$$

The system is consistent and so has a solution. Hence W is spanned by u and v. (Observe that we do not need to solve for x and y; it is only necessary to know that a solution exists.)

7.95 Show that $u = (3, 2, 0)$ and $v = (1, 1, 2)$ span W.

l The vectors u and v cannot span W since v does not belong to W. In other words, $W \neq$ span(u, v).

Problems 7.96–7.97 refer to the vector space V of all polynomials (in t).

7.96 Show that the polynomials $1, t, t^2, t^3, \ldots$ span V.

▮ Any polynomial $f(t)$ in V is a linear combination of 1 and powers of t. Hence $V = \text{span}(1, t, t^2, t^3, \ldots)$.

7.97 Show that no finite set S of polynomials in V can span V.

▮ Any finite set S of polynomials contains one of maximum degree, say m. Then $\text{span}(S)$ cannot contain polynomials of degree greater than m. Accordingly, $V \neq \text{span}(S)$, for any finite set S.

7.98 Suppose $u_1, u_2, \ldots, u_m$ span V. Show that, for any vector $w \in V$, the vectors $u_1, u_2, \ldots, u_m, w$ span V.

▮ *Method 1.* Let $v \in V$. Since the u_i span V, there exist scalars $a_1, \ldots, a_m$ such that $v = a_1 u_1 + \cdots + a_m u_m$. Then $v = a_1 u_1 + a_2 u_2 + \cdots + a_m u_m + 0w$. Thus $u_1, u_2, \ldots, u_m, w$ span V.

Method 2. By Problem 7.86, $\text{span}(u_i) \subseteq \text{span}(u_i, w)$. Hence $V = \text{span}(u_i) \subseteq \text{span}(u_i, w) \subseteq V$. Thus no inclusion can be proper and so $\text{span}(u_i, w) = V$.

7.99 Suppose $u_1, u_2, \ldots, u_m$ span V. Suppose, for $k > 1$, the vector u_k is a linear combination of the preceding vectors $u_1, u_2, \ldots, u_{k-1}$. Show that the u_i without u_k span V, i.e., show $\text{span}(u_i, \ldots, u_{k-1}, u_{k+1}, \ldots, u_m) = V$.

▮ Let $v \in V$. Since the u_i span V, there exist scalars $a_1, \ldots, a_m$ such that $v = a_1 u_1 + \cdots + a_m u_m$. Since u_k is a linear combination of $u_1, \ldots, u_{k-1}$, there exist scalars $b_1, \ldots, b_{k-1}$ such that $u_k = b_1 u_1 + \cdots + b_{k-1} u_{k-1}$. Thus

$$
\begin{aligned}
v &= a_1 u_1 + \cdots + a_k u_k + \cdots + a_m u_m \\
&= a_1 u_1 + \cdots + a_k(b_1 u_1 + \cdots + b_{k-1} u_{k-1}) + \cdots + a_m u_m \\
&= (a_1 + a_k b_1) u_1 + \cdots + (a_{k-1} + a_k b_{k-1}) u_{k-1} + a_{k+1} u_{k+1} + \cdots + a_m u_m
\end{aligned}
$$

Hence $\text{span}(u_1, \ldots, u_{k-1}, u_{k+1}, \ldots, u_m) = V$.

7.100 Let $W_1, W_2, \ldots$ be subspaces of a vector space V for which $W_1 \subset W_2 \subset \cdots$. Let $W = W_1 \cup W_2 \cup \cdots$. Show that W is a subspace of V.

▮ The zero vector $0 \in W_1$; hence $0 \in W$. Suppose $u, v \in W$. Then there exist j_1 and j_2 such that $u \in W_{j_1}$ and $v \in W_{j_2}$. Let $j = \max(j_1, j_2)$. Then $W_{j_1} \subseteq W_j$ and $W_{j_2} \subseteq W_j$, and so $u, v \in W_j$. But W_j is a subspace; hence $u + v \in W_j$ and, for any scalar k, the multiple $ku \in W_j$. Since $W_j \subseteq W$, we have $u + v, ku \in W$. Thus W is a subspace of V.

7.101 In the preceding problem, suppose S_i spans W_i for $i = 1, 2, \ldots$. Show that $S = S_1 \cup S_2 \cup \cdots$ spans W.

▮ Let $v \in W$. Then there exists j such that $v \in W_j$. Then $v \in \text{span}(S_j) \subseteq \text{span}(S)$. Thus $W \subseteq \text{span}(S)$. But $S \subseteq W$ and W is a subspace; hence $\text{span}(S) \subseteq W$. Both inclusions give $\text{span}(S) = W$, i.e., S spans W.

7.5 ROW SPACE OF A MATRIX

7.102 Define the row space of a matrix A, denoted by $\text{rowsp}(A)$.

▮ Let A be an arbitrary $m \times n$ matrix over a field K:

$$
A = \begin{pmatrix}
a_{11} & a_{12} & \cdots & a_{1n} \\
a_{21} & a_{22} & \cdots & a_{2n} \\
\cdots\cdots\cdots\cdots\cdots\cdots \\
a_{m1} & a_{m2} & \cdots & a_{mn}
\end{pmatrix}
$$

The rows of A, $R_1 = (a_{11}, a_{12}, \ldots, a_{1n}), \ldots, R_m = (a_{m1}, a_{m2}, \ldots, a_{mn})$, viewed as vectors in K^n, span a subspace of K^n called the *row space* of A. That is, $\text{rowsp}(A) = \text{span}(R_1, R_2, \ldots, R_m)$.

7.103 Define the column space of a matrix A, denoted by $\text{colsp}(A)$.

▮ The columns, $C_1, C_2, \ldots, C_n$, of an $m \times n$ matrix A over a field K, viewed as vectors in K^m, span a subspace of K^m called the column space of A. That is, $\text{colsp}(A) = \text{span}(C_1, C_2, \ldots, C_n)$. Alternatively, $\text{colsp}(A) = \text{rowsp}(A^T)$.

Theorem 7.8: Row equivalent matrices have the same row space.

7.104 Prove Theorem 7.8.

▌ Suppose we apply an elementary row operation on a matrix A: (i) $R_i \leftrightarrow R_j$, (ii) $R_i \to kR_i$, $k \neq 0$, or (iii) $R_i \to kR_j + R_i$ and obtain a matrix B. Then each row of B is clearly a row of A or a linear combination of rows of A. Hence the row space of B is contained in the row space of A. On the other hand, we can apply the inverse elementary row operation on B and obtain A; hence the row space of A is contained in the row space of B. Accordingly, A and B have the same row space. Thus any sequence of elementary row operations produces a matrix with the same row space. Accordingly, row equivalent matrices have the same row space.

7.105 Determine whether the following matrices have the same row space:

$$A = \begin{pmatrix} 1 & 1 & 5 \\ 2 & 3 & 13 \end{pmatrix} \qquad B = \begin{pmatrix} 1 & -1 & -2 \\ 3 & -2 & -3 \end{pmatrix} \qquad C = \begin{pmatrix} 1 & -1 & -1 \\ 4 & -3 & -1 \\ 3 & -1 & 3 \end{pmatrix}$$

▌ Matrices have the same row space if and only if their row canonical forms have the same nonzero rows; hence row reduce each matrix to row canonical form:

$$A = \begin{pmatrix} 1 & 1 & 5 \\ 2 & 3 & 13 \end{pmatrix} \sim \begin{pmatrix} 1 & 1 & 5 \\ 0 & 1 & 3 \end{pmatrix} \sim \begin{pmatrix} 1 & 0 & 2 \\ 0 & 1 & 3 \end{pmatrix}$$

$$B = \begin{pmatrix} 1 & -1 & -2 \\ 3 & -2 & -3 \end{pmatrix} \sim \begin{pmatrix} 1 & -1 & -2 \\ 0 & 1 & 3 \end{pmatrix} \sim \begin{pmatrix} 1 & 0 & 1 \\ 0 & 1 & 3 \end{pmatrix}$$

$$C = \begin{pmatrix} 1 & -1 & -1 \\ 4 & -3 & -1 \\ 3 & -1 & 3 \end{pmatrix} \sim \begin{pmatrix} 1 & -1 & -1 \\ 0 & 1 & 3 \\ 0 & 2 & 6 \end{pmatrix} \sim \begin{pmatrix} 1 & -1 & -1 \\ 0 & 1 & 3 \\ 0 & 0 & 0 \end{pmatrix} \sim \begin{pmatrix} 1 & 0 & 2 \\ 0 & 1 & 3 \\ 0 & 0 & 0 \end{pmatrix}$$

Since the nonzero rows of the reduced form of A and of the reduced form of C are the same, A and C have the same row space. On the other hand, the nonzero rows of the reduced form of B are not the same as the others, and so B has a different row space.

7.106 Determine which of the following matrices have the same row space:

$$A = \begin{pmatrix} 1 & -2 & -1 \\ 3 & -4 & 5 \end{pmatrix} \qquad B = \begin{pmatrix} 1 & -1 & 2 \\ 2 & 3 & -1 \end{pmatrix} \qquad C = \begin{pmatrix} 1 & -1 & 3 \\ 2 & -1 & 10 \\ 3 & -5 & 1 \end{pmatrix}$$

▌ Row reduce each matrix to row canonical form:

$$A \sim \begin{pmatrix} 1 & -2 & -1 \\ 0 & 2 & 8 \end{pmatrix} \sim \begin{pmatrix} 1 & -2 & -1 \\ 0 & 1 & 4 \end{pmatrix} \sim \begin{pmatrix} 1 & 0 & 7 \\ 0 & 1 & 4 \end{pmatrix}$$

$$B \sim \begin{pmatrix} 1 & -1 & 2 \\ 0 & 5 & -5 \end{pmatrix} \sim \begin{pmatrix} 1 & -1 & 2 \\ 0 & 1 & -1 \end{pmatrix} \sim \begin{pmatrix} 1 & 0 & 1 \\ 0 & 1 & -1 \end{pmatrix}$$

$$C \sim \begin{pmatrix} 1 & -1 & 3 \\ 0 & 1 & 4 \\ 0 & -2 & -8 \end{pmatrix} \sim \begin{pmatrix} 1 & -1 & 3 \\ 0 & 1 & 4 \\ 0 & 0 & 0 \end{pmatrix} \sim \begin{pmatrix} 1 & 0 & 7 \\ 0 & 1 & 4 \\ 0 & 0 & 0 \end{pmatrix}$$

The matrices A and C have the same row space since they have the same row canonical forms (excluding zero rows). B does not have the same row space as A or C.

7.107 Determine whether the following matrices have the same column space:

$$A = \begin{pmatrix} 1 & 3 & 5 \\ 1 & 4 & 3 \\ 1 & 1 & 9 \end{pmatrix} \qquad B = \begin{pmatrix} 1 & 2 & 3 \\ -2 & -3 & -4 \\ 7 & 12 & 17 \end{pmatrix}$$

▌ Observe that A and B have the same column space if and only if the transposes A^T and B^T have the same row space. Thus reduce A^T and B^T to row canonical form:

$$A^T = \begin{pmatrix} 1 & 1 & 1 \\ 3 & 4 & 1 \\ 5 & 3 & 9 \end{pmatrix} \sim \begin{pmatrix} 1 & 1 & 1 \\ 0 & 1 & -2 \\ 0 & -2 & 4 \end{pmatrix} \sim \begin{pmatrix} 1 & 1 & 1 \\ 0 & 1 & -2 \\ 0 & 0 & 0 \end{pmatrix} \sim \begin{pmatrix} 1 & 0 & 3 \\ 0 & 1 & -2 \\ 0 & 0 & 0 \end{pmatrix}$$

$$B^T = \begin{pmatrix} 1 & -2 & 7 \\ 2 & -3 & 12 \\ 3 & -4 & 17 \end{pmatrix} \sim \begin{pmatrix} 1 & -2 & 7 \\ 0 & 1 & -2 \\ 0 & 2 & -4 \end{pmatrix} \sim \begin{pmatrix} 1 & -2 & 7 \\ 0 & 1 & -2 \\ 0 & 0 & 0 \end{pmatrix} \sim \begin{pmatrix} 1 & 0 & 3 \\ 0 & 1 & -2 \\ 0 & 0 & 0 \end{pmatrix}$$

Since A^T and B^T have the same row space, A and B have the same column space.

7.108 Let $U = \text{span}(u_1, u_2, u_3)$ and $W = \text{span}(v_1, v_2)$ be subspaces of $\mathbf{R}^4$ where $u_1 = (1, 2, -1, 3)$, $u_2 = (2, 4, 1, -2)$, $u_3 = (3, 6, 3, -7)$, $v_1 = (1, 2, -4, 11)$, $v_2 = (2, 4, -5, 14)$. Show that $U = W$.

▮ **Method 1.** Show that each u_i is a linear combination of v_1 and v_2, and show that each v_i is a linear combination of u_1, u_2, and u_3. Observe that we have to show that six systems of linear equations are consistent.

Method 2. Form the matrix A whose rows are the u_i, and row reduce A to row canonical form:

$$A = \begin{pmatrix} 1 & 2 & -1 & 3 \\ 2 & 4 & 1 & -2 \\ 3 & 6 & 3 & -7 \end{pmatrix} \sim \begin{pmatrix} 1 & 2 & -1 & 3 \\ 0 & 0 & 3 & -8 \\ 0 & 0 & 6 & -16 \end{pmatrix} \sim \begin{pmatrix} 1 & 2 & 0 & \frac{1}{3} \\ 0 & 0 & 1 & -\frac{8}{3} \end{pmatrix}$$

Now form the matrix B whose rows are v_1 and v_2, and row reduce B to row canonical form:

$$B = \begin{pmatrix} 1 & 2 & -4 & 11 \\ 2 & 4 & -5 & 14 \end{pmatrix} \sim \begin{pmatrix} 1 & 2 & -4 & 11 \\ 0 & 0 & 3 & -8 \end{pmatrix} \sim \begin{pmatrix} 1 & 2 & 0 & \frac{1}{3} \\ 0 & 0 & 1 & -\frac{8}{3} \end{pmatrix}$$

Since the nonzero rows of the reduced matrices are identical, the row spaces of A and B are equal and so $U = W$.

7.109 Let $U = \text{span}(u_1, u_2, u_3)$ and $W = \text{span}(v_1, v_2, v_3)$ be subspaces of $\mathbf{R}^3$ where $u_1 = (1, 1, -1)$, $u_2 = (2, 3, -1)$, $u_3 = (3, 1, -5)$, $v_1 = (1, -1, -3)$, $v_2 = (3, -2, -8)$, and $v_3 = (2, 1, -3)$. Show $U = W$.

▮ Form the matrix A whose rows are the u_i, and row reduce A to row canonical form:

$$A = \begin{pmatrix} 1 & 1 & -1 \\ 2 & 3 & -1 \\ 3 & 1 & -5 \end{pmatrix} \sim \begin{pmatrix} 1 & 1 & -1 \\ 0 & 1 & 1 \\ 0 & -2 & -2 \end{pmatrix} \sim \begin{pmatrix} 1 & 0 & -2 \\ 0 & 1 & 1 \\ 0 & 0 & 0 \end{pmatrix}$$

Next form the matrix B whose rows are the v_i, and row reduce B to row canonical form:

$$B = \begin{pmatrix} 1 & -1 & -3 \\ 3 & -2 & -8 \\ 2 & 1 & -3 \end{pmatrix} \sim \begin{pmatrix} 1 & -1 & -3 \\ 0 & 1 & 1 \\ 0 & 3 & 3 \end{pmatrix} \sim \begin{pmatrix} 1 & 0 & -2 \\ 0 & 1 & 1 \\ 0 & 0 & 0 \end{pmatrix}$$

Since A and B have the same row canonical form, the row spaces of A and B are equal and so $U = W$.

7.110 Consider an arbitrary matrix $A = (a_{ij})$. Suppose $u = (b_1, \ldots, b_n)$ is a linear combination of the rows $R_1, \ldots, R_m$ of A; say $u = k_1 R_1 + \cdots + k_m R_m$. Show that, $\forall i$, $b_i = k_1 a_{1i} + k_2 a_{2i} + \cdots + k_m a_{mi}$ where $a_{1i}, \ldots, a_{mi}$ are the entries of the ith column of A.

▮ We have $u = k_1 R_1 + \cdots + k_m R_m$; hence $(b_1, \ldots, b_n) = k_1(a_{11}, \ldots, a_{1n}) + \cdots + k_m(a_{m1}, \ldots, a_{mn})$ $= (k_1 a_{11} + \cdots + k_m a_{m1}, \ldots, k_1 a_{m1} + \cdots + k_m a_{mn})$. Setting corresponding components equal to each other, we obtain the desired result.

7.111 Let $A = (a_{ij})$ be an echelon matrix with leading nonzero entries $a_{1j_1}, a_{2j_2}, \ldots, a_{rj_r}$, and let $B = (b_{ij})$ be an echelon matrix with leading nonzero entries $b_{1k_1}, b_{2k_2}, \ldots, b_{sk_s}$:

$$A = \begin{pmatrix} a_{1j_1} & * & * & * & * & * & * \\ & & a_{2j_2} & * & * & * & * \\ & & \cdots & \cdots & \cdots & \cdots & \cdots \\ & & & & a_{rj_r} & * & * \end{pmatrix} \qquad B = \begin{pmatrix} b_{1k_1} & * & * & * & * & * & * \\ & & b_{2k_2} & * & * & * & * \\ & & \cdots & \cdots & \cdots & \cdots & \cdots \\ & & & & b_{sk_s} & * & * \end{pmatrix}$$

Suppose A and B have the same row space. Show that the leading nonzero entries of A and of B are in the same position, i.e., $j_1 = k_1$, $j_2 = k_2, \ldots, j_r = k_r$, and $r = s$.

▌ Clearly $A = 0$ if and only if $B = 0$, and so we need only prove the theorem when $r \geq 1$ and $s \geq 1$. We first show that $j_1 = k_1$. Suppose $j_1 < k_1$. Then the j_1th column of B is zero. Since the first row of A is in the row space of B, we have by the preceding problem, $a_{1j_1} = c_1 0 + c_2 0 + \cdots + c_m 0 = 0$ for scalars c_i. But this contradicts the fact that the element $a_{1j_1} \neq 0$. Hence $j_1 \geq k_1$, and similarly $k_1 \geq j_1$. Thus $j_1 = k_1$.

Now let A' be the submatrix of A obtained by deleting the first row of A, and let B' be the submatrix of B obtained by deleting the first row of B. We prove that A' and B' have the same row space. The theorem will then follow by induction since A' and B' are also echelon matrices.

Let $R = (a_1, a_2, \ldots, a_n)$ be any row of A' and let $R_1, \ldots, R_m$ be the rows of B. Since R is in the row space of B, there exist scalars $d_1, \ldots, d_m$ such that $R = d_1 R_1 + d_2 R_2 + \cdots + d_m R_m$. Since A is in echelon form and R is not the first row of A, the j_1th entry of R is zero: $a_i = 0$ for $i = j_1 = k_1$. Furthermore, since B is in echelon form, all the entries in the k_1th column of B are 0 except the first: $b_{1k_1} \neq 0$, but $b_{2k_1} = 0, \ldots, b_{mk_1} = 0$. Thus $0 = a_{k_1} = d_1 b_{1k_1} + d_2 0 + \cdots + d_m 0 = d_1 b_{1k_1}$. Now $b_{1k_1} \neq 0$ and so $d_1 = 0$. Thus R is a linear combination of $R_2, \ldots, R_m$ and so is in the row space of B'. Since R was any row of A', the row space of A' is contained in the row space of B'. Similarly, the row space of B' is contained in the row space of A'. Thus A' and B' have the same row space, and so the theorem is proved.

Theorem 7.9: Let $A = (a_{ij})$ and $B = (b_{ij})$ be echelon matrices in row canonical form. Then A and B have the same row space if and only if they have the same nonzero rows.

7.112 Prove Theorem 7.9.

▌ Obviously, if A and B have the same nonzero rows then they have the same row space. Thus we only have to prove the converse.

Suppose A and B have the same row space, and suppose $R \neq 0$ is the ith row of A. Then there exist scalars $c_1, \ldots, c_s$ such that

$$R = c_1 R_1 + c_2 R_2 + \cdots + c_s R_s \tag{1}$$

where the R_i are the nonzero rows of B. The theorem is proved if we show that $R = R_i$, that is, $c_i = 1$ but $c_k = 0$ for $k \neq i$.

Let a_{ij_i} be the leading nonzero entry of R. By (1) and Problem 7.110,

$$a_{ij_i} = c_1 b_{1j_i} + c_2 b_{2j_i} + \cdots + c_s b_{sj_i} \tag{2}$$

By Problem 7.111, b_{ij_i} is a leading nonzero entry of B and, since B is row reduced, it is the only nonzero entry in the j_ith column of B. Thus from (2) we obtain $a_{ij_i} = c_i b_{ij_i}$. However, $a_{ij_i} = 1$ and $b_{ij_i} = 1$ since A and B are row reduced; hence $c_i = 1$.

Now suppose $k \neq i$, and b_{kj_k} is the distinguished entry in R_k. By (1) and Problem 7.110,

$$a_{ij_k} = c_1 b_{1j_k} + c_2 b_{2j_k} + \cdots + c_s b_{sj_k} \tag{3}$$

Since B is row reduced, b_{kj_k} is the only nonzero entry in the j_kth column of B; hence by (3), $a_{ij_k} = c_k b_{kj_k}$. By Problem 7.111, a_{kj_k} is a leading nonzero entry of A and, since A is row reduced, $a_{ij_k} = 0$. Thus $c_k b_{kj_k} = 0$ and, since $b_{kj_k} = 1$, $c_k = 0$. Accordingly $R = R_i$ and the theorem is proved.

Theorem 7.10: Let A be any matrix. Then A is row equivalent to a unique matrix in row canonical form.

7.113 Prove Theorem 7.10.

▌ Suppose A is row equivalent to matrices A_1 and A_2 where A_1 and A_2 are in row canonical form. By Theorem 7.8, $\text{rowsp}(A) = \text{rowsp}(A_1)$ and $\text{rowsp}(A) = \text{rowsp}(A_2)$; hence $\text{rowsp}(A_1) = \text{rowsp}(A_2)$. Since A_1 and A_2 are in row canonical form, $A_1 = A_2$ by Theorem 7.9.

Theorem 7.11: Matrices A and B have the same row space if and only if their row canonical forms have the same nonzero rows.

7.114 Prove Theorem 7.11.

❚ Let A_1 and B_1 be the row canonical forms of A and B, respectively. Suppose A and B have the same row space. Then $\text{rowsp}(A_1) = \text{rowsp}(A) = \text{rowsp}(B) = \text{rowsp}(B_1)$. By Theorem 7.9, A_1 and B_1 have the same nonzero rows. Conversely, suppose A_1 and B_1 have the same nonzero rows. Then $\text{rowsp}(A) = \text{rowsp}(A_1) = \text{rowsp}(B_1) = \text{rowsp}(B)$. Thus the theorem is proved.

7.115 Let R be a row vector and B a matrix for which RB is defined. Show that RB is a linear combination of the rows of B.

❚ Suppose $R = (a_1, a_2, \ldots, a_m)$ and $B = (b_{ij})$. Let $B_1, \ldots, B_m$ denote the rows of B and $B^1, \ldots, B^n$ its columns. Then

$$
\begin{aligned}
RB &= (R \cdot B^1, R \cdot B^2, \ldots, R \cdot B^n) \\
&= (a_1 b_{11} + a_2 b_{21} + \cdots + a_m b_{m1}, a_1 b_{12} + a_2 b_{22} + \cdots + a_m b_{m2}, \ldots, a_1 b_{1n} + a_2 b_{2n} + \cdots + a_m b_{mn}) \\
&= a_1(b_{11}, b_{12}, \ldots, b_{1n}) + a_2(b_{21}, b_{22}, \ldots, b_{2n}) + \cdots + a_m(b_{m1}, b_{m2}, \ldots, b_{mn}) \\
&= a_1 B_1 + a_2 B_2 + \cdots + a_m B_m
\end{aligned}
$$

Thus RB is a linear combination of the rows of B, as claimed.

Theorem 7.12: Let A and B be matrices such that the product AB is defined. Then the row space of AB is contained in the row space of B.

7.116 Prove Theorem 7.12.

❚ The rows of AB are $R_i B$ where R_i is the ith row of A. Hence by the above result each row of AB is in the row space of B. Thus the row space of AB is contained in the row space of B.

7.117 Show that $\text{colsp}(AB) \subseteq \text{colsp}(A)$.

❚ Using Theorem 7.12, we have

$$\text{colsp}(AB) = \text{rowsp}((AB)^T) = \text{rowsp}(B^T A^T) \subseteq \text{rowsp}(A^T) = \text{colsp}(A)$$

7.118 Suppose P is a nonsingular (invertible) matrix. Show that $\text{rowsp}(PA) = \text{rowsp}(A)$.

❚ Using Theorem 7.12, we have $\text{rowsp}(A) = \text{rowsp}(IA) = \text{rowsp}(P^{-1}PA) \subseteq \text{rowsp}(PA) \subseteq \text{rowsp}(A)$. Thus no inclusion can be proper, and so $\text{rowsp}(PA) = \text{rowsp}(A)$. [Alternatively, PA is row equivalent to A and hence PA and A have the same row space by Theorem 7.8.]

7.6 SUMS AND DIRECT SUMMS

7.119 Suppose U and W are subsets of a vector space V. Define $U + W$.

❚ $U + W$ consists of all sums $u + w$ where $u \in U$ and $w \in W$:

$$U + W = \{u + w : u \in U, w \in W\}$$

7.120 Suppose U and W are subspaces of a vector space V. Show that $U + W$ is a subspace of V.

❚ Since U and W are subspaces, $0 \in U$ and $0 \in W$. Hence $0 = 0 + 0 \in U + W$. Suppose $v, v' \in U + W$. Then there exist $u, u' \in U$ and $w, w' \in W$ such that $v = u + w$ and $v' = u' + w'$. Since U and W are subspaces, $u + u' \in U$ and $w + w' \in W$ and, for any scalar k, $ku \in U$ and $kw \in W$. Accordingly, $v + v' = (u + w) + (u' + w') = (u + u') + (w + w') \in U + W$ and, for any scalar k, $kv = k(u + w) = ku + kw \in U + W$. Thus $U + W$ is a subspace of V.

7.121 Let V be the vector space of 2 by 2 matrices over **R**. Let U consist of those matrices in V whose second row is zero, and let W consist of those matrices in V whose second column is zero:

$$U = \left\{ \begin{pmatrix} a & b \\ 0 & 0 \end{pmatrix} : a, b \in \mathbf{R} \right\} \qquad W = \left\{ \begin{pmatrix} a & 0 \\ c & 0 \end{pmatrix} : a, c \in \mathbf{R} \right\}$$

Describe $U + W$ and $U \cap W$.

▮ $U + W$ consists of those matrices whose lower right entry is 0, and $U \cap W$ consists of those matrices whose second row and second column are zero:

$$U + W = \left\{ \begin{pmatrix} a & b \\ c & 0 \end{pmatrix} : a, b, c \in \mathbf{R} \right\} \quad \text{and} \quad U \cap W = \left\{ \begin{pmatrix} a & 0 \\ 0 & 0 \end{pmatrix} : a \in \mathbf{R} \right\}$$

Problems 7.122–7.124 refer to subspaces U and W of a vector space V.

7.122   Show that U and W are contained in $U + W$.

▮ Let $u \in U$. By hypothesis W is a subspace of V and so $0 \in W$. Hence $u = u + 0 \in U + W$. Accordingly, U is contained in $U + W$. Similarly, W is contained in $U + W$.

7.123   Show that $U + W$ is the smallest subspace of V containing U and W, that is, show that $U + W = \text{span}(U, W)$.

▮ Since $U + W$ is a subspace of V containing both U and W, it must also contain the linear span of U and W; i.e., $\text{span}(U, W) \subseteq U + W$.
On the other hand, if $v \in U + W$ then $v = u + w = 1u + 1w$ where $u \in U$ and $w \in W$; hence v is a linear combination of elements in $U \cup W$ and so belongs to $\text{span}(U, W)$. Thus $U + W \subseteq \text{span}(U, W)$. Both inclusions gives us the required result.

7.124   Show that $W + W = W$.

▮ Since W is a subspace of V, we have that W is closed under vector addition; hence $W + W \subseteq W$. By Problem 7.122, $W \subseteq W + W$. Hence $W + W = W$.

7.125   Give an example of a subset S of $\mathbf{R}^2$ such that $S + S \subset S$ (properly contained).

▮ Let $S = \{(0, 5), (0, 6), (0, 7), \ldots\}$. Then $S + S \subset S$.

7.126   Give an example of a subset S of $\mathbf{R}^2$ such that $S \subset S + S$ (properly contained).

▮ Let $S = \{(0, 0), (0, 1)\}$. Then $S \subset S + S$.

7.127   Give an example of a subset S of $\mathbf{R}^2$ which is not a subspace of $\mathbf{R}^2$ but for which $S + S = S$.

▮ Let $S = \{(0, 0), (0, 1), (0, 2), (0, 3), \ldots\}$. Then $S + S = S$.

7.128   Suppose U and W are subspaces of a vector space V such that $U = \text{span}(S)$ and $W = \text{span}(T)$. Show that $U + W = \text{span}(S \cup T)$.

▮ Since $S \subseteq U \subseteq U + W$ and $T \subseteq W \subseteq U + W$, we have $S \cup T \subseteq U + W$. Hence $\text{span}(S \cup T) \subseteq U + W$. Now suppose $v \in U + W$. Then $v = u + W$ where $u \in U$ and $w \in W$. Since $U = \text{span}(S)$ and $W = \text{span}(T)$, $u = a_1 u_1 + \cdots + a_r u_r$ and $w = b_1 w_1 + \cdots + b_s w_s$ where a_i, $b_j \in K$, $u_i \in S$, and $w_i \in T$. Then $v = u + w = a_1 u_1 + \cdots + a_r u_r + b_1 w_1 + \cdots + b_s w_s$. Thus $U + W \subseteq \text{span}(S \cup T)$. Both inclusions yield $U + W = \text{span}(S \cup T)$.

7.129   Suppose U and W are subspaces of V. Show that $V = U + W$ if every $v \in V$ can be written in the form $v = u + w$ where $u \in U$ and $w \in W$.

▮ Suppose, for any $v \in V$, we have $v = u + w$ where $u \in U$ and $w \in W$. Then $v \in U + W$ and so $V \subseteq U + W$. Since U and W are subspaces of V, we have $U + W \subseteq V$. Both inclusions imply $V = U + W$.

7.130   Define the direct sum $V = U \oplus W$.

▮ The vector space V is said to be the *direct sum* of its subspaces U and W, denoted by $V = U \oplus W$ if every vector $v \in V$ can be written in one and only one way as $v = u + w$ where $u \in U$ and $w \in W$.

Theorem 7.13: The vector space V is the direct sum of its subspaces U and W if and only if (i) $V = U + W$ and (ii) $U \cap W = \{0\}$.

7.131 Prove Theorem 7.13.

❚ Suppose $V = U \oplus W$. Then any $v \in V$ can be uniquely written in the form $v = u + w$ where $u \in U$ and $w \in W$. Thus, in particular, $V = U + W$. Now suppose $v \in U \cap W$. Then

$$v = v + 0 \qquad \text{where } v \in U, 0 \in W \qquad\qquad (1)$$

and

$$v = 0 + v \qquad \text{where } 0 \in U, v \in W \qquad\qquad (2)$$

Since such a sum for v must be unique, $v = 0$. Accordingly, $U \cap W = \{0\}$.

On the other hand, suppose $V = U + W$ and $U \cap W = \{0\}$. Let $v \in V$. Since $V = U + W$, there exist $u \in U$ and $w \in W$ such that $v = u + w$. We need to show that such a sum is unique. Suppose also that $v = u' + w'$ where $u' \in U$ and $w' \in W$. Then $u + w = u' + w'$ and so $u - u' = w' - w$. But $u - u' \in U$ and $w' - w \in W$; hence by $U \cap W = \{0\}$, $u - u' = 0$, $w' - w = 0$ and so $u = u'$, $w = w'$. Thus such a sum for $v \in V$ is unique and $V = U \oplus W$.

7.132 In the vector space $\mathbf{R}^3$, let U be the xy plane and let W be the yz plane:

$$U = \{(a, b, 0): a, b \in \mathbf{R}\} \qquad \text{and} \qquad W = \{(0, b, c): b, c \in \mathbf{R}\}$$

Then $\mathbf{R}^3 = U + W$ since every vector in $\mathbf{R}^3$ is the sum of a vector in U and a vector in W. Show that $\mathbf{R}^3$ is not the direct sum of U and W.

❚ Show that a vector $v \in \mathbf{R}^3$ can be written in more than one way as the sum of a vector in U and a vector in W, e.g., $(3, 5, 7) = (3, 1, 0) + (0, 4, 7)$ and also $(3, 5, 7) = (3, -4, 0) + (0, 9, 7)$. Alternatively, $(0, 1, 0) \in U \cap W$; hence $U \cap W \neq \{0\}$. Thus $\mathbf{R}^3 \neq U \oplus W$.

7.133 In $\mathbf{R}^3$, let U be the xy plane and let W be the z axis:

$$U = \{(a, b, 0): a, b \in \mathbf{R}\} \qquad \text{and} \qquad W = \{(0, 0, c): c \in \mathbf{R}\}$$

Show that $\mathbf{R}^3 = U \oplus W$.

❚ Any vector $(a, b, c) \in \mathbf{R}^3$ can be written as the sum of a vector in U and a vector in V in one and only one way:

$$(a, b, c) = (a, b, 0) + (0, 0, c)$$

Accordingly, $\mathbf{R}^3$ is the direct sum of U and W, that is, $\mathbf{R}^3 = U \oplus W$.

7.134 Let U and W be the subspaces of $\mathbf{R}^3$ defined by

$$U = \{(a, b, c): a = b = c\} \qquad \text{and} \qquad W = \{(0, b, c)\}$$

(Note that W is the yz plane.) Show that $\mathbf{R}^3 = U \oplus W$.

❚ Note first that $U \cap W = \{0\}$, for $v = (a, b, c) \in U \cap W$ implies that $a = b = c$ and $a = 0$ which implies $a = 0$, $b = 0$, $c = 0$, i.e., $v = (0, 0, 0)$.

We also claim that $\mathbf{R}^3 = U + W$. For if $v = (a, b, c) \in \mathbf{R}^3$, then $v = (a, a, a) + (0, b - a, c - a)$ where $(a, a, a) \in U$ and $(0, b - a, c - a) \in W$. Both conditions, $U \cap W = \{0\}$ and $\mathbf{R}^3 = U + W$, imply $\mathbf{R}^3 = U \oplus W$.

7.135 Let V be the vector space of n-square matrices over a field $\mathbf{R}$. Let U and W be the subspaces of symmetric and antisymmetric matrices, respectively. Show that $V = U \oplus W$. (The matrix M is symmetric iff $M = M^T$, and antisymmetric iff $M^T = -M$.)

❚ We first show that $V = U + W$. Let A be any arbitrary n-square matrix. Note that $A = \frac{1}{2}(A + A^T) + \frac{1}{2}(A - A^T)$. We claim that $\frac{1}{2}(A + A^T) \in U$ and that $\frac{1}{2}(A - A^T) \in W$. For $(\frac{1}{2}(A + A^T))^T =$

$\frac{1}{2}((A + A^T)^T = \frac{1}{2}(A^T + A^{TT}) = \frac{1}{2}(a + A^T)$,   that is, $\frac{1}{2}(A + A^T)$ is symmetric.   Furthermore, $(\frac{1}{2}(A - A^T))^T = \frac{1}{2}(A - A^T)^T = \frac{1}{2}(A^T - A) = -\frac{1}{2}(A - A^T)$,   that is, $\frac{1}{2}(A - A^T)$ is antisymmetric.

We next show that $U \cap W = \{0\}$.   Suppose $M \in U \cap W$.   Then $M = M^T$ and $M^T = -M$ which implies $M = -M$ or $M = 0$.   Hence $U \cap W = \{0\}$.   Accordingly, $V = U \oplus W$.

7.136   Show $S + T = T + S$ for any subsets S, T of a vector space V.

▌ Since $u + w = w + u$ for any vectors $u, w \in V$,   we have $S + T = \{u + w : u \in S, w \in T\} = \{w + u : u \in S, w \in T\} = T + S$.

7.137   Show $(S_1 + S_2) + S_3 = S_1 + (S_2 + S_3)$ for any subsets S_1, S_2, S_3 of a vector space V.

▌ Since $(u + v) + w = u + (v + w)$ for any vectors $u, v, w \in V$,   we have

$$(S_1 + S_2) + S_3 = \{(u + v) + w : u \in S_1, v \in S_2, w \in S_3\}$$
$$= \{u + (v + w) : u \in S_1, v \in S_2, W \in S_3\} = S_1 + (S_2 + S_3)$$

7.138   Show that $S + V = V + S = V$ for any subset S of a vector space V.

▌ Since $S \subseteq V$ and $V \subseteq V$,   we have $S + V \subseteq V$.   Consider any vector $v \in V$.   Let $s \in S$.   Then $v - s \in V$.   Since $v = s + (v - s)$,   we have $v \in S + V$.   Hence $V \subseteq S + V$.   Both inclusions yield $S + V = V$.   By Problem 7.136, $V + S = S + V = V$.

7.139   Show $S + \{0\} = \{0\} + S = S$ for any subset S of a vector space V.

▌ For any $u \in S$,   we have $u + 0 = 0 + u = u$.   Hence $S + \{0\} = \{u + 0 : u \in S\} = \{u : u \in S\} = S$. Thus $S + \{0\} = S$.   By Problem 7.136, $\{0\} + S = S + \{0\} = S$.

7.140   Suppose U, V, and W are subspaces of a vector space.   Prove that $(U \cap V) + (U \cap W) \subseteq U \cap (V + W)$.

▌ Let $u \in (U \cap V) + (U \cap W)$.   Then $u = u_1 + u_2$ where $u_1 \in U \cap V$ and $u_2 \in U \cap W$.   Hence $u_1, u_2 \in U$.   Since U is a subspace, $u = u_1 + u_2 \in U$.   Also, $u_1 \in V$ and $u_2 \in W$,   therefore, $u = u_1 + u_2 \in V + W$.   Thus $u \in U \cap (V + W)$ and so $(U \cap V) + (U \cap W) \subseteq U \cap (V + W)$.

7.141   Find subspaces U, V, W of $\mathbf{R}^2$ such that $(U \cap V) + (U \cap W) \neq U \cap (V + W)$.

▌ Let $U = \{(a, b : a = b\}$ (the line $y = x$), $V = \{(a, 0)\}$ (the x axis) and $W = \{(0, b)\}$ (the y axis). Then $U \cap V = \{0\}$ and $U \cap W = \{0\}$,   and hence $(U \cap V) + (U \cap W) = \{0\}$.   On the other hand, $\mathbf{R}^2 = V + W$,   so $U \cap (V + W) = U \cap \mathbf{R}^2 = U \neq \{0\} = (U \cap V) + (U \cap W)$.

7.142   Let V be the vector space of n-square matrices over a field K.   Let U be the subspace of upper triangular matrices and W the subspace of lower triangular matrices.   Note $V = U + W$.   Show that $V \neq U \oplus W$.

▌ $U \cap W \neq \{0\}$ since $U \cap W$ consists of all the diagonal matrices.   Thus the sum cannot be direct.

7.143   Let V be the vector space of all functions from the real field $\mathbf{R}$ into $\mathbf{R}$.   Let U be the subspace of even functions and W the subspace of odd functions.   Show that $V = U \oplus W$.   [Recall that f is even iff $f(-x) = f(x)$ and f is odd iff $f(-x) = -f(x)$.]

▌ $f(x) = \frac{1}{2}(f(x) + f(-x)) + \frac{1}{2}(f(x) - f(-x))$,   where $\frac{1}{2}(f(x) + f(-x))$ is even and $\frac{1}{2}(f(x) - f(-x))$ is odd.   Thus $V = U + W$.   Suppose $f \in U \cap W$.   Then $f(x) = f(-x)$ and $f(x) = -f(-x)$.   Hence $f(-x) = -f(-x)$ and so $f(-x) = 0$.   Thus, for every $x \in \mathbf{R}$, $f(x) = 0$.   Therefore $f = 0$,   the zero function.   Thus $U \cap W = \{0\}$ and so $V = U \oplus W$.

7.144   Suppose $W_1, W_2, \ldots, W_r$ are subspaces of a vector space V.   Discuss the difference between $V = W_1 + W_2 + \cdots + W_r$ and $V = W_1 \oplus W_2 \oplus \cdots \oplus W_r$.

▌ Suppose each $v \in V$ can be written as a sum in the form $v = w_1 + w_2 + \cdots + w_r$ where $w_i \in W_i$. Then $V = W_1 + W_2 + \cdots + W_r$.   If such a sum for v is unique, then the sum is direct, i.e., $V = W_1 \oplus W_2 \oplus \cdots \oplus W_r$.

7.145   Let W_1, W_2, W_3 be the x, y, and z axis, respectively, in $\mathbf{R}^3$.   Show that $\mathbf{R}^3 = W_1 \oplus W_2 \oplus W_3$.

▌ Any vector $(a, b, c) \in \mathbf{R}^3$ can be written uniquely as a sum of a vector in W_1, a vector in W_2, and a vector in W_3 as follows:

$$(a, b, c) = (a, 0, 0) + (0, b, 0) + (0, 0, c)$$

Thus $\mathbf{R}^3 = W_1 \oplus W_2 \oplus W_3$.

7.146 Suppose $W_1, W_2, \ldots, W_r$ are subspaces of V such that $V = W_1 + W_2 + \cdots + W_r$. Suppose $0 \in V$ can be written uniquely as a sum $0 = w_1 + w_2 + \cdots + w_r$, where $w_i \in W_i$. Show that $V = W_1 \oplus W_2 \oplus \cdots \oplus W_r$, i.e., that the sum is direct.

  ▮ Since $0 = 0_1 + \cdots + 0_r$ where 0_i is the zero vector in W_i, this is the unique sum for $0 \in V$. Let $v \in V$ and suppose $v = u_1 + u_2 + \cdots + u_r$ and $v = w_1 + w_2 + \cdots + w_r$, where $u_i, w_i \in W_i$. Then $0 = v - v = (u_1 - w_1) + (u_2 - w_2) + \cdots + (u_r - v_r)$ where $u_i - v_i \in W_i$. Since such a sum for 0 is unique, $u_i - v_i = 0$, for every i, and hence $u_i = v_i$, for every i. Thus such a sum for v is also unique, and so $V = W_1 \oplus W_2 \oplus \cdots \oplus W_r$.

7.147 Suppose W_1, W_2, and W_3 are the x axis, the y axis, and the line $y = x$, respectively, in the plane $\mathbf{R}^2$. Show that $\mathbf{R}^2 \neq W_1 \oplus W_2 \oplus W_3$ (even though $\mathbf{R}^2 = W_1 + W_2 + W_3$ and $W_i \cap W_j = \{0\}$ for $i \neq j$).

  ▮ Show that a vector $v \in \mathbf{R}^2$, say $v = (0, 0)$, can be written in more than one way as a sum of a vector in W_1, a vector in W_2, and a vector in W_3:

$$(0, 0) = (0, 0) + (0, 0) + (0, 0) = (1, 0) + (0, 1) + (-1, -1)$$

Thus $\mathbf{R}^2 \neq W_1 \oplus W_2 \oplus W_3$.

7.148 Let U and W be vector spaces over a field K. Define the external direct sum of U and W.

  ▮ Let V be the set of ordered pairs (u, w) where u belongs to U and w to W: $V = \{(u, w): u \in U, w \in W\}$. Then V is a vector space over K with addition in V and scalar multiplication on V defined by

$$(u, w) + (u', w') = (u + u', w + w') \qquad \text{and} \qquad k(u, w) = (ku, kw)$$

where $u, u' \in U$, $w, w' \in W$ and $k \in K$. (This space V is called the *external direct sum* of U and W.)

  Problems 7.149–7.152 refer to a vector space V which is the external direct sum (see Problem 7.148) of vector spaces U and W over a field K. Also, suppose 0_1 and 0_2 are, respectively, the zero vectors of U and W.

7.149 Show that $\theta = (0_1, 0_2)$ is the zero vector of V.

  ▮ Let $v \in V$, say $v = (u, w)$ where $u \in U$ and $w \in W$. Then $v + \theta = (u, w) + (0_1, 0_2) = (u + 0_1, w + 0_2) = (u, w) = v$. Similarly, $\theta + v = v$. Hence θ is the zero vector in V.

7.150 Find the negative $-v$ of a vector $v \in V$.

  ▮ Suppose $v = (u, w)$ where $u \in U$ and $w \in W$. Then $-v = (-u, -w)$ since we have $(-u, -w) + (u, w) = (0_1, 0_2) = \theta$.

7.151 Let $\hat{U} = \{v \in V : v = (u, 0_2)\}$ and $\hat{W} = \{v \in V : v = (0_1, w)\}$. Show that $\hat{U}$ and $\hat{W}$ are subspaces of V.

  ▮ First we have $\theta = (0_1, 0_2) \in \hat{U}$. Suppose $v_1, v_2 \in \hat{U}$; say $v_1 = (u_1, 0_2)$ and $v_2 = (u_2, 0_2)$. Then

$$v_1 + v_2 = (u_1, 0_2) + (u_2, 0_2) = (u_1 + u_2, 0_2)$$
$$kv_1 = k(u_1, 0_2) = (ku_1, k0_2) = (ku_1, 0_2)$$

where $k \in K$. Thus $v_1 + v_2$ and kv_1 belong to $\hat{U}$, and so $\hat{U}$ is a subspace of V. Similarly, $\hat{W}$ is a subspace of V.

7.152 Show that $V = \hat{U} \oplus \hat{W}$.

  ▮ Let $v \in V$. Then $v = (u, w)$ where $u \in U$ and $w \in W$. Hence $v = (u, 0_2) + (0_1, w) \in \hat{U} + \hat{W}$. Thus $V = \hat{U} + \hat{W}$. On the other hand, suppose $v = (u, w) \in \hat{U} + \hat{W}$. Then $w = 0_2$ since $v \in \hat{U}$ and $u = 0_1$ since $v \in \hat{W}$. Thus $v = \theta$ and so $\hat{U} \cap \hat{W} = \{0\}$. Accordingly $V = \hat{U} \oplus \hat{W}$.

CHAPTER 8
Linear Dependence, Basis, Dimension

The notation in this chapter is the same as in the preceding chapter; i.e., V denotes a vector space, K the field of scalars, u, v, w vectors in V, and a, b, c, k scalars in K [with or without subscripts].

8.1 ELEMENTARY PROPERTIES OF LINEAR DEPENDENCE AND INDEPENDENCE

8.1   Define linear dependence and linear independence.

▌ Let V be a vector space over a field K. The vectors $v_1, \ldots, v_m \in V$ are said to be *linearly dependent over K*, or simply *dependent*, if there exist scalars $a_1, \ldots, a_m \in K$, not all of them 0, such that

$$a_1 v_1 + a_2 v_2 + \cdots + a_m v_m = 0 \qquad\qquad (1)$$

Otherwise, the vectors are said to be *linearly independent over K*, or simply *independent*.

    Remark: Observe that the relation *(1)* will always hold if the a's are all 0. If this relation holds only in this case, i.e.,

$$a_1 v_1 + a_2 v_2 + \cdots + a_m v_m = 0 \qquad \text{implies} \qquad a_1 = 0, \ldots, a_m = 0$$

then the vectors are linearly independent. On the other hand, if the relation *(1)* also holds when one of the a's is not 0, then the vectors are linearly dependent.

8.2   Show that if 0 is one of the vectors $v_1, \ldots, v_m$, say $v_1 = 0$, then the vectors must be linearly dependent.

▌ We have $1v_1 + 0v_2 + \cdots + 0v_m = 1 \cdot 0 + 0 + \cdots + 0 = 0$ and the coefficient of v_1 is not 0.

8.3   Show that any nonzero vector v is, by itself, linearly independent.

▌ Suppose $kv = 0$, but $v \neq 0$. Then $k = 0$. Hence v is linearly independent.

8.4   Suppose $m > 1$. Show that the vectors $v_1, \ldots, v_m$ are linearly dependent if and only if one of them is a linear combination of the others.

▌ Suppose, say, v_i is a linear combination of the others:

$$v_i = a_1 v_1 + \cdots + a_{i-1} v_{i-1} + a_{i+1} v_{i+1} + \cdots + a_m v_m$$

Then by adding $-v_i$ to both sides, we obtain $a_1 v_1 + \cdots + a_{i-1} v_{i-1} - v_i + a_{i+1} v_{i+1} + \cdots + a_m v_m = 0$, where the coefficient of v_i is not 0; hence the vectors are linearly dependent. Conversely, suppose the vectors are linearly dependent, say, $b_1 v_1 + \cdots + b_j v_j + \cdots + b_m v_m = 0$, where $b_j \neq 0$. Then $v_j = -b_j^{-1} b_1 v_1 - \cdots - b_j^{-1} b_{j-1} v_{j-1} - b_j^{-1} b_{j+1} v_{j-1} - \cdots - b_j^{-1} b_m v_m$ and so v_j is a linear combination of the other vectors.

8.5   Define a dependent or independent set of vectors.

▌ A set $\{v_1, \ldots, v_m\}$ is called a *dependent* or *independent set* according as the vectors $v_1, \ldots, v_m$ are linearly dependent or independent. An infinite set S of vectors is linearly dependent if there exist vectors $u_1, \ldots, u_k$ in S which are linearly dependent; otherwise S is linearly independent. The empty set $\emptyset$ is defined to be linearly independent.

8.6   Show that if two of the vectors $v_1, \ldots, v_m$ are equal, say $v_1 = v_2$, then the vectors are linearly dependent.

▌ We have $v_1 - v_2 + 0v_3 + \cdots + 0v_m = 0$ and the coefficient of v_1 is not 0.

8.7   Show that two vectors v_1 and v_2 are dependent if and only if one of them is a multiple of the other.

❙ Suppose v_1 and v_2 are dependent. Then there exist scalars a, b, not both 0, such that $av_1 + bv_2 = 0$. Say $a \neq 0$. Then $v_1 = (-b/a)v_2$. Conversely, suppose $v_1 = kv_2$. Then $v_1 - kv_2 = 0$ where 1 is the coefficient of v_1, and hence v_1 and v_2 are dependent. [Note that this is a special case of Problem 8.4.]

8.8 Describe geometrically the linear dependence of two vectors and of three vectors in real space $\mathbf{R}^3$.

❙ Two vector u and v in $\mathbf{R}^3$ are dependent if and only if they lie on the same line through the origin. Three vectors u, v, and w in $\mathbf{R}^3$ are dependent if and only if they lie on the same plane through the origin.

8.9 Show that if the set $\{v_1, \ldots, v_m\}$ is dependent, then any rearrangement of the vectors $\{v_{i_1}, v_{i_2}, \ldots, v_{i_m}\}$ is also dependent. [Thus, if $v_1, \ldots, v_m$ are independent, then any rearrangement is also independent.]

❙ Suppose $v_1, v_2, \ldots, v_m$ are dependent. Then there exist scalars, $a_1, a_2, \ldots, a_m$, not all 0, such that $a_1 v_1 + a_2 v_2 + \cdots + a_m v_m = 0$. Then $a_{i_1} v_{i_1} + a_{i_2} v_{i_2} + \cdots + a_{i_m} v_{i_m} = 0$ and some $a_{i_j} \neq 0$. Thus $v_{i_1}, v_{i_2}, \ldots, v_{i_m}$ are dependent.

8.10 Suppose $S = \{v_1, \ldots, v_m\}$ contains a dependent subset, say $\{v_1, \ldots, v_r\}$. Show that S is also dependent. Hence every subset of an independent set is independent.

❙ Since $\{v_1, \ldots, v_r\}$ is dependent, there exist scalars $a_1, \ldots, a_r$, not all 0, such that $a_1 v_1 + a_2 v_2 + \cdots + a_r v_r = 0$. Hence there exist scalars $a_1, \ldots, a_r, 0, \ldots, 0$, not all 0, such that $a_1 v_1 + \cdots + a_r v_r + 0v_{r+1} + \cdots + 0v_m = 0$. Accordingly, S is dependent.

8.11 Suppose $\{v_1, \ldots, v_m\}$ is independent, but $\{v_1, \ldots, v_m, w\}$ is dependent. Show that w is a linear combination of the v_1.

❙ Since $\{v_1, \ldots, v_m, w\}$ is dependent, there exist scalars $a_1, \ldots, a_m, b$, not all 0, such that $a_1 v_1 + \cdots + a_m v_m + bw = 0$. If $b = 0$, then one of the a_i is not zero and $a_1 v_1 + \cdots + a_m v_m = 0$. But this contradicts the hypothesis that $\{v_1, \ldots, v_m\}$ is independent. Accordingly, $b \neq 0$ and so $w = b^{-1}(-a_1 v_1 - \cdots - a_m v_m) = -b^{-1} a_1 v_1 - \cdots - b^{-1} a_m v_m$. That is, w is a linear combination of the v_i.

8.12 Suppose $A_1, A_2, \ldots$ are linearly independent sets of vectors, and that $A_1 \subset A_2 \subset \cdots$. Show that the union $A = A_1 \cup A_2 \cup \cdots$ is also linearly independent.

❙ Suppose A is linearly dependent. Then there exist vectors $v_1, \ldots, v_n \in A$ and scalars $a_1, \ldots, a_n \in K$, not all of them 0, such that

$$a_1 v_1 + a_2 v_2 + \cdots + a_n v_n = 0 \qquad (1)$$

Since $A = \cup A_i$ and the $v_i \in A$, there exist sets $A_{i_1}, \ldots, A_{i_n}$ such that

$$v_1 \in A_{i_1} \qquad v_2 \in A_{i_2} \qquad \ldots \qquad v_n \in A_{i_n}$$

Let k be the maximum index of the sets A_{i_j}: $k = \max(i_1, \ldots, i_n)$. It follows then, since $A_1 \subset A_2 \subset \cdots$, that each A_{i_j} is contained in A_k. Hence $v_1, v_2, \ldots, v_n \in A_k$ and so, by (1), A_k is linearly dependent, which contradicts our hypothesis. Thus A is linearly independent.

8.2 LINEAR DEPENDENCE OF VECTORS

8.13 Determine whether u and v are linearly dependent where (a) $u = (3, 4)$, $v = (1, -3)$; (b) $u = (2, -3)$, $v = (6, -9)$.

❙ Two vectors u and v are dependent if and only if one is a multiple of the other. (a) No; neither is a multiple of the other. (b) Yes; for $v = 3u$.

8.14 Determine whether u and v are linearly dependent where (a) $u = (4, 3, -2)$, $v = (2, -6, 7)$; (b) $u = (-4, 6, -2)$, $v = (2, -3, 1)$.

❙ (a) No, since neither is a multiple of the other. (b) Yes; for $u = -2v$.

8.15   Determine whether the matrices A and B are dependent where

$$(a) \quad A = \begin{pmatrix} 1 & -2 & 4 \\ 3 & 0 & -1 \end{pmatrix}, \quad B = \begin{pmatrix} 2 & -4 & 8 \\ 6 & 0 & -2 \end{pmatrix} \qquad (b) \quad A = \begin{pmatrix} 1 & 2 & -3 \\ 6 & -5 & 4 \end{pmatrix}, \quad B = \begin{pmatrix} 6 & -5 & 4 \\ 1 & 2 & -3 \end{pmatrix}$$

▌ (a) Yes; for $B = 2A$.   (b) No; since neither is a multiple of the other.

8.16   Determine whether the polynomials u are v are dependent where (a) $u = 2 - 5t + 6t^2 - t^3$, $v = 3 + 2t - 4t^2 + 5t^3$;   (b) $u = 1 - 3t + 2t^2 - 3t^3$, $v = -3 + 9t - 6t^2 + 9t^3$.

▌ (a) No, since neither is a multiple of the other.   (b) Yes; for $v = -3u$.

Problems 8.17–8.20 refer to vectors in real space $\mathbf{R}^3$.

8.17   Determine whether or not the vectors $(1, -2, 1)$, $(2, 1, -1)$, $(7, -4, 1)$ are linearly dependent.

▌ *Method 1.*   Set a linear combination of the vectors equal to the zero vector using unknown scalars x, y, and z:

$$x(1, -2, 1) + y(2, 1, -1) + z(7, -4, 1) = (0, 0, 0)$$

Then
$$(x, -2x, x) + (2y, y, -y) + (7z, -4z, z) = (0, 0, 0)$$
or
$$(x + 2y + 7z, -2x + y - 4z, x - y + z) = (0, 0, 0)$$

Set corresponding components equal to each other to obtain the equivalent homogeneous system, and reduce to echelon form:

$$\begin{array}{ccc}
\begin{aligned}
x + 2y + 7z &= 0 \\
-2x + y - 4z &= 0 \\
x - y + z &= 0
\end{aligned}
& \text{or} \quad
\begin{aligned}
x + 2y + 7z &= 0 \\
5y + 10z &= 0 \\
-3y - 6z &= 0
\end{aligned}
& \text{or} \quad
\begin{aligned}
x + 2y + 7z &= 0 \\
y + 2z &= 0
\end{aligned}
\end{array}$$

The system, in echelon form, has only two nonzero equations in the three unknowns; hence the system has a nonzero solution.   Thus the original vectors are linearly dependent.

Method 2.   Form the matrix whose rows are the given vectors, and reduce to echelon form using the elementary row operations:

$$\begin{pmatrix} 1 & -2 & 1 \\ 2 & 1 & -1 \\ 7 & -4 & 1 \end{pmatrix} \quad \text{to} \quad \begin{pmatrix} 1 & -2 & 1 \\ 0 & 5 & -3 \\ 0 & 10 & -6 \end{pmatrix} \quad \text{to} \quad \begin{pmatrix} 1 & -2 & 1 \\ 0 & 5 & -3 \\ 0 & 0 & 0 \end{pmatrix}$$

Since the echelon matrix has a zero row, the vectors are dependent.

8.18   Determine whether $(1, -3, 7)$, $(2, 0, -6)$, $(3, -1, -1)$, $(2, 4, -5)$ are linearly dependent.

▌ Yes, since any $n + 1$ (or more) vectors in K^n are automatically dependent.

8.19   Determine whether $(1, 2, -3)$, $(1, -3, 2)$, $(2, -1, 5)$ are linearly dependent.

▌ Form the matrix whose rows are given vectors and row reduce the matrix to echelon form:

$$\begin{pmatrix} 1 & 2 & -3 \\ 1 & -3 & 2 \\ 2 & -1 & 5 \end{pmatrix} \quad \text{to} \quad \begin{pmatrix} 1 & 2 & -3 \\ 0 & -5 & 5 \\ 0 & -5 & 11 \end{pmatrix} \quad \text{to} \quad \begin{pmatrix} 1 & 2 & -3 \\ 0 & -5 & 5 \\ 0 & 0 & 6 \end{pmatrix}$$

Since the echelon matrix has no zero rows, the vectors are independent.

8.20   Determine whether $(2, -3, 7)$, $(0, 0, 0)$, $(3, -1, -4)$ are linearly dependent.

▌ Yes, since $0 = (0, 0, 0)$ is one of the vectors.

8.21   Let V be the vector space of 2×2 matrices over $\mathbf{R}$.   Determine whether the matrices $A, B, C \in V$ are dependent where

$$A = \begin{pmatrix} 1 & 1 \\ 1 & 1 \end{pmatrix} \qquad B = \begin{pmatrix} 1 & 0 \\ 0 & 1 \end{pmatrix} \qquad C = \begin{pmatrix} 1 & 1 \\ 0 & 0 \end{pmatrix}$$

▌ Set a linear combination of the matrices A, B, and C equal to the zero matrix using unknown scalars x, y, and z; i.e., set $xA + yB + zC = 0$. Thus

$$x\begin{pmatrix} 1 & 1 \\ 1 & 1 \end{pmatrix} + y\begin{pmatrix} 1 & 0 \\ 0 & 1 \end{pmatrix} + z\begin{pmatrix} 1 & 1 \\ 0 & 0 \end{pmatrix} = \begin{pmatrix} 0 & 0 \\ 0 & 0 \end{pmatrix}$$

or

$$\begin{pmatrix} x & x \\ x & x \end{pmatrix} + \begin{pmatrix} y & 0 \\ 0 & y \end{pmatrix} + \begin{pmatrix} z & z \\ 0 & 0 \end{pmatrix} = \begin{pmatrix} 0 & 0 \\ 0 & 0 \end{pmatrix}$$

or

$$\begin{pmatrix} x+y+z & x+z \\ x & x+y \end{pmatrix} = \begin{pmatrix} 0 & 0 \\ 0 & 0 \end{pmatrix}$$

Set corresponding entries equal to each other to obtain the equivalent homogeneous system of equations:

$$\begin{aligned} x + y + z &= 0 \\ x + \quad\; z &= 0 \\ x \quad\quad\quad &= 0 \\ x + y \quad\;\; &= 0 \end{aligned}$$

Solving the above system we obtain only the zero solution, $x = 0$, $y = 0$, $z = 0$. We have shown that $xA + yB + zC$ implies $x = 0$, $y = 0$, $z = 0$; hence the matrices A, B, and C are linearly independent.

8.22 Determine whether the matrices A, B, C are dependent where

$$A = \begin{pmatrix} 1 & 2 \\ 3 & 1 \end{pmatrix} \qquad B = \begin{pmatrix} 3 & -1 \\ 2 & 2 \end{pmatrix} \qquad C = \begin{pmatrix} 1 & -5 \\ -4 & 0 \end{pmatrix}$$

▌ Set a linear combination of the matrices A, B, and C equal to the zero vector using unknown scalars x, y, and z; i.e., set $xA + yB + zC = 0$. Thus

$$x\begin{pmatrix} 1 & 2 \\ 3 & 1 \end{pmatrix} + y\begin{pmatrix} 3 & -1 \\ 2 & 2 \end{pmatrix} + z\begin{pmatrix} 1 & -5 \\ -4 & 0 \end{pmatrix} = \begin{pmatrix} 0 & 0 \\ 0 & 0 \end{pmatrix}$$

or

$$\begin{pmatrix} x & 2x \\ 3x & x \end{pmatrix} + \begin{pmatrix} 3y & -y \\ 2y & 2y \end{pmatrix} + \begin{pmatrix} z & -5z \\ -4z & 0 \end{pmatrix} = \begin{pmatrix} 0 & 0 \\ 0 & 0 \end{pmatrix}$$

or

$$\begin{pmatrix} x+3y+z & 2x-y-5z \\ 3x+2y-4z & x+2y \end{pmatrix} = \begin{pmatrix} 0 & 0 \\ 0 & 0 \end{pmatrix}$$

Set corresponding entries equal to each other to obtain the equivalent homogeneous system of linear equations and reduce to echelon form:

$$\begin{aligned} x + 3y + z &= 0 \\ 2x - y - 5z &= 0 \\ 3x + 2y - 4z &= 0 \\ x + 2y \quad\;\; &= 0 \end{aligned} \qquad \text{or} \qquad \begin{aligned} x + 3y + z &= 0 \\ -7y - 7z &= 0 \\ -7y - 7z &= 0 \\ -y - z &= 0 \end{aligned}$$

or finally

$$\begin{aligned} x + 3y + z &= 0 \\ y + z &= 0 \end{aligned}$$

The system in echelon form has a free variable and hence a nonzero solution, for example, $x = 2$, $y = -1$, $z = 1$. We have shown that $xA + yB + zC = 0$ does not imply that $x = 0$, $y = 0$, $z = 0$; hence the matrices are linearly dependent.

8.23 Let V be the vector space of polynomials of degree 3 over **R**. Determine whether $u = t^3 - 3t^2 + 5t + 1$, $v = t^3 - t^2 + 8t + 2$, $w = 2t^3 + 4t^2 + 9t + 5$.

▌ Set a linear combination of the polynomials u, v, and w equal to the zero polynomial using unknown scalars x, y, and z; i.e., set $xu + yv + zw = 0$. Thus

$$x(t^2 - 3t^2 + 5t + 1) + y(t^3 - t^2 + 8t + 2) + z(2t^3 - 4t^2 + 9t + 5) = 0$$

or

$$xt^3 - 3xt^3 + 5xt + x + yt^3 - yt^2 + 8yt + 2y + 2zt^3 - 4zt^2 + 9zt + 5z = 0$$

or

$$(x + y + 2z)t^3 + (-3x - y - 4z)t^2 + (5x + 8y + 9z)t + (x + 2y + 5z) = 0$$

The coefficients of the powers of t must each be 0:

$$x + y + 2z = 0$$
$$-3x - y - 4z = 0$$
$$5x + 8y + 9z = 0$$
$$x + 2y + 5z = 0$$

Solving the above homogeneous system, we obtain only the zero solution: $x = 0$, $y = 0$, $z = 0$; hence u, v, and w are independent.

8.24 Determine whether the polynomials u, v, w are dependent where $u = t^3 + 4t^2 - 2t + 3$, $v = t^3 + 6t^2 - t + 4$, $w = 3t^3 + 8t^2 - 8t + 7$.

▎ Set a linear combination of the polynomials u, v, and w equal to the zero polynomial using unknown scalars x, y, and z; i.e., set $xu + yv + zw = 0$. Thus

$$x(t^3 + 4t^2 - 2t + 3) + y(t^3 + 6t^2 - t + 4) + z(3t^3 + 8t^2 - 8t + 7) = 0$$

or

$$xt^3 + 4xt^2 - 2xt + 3x + yt^3 + 6yt^2 - yt + 4y + 3zt^3 + 8zt^2 - 8zt + 7z = 0$$

or

$$(x + y + 3z)t^3 + (4x + 6y + 8z)t^2 + (-2x - y - 8z)t + (3x + 4y + 7z) = 0$$

Set the coefficients of the powers of t each equal to 0 and reduce the system to echelon form:

$$
\begin{array}{ll}
\begin{aligned}
x + y + 3z &= 0 \\
4x + 6y + 8z &= 0 \\
-2x - y - 8z &= 0 \\
3x + 4y + 7z &= 0
\end{aligned}
\quad \text{or} \quad
&\begin{aligned}
x + y + 3z &= 0 \\
2y - 4z &= 0 \\
y - 2z &= 0 \\
y - 2z &= 0
\end{aligned}
\end{array}
$$

or finally

$$
\begin{aligned}
x + y + 3z &= 0 \\
y - 2z &= 0
\end{aligned}
$$

The system in echelon form has a free variable and hence a nonzero solution. Thus $xu + yv + zw = 0$ does not imply that $x = 0$, $y = 0$, $z = 0$; hence the polynomials are linearly dependent.

8.25 Let V be the vector space of functions from **R** into **R**. Show that $f, g, h \in V$ are linearly independent where $f(t) = e^{2t}$, $g(t) = t^2$, $h(t) = t$.

▎ Set a linear combination of the functions equal to the zero function 0 using unknown scalars x, y, and z: $xf + yg + zh = 0$; and then show that $x = 0$, $y = 0$, $z = 0$. We emphasize that $xf + yg + zh = 0$ means that, for every value of t, $xf(t) + yg(t) + zh(t) = 0$.

In the equation $xe^{2t} + yt^2 + zt = 0$, substitute

$$
\begin{array}{lll}
t = 0 & \text{to obtain} & xe^0 + y0 + z0 = 0 \quad \text{or} \quad x = 0 \\
t = 1 & \text{to obtain} & xe^2 + y + z = 0 \\
t = 2 & \text{to obtain} & xe^4 + 4y + 2z = 0
\end{array}
$$

Solve the system $\begin{cases} x = 0 \\ xe^2 + y + z = 0 \\ xe^4 + 4y + 2z = 0 \end{cases}$ to obtain only the zero solution: $x = 0$, $y = 0$, $z = 0$. Hence f, g, and h are independent.

8.26 Show that the functions $f(t) = \sin t$, $g(t) = \cos t$, $h(t) = t$ are linearly independent.

▎ Form the functional equation $xf + yg + zh = 0$, that is, $x \sin t + y \cos t + zt = 0$, using unknowns x, y, z, and then show $x = 0$, $y = 0$, $z = 0$.

Method 1. In the equation $x \sin t + y \cos t + zt = 0$, substitute

$$
\begin{array}{lll}
t = 0 & \text{to obtain} & x \cdot 0 + y \cdot 1 + z \cdot 0 = 0 \quad \text{or} \quad y = 0 \\
t = \pi/2 & \text{to obtain} & x \cdot 1 + y \cdot 0 + z(\pi/2) = 0 \quad \text{or} \quad x + (\pi/2)z = 0 \\
t = \pi & \text{to obtain} & x \cdot 0 + y(-1) + z \cdot \pi = 0 \quad \text{or} \quad -y + \pi z = 0
\end{array}
$$

Solve the system $\begin{cases} y = 0 \\ x + (\pi/2)z = 0 \\ -y + \pi z = 0 \end{cases}$ to obtain only the zero solution: $x = 0$, $y = 0$, $z = 0$. Hence f, g, and h are independent.

Method 2. Take the first, second, and third derivatives of $x \sin t + y \cos t + zt = 0$ with respect to t to get

$$x \cos t - y \sin t + z = 0 \qquad (1)$$
$$-x \sin t - y \cos t = 0 \qquad (2)$$
$$-x \cos t + y \sin t = 0 \qquad (3)$$

Add *(1)* and *(3)* to obtain $z = 0$. Multiply *(2)* by $\sin t$ and *(3)* by $\cos t$, and then add:

$$\begin{array}{ll} \sin t \times (2): & -x \sin^2 t - y \sin t \cos t = 0 \\ \cos t \times (3): & -x \cos^2 t + y \sin t \cos t = 0 \\ \hline & -x(\sin^2 t + \cos^2 t) = 0 \qquad \text{or} \qquad x = 0 \end{array}$$

Last, multiply *(2)* by $-\cos t$ and *(3)* by $\sin t$; and then add to obtain

$$y(\cos^2 t + \sin^2 t) = 0 \qquad \text{or} \qquad y = 0$$

Since $\qquad x \sin t + y \cos t + zt = 0 \qquad$ implies $\qquad x = 0, y = 0, z = 0$

f, g, and h are independent.

8.27 Show that the vectors $v = (1 + i, 2i)$ and $w = (1, 1 + i)$ in $\mathbf{C}^2$ are linearly dependent over the complex field $\mathbf{C}$ but are linearly independent over the real field $\mathbf{R}$.

▌ Recall that two vectors are dependent iff one is a multiple of the other. Since the first coordinate of w is 1, v can be a multiple of w iff $v = (1 + i)w$. But $1 + i \notin \mathbf{R}$; hence v and w are independent over $\mathbf{R}$. Since $(1 + i)w = (1 + i)(1, 1 + i) = (1 + i, 2i) = v$ and $1 + i \in \mathbf{C}$, they are dependent over $\mathbf{C}$.

8.28 Let u, v, and w be independent vectors. Show that $u + v$, $u - v$, and $u - 2v + w$ are also independent.

▌ Suppose $x(u + v) + y(u - v) + z(u - 2v + w) = 0$ where x, y, and z are scalars. Then $xu + xv + yu - yv + zu - 2zv + zw = 0$ or $(x + y + z)u + (x - y - 2z)v + zw = 0$. But u, v, and w are linearly independent; hence the coefficients in the above relation are each 0:

$$\begin{array}{r} x + y + \ z = 0 \\ x - y - 2z = 0 \\ z = 0 \end{array}$$

The only solution to the above system is $x = 0$, $y = 0$, $z = 0$. Hence $u + v$, $u - v$, and $u - 2v + w$ are independent.

8.3 THEOREMS ON BASES AND DIMENSION

8.29 Define a basis of a vector space V.

▌ A sequence of vectors $\{u_1, u_2, \ldots, u_n\}$ is a basis of V if (1) $u_1, u_2, \ldots, u_n$ are linearly independent and (2) $u_1, u_2, \ldots, u_n$ span V.

8.30 Define the dimension of a vector space V.

▌ A vector space V is said to be of finite dimension n or to be n-dimensional, written $\dim V = n$, if V contains a basis with n elements. [This definition of dimension is well-defined by Theorem 8.4 which states that any two bases have the same number of elements.]

The vector space $\{0\}$ is defined to have dimension 0. [In a certain sense this agrees with the above definition since, by definition, $\emptyset$ is independent and generates $\{0\}$.] When a vector space is not of finite dimension, it is said to be of *infinite dimension*.

Lemma 8.1: The nonzero vectors $v_1, \ldots, v_m$ are linearly dependent if and only if one of them, say v_i, is a linear combination of the preceding vectors:

$$v_i = a_1 v_1 + \cdots + a_{i-1} v_{i-1}$$

8.31 Prove Lemma 8.1.

∎ Suppose $v_i = a_1 v_1 + \cdots + a_{i-1} v_{i-1}$. Then $a_1 v_1 + \cdots + a_{i-1} v_{i-1} - v_i + 0 v_{i+1} + \cdots + 0 v_m = 0$ and the coefficient of v_i is not 0. Hence the v_i are linearly dependent.

Conversely, suppose the v_i are linearly dependent. Then there exist scalars $a_1, \ldots, a_m$, not all 0, such that $a_1 v_1 + \cdots + a_m v_m = 0$. Let k be the largest integer such that $a_k \neq 0$. Then

$$a_1 v_1 + \cdots + a_k v_k + 0 v_{k+1} + \cdots + 0 v_m = 0 \qquad \text{or} \qquad a_1 v_1 + \cdots + a_k v_k = 0$$

Suppose $k = 1$; then $a_1 v_1 = 0$, $a_1 \neq 0$, and so $v_1 = 0$. But the v_i are nonzero vectors; hence $k > 1$ and $v_k = -a_k^{-1} a_1 v_1 - \cdots - a_k^{-1} a_{k-1} v_{k-1}$. That is, v_k is a linear combination of the preceding vectors.

Theorem 8.2: The nonzero rows $R_1, \ldots, R_n$ of a matrix in echelon form are linearly independent.

8.32 Prove Theorem 8.2.

∎ Suppose $\{R_n, R_{n-1}, \ldots, R_1\}$ is dependent. Then one of the rows, say R_m, is a linear combination of the preceding rows:

$$R_m = a_{m+1} R_{m+1} + a_{m+2} R_{m+2} + \cdots + a_n R_n \qquad\qquad (1)$$

Now suppose the kth component of R_m is its first nonzero entry. Then, since the matrix is in echelon form, the kth components of $R_{m+1}, \ldots, R_n$ are all 0, and so the kth component of (1) is $a_{m+1} \cdot 0 + a_{m+2} \cdot 0 + \cdots + a_n \cdot 0 = 0$. But this contradicts the assumption that the kth component of R_m is not 0. Thus $R_1, \ldots, R_n$ are independent.

8.33 Suppose $\{v_1, \ldots, v_m\}$ spans a vector space V and suppose $w \in V$. Show that $\{w, v_1, \ldots, v_m\}$ is linearly dependent and spans V.

∎ The vector w is a linear combination of the v_i since $\{v_i\}$ spans V. Accordingly, $\{w, v_1, \ldots, v_m\}$ is linearly dependent. Clearly, w with the v_i span V since the v_i by themselves span V. That is, $\{w, v_1, \ldots, v_m\}$ spans V.

8.34 Suppose $\{v_1, \ldots, v_m\}$ spans a vector space V and suppose v_i is a linear combination of the preceding vectors. Show that $\{v_1, \ldots, v_{i-1}, v_{i+1}, \ldots, v_m\}$ spans V.

∎ Suppose $v_i = k_1 v_1 + \cdots + k_{i-1} v_{i-1}$. Let $u \in V$. Since $\{v_i\}$ spans V, u is a linear combination of the v_i, say, $u = a_1 v_1 + \cdots + a_m v_m$. Substituting for v_i, we obtain

$$u = a_1 v_1 + \cdots + a_{i-1} v_{i-1} + a_i(k_1 v_1 + \cdots + k_{i-1} v_{i-1}) + a_{i+1} v_{i+1} + \cdots + a_m v_m$$
$$= (a_1 + a_i k_1) v_1 + \cdots + (a_{i-1} + a_i k_{i-1}) v_{i-1} + a_{i+1} v_{i+1} + \cdots + a_m v_m$$

Thus $\{v_1, \ldots, v_{i-1}, v_{i+1}, \ldots, v_m\}$ spans V. In other words, we can delete v_i from the spanning set and still retain a spanning set.

Lemma 8.3 ("Replacement" Lemma): Suppose $\{v_1, \ldots, v_n\}$ spans a vector space V and $\{w_1, \ldots, w_m\}$ is linearly independent. Then $m \leq n$ and V is spanned by a set of the form $\{w_1, \ldots, w_m, v_{i_1}, \ldots, v_{i_{n-m}}\}$. Thus, in particular, any $n + 1$ or more vectors in V are linearly dependent.

8.35 Prove Lemma 8.3.

▌ It suffices to prove the theorem in the case that the v_i are all not 0.   (Prove!)   Since the $\{v_i\}$ generates V, we have, by Problem 8.33, that

$$\{w_1, v_1, \ldots, v_n\} \tag{1}$$

is linearly dependent and also generates V.   By Lemma 8.1, one of the vectors in *(1)* is a linear combination of the preceding vectors.   This vector cannot be w_1, so it must be one of the v's, say v_j. Thus by the preceding problem we can delete v_j from the generating set *(1)* and obtain the generating set

$$\{w_1, v_1, \ldots, v_{j-1}, v_{j+1}, \ldots, v_n\} \tag{2}$$

Now we repeat the argument with the vector w_2.   That is, since *(2)* generates V, the set

$$\{w_1, w_2, v_1, \ldots, v_{j-1}, v_{j+1}, \ldots, v_n\} \tag{3}$$

is linearly dependent and also generates V.   Again by Lemma 8.1, one of the vectors in *(3)* is a linear combination of the preceding vectors.   We emphasize that this vector cannot be w_1 or w_2 since $\{w_1, \ldots, w_m\}$ is independent; hence it must be one of the v's, say v_k.   Thus by the preceding problem we can delete v_k from the generating set *(3)* and obtain the generating set $\{w_1, w_2, v_1, \ldots, v_{j-1}, v_{j+1}, \ldots, v_{k-1}, v_{k+1}, \ldots, v_n\}$.

We repeat the argument with w_3 and so forth. At each step we are able to add one of the w's and delete one of the v's in the generating set.   If $m \leq n$,   then we finally obtain a generating set of the required form:

$$\{w_1, \ldots, w_m, v_{i_1}, \ldots, v_{i_{n-m}}\}$$

Last, we show that $m > n$ is not possible.   Otherwise, after n of the above steps, we obtain the generating set $\{w_1, \ldots, w_n\}$.   This implies that w_{n+1} is a linear combination of $w_1, \ldots, w_n$ which contradicts the hypothesis that $\{w_i\}$ is linearly independent.

Theorem 8.4: Let V be a finite-dimensional vector space.   Then every basis of V has the same number of vectors.

8.36   Prove Theorem 8.4 (a basic result of linear algebra).

▌ Suppose $\{e_1, e_2, \ldots, e_n\}$ is a basis of V and $\{f_1, f_2, \ldots\}$ is another basis of V.   Since $\{e_i\}$ generates V, the basis $\{f_1, f_2, \ldots\}$ must contain n or less vectors, or else it is dependent by the preceding problem.   On the other hand, if the basis $\{f_1, f_2, \ldots\}$ contains less than n vectors, then $\{e_1, \ldots, e_n\}$ is dependent by the preceding problem.   Thus the basis $\{f_1, f_2, \ldots\}$ contains exactly n vectors, and so the theorem is true.

8.37   Define a maximal independent subset of a set S of vectors in V.

▌ A subset $\{v_1, \ldots, v_m\}$ of S is a maximal independent subset of S if it is independent and if, for any $w \in S$,   the set $\{v_1, \ldots, v_m, w\}$ is dependent.

Theorem 8.5: Suppose $\{v_1, \ldots, v_m\}$ is a maximal independent subset of a set S where S spans a vector space V.   Then $\{v_1, \ldots, v_m\}$ is a basis of V.

8.38   Prove Theorem 8.5.

▌ Suppose $w \in S$.   Then, since $\{v_i\}$ is a maximal independent subset of S, $\{v_1, \ldots, v_m, w\}$ is dependent. Thus w is a linear combination of the v_i, that is, $w \in \text{span}(v_i)$.   Hence $S \subseteq \text{span}(v_i)$.   This leads to $V = \text{span}(S) \subseteq \text{span}(v_i) \subseteq V$.   Thus $\{v_i\}$ spans V and, since it is independent, it is a basis of V.

8.39   Suppose V is generated by a finite set S.   Show that V is of finite dimension and, in particular, a subset of S is a basis of V.

▌ *Method 1.*   Of all the independent subsets of S, and there is a finite number of them since S is finite, one of them is maximal.   By the preceding problem this subset of S is a basis of V.

Method 2. If S is independent, it is a basis of V. If S is dependent, one of the vectors is a linear combination of the preceding vectors. We may delete this vector and still retain a generating set. We continue this process until we obtain a subset which is independent and generates V, i.e., is a basis of V.

8.40 Consider a finite sequence of vectors $S = \{v_1, v_2, \ldots, v_n\}$. Let T be the sequence of vectors obtained from S by one of the following "elementary operations": (i) interchange two vectors, (ii) multiply a vector by a nonzero scalar, (iii) add a multiple of one vector to another. Show that S and T generate the same space W. Also show that T is independent if and only if S is independent.

■ Observe that, for each operation, the vectors in T are linear combinations of vectors in S. On the other hand, each operation has an inverse of the same type (Prove!); hence the vectors in S are linear combinations of vectors in T. Thus S and T generate the same space W. Also, T is independent if and only if $\dim W = n$, and this is true iff S is also independent.

8.41 Let $A = (a_{ij})$ and $B = (b_{ij})$ be row equivalent $m \times n$ matrices over a field K, and let $v_1, \ldots, v_n$ be any vectors in a vector space V over K. Let

$$
\begin{aligned}
u_1 &= a_{11}v_1 + a_{12}v_2 + \cdots + a_{1n}v_n \qquad\qquad w_1 = b_{11}v_1 + b_{12}v_2 + \cdots + b_{1n}v_n \\
u_2 &= a_{21}v_1 + a_{22}v_2 + \cdots + a_{2n}v_n \qquad\qquad w_2 = b_{21}v_1 + b_{22}v_2 + \cdots + b_{2n}v_n \\
&\,\cdots\cdots\cdots\cdots\cdots\cdots\cdots\cdots\cdots\cdots\qquad\qquad\quad\cdots\cdots\cdots\cdots\cdots\cdots\cdots\cdots\cdots\cdots \\
u_m &= a_{m1}v_1 + a_{m2}v_2 + \cdots + a_{mn}v_n \qquad\quad w_m = b_{m1}v_1 + b_{m2}v_2 + \cdots + b_{mn}v_n
\end{aligned}
$$

Show that $\{u_i\}$ and $\{w_i\}$ generate the same space.

■ Applying an "elementary operation" of the preceding problem to $\{u_i\}$ is equivalent to applying an elementary row operation to the matrix A. Since A and B are row equivalent, B can be obtained from A by a sequence of elementary row operations; hence $\{w_i\}$ can be obtained from $\{u_i\}$ by the corresponding sequence of operations. Accordingly, $\{u_i\}$ and $\{w_i\}$ generate the same space.

Theorem 8.6: Let $v_1, \ldots, v_n$ belong to a vector space V over a field K. Let

$$
\begin{aligned}
w_1 &= a_{11}v_1 + a_{12}v_2 + \cdots + a_{1n}v_n \\
w_2 &= a_{21}v_1 + a_{22}v_2 + \cdots + a_{2n}v_n \\
&\,\cdots\cdots\cdots\cdots\cdots\cdots\cdots\cdots\cdots\cdots \\
w_n &= a_{n1}v_1 + a_{n2}v_2 + \cdots + a_{nn}v_n
\end{aligned}
$$

where $a_{ij} \in K$. Let P be the n-square matrix of coefficients, i.e., let $P = (a_{ij})$.
(i) Suppose P is invertible. Then $\{w_i\}$ and $\{v_i\}$ span the same space; hence $\{w_i\}$ is independent if and only if $\{v_i\}$ is independent.
(ii) Suppose P is not invertible. Then $\{w_i\}$ is dependent.
(iii) Suppose $\{w_i\}$ is independent. Then P is invertible.

8.42 Prove (i) of Theorem 8.6: Suppose P is invertible. Then $\mathrm{span}(w_i) = \mathrm{span}(v_i)$; hence $\{w_i\}$ is independent if and only if $\{v_i\}$ is independent.

■ Since P is invertible, it is row equivalent to the identity matrix I. Hence by the preceding problem $\{w_i\}$ and $\{v_i\}$ generate the same space. Thus one is independent if and only if the other is.

8.43 Prove (ii) of Theorem 8.6: Suppose P is not invertible. Then $\{w_i\}$ is dependent.

■ Since P is not invertible, it is row equivalent to a matrix with a zero row. This means that $\{w_i\}$ generates a space which has a generating set of less than n elements. Thus $\{w_i\}$ is dependent.

8.44 Prove (iii) of Theorem 8.6: Suppose $\{w_i\}$ is independent. Then P is invertible.

■ This is the contrapositive of the statement of (ii) and so it follows from (ii).

8.45 Let K be a subfield of a field L and L a subfield of a field E: that is, $K \subset L \subset E$. [Hence K is a subfield of E.] Suppose that E is of dimension n over L and L is of dimension m over K. Show that E is of dimension mn over K.

▮ Suppose $\{v_1, \ldots, v_n\}$ is a basis of E over L and $\{a_1, \ldots, a_m\}$ is a basis of L over K. We claim that $\{a_i v_j: i = 1, \ldots, m, j = 1, \ldots, n\}$ is a basis of E over K. Note that $\{a_i v_j\}$ contains mn elements.

Let w be any arbitrary element in E. Since $\{v_1, \ldots, v_n\}$ generates E over L, w is a linear combination of the v_i with coefficients in L:

$$w = b_1 v_1 + b_2 v_2 + \cdots + b_n v_n \qquad b_i \in L \tag{1}$$

Since $\{a_1, \ldots, a_m\}$ generates L over K, each $b_i \in L$ is a linear combination of the a_j with coefficients in K:

$$b_1 = k_{11} a_1 + k_{12} a_2 + \cdots + k_{1m} a_m$$
$$b_2 = k_{21} a_1 + k_{22} a_2 + \cdots + k_{2m} a_m$$
$$\cdots\cdots\cdots\cdots\cdots\cdots\cdots\cdots\cdots\cdots\cdots\cdots$$
$$b_n = k_{n1} a_1 + k_{n2} a_2 + \cdots + k_{nm} a_m$$

where $k_{ij} \in K$. Substituting in (1), we obtain

$$w = (k_{11} a_1 + \cdots + k_{1m} a_m) v_1 + (k_{21} a_1 + \cdots + k_{2m} a_m) v_2 + \cdots + (k_{n1} a_1 + \cdots + k_{nm} a_m) v_n$$
$$= k_{11} a_1 v_1 + \cdots + k_{1m} a_m v_1 + k_{21} a_1 v_2 + \cdots + k_{2m} a_m v_2 + \cdots + k_{n1} a_1 v_n + \cdots + k_{nm} a_m v_n$$
$$= \sum_{i,j} k_{ji} (a_i v_j)$$

where $k_{ji} \in K$. Thus w is a linear combination of the $a_i v_j$ with coefficients in K; hence $\{a_i v_i\}$ generates E over K.

The proof is complete if we show that $\{a_i v_j\}$ is linearly independent over K. Suppose, for scalars $x_{ji} \in K$, $\sum_{i,j} x_{ji} (a_i v_j) = 0$; that is,

$$(x_{11} a_1 v_1 + x_{12} a_2 v_1 + \cdots + x_{1m} a_m v_1) + \cdots + (x_{n1} a_1 v_n + x_{n2} a_2 v_n + \cdots + x_{nm} a_m v_n) = 0$$
or
$$(x_{11} a_1 + x_{12} a_2 + \cdots + x_{1m} a_m) v_1 + \cdots + (x_{n1} a_1 + x_{n2} a_2 + \cdots + x_{nm} a_m) v_n = 0$$

Since $\{v_1, \ldots, v_n\}$ is linearly independent over L and since the above coefficients of the v_i belong to L, each coefficient must be 0:

$$x_{11} a_1 + x_{12} a_2 + \cdots + x_{1m} a_m = 0, \quad \ldots, \quad x_{n1} a_1 + x_{n2} a_2 + \cdots + x_{nm} a_n = 0$$

But $\{a_1, \ldots, a_m\}$ is linearly independent over K; hence since the $x_{ji} \in K$,

$$x_{11} = 0, \quad x_{12} = 0, \quad \ldots, \quad x_{1m} = 0, \quad \ldots, \quad x_{n1} = 0, \quad x_{n2} = 0, \quad \ldots, \quad x_{nm} = 0$$

Accordingly, $\{a_i v_j\}$ is linearly independent over K and the theorem is proved.

8.4 BASES AND DIMENSION

8.46 What is meant by the usual basis of the vector space $\mathbf{R}^n$?

▮ Consider the following n vectors in $\mathbf{R}^n$:

$$e_1 = (1, 0, 0, \ldots, 0, 0), \quad e_2 = (0, 1, 0, \ldots, 0, 0), \quad \ldots, \quad e_n = (0, 0, \ldots, 0, 1)$$

These vectors are linearly independent and span $\mathbf{R}^n$. [See Problem 8.49.] Thus the vectors form a basis of $\mathbf{R}^n$ called the *usual basis* of $\mathbf{R}^n$.

8.47 Show that $\dim \mathbf{R}^n = n$.

▮ The above usual basis of $\mathbf{R}^n$ has n vectors; hence $\dim \mathbf{R}^n = n$.

8.48 Let U be the vector space of all 2×3 matrices over a field K. Show that $\dim U = 6$.

▮ The following six matrices,

$$\begin{pmatrix} 1 & 0 & 0 \\ 0 & 0 & 0 \end{pmatrix} \quad \begin{pmatrix} 0 & 1 & 0 \\ 0 & 0 & 0 \end{pmatrix} \quad \begin{pmatrix} 0 & 0 & 1 \\ 0 & 0 & 0 \end{pmatrix} \quad \begin{pmatrix} 0 & 0 & 0 \\ 1 & 0 & 0 \end{pmatrix} \quad \begin{pmatrix} 0 & 0 & 0 \\ 0 & 1 & 0 \end{pmatrix} \quad \begin{pmatrix} 0 & 0 & 0 \\ 0 & 0 & 1 \end{pmatrix}$$

are linearly independent and span U, and hence form a basis of U. [See Problem 8.49.] Thus $\dim U = 6$.

8.49   Let V be the vector space of $m \times n$ matrices over a field K. Let $E_{ij} \in V$ be the matrix with 1 as the ij-entry and 0 elsewhere. Show that $\{E_{ij}\}$ is a basis of V. Thus $\dim V = mn$. [This basis is called the usual basis of V.]

ǀ   We need to show that $\{E_{ij}\}$ spans V and is independent. Let $A = (a_{ij})$ be any matrix in V. Then $A = \sum_{i,j} a_{ij} E_{ij}$. Hence $\{E_{ij}\}$ spans V.

Now suppose that $\sum_{i,j} x_{ij} E_{ij} = 0$ where the x_{ij} are scalars. The ij-entry of $\sum_{i,j} x_{ij} E_{ij}$ is x_{ij}, and the ij-entry of 0 is 0. Thus $x_{ij} = 0$, $i = 1, \ldots, m$, $j = 1, \ldots, n$. Accordingly the matrices E_{ij} are independent. Thus $\{E_{ij}\}$ is a basis of V.

Remark: Viewing a vector in K^n as a $1 \times n$ matrix, we have shown by the above result that the usual basis of $\mathbf{R}^n$ defined in Problem 8.46 is a basis of $\mathbf{R}^n$.

Theorem 8.7:   Suppose $\dim V = n$; say $\{e_1, \ldots, e_n\}$ is a basis of V. Then

(i)   Any set of $n + 1$ or more vectors is linearly dependent.
(ii)   Any linearly independent set is part of a basis.
(iii)   A linearly independent set with n elements is a basis.

8.50   Prove (i) of Theorem 8.7: Any set of $n + 1$ or more vectors is linearly dependent.

ǀ   Since $\{e_1, \ldots, e_n\}$ generates V, any $n + 1$ or more vectors is dependent by Lemma 8.3.

8.51   Prove (ii) of Theorem 8.7: Any linearly independent set is part of a basis.

ǀ   Suppose $\{v_1, \ldots, v_r\}$ is independent. By Lemma 8.3, V is generated by a set of the form $S = \{v_1, \ldots, v_r, e_{i_1}, \ldots, e_{i_{n-r}}\}$. By the preceding problem, a subset of S is a basis. But S contains n elements and every basis of V contains n elements. Thus S is a basis of V and contains $\{v_1, \ldots, v_r\}$ as a subset.

8.52   Prove (iii) of Theorem 8.7: A linearly independent set with n elements is a basis.

ǀ   By (ii), an independent set T with n elements is part of a basis. But every basis of V contains n elements. Thus, T is a basis.

8.53   Show that the following four vectors form a basis of $\mathbf{R}^4$: $(1, 1, 1, 1)$, $(0, 1, 1, 1)$, $(0, 0, 1, 1)$, $(0, 0, 0, 1)$.

ǀ   The vectors form a matrix in echelon form, and so the vectors are linearly independent. Furthermore, since $\dim \mathbf{R}^4 = 4$, they form a basis of $\mathbf{R}^4$.

8.54   Determine whether or not each of the following form a basis of $\mathbf{R}^3$: (a) $(1, 1, 1)$ and $(1, -1, 5)$; (b) $(1, 2, 3)$, $(1, 0, -1)$, $(3, -1, 0)$, and $(2, 1, -2)$.

ǀ   A basis of $\mathbf{R}^3$ must contain exactly three elements, since $\dim \mathbf{R}^3 = 3$. Therefore, neither the vectors in (a) nor the vectors in (b) form a basis of $\mathbf{R}^3$.

8.55   Determine whether the vectors $(1, 1, 1)$, $(1, 2, 3)$, $(2, -1, 1)$ form a basis of $\mathbf{R}^3$.

ǀ   The three vectors form a basis if and only if they are independent. Thus form the matrix whose rows are the given vectors and row reduce to echelon form:

$$\begin{pmatrix} 1 & 1 & 1 \\ 1 & 2 & 3 \\ 2 & -1 & 1 \end{pmatrix} \quad \text{to} \quad \begin{pmatrix} 1 & 1 & 1 \\ 0 & 1 & 2 \\ 0 & -3 & -1 \end{pmatrix} \quad \text{to} \quad \begin{pmatrix} 1 & 1 & 1 \\ 0 & 1 & 2 \\ 0 & 0 & 5 \end{pmatrix}$$

The echelon matrix has no zero rows; hence the three vectors are independent and so form a basis for $\mathbf{R}^3$.

8.56   Determine whether $(1, 1, 2)$, $(1, 2, 5)$, $(5, 3, 4)$ form a basis of $\mathbf{R}^3$.

ǀ   Form the matrix whose rows are the given vectors and row reduce to echelon form:

$$\begin{pmatrix} 1 & 1 & 2 \\ 1 & 2 & 5 \\ 5 & 3 & 4 \end{pmatrix} \quad \text{to} \quad \begin{pmatrix} 1 & 1 & 2 \\ 0 & 1 & 3 \\ 0 & -2 & -6 \end{pmatrix} \quad \text{to} \quad \begin{pmatrix} 1 & 1 & 2 \\ 0 & 1 & 3 \\ 0 & 0 & 0 \end{pmatrix}$$

The echelon matrix has a zero row, i.e., only two nonzero rows; hence the three vectors are dependent and so do not form a basis for $\mathbf{R}^3$.

Problems 8.57–8.59 refer to the vector space V of polynomials in t of degree $\leq n$.

8.57 Show that $\{1, t, t^2, \ldots, t^n\}$ is a basis of V; hence $\dim V = n + 1$.

▮ Clearly each polynomial in V is a linear combination of $1, t, \ldots, t^{n-1}$ and t^n. Furthermore, $1, t, \ldots, t^{n-1}$ and t^n are independent since none is a linear combination of the preceding polynomials. Thus $\{1, t, \ldots, t^n\}$ is a basis of V.

8.58 Show that $\{1, t - 1, (t - 1)^2, \ldots, (t - 1)^n\}$ is a basis of V.

▮ [Since $\dim V = n + 1$, any $n + 1$ independent polynomials form a basis of V.] Now each polynomial in the sequence $1, 1 - t, \ldots, (1 - t)^n$ is of degree higher than the preceding ones and so is not a linear combination of the preceding ones. Thus the $n + 1$ polynomials $1, 1 - t, \ldots, (1 - t)^n$ are independent and so form a basis of V.

8.59 Determine whether or not $\{1 + t, t + t^2, t^2 + t^3, \ldots, t^{n-1} + t^n\}$ is a basis of V.

▮ The polynomials are linearly independent since each one is of degree higher than the preceding ones. However, the set contains only n elements and $\dim V = n + 1$; hence it is not a basis of V.

8.60 Let V be the vector space of 2×2 symmetric matrices over K. Show that $\dim V = 3$. [Recall that $A = (a_{ij})$ is symmetric iff $A = A^T$ or, equivalently, $a_{ij} = a_{ji}$.]

▮ An arbitrary 2×2 symmetric matrix is of the form $A = \begin{pmatrix} a & b \\ b & c \end{pmatrix}$ where $a, b, c \in K$. [Note that there are three "variables."] Setting (i) $a = 1$, $b = 0$, $c = 0$; (ii) $a = 0$, $b = 1$, $c = 0$; and (iii) $a = 0$, $b = 0$, $c = 1$, we obtain the respective matrices

$$E_1 = \begin{pmatrix} 1 & 0 \\ 0 & 0 \end{pmatrix} \qquad E_2 = \begin{pmatrix} 0 & 1 \\ 1 & 0 \end{pmatrix} \qquad E_3 = \begin{pmatrix} 0 & 0 \\ 0 & 1 \end{pmatrix}$$

We show that $\{E_1, E_2, E_3\}$ is a basis of V, i.e., that it (1) generates V and (2) is independent.
(1) For the above arbitrary matrix A in V, we have

$$A = \begin{pmatrix} a & b \\ b & c \end{pmatrix} = aE_1 + bE_2 + cE_3$$

Thus $\{E_1, E_2, E_3\}$ generates V.
(2) Suppose $xE_1 + yE_2 + zE_3 = 0$, where x, y, z are unknown scalars. That is, suppose

$$x\begin{pmatrix} 1 & 0 \\ 0 & 0 \end{pmatrix} + y\begin{pmatrix} 0 & 1 \\ 1 & 0 \end{pmatrix} + z\begin{pmatrix} 0 & 0 \\ 0 & 1 \end{pmatrix} = \begin{pmatrix} 0 & 0 \\ 0 & 0 \end{pmatrix} \qquad \text{or} \qquad \begin{pmatrix} x & y \\ y & z \end{pmatrix} = \begin{pmatrix} 0 & 0 \\ 0 & 0 \end{pmatrix}$$

Setting corresponding entries equal to each other, we obtain $x = 0$, $y = 0$, $z = 0$. In other words, $xE_1 + yE_2 + zE_3 = 0$ implies $x = 0$, $y = 0$, $z = 0$. Accordingly, $\{E_1, E_2, E_3\}$ is independent. Thus $\{E_1, E_2, E_3\}$ is a basis of V and so the dimension of V is 3.

8.61 Let W be the vector space of 3×3 symmetric matrices over K. Show that $\dim W = 6$ by exhibiting a basis of W. [Recall that $A = (a_{ij})$ is symmetric iff $a_{ij} = a_{ji}$.]

▮ The following six matrices form a basis of W:

$$\begin{pmatrix} 1 & 0 & 0 \\ 0 & 0 & 0 \\ 0 & 0 & 0 \end{pmatrix} \quad \begin{pmatrix} 0 & 0 & 0 \\ 0 & 1 & 0 \\ 0 & 0 & 0 \end{pmatrix} \quad \begin{pmatrix} 0 & 0 & 0 \\ 0 & 0 & 0 \\ 0 & 0 & 1 \end{pmatrix} \quad \begin{pmatrix} 0 & 1 & 0 \\ 1 & 0 & 0 \\ 0 & 0 & 0 \end{pmatrix} \quad \begin{pmatrix} 0 & 0 & 1 \\ 0 & 0 & 0 \\ 1 & 0 & 0 \end{pmatrix} \quad \begin{pmatrix} 0 & 0 & 0 \\ 0 & 0 & 1 \\ 0 & 1 & 0 \end{pmatrix}$$

8.62 What is the dimension of the vector space U of $n \times n$ symmetric matrices over a field K?

▮ As indicated by Problem 8.61, each element on or above the diagonal corresponds to a basis element; hence $\dim U = n + (n - 1) + \cdots + 2 + 1 = \frac{1}{2}n(n + 1)$.

8.63 Let W be the vector space of 3×3 antisymmetric matrices over K. Show that $\dim W = 3$ by exhibiting a basis of W. [Recall that $A = (a_{ij})$ is antisymmetric iff $a_{ij} = -a_{ji}$.]

▋ The following three matrices form a basis of W:

$$\begin{pmatrix} 0 & 1 & 0 \\ -1 & 0 & 0 \\ 0 & 0 & 0 \end{pmatrix} \qquad \begin{pmatrix} 0 & 0 & 1 \\ 0 & 0 & 0 \\ -1 & 0 & 0 \end{pmatrix} \qquad \begin{pmatrix} 0 & 0 & 0 \\ 0 & 0 & 1 \\ 0 & -1 & 0 \end{pmatrix}$$

8.64 What is the dimension of the vector space U of $n \times n$ antisymmetric matrices over a field K?

▋ As indicated by Problem 8.63, each element above the diagonal corresponds to a basis element; hence $\dim U = (n-1) + (n-2) + \cdots + 2 + 1 = \frac{1}{2}n(n-1)$.

8.65 Show that the complex field $\mathbf{C}$ is a vector space of dimension 2 over the real field $\mathbf{R}$.

▋ We claim that $\{1, i\}$ is a basis of $\mathbf{C}$ over $\mathbf{R}$. For if $v \in \mathbf{C}$, then $v = a + bi = a \cdot 1 + b \cdot i$ where $a, b \in \mathbf{R}$; i.e., $\{1, i\}$ generates $\mathbf{C}$ over $\mathbf{R}$. Furthermore, if $x \cdot 1 + y \cdot i = 0$ or $x + yi = 0$, where $x, y \in \mathbf{R}$, then $x = 0$ and $y = 0$; i.e., $\{1, i\}$ is linearly independent over $\mathbf{R}$. Thus $\{1, i\}$ is a basis of $\mathbf{C}$ over $\mathbf{R}$, and so $\mathbf{C}$ is of dimension 2 over $\mathbf{R}$.

8.66 Show that the real field $\mathbf{R}$ is a vector space of infinite dimension over the rational field $\mathbf{Q}$.

▋ We claim that, for any n, $\{1, \pi, \pi^2, \ldots, \pi^n\}$ is linearly independent over $\mathbf{Q}$. For suppose $a_0 1 + a_1 \pi + a_2 \pi^2 + \cdots + a_n \pi^n = 0$, where the $a_i \in \mathbf{Q}$, and not all the a_i are 0. Then π is a root of the following nonzero polynomial over $\mathbf{Q}$: $a_0 + a_1 x + a_2 x^2 + \cdots + a_n x^n$. But it can be shown that π is a transcendental number, i.e., that π is not a root of any nonzero polynomial over $\mathbf{Q}$. Accordingly, the $n + 1$ real numbers $1, \pi, \pi^2, \ldots, \pi^2$ are linearly independent over $\mathbf{Q}$. Thus for any finite n, $\mathbf{R}$ cannot be of dimension n over $\mathbf{Q}$, i.e., $\mathbf{R}$ is of infinite dimension over $\mathbf{Q}$.

8.67 Let V be the vector space of ordered pairs of complex numbers over the real field $\mathbf{R}$. Show that V is of dimension 4.

▋ We claim that the following is a basis of V: $B = \{(1, 0), (i, 0), (0, 1), (0, i)\}$. Suppose $v \in V$. Then $v = (z, w)$ where z, w are complex numbers, and so $v = (a + bi, c + di)$ where a, b, c, d are real numbers. Then $v = a(1, 0) + b(i, 0) + c(0, 1) + d(0, i)$. Thus B generates V.

The proof is complete if we show that B is independent. Suppose $x_1(1, 0) + x_2(i, 0) + x_3(0, 1) + x_4(0, i) = 0$ where $x_1, x_2, x_3, x_4 \in R$. Then

$$(x_1 + x_2 i, x_3 + x_4 i) = (0, 0) \qquad \text{and so} \qquad \begin{cases} x_1 + x_2 i = 0 \\ x_3 + x_4 i = 0 \end{cases}$$

Accordingly $x_1 = 0$, $x_2 = 0$, $x_3 = 0$, $x_4 = 0$ and so B is independent.

8.68 Suppose $\dim V = n$. Show that a generating set with n elements is a basis.

▋ Suppose $u_1, u_2, \ldots, u_n$ span V and the vectors are linearly dependent. Then one of them is a linear combination of the others and so may be deleted from the spanning set. Hence V is spanned by $n - 1$ vectors. This is impossible since $\dim V = n$. Thus the u_i are linearly independent and hence form a basis of V.

8.5 DIMENSION AND SUBSPACES

Theorem 8.8: Let W be a subspace of an n-dimensional vector space V. Then $\dim W \leq n$. In particular, if $\dim W = n$, then $W = V$.

8.69 Prove Theorem 8.8 which gives the basic relationship between the dimension of a vector space V and the dimension of a subspace W of V.

▋ Since V is of dimension n, any $n + 1$ or more vectors are linearly dependent. Furthermore, since a basis of W consists of linearly independent vectors, it cannot contain more than n elements. Accordingly, $\dim W \leq n$.

In particular, if $\{w_1, \ldots, w_n\}$ is a basis of W, then since it is an independent set with n elements it is also a basis of V. Thus $W = V$ when $\dim W = n$.

8.70 Let W be a subspace of real space $\mathbf{R}^3$. Give a geometrical description of W in terms of its dimension.

   ▮ Since $\dim \mathbf{R}^3 = 3$, the dimension of W can only be $0, 1, 2,$ or 3. The following cases apply:
(i) $\dim W = 0$,   then $W = \{0\}$, a point
(ii) $\dim W = 1$,   then W is a line through the origin
(iii) $\dim W = 2$,   then W is a plane through the origin
(iv) $\dim W = 3$,   then W is the entire space $\mathbf{R}^3$

8.71 Find the dimension of the subspace W of $\mathbf{R}^4$ spanned by (a) $(1, -2, 3, -1)$ and $(1, 1, -2, 3)$, (b) $(3, -6, 3, -9)$ and $(-2, 4, -2, 6)$.

   ▮ Two nonzero vectors span a space W of dimension 2 if they are independent, and of dimension 1 if they are dependent. Recall that two vectors are dependent if and only if one is a multiple of the other. Thus (a) $\dim W = 2$, (b) $\dim W = 1$.

8.72 Let W be the subspace of $\mathbf{R}^4$ generated by the vectors $(1, -2, 5, -3)$, $(2, 3, 1, -4)$, and $(3, 8, -3, -5)$. Find a basis and the dimension of W.

   ▮ Form the matrix A whose rows are the given vectors and row reduce A to an echelon form:

$$A = \begin{pmatrix} 1 & -2 & 5 & -3 \\ 2 & 3 & 1 & -4 \\ 3 & 8 & -3 & -5 \end{pmatrix} \quad \text{to} \quad \begin{pmatrix} 1 & -2 & 5 & -3 \\ 0 & 7 & -9 & 2 \\ 0 & 14 & -18 & 4 \end{pmatrix} \quad \text{to} \quad \begin{pmatrix} 1 & -2 & 5 & -3 \\ 0 & 7 & -9 & 2 \\ 0 & 0 & 0 & 0 \end{pmatrix}$$

The nonzero rows $(1, -2, 5, -3)$ and $(0, 7, -9, 2)$ of the echelon matrix form a basis of the row space of A which is W. Thus, in particular, $\dim W = 2$.

8.73 Extend the basis of W in Problem 8.72 to a basis of the whole space $\mathbf{R}^4$.

   ▮ We seek four independent vectors which include the above two vectors. The vectors $(1, -2, 5, -3)$, $(0, 7, -9, 2)$, $(0, 0, 1, 0)$, and $(0, 0, 0, 1)$ are independent (since they form an echelon matrix), and so they form a basis of $\mathbf{R}^4$ which is an extension of the basis of W.

8.74 Let W be the subspace of $\mathbf{R}^3$ defined by $W = \{(a, b, c): a + b + c = 0\}$. Find a basis and dimension of W.

   ▮ Note $W \neq \mathbf{R}^3$ since, for example, $(1, 2, 3) \notin W$. Thus $\dim W < 3$. Note $u_1 = (1, 0, -1)$ and $u_2 = (0, 1, -1)$ are two independent vectors in W. Thus $\dim W = 2$ and so u_1 and u_2 form a basis of W.

8.75 Let W be the subspace of $\mathbf{R}^3$ defined by $W = \{(a, b, c): a = b = c\}$. Find a basis and dimension of W.

   ▮ The vector $u = (1, 1, 1) \in W$. Any vector $w \in W$ has the form $w = (k, k, k)$. Hence $w = ku$. Thus u spans W and $\dim W = 1$.

8.76 Let W be the subspace of $\mathbf{R}^3$ defined by $W = \{(a, b, c): c = 3a\}$. Find a basis and dimension of W.

   ▮ $W \neq \mathbf{R}^3$ since, for example $(1, 1, 1) \notin W$. Thus $\dim W < 3$. The vectors $u_1 = (1, 0, 3)$ and $u_2 = (0, 1, 0)$ belong to W and are linearly independent. Thus $\dim W = 2$ and u_1, u_2 form a basis of W.

8.77 Find a basis and the dimension of the subspace W of $\mathbf{R}^4$ spanned by $(1, 4, -1, 3)$, $(2, 1, -3, -1)$, and $(0, 2, 1, -5)$.

   ▮ Reduce to echelon form the matrix whose rows are the given vectors:

$$\begin{pmatrix} 1 & 4 & -1 & 3 \\ 2 & 1 & -3 & -1 \\ 0 & 2 & 1 & -5 \end{pmatrix} \quad \text{to} \quad \begin{pmatrix} 1 & 4 & -1 & 3 \\ 0 & -7 & -1 & -7 \\ 0 & 2 & 1 & -5 \end{pmatrix} \quad \text{to} \quad \begin{pmatrix} 1 & 4 & -1 & 3 \\ 0 & -7 & -1 & -7 \\ 0 & 0 & 5 & -49 \end{pmatrix}$$

The nonzero rows in the echelon matrix form a basis of W; hence $\dim W = 3$. In particular, this means the original three vectors are linearly independent and also form a basis for W.

8.78 Find a basis and the dimension of the subspace W of $\mathbf{R}^4$ spanned by $(1, -4, -2, 1)$, $(1, -3, -1, 2)$, and $(3, -8, -2, 7)$.

▮ Reduce to echelon form the matrix whose rows are the given vectors:

$$\begin{pmatrix} 1 & -4 & -2 & 1 \\ 1 & -3 & -1 & 2 \\ 3 & -8 & -2 & 7 \end{pmatrix} \quad \text{to} \quad \begin{pmatrix} 1 & -4 & -2 & 1 \\ 0 & 1 & 1 & 1 \\ 0 & 4 & 4 & 4 \end{pmatrix} \quad \text{to} \quad \begin{pmatrix} 1 & -4 & -2 & 1 \\ 0 & 1 & 1 & 1 \\ 0 & 0 & 0 & 0 \end{pmatrix}$$

The nonzero rows in the echelon matrix, $(1, -4, -2, 1)$ and $(0, 1, 1, 1)$, form a basis of W and so $\dim W = 2$. In particular, this means that the original three vectors were linearly dependent.

8.79 Let W be the subspace of $\mathbf{R}^5$ spanned by $u_1 = (1, 2, -1, 3, 4)$, $u_2 = (2, 4, -2, 6, 8)$, $u_3 = (1, 3, 2, 2, 6)$, $u_4 = (1, 4, 5, 1, 8)$, $u_5 = (2, 7, 3, 3, 9)$. Find a subset of the vectors which form a basis of W.

▮ *Method 1.* Find the first vector in the sequence u_1, u_2, u_3, u_4, u_5 which is a linear combination of the preceding vectors and then eliminate this vector from the spanning set. Repeat this process until an independent set of vectors remain. This independent set of vectors is then a basis of W.

Method 2. Form the matrix whose rows are the given vectors and reduce the matrix to an "echelon" form but without interchanging any zero rows:

$$\begin{pmatrix} 1 & 2 & -1 & 3 & 4 \\ 2 & 4 & -2 & 6 & 8 \\ 1 & 3 & 2 & 2 & 6 \\ 1 & 4 & 5 & 1 & 8 \\ 2 & 7 & 3 & 3 & 9 \end{pmatrix} \quad \text{to} \quad \begin{pmatrix} 1 & 2 & -1 & 3 & 4 \\ 0 & 0 & 0 & 0 & 0 \\ 0 & 1 & 3 & -1 & 2 \\ 0 & 2 & 6 & -2 & 4 \\ 0 & 3 & 5 & -3 & 1 \end{pmatrix} \quad \text{to} \quad \begin{pmatrix} 1 & 2 & -1 & 3 & 4 \\ 0 & 0 & 0 & 0 & 0 \\ 0 & 1 & 3 & -1 & 2 \\ 0 & 0 & 0 & 0 & 0 \\ 0 & 0 & -4 & 0 & -5 \end{pmatrix}$$

The nonzero rows are the first, third, and fifth rows; hence u_1 u_3, u_5 form a basis of W.

Problems 8.80–8.81 refer to the vector space V of polynomials over $\mathbf{R}$.

8.80 Find the dimension of the subspace W of V spanned by (a) $t^3 + 2t^2 + 3t + 1$ and $2t^3 + 4t^2 + 6t + 2$, (b) $t^3 - 2t^2 + 5$ and $t^2 + 3t - 4$.

▮ Dim $W = 1$ or 2 according as the vectors are dependent or independent, and two vectors are dependent iff one is a multiple of the other. Thus (a) $\dim W = 2$, (b) $\dim W = 1$.

8.81 Find a basis and dimension of the subspace W of V spanned by the polynomials $v_1 = t^3 - 2t^2 + 4t + 1$, $v_2 = 2t^3 - 3t^2 + 9t - 1$, $v_3 = t^3 + 6t - 5$, $v_4 = 2t^3 - 5t^2 + 7t + 5$.

▮ The coordinate vectors (see Section 8.9) of the given polynomials relative to the basis $\{t^3, t^2, t, 1\}$ are, respectively, $[v_1] = (1, -2, 4, 1)$, $[v_2] = (2, -3, 9, -1)$, $[v_3] = (1, 0, 6, -5)$, $[v_4] = (2, -5, 7, 5)$. Form the matrix whose rows are the above coordinate vectors and row reduce to echelon form:

$$\begin{pmatrix} 1 & -2 & 4 & 1 \\ 2 & -3 & 9 & -1 \\ 1 & 0 & 6 & -5 \\ 2 & -5 & 7 & 5 \end{pmatrix} \quad \text{to} \quad \begin{pmatrix} 1 & -2 & 4 & 1 \\ 0 & 1 & 1 & -3 \\ 0 & 2 & 2 & -6 \\ 0 & -1 & -1 & 3 \end{pmatrix} \quad \text{to} \quad \begin{pmatrix} 1 & -2 & 4 & 1 \\ 0 & 1 & 1 & -3 \\ 0 & 0 & 0 & 0 \\ 0 & 0 & 0 & 0 \end{pmatrix}$$

The nonzero rows $(1, -2, 4, 1)$ and $(0, 1, 1, -3)$ of the echelon matrix form a basis of the space generated by the coordinate vectors, and so the corresponding polynomials $t^3 - 2t^2 + 4t + 1$ and $t^2 + t - 3$ form a basis of W. Thus $\dim W = 2$.

8.82 Let V be the vector space of functions from $\mathbf{R}$ into $\mathbf{R}$. Find a basis and dimension of the subspace W of V spanned by the functions $f(t) = \sin t$, $g(t) = \cos t$, $h(t) = t$.

▮ By Problem 8.26, f, g, and h are linearly independent. Thus $\{f, g, h\}$ is a basis of W and $\dim W = 3$.

Problems 8.83–8.84 refer to the vector space V of real 2×2 matrices.

8.83 Find the dimension of the subspace W of V spanned by

(a) $\begin{pmatrix} 1 & 2 \\ 1 & 2 \end{pmatrix}$ and $\begin{pmatrix} 1 & 1 \\ 2 & 2 \end{pmatrix}$ (b) $\begin{pmatrix} 1 & 1 \\ -1 & -1 \end{pmatrix}$ and $\begin{pmatrix} -3 & -3 \\ 3 & 3 \end{pmatrix}$

∎ (a) dim $W = 2$ since neither matrix is a multiple of the other; (b) dim $W = 1$ since the matrices are multiples of each other.

8.84 Find the dimension and a basis of the subspace W of V spanned by

$$A = \begin{pmatrix} 1 & 2 \\ -1 & 3 \end{pmatrix} \qquad B = \begin{pmatrix} 2 & 5 \\ 1 & -1 \end{pmatrix} \qquad C = \begin{pmatrix} 5 & 12 \\ 1 & 1 \end{pmatrix} \qquad D = \begin{pmatrix} 3 & 4 \\ -2 & 5 \end{pmatrix}$$

∎ The coordinate vectors [see Section 8.9] of the given matrices relative to the usual basis of V:

$$E_1 = \begin{pmatrix} 1 & 0 \\ 0 & 0 \end{pmatrix} \qquad E_2 = \begin{pmatrix} 0 & 1 \\ 0 & 0 \end{pmatrix} \qquad E_3 = \begin{pmatrix} 0 & 0 \\ 1 & 0 \end{pmatrix} \qquad E_4 = \begin{pmatrix} 0 & 0 \\ 0 & 1 \end{pmatrix}$$

are as follows:

$$[A] = [1, 2, -1, 3] \qquad [B] = [2, 5, 1, -1] \qquad [C] = [5, 12, 1, 1] \qquad [D] = [3, 4, -2, 5]$$

Reduce to echelon form the matrix whose rows are the coordinate vectors:

$$\begin{pmatrix} 1 & 2 & -1 & 3 \\ 2 & 5 & 1 & -1 \\ 5 & 12 & 1 & 1 \\ 3 & 4 & -2 & 5 \end{pmatrix} \quad \text{to} \quad \begin{pmatrix} 1 & 2 & -1 & 3 \\ 0 & 1 & 3 & -7 \\ 0 & 2 & 6 & -14 \\ 0 & -2 & 1 & -4 \end{pmatrix} \quad \text{to} \quad \begin{pmatrix} 1 & 2 & -1 & 3 \\ 0 & 1 & 3 & -7 \\ 0 & 0 & 0 & 0 \\ 0 & 0 & 7 & -18 \end{pmatrix}$$

The nonzero rows are linearly independent, hence the corresponding matrices $\begin{pmatrix} 1 & 2 \\ -1 & 3 \end{pmatrix}$, $\begin{pmatrix} 0 & 1 \\ 3 & -7 \end{pmatrix}$,

and $\begin{pmatrix} 0 & 0 \\ 7 & -18 \end{pmatrix}$ form a basis of W and dim $W = 3$. [Note also that the matrices A, B, and D form a basis of W.]

8.6 RANK OF A MATRIX

8.85 Define the rank of a matrix A.

∎ The rank, or row rank, of a matrix A, denoted by rank(A), is equal to the maximum number of linearly independent rows or, equivalently, is equal to the dimension of the row space of A.

8.86 Find the rank of $A = \begin{pmatrix} 1 & 2 & 0 & -1 \\ 2 & 6 & -3 & -3 \\ 3 & 10 & -6 & -5 \end{pmatrix}$.

∎ Row reduce A to an echelon form:

$$A \quad \text{to} \quad \begin{pmatrix} 1 & 2 & 0 & -1 \\ 0 & 2 & -3 & -1 \\ 0 & 4 & -6 & -2 \end{pmatrix} \quad \text{to} \quad \begin{pmatrix} 1 & 2 & 0 & -1 \\ 0 & 2 & -3 & -1 \\ 0 & 0 & 0 & 0 \end{pmatrix}$$

The nonzero rows of the echelon matrix form a basis of the row space of A and hence the dimension of the row space of A is two. Thus rank(A) = 2.

8.87 Find the rank of $B = \begin{pmatrix} 1 & 3 & 1 & -2 & -3 \\ 1 & 4 & 3 & -1 & -4 \\ 2 & 3 & -4 & -7 & -3 \\ 3 & 8 & 1 & -7 & -8 \end{pmatrix}$.

∎ Row reduce B to an echelon form:

$$B \quad \text{to} \quad \begin{pmatrix} 1 & 3 & 1 & -2 & -3 \\ 0 & 1 & 2 & 1 & -1 \\ 0 & -3 & -6 & -3 & 3 \\ 0 & -1 & -2 & -1 & 1 \end{pmatrix} \quad \text{to} \quad \begin{pmatrix} 1 & 3 & 1 & -2 & -3 \\ 0 & 1 & 2 & 1 & -1 \\ 0 & 0 & 0 & 0 & 0 \\ 0 & 0 & 0 & 0 & 0 \end{pmatrix}$$

Since the echelon matrix has two nonzero rows, rank(B) = 2.

8.88 Define the column rank of a matrix A.

▌ The column rank of A is equal to the maximum number of linearly independent columns of A or, equivalently, is equal to the dimension of the column space of A.

Theorem 8.9: The row rank and the column rank of any matrix are equal.

8.89 Prove Theorem 8.9 [which justifies the use of the word rank by itself].

▌ Let A be an arbitrary $m \times n$ matrix:

$$A = \begin{pmatrix} a_{11} & a_{22} & \cdots & a_{1n} \\ a_{21} & a_{22} & \cdots & a_{2n} \\ \cdots\cdots\cdots\cdots\cdots\cdots \\ a_{m1} & a_{m2} & \cdots & a_{mn} \end{pmatrix}$$

Let $R_1, R_2, \ldots, R_m$ denote its rows:

$$R_1 = (a_{11}, a_{12}, \ldots, a_{1n}), \ldots, R_m = (a_{m1}, a_{m2}, \ldots, a_{mn})$$

Suppose the row rank is r and that the following r vectors form a basis for the row space:

$$S_1 = (b_{11}, b_{12}, \ldots, b_{1n}), S_2 = (b_{21,}\ b_{22}, \ldots, b_{2n}), \ldots, S_r = (b_{r1}, b_{r2}, \ldots, b_{rn})$$

Then each of the row vectors is a linear combination of the S_i:

$$R_1 = k_{11}S_1 + k_{12}S_2 + \cdots + k_{1r}S_\tau$$
$$R_2 = k_{21}S_1 + k_{22}S_2 + \cdots + k_{2r}S_r$$
$$\cdots\cdots\cdots\cdots\cdots\cdots\cdots\cdots$$
$$R_m = k_{m1}S_1 + k_{m2}S_2 + \cdots + k_{mr}S_r$$

where the k_{ij} are scalars. Setting the ith components of each of the above vector equations equal to each other, we obtain the following system of equations, each valid for $i = 1, \ldots, n$:

$$a_{1i} = k_{11}b_{1i} + k_{12}b_{2i} + \cdots + k_{1r}b_{ri}$$
$$a_{2i} = k_{21}b_{1i} + k_{22}b_{2i} + \cdots + k_{2bri}$$
$$\cdots\cdots\cdots\cdots\cdots\cdots\cdots\cdots$$
$$a_{mi} = k_{m1}b_{1i} + k_{m2}b_{2i} + \cdots + k_{mr}b_{ri}$$

Thus for $i = 1, \ldots, n$:

$$\begin{pmatrix} a_{1i} \\ a_{2i} \\ \vdots \\ a_{mi} \end{pmatrix} = b_{1i}\begin{pmatrix} k_{11} \\ k_{21} \\ \vdots \\ k_{m1} \end{pmatrix} + b_{2i}\begin{pmatrix} k_{12} \\ k_{22} \\ \vdots \\ k_{m2} \end{pmatrix} + \cdots + b_{ri}\begin{pmatrix} k_{1r} \\ k_{2r} \\ \vdots \\ k_{mr} \end{pmatrix}$$

In other words, each of the columns of A is a linear combination of the r vectors

$$\begin{pmatrix} k_{11} \\ k_{21} \\ \vdots \\ k_{m1} \end{pmatrix}, \begin{pmatrix} k_{12} \\ k_{22} \\ \vdots \\ k_{m2} \end{pmatrix}, \ldots, \begin{pmatrix} k_{1r} \\ k_{2r} \\ \vdots \\ k_{mr} \end{pmatrix}$$

Thus the column space of the matrix A has dimension at most r, i.e., column rank $\leq r$.
Hence, column rank $\leq$ row rank.

Similarly (or considering the transpose matrix A^T) we obtain row rank $\leq$ column rank. Thus the row rank and column rank are equal.

8.90 Find the rank of $A = \begin{pmatrix} 1 & 2 & -3 \\ 2 & 1 & 0 \\ -2 & -1 & 3 \\ -1 & 4 & -2 \end{pmatrix}$.

▮ Since row rank equals column rank, it is easier to form the transpose of A and then row reduce to echelon form:

$$\begin{pmatrix} 1 & 2 & -2 & -1 \\ 2 & 1 & -1 & 4 \\ -3 & 0 & 3 & -2 \end{pmatrix} \quad\text{to}\quad \begin{pmatrix} 1 & 2 & -2 & -1 \\ 0 & -3 & 3 & 6 \\ 0 & 6 & -6 & -5 \end{pmatrix} \quad\text{to}\quad \begin{pmatrix} 1 & 2 & -2 & -1 \\ 0 & -3 & 3 & 6 \\ 0 & 0 & 3 & 7 \end{pmatrix}$$

Thus $\operatorname{rank}(A) = 3$.

8.91 Find the rank of $B = \begin{pmatrix} 1 & 3 \\ 0 & -2 \\ 5 & -1 \\ -2 & 3 \end{pmatrix}$.

▮ The two columns are linearly independent since one is not a multiple of the other. Thus $\operatorname{rank}(B) = 2$.

8.92 Let A and B be arbitrary matrices for which the product AB is defined. Show that $\operatorname{rank}(AB) \leq \operatorname{rank}(B)$ and $\operatorname{rank}(AB) \leq \operatorname{rank}(A)$.

▮ The row space of AB is contained in the row space of B; hence $\operatorname{rank}(AB) \leq \operatorname{rank}(B)$. Furthermore, the column space of AB is contained in the column space of A; hence $\operatorname{rank}(AB) \leq \operatorname{rank}(A)$.

8.93 Let A be an n-square matrix. Show that A is invertible if and only if $\operatorname{rank}(A) = n$.

▮ Note that the rows of the n-square identity matrix I_n are linearly independent since I_n is in echelon form; hence $\operatorname{rank}(I_n) = n$. Now if A is invertible then A is row equivalent to I_n; hence $\operatorname{rank}(A) = n$. But if A is not invertible then A is row equivalent to a matrix with a zero row; hence $\operatorname{rank}(A) < n$. That is, A is invertible if and only if $\operatorname{rank}(A) = n$.

8.94 Define the determinantal rank of a matrix A.

▮ The determinantal rank of A is the order of the largest square submatrix of A [obtained by deleting rows and columns of A] whose determinant is not zero.

8.95 Show that the determinantal rank of a matrix A is equal to $\operatorname{rank}(A)$.

▮ Let d equal the determinantal rank of A. Hence there is a d-square submatrix of A with nonzero determinant. Thus the d rows in the submatrix are linearly independent. Hence $d \leq \operatorname{rank}(A)$. On the other hand, let B be an echelon matrix row equivalent to A. Choose the submatrix of B which contains the nonzero rows of B and the columns which contain the leading nonzero entries in the rows. This gives an upper triangular $\operatorname{rank}(A)$-square submatrix with nonzero determinant. Hence $\operatorname{rank}(A) \leq d$. Thus $d = \operatorname{rank}(A)$.

8.7 APPLICATIONS TO LINEAR EQUATIONS

8.96 Suppose a homogeneous system $AX = 0$ is in echelon form. Moreover, suppose the system has n unknowns and r (nonzero) linear equations. Give a method to obtain a basis for the solution space W of the system.

▮ The system has $n - r$ free variables $x_{i_1}, x_{i_2}, \ldots, x_{i_{n-r}}$. Find the solution v_j obtained by setting $x_{i_j} = 1$ (or any nonzero constant) and the remaining free variables equal to 0. Then the solutions $v_1, v_2, \ldots, v_{n-r}$ form a basis of W and so $\dim W = n - r$.

8.97 Find the dimension and a basis of the solution space W of the system

$$\begin{aligned} x + 2y + 2z - s + 3t &= 0 \\ x + 2y + 3z + s + t &= 0 \\ 3x + 6y + 8z + s + 5t &= 0 \end{aligned}$$

▮ Reduce the system to echelon form:

$$\begin{aligned} x + 2y + 2z - s + 3t &= 0 \\ z + 2s - 2t &= 0 \\ 2z + 4s - 4t &= 0 \end{aligned} \qquad\text{or}\qquad \begin{aligned} x + 2y + 2z - s + 3t &= 0 \\ z + 2s - 2t &= 0 \end{aligned}$$

The system in echelon form has two (nonzero) equations in five unknowns; and hence the system has $5 - 2 = 3$ free variables which are y, s, and t. Thus $\dim W = 3$. To obtain a basis for W, set

(i) $y = 1$, $s = 0$, $t = 0$ to obtain the solution $v_1 = (-2, 1, 0, 0, 0)$

(ii) $y = 0$, $s = 1$, $t = 0$ to obtain the solution $v_2 = (5, 0, -2, 1, 0)$

(iii) $y = 0$, $s = 0$, $t = 1$ to obtain the solution $v_3 = (-7, 0, 2, 0, 1)$

The set $\{v_1, v_2, v_3\}$ is a basis of the soution space W.

8.98   Find the dimension and a basis of the solution space W of the system

$$\begin{aligned} x + 2y + z - 3t &= 0 \\ 2x + 4y + 4z - t &= 0 \\ 3x + 6y + 7z + t &= 0 \end{aligned}$$

❚   Reduce the system to echelon form:

$$\begin{aligned} x + 2y + z - 3t &= 0 \\ 2z + 5t &= 0 \\ 4z + 10t &= 0 \end{aligned} \qquad \text{or} \qquad \begin{aligned} x + 2y + z - 3t &= 0 \\ 2z + 5t &= 0 \end{aligned}$$

The free variables are y and t and $\dim W = 2$. Set

(i) $y = 1$, $z = 0$ to obtain the solution $u_1 = (-2, 1, 0, 0)$

(ii) $y = 0$, $t = 2$ to obtain the solution $u_2 = (11, 0, -5, 2)$

Then $\{u_1, u_2\}$ is a basis of W. [We could have chosen $y = 0$, $t = 1$ in (ii), but such a choice would introduce fractions into the solution.]

8.99   Find the dimension and a basis of the solution space W of the system

$$\begin{aligned} x + 2y - 4z + 3r - s &= 0 \\ x + 2y - 2z + 2r + s &= 0 \\ 2x + 4y - 2z + 3r + 4s &= 0 \end{aligned}$$

❚   Reduce the system to echelon form:

$$\begin{aligned} x + 2y - 4z + 3r - s &= 0 \\ 2z - r + 2s &= 0 \\ 6z - 3r + 6s &= 0 \end{aligned} \qquad \text{and then} \qquad \begin{aligned} x + 2y - 4z + 3r - s &= 0 \\ 2z - r + 2s &= 0 \end{aligned}$$

There are five unknowns and two (nonzero) equations in echelon form; hence there are $5 - 2 = 3$ free variables, y, r, and s. Thus $\dim W = 3$. Set

(i) $y = 1$, $r = 0$, $s = 0$ to obtain the solution $v_1 = (-2, 1, 0, 0, 0)$

(ii) $y = 0$, $r = 2$, $s = 0$ to obtain the solution $v_2 = (-2, 0, 1, 2, 0)$

(iii) $y = 0$, $r = 0$, $s = 1$ to obtain the solution $v_3 = (-3, 0, -1, 0, 1)$

The set $\{v_1, v_2, v_3\}$ is a basis of the solution space W.

8.100   Find the dimension and a basis of the solution space W of the system $x + 2y - 3z = 0$, $2x + 5y + z = 0$, $x - y + 2z = 0$.

❚   Reduce the system to echelon form:

$$\begin{aligned} x + 2y - 3z &= 0 \\ y + 7z &= 0 \\ -3y + 5z &= 0 \end{aligned} \qquad \text{and then} \qquad \begin{aligned} x + 2y - 3z &= 0 \\ y + 7z &= 0 \\ 26z &= 0 \end{aligned}$$

The echelon system is in triangular form and hence has no free variables. Thus 0 is the only solution, that is, $W = \{0\}$. Accordingly, $\dim W = 0$.

8.101   Find a homogeneous system whose solution set W is generated by $\{(1, -2, 0, 3), (1 -1, -1, 4), (1, 0, -2, 5)\}$.

❚   Let $v = (x, y, z, t)$. Form the matrix M whose first rows are the given vectors and whose last row is v; and then row reduce to echelon form:

$$M = \begin{pmatrix} 1 & -2 & 0 & 3 \\ 1 & -1 & -1 & 4 \\ 1 & 0 & -2 & 5 \\ x & y & z & t \end{pmatrix} \quad \text{to} \quad \begin{pmatrix} 1 & -2 & 0 & 3 \\ 0 & -1 & -1 & 1 \\ 0 & -2 & -2 & 2 \\ 0 & 2x+y & z & -3x+t \end{pmatrix} \quad \text{to} \quad \begin{pmatrix} 1 & -2 & 0 & 3 \\ 0 & 1 & -1 & 1 \\ 0 & 0 & 2x+y+z & -5x-y+t \\ 0 & 0 & 0 & 0 \end{pmatrix}$$

The original first three rows show that W has dimension 2. Thus $v \in W$ if and only if the additional row does not increase the dimension of the row space. Hence we set the last two entries in the third row on the right equal to 0 to obtain the required homogeneous system

$$\begin{aligned} 2x + y + z &= 0 \\ 5x + y - t &= 0 \end{aligned}$$

Problems 8.102–8.104 refer to the following subspaces of $\mathbf{R}^4$:

$$U = \{(a, b, c, d): b + c + d = 0\} \qquad W = \{(a, b, c, d): a + b = 0, c = 2d\}$$

8.102 Find the dimension and a basis of U.

❚ Find a basis of the set of solutions (a, b, c, d) of the equation $b + c + d = 0$ or $0 \cdot a + b + c + d = 0$. The free variables are a, c, and d. Set (1) $a = 1$, $c = 0$, $d = 0$, (2) $a = 0$, $c = 1$, $d = 0$, and (3) $a = 0$, $c = 0$, $d = 1$ to obtain the respective solutions

$$v_1 = (1, 0, 0, 0) \qquad v_2 = (0, -1, 1, 0) \qquad v_3 = (0, -1, 0, 1)$$

The set $\{v_1, v_2, v_3\}$ is a basis of U and $\dim U = 3$.

8.103 Find the dimension and a basis of W.

❚ We seek a basis of the set of solutions (a, b, c, d) of the system

$$\begin{aligned} a + b &= 0 \\ c &= 2d \end{aligned} \quad \text{or} \quad \begin{aligned} a + b &= 0 \\ c - 2d &= 0 \end{aligned}$$

The free variables are b and d. Set (1) $b = 1$, $d = 0$ to obtain the solution $v_1 = (-1, 1, 0, 0)$ and (2) $b = 0$, $d = 1$ to obtain the solution $v_2 = (0, 0, 2, 1)$. The set $\{v_1, v_2\}$ is a basis of W and $\dim W = 2$.

8.104 Find the dimension and a basis of $U \cap W$.

❚ $U \cap W$ consists of those vectors (a, b, c, d) which satisfy the conditions defining U and the conditions defining W, i.e., the three equations

$$\begin{aligned} b + c + d &= 0 \\ a + b &= 0 \\ c &= 2d \end{aligned} \quad \text{or} \quad \begin{aligned} a + b &= 0 \\ b + c + d &= 0 \\ c - 2d &= 0 \end{aligned}$$

The free variable is d. Set $d = 1$ to obtain the solution $v = (3, -3, 2, 1)$. Thus $\{v\}$ is a basis of $U \cap W$ and $\dim(U \cap W) = 1$.

8.105 Let $x_{i_1}, x_{i_2}, \ldots, x_{i_k}$ be the free variables of a homogeneous system of linear equations with n unknowns. Let v_j be the solution for which $x_{i_j} = 1$ and all other free variables $= 0$. Show that the solutions $v_1, v_2, \ldots, v_k$ are linearly independent.

❚ Let A be the matrix whose rows are the v_i, respectively. We interchange column 1 and column i_1, then the column 2 and column $i_2, \ldots$, and then column k and column i_k; and obtain the $k \times n$ matrix

$$B = (I, C) = \begin{pmatrix} 1 & 0 & 0 & \cdots & 0 & 0 & c_{1,k+1} & \cdots & c_{1n} \\ 0 & 1 & 0 & \cdots & 0 & 0 & c_{2,k+1} & \cdots & c_{2n} \\ \multicolumn{9}{c}{\dotfill} \\ 0 & 0 & 0 & \cdots & 0 & 1 & c_{k,k+1} & \cdots & c_{kn} \end{pmatrix}$$

The above matrix B is in echelon form and so its rows are independent; hence $\operatorname{rank}(B) = k$. Since A and B are column equivalent, they have the same rank, i.e., $\operatorname{rank}(A) = k$. But A has k rows; hence these rows, i.e., the v_i, are linearly independent as claimed.

Theorem 8.10: The system of linear equations $AX = B$ has a solution if and only if the coefficient matrix A and the augmented matrix (A, B) have the same rank.

8.106 Prove Theorem 8.10 which refers to a system of m linear equations in n unknowns $x_1, \ldots, x_n$ over a field K:

$$
\begin{aligned}
a_{11}x_1 + a_{12}x_2 + \cdots + a_{1n}x_n &= b_1 \\
a_{21}x_1 + a_{22}x_2 + \cdots + a_{2n}x_n &= b_2 \\
&\cdots\cdots\cdots\cdots\cdots\cdots\cdots \\
a_{m1}x_1 + a_{m2}x_2 + \cdots + a_{mn}x_n &= b_m
\end{aligned}
$$

or the equivalent matrix equation $AX = B$ where $A = (a_{ij})$ is the coefficient matrix and where $X = (x_i)$ and $B = (b_i)$ are the column vectors consisting of the unknowns and of the constants, respectively.

❚ The above system is equivalent to the following vector equation:

$$
x_1 \begin{pmatrix} a_{11} \\ a_{21} \\ \vdots \\ a_{m1} \end{pmatrix} + x_2 \begin{pmatrix} a_{12} \\ a_{22} \\ \vdots \\ a_{m2} \end{pmatrix} + \cdots + x_n \begin{pmatrix} a_{1n} \\ a_{2n} \\ \vdots \\ a_{mn} \end{pmatrix} = \begin{pmatrix} b_1 \\ b_2 \\ \vdots \\ b_m \end{pmatrix}
$$

Thus the system $AX = B$ has a solution if and only if the column vector B is a linear combination of the columns of A. Thus $AX = B$ has a solution if and only if the augmented matrix

$$
(A, B) = \begin{pmatrix} a_{11} & a_{12} & \cdots & a_{1n} & b_1 \\ a_{21} & a_{22} & \cdots & a_{2n} & b_2 \\ \cdots\cdots\cdots\cdots\cdots\cdots\cdots\cdots \\ a_{m1} & a_{m2} & \cdots & a_{mn} & b_m \end{pmatrix}
$$

has the same column space as A. Thus $AX = B$ has a solution if and only if $\operatorname{rank}(A, B) = \operatorname{rank}(A)$.

8.8 SUMS, DIRECT SUMS, INTERSECTIONS

Theorem 8.11: Let U and W be finite-dimensional subspaces of a vector space V. Then $\dim(U + W) = \dim U + \dim W - \dim(U \cap W)$. [Thus $\dim(U \cap W) = \dim U + \dim W - \dim(U + W)$.]

8.107 Prove Theorem 8.11 which gives the relationship between the dimension of a sum and its subspaces.

❚ Observe that $U \cap W$ is a subspace of both U and W. Suppose $\dim U = m$, $\dim W = n$, and $dim(U \cap W) = r$. Suppose $\{v_1, \ldots, v_r\}$ is a basis of $U \cap W$. We can extend $\{v_i\}$ to a basis of U and to a basis of W; say, $\{v_1, \ldots, v_r, u_1, \ldots, u_{m-r}\}$ and $\{v_1, \ldots, v_r, w_1, \ldots, w_{n-r}\}$ are bases of U and W, respectively. Let $B = \{v_1, \ldots, v_r, u_1, \ldots, u_{m-r}, w_1, \ldots, w_{n-r}\}$. Note that B has exactly $m + n - r$ elements. Thus the theorem is proved if we can show that B is a basis of $U + W$. Since $\{v_i, u_j\}$ generates U and $\{v_i, w_k\}$ generates W, the union $B = \{v_i, u_j, w_k\}$ generates $U + W$. Thus it suffices to show that B is independent.

Suppose

$$
a_1 v_1 + \cdots + a_r v_r + b_1 u_1 + \cdots + b_{m-r} u_{m-r} + c_1 w_1 + \cdots + c_{n-r} w_{n-r} = 0 \tag{1}
$$

where a_i, b_j, c_k are scalars. Let

$$
v = a_1 v_1 + \cdots + a_r v_r + b_1 u_1 + \cdots + b_{m-r} u_{m-r} \tag{2}
$$

By *(1)*, we also have that

$$
v = -c_1 w_1 - \cdots - c_{n-r} w_{n-r} \tag{3}
$$

Since $\{v_i, u_j\} \subset U$, $v \in U$ by *(2)*; and since $\{w_k\} \subset W$, $v \in W$ by *(3)*. Accordingly, $v \in U \cap W$. Now $\{v_i\}$ is a basis of $U \cap W$ and so there exist scalars $d_1, \ldots, d_r$ for which $v = d_1 v_1 + \cdots + d_r v_r$. Thus by *(3)* we have $d_1 v_1 + \cdots + d_r v_r + c_1 w_1 + \cdots + c_{n-r} w_{n-r} = 0$. But $\{v_i, w_k\}$ is a basis of W and so is independent. Hence the above equation forces $c_1 = 0, \ldots, c_{n-r} = 0$. Substituting this into *(1)*, we obtain

$a_1 v_1 + \cdots + a_r v_r + b_1 u_1 + \cdots + b_{m-r} u_{m-r} = 0$. But $\{v_i, u_j\}$ is a basis of U and so is independent. Hence the above equation forces $a_1 = 0, \ldots, \quad a_r = 0, \quad b_1 = 0, \ldots, \quad b_{m-r} = 0$.

Since equation *(1)* implies that the a_i, b_j, and c_k are all 0, $B = \{v_i, u_j, w_k\}$ is independent and the theorem is proved.

8.108 Suppose U and W are distinct four-dimensional subspaces of a vector space V of dimension 6. Find the possible dimensions of $U \cap W$.

❚ Since U and W are distinct, $U + W$ properly contains U and W; hence $\dim(U + W) > 4$. But $\dim(U + W)$ cannot be greater than 6, since $\dim V = 6$. Hence we have two possibilities: (i) $\dim(U + W) = 5$ or (ii) $\dim(U + W) = 6$. By Theorem 8.11, $\dim(U \cap W) = \dim U + \dim W - \dim(U + W) = 8 - \dim(U + W)$. Thus (i) $\dim(U \cap W) = 3$ or (ii) $\dim(U \cap W) = 2$.

8.109 Suppose U and W are two-dimensional subspaces of $\mathbf{R}^3$. Show that $U \cap W \neq \{0\}$. In particular, find the possible dimensions of $U \cap W$.

❚ Suppose $U = W$. Then $U \cap W = U = W$ and hence $\dim(U \cap W) = 2$. Suppose $U \neq W$. Then $U + W$ properly contains U (and W). Hence $\dim(U + W) > \dim U = 2$. But $U + W \subseteq \mathbf{R}^3$ which has dimension 3. Therefore $\dim(U + W) = 3$. Thus, by Theorem 8.11, $\dim(U \cap W) = \dim U + \dim W - \dim(U + W) = 2 + 2 - 3 = 1$. That is, $U \cap W$ is a line through the origin.

Remark: The above agrees with the well-known result in solid geometry that the intersection of two distinct planes is a line.

Problems 8.110–8.113 refer to the following subspaces of $\mathbf{R}^4$:

$$U = \text{span}\{(1, 1, 0, -1), (1, 2, 3, 0), (2, 3, 3, -1)\} \qquad W = \text{span}\{(1, 2, 2, -2), (2, 3, 2, -3), (1, 3, 4, -3)\}$$

8.110 Find a basis and the dimension of $U + W$.

❚ $U + W$ is the space spanned by all six vectors. Hence form the matrix whose rows are the given six vectors and then row reduce to echelon form:

$$
\begin{pmatrix} 1 & 1 & 0 & -1 \\ 1 & 2 & 3 & 0 \\ 2 & 3 & 3 & -1 \\ 1 & 2 & 2 & -2 \\ 2 & 3 & 2 & -3 \\ 1 & 3 & 4 & -3 \end{pmatrix}
\text{ to }
\begin{pmatrix} 1 & 1 & 0 & -1 \\ 0 & 1 & 3 & 1 \\ 0 & 1 & 3 & 1 \\ 0 & 1 & 2 & -1 \\ 0 & 1 & 2 & -1 \\ 0 & 2 & 4 & -2 \end{pmatrix}
\text{ to }
\begin{pmatrix} 1 & 1 & 0 & -1 \\ 0 & 1 & 3 & 1 \\ 0 & 1 & 2 & -1 \\ 0 & 0 & 0 & 0 \\ 0 & 0 & 0 & 0 \\ 0 & 0 & 0 & 0 \end{pmatrix}
$$

$$
\text{ to }
\begin{pmatrix} 1 & 1 & 0 & -1 \\ 0 & 1 & 3 & 1 \\ 0 & 0 & -1 & -2 \\ 0 & 0 & 0 & 0 \\ 0 & 0 & 0 & 0 \\ 0 & 0 & 0 & 0 \end{pmatrix}
$$

The nonzero rows of the echelon matrix, $(1, 1, 0, -1)$, $(0, 1, 3, 1)$, and $(0, 0, -1, -2)$, form a basis of $U + W$, and so $\dim(U + W) = 3$.

8.111 Find a basis and the dimension of U.

❚ Reduce to echelon form the matrix whose rows span U:

$$
\begin{pmatrix} 1 & 1 & 0 & -1 \\ 1 & 2 & 3 & 0 \\ 2 & 3 & 3 & -1 \end{pmatrix}
\text{ to }
\begin{pmatrix} 1 & 1 & 0 & -1 \\ 0 & 1 & 3 & 1 \\ 0 & 1 & 3 & 1 \end{pmatrix}
\text{ to }
\begin{pmatrix} 1 & 1 & 0 & -1 \\ 0 & 1 & 3 & 1 \\ 0 & 0 & 0 & 0 \end{pmatrix}
$$

The two nonzero rows of the echelon matrix, $(1, 1, 0, -1)$ and $(0, 1, 3, 1)$, form a basis of U and so $\dim U = 2$.

8.112   Find a basis and the dimension of W.

❚   Reduce to echelon form the matrix whose rows span W.

$$\begin{pmatrix} 1 & 2 & 2 & -2 \\ 2 & 3 & 2 & -3 \\ 1 & 3 & 4 & -3 \end{pmatrix} \quad \text{to} \quad \begin{pmatrix} 1 & 2 & 2 & -2 \\ 0 & -1 & -2 & 1 \\ 0 & 1 & 2 & -1 \end{pmatrix} \quad \text{to} \quad \begin{pmatrix} 1 & 2 & 2 & -2 \\ 0 & -1 & -2 & 1 \\ 0 & 0 & 0 & 0 \end{pmatrix}$$

The two nonzero rows of the echelon matrix, $(1, 2, 2, -2)$ and $(0, -1, -2, 1)$, form a basis of W and so $\dim W = 2$.

8.113   Find the dimension of $U \cap W$.

❚   Use Theorem 8.11: $\dim(U \cap W) = \dim U + \dim W - \dim(U + W) = 2 + 2 - 3 = 1$.   [Note that Theorem 8.11 does not help us to find a basis of $U \cap W$, just its dimension.   (See Problems 8.114–8.117)]

Problems 8.114–8.117 refer to the following subspaces of $\mathbf{R}^5$:

$$U = \text{span}\{(1, 3, -2, 2, 3), (1, 4, -3, 4, 2), (2, 3, -1, -2, 9)\}$$
$$W = \text{span}\{(1, 3, 0, 2, 1), (1, 5, -6, 6, 3), (2, 5, 3, 2, 1)\}$$

8.114   Find a basis and the dimension of $U + W$.

❚ $U + W$ is the space generated by all six vectors.   Hence form the matrix whose rows are the six vectors and then row reduce to echelon form:

$$\begin{pmatrix} 1 & 3 & -2 & 2 & 3 \\ 1 & 4 & -3 & 4 & 2 \\ 2 & 3 & -1 & -2 & 9 \\ 1 & 3 & 0 & 2 & 1 \\ 1 & 5 & -6 & 6 & 3 \\ 2 & 5 & 3 & 2 & 1 \end{pmatrix} \quad \text{to} \quad \begin{pmatrix} 1 & 3 & -2 & 2 & 3 \\ 0 & 1 & -1 & 2 & -1 \\ 0 & -3 & 3 & -6 & 3 \\ 0 & 0 & 2 & 0 & -2 \\ 0 & 2 & -4 & 4 & 0 \\ 0 & -1 & 7 & -2 & -5 \end{pmatrix} \quad \text{to} \quad \begin{pmatrix} 1 & 3 & -2 & 2 & 3 \\ 0 & 1 & -1 & 2 & -1 \\ 0 & 0 & 0 & 0 & 0 \\ 0 & 0 & 2 & 0 & -2 \\ 0 & 0 & -2 & 0 & 2 \\ 0 & 0 & 6 & 0 & -6 \end{pmatrix}$$

$$\text{to} \quad \begin{pmatrix} 1 & 3 & -2 & 2 & 3 \\ 0 & 1 & -1 & 2 & -1 \\ 0 & 0 & 2 & 0 & -2 \\ 0 & 0 & 0 & 0 & 0 \\ 0 & 0 & 0 & 0 & 0 \\ 0 & 0 & 0 & 0 & 0 \end{pmatrix}$$

The set of nonzero rows of the echelon matrix, $\{(1, 3, -2, 2, 3), (0, 1, -1, 2, -1), (0, 0, 2, 0, -2)\}$ is a basis of $U + W$;   thus $\dim(U + W) = 3$.

8.115   Find a homogeneous system whose solution space is U.

❚   Form the matrix whose first three rows span U and whose last row is (x, y, z, s, t) and then row reduce to an echelon form:

$$\begin{pmatrix} 1 & 3 & -2 & 2 & 3 \\ 1 & 4 & -3 & 4 & 2 \\ 2 & 3 & -1 & -2 & 9 \\ x & y & z & s & t \end{pmatrix} \quad \text{to} \quad \begin{pmatrix} 1 & 3 & -2 & 2 & 3 \\ 0 & 1 & -1 & 2 & -1 \\ 0 & -3 & 3 & -6 & 3 \\ 0 & -3x + y & 2x + z & -2x + s & -3x + t \end{pmatrix}$$

$$\text{to} \quad \begin{pmatrix} 1 & 3 & -2 & 2 & 3 \\ 0 & 1 & -1 & 2 & -1 \\ 0 & 0 & -x + y + z & 4x - 2y + s & -6x + y + t \\ 0 & 0 & 0 & 0 & 0 \end{pmatrix}$$

Set the entries of the third row equal to 0 to obtain the homogeneous system whose solution space is U:

$$-x + y + z = 0 \qquad 4x - 2y + s = 0 \qquad -6x + y + t = 0$$

8.116   Find a homogeneous system whose solution space is W.

❚ Form the matrix whose first rows span W and whose last row is (x, y, z, s, t) and then row reduce to an echelon form:

$$\begin{pmatrix} 1 & 3 & 0 & 2 & 1 \\ 1 & 5 & -6 & 6 & 3 \\ 2 & 5 & 3 & 2 & 1 \\ x & y & z & s & t \end{pmatrix} \quad \text{to} \quad \begin{pmatrix} 1 & 3 & 0 & 2 & 1 \\ 0 & 2 & -6 & 4 & 2 \\ 0 & -1 & 3 & -2 & -1 \\ 0 & -3x+y & z & -2x+s & -x+t \end{pmatrix}$$

$$\text{to} \quad \begin{pmatrix} 1 & 3 & 0 & 2 & 1 \\ 0 & 1 & -3 & 2 & 1 \\ 0 & 0 & -9x+3y+z & 4x-2y+s & 2x-y+t \\ 0 & 0 & 0 & 0 & 0 \end{pmatrix}$$

Set the entries of the third row equal to 0 to obtain the homogeneous system whose solution space is W:

$$-9x + 3y + z = 0 \qquad 4x - 2y + s = 0 \qquad 2x - y + t = 0$$

8.117 Find a basis and the dimension of $U \cap W$.

❚ Combine both of the above systems to obtain a homogeneous system whose solution space is $U \cap W$, and then solve:

$$\begin{cases} x + y + z & = 0 \\ 4x - 2y + s & = 0 \\ -6x + y + t & = 0 \\ -9x + 3y + z & = 0 \\ 4x - 2y + s & = 0 \\ 2x - y + t & = 0 \end{cases} \quad \text{or} \quad \begin{cases} -x + y + z & = 0 \\ 2y + 4z + s & = 0 \\ -5y - 6z + t & = 0 \\ -6y - 8z & = 0 \\ 2y + 4z + s & = 0 \\ y + 2z + t & = 0 \end{cases}$$

$$\text{or} \quad \begin{cases} -x + y + z & = 0 \\ 2y + 4z + s & = 0 \\ 8z + 5s + 2t & = 0 \\ 4z + 3s & = 0 \\ s - 2t & = 0 \end{cases} \quad \text{or} \quad \begin{cases} -x + y + z & = 0 \\ 2y + 4z + s & = 0 \\ 8z + 5s + 2t & = 0 \\ s - 2t & = 0 \end{cases}$$

There is one free variable, which is t; hence $\dim(U \cap W) = 1$. Setting $t = 2$, we obtain the solution $x = 1$, $y = 4$, $z = -3$, $s = 4$, $t = 2$. Thus $\{(1, 4, -3, 4, 2)\}$ is a basis of $U \cap W$.

8.118 Suppose U and W are subspaces of V and that $\dim U = 4$, $\dim W = 5$, and $\dim V = 7$. Find the possible dimensions of $U \cap W$.

❚ Since $W \subseteq U + W \subseteq V$, where $\dim W = 5$ and $\dim V = 7$, the only possible dimension of $U + W$ is 5, 6, or 7. But, by Theorem 8.11, $\dim(U \cap W) = \dim U + \dim W - \dim(U + W) = 9 - \dim(U + W)$. Since $\dim(U + W) = 5$, 6, or 7, we have $\dim(U \cap W) = 4$, 3, or 2.

8.119 Suppose V is the direct sum of its subspaces U and W, i.e., suppose $V = U \oplus W$. Show that $\dim V = \dim U + \dim W$.

❚ Since $V = U \oplus W$, we have $V = U + W$ and $U \cap W = \{0\}$. Thus,

$$\dim V = \dim U + \dim W - \dim(U \cap W) = \dim U + \dim W - 0$$
$$= \dim U + \dim W$$

8.120 Let U and W be subspaces of $\mathbf{R}^3$ for which $\dim U = 1$, $\dim W = 2$, and $U \nsubseteq W$. Show that $\mathbf{R}^3 = U \oplus W$.

❚ Since $U \nsubseteq W$, the intersection $U \cap W$ is properly contained in U and $U + W$ properly contains W. Thus $\dim(U \cap W) < \dim U = 1$. Hence $\dim(U \cap W) = 0$ and so $U \cap W = \{0\}$. Also, $\dim(U + W) > \dim W = 2$, but $U + W \subseteq \mathbf{R}^3$ where $\dim \mathbf{R}^3 = 3$. Thus $\dim(U + W) = 3$ and so $\mathbf{R}^3 = U + W$. The conditions $\mathbf{R}^3 = U + W$ and $U \cap W = \{0\}$ imply that $\mathbf{R}^3 = U \oplus W$.

8.121 Suppose $V = U_1 \oplus U_2 \oplus \cdots \oplus U_r$, and suppose, for $i = 1, \ldots, r$, $B_i = \{u_{j1}, u_{i2}, \ldots, u_{in_j}\}$ is contained in U_i and is linearly independent. Show that the union $B = B_1 \cup B_2 \cup \cdots \cup B_r$ is linearly independent.

▮ Suppose

$$\sum_{j_1=1}^{n_1} a_{1j_1}u_{1j_1} + \sum_{j_2=1}^{n_2} a_{2j_2}u_{2j_2} + \cdots + \sum_{j_r=1}^{n_r} a_{rj_r}u_{rj_r} = 0 \qquad (1)$$

where the a_{ij_i} are scalars. Each $\sum_{j_i} a_{ij_i}u_{ij_i}$ belongs to U_i. Since V is the direct sum of the U_i, the sum (1) for 0 is unique. Thus, for $i = 1, \ldots, r$, $\sum_{j_i} a_{ij_i}u_{ij_i} = 0$. But B_i is linearly independent. Hence, for $i = 1, \ldots, r$, we have $a_{i1} = 0, a_{i2} = 0, \ldots, a_{in_i} = 0$. In other words, every scalar in (1) is equal to 0. Thus B is linearly independent.

8.122 Suppose $V = U_1 \oplus U_2 \oplus \cdots \oplus U_r$, and suppose, for $i = 1, \ldots, r$, B_i is a basis of U_i. Show that $B = \cup B_i$ is a basis of V.

▮ By Problem 8.121, B is linearly independent since each B_i is linearly independent. Suppose $v \in V$. Then $v = u_1 + \cdots + u_r$ where $u_i \in U_i$. Then u_i is a linear combination of the vectors in B_i. Hence v is a linear combination of the vectors in B. Thus B spans V. Since B is linearly independent and spans V, B is a basis of V.

8.123 Let $V = U_1 \oplus U_2 \oplus \cdots \oplus U_r$, where $\dim U_i = n_i$. Prove $\dim V = \dim U_1 + \dim U_2 + \cdots + \dim U_r$.

▮ Let B_i be a basis of U_i. Hence B_i has n_i elements. Thus $B = \cup B_i$ has $n_1 + \cdots + n_r$ elements. By Problem 8.122, B is a basis of V. Hence $\dim V = n_1 + \cdots + n_r = \dim U_1 + \cdots + \dim U_r$.

8.124 Suppose $\{u_1, \ldots, u_r, w_1, \ldots, w_s\}$ is a linearly independent subset of a vector space V. Show that $\mathrm{span}(u_i) \cap \mathrm{span}(w_j) = \{0\}$.

▮ Suppose $v \in \mathrm{span}(u_i) \cap \mathrm{span}(w_j)$. Then there exist scalars a_i and scalars b_j such that $v = a_1u_1 + \cdots + a_ru_r = b_1w_1 + \cdots + b_sw_s$. Hence $a_1u_1 + \cdots + a_ru_r - b_1w_1 - \cdots - b_sw_s = 0$. But $\{u_i, w_j\}$ is linearly independent. Hence each $a_i = 0$ and each $b_j = 0$. Therefore $v = 0$.

8.125 Let U be a subspace of a vector space V of finite dimension. Show that there exists a subspace W of V such that $V = U \oplus W$.

▮ Let $\{u_1, \ldots, u_r\}$ be a basis of U. Since $\{u_i\}$ is linearly independent, it can be extended to a basis of V, say, $\{u_1, \ldots, u_r, w_1, \ldots, w_s\}$. Let W be the space generated by $\{w_1, \ldots, w_s\}$. Since $\{u_i, w_j\}$ spans V, we have $V = U + W$. On the other hand, by Problem 8.124, $U \cap W = \{0\}$. Accordingly, $V = U \oplus W$.

8.126 Suppose B is a linearly independent subset of V. Let $[B_1, B_2, \ldots, B_r]$ be a partition of B. Show that $\mathrm{span}(B) = \mathrm{span}(B_1) \oplus \mathrm{span}(B_2) \oplus \cdots \oplus \mathrm{span}(B_r)$.

▮ Since $B = \cup_i B_i$ and each $B_i \subseteq B$, we have $\mathrm{span}(B) = \mathrm{span}(\cup_i B_i) \subseteq \Sigma_i \mathrm{span}(B_i) \subseteq \mathrm{span}(B)$. Hence $\mathrm{span}(B) = \Sigma_i \mathrm{span}(B_i)$. Suppose

$$0 = \sum a_{1j_1}u_{1j_1} + \sum a_{2j_2}u_{2j_2} + \cdots + \sum a_{rj_r}u_{rj_r} \qquad (1)$$

where a_{ij_i} are scalars and the $u_{ij_i} \in B_i$. Since B is linearly independent, each $a_{ij_i} = 0$ in (1). Thus 0 can only be written uniquely as $0 = 0 + 0 + \cdots + 0$. Therefore, $\mathrm{span}(B) = \mathrm{span}(B_1) \oplus \cdots \oplus \mathrm{span}(B_r)$.

8.127 Suppose $V = U_1 + U_2 + \cdots + U_r$ and $\dim V = \dim U_1 + \dim U_2 + \cdots + \dim U_r$. Show that $V = U_1 \oplus U_2 \oplus \cdots \oplus U_r$.

▮ Suppose $\dim V = n$. Let B_i be a basis for U_i. Then $B = \cup_i B_i$ has n elements and spans V. Thus B is a basis for V. By Problem 8.126, $V = U_1 \oplus U_2 \oplus \cdots \oplus U_r$.

8.9 COORDINATES

8.128 Define the coordinates of a vector v in a vector space V over a field K where $\dim V = n$.

▮ Let $\{e_1, \ldots, e_n\}$ be a basis of V. Since $\{e_i\}$ spans V, the vector v is a linear combination of the e_i:

$$v = a_1e_1 + a_2e_2 + \cdots + a_ne_n \qquad a_i \in K$$

Since the e_i are independent, such a representation is unique (Problem 8.129). Thus the n scalars

$a_1, \ldots, a_n$ are completely determined by the vector v and the basis $\{e_i\}$. We call these scalars the *coordinates* of v in $\{e_i\}$, and we call the n-tuple $(a_1, \ldots, a_n)$ the *coordinate vector* of v relative to $\{e_i\}$ and denote it by $[v]_e$ or simply $[v]$:

$$[v]_e = (a_1, a_2, \ldots, a_n)$$

8.129 Let $v_1, v_2, \ldots, v_m$ be independent vectors, and suppose u is a linear combination of the v_i, say $u = a_1 v_1 + a_2 v_2 + \cdots + a_m v_m$ where the a_i are scalars. Show that the above representation of u is unique.

▌ Suppose $u = b_1 v_1 + b_2 v_2 + \cdots + b_m v_m$ where the b_i are scalars. Subtracting,

$$0 = u - u = (a_1 - b_1)v_1 + (a_2 - b_2)v_2 + \cdots + (a_m - b_m)v_m$$

But the v_i are linearly independent; hence the coefficients in the above relation are each 0:

$$a_1 - b_1 = 0, \quad a_2 - b_2 = 0, \ldots, a_m - b_m = 0$$

Hence $a_1 = b_1, a_2 = b_2, \ldots, a_m = b_m$ and so the above representation of u as a linear combination of the v_i is unique.

Problems 8.130–8.131 refer to the vector $v = (3, 1, -4)$ in $\mathbf{R}^3$.

8.130 Find the coordinate vector of v relative to the basis $f_1 = (1, 1, 1)$, $f_2 = (0, 1, 1)$, $f_3 = (0, 0, 1)$.

▌ Set v as a linear combination of the f_i using the unknowns x, y, and z; i.e., set $v = xf_1 + yf_2 + zf_3$:

$$
\begin{aligned}
(3, 1, -4) &= x(1, 1, 1) + y(0, 1, 1) + z(0, 0, 1) \\
&= (x, x, x) + (0, y, y) + (0, 0, z) \\
&= (x, x + y, x + y + z)
\end{aligned}
$$

Then set the corresponding components equal to each other to obtain the equivalent system of equations

$$
\begin{aligned}
x & & & = 3 \\
x + y & & & = 1 \\
x + y + z & & & = -4
\end{aligned}
$$

having solution $x = 3$, $y = -2$, $z = -5$. Thus $[v]_f = [3, -2, -5]$.

8.131 Find the coordinate vector of v relative to the usual basis $e_1 = (1, 0, 0)$, $e_2 = (0, 1, 0)$, $e_3 = (0, 0, 1)$.

▌ Set $v = xe_1 + ye_2 + ze_3$ using unknowns x, y, z: $(3, 1, -4) = x(1, 0, 0) + y(0, 1, 0) + z(0, 0, 1) = (x, y, z)$. Set corresponding components equal to each other to obtain $x = 3$, $y = 1$, $z = -4$, Thus $[v]_e = [3, 1, -4]$. In other words, relative to the usual basis, $[v]_e$ has the same components as v. [The next problem shows that this result is true in general.]

8.132 Let v be a vector in K^n. Show that the coordinate vector $[v]$ relative to the usual basis $e_1 = (1, 0, \ldots, 0)$, $e_2 = (0, 1, 0, \ldots, 0)$, $\ldots$, $e_n = (0, 0, \ldots, 0, 1)$ has the same components as v.

▌ Suppose $v = (a_1, a_2, \ldots, a_n)$. Then $v = x_1 e_1 + x_2 e_2 + \cdots + x_n e_n = (x_1, x_2, \ldots, x_n)$. Thus $x_1 = a_1$, $x_2 = a_2, \ldots$, $x_n = a_n$. Accordingly, $[v] = [a_1, a_2, \ldots, a_n]$.

8.133 Let V be the vector space of polynomials with degree ≤ 2:

$$V = \{at^2 + bt + c: a, b, c \in \mathbf{R}\}$$

The polynomials $e_1 = 1$, $e_2 = t - 1$, and $e_3 = (t - 1)^2 = t^2 - 2t + 1$ form a basis for V. Let $v = 2t^2 - 5t + 6$. Find $[v]_e$, the coordinate vector of v relative to the basis $\{e_1, e_2, e_3\}$.

▌ Set v as a linear combination of the e_i using the unknowns x, y, and z, i.e., set $v = xe_1 + ye_2 + ze_3$:

$$
\begin{aligned}
2t^2 - 5t + 6 &= x(1) + y(t - 1) + z(t^2 - 2t + 1) \\
&= x + yt - y + zt^2 - 2zt + z \\
&= zt^2 + (y - 2z)t + (x - y + z)
\end{aligned}
$$

Then set the coefficients of the same powers of t equal to each other:

$$\begin{aligned} x - y + z &= 6 \\ y - 2z &= -5 \\ z &= 2 \end{aligned}$$

The solution of the above system is $x = 3$, $y = -1$, $z = 2$. Thus $v = 3e_1 - e_2 + 2e_3$, and so $[v]_e = [3, -1, 2]$.

Problems 8.134–8.137 refer to the basis $u_1 = (2, 1)$, $u_2 = (1, -1)$ of $\mathbf{R}^2$.

8.134 Find the coordinate vector $[v]$ of $v = (2, 3)$.

▌ Set $v = xu_1 + yu_2$ to obtain $(2, 3) = x(2, 1) + y(1, -1) = (2x + y, x - y)$. Set corresponding components equal to each other to obtain the equations $2x + y = 2$ and $x - y = 3$. Solve to obtain $x = \frac{5}{3}$, $y = -\frac{4}{3}$. Thus $[v] = [\frac{5}{3}, -\frac{4}{3}]$.

8.135 Find the coordinate vector $[u]$ where $u = (4, -1)$.

▌ Set $u = xu_1 + yu_2$ to obtain $(4, -1) = (2x + y, x - y)$. Solve $2x + y = 4$ and $x - y = -1$ to get $x = 1$, $y = 2$. Hence $[u] = [1, 2]$.

8.136 Find the coordinate vector $[w]$ where $w = (3, -3)$.

▌ Set $w = xu_1 + yu_2$ to obtain $(3, -3) = (2x + y, x - y)$. Solve $2x + y = 3$ and $x - y = -3$ to get $x = 0$, $y = 3$. Thus $[w] = [0, 3]$.

8.137 Find the coordinate vector $[v]$ where $v = (a, b)$.

▌ Set $v = xu_1 + yu_2$ to obtain $(a, b) = (2x + y, x - y)$. Solve $2x + y = a$ and $x - y = b$ to obtain $x = (a + b)/3$, $y = (a - 2b)/3$. Thus $[v] = [(a + b)/3, (a - 2b)/3]$.

Problems 8.138–8.139 refer to the basis $\{(1, 1, 1), (1, 1, 0), (1, 0, 0)\}$ of $\mathbf{R}^3$.

8.138 Find the coordinates of the vector $v = (4, -3, 2)$.

▌ Set v as a linear combination of the basis vectors using unknown scalars x, y, and z:

$$v = x(1, 1, 1) + y(1, 1, 0) + z(1, 0, 0)$$

and then solve for the solution vector (x, y, z). [The solution is unique since the basis vectors are linearly independent.]

$$(4, -3, 2) = x(1, 1, 1) + y(1, 1, 0) + z(1, 0, 0) = (x + y + z, x + y, x)$$

Set corresponding components equal to each other to obtain the system $x + y + z = 4$, $x + y = -3$, $x = 2$. Substitute $x = 2$ into the second equation to obtain $y = -5$; then put $x = 2$, $y = -5$ into the first equation to obtain $z = 7$. Thus $x = 2$, $y = -5$, $z = 7$ is the unique solution to the system. Thus $[v] = [2, -5, 7]$.

8.139 Find the coordinate vector $[w]$ where $w = (a, b, c)$.

▌ Set w as a linear combination of the basis vectors:

$$(a, b, c) = x(1, 1, 1) + y(1, 1, 0) + z(1, 0, 0) = (x + y + z, x + y, x)$$

Then $x + y + z = a$, $x + y = b$, $x = c$. Solve to get $x = c$, $y = b - c$, $z = a - b$. Thus $[w] = [c, b - c, a - b]$.

Problems 8.140–8.141 refer to the matrix $A = \begin{pmatrix} 2 & 3 \\ 4 & -7 \end{pmatrix}$ in the vector space V of real 2×2 matrices.

8.140 Find the coordinate vector $[A]_B$ of the matrix A relative to the basis

$$B = \left\{ \begin{pmatrix} \bar{1} & 1 \\ 1 & 1 \end{pmatrix}, \begin{pmatrix} 0 & -1 \\ 1 & 0 \end{pmatrix}, \begin{pmatrix} 1 & -1 \\ 0 & 0 \end{pmatrix}, \begin{pmatrix} 1 & 0 \\ 0 & 0 \end{pmatrix} \right\}$$

❚ Set A as a linear combination of the matrices in the basis using unknown scalars x, y, z, t:

$$A = \begin{pmatrix} 2 & 3 \\ 4 & -7 \end{pmatrix} = x\begin{pmatrix} 1 & 1 \\ 1 & 1 \end{pmatrix} + y\begin{pmatrix} 0 & -1 \\ 1 & 0 \end{pmatrix} + z\begin{pmatrix} 1 & -1 \\ 0 & 0 \end{pmatrix} + t\begin{pmatrix} 1 & 0 \\ 0 & 0 \end{pmatrix}$$

$$= \begin{pmatrix} x & x \\ x & x \end{pmatrix} + \begin{pmatrix} 0 & -y \\ y & 0 \end{pmatrix} + \begin{pmatrix} z & -z \\ 0 & 0 \end{pmatrix} + \begin{pmatrix} t & 0 \\ 0 & 0 \end{pmatrix}$$

$$= \begin{pmatrix} x+z+t & x-y-z \\ x+y & x \end{pmatrix}$$

Set corresponding entries equal to each other to obtain the system $x+z+t=2$, $x-y-z=3$, $x+y=4$, $x = -7$ from which $x=-7$, $y=11$, $z=-21$, $t=30$. Thus $[A] = [-7, 11, -21, 30]$. [Note that the coordinate vector of A must be a vector in $\mathbf{R}^4$ since $\dim V = 4$.]

8.141 Find the coordinate vector $[A]_E$ of the matrix A relative to the usual basis of V; that is, the basis

$$E = \left\{ \begin{pmatrix} 1 & 0 \\ 0 & 0 \end{pmatrix}, \begin{pmatrix} 0 & 1 \\ 0 & 0 \end{pmatrix}, \begin{pmatrix} 0 & 0 \\ 1 & 0 \end{pmatrix}, \begin{pmatrix} 0 & 0 \\ 0 & 1 \end{pmatrix} \right\}$$

❚

$$\begin{pmatrix} 2 & 3 \\ 4 & -7 \end{pmatrix} = x\begin{pmatrix} 1 & 0 \\ 0 & 0 \end{pmatrix} = y\begin{pmatrix} 0 & 1 \\ 0 & 0 \end{pmatrix} = z\begin{pmatrix} 0 & 0 \\ 1 & 0 \end{pmatrix} = t\begin{pmatrix} 0 & 0 \\ 0 & 1 \end{pmatrix} = \begin{pmatrix} x & y \\ z & t \end{pmatrix}$$

Thus $x=2$, $y=3$, $z=4$, $t=-7$. Hence $[A]_E = [2, 3, 4, -7]$, whose components are the elements of A written row by row.

Remark: The above result is true in general, i.e., if A is any $m \times n$ matrix in the vector space V of $m \times n$ matrices over a field K, then the coordinate vector $[A]$ of A relative to the usual basis of V is the mn coordinate vector in K^{mn} whose components are the elements of A written row by row.

8.142 Determine whether the following matrices are dependent or independent:

$$A = \begin{pmatrix} 1 & 2 & -3 \\ 4 & 0 & 1 \end{pmatrix} \qquad B = \begin{pmatrix} 1 & 3 & -4 \\ 6 & 5 & 4 \end{pmatrix} \qquad C = \begin{pmatrix} 3 & 8 & -11 \\ 16 & 10 & 9 \end{pmatrix}$$

❚ The coordinate vectors of the above matrices relative to the usual basis are as follows:

$$[A] = [1, 2, -3, 4, 0, 1] \qquad [B] = [1, 3, -4, 6, 5, 4] \qquad [C] = [3, 8, -11, 16, 10, 9]$$

Form the matrix M whose rows are the above coordinate vectors:

$$M = \begin{pmatrix} 1 & 2 & -3 & 4 & 0 & 1 \\ 1 & 3 & -4 & 6 & 5 & 4 \\ 3 & 8 & -11 & 16 & 10 & 9 \end{pmatrix}$$

Row reduce M to echelon form:

$$M \quad \text{to} \quad \begin{pmatrix} 1 & 2 & -3 & 4 & 0 & 1 \\ 0 & 1 & -1 & 2 & 5 & 3 \\ 0 & 2 & -2 & 4 & 10 & 6 \end{pmatrix} \quad \text{to} \quad \begin{pmatrix} 1 & 2 & -3 & 4 & 0 & 1 \\ 0 & 1 & -1 & 2 & 5 & 3 \\ 0 & 0 & 0 & 0 & 0 & 0 \end{pmatrix}$$

Since the echelon matrix has only two nonzero rows, the coordinate vectors $[A]$, $[B]$, and $[C]$ generate a space of dimension 2 and so are dependent. Accordingly, the original matrices A, B, and C are dependent.

8.143 Let W be the vector space of 2×2 symmetric matrices over $\mathbf{R}$. [See Problem 8.60.] Find the coordinate vector of the matrix $A = \begin{pmatrix} 4 & -11 \\ -11 & -7 \end{pmatrix}$ relative to the basis $\left\{ \begin{pmatrix} 1 & -2 \\ -2 & 1 \end{pmatrix}, \begin{pmatrix} 2 & 1 \\ 1 & 3 \end{pmatrix}, \begin{pmatrix} 4 & -1 \\ -1 & -5 \end{pmatrix} \right\}$.

❚ Set A as a linear combination of the matrices in the basis using unknown scalars x, y, and z:

$$A = \begin{pmatrix} 4 & -11 \\ -11 & -7 \end{pmatrix} = x\begin{pmatrix} 1 & -2 \\ -2 & 1 \end{pmatrix} + y\begin{pmatrix} 2 & 1 \\ 1 & 3 \end{pmatrix} + z\begin{pmatrix} 4 & -1 \\ -1 & -5 \end{pmatrix} = \begin{pmatrix} x + 2y + 4z & -2x + y - z \\ -2x + y - z & x + 3y - 5z \end{pmatrix}$$

Set corresponding entries equal to each other to obtain the equivalent system of linear equations and reduce to echelon form:

$$\begin{array}{r} x + 2y + 4z = 4 \\ -2x + y - z = -11 \\ 2x + y - z = -11 \\ x + 3y - 5z = -7 \end{array} \quad \text{or} \quad \begin{array}{r} x + 2y + 4z = 4 \\ 5y + 7z = -3 \\ y - 9z = -11 \end{array} \quad \text{or} \quad \begin{array}{r} x + 2y + 4z = 4 \\ 5y + 7z = -3 \\ 52z = 52 \end{array}$$

We obtain $z = 1$ from the third equation, then $y = -2$ from the second equation, and then $x = 4$ from the first equation. Thus the solution of the system is $x = 4$, $y = -2$, $z = 1$; hence $[A] = [4, -2, 1]$. [Since $\dim W = 3$ by Problem 8.60, the coordinate vector of A must be a vector in $\mathbf{R}^3$.]

8.144 Let $\{e_1, e_2, e_3\}$ and $\{f_1, f_2, f_3\}$ be bases of a vector space V (of dimension 3). Suppose

$$\begin{array}{l} e_1 = a_1 f_1 + a_2 f_2 + a_3 f_3 \\ e_2 = b_1 f_1 + b_2 f_2 + b_3 f_3 \\ e_3 = c_1 f_1 + c_2 f_2 + c_3 f_3 \end{array} \quad \text{and} \quad P = \begin{pmatrix} a_1 & a_2 & a_3 \\ b_1 & b_2 & b_3 \\ c_1 & c_2 & c_3 \end{pmatrix}$$

$$(1)$$

Here P is the matrix whose rows are the coordinate vectors of e_1, e_2, and e_3, respectively, relative to the basis $\{f_i\}$. Show that, for any vector $v \in V$, $[v]_e P = [v]_f$. That is, multiplying the coordinate vector of v relative to the basis $\{e_i\}$ by the matrix P, we obtain the coordinate vector of v relative to the basis $\{f_i\}$. [The matrix P is frequently called the change of basis matrix.]

❚ Suppose $v = re_1 + se_2 + te_3$; then $[v]_e = (r, s, t)$. Using (1), we have

$$\begin{array}{l} v = r(a_1 f_1 + a_2 f_2 + a_3 f_3) + s(b_1 f_1 + b_2 f_2 + b_3 f_3) + t(c_1 f_1 + c_2 f_2 + c_3 f_3) \\ = (ra_1 + sb_1 + tc_1)f_1 + (ra_2 + sb_2 + tc_2)f_2 + (ra_3 + sb_3 + tc_3)f_3 \end{array}$$

Hence

$$[v]_f = (ra_1 + sb_1 + tc_1, ra_2 + sb_2 + tc_2, ra_3 + sb_3 + tc_3)$$

On the other hand,

$$[v]_e P = (r, s, t)\begin{pmatrix} a_1 & a_2 & a_3 \\ b_1 & b_2 & b_3 \\ c_1 & c_2 & c_3 \end{pmatrix}$$

$$= (ra_1 + sb_1 + tc_1, ra_2 + sb_2 + tc_2, ra_3 + sb_3 + tc_3)$$

Accordingly, $[v]_e P = [v]_f$.

Remark: In Chapters 9–11 we shall write coordinate vectors as column vectors rather than as row vectors. Then, by above,

$$Q[v]_e = \begin{pmatrix} a_1 & b_1 & c_1 \\ a_2 & b_2 & c_2 \\ a_3 & b_3 & c_3 \end{pmatrix}\begin{pmatrix} r \\ s \\ t \end{pmatrix} = \begin{pmatrix} ra_1 + sb_1 + tc_1 \\ ra_2 + sb_2 + tc_2 \\ ra_3 + sb_3 + tc_3 \end{pmatrix} = [v]_f$$

where Q is the matrix whose columns are the coordinate vectors of e_1, e_2, and e_3, respectively, relative to the basis $\{f_i\}$. Note that Q is the transpose of P and that Q appears on the left of the column vector $[v]_e$ whereas P appears on the right of the row vector $[v]_e$.

Problems 8.145–8.150 refer to the basis $B = \{1, 1 - t, (1 - t)^2, (1 - t)^3\}$ of the vector space V of polynomials in t of degree ≤ 3 and the polynomials:

$$u = 2 - 3t + t^2 + 2t^3 \qquad w = 3 - 2t - t^2 \qquad v = a + bt + ct^2 + dt^3$$

8.145 Find the coordinate vector $[u]$ relative to the basis B of V.

$\blacksquare$ Set u as a linear combination of the basis vectors using unknowns x, y, z, s:

$$\begin{aligned} u = 2 - 3t + t^2 + 2t^3 &= x(1) + y(1 - t) + z(1 - t)^2 + s(1 - t)^3 \\ &= x(1) + y(1 - t) + z(1 - 2t + t^2) + s(1 - 3t + 3t^2 - t^3) \\ &= x + y - yt + z - 2zt + zt^2 + s - 3st + 3st^2 - st^3 \\ &= (x + y + z + s) + (-y - 2z - 3s)t + (z + 3s)t^2 + (-s)t^3 \end{aligned}$$

Then set the coefficients of the same powers of t equal to each other:

$$x + y + z + s = 2 \qquad -y - 2z - 3s = -3 \qquad z + 3s = 1 \qquad -s = 2$$

The solution is $x = 2$, $y = -5$, $z = 7$, $s = -2$. Thus $[u] = [2, -5, 7, -2]$.

8.146 Find the coordinate vector $[u]$ relative to the basis $\{1, t, t^2, t^3\}$ of V.

$\blacksquare$ The basis consists of the powers of t; hence simply write down the corresponding coefficients to obtain $[u] = [2, -3, 1, 2]$.

8.147 Find the coordinate vector $[w]$ relative to the above basis B of V.

$\blacksquare$ Set w as a linear combination of the basis vectors using unknowns x, y, z, s:

$$w = 3 - 2t - t^2 = x(1) + y(1 - t) + z(1 - t)^2 + s(1 - t)^3 = (x + y + z + s) + (-y - 2z - 3s)t + (z + 3s)t^2 + (-s)t^3$$

Then set the coefficients of the same powers of t equal to each other:

$$x + y + z + s = 3 \qquad -y - 2z - 3s = -2 \qquad z + 3s = -1 \qquad -s = 0$$

The solution is $x = 0$, $y = 4$, $z = -1$, $s = 0$. Thus $[w] = [0, 4, -1, 0]$.

8.148 Find the coordinate vector $[w]$ relative to the basis $\{t^3, t^2, t, 1\}$ of V.

$\blacksquare$ The basis consists of the powers of t, hence write down the corresponding coefficients to obtain $[w] = [0, -1, -2, 3]$.

8.149 Find the coordinate vector $[v]$ relative to the above basis B of V.

$\blacksquare$ Set v as a linear combination of the basis vectors using unknowns x, y, z, s:

$$\begin{aligned} v = a + bt + ct^2 + dt^3 &= x(1) + y(1 - t) + z(1 - t)^2 + s(1 - t)^3 \\ &= (x + y + z + s) + (-y - 2z - 3s)t + (z + 3s)t^2 + (-s)t^3 \end{aligned}$$

Then set the coefficients of the same powers of t equal to each other:

$$x + y + z + s = a \qquad -y - 2z - 3s = b \qquad z + 3s = c \qquad -s = d$$

The solution is $x = a + b + c + d$, $y = -b - 2c - 3d$, $z = c + 3d$, $s = -d$. Thus

$$[v] = [a + b + c + d, \quad -b - 2c - 3d, \quad c + 3d, \quad -d]$$

8.150 Find the coordinate vector $[v]$ relative (a) to the basis $\{1, t, t^2, t^3\}$ of V, (b) to the basis $\{t^3, t^2, t, 1\}$ of V. [The bases are distinct since the orders of the elements are different.]

$\blacksquare$ In each case, write down the corresponding coefficients of the basis vectors: (a) $[v] = [d, c, b, a]$, (b) $[v] = [a, b, c, d]$.

8.151 Show that the coordinate vector of $0 \in V$ relative to any basis of V is always the zero n-tuple, i.e., $[0] = [0, 0, \ldots, 0]$.

$\blacksquare$ Let $\{u_1, u_2, \ldots, u_n\}$ be a basis of V. Suppose $0 = a_1 u_1 + \cdots + a_n u_n$. Since the u_i are linearly independent, $a_1 = 0, \ldots, a_n \equiv 0$. Hence $[0] = [0, 0, \ldots, 0]$.

8.152 Suppose V and V' are vector spaces over the same field K. Define an isomorphism between V and V', and define isomorphic vector spaces.

    ▌ Suppose $f: V \to V'$ is a one-to-one correspondence, i.e., suppose f is one-to-one and onto. Then f is called an isomorphism between V and V' if f "preserves" the vector space operations of vector addition and scalar multiplication; i.e., for every $v, w \in V$ and for any scalar $k \in K$, $f(v + w) = f(v) + f(w)$ and $f(kv) = kf(v)$. In such a case, we say that V and V' are isomorphic vector spaces and write $V \simeq V'$.

Theorem 8.12: Let V be an n-dimensional vector space over a field K. Then V and K^n are isomorphic.

8.153 Prove Theorem 8.12.

    ▌ Let $\{e_1, e_2, \ldots, e_n\}$ be a basis of V. Then to each vector $v \in V$, there corresponds the n-tuple $[v]_e$ in K^n. On the other hand, for any vector $(a_1, a_2, \ldots, a_n) \in K^n$, there exists a vector in V of the form $a_1 e_1 + \cdots + a_n e_n$. Thus the basis $\{e_i\}$ determines a one-to-one correspondence between the vectors in V and the n-tuples in K^n. Observe also that if

and

$$v = a_1 e_1 + \cdots + a_n e_n \qquad \text{corresponds to} \qquad (a_1, \ldots, a_n)$$

$$w = b_1 e_1 + \cdots + b_n e_n \qquad \text{corresponds to} \qquad (b_1, \ldots, b_n)$$

then

$$v + w = (a_1 + b_1) e_1 + \cdots + (a_n + b_n) e_n \qquad \text{corresponds to} \qquad (a_1, \ldots, a_n) + (b_1, \ldots, b_n)$$

and, for any scalar $k \in K$,

$$kv = (ka_1) e_1 + \cdots + (ka_n) e_n \qquad \text{corresponds to} \qquad k(a_1, \ldots, a_n)$$

That is,

$$[v + w]_e = [v]_e + [w]_e \qquad \text{and} \qquad [kv]_e = k[v]_e$$

Thus the above one-to-one correspondence between V and K^n preserves the vector space operations of vector addition and scalar multiplication. Thus V and K are isomorphic, written $V \simeq K^n$.

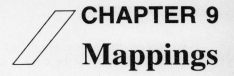

CHAPTER 9
Mappings

This chapter considers mappings and functions on arbitrary sets, not necessarily vector spaces. The concepts discussed in this chapter may be viewed as preliminary to the next chapter which discusses linear mappings on vector spaces.

9.1 MAPPINGS, FUNCTIONS

9.1 Define a mapping from a set A into a set B.

▮ Suppose that to each element of A there is assigned a unique element of B; the collection of such assignments is called a *mapping* (or *map*) from A into B. We denote a mapping f from A into B by $f: A \to B$. We write $f(a)$, read "f of a," for the element of B that f assigns to $a \in A$; it is called the *value* of f at a or the *image* of a under f.

Remark: The term function is used synonymously with the word mapping, although some texts reserve the word function for a real-valued or complex-valued mapping, i.e., which maps a set into **R** or **C**.

9.2 What is the domain of a mapping $f: A \to B$?

▮ The set A is the domain of f.

9.3 What is the co-domain of a mapping $f: A \to B$?

▮ The set B is the co-domain of f.

9.4 Define the image of a mapping $f: A \to B$.

▮ The set of all image values of f is called the *image* (or *range*) of f and is denoted by Im f or $f(A)$. That is,

$$\text{Im } f = \{b \in B: \text{ there exists } a \in A \text{ for which } f(a) = b\}$$

[Observe that Im f is a subset (perhaps a proper subset) of B.]

9.5 Let $f: A \to B$ and let S be a subset of A. Define the image of S under f, denoted by $f(S)$.

▮ Here $f(S) = \{f(a): a \in S\} = \{b \in B: \exists a \in S \text{ such that } f(a) = b\}$. In other words, $f(S)$ consists of all images of the elements in S.

9.6 Let $f: A \to B$ and let T be a subset of B. Define the inverse image or preimage of T under f, denoted by $f^{-1}(T)$.

▮ Here $f^{-1}(T) = \{a \in A: f(a) \in T\}$. In other words, $f^{-1}(T)$ consists of the elements of A each of whose images belongs to T.

9.7 Define the equality of functions.

▮ Two functions $f: A \to B$ and $g: A \to B$ are defined to be equal, written $f = g$, if $f(a) = g(a)$ for every $a \in A$. The negation of $f = g$ is written $f \neq g$ and is the statement: There exists an $a \in A$ for which $f(a) \neq g(a)$.

9.8 Define the graph of a function $f: A \to B$.

▮ To each function $f: A \to B$ there corresponds the subset of $A \times B$ given by $\{(a, f(a)): a \in A\}$.

We call this set the *graph* of f. We note that two functions $f: A \to B$ and $g: A \to B$ are equal if and only if they have the same graph. Thus we do not distinguish between a function and its graph.

Problems 9.9–9.14 refer to the mapping f from $A = \{a, b, c, d\}$ into $B = \{x, y, z, w\}$ defined by Fig. 9-1.

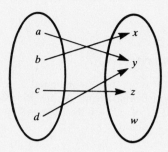

Fig. 9-1

9.9   Find the image of each element of A.

▮   The arrow indicates the image of an element. Thus $f(a) = y$, $f(b) + x$, $f(c) = z$, and $f(d) = y$.

9.10   Find the image of f.

▮   The image $f(A)$ of f consists of all image values. Only x, y, and z appear as image values; hence $f(A) = \{x, y, z\}$.

9.11   Find $f(S)$ where $S = \{a, b, d\}$.

▮ $f(S) + f(\{a, b, d\}) = \{f(a), f(b), f(d)\} = \{y, x, y\} = \{x, y\}$

9.12   Find $f^{-1}(T)$ where $T = \{y, z\}$.

▮   The elements a, c, and d have images in T, hence $f^{-1}(T) = \{a, c, d\}$.

9.13   Find $f^{-1}(w)$.

▮   No element has the image w under f; hence $f^{-1}(w) = \emptyset$, the empty set.

9.14   Find the graph of f, i.e., write f as a set of ordered pairs.

▮   The ordered pairs $(a, f(a))$, where $a \in A$, form the graph of f. Thus $f = \{(a, y), (b, x), (c, z), (d, y)\}$.

Problems 9.15–9.17 refer to the set $A = \{1, 2, 3, 4, 5\}$ and the function $f: A \to A$ defined by Fig. 9-2.

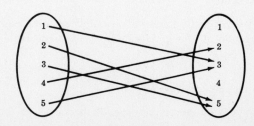

Fig. 9-2

9.15   Find the image of each element of A.

▌   The arrow indicates the image of an element; thus $f(1) = 3$, $f(2) = 5$, $f(3) = 5$, $f(4) = 2$, $f(5) = 3$.

9.16   Find the image $f(A)$ of the function f.

▌   The image $f(A)$ of f consists of all the image values.   Now only 2, 3, and 5 appear as the image of any elements of A; hence $f(A) = \{2, 3, 5\}$.

9.17   Find the graph of f, i.e., write f as a set of ordered pairs.

▌   The ordered pairs $(a, f(a))$, where $a \in A$,   form the graph of f.   Thus $f = \{(1, 3), (2, 5), (3, 5), (4, 2), (5, 3)\}$.

Problems 9.18–9.20 refer to sets $A = \{a, b, c\}$ and $B = \{x, y, z\}$ and Fig. 9-3.

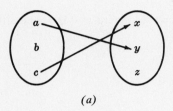 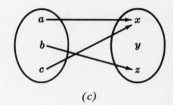

(a)   (b)   (c)

Fig. 9-3

9.18   Does Fig. 9-3(a) define a function from A into B?

▌   No.   There is nothing assigned to the element $b \in A$.

9.19   Does Fig. 9-3(b) define a function from A into B?

▌   No.   Two elements, x and z, are assigned to $c \in A$.

9.20   Does Fig. 9-3(c) define a function from A into B?

▌   Yes, since each element of A is assigned a unique element of B.

9.21   Let f be a subset of $A \times B$.   When does f define a function from A into B?

▌   A subset f of $A \times B$ is a function $f: A \to B$ if and only if each $a \in A$ appears as the first coordinate in exactly one ordered pair in f.

Problems 9.22–9.24 refer to the set $X = \{1, 2, 3, 4\}$ and the following subsets of $X \times X$:

$$f = \{(2, 3), (1, 4), (2, 1), (3, 2), (4, 4)\} \qquad g = \{(3, 1), (4, 2), (1, 1)\}$$
$$h = \{(2, 1), (3, 4), (1, 4), (2, 1), (4, 4)\}$$

9.22   Does f define a function $f: X \to X$?

▌   No.   Two different ordered pairs $(2, 3)$ and $(2, 1)$ in f have the same number 2 as their first coordinate.

9.23   Does g define a function $g: X \to X$?

▌   No.   The element $2 \in X$ does not appear as the first coordinate in any ordered pair in g.

9.24   Does h define a function $h: X \to X$?

▮ Yes. Although $2 \in X$ appears as the first coordinate in two ordered pairs in h, these two ordered pairs are equal.

9.25 Let A be the set of students in a school. Determine which of the following assignments defines a mapping on A:
(a) To each student assign his or her age
(b) To each student assign his or her teacher
(c) To each student assign his or her sex
(d) To each student assign his or her spouse.

▮ A collection of assignments is a mapping on A providing each element $a \in A$ is assigned exactly one element. Thus
(a) Yes, because each student has one and only one age.
(b) Yes, if each student has only one teacher; no, if any student has more than one teacher.
(c) Yes.
(d) No, if some student is not married.

9.26 Let $f: \mathbf{R} \to \mathbf{R}$ be the mapping which assigns to each real number x its square x^2. Describe different ways of defining f.

▮ The function f may be described by any of the following:

$$f(x) = x^2 \quad \text{or} \quad x \mapsto x^2 \quad \text{or} \quad y = x^2$$

Here the barred arrow $\mapsto$ is read "goes into." In the last notation, x is called the *independent variable* and y is called the *dependent variable* since the value of y will depend on the value that x takes.

Remark: Whenever a function f is given by a formula using the independent variable x, as above, we assume unless otherwise stated or implied that f is a function from $\mathbf{R}$ [or the largest subset of $\mathbf{R}$ for which f has meaning] into $\mathbf{R}$. [See Section 9.2.]

Problems 9.27–9.30 refer to the above function $f(x) = x^2$.

9.27 Find the value of f at 5, -4, and 0.

▮ $f(5) = 5^2 = 25$, $f(-4) = (-4)^2 = 16$, $f(0) = 0^2 = 0$.

9.28 Find (a) $f(y + 2)$, (b) $f(x + h)$.

▮ (a) $f(y + 2) = (y + 2)^2 = y^2 + 4y + 4$, (b) $f(x + h) = (x + h)^2 = x^2 + 2xh + h^2$.

9.29 Find $[f(x + h) - f(x)]/h$.

▮ $[f(x + h) - f(x)]/h = (x^2 + 2xh + h^2 - x^2)/h = (2xh + h^2)/h = 2x + h$.

9.30 Find Im f, the image of f.

▮ Every nonnegative real number a is the square of $\sqrt{a}$ and the square of any number cannot be negative. Hence Im $f = \{x : x \geq 0\}$, i.e., the set of nonnegative real numbers.

9.31 Find the number of functions from $X = \{a, b\}$ into $Y = \{1, 2, 3\}$.

▮ There are three choices for the image of a, and there are three choices for the image of b; hence there are $3 \cdot 3 = 3^2 = 9$ possible functions from X into Y.

9.32 Suppose X has $|X|$ elements and Y has $|Y|$ elements. Show that there are $|Y|^{|X|}$ functions from X into Y. [For this reason, one frequently writes Y^X for the collection of all functions from X into Y.]

▮ There are $|Y|$ choices for the image of each of the $|X|$ elements of X; hence there are $|Y|^{|X|}$ possible functions from X into Y.

9.33 Let A be any nonempty set. Define the identity mapping on A.

▮ The identity mapping on A, denoted by 1_A, is the mapping defined by $1_A(x) = x$ for every $x \in A$.

9.34 Let $A = \{1, 2, 3, \ldots, 9\}$. Find $1_A(3)$, $1_A(6)$, and $1_A(9)$.

▮ Under the identity map, the image of an element is the element itself; so $1_A(3) + 3$, $1_A(6) = 6$, $1_A(8) = 8$.

9.35 Define a constant map.

▮ Let f be a function with domain A. Then f is a constant map if every $a \in A$ is assigned the same element.

9.36 Given sets A and B, now many constant maps are there from A into B?

▮ Each $b \in B$ defines the constant map $f(x) = b$ for every $x \in A$. Hence there are $|B|$ constant maps where $|B|$ denotes the number of elements in B.

9.37 Let S be a subset of A and let $f: A \to B$. Define the restriction of f to S.

▮ The restriction of f to S is the mapping $\hat{f}: S \to B$ defined by $\hat{f}(s) = f(s)$ for every $s \in S$. One usually writes $f|_S$ to denote the restriction of f to S.

9.38 Let $f: \mathbf{R} \to \mathbf{R}$ be defined by $f(x) = x^2$. Let $\hat{f}: \mathbf{N} \to \mathbf{R}$ be the restriction of f to $\mathbf{N}$, i.e., let $\hat{f} = f|_\mathbf{N}$. Find $f(4)$, $f(-3)$, and $f(\frac{1}{2})$.

▮ By definition, $\hat{f}(n) = f(n)$ for every $n \in \mathbf{N}$. Thus $\hat{f}(4) = f(4) = 4^2 = 16$ and $\hat{f}(-3) = f(-3) = (-3)^2 = 9$. However, $\hat{f}(\frac{1}{2})$ is not defined since $\frac{1}{2}$ is not in the domain of $\hat{f}$.

9.2 REAL-VALUED FUNCTIONS

This section covers real-valued functions, i.e., functions f which map sets into $\mathbf{R}$. Frequently, the domain of f is $\mathbf{R}$ itself or a subset of $\mathbf{R}$ and hence can be plotted in the coordinate plane $\mathbf{R} \times \mathbf{R} = \mathbf{R}^2$. We also use the following notation for intervals from a to b where a and b are real numbers such that $a < b$:

$[a, b] = \{x : a \le x \le b\}$, called the closed interval from a to b
$[a, b) = \{x : a \le x < b\}$, called a half-open interval from a to b
$(a, b] = \{x : a < x \le b\}$, called a half-open interval from a to b
$(a, b) = \{x : a < x < b\}$, called the open interval from a to b

9.39 What is the domain D of a real-valued function $f(x)$ [where x is a real variable] when $f(x)$ is given by a formula?

▮ The domain D consists of the largest subset of $\mathbf{R}$ for which $f(x)$ has meaning and is real, unless otherwise specified.

9.40 Find the domain D of the function $f(x) = 1/(x - 2)$.

▮ f is not defined for $x - 2 = 0$, i.e., for $x = 2$; hence $D = \mathbf{R} \backslash \{2\}$.

9.41 Find the domain D of the function $g(x) = x^2 - 3x - 4$.

▮ g is defined for every real number; hence $D = \mathbf{R}$.

9.42 Find the domain D of the function $h(x) = \sqrt{25 - x^2}$.

▮ h is not defined when $25 - x^2$ is negative; hence $D = [-5, 5] = \{x : -5 \le x \le 5\}$.

9.43 Find the domain D of the function $f(x) = x^2$ where $0 \le x \le 2$.

▮ Although the formula for f is meaningful for every real number, the domain of f is explicitly given as $D = \{x : 0 \le x \le 2\}$.

Problems 9.44–9.49 refer to the following functions from **R** into **R**:
(i) To each number let f assign its cube.
(ii) To each number let g assign the number 5.
(iii) To each positive number let h assign its square, and to each nonpositive number let h assign the number 6.

9.44 Use a formula to define f.

▮ Since f assigns to any number x its cube x^3, we can define f by $f(x) = x^3$.

9.45 Find the value of f at 4, -2, and 0.

▮ $f(4) = 4^3 = 64$, $f(-2) = (-2)^3 = -8$, $f(0) = 0^3 = 0$.

9.46 Use a formula to define g.

▮ Since g assigns 5 to any number x, we can define g by $g(x) = 5$.

9.47 Find the image of 4, -2, and 0 under g.

▮ The image of every number is 5, so $g(4) = 5$, $g(-2) = 5$, $g(0) = 5$.

9.48 Use a formula to define h.

▮ Two different rules are used to define h as follows:

$$h(x) = \begin{cases} x^2 & \text{if } x > 0 \\ 6 & \text{if } x \le 0 \end{cases}$$

9.49 Find $h(4)$, $h(-2)$, and $h(0)$.

▮ Since $4 > 0$, $h(4) = 4^2 = 16$. On the other hand, $-2, 0 \le 0$ and so $h(-2) = 6$, $h(0) = 6$.

Problems 9.50–9.54 refer to the function $f : \mathbf{R} \to \mathbf{R}$ defined by $f(x) = x^3$.

9.50 Find $f(3)$ and $f(-5)$.

▮ $f(3) = 3^3 = 27$, $f(-5) = (-5)^3 = -125$.

9.51 Find $f(y)$ and $f(y + 1)$.

▮ $f(y) = (y)^3 = y^3$, $f(y + 1) = (y + 1)^3 = y^3 + 3y^2 + 3y + 1$.

9.52 Find $f(x + h)$.

▮ $f(x + h) = (x + h)^3 = x^3 + 3x^2h + 3xh^2 + h^3$.

9.53 Find $[f(x + h) - f(x)]/h$.

▮ $[f(x + h) - f(x)]/h = (x^3 + 3x^2h + 3xh^2 + h^3 - x^3)/h = (3x^2h + 3xh^2 + h^3)/h = 3x^2 + 3xh + h^2$.

9.54 Sketch the graph of f.

▮ Since f is a polynomial function, it can be sketched by first plotting some points of its graph and then drawing a smooth curve through these points as in Fig. 9-4.

x	$f(x)$
-3	-27
-2	-8
-1	-1
0	0
1	1
2	8
3	27

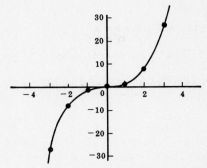

Graph of $f(x) = x^3$ **Fig. 9-4**

9.55   Sketch the graph of $f(x) = 3x - 2$.

❚ Since f is linear, only two points (three as a check) are needed to sketch its graph. Set up a table with three values of x, say, $x = -2, 0, 2$ and find the corresponding values of $f(x)$:

$$f(-2) = 3(-2) - 2 = -8 \qquad f(0) = 3(0) - 2 = -2 \qquad f(2) = 3(2) - 2 = 4$$

Draw the line through these points as in Fig. 9-5.

x	$f(x)$
-2	-8
0	-2
2	4

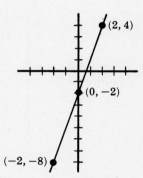

Graph of f   **Fig. 9-5**

Problems 9.56–9.58 refer to the function $g(x) = x^2 + x - 6$.

9.56   Sketch the graph of g.

❚ Set up a table of values for x and then find the corresponding values of the function. Plot the points in a coordinate diagram and then draw a smooth continuous curve through the points as in Fig. 9-6.

x	$g(x)$
-4	6
-3	0
-2	-4
-1	-6
0	-6
1	-4
2	0
3	6

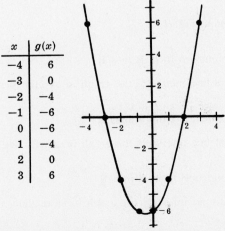

Graph of g   **Fig. 9-6**

9.57   Find $g^{-1}(14)$.

▌ Set $g(x) = 14$ and solve for x:

$$x^2 + x - 6 = 14 \quad \text{or} \quad x^2 + x - 20 = 0 \quad \text{or} \quad (x + 5)(x - 4) = 0$$

Thus $x = -5$ and $x = 4$. In other words, $g^{-1}(-4) = -5, 4$.

9.58   Find $g^{-1}(-8)$.

▌ Set $g(x) = -8$ and solve for x: $x^2 + x - 6 = -8$ or $x^2 + x + 2 = 0$. Using the quadratic formula, the discriminant $D = b^2 - 4ac = 1^2 - 4 \cdot 1 \cdot 2 = -7$ is negative and hence there are no real solutions. Thus $g^{-1}(-8) = \varnothing$, the empty set.

9.59   Sketch the graph of $h(x) = x^3 - 3x^2 - x + 3$.

▌ Draw a smooth continuous curve through some of the points of the graph of h as in Fig. 9-7.

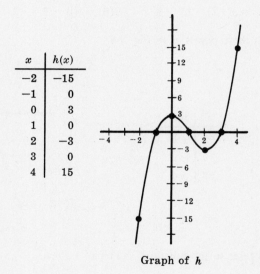

x	$h(x)$
−2	−15
−1	0
0	3
1	0
2	−3
3	0
4	15

Graph of h **Fig. 9-7**

9.60   Let f be a subset of $\mathbf{R} \times \mathbf{R}$. State a geometrical condition for f to be a function from $\mathbf{R}$ into $\mathbf{R}$.

▌ The graph f is a function from $\mathbf{R}$ into $\mathbf{R}$ if every vertical line intersects the graph in exactly one point.

Problems 9.61–9.66 refer to Fig. 9-8.

9.61   Does Fig. 9-8(a) define a function from $\mathbf{R}$ into $\mathbf{R}$?

▌ Yes, since every vertical line intersects the graph in exactly one point.

9.62   Does Fig. 9-8(b) define a function from $\mathbf{R}$ into $\mathbf{R}$?

▌ Yes, since every vertical line intersects the graph in exactly one point.

9.63   Does Fig. 9-8(c) define a function from $\mathbf{R}$ into $\mathbf{R}$?

▌ No, since some vertical lines intersect the graph in more than one point.

9.64   Does Fig. 9-8(d) define a function from $\mathbf{R}$ into $\mathbf{R}$?

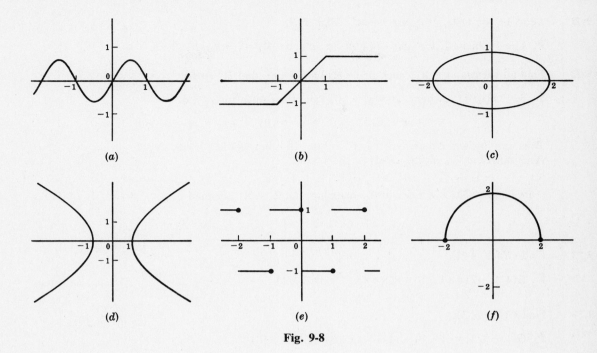

Fig. 9-8

▮ No, since some vertical lines intersect the graph in more than one point or do not intersect the graph at all.

9.65 Does Fig. 9-8(e) define a function from **R** into **R**?

▮ Yes, since every vertical line intersects the graph in exactly one point.

9.66 Does Fig. 9-8(f) define a function from **R** into **R**?

▮ No; however the graph does define a function from D into **R** where $D = \{x: -2 \leq x \leq 2\}$.

9.3 VECTOR-VALUED MAPPINGS

This section considers mappings from one vector space V into another vector space V'. [Special types of such mappings, called linear maps (Chapter 10), are the main subject matter of linear algebra.]

Problems 9.67–9.72 refer to the mapping $F: \mathbf{R}^3 \to \mathbf{R}^2$ defined by $F(x, y, z) = (yz, x^2)$.

9.67 Find $F(2, 3, 4)$.

▮ Substitute in the formula for F to get $F(2, 3, 4) = (3 \cdot 4, 2^2) = (12, 4)$.

9.68 Find $F(5, -2, 7)$.

▮ $F(5, -2, 7) = (-2 \cdot 7, 5^2) = (-14, 25)$.

9.69 Find $F(3, -5)$.

▮ The domain of F is not $\mathbf{R}^2$, so $F(3, -5)$ is not defined.

9.70 Find $F(a, a, a)$.

▮ $F(a, a, a) = (a \cdot a, a^2) = (a^2, a^2)$.

9.71   Let S be the line $x = y = z$ in $\mathbf{R}^3$. Find $F(S)$.

    ∎   Use Problem 9.70 to get $F(S) = \{(a^2, a^2): a \in \mathbf{R}\} = \{(b, b): b \geq 0\}$.

9.72   Find all vectors $v \in \mathbf{R}^3$ such that $F(v) = 0$, i.e., find $F^{-1}(0, 0)$.

    ∎   Set $F(v) = 0$ where $v = (x, y, z)$, and then solve for x, y, z.

$$F(x, y, z) = (yz, x^2) = (0, 0) \quad \text{or} \quad yz = 0 \quad \text{and} \quad x^2 = 0$$

Thus $x = 0$ and either $y = 0$ or $z = 0$. In other words, $x = 0$, $y = 0$ or $x = 0$, $z = 0$. Accordingly, v lies on the z axis or the y axis.

    Problems 9.73–9.78 refer to the mapping $G: \mathbf{R}^3 \to \mathbf{R}^2$ defined by

$$G(x, y, z) = (x + 2y - 4z, 2x + 3y + z)$$

9.73   Find $G(4, 5, -2)$.

    ∎ $G(4, 5, -2) = (4 + 10 + 8, 8 + 15 - 2) = (22, 21)$.

9.74   Find $G(1, -5, 3)$.

    ∎ $G(1, -5, 3) = (1 - 10 - 12, 2 - 15 + 3) = (-21, -10)$.

9.75   Find $G(0)$ [where $0 = (0, 0, 0)$].

    ∎ $G(0) = G(0, 0, 0) = (0 + 0 + 0, 0 + 0 + 0) = (0, 0) = 0$.

9.76   Find $G(a, a, a)$.

    ∎ $G(a, a, a) = (a + 2a - 4a, 2a + 3a + a) = (-a, 6a)$.

9.77   Find $G(14, -9, -1)$.

    ∎ $G(14, -9, -1) = (14 - 18 + 4, 28 - 27 - 1) = (0, 0)$.

9.78   Find $G^{-1}(3, 4)$.

    ∎   Set $G(x, y, z) = (3, 4)$ to get the system

$$\begin{array}{lll} x + 2y - 4z = 3 & x + 2y - 4z = \ \ 3 & x + 2y - 4z = 3 \\ 2x + 3y + \ z = 4 \quad \text{or} & \ \ - \ y + 9z = -2 \quad \text{or} & y - 9z = 2 \end{array}$$

Here z is a free variable. Set $z = a$ to obtain the general solution:

$$x = -14a - 1 \qquad y = 9a + 2 \qquad z = a$$

In other words, $G^{-1}(3, 4) = \{(-14a - 1, 9a + 2, a)\}$, where $a \in \mathbf{R}$.

    Problems 9.79–9.84 refer to the mapping $H: \mathbf{R} \to \mathbf{R}^3$ defined by $H(t) = (2t, t^2, 3t + 5)$. [Such a mapping from $\mathbf{R}$ into $\mathbf{R}^n$ is called a curve in $\mathbf{R}^n$. This curve is sometimes presented in the form $x = 2t$, $y = t^2$, $z = 3t + 5$.]

9.79   Find $H(0)$.

    ∎ $H(0) = (0, 0^2, 0 + 5) = (0, 0, 5)$.

9.80   Find $H(2)$.

    ∎ $H(2) = (4, 4, 6 + 5) = (4, 4, 11)$.

9.81   Find $H(1, 2, 3)$.

▮   The domain of H is $\mathbf{R}$, hence $H(1, 2, 3)$ is not defined.

9.82   Find $H^{-1}(8)$.

▮   The co-domain of H is $\mathbf{R}^3$, so $H^{-1}(8)$ is not defined.

9.83   Find $H^{-1}(v)$ where $v = (6, 9, 14)$.

▮   Set $H(t) = v$ and solve for t:

$$(2t, t^2, 3t + 5) = (6, 9, 14) \qquad \text{or} \qquad 2t = 6, \quad t^2 = 9, \quad 3t + 5 = 14$$

This gives $t = 3$. Thus $H^{-1}(v) = 3$.

9.84   Find $H^{-1}(v)$ where $v = (8, 4, 20)$.

▮   Set $H(t) = v$ and solve for t:

$$(2t, t^2, 3t + 5) = (8, 4, 20) \qquad \text{or} \qquad 2t = 8, \quad t^2 = 4, \quad 3t + 5 = 20$$

There is no single value of t which is a solution of all three equations. Thus $H^{-1}(v) = \varnothing$, the empty set.

Problems 9.85–9.88 refer to the map $F: \mathbf{R}^2 \to \mathbf{R}^2$ defined by $F(x, y) = (3y, 2x)$.

9.85   Find $F(4, -5)$.

▮ $F(4, -5) = (3 \cdot (-5), 2 \cdot 4) = (-15, 8)$.

9.86   Find $F^{-1}(6, -8)$.

▮   Set $F(x, y) = (6, -8)$ and solve for x and y:

$$(3y, 2x) = (6, -8) \qquad \text{or} \qquad 3y = 6, \quad 2x = -8 \qquad \text{or} \qquad y = 2, \quad x = -4$$

Hence $F^{-1}(6, -8) = (-4, 2)$.

9.87   Let S be the unit circle in $\mathbf{R}^2$, i.e., the solution set of $x^2 + y^2 = 1$. Describe $F(S)$.

▮   Let (a, b) be an element of $F(S)$. Then there exists $(x, y) \in S$ such that $F(x, y) = (a, b)$. Hence:

$$(3y, 2x) = (a, b) \qquad \text{or} \qquad 3y = a, \quad 2x = b \qquad \text{or} \qquad y = \frac{a}{3}, \quad x = \frac{b}{2}$$

Since $(x, y) \in S$, that is, $x^2 + y^2 = 1$, we have

$$\left(\frac{b}{2}\right)^2 + \left(\frac{a}{3}\right)^2 = 1 \qquad \text{or} \qquad \frac{a^2}{9} + \frac{b^2}{4} = 1$$

Thus $F(S)$ is an ellipse.

9.88   Find $F^{-1}(S)$ where S is the unit circle in $\mathbf{R}^2$.

▮   Let $F(x, y) = (a, b)$ where $(a, b) \in S$. Then $(3y, 2x) = (a, b)$ or $3y = a$, $2x = b$. Since (a, b) is in S, we have $a^2 + b^2 = 1$. Thus $(3y)^2 + (2x)^2 = 1$. Accordingly, $F^{-1}(S)$ is the ellipse $4x^2 + 9y^2 = 1$.

Problems 9.89–9.90 refer to the real 2×3 matrix $A = \begin{pmatrix} 1 & -3 & 5 \\ 2 & 4 & -1 \end{pmatrix}$.

9.89 If vectors in $\mathbf{R}^2$ and $\mathbf{R}^3$ are viewed as row vectors, then A determines a mapping $f: \mathbf{R}^2 \to \mathbf{R}^3$ defined by $f(v) = vA$. Find $f(v)$ where $v = (2, -3)$.

▮ $f(v) = vA = (2, -3)\begin{pmatrix} 1 & -3 & 5 \\ 2 & 4 & -1 \end{pmatrix} = (2 - 6, -6 - 12, 10 + 3) = (-4, -18, 13).$

9.90 If vectors in $\mathbf{R}^2$ and $\mathbf{R}^3$ are viewed as column vectors, then A determines a mapping $g: \mathbf{R}^3 \to \mathbf{R}^2$ defined by $g(v) = Av$. Find $g(v)$ where $v = (3, 1, -2)$. [For notational convenience, column vectors are frequently presented as rows.]

▮ $g(v) = Av = \begin{pmatrix} 1 & -3 & 5 \\ 2 & 4 & -1 \end{pmatrix}\begin{pmatrix} 3 \\ 1 \\ -2 \end{pmatrix} = \begin{pmatrix} -10 \\ 12 \end{pmatrix}.$

Remark: Problems 9.89 and 9.90 indicate that any $m \times n$ matrix A over a field K may be viewed as a mapping from K^m to K^n or as a mapping from K^n to K^m depending on whether the vectors are viewed as rows or as columns. Unless otherwise specified or implied, we will assume that A is a mapping from K^m to K^n as in Problem 9.90 and so vectors will be viewed as columns rather than as rows. Moreover, we will usually denote this mapping by A, the same symbol used for the matrix.

Problems 9.91–9.92 refer to the real matrix $B = \begin{pmatrix} 1 & 2 \\ 4 & 3 \end{pmatrix}$.

9.91 Find $B(v)$ where $v = (3, -2)$.

▮ Since we view v as a column vector,

$$B(v) = \begin{pmatrix} 1 & 2 \\ 4 & 3 \end{pmatrix}\begin{pmatrix} 3 \\ -2 \end{pmatrix} = \begin{pmatrix} 3 - 4 \\ 12 - 6 \end{pmatrix} = \begin{pmatrix} -1 \\ 6 \end{pmatrix}$$

That is, $B(v) = (-1, 6)$.

9.92 Find $B^{-1}(w)$ where $w = (-3, 8)$.

▮ Set $B(v) = w$ where $v = (x, y)$ and solve for x and y:

$$\begin{pmatrix} 1 & 2 \\ 4 & 3 \end{pmatrix}\begin{pmatrix} x \\ y \end{pmatrix} = \begin{pmatrix} -3 \\ 8 \end{pmatrix} \quad \text{or} \quad \begin{pmatrix} x + 2y \\ 4x + 3y \end{pmatrix} = \begin{pmatrix} -3 \\ 8 \end{pmatrix} \quad \text{or} \quad \begin{matrix} x + 2y = -3 \\ 4x + 3y = 8 \end{matrix}$$

The solution of the system is $x = 5$, $y = -4$. Thus $B^{-1}(w) = (5, -4)$.

Remark: Let V be the vector space of polynomials in the variable t over the real field $\mathbf{R}$. Then the derivative defines a mapping $D: V \to V$ where, for any polynomial $f \in V$, we let $D(f) = df/dt$.

Problems 9.93–9.96 refer to the above derivative map $D: V \to V$ where V is the vector space of real polynomials in the variable t.

9.93 Find $D(3t^2 - 5t + 2)$.

▮ Take the derivative: $D(3t^2 - 5t + 2) = 6t - 5$.

9.94 Find $D(at^3 + bt^2 + ct + d)$.

▮ Take the derivative: $D(at^3 + bt^2 + ct + d) = 3at^2 + 2bt + c$.

9.95   Find $D^{-1}(g)$ where $g(t) = 6t^2 + 8t - 5$.

▌ Take the antiderivative (integral) of g to obtain $D^{-1}(g) = 2t^3 + 4t^2 - 5t + C$ where C is the constant of integration.

9.96   Find Im D, the image of D.

▌ Every polynomial $g \in V$ is the derivative of a polynomial; hence Im $D = V$.

Remark: Let V be the vector space of polynomials in t over $\mathbf{R}$. Then the integral from, say, 0 to 1 defines a mapping $I: V \to \mathbf{R}$ where, for any polynomial $f \in V$, we let $I(f) = \int_0^1 f(t)\, dt$.

Problems 9.97–9.98 refer to the above integral map $I: V \to \mathbf{R}$.

9.97   Find $I(f)$ where $f(t) = 3t^2 - 5t + 2$.

▌ $I(f) = \int_0^1 (3t^2 - 5t + 2)\, dt = \frac{1}{2}$.

9.98   Find $I(g)$ where $g(t) = at^3 + bt^2 + ct + d$.

▌ $I(g) = \int_0^1 (at^3 + bt^2 + ct + d)\, dt = a/4 + b/3 + c/2 + d$.

9.4   COMPOSITION OF MAPPINGS

9.99   Consider two mappings $f: A \to B$ and $g: B \to C$. Define the composition mapping of f and g.

▌ Let $a \in A$; then $f(a) \in B$, the domain of g. Hence we can obtain the image of $f(a)$ under the mapping g; that is, we can obtain $g(f(a))$. This mapping from A into C is called the *composition* or *product* of f and g and is denoted by $g \circ f$. In other words, $(g \circ f): A \to C$ is the mapping defined by $(g \circ f)(a) = g(f(a))$.

Remark: Let $F: A \to B$. Some texts write aF instead of $F(a)$ for the image of $a \in A$ under F. With this notation, the composition of functions $F: A \to B$ and $G: B \to C$ is denoted by $F \circ G$ and not by $G \circ F$ as used in this text.

Problems 9.100–9.103 refer to mappings $f: A \to B$ and $g: B \to C$ defined by Fig. 9-9.

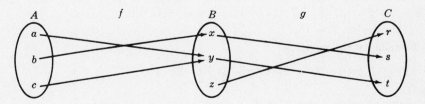

Fig. 9-9

9.100   Find the composition mapping $(g \circ f): A \to C$.

▌ We use the definition of the composition mapping to compute

$$(g \circ f)(a) = g(f(a)) = g(y) = t$$
$$(g \circ f)(b) = g(f(b)) = g(x) = s$$
$$(g \circ f)(c) = g(f(c)) = g(y) = t$$

Observe that we arrive at the same answer if we "follow the arrows" in the diagram:

$$a \rightarrow y \rightarrow t, \qquad b \rightarrow x \rightarrow s, \qquad c \rightarrow y \rightarrow t$$

9.101 Find the image of f and of g.

▎ By the diagram, the image values under the mapping f are x and y and the image values under g are r, s, and t; hence $\text{Im } f = (x, y\}$ and $\text{Im } g = \{r, s, t\}$.

9.102 Find the image of the composition mapping $g \circ f$.

▎ By problem 9.100, the image values under the composition mapping $g \circ f$ are t and s: hence $\text{Im } g \circ f = \{s, t\}$. Note that the images of g and $g \circ f$ are different.

9.103 Find the composition mapping $f \circ g$.

▎ The composition $f \circ g$ is not defined since the domain of f is not the co-domain of g.

Problems 9.104–9.110 refer to the mappings $f: \mathbf{R} \rightarrow \mathbf{R}$ and $g: \mathbf{R} \rightarrow \mathbf{R}$ defined by $f(x) = 2x + 1$ and $g(x) = x^2 - 2$.

9.104 Find: (a) $(g \circ f)(4)$ and (b) $(f \circ g)(4)$.

▎ (a) $f(4) = 2 \cdot 4 + 1 = 9$. Hence $(g \circ f)(4) = f(f(4)) = g(9) = 9^2 - 2 = 79$. (b) $g(4) = 4^2 - 2 = 14$. Hence $(f \circ g)(4) = f(g(4)) = f(14) = 2 \cdot 14 + 1 = 29$. [Note that $f \circ g \neq g \circ f$ since they differ on $x = 4$.]

9.105 Find $(g \circ f)(a + 2)$.

▎ $f(a + 2) = 2(a + 2) + 1 = 2a + 5$. Hence

$$(g \circ f)(a + 2) = g(f(a + 2)) = g(2a + 5) = (2a + 5)^2 - 2 = 4a^2 + 20a + 23$$

9.106 Find $(f \circ g)(a + 2)$.

▎ $g(a + 2) = (a + 2)^2 - 2 = a^2 + 4a + 2$. Hence

$$(f \circ g)(a + 2) = f(g(a + 2)) = f(a^2 + 4a + 2) = 2(a^2 + 4a + 2) + 1 = 2a^2 + 8a + 5$$

9.107 Find a formula for the mapping $g \circ f$.

▎ Compute the formula for $g \circ f$ as follows:

$$(g \circ f)(x) = g(f(x)) = g(2x + 1) = (2x + 1)^2 - 2 = 4x^2 + 4x - 1$$

Observe that the same answer can be found by writing $y = f(x) = 2x + 1$ and $z = g(y) = y^2 - 2$, and then eliminating y: $z = y^2 - 2 = (2x + 1)^2 - 2 = 4x^2 + 4x - 1$.

9.108 Find a formula for the mapping $f \circ g$.

▎ $(f \circ g)(x) = f(g(x)) = f(x^2 - 2) = 2(x^2 - 2) + 1 = 2x^2 - 3$.

9.109 Find a formula for the mapping $f \circ f$ [sometimes denoted by f^2].

▎ $(f \circ f)(x) = f(f(x)) = f(2x + 1) = 2(2x + 1) + 1 = 4x + 3$.

9.110 Find a formula for the mapping $g \circ g$.

▎ $(g \circ g)(x) = g(g(x)) = g(x^2 - 2) = (x^2 - 2)^2 - 2 = x^4 - 4x^2 + 2$.

9.111 Consider an arbitrary mapping $f: A \rightarrow B$. When is $f \circ f$ defined?

▌ The composition $f \circ f$ is defined when the domain of f is equal to the co-domain of f, i.e., when $A = B$.

Problems 9.112–9.118 refer to the mappings $f: \mathbf{R}^2 \to \mathbf{R}^2$ and $g: \mathbf{R}^2 \to \mathbf{R}$ defined by $f(x, y) = (x^2 + 1, x + y)$ and $g(x, y) = 2x + 3y$.

9.112 Find: (a) $f(1, 4)$, (b) $g(1, 4)$.

▌ (a) $f(1, 4) = (1^2 + 1, 1 + 4) = (2, 5)$. (b) $g(1, 4) = 2 \cdot 1 + 3 \cdot 4 = 2 + 12 = 14$. [Note that the image of a vector under f is a vector in $\mathbf{R}^2$ whereas the image of a vector under g is an element in $\mathbf{R}$.]

9.113 Find $(g \circ f)(2, 3)$.

▌ First compute $f(2, 3) = (2^2 + 1, 2 + 3) = (5, 5)$. Then

$$(g \circ f)(2, 3) = g(f(2, 3)) = g(5, 5) = 2 \cdot 5 + 3 \cdot 5 = 25.$$

9.114 Find $(f \circ g)(2, 3)$.

▌ The composition $f \circ g$ is not defined since the co-domain $\mathbf{R}$ of g is not the domain of f. Hence $(f \circ g)(2, 3)$ does not exist.

9.115 Find $(f \circ f)(3, 1)$.

▌ First compute $f(3, 1) = (3^2 + 1, 3 + 1) = (10, 4)$. Then compute $f(f(3, 1)) = f(10, 4) = (10^2 + 1, 10 + 4) = (101, 14)$. Thus $(f \circ f)(3, 1) = (101, 14)$.

9.116 Find $(g \circ g)(3, 1)$.

▌ The composition $g \circ g$ is not defined since the co-domain $\mathbf{R}$ of g is not the domain of g.

9.117 Find $(f \circ f \circ f)(v)$ [or $f^3(v)$] where $v = (2, 5)$.

▌ First compute $f(v) = f(2, 5) = (2^2 + 1, 2 + 5) = (5, 7)$. Then compute $f(f(v)) = f(5, 7) = (5^2 + 1, 5 + 7) = (26, 12)$. Last, compute $f(f(f(v))) = f(26, 12) = (26^2 + 1, 26 + 12) = (677, 38)$. Thus $f^3(v) = (677, 38)$.

9.118 Find a formula for $f \circ f$.

▌ $(f \circ f)(x, y) = f(f(x, y)) = f(x^2 + 1, x + y) = [(x^2 + 1)^2 + 1, (x^2 + 1) + (x + y)]$
$\qquad\qquad = (x^4 + 2x^2 + 2, x^2 + x + 1 + y)$.

9.119 Show that $1_B \circ f = f$ for any map $f: A \to B$. [Here $1_B: B \to B$ is the identity map on B, that is, $1_B(b) = b$ for every $b \in B$.]

▌ $(1_B \circ f)(a) = 1_B(f(a)) = f(a)$, for every $a \in A$. Thus $1_B \circ f = f$.

9.120 Show that $f \circ 1_A = f$ for any map $f: A \to B$. [Here $1_A: A \to A$ is the identity map on A.]

▌ $(f \circ 1_A)(a) = f(1_A(a)) = f(a)$, for every $a \in A$. Thus $f \circ 1_B = f$.

Theorem 9.1: Let $f: A \to B$, $g: B \to C$, and $h: C \to D$. Then $h \circ (g \circ f) = (h \circ g) \circ f$.

9.121 Prove Theorem 9.1 which states that composition of mappings satisfies the associative law.

▌ Consider any element $a \in A$. Then

and
$$(h \circ (g \circ f))(a) = h((g \circ f)(a)) = h(g(f(a)))$$
$$((h \circ g) \circ f)(a) = (h \circ g)(f(a)) = h(g(f(a)))$$

Thus $(h \circ (g \circ f))(a) = ((h \circ g) \circ f)(a)$ for every $a \in A$ and so $h \circ (g \circ f) = (h \circ g) \circ f$.

9.122 Define a diagram of maps.

❚ A directed graph in which the vertices are sets and the edges denote maps between the sets is called a diagram of maps.

Problems 9.123–9.126 refer to maps $f: A \to B$, $g: B \to A$, $h: C \to B$, $F: B \to C$, and $G: A \to C$ which are pictured in the diagram of maps in Fig. 9-10.

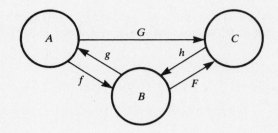

Fig. 9-10

9.123 Is $g \circ f$ defined? If so, what is its domain and co-domain?

❚ Since f goes from A to B and g goes from B to A, $g \circ f$ is defined and A is its domain and co-domain.

9.124 Is $h \circ f$ defined? If so, what is its domain and co-domain?

❚ Note h does not "follow" f in the diagram, i.e., the co-domain B of f is not the domain of h. Hence $h \circ f$ is not defined.

9.125 Is $F \circ h \circ G$ defined? If so, what is its domain and co-domain?

❚ The arrows representing G, h, and F do follow each other in the diagram and go from A to C to B to C. Thus $F \circ h \circ G$ is defined with domain A and co-domain C. [We emphasize that mappings are "read" from right to left.]

9.126 Is $G \circ F \circ h$ defined? If so, what is its domain and co-domain?

❚ F follows h in the diagram but G does not follow F, i.e., the co-domain C of F is not the domain of G. Hence $G \circ F \circ h$ is not defined.

9.127 Define a commutative diagram of maps.

❚ A diagram of maps is commutative if any two paths with the same initial and terminal vertices are equal.

Problems 9.128–9.132 refer to the commutative diagram of maps in Fig. 9-11.

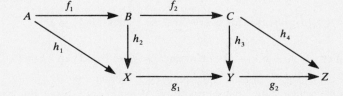

Fig. 9-11

9.128 Represent $h_2 \circ f_1$ by a single map.

❚ The composition map $h_2 \circ f_1$ goes from A to B to X. Since the diagram is commutative, $h_2 \circ f_1 = h_1$.

9.129 Represent $h_3 \circ f_2$ in as many ways as possible.

▌ The map $h_3 \circ f_2$ goes from B to C to Y. The only other path from B to Y is the map $g_1 \circ h_2$.

9.130 Represent the map $g_2 \circ h_3$ by a single map.

▌ The map $g_2 \circ h_3$ goes from C to Y to Z. The map h_4 goes from C to Z. Since the diagram is commutative, $g_2 \circ h_3 = h_4$.

9.131 Represent the map $g_1 \circ h_3$ by a single map.

▌ The map $g_1 \circ h_3$ is not defined since the co-domain Y of h_3 is not the domain of g_1.

9.132 Represent the map $g_2 \circ h_3 \circ f_2 \circ f_1$ in as many ways as possible.

▌ The map $g_2 \circ h_3 \circ f_2 \circ f_1$ goes from A to B to C to Y to Z. There are three other paths from A to Z:
(i) $g_2 \circ g_1 \circ h_1$, (ii) $g_2 \circ g_1 \circ h_2 \circ f_1$, and (iii) $h_4 \circ f_2 \circ f_1$.

9.133 Figure 9-12 defines maps $f: A \to B$, $g: B \to C$, and $h: C \to D$. Find the composition map $h \circ g \circ f$.

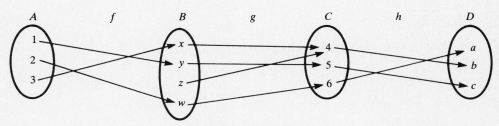

Fig. 9-12

▌ Follow the arrows from A to B to C to D as follows:

$$
\begin{array}{lll}
1 \to y \to 5 \to c & \text{hence} & (h \circ g \circ f)(1) = c \\
2 \to w \to 6 \to a & \text{hence} & (h \circ g \circ f)(2) = a \\
3 \to x \to 4 \to b & \text{hence} & (h \circ g \circ f)(3) = b
\end{array}
$$

9.5 ONE-TO-ONE, ONTO, AND INVERTIBLE MAPPINGS

9.134 Define a one-to-one or injective mapping.
▌ A mapping $f: A \to B$ is said to be one-to-one (or 1–1) or injective if different elements of A have distinct images; i.e., if $a \neq a'$ implies $f(a) \neq f(a')$ or, equivalently, if $f(a) = f(a')$ implies $a = a'$.

9.135 Define an onto or surjective mapping.

▌ A mapping $f: A \to B$ is said to be *onto* (or f maps A *onto* B) or *surjective* if every $b \in B$ is the image of at least one $a \in A$.

9.136 Define a one-to-one correspondence or bijective mapping.

▌ A mapping $f: A \to B$ is called a one-to-one correspondence between A and B or a bijective mapping if f is both one-to-one and onto.

Problems 9.137–9.145 refer to the maps $f: A \to B$, $g: B \to C$, and $h: C \to D$ in Fig. 9-12.

9.137 Is f one-to-one?

▌ Yes, since 1, 2, and 3 have distinct images.

9.138 Is f an onto map?

⫽ No, since z has no preimage under f.

9.139 Is f a one-to-one correspondence?

⫽ No, since f is not an onto map.

9.140 Is g one-to-one?

⫽ No, since x and z have the same image 4.

9.141 Is g an onto map?

⫽ Yes, since every element of C has a preimage.

9.142 Is g a one-to-one correspondence?

⫽ No, since g is not one-to-one.

9.143 Is h one-to-one?

⫽ Yes, since 4, 5, and 6 have distinct images.

9.144 Is h an onto map?

⫽ Yes, since a, b, and c have preimages.

9.145 Is h a one-to-one correspondence?

⫽ Yes, since h is both one-to-one and onto.

9.146 State a geometrical condition for a function $f: \mathbf{R} \rightarrow \mathbf{R}$ to be one-to-one.

⫽ A function $f: \mathbf{R} \rightarrow \mathbf{R}$ is one-to-one if no horizontal line contains more than one point of f.

9.147 State a geometrical condition for a function $g: \mathbf{R} \rightarrow \mathbf{R}$ to be an onto function.

⫽ A function $g: \mathbf{R} \rightarrow \mathbf{R}$ is an onto function if every horizontal line contains at least one point of g.

9.148 State a geometrical condition for a function $h: \mathbf{R} \rightarrow \mathbf{R}$ to be a one-to-one correspondence.

⫽ A function $h: \mathbf{R} \rightarrow \mathbf{R}$ is a one-to-one correspondence if every horizontal line contains exactly one point of h.

Problems 9.149–9.157 refer to the functions $f(x) = 2^x$, $g(x) = x^3 - x$, and $h(x) = x^2$ whose graphs appear in Fig. 9-13.

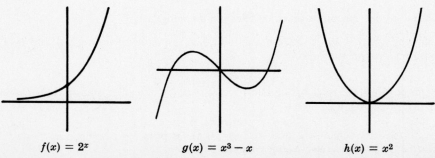

$f(x) = 2^x$ $g(x) = x^3 - x$ $h(x) = x^2$ **Fig. 9-13**

9.149   Is f one-to-one?

❙   Yes, since no horizontal line contains more than one point of f.

9.150   Is f onto?

❙   No, since some horizontal lines (those below the y axis) contain no point of f.

9.151   Is f bijective?

❙   No, since f is not surjective, i.e., f is not an onto function.

9.152   Is g injective (i.e., one-to-one)?

❙   No, since some horizontal lines contain more than one point of g, for example, $y = 0$ contains three points of g. In other words, $g(1) = g(0) = g(-1) = 0$, so g is not one-to-one.

9.153   Is g surjective (i.e., an onto function)?

❙   Yes, since every horizontal line contains at least one point of g.

9.154   Is g bijective?

❙   No, since g is not injective.

9.155   Is h one-to-one?

❙   No, for example, $h(2) = h(-2) = 4$. That is, the horizontal line $y = 4$ contains two points of h.

9.156   Is h an onto function?

❙   No, for example, -16 has no preimage. That is, the horizontal line $y = -16$ contains no points of h.

9.157   Is h bijective?

❙   No, as h is neither one-to-one nor onto.

9.158   Suppose $f: A \to B$ and $g: B \to C$ are one-to-one. Show that $g \circ f: A \to C$ is one-to-one.

❙   Suppose $(g \circ f)(x) = (g \circ f)(y)$. Then $g(f(x)) = g(f(y))$. Since g is one-to-one, $f(x) = f(y)$. Since f is one-to-one, $x = y$. We have proven that $(g \circ f)(x) = (g \circ f)(y)$ implies $x = y$; hence $g \circ f$ is one-to-one.

9.159   Suppose $f: A \to B$ and $g: B \to C$ are onto mappings. Show that $g \circ f: A \to C$ is an onto mapping.

❙   Suppose $c \in C$. Since g is onto, there exists $b \in B$ for which $g(b) = c$. Since f is onto, there exists $a \in A$ for which $f(a) = b$. Thus $(g \circ f)(a) = g(f(a)) = g(b) = c$; hence $g \circ f$ is onto.

9.160   Given $f: A \to B$ and $g: B \to C$, show that if $g \circ f$ is one-to-one, then f is one-to-one.

❙   Suppose f is not one-to-one. Then there exists distinct elements $x, y \in A$ for which $f(x) = f(y)$. Thus $(g \circ f)(x) = g(f(x)) = g(f(y)) = (g \circ f)(y)$; hence $g \circ f$ is not one-to-one. Therefore, if $g \circ f$ is one-to-one, then f must be one-to-one.

9.161   Given $f: A \to B$ and $g: B \to C$, show that if $g \circ f$ is onto, then g is onto.

❙   If $a \in A$, then $(g \circ f)(a) = g(f(a)) \in g(B)$; hence $(g \circ f)(A) \subset g(B)$. Suppose g is not onto. Then $g(B)$ is properly contained in C and so $(g \circ f)(A)$ is properly contained in C; thus $g \circ f$ is not onto. Accordingly if $g \circ f$ is onto, then g must be onto.

9.162 Define an invertible mapping.

▐ A mapping $f: A \to B$ is said to be *invertible* if there exists a mapping $g: B \to A$ such that $f \circ g = 1_B$ and $g \circ f = 1_A$ (where 1_A and 1_B are the identity maps). In such a case, the map g is called the *inverse* of f and is denoted by f^{-1}. Alternatively, f is invertible if the inverse relation f^{-1} is a mapping from B to A. [By Problem 9.163, f has an inverse if and only if f is both one-to-one and onto. Also, if $b \in B$ then $f^{-1}(b) = a$ where a is the unique element of A for which $f(a) = b$.]

9.163 Prove that a mapping $f: A \to B$ has an inverse if and only if it is one-to-one and onto.

▐ Suppose f has an inverse, i.e., there exists a function $f^{-1}: B \to A$ for which $f^{-1} \circ f = 1_A$ and $f \circ f^{-1} = 1_B$. Since 1_A is one-to-one, f is one-to-one by Problem 9.160; and since 1_B is onto, f is onto by Problem 9.161. That is, f is both one-to-one and onto.

Now suppose f is both one-to-one and onto. Then each $b \in B$ is the image of a unique element in A, say $\hat{b}$. Thus if $f(a) = b$, then $a = \hat{b}$; hence $f(\hat{b}) = b$. Now let g denote the mapping from B to A defined by $b \mapsto \hat{b}$. We have
(i) $(g \circ f)(a) = g(f(a)) = g(b) = \hat{b} = a$, for every $a \in A$; hence $g \circ f = 1_A$.
(ii) $(f \circ g)(b) = f(g(b)) = f(\hat{b}) = b$, for every $b \in B$; hence $f \circ g = 1_B$.
Accordingly, f has an inverse. Its inverse is the mapping g.

9.164 Let $f: \mathbf{R} \to \mathbf{R}$ be defined by $f(x) = 2x - 3$. Now f is one-to-one and onto; hence f has an inverse mapping f^{-1}. Find a formula for f^{-1}.

▐ Let y be the image of x under the mapping f, i.e., set $y = 2x - 3$. Interchange x and y to obtain $x = 2y - 3$ which is the inverse relation f^{-1}. Solve for y in terms of x to get $y = (x + 3)/2$. Thus the formula defining the inverse mapping is $f^{-1}(x) = (x + 3)/2$.

9.165 Find a formula for the inverse of $g(x) = x^2 - 1$.

▐ Set $y = x^2 - 1$. Interchange x and y to get $x = y^2 - 1$. Solve for y to get $y = \pm\sqrt{x + 1}$. The inverse of g does not exist unless the domain of g^{-1} is restricted to $x \geq 1$. Also, assume only the positive value of $\sqrt{x + 1}$. Then $g^{-1}(x) = \sqrt{x + 1}$.

9.166 Let $P = \{A_i\}$ be a partition of a set S. Define the natural (or canonical) map f from S into P.

▐ Let $s \in S$. Since P is a partition of S, there exists a unique index i_0 such that $s \in A_{i_0}$. Define $f: S \to P$ by $f(s) = A_{i_0}$. This is the natural map.

9.167 Consider the natural map $f: S \to P$ in Problem 9.166. Show that the natural map is an onto mapping.

▐ Let $A_i \in P$. Then $A_i \neq \varnothing$. Thus there exists $s \in S$ such that $s \in A_i$. Then $f(s) = A_i$. Thus f is an onto mapping.

9.168 Let S be subset of a set A. Define the inclusion mapping from S into A.

▐ The inclusion mapping from S into A, denote d by $i: S \hookrightarrow A$, is defined by $i(s) = s$ for every $s \in S$. [Note that the inclusion mapping is the identity mapping when $S = A$.]

9.169 Let $i: S \hookrightarrow A$ be the inclusion map in Problem 9.168. Show that i is one-to-one.

▐ Suppose $i(x) = i(y)$. Note $i(x) = x$ and $i(y) = y$. Hence $x = y$. Thus i is one-to-one.

CHAPTER 10

Linear Mappings

All vector spaces are assumed to be over the same field K unless otherwise stated or implied.

10.1 LINEAR MAPPINGS

10.1 Define a linear mapping.

▌ Let V and U be vector spaces over the same field K. A mapping $F: V \rightarrow U$ is called a *linear mapping* [or *linear transformation* or *vector space homomorphism*] if it satisfies the following two conditions:
(1) For any $v, w \in V$, $F(v + w) = F(v) + F(w)$.
(2) For any $k \in K$ and any $v \in V$, $F(kv) = kF(v)$.
In other words, $F: V \rightarrow U$ is linear if it "preserves" the two basic operations of a vector space, that of vector addition and that of scalar multiplication.

10.2 Suppose $F: V \rightarrow U$ is linear. Show that $F(0) = 0$.

▌ Substitute $k = 0$ into $F(kv) = kF(v)$ to get $F(0) = 0$.

10.3 Suppose $F: V \rightarrow U$ is linear. Show that $F(-u) = -F(u)$.

▌ Using $F(ku) = kF(u)$, we have $F(-u) = F[(-1)u] = (-1)F(u) = -F(u)$.

10.4 Show that $F: V \rightarrow U$ is linear if and only if, for any scalars $a, b \in K$ and any vectors $v, w \in V$,

$$F(av + bw) = F(av) + F(bw) = aF(v) + bF(w) \tag{1}$$

▌ Suppose F is linear, then $F(av + bw) = F(av) + F(bw) = aF(v) + bF(w)$. Conversely, suppose *(1)* holds. For $a = 1$ and $b = 1$ we get $F(v + w) = F(v) + F(w)$, and for $b = 0$ we get $F(av) = aF(v)$; hence F is linear.

Remark: The condition $F(av + bw) = aF(v) + bF(w)$ completely characterizes linear mappings and is sometimes used as its definition.

10.5 Suppose $F: V \rightarrow U$ is linear. Show that, for any $a_i \in K$ and any $v_i \in V$,

$$F(a_1 v_1 + a_2 v_2 + \cdots + a_n v_n) = a_1 F(v_1) + a_2 F(v_2) + \cdots + a_n F(v_n)$$

▌ Since F is linear, the condition holds for $n = 1$ and $n = 2$. Suppose $n > 2$. Then, by induction, $F(a_1 v_1 + a_2 v_2 + \cdots + a_n v_n) = F(a_1 v_1) + F(a_1 v_2 + \cdots + a_n v_n) = a_1 F(v_1) + a_2 F(v_2) + \cdots + a_n F(v_n)$. [This condition will be used frequently throughout the text.]

10.6 Let A be any $m \times n$ matrix over a field K. As noted previously, A determines a mapping $T: K^n \rightarrow K^m$ by the assignment $v \mapsto Av$. [Here the vectors in K^n in K^m are written as columns.] Show that T is linear.

▌ By properties of matrices, $T(v + w) = A(v + w) = Av + Aw = T(v) + T(w)$ and $T(kv) = A(kv) = kAv = kT(v)$, where $v, w \in K^n$ and $k \in K$. Thus T is linear.

Remark: The above type of linear mapping shall occur again and again. In fact, in the next chapter we show that every linear mapping from one finite-dimensional vector space into another can be represented as a linear mapping of the above type.

10.7 Let $F: \mathbf{R}^3 \rightarrow \mathbf{R}^3$ be the "projection" mapping into the xy plane, i.e., $F(x, y, z) = (x, y, 0)$. Show that F is linear.

▌ Let $v = (a, b, c)$ and $w = (a', b', c')$. Then

$$F(v + w) = F(a + a', b + b', c + c') = (a + a', b + b', 0) = (a, b, 0) + (a', b', 0) = F(v) + F(w)$$

and, for any $k \in \mathbf{R}$, $F(kv) = F(ka, kb, kc) = (ka, kb, 0) = k(a, b, 0) = kF(v)$. That is, F is linear.

10.8 Let $F: \mathbf{R}^2 \to \mathbf{R}^2$ be the "translation" mapping defined by $F(x, y) = (x + 1, y + 2)$. Show that F is not linear.

▌ Observe that $F(0) = F(0, 0) = (1, 2) \neq 0$. That is, the zero vector is not mapped onto the zero vector. Hence F is not linear.

10.9 Let $F: V \to U$ be the mapping which assigns $0 \in U$ to every $v \in V$. Show that F is linear.

▌ For any $v, w \in V$ and any $k \in K$, $F(v + w) = 0 = 0 + 0 = F(v) + F(w)$ and $F(kv) = 0 = k0 = kF(v)$. Thus F is linear. We call F the *zero mapping* and shall usually denote it by 0.

10.10 Consider the identity mapping $I: V \to V$ which maps each $v \in V$ into itself. Show that I is linear.

▌ For any $v, w \in V$ and any $a, b \in K$, we have $I(av + bw) = av + bw = aI(v) + bI(w)$. Thus I is linear.

Problems 10.11–10.12 refer to the vector space V of polynomials in the variable t over the real field $\mathbf{R}$.

10.11 Let $D: V \to V$ be the differential mapping $D(v) = dv/dt$. Show that D is linear.

▌ It is proven in calculus that

$$\frac{d(u + v)}{dt} = \frac{du}{dt} + \frac{dv}{dt} \quad \text{and} \quad \frac{d(ku)}{dt} = k\frac{du}{dt}$$

i.e., $D(u + v) = D(u) + D(v)$ and $D(ku) = kD(u)$. Thus D is linear.

10.12 Let $I: V \to \mathbf{R}$ be the integral mapping $I(v) = \int_0^1 v(t)\, dt$. Show that I is linear.

▌ It is proven in calculus that

$$\int_0^1 (u(t) + v(t))dt = \int_0^1 u(t)dt + \int_0^1 v(t)dt$$

and

$$\int_0^1 ku(t)dt = k\int_0^1 u(t)dt$$

i.e., $I(u + v) = I(u) + I(v)$ and $I(ku) = kI(u)$. Thus I is linear.

10.13 Consider the mapping $F: \mathbf{R}^2 \to \mathbf{R}^2$ defined by $F(x, y) = (x + y, x)$. Show that F is linear.

▌ Let $v = (a, b)$ and $w = (a', b')$; hence $v + w = (a + a', b + b')$ and $kv = (ka, kb)$. We have $F(v) = (a + b, a)$ and $F(w) = (a' + b', a')$. Thus

$$F(v + w) = F(a + a', b + b') = (a + a' + b + b', a + a') = (a + b, a) + (a' + b', a') = F(v) + F(w)$$

and $F(kv) = F(ka, kb) = (ka + kb, ka) = k(a + b, a) = kF(v)$. Since v, w, and k were arbitrary, F is linear.

10.14 Consider $F: \mathbf{R}^3 \to \mathbf{R}$ defined by $F(x, y, z) = 2x - 3y + 4z$. Show that F is linear.

▌ Let $v = (a, b, c)$ and $w = (a', b', c')$; hence

$$v + w = (a + a', b + b', c + c') \quad \text{and} \quad kv = (ka, kb, kc) \qquad k \in \mathbf{R}$$

We have $F(v) = 2a - 3b + 4c$ and $F(w) = 2a' - 3b' + 4c'$. Thus $F(v + w) = F(a + a', b + b', c + c') = 2(a + a') - 3(b + b') + 4(c + c') = (2a - 3b + 4c) + (2a' - 3b' + 4c') = F(v) + F(w)$ and $F(kv) = F(ka, kb, kc) = 2ka - 3kb + 4kc = k(2a - 3b + 4c) = kF(v)$. Accordingly, F is linear.

10.15   Consider $F: \mathbf{R}^2 \to \mathbf{R}$ defined by $F(x, y) = xy$. Show that F is not linear.

❚   Let $v = (1, 2)$ and $w = (3, 4)$; then $v + w = (4, 6)$. We have $F(v) = 1 \cdot 2 = 2$ and $F(w) = 3 \cdot 4 = 12$. Hence $F(v + w) = F(4, 6) = 4 \cdot 6 = 24 \neq F(v) + F(w)$. Accordingly, F is not linear.

10.16   Consider $F: \mathbf{R}^2 \to \mathbf{R}^3$ defined by $F(x, y) = (x + 1, 2y, x + y)$. Show that F is not linear.

❚   Since $F(0, 0) = (1, 0, 0) \neq (0, 0, 0)$, F cannot be linear.

10.17   Consider $F: \mathbf{R}^3 \to \mathbf{R}^2$ defined by $F(x, y, z) = (|x|, 0)$. Show that F is not linear.

❚   Let $v = (1, 2, 3)$ and $k = -3$; hence $kv = (-3, -6, -9)$. We have $F(v) = (1, 0)$ and so $kF(v) = -3(1, 0) = (-3, 0)$. Then $F(kv) = F(-3, -6, -9) = (3, 0) \neq kF(v)$ and hence F is not linear.

10.18   Consider $F: \mathbf{R}^2 \to \mathbf{R}^2$ defined by $F(x, y) = (2x - y, x)$. Show that F is linear.

❚   Let $u = (a, b)$ and $v = (a', b')$. Then $u + v = (a + a', b + b')$ and $k(u) = (ka, kb)$. We have $F(u) = (2a - b, a)$ and $F(v) = (2a' - b', a')$. Thus

$$F(u + v) = F(a + a', b + b') = [2(a + a') - (b + b'), \; a + a'] = (2a - b, a) + (2a' - b', a') = F(u) + F(v)$$

and $F(ku) = F(ka, kb) = (2ka - kb, ka) = k(2a - b, a) = kF(u)$. Thus F is linear.

10.19   Consider $F: \mathbf{R}^2 \to \mathbf{R}$ defined by $F(t) = (2t, 3t)$. Show that F is linear.

❚ $$F(t_1 + t_2) = [2(t_1 + t_2), 3(t_1 + t_2)] = [2t_1 + 2t_2, 3t_1 + 3t_2] = (2t_1, 3t_1) + (2t_2, 3t_2) = F(t_1) + F(t_2)$$

and

$$F(kt) = (2kt, 3kt) = k(2t, 3t) = kF(t)$$

Thus F is linear.

10.20   Consider $F: \mathbf{R}^2 \to \mathbf{R}^2$ defined by $F(x, y) = (x^2, y^2)$. Show that F is not linear.

❚   Let $u = (1, 2)$ and $k = 3$. Then $ku = (3, 6)$. We have $F(u) = (1, 4)$ and so $kF(u) = (3, 12)$. Hence $F(ku) = F(3, 6) = (9, 16) \neq kF(u)$. Thus F is not linear.

10.21   Consider $F: \mathbf{R}^3 \to \mathbf{R}^2$ defined by $F(x, y, z) = (x + 1, y + z)$. Show that F is not linear.

❚ $F(0) = F(0, 0, 0) = (0 + 1, 0 + 0) = (1, 0) \neq (0, 0)$. Thus F is not linear.

10.22   Consider $F: \mathbf{R}^2 \to \mathbf{R}$ defined by $F(x, y) = |x + y|$. Show that F is not linear.

❚   Let $u = (1, 2)$ and $k = -3$; so $ku = (-3, -6)$. We have $F(u) = 1 + 2 = 3$, hence $kF(u) = (-3) \cdot 3 = -9$. Thus $F(ku) = F(-3, -6) = -3 - 6 = -9 = 9 \neq kF(u)$. Accordingly, F is not linear.

Problems 10.23–10.25 refer to the vector space V of n-square matrices over a field K and an arbitrary matrix M in V.

10.23   Let $T: V \to V$ be defined by $T(A) = AM + MA$, where $A \in V$. Show that T is linear.

❚   For any $A, B \in V$ and any $k \in K$, we have $T(A + B) = (A + B)M + M(A + B) = AM + BM + MA + MB = (AM + MA) + (BM + MB) = T(A) + T(B)$ and $T(kA) = (kA)M + M(kA) = k(AM) + k(MA) = k(AM + MA) = kT(A)$. Accordingly, T is linear.

10.24   Let $T: V \to V$ be defined by $T(A) = M + A$ where $A \in V$. Show that T is linear if and only if $M = 0$.

❚   If $M = 0$, then $T(A) = A$, that is, T is the identity map; hence T is linear. On the other hand, suppose $M \neq 0$. Then $T(0) = M + 0 = M \neq 0$; and so T is not linear.

10.25 Let $T: V \to V$ be defined by $T(A) = MA$ where $A \in V$. Show that T is linear.

▌ For any $A, B \in V$ and any $a, b \in K$, we have $T(aA + bB) = M(aA + bB) = aMA + bMB = aT(A) + bT(B)$. Thus T is linear.

10.26 Let V be the vector space of polynomials in t over K. Show that the mapping $T: V \to V$ is linear where $T(a_0 + a_1 t + \cdots + a_n t^n) = a_0 t + a_1 t^2 + \cdots + a_n t^{n+1}$.

▌ Note T multiplies a polynomial $f(t)$ by t, that is, $T(f(t)) = tf(t)$. Hence $T(f(t) + g(t)) = t(f(t) + g(t)) = tf(t) + tg(t) = T(f(t)) + T(g(t))$ and, for any scalar $k \in K$, $T(kf(t)) = t(kf(t)) = k(tf(t)) = kT(f(t))$. Thus T is linear.

Problems 10.27–10.28 refer to the conjugate mapping $T: \mathbf{C} \to \mathbf{C}$ on the complex field $\mathbf{C}$. That is, $T(z) = \bar{z}$ where $z \in \mathbf{C}$, or $T(a + bi) = a - bi$ where $a, b \in \mathbf{R}$.

10.27 Show that T is not linear if $\mathbf{C}$ is viewed as a vector space over itself.

▌ Let $u = 3 + 4i$ and $k = 2 - i$. Then $ku = (2 - i)(3 + 4i) = 10 + 5i$ and $T(ku) = 10 - 5i$. However, $kT(u) = (2 - i)(3 - 4i) = 2 - 11i \neq T(ku)$. Thus T is not linear.

10.28 Show that T is linear if $\mathbf{C}$ is viewed as a vector space over the real field $\mathbf{R}$.

▌ Let $z = a + bi$ and $w = c + di$ where $a, b, c, d \in \mathbf{R}$. Then $z + w = (a + c) + (b + d)i$. Then $T(z + w) = (a + c) - (b + d)i = (a - bi) + (c - di) = T(z) + T(w)$. Also, for $k \in \mathbf{R}$, we have $kz = ka + kbi$. Hence $T(kz) = ka - kbi = k(a - bi) = kT(z)$. Thus T is linear.

10.2 PROPERTIES OF LINEAR MAPPINGS

Theorem 10.1: Let V and U be vector spaces over a field K. Let $\{v_1, \ldots, v_n\}$ be a basis of V and let $u_1, \ldots, u_n$ be any arbitrary vectors in U. Then there exists a unique linear mapping $F: V \to U$ such that $F(v_1) = u_1, F(v_2) = u_2, \ldots, F(v_n) = u_n$.

This section uses the Theorem 10.1 whose proof appears in Problems 10.43–10.45.

10.29 Show that there is a unique linear map $F: \mathbf{R}^2 \to \mathbf{R}^2$ for which $F(1, 2) = (2, 3)$ and $F(0, 1) = (1, 4)$.

▌ Since $(1, 2)$ and $(0, 1)$ form a basis of $\mathbf{R}^2$, such a linear map F exists and is unique by Theorem 10.1.

Problems 10.30–10.32 refer to the linear map F in Problem 10.29.

10.30 Find a formula for F, i.e., find $F(a, b)$.

▌ Write (a, b) as a linear combination of $(1, 2)$ and $(0, 1)$ using unknowns x and y:

$$(a, b) = x(1, 2) + y(0, 1) = (x, 2x + y) \qquad \text{so} \quad a = x, b = 2x + y$$

Solve for x and y in terms of a and b to get $x = a$, $y = -2a + b$. Then $F(a, b) = xF(1, 2) + yF(0, 1) = a(2, 3) + (-2a + b)(1, 4) = (b, -5a + 4b)$.

10.31 Find $F(5, 6)$.

▌ Use the formula for F to get $F(5, 6) = (6, -25 + 24) = (6, -1)$.

10.32 Find $F^{-1}(-2, 7)$.

▌ Set $F(a, b) = (-2, 7)$ and solve for a and b. We get $(b, -5a + 4b) = (-2, 7)$ so $b = -2$, and $-5a + 4b = 7$. Then $a = -3$, $b = -2$. Thus $F^{-1}(-2, 7) = (-3, -2)$.

10.33 Show there is a unique linear map $T: \mathbf{R}^2 \to \mathbf{R}^2$ for which $T(3, 1) = (2, -4)$ and $T(1, 1) = (0, 2)$.

▮ Since $(3, 1)$ and $(1, 1)$ are linearly independent, they form a basis for $\mathbf{R}^2$; hence such a linear map T exists and is unique by Theorem 10.1.

Problems 10.34–10.36 refer to the linear map T in Problem 10.33.

10.34 Find a formula for T.

▮ First write (a, b) as a linear combination of $(3, 1)$ and $(1, 1)$ using unknown scalars x and y:

$$(a, b) = x(3, 1) + y(1, 1)$$

Hence

$$(a, b) = (3x, x) + (y, y) = (3x + y, x + y) \qquad \text{and so} \qquad \begin{cases} 3x + y = a \\ x + y = b \end{cases}$$

Solving for x and y in terms of a and b, $x = \frac{1}{2}a - \frac{1}{2}b$ and $y = -\frac{1}{2}a + \frac{3}{2}b$. Therefore, $T(a, b) = xT(3, 1) + yT(1, 1) = x(2, -4) + y(0, 2) = (2x, -4x) + (0, 2y) = (2x, -4x + 2y) = (a - b, 5b - 3a)$.

10.35 Find $T(7, 4)$.

▮ Use the formula for T to get $T(7, 4) = (7 - 4, 20 - 21) = (3, -1)$.

10.36 Find $T^{-1}(5, -3)$.

▮ Set $T(a, b) = (5, -3)$ and solve for a and b. We get $(a - b, -3a + 5b) = (5, -3)$ so $a - b = 5$, $-3a + 5b = -3$. Then $a = 11$, $b = 6$. Thus $F^{-1}(5, -3) = (11, 6)$.

10.37 Show there is a unique linear map $T: \mathbf{R}^2 \to \mathbf{R}$ for which $T(1, 1) = 3$ and $T(0, 1) = -2$.

▮ Since $\{(1, 1), (0, 1)\}$ is a basis of $\mathbf{R}^2$, such a linear mapping exists and is unique by Theorem 10.1.

Problems 10.38–10.41 refer to the linear map T in Problem 10.37.

10.38 Find a formula for T.

▮ First we write (a, b) as a linear combination of $(1, 1)$ and $(0, 1)$ using unknown scalars x and y:

$$(a, b) = x(1, 1) + y(0, 1)$$

Then $(a, b) = (x, x) + (0, y) = (x, x + y)$ and so $x = a$, $x + y = b$. Solving for x and y in terms of a and b, we obtain $x = a$ and $y = b - a$. Therefore, $T(a, b) = T(x(1, 1) + y(0, 1)) = xT(1, 1) + yT(0, 1) = a(3) + (b - a)(-2) = 5a - 2b$.

10.39 Find $T(8, 2)$ and $T(-4, 6)$.

▮ Use the formula for T to get $T(8, 2) = 40 - 4 = 36$ and $T(-4, 6) = -20 - 12 = -32$.

10.40 Find $T^{-1}(6)$.

▮ Set $T(a, b) = 6$ to get $5a - 2b = 6$. Here b is a free variable. Set $b = t$ where t is a parameter to get the solution $a = (2t + 6)/5$, $b = t$. Thus $T^{-1}(6) = \{((2t + 6)/5, t): t \in \mathbf{R}\}$.

10.41 Is T one-to-one?

▮ No, since $T^{-1}(6)$ has more than one element, e.g., $T(\frac{6}{5}, 0) = 6$ and $T(\frac{8}{5}, 0) = 6$.

10.42   Is there a linear map $T: \mathbf{R}^2 \to \mathbf{R}^2$ for which $T(2, 2) = (8, -6)$ and $T(5, 5) = (3, -2)$?

▌ Theorem 10.1 does not apply here since $(2, 2)$ and $(5, 5)$ are linearly dependent and hence do not form a basis of $\mathbf{R}^2$.   Note $(5, 5) = \frac{5}{2}(2, 2)$.   If T is linear then $T(5, 5) = T(\frac{5}{2}(2, 2)) = \frac{5}{2}T(2, 2) = \frac{5}{2}(8, -6) = (20, -15)$.   But $T(5, 5) = (3, -2)$ is given.   Thus such a linear map T does not exist.

Problems 10.43–10.45 refer to the proof of Theorem 10.1 which consists of three steps:
(1)   Define the mapping $F: V \to U$ such that $F(v_i) = u_1$, $i = 1, \ldots, n$.
(2)   Show that F is linear.
(3)   Show that F is unique.

10.43   **Step (1):**   Define the required map $F: V \to U$ such that $F(v_i) = u_i$.

▌ Let $v \in V$.   Since $\{v_1, \ldots, v_n\}$ is a basis of V, there exist unique scalars $a_1, \ldots, a_n \in K$ for which $v = a_1 v_1 + a_2 v_2 + \cdots + a_n v_n$.   We define $F: V \to U$ by $F(v) = a_1 u_1 + a_2 u_2 + \cdots + a_n u_n$.   [Since the a_i are unique, the mapping F is well-defined.]   Now, for $i = 1, \ldots, n$, $v_i = 0 v_1 + \cdots + 1 v_i + \cdots + 0 v_n$.   Hence $F(v_i) = 0 u_1 + \cdots + 1 u_i + \cdots + 0 u_n = u_i$.   Thus the first step of the proof is complete.

10.44   **Step (2):**   Show F is linear.

▌ Suppose $v = a_1 v_1 + a_2 v_2 + \cdots + a_n v_n$ and $w = b_1 v_1 + b_2 v_2 + \cdots + b_n v_n$.   Then $v + w = (a_1 + b_1)v_1 + (a_2 + b_2)v_2 + \cdots + (a_n + b_n)v_n$ and, for any $K \in K$, $kv = k a_1 v_1 + k a_2 v_2 + \cdots + k a_n v_n$.   By definition of the mapping F, $F(v) = a_1 u_1 + a_2 u_2 + \cdots + a_n u_n$ and $F(w) = b_1 u_1 + b_2 u_2 + \cdots + b_n u_n$. Hence

$$F(v + w) = (a_1 + b_1)u_1 + (a_2 + b_2)u_2 + \cdots + (a_n + b_n)u_n$$
$$= (a_1 u_1 + a_2 u_2 + \cdots + a_n u_n) + (b_1 u_1 + b_2 u_2 + \cdots + b_n u_n)$$
$$= F(v) + F(w)$$

and

$$F(k) = k(a_1 u_1 + a_2 u_2 + \cdots + a_n u_n) = kF(v)$$

Thus F is linear.

10.45   **Step (3):**   Show F is unique.

▌ Suppose $G: V \to U$ is linear and $G(v_i) = u_i$, $i = 1, \ldots, n$.   If $v = a_1 v_1 + a_2 v_2 + \cdots + a_n v_n$,   then

$$G(v) = G(a_1 v_1 + a_2 v_2 + \cdots + a_n v_n) = a_1 G(v_1) + a_2 G(v_2) + \cdots + a_n G(v_n)$$
$$= a_1 u_1 + a_2 u_2 + \cdots + a_n u_n = F(v)$$

Since $G(v) = F(v)$ for every $v \in V$, $G = F$.   Thus F is unique and the theorem is proved.

10.46   Suppose the linear mapping $F: V \to U$ is one-to-one and onto.   Show that the inverse mapping $F^{-1}: U \to V$ is also linear.

▌ Suppose $u, u' \in U$.   Since F is one-to-one and onto, there exist unique vectors $v, v' \in V$ for which $F(v) = u$ and $F(v') = u'$.   Since F is linear, we also have $F(v + v') = F(v) + F(v') = u + u'$ and $F(kv) = kF(v) = ku$.   By definition of the inverse mapping, $F^{-1}(u) = v$, $F^{-1}(u') = v'$, $F^{-1}(u + u') = v + v'$,   and $F^{-1}(ku) = kv$. Then $F^{-1}(u + u') = v + v' = F^{-1}(u) + F^{-1}(u')$ and $F^{-1}(ku) = kv = kF^{-1}(u)$ and thus F^{-1} is linear.

10.47   Suppose $F: V \to U$ and $G: U \to W$ are linear mappings.   Show that the composition mapping $G \circ F: V \to W$ is linear.   [Recall that $G \circ F$ is defined by $(G \circ F)(v) = G(F(v))$.]

▌ For any vectors $v, w \in V$ and any scalars $a, b \in K$, $(G \circ F)(av + bw) = G(F(av + bw)) = G(aF(v) + bF(w)) = aG(F(v)) + bG(F(w)) = a(G \circ F)(v) + b(G \circ F)(w)$.   Thus $G \circ F$ is linear.

10.48 Let $\{e_1, e_2, e_3\}$ be a basis of V and $\{f_1, f_2\}$ a basis of U. Let $T: V \to U$ be linear. Furthermore, suppose

$$\begin{array}{ll} T(e_1) = a_1 f_1 + a_2 f_2 \\ T(e_2) = b_1 f_1 + b_2 f_2 & \text{and} \qquad A = \begin{pmatrix} a_1 & b_1 & c_1 \\ a_2 & b_2 & c_2 \end{pmatrix} \\ T(e_3) = c_1 f_1 + c_2 f_2 \end{array}$$

Show that, for any $v \in V$, $A[v]_e = [T(v)]_f$ where the vectors in K^2 and K^3 are written as column vectors.

▮ Suppose $v = k_1 e_1 + k_2 e_2 + k_3 e_3$; then $[v]_e = \begin{pmatrix} k_1 \\ k_2 \\ k_3 \end{pmatrix}$. Also,

$$\begin{aligned} T(v) &= k_1 T(e_1) + k_2 T(e_2) + k_3 T(e_3) \\ &= k_1(a_1 f_1 + a_2 f_2) + k_2(b_1 f_1 + b_2 f_2) + k_3(c_1 f_1 + c_2 f_2) \\ &= (a_1 k_1 + b_1 k_2 + c_1 k_3) f_1 + (a_2 k_1 + b_2 k_2 + c_2 k_3) f_2 \end{aligned}$$

Accordingly,

$$[T(v)]_f = \begin{pmatrix} a_1 k_1 + b_1 k_2 + c_1 k_3 \\ a_2 k_1 + b_2 k_2 + c_2 k_3 \end{pmatrix}$$

Computing, we obtain

$$A[v]_e = \begin{pmatrix} a_1 & b_1 & c_1 \\ a_2 & b_2 & c_2 \end{pmatrix} \begin{pmatrix} k_1 \\ k_2 \\ k_3 \end{pmatrix} = \begin{pmatrix} a_1 k_1 + b_1 k_2 + c_1 k_3 \\ a_2 k_1 + b_2 k_2 + c_2 k_3 \end{pmatrix} = [T(v)]_f$$

10.49 Let $T: V \to U$ be linear, and suppose $v_1, \ldots, v_n \in V$ have the property that their images $T(v_1), \ldots, T(v_n)$ are linearly independent. Show that the vectors $v_1, \ldots, v_n$ are also linearly independent.

▮ Suppose that, for scalars $a_1, \ldots, a_n$, $a_1 v_1 + a_2 v_2 + \cdots + a_n v_n = 0$. Then

$$0 = T(0) = T(a_1 v_1 + a_2 v_2 + \cdots + a_n v_n) = a_1 T(v_1) + a_2 T(v_2) + \cdots + a_n T(v_n)$$

Since the $T(v_i)$ are linearly independent, all the $a_i = 0$. Thus $v_1, \ldots, v_n$ are linearly independent.

10.3 KERNEL AND IMAGE OF A LINEAR MAPPING

10.50 Let $F: V \to U$ be a linear mapping. Define the kernel of F.

▮ The *kernel* of F, written Ker F, is the set of elements in V which map into $0 \in U$:

$$\text{Ker } F = \{v \in V: F(v) = 0\}$$

10.51 Let $F: V \to U$ be a linear mapping. Define the image of F.

▮ The *image* of F, written Im F, is the set of image points in U:

$$\text{Im } F = \{u \in U: \exists v \in V \text{ for which } F(v) = u\}$$

10.52 Let $F: \mathbf{R}^3 \to \mathbf{R}^3$ be the projection mapping into the xy plane, i.e., defined by $F(x, y, z) = (x, y, 0)$. Find the kernel of F.

▮ The points on the z axis, and only these points, map into the zero vector $0 = (0, 0, 0)$. Thus Ker $F = \{(0, 0, c): c \in \mathbf{R}\}$.

10.53 Find the image of the projection mapping $F(x, y, z) = (x, y, 0)$ in Problem 10.52.

▮ The image of F consists precisely of those points in the xy plane: Im $F = \{(a, b, 0): a, b \in \mathbf{R}\}$.

10.54 Let $F: \mathbf{R}^3 \to \mathbf{R}^3$ be the linear mapping which rotates a vector about the z axis through an angle θ:

$$F(x, y, z) = (x \cos \theta - y \sin \theta, x \sin \theta + y \cos \theta, z)$$

Find the kernel of F.

▮ Under a rotation, the length of a vector does not change. Thus only the zero vector is mapped into the zero vector; hence Ker $F = \{0\}$. [In other words, setting $F(x, y, z) = (0, 0, 0)$ yields only $x = 0$, $y = 0$, $z = 0$.]

10.55 Find the image of the rotation map F in Problem 10.54.

▮ Since one can always rotate back by an angle $-\theta$, every vector $v \in \mathbf{R}^3$ is in the image of F; that is, Im $F = \mathbf{R}^3$.

Problems 10.56–10.60 refer to the vector space V of real polynomials in the variable t and the third derivative map $D^3 : V \to V$, that is, $D^3(f) = d^3f/dt^3$. [Frequently, one uses D for the first derivative, D^2 for the second derivative, and so on.]

10.56 Find $D^3(f)$ where $f(t) = t^4 - 2t^3 + 5t^2 - 6t + 9$.

▮ Take the derivative three times:

$$\frac{df}{dt} = 4t^3 - 6t^2 + 10t - 6 \qquad \frac{d^2f}{dt^2} = 12t^2 - 12t + 10 \qquad D^3(f) = \frac{d^3f}{dt^3} = 24t - 12$$

10.57 Find $D^3(g)$ where $g(t) = at^2 + bt + c$.

▮

$$\frac{dg}{dt} = 2at + b \qquad \frac{d^2g}{dt^2} = 2a \qquad D^3(g) = \frac{d^3g}{dt^3} = 0$$

10.58 Find the kernel of D^3.

▮ The third derivative of any polynomial of degree two or less equals zero and those of higher degree are not zero. Thus Ker $D^3 = \{f \in V : \deg f \leq 2\}$.

10.59 Find the preimage of $h(t) = t^3$ [denoted by $D^{-3}(h)$].

▮ Integrate three times:

$$D^{-1}(h) = \frac{t^4}{4} + C_1 \qquad D^{-2}(h) = \frac{t^5}{20} + C_1 t + C_2 \qquad D^{-3}(h) = \frac{t^6}{120} + \frac{C_1 t^2}{2} + C_2 t + C_3 = \frac{t^6}{120} + at^2 + bt + c$$

10.60 Find the image of D^3.

▮ Given any polynomial $f(t)$, one can integrate three times to obtain a polynomial $F(t)$ such that d^3F/dt^3 yields $f(t)$. Thus the image of D^3 contains every polynomial $f(t)$, that is, Im $D^3 = V$.

10.61 Suppose $F : V \to U$ is a linear mapping. Show that the kernel of F is a subspace of V.

▮ Since $F(0) = 0$, $0 \in$ Ker F. Now suppose $v, w \in$ Ker F and $a, b \in K$. Since v and w belong to the kernel of F, $F(v) = 0$ and $F(w) = 0$. Thus $F(av + bw) = aF(v) + bF(w) = a0 + b0 = 0$ and so $av + bw \in$ Ker F. Thus the kernel of F is a subspace of V.

10.62 Suppose $F : V \to U$ is a linear mapping. Show that the image of F is a subspace of U.

▮ Since $F(0) = 0$, $0 \in$ Im F. Now suppose $u, u' \in$ Im F and $a, b \in K$. Since u and u' belong to the image of F, there exist vectors $v, v' \in V$ such that $F(v) = u$ and $F(v') = u'$. Then $F(av + bv') = aF(v) + bF(v') = au + bu' \in$ Im F. Thus the image of F is a subspace of U.

10.63 Suppose the vectors $v_1, \ldots, v_n$ span V and that $F : V \to U$ is linear. Show that the vectors $F(v_1), \ldots, F(v_n) \in U$ span Im F.

❚ Suppose $u \in \operatorname{Im} F$; then $F(v) = u$ for some vector $v \in V$. Since $v_1, \ldots, v_n$ span V and since $v \in V$, there exist scalars $a_1, \ldots, a_n$ for which $v = a_1 v_1 + a_2 v_2 + \cdots + a_n v_n$. Accordingly,

$$u = F(v) = F(a_1 v_1 + a_2 v_2 + \cdots + a_n v_n) = a_1 F(v_1) + a_2 F(v_2) + \cdots + a_n F(v_n)$$

Thus the vectors $F(v_1), \ldots, F(v_n)$ span $\operatorname{Im} F$.

10.64 Let $A = \begin{pmatrix} a_1 & a_2 & a_3 \\ b_1 & b_2 & b_3 \\ c_1 & c_2 & c_3 \\ d_1 & d_2 & d_3 \end{pmatrix}$ be an arbitrary 4×3 matrix over a field K. [Recall that we view A as a linear

mapping $A: K^3 \to K^4$.] Show that the image of A is precisely the column space of A.

❚ Let e_1, e_2, e_3 be the usual basis vectors of K^3. Since e_1, e_2, e_3 span K^3, their values Ae_1, Ae_2, Ae_3 under A span the image of A. But the vectors $Ae_1, Ae_2,$ and Ae_3 are the columns of A:

$$Ae_1 = \begin{pmatrix} a_1 & a_2 & a_3 \\ b_1 & b_2 & b_3 \\ c_1 & c_2 & c_3 \\ d_1 & d_2 & d_3 \end{pmatrix} \begin{pmatrix} 1 \\ 0 \\ 0 \end{pmatrix} = \begin{pmatrix} a_1 \\ b_1 \\ c_1 \\ d_1 \end{pmatrix} \qquad Ae_2 = \begin{pmatrix} a_1 & a_2 & a_3 \\ b_1 & b_2 & b_3 \\ c_1 & c_2 & c_3 \\ d_1 & d_2 & d_3 \end{pmatrix} \begin{pmatrix} 0 \\ 1 \\ 0 \end{pmatrix} = \begin{pmatrix} a_2 \\ b_2 \\ c_2 \\ d_2 \end{pmatrix} \qquad Ae_3 = \begin{pmatrix} a_1 & a_2 & a_3 \\ b_1 & b_2 & b_3 \\ c_1 & c_2 & c_3 \\ d_1 & d_2 & d_3 \end{pmatrix} \begin{pmatrix} 0 \\ 0 \\ 1 \end{pmatrix} = \begin{pmatrix} a_3 \\ b_3 \\ c_3 \\ d_3 \end{pmatrix}$$

Thus the image of A is precisely the column space of A.

Remark: We emphasize that if A is any $m \times n$ matrix over a field K, then we view A as a linear map $A: K^n \to K^m$ where vectors are written as columns. In such a case, the image of A is the column space of A. On the other hand, some texts view the matrix A as a linear map $A: K^m \to K^n$ where vectors are written as rows, and there the image of A is the row space of A.

10.65 Suppose V has finite dimension and $F: V \to U$ is linear. Show that $\operatorname{Im} F$ has finite dimension, and $\dim(\operatorname{Im} F) \leq \dim V$.

❚ Suppose $\dim V = n$ and $\dim(\operatorname{Im} F) > \dim V$. Then there exist vectors $w_1, w_2, \ldots, w_{n+1} \in \operatorname{Im} F$ which are linearly independent. Let $v_1, v_2, \ldots, v_{n+1}$ be vectors in V such that $F(v_i) = w_i$. Suppose $a_1 v_1 + \cdots + a_{n+1} v_{n+1} = 0$. Then $0 = F(0) = F(a_1 v_1 + \cdots + a_{n+1} v_{n+1}) = a_1 F(v_1) + \cdots + a_{n+1} F(v_{n+1}) = a_1 w_1 + \cdots + a_{n+1} w_{n+1}$. Since the w_i are linearly independent, $a_1 = 0, \ldots, a_{n+1} = 0$. Thus $v_1, v_2, \ldots, v_{n+1}$ are linearly independent. This contradicts the fact that $\dim V = n$. Thus $\dim(\operatorname{Im} F) \leq \dim V$.

Theorem 10.2: Let V be of finite dimension and let $F: V \to U$ be a linear mapping. Then

$$\dim V = \dim(\operatorname{Ker} F) + \dim(\operatorname{Im} F)$$

[That is, the sum of the dimensions of the image and kernel of a linear mapping is equal to the dimension of its domain.]

10.66 Prove Theorem 10.2.

❚ Suppose $\dim(\operatorname{Ker} F) = r$ and $\{w_1, \ldots, w_r\}$ is a basis of $\operatorname{Ker} F$, and suppose $\dim(\operatorname{Im} F) = s$ and $\{u_1, \ldots, u_s\}$ is a basis of $\operatorname{Im} F$. [By Problem 10.65, $\operatorname{Im} F$ has finite dimension.] Since $u_i \in \operatorname{Im} F$, there exist vectors $v_1, \ldots, v_s$ in V such that $F(v_1) = u_1, \ldots, F(v_s) = u_s$. We claim that the set $B = \{w_1, \ldots, w_r, v_1, \ldots, v_3\}$ is a basis of V, i.e., (i) B spans V and (ii) B is linearly independent. Once we prove (i) and (ii), then $\dim V = r + s = \dim(\operatorname{Ker} F) + \dim(\operatorname{Im} F)$.

(i) B spans V. Let $v \in V$. Then $F(v) \in \operatorname{Im} F$. Since the u_j span $\operatorname{Im} F$, there exist scalars $a_1, \ldots, a_s$ such that $F(v) = a_1 u_1 + \cdots + a_s u_s$. Set $\hat{v} = a_1 v_1 + \cdots + a_s v_s - v$. Then $F(\hat{v}) = F(a_1 v_1 + \cdots + a_s v_s - v) = a_1 F(v_1) + \cdots + a_s F(v_s) - F(v) = a_1 u_1 + \cdots + a_s u_s - F(v) = 0$. Thus $v \in \operatorname{Ker} F$. Since the w_i span $\operatorname{Ker} F$, there exist scalars $b_1, \ldots, b_r$ such that $\hat{v} = b_1 w_1 + \cdots + b_r w_r = a_1 v_1 + \cdots + a_s v_s - v$. Accordingly, $v = a_1 v_1 + \cdots + a_s v_s - b_1 w_1 - \cdots - b_r w_r$. Thus B spans V.

(ii) B is linearly independent. Suppose

$$x_1 w_1 + \cdots + x_r w_r + y_1 v_1 + \cdots + y_s v_s = 0 \tag{1}$$

where the $x_i, y_j \in K$. Then

$$0 = F(0) = F(x_1 w_1 + \cdots + x_r w_r + y_1 v_1 + \cdots + y_s v_s) = x_1 F(w_1) + \cdots + x_r F(w_r) + y_1 F(v_1) + \cdots + y_s F(v_s) \tag{2}$$

But $F(w_i) = 0$ since $w_i \in \operatorname{Ker} F$ and $F(v_j) = u_j$. Substitution in *(2)* gives $y_1 u_1 + \cdots + y_s u_s = 0$. Since the u_j are linearly independent, each $y_j = 0$. Substitution in *(1)* gives $x_1 w_1 + \cdots + x_r w_r = 0$. Since the w_i are linearly independent, each $x_i = 0$. Thus B is linearly independent.

10.67 Define the rank of a linear map $F: V \to U$.

∥ The *rank* of F is defined to be the dimension of its image, i.e., $\operatorname{rank}(F) = \dim(\operatorname{Im} F)$.

10.68 Define the nullity of a linear map $F: V \to U$.

∥ The *nullity* of F is defined to be the dimension of its kernel, i.e., $\operatorname{nullity}(F) = \dim(\operatorname{Ker} F)$.

10.69 Restate Theorem 10.2 using the above terminology.

∥ **Theorem 10.2:** Let $F: V \to U$ be linear where V has finite dimension. Then $\operatorname{rank}(F) + \operatorname{nullity}(F) = \dim(\operatorname{Dom} F)$ [where $\operatorname{Dom} F$ denotes the domain V of F].

10.70 The rank of a matrix A was originally defined to be the dimension of its column space and of its row space. How is this definition related to the definition of rank in Problem 10.67?

∥ Both definitions give the same value since the image of A is precisely its column space.

10.71 Let $F: V \to U$ and $G: U \to W$ be linear. Show that $\operatorname{rank}(G \circ F) \leq \operatorname{rank} G$.

∥ Since $F(V) \subset U$, we also have $G(F(V)) \subset G(U)$ and so $\dim G(F(V)) \leq \dim G(U)$. Then $\operatorname{rank}(G \circ F) = \dim((G \circ F)(V)) = \dim(G(F(V))) \leq \dim G(U) = \operatorname{rank} G$.

10.72 Let $F: V \to U$ and $G: U \to W$ be linear. Show that $\operatorname{rank}(G \circ F) \leq \operatorname{rank} F$.

∥ We have $\dim(G(F(V))) \leq \dim F(V)$. Hence $\operatorname{rank}(G \circ F) = \dim((G \circ F)(V)) = \dim(G(F(V))) \leq \dim F(V) = \operatorname{rank} F$.

10.73 Suppose $f: V \to U$ is linear with kernel W and $f(v) = u$. Show that $v + W = \{v + w : w \in W\}$ is the preimage of u, that is, $f^{-1}(u) = v + W$. [The set $v + W$ is called a coset of W.]

∥ We must prove that (i) $f^{-1}(u) \subset v + W$ and (ii) $v + W \subset f^{-1}(u)$. We first prove (i). Suppose $v' \in f^{-1}(u)$. Then $f(v') = u$ and so $f(v' - v) = f(v') - f(v) = u - u = 0$, that is, $v' - v \in W$. Thus $v' = v + (v' - v) \in v + W$ and hence $f^{-1}(u) \subset v + W$.

Now we prove (ii). Suppose $v' \in v + W$. Then $v' = v + w$ where $w \in W$. Since W is the kernel of f, $f(w) = 0$. Accordingly, $f(v') = f(v + w) = f(v) + f(w) = f(v) + 0 = f(v) = u$. Thus $v' \in f^{-1}(u)$ and so $v + W \subset f^{-1}(u)$.

10.4 COMPUTING THE KERNEL AND IMAGE OF LINEAR MAPPINGS

10.74 Let $F: \mathbf{R}^4 \to \mathbf{R}^3$ be the linear mapping defined by $F(x, y, s, t) = (x - y + s + t, x + 2s - t, x + y + 3s - 3t)$. Find a basis and the dimension of the image U of F.

∥ Find the image of the usual basis vectors of $\mathbf{R}^4$:

$$F(1, 0, 0, 0) = (1, 1, 1) \qquad F(0, 1, 0, 0) = (-1, 0, 1) \qquad F(0, 0, 1, 0) = (1, 2, 3)$$
$$F(0, 0, 0, 1) = (1, -1, -3)$$

The image vectors span U; hence form the matrix whose rows are these image vectors and row reduce to echelon form:

$$\begin{pmatrix} 1 & 1 & 1 \\ -1 & 0 & 1 \\ 1 & 2 & 3 \\ 1 & -1 & -3 \end{pmatrix} \quad \text{to} \quad \begin{pmatrix} 1 & 1 & 1 \\ 0 & 1 & 2 \\ 0 & 1 & 2 \\ 0 & -2 & -4 \end{pmatrix} \quad \text{to} \quad \begin{pmatrix} 1 & 1 & 1 \\ 0 & 1 & 2 \\ 0 & 0 & 0 \\ 0 & 0 & 0 \end{pmatrix}$$

Thus $\{(1, 1, 1), (0, 1, 2)\}$ is a basis of U; hence   dim $U = 2$.

10.75 Find a basis and the dimension of the kernel W of the map F in Problem 10.74.

▌ Set $F(v) = 0$ where $v = (x, y, z, t)$:

$$F(x, y, s, t) = (x - y + s + t, \ x + 2s - t, \ x + y + 3s - 3t) = (0, 0, 0)$$

Set corresponding components equal to each other to form the following homogeneous system whose solution space is the kernel W of F:

$$\begin{array}{ll}
x - y + s + t = 0 & \\
x \quad\quad + 2s - t = 0 \quad \text{or} \\
x + y + 3s - 3t = 0 &
\end{array}
\quad
\begin{array}{ll}
x - y + s + t = 0 & \\
y + s - 2t = 0 \quad \text{or} \\
2y + 2s - 4t = 0 &
\end{array}
\quad
\begin{array}{l}
x - y + s + t = 0 \\
y + s - 2t = 0
\end{array}$$

The free variables are s and t; hence   dim $W = 2$.   Set
(a) $s = -1$, $t = 0$ to obtain the solution $(2, 1, -1, 0)$
(b) $s = 0$, $t = 1$,   to obtain the solution $(1, 2, 0, 1)$
Thus $\{(2, 1, -1, 0), (1, 2, 0, 1)\}$ is a basis of W. [Observe that   dim U + dim $W = 2 + 2 = 4$,   which is the dimension of the domain $\mathbf{R}^4$ of F.]

10.76 Let $T: \mathbf{R}^3 \to \mathbf{R}^3$ be the linear mapping defined by $T(x, y, z) = (x + 2y - z, \ y + z, \ x + y - 2z)$.   Find a basis and the dimension of the image U of T.

▌ Find the image of vectors which span the domain $\mathbf{R}^3$:

$$T(1, 0, 0) = (1, 0, 1) \qquad T(0, 1, 0) = (2, 1, 1) \qquad T(0, 0, 1) = (-1, 1, -2)$$

The images span the image U of T; hence form the matrix whose rows are the image vectors and row reduce to echelon form:

$$\begin{pmatrix} 1 & 0 & 1 \\ 2 & 1 & 1 \\ -1 & 1 & -2 \end{pmatrix} \quad \text{to} \quad \begin{pmatrix} 1 & 0 & 1 \\ 0 & 1 & -1 \\ 0 & 1 & -1 \end{pmatrix} \quad \text{to} \quad \begin{pmatrix} 1 & 0 & 1 \\ 0 & 1 & -1 \\ 0 & 0 & 0 \end{pmatrix}$$

Thus $\{(1, 0, 1), (0, 1, -1)\}$ is a basis of U, and so   dim $U = 2$.

10.77 Find a basis and the dimension of the kernel W of the map T in Problem 10.76.

▌ Set $T(v) = 0$ where $v = (x, y, z)$: $T(x, y, z) = (x + 2y - z, \ y + z, \ x + y - 2z) = (0, 0, 0)$.   Set corresponding components equal to each other to form the homogeneous system whose solution space is the kernel W of T:

$$\begin{array}{ll}
x + 2y - z = 0 & \\
y + z = 0 \quad \text{or} \\
x + y - 2z = 0 &
\end{array}
\quad
\begin{array}{ll}
x + 2y - z = 0 & \\
y + z = 0 \quad \text{or} \\
-y - z = 0 &
\end{array}
\quad
\begin{array}{l}
x + 2y - z = 0 \\
y + z = 0
\end{array}$$

The only free variable is z; hence   dim $W = 1$.   Let $z = 1$;   then $y = -1$ and $x = 3$.   Thus $\{(3, -1, 1)\}$ is a basis of W. [Observe that   dim U + dim $W = 2 + 1 = 3$,   which is the dimension of the domain $\mathbf{R}^3$ of T.]

10.78 Let $F: \mathbf{R}^3 \to \mathbf{R}^4$ be defined by $F(x, y, z) = (x + y + z, \ x + 2y - 3z, \ 2x + 3y - 2z, \ 3x + 4y - z)$.   Find a basis and the dimension of the image of F.

▌ First find the image of vectors which span the domain $\mathbf{R}^3$ of F:

$$F(1, 0, 0) = (1, 1, 2, 3) \qquad F(0, 1, 0) = (1, 2, 3, 4) \qquad F(0, 0, 1) = (1, -3, -2, -1)$$

[The three image vectors span Im F.] Form the matrix whose rows are the image vectors and row reduce to echelon form:

$$\begin{pmatrix} 1 & 1 & 2 & 3 \\ 1 & 2 & 3 & 4 \\ 1 & -3 & -2 & -1 \end{pmatrix} \quad \text{to} \quad \begin{pmatrix} 1 & 1 & 2 & 3 \\ 0 & 1 & 1 & 1 \\ 0 & -4 & -4 & -4 \end{pmatrix} \quad \text{to} \quad \begin{pmatrix} 1 & 1 & 2 & 3 \\ 0 & 1 & 1 & 1 \\ 0 & 0 & 0 & 0 \end{pmatrix}$$

Thus $\{(1, 1, 2, 3), (0, 1, 1, 1)\}$ is a basis of Im F and $\dim(\text{Im } F) = 2$.

10.79 Find a basis and the dimension of the kernel of the map F in Problem 10.78.

▮ Set $F(v) = 0$ where $v = (x, y, z)$ and solve the homogeneous system: $F(x, y, z) = (x + y + z, x + 2y - 3z, 2x + 3y - 2z, 3x + 4y - z) = (0, 0, 0, 0)$. Thus

$$\begin{array}{c} x + y + z = 0 \\ x + 2y - 3z = 0 \\ 2x + 3y - 2z = 0 \\ 3x + 4y - z = 0 \end{array} \quad \text{or} \quad \begin{array}{c} x + y + z = 0 \\ y - 4z = 0 \\ y - 4z = 0 \\ y - 4z = 0 \end{array} \quad \text{or} \quad \begin{array}{c} x + y + z = 0 \\ y - 4z = 0 \end{array}$$

The only free variable is z, so $\dim(\text{Ker } F) = 1$. Set $z = 1$ and get $y = 4$ and $x = -5$. Thus $\{(-5, 4, 1)\}$ is a basis of Ker F.

Problems 10.80–10.85 refer to the matrix maps $A : \mathbf{R}^4 \to \mathbf{R}^3$ and $B : \mathbf{R}^3 \to \mathbf{R}^3$ defined by the matrices

$$A = \begin{pmatrix} 1 & 2 & 3 & 1 \\ 1 & 3 & 5 & -2 \\ 3 & 8 & 13 & -3 \end{pmatrix} \quad \text{and} \quad B = \begin{pmatrix} 1 & 2 & 5 \\ 3 & 5 & 13 \\ -2 & -1 & -4 \end{pmatrix}$$

10.80 Find the dimension and a basis for the image of A.

▮ The column space of A is equal to Im A. Thus reduce A^T to echelon form:

$$A^T = \begin{pmatrix} 1 & 1 & 3 \\ 2 & 3 & 8 \\ 3 & 5 & 13 \\ 1 & -2 & -3 \end{pmatrix} \quad \text{to} \quad \begin{pmatrix} 1 & 1 & 3 \\ 0 & 1 & 2 \\ 0 & 2 & 4 \\ 0 & -3 & -6 \end{pmatrix} \quad \text{to} \quad \begin{pmatrix} 1 & 1 & 3 \\ 0 & 1 & 2 \\ 0 & 0 & 0 \\ 0 & 0 & 0 \end{pmatrix}$$

Thus $\{(1, 1, 3), (0, 1, 2)\}$ is a basis of Im A and $\dim(\text{Im } A) = 2$.

10.81 Find the dimension of the kernel of the matrix map A.

▮ The domain of A is $\mathbf{R}^4$; hence $\dim(\text{Dom } A) = 4$. By Theorem 10.2, $\dim(\text{Ker } A) = \dim(\text{Dom } A) - \dim(\text{Im } A) = 4 - 2 = 2$.

10.82 Find a basis of the kernel of the matrix map A.

▮ Set $0 = A(v)$ where $v = (x, y, z, t)$, and solve the homogeneous system:

$$\begin{pmatrix} 0 \\ 0 \\ 0 \end{pmatrix} = \begin{pmatrix} 1 & 2 & 3 & 1 \\ 1 & 3 & 5 & -2 \\ 3 & 8 & 13 & -3 \end{pmatrix} \begin{pmatrix} x \\ y \\ z \\ t \end{pmatrix} = \begin{pmatrix} x + 2y + 3z + t \\ x + 3y + 5z - 2t \\ 3x + 8y + 13z - 3t \end{pmatrix} \quad \text{or} \quad \begin{array}{c} x + 2y + 3z + t = 0 \\ x + 3y + 5z - 2t = 0 \\ 3x + 8y + 13z - 3t = 0 \end{array}$$

The matrix of coefficients of the homogeneous system is the given matrix A. Reduce A to echelon form:

$$A \quad \text{to} \quad \begin{pmatrix} 1 & 2 & 3 & 1 \\ 0 & 1 & 2 & -3 \\ 0 & 2 & 4 & -6 \end{pmatrix} \quad \text{to} \quad \begin{pmatrix} 1 & 2 & 3 & 1 \\ 0 & 1 & 2 & -3 \\ 0 & 0 & 0 & 0 \end{pmatrix} \quad \text{or} \quad \begin{array}{c} x + 2y + 3z + t = 0 \\ y + 2z - 3t = 0 \end{array}$$

The free variables are z and t. Set (i) $z = 1$, $t = 0$ to get the solution $(1, -2, 1, 0)$ and (ii) $z = 0$, $t = 1$ to get the solution $(-7, 3, 0, 1)$. Thus $(1, -2, 1, 0)$ and $(-7, 3, 0, 1)$ form a basis for Ker A. [As expected, $\dim(\text{Ker } A) = 2$.]

10.83 Find the dimension and a basis of the kernel of the matrix map B.

▮ Reduce B to echelon form to get the homogeneous system corresponding to Ker B:

$$B = \begin{pmatrix} 1 & 2 & 5 \\ 3 & 5 & 13 \\ -2 & -1 & -4 \end{pmatrix} \quad \text{to} \quad \begin{pmatrix} 1 & 2 & 5 \\ 0 & -1 & -2 \\ 0 & 3 & 6 \end{pmatrix} \quad \text{to} \quad \begin{pmatrix} 1 & 2 & 5 \\ 0 & 1 & 2 \\ 0 & 0 & 0 \end{pmatrix} \quad \text{or} \quad \begin{matrix} x + 2y + 5z = 0 \\ y + 2z = 0 \end{matrix}$$

There is one free variable z so $\dim(\text{Ker } B) = 1$. Set $z = 1$ to get the solution $(-1, -2, 1)$ which forms a basis of Ker B.

10.84 Find the dimension of the image of the matrix map B.

▮ The domain of B is $\mathbf{R}^3$, so $\dim(\text{Dom } B) = 3$. Thus $\dim(\text{Im } B) = 3 - 1 = 2$.

10.85 Find a basis of the image of B.

▮ Reduce B^T to echelon form:

$$B^T = \begin{pmatrix} 1 & 3 & -2 \\ 2 & 5 & -1 \\ 5 & 13 & -4 \end{pmatrix} \quad \text{to} \quad \begin{pmatrix} 1 & 3 & -2 \\ 0 & -1 & 3 \\ 0 & -2 & 6 \end{pmatrix} \quad \text{to} \quad \begin{pmatrix} 1 & 3 & -2 \\ 0 & 1 & -3 \\ 0 & 0 & 0 \end{pmatrix}$$

Thus $(1, 3, -2)$ and $(0, 1, -3)$ form a basis of Im B. [As expected, the basis consists of two vectors.]

10.86 Find a linear map $F: \mathbf{R}^3 \to \mathbf{R}^4$ whose image is spanned by $(1, 2, 0, -4)$ and $(2, 0, -1, -3)$.

▮ Consider the usual basis of $\mathbf{R}^3$: $e_1 = (1, 0, 0)$, $e_2 = (0, 1, 0)$, $e_3 = (0, 0, 1)$. Set $F(e_1) = (1, 2, 0, -4)$, $F(e_2) = (2, 0, -1, -3)$, and $F(e_3) = (0, 0, 0, 0)$. By Theorem 10.1, such a linear map F exists and is unique. Furthermore, the image of F is spanned by the $F(e_i)$; hence F has the required property. We find a general formula for $F(x, y, z)$:

$$\begin{aligned} F(x, y, z) &= F(xe_1 + ye_2 + ze_3) = xF(e_1) + yF(e_2) + zF(e_3) \\ &= x(1, 2, 0, -4) + y(2, 0, -1, -3) + z(0, 0, 0, 0) \\ &= (x + 2y, 2x, -y, -4x - 3y) \end{aligned}$$

10.87 Find a matrix map $A: \mathbf{R}^3 \to \mathbf{R}^4$ whose image is spanned by the above vectors $(1, 2, 0, -4)$ and $(2, 0, -1, -3)$.

▮ Form a 4×3 matrix A whose columns consist only of the given vectors; say,

$$A = \begin{pmatrix} 1 & 2 & 2 \\ 2 & 0 & 0 \\ 0 & -1 & -1 \\ -4 & -3 & -3 \end{pmatrix}$$

Recall that A determines a linear map $A: \mathbf{R}^3 \to \mathbf{R}^4$ whose image is spanned by the columns of A. Thus A satisfies the required condition.

Problems 10.88–10.91 refer to the vector space V of real polynomials $f(t)$ of degree 10 or less and the linear map $D^4: V \to V$ defined by d^4f/dt^4, i.e., the fourth derivative.

10.88 What is the dimension of V?

▮ Any polynomial $f(t)$ in V has degree 10 or less; hence the eleven polynomials $1, t, t^2, \ldots, t^{10}$ form a basis of V. Thus $\dim V = 11$.

10.89 Find the dimension and a basis of Ker D^4.

▮ Ker D^4 consists of those polynomials of degree 3 or less. Thus $\{1, t, t^2, t^3\}$ is a basis of Ker D^4 and $\dim(\text{Ker } D^4) = 4$.

10.90 What is the dimension of Im D^4?

▌ By Theorem 10.2, $\dim(\text{Im } D^4) = \dim(\text{Dom } D^4) - \dim(\text{Ker } D^4) = 11 - 4 = 7$.

10.91 Find a basis of Im D^4.

▌ The fourth derivative of polynomials of degree 10 or less will yield precisely the polynomials of degree 6 or less. Thus $1, t, t^2, t^3, t^4, t^5, t^6$ form a basis of Im D^4.

10.5 SINGULAR AND NONSINGULAR LINEAR MAPPINGS, ISOMORPHISMS

10.92 Define singular and nonsingular linear mappings.

▌ A linear mapping $F: V \to U$ is said to be *singular* if the image of some nonzero vector under F is 0, i.e., if there exists $v \in V$ for which $v \neq 0$ but $F(v) = 0$. Thus $F: V \to U$ is *nonsingular* if only $0 \in V$ maps into $0 \in U$ or, equivalently, if its kernel consists only of the zero vector: Ker $F = \{0\}$.

10.93 Let $F: \mathbf{R}^3 \to \mathbf{R}^3$ be the projection mapping into the xy plane that is defined $F(x, y, z) = (x, y, 0)$. Is F singular or nonsingular?

▌ F is singular since nonzero vectors on the z axis map into 0.

10.94 Let $F: \mathbf{R}^3 \to \mathbf{R}^3$ be the linear mapping which rotates a vector about the z axis through an angle θ:

$$F(x, y, z) = (x \cos \theta - y \sin \theta, x \sin \theta + y \cos \theta, z)$$

Is F singular or nonsingular?

▌ Since the length of a vector does not change under a rotation, only the zero vector is mapped into the zero vector. Thus the rotation map F is nonsingular.

10.95 Let $F: \mathbf{R}^2 \to \mathbf{R}^2$ be defined by $F(x, y) = (x - y, x - 2y)$. Is F nonsingular? If not, find $v \neq 0$ such that $F(v) = 0$.

▌ Find Ker F by setting $F(v) = 0$ where $v = (x, y)$:

$$(x - y, x - 2y) = (0, 0) \quad \text{or} \quad \begin{matrix} x - y = 0 \\ x - 2y = 0 \end{matrix} \quad \text{or} \quad \begin{matrix} x - y = 0 \\ -y = 0 \end{matrix}$$

The only solution is $x = 0$, $y = 0$; hence F is nonsingular.

10.96 Let $G: \mathbf{R}^2 \to \mathbf{R}^2$ be defined by $G(x, y) = (2x - 4y, 3x - 6y)$. Is G nonsingular? If not, find $v \neq 0$ such that $G(v) = 0$.

▌ Set $G(x, y) = (0, 0)$ to find Ker G:

$$(2x - 4y, 3x - 6y) = (0, 0) \quad \text{or} \quad \begin{matrix} 2x - 4y = 0 \\ 3x - 6y = 0 \end{matrix} \quad \text{or} \quad x - 2y = 0$$

The system has nonzero solutions, i.e., y is a free variable; hence G is singular. Let $y = 1$ to obtain the solution $v = (-2, 1)$ which is a nonzero vector such that $G(v) = 0$.

10.97 Let $H: \mathbf{R}^3 \to \mathbf{R}^3$ be defined by $H(x, y, z) = (x + y - 2z, x + 2y + z, 2x + 2y - 3z)$. Is H nonsingular? If not, find $v \neq 0$ such that $H(v) = 0$.

▌ Set $H(x, y, z) = (0, 0, 0)$:

$$(x + y - 2z, x + 2y + z, 2x + 2y - 3z) = (0, 0, 0) \quad \text{or} \quad \begin{matrix} x + y - 2z = 0 \\ x + 2y + z = 0 \\ 2x + 2y - 3z = 0 \end{matrix} \quad \text{or} \quad \begin{matrix} x + y - 2z = 0 \\ y + 3z = 0 \\ z = 0 \end{matrix}$$

The echelon system is in triangular form so the only solution is $x = 0$, $y = 0$, $z = 0$. Thus H is nonsingular.

10.98 Let $F: \mathbf{R}^3 \to \mathbf{R}^3$ be defined by $F(x, y, z) = (x + y + z, x + 2y - z, 3x + 5y - z)$. Is H nonsingular? If not, find $v \neq 0$ such that $F(v) = 0$.

▌ Set $F(x, y, z) = (0, 0, 0)$ to get the homogeneous system:

$$
\begin{array}{lll}
\begin{aligned}
x + \ y + z &= 0 \\
x + 2y - z &= 0 \\
3x + 5y - z &= 0
\end{aligned}
\quad \text{or} \quad
\begin{aligned}
x + y + \ z &= 0 \\
y - 2z &= 0 \\
2y - 4z &= 0
\end{aligned}
\quad \text{or} \quad
\begin{aligned}
x + y \ z &= 0 \\
y - 2z &= 0
\end{aligned}
\end{array}
$$

Since z is a free variable, the system has a nonzero solution, so F is singular. Set $z = 1$ to obtain the nonzero solution $v = (-3, 2, 1)$ such that $F(v) = 0$.

Problems 10.99–10.100 refer to the vector space V of real polynomials (in the variable t).

10.99 Let $D^n: V \to V$ be the nth-derivative mapping, i.e., $D^n(f) = d^n f / dt^n$ (where $n > 0$). Is D^n singular or nonsingular?

▌ Since the derivative of a nonzero constant polynomial $f(t) = k$ (where $k \neq 0$) is zero, D^n is singular for every n.

10.100 Let $G: V \to V$ be the linear mapping which multiplies a polynomial by t, that is, $G(f(t)) = tf(t)$. Is G singular or nonsingular?

▌ If $f(t) \neq 0$, then $tf(t) \neq 0$; hence G is nonsingular.

10.101 Suppose a linear map $F: V \to U$ is one-to-one. Show that F is nonsingular.

▌ Since F is linear, $F(0) = \{0\}$. Since F is one-to-one, only $0 \in V$ can map into $0 \in U$, that is, $\operatorname{Ker} F = \{0\}$. Thus F is nonsingular.

10.102 Suppose $F: V \to U$ is nonsingular. Show that F is one-to-one.

▌ Suppose $F(v) = F(w)$; then $F(v - w) = F(v) - F(w) = 0$. Thus $v - w \in \operatorname{Ker} F$. But F is nonsingular, i.e., $\operatorname{Ker} F = 0$; hence $v - w = 0$ and so $v = w$. Thus $F(v) = F(w)$ implies $v = w$, That is, F is one-to-one.

10.103 Give an example of a nonlinear map $F: V \to U$ such that $F^{-1}(0) = \{0\}$ but F is not one-to-one.

▌ Let $F: \mathbf{R} \to \mathbf{R}$ be defined by $F(x) = x^2$. Then $F^{-1}(0) = \{0\}$, but $F(2) = F(-2) = 4$, that is, F is not one-to-one.

10.104 Suppose $F: V \to U$ is linear and that V is of finite dimension. Show that V and the image of F have the same dimension if and only if F is nonsingular.

▌ By Theorem 10.2, $\dim V = \dim(\operatorname{Im} F) + \dim(\operatorname{Ker} F)$. Hence V and $\operatorname{Im} F$ have the same dimension if and only if $\dim(\operatorname{Ker} F) = 0$ or $\operatorname{Ker} F = \{0\}$, i.e., if and only if F is nonsingular.

10.105 Determine all nonsingular linear mappings $T: \mathbf{R}^4 \to \mathbf{R}^3$.

▌ Since $\dim \mathbf{R}^3$ is less than $\dim \mathbf{R}^4$, we have $\dim(\operatorname{Im} T)$ is less than the dimension of the domain $\mathbf{R}^4$ of T. Accordingly, no linear mapping $T: \mathbf{R}^4 \to \mathbf{R}^3$ can be nonsingular.

10.106 Let A be an n-square matrix over a field K [which defines a linear mapping $A: K^n \to K^n$]. The matrix A is said to be nonsingular if $\det(A) \neq 0$. Show that both definitions of nonsingularity agree.

▌ We have that $\det(A) \neq 0$ if and only if the homogeneous system $Ax = 0$ has only the zero solution if and only if $\operatorname{Ker} A = \{0\}$. Thus both definitions agree.

10.107 Show that if $F: V \to U$ is linear and maps independent sets into independent sets then F is nonsingular.

❚ Suppose $v \in V$ is nonzero, then $\{v\}$ is independent. Then $\{F(v)\}$ is independent and so $F(v) \neq 0$. Accordingly, F in nonsingular.

Theorem 10.3: Suppose a linear mapping $F: V \to U$ is nonsingular. Then the image of any linearly independent set is linearly independent.

10.108 Prove Theorem 10.3.

❚ Suppose $v_1, v_2, \ldots, v_n$ are linearly independent vectors in V. We claim that the vectors $F(v_1), F(v_2), \ldots, F(v_n)$ are also linearly independent. Suppose $a_1 F(v_1) + a_2 F(v_2) + \cdots + a_n F(v_n) = 0$, where $a_i \in K$. Since F is linear, $F(a_1 v_1 + a_2 v_2 + \cdots + a_n v_n = 0$; hence $a_1 v_1 + a_2 v_2 + \cdots + a_n v_n$ belongs to Ker F. But F is nonsingular, i.e., Ker $F = \{0\}$; hence $a_1 v_1 + a_2 v_2 + \cdots + a_n v_n = 0$. Since the v_i are linearly independent, all the a_i are 0. Accordingly, the $F(v_i)$ are linearly independent. Thus the theorem is proved.

10.109 Suppose V has finite dimension and $\dim V = \dim U$. Show that a linear map $F: V \to U$ is nonsingular if and only if F is surjective, i.e., maps V onto U.

❚ By Theorem 10.2, $\dim(V) = \dim(\text{Ker } F) + \dim(\text{Im } F)$. Thus F is surjective iff Im $F = U$ iff $\dim(\text{Im } F) = \dim U = \dim V$ iff $\dim(\text{Ker } F) = 0$ iff F is nonsingular.

10.110 Give an example of a linear map $F: V \to V$ which is onto but not nonsingular. [By Problem 10.109, V cannot have finite dimension.]

❚ Let V be the vector space of polynomials $f(t)$. Let $D: V \to V$ be the derivative mapping, i.e., $D(f) = df/dt$. Then D is onto, but not nonsingular.

10.111 Give an example of a linear map $G: V \to V$ which is nonsingular, but not onto. [By Problem 10.109, V cannot have finite dimension.]

❚ Let V be the vector space of polynomials $f(t)$. Let $G: V \to V$ be the linear map which multiplies a polynomial by t, that is, $G(f(t)) = tf(t)$. Then G is nonsingular but not onto.

10.112 Define a vector space isomorphism.

❚ A mapping $F: V \to U$ is called an isomorphism if F is linear and if F is bijective, i.e., if F is one-to-one and onto. [In such a case, F has an inverse $F^{-1}: U \to V$ and so F is said to be *invertible*.]

10.113 Define isomorphic vector spaces.

❚ A vector space V is said to be isomorphic to a vector space U, written $V \simeq U$, if there is an isomorphism $F: V \to U$.

10.114 Suppose V is a vector space over a field K and $\dim V = n$. Show that $V \simeq K^n$.

❚ Let $\{w_1, w_2, \ldots, w_n\}$ be a basis of V. Let $[v]$ denote the coordinates of $v \in V$ relative to the given basis. Then the map $F: V \to K^n$ defined by $F(v) = [v]$ is an isomorphism. Thus $V \simeq K^n$.

Theorem 10.4: Suppose V has finite dimension and $\dim V = \dim U$. Suppose $F: V \to U$ is linear. Then F is an isomorphism if and only if F is nonsingular.

10.115 Prove Theorem 10.4.

❚ If F is an isomorphism then only 0 maps to 0 so F is nonsingular. Suppose F is nonsingular. Then $\dim(\text{Ker } F) = 0$. By Theorem 10.2, $\dim V = \dim(\text{Ker } F) + \dim(\text{Im } F)$. Thus $\dim U = \dim V = \dim(\text{Im } F)$. Since U has finite dimension, Im $F = U$ and so F is surjective. Thus F is both one-to-one and onto, i.e., F is an isomorphism.

10.116 The linear map $F: \mathbf{R}^2 \to \mathbf{R}^2$ defined by $F(x, y) = (x - y, x - 2y)$ is nonsingular (Problem 10.95). Find a formula for F^{-1}.

▌ Set $F(x, y) = (a, b)$ [so $F^{-1}(a, b) = (x, y)$]:

$$(x - y, x - 2y) = (a, b) \quad \text{or} \quad \begin{matrix} x - y = a \\ x - 2y = b \end{matrix} \quad \text{or} \quad \begin{matrix} x - y = a \\ -y = b - a \end{matrix}$$

Solve for x and y in terms of a and b to get $x = 2a - b$, $y = a - b$. Thus $F^{-1}(a, b) = (2a - b, a - b)$ [or, equivalently, $F^{-1}(x, y) = (2x - y, x - y)$].

10.117 The linear map $G: \mathbf{R}^2 \to \mathbf{R}^3$ defined by $F(x, y) = (x + y, x - 2y, 3x + y)$ is nonsingular. Find a formula for F^{-1}.

▌ Although G is nonsingular, it is not invertible since $\mathbf{R}^2$ and $\mathbf{R}^3$ have different dimensions. [Thus Theorem 10.4 does not apply.] Accordingly, F^{-1} does not exist.

10.118 The linear map $H: \mathbf{R}^3 \to \mathbf{R}^3$ defined by $H(x, y, z) = (x + y - 2z, x + 2y + z, 2x + 2y - 3z)$ is nonsingular (Problem 10.97). Find a formula for H^{-1}.

▌ Set $H(x, y, z) = (a, b, c)$ and then solve for x, y, z in terms of a, b, c:

$$\begin{matrix} x + y - 2z = a \\ x + 2y + z = b \\ 2x + 2y - 3z = c \end{matrix} \quad \text{or} \quad \begin{matrix} x + y - 2z = a \\ y + 3z = b - a \\ z = c - 2a \end{matrix}$$

Solving for x, y, z yields $x = -8a - b + 5c$, $y = 5a + b - 3c$, $z = -2a + c$. Thus $H^{-1}(a, b, c) = (-8a - b + 5c, \; 5a + b - 3c, \; -2a + c)$ or, replacing a, b, c by x, y, z, respectively, $H^{-1}(x, y, z) = (-8x - y + 5z, \; 5x + y - 3z, \; -2x + z)$.

Problems 10.119–10.121 show that the relation $V \simeq U$ of isomorphism of vector spaces is an equivalence relation, i.e., is reflexive, symmetric, and transitive.

10.119 Show that $\simeq$ is reflexive, i.e., $V \simeq V$ for any vector space V.

▌ The identity map $1_V: V \to V$ is linear and bijective, i.e., an isomorphism. Hence $V \simeq V$ for any vector space V.

10.120 Show that $\simeq$ is symmetric, i.e., if $V \simeq U$ then $U \simeq V$.

▌ Suppose $V \simeq U$ and $F: V \to U$ is an isomorphism. By Problem 10.30, F^{-1} is also linear. Also F^{-1} is bijective. Thus $F^{-1}: U \to V$ is an isomorphism, and so $U \simeq V$.

10.121 Show that $\simeq$ is transitive, i.e., if $V \simeq U$ and $U \simeq W$, then $V \simeq W$.

▌ Suppose $V \simeq U$ and $U \simeq W$, say $F: V \to U$ and $G: U \to W$ are isomorphisms. Since F and G are bijective, so is the composition $G \circ F$. By Problem 10.47, $G \circ F$ is linear since F and G are linear. Thus $G \circ F: V \to W$ is an isomorphism, and so $V \simeq W$.

10.6 APPLICATIONS TO GEOMETRY, CONVEX SETS

This section assumes that all vector spaces are over the real field $\mathbf{R}$.

10.122 Let v and w be elements of V. The line segment L from v to $v + w$ is defined to be the set of vectors $v + tw$ for $0 \le t \le 1$. [See Fig. 10.1.] Describe the point: (a) midway between v and $v + w$, (b) one-third the way from v to $v + w$, (c) three-fourths the way from v to $v + w$.

▌ (a) Set $t = \frac{1}{2}$ to get the point $v + \frac{1}{2}w$ which is midway between v and $v + w$.
(b) Set $t = \frac{1}{3}$ to get the point $v + \frac{1}{3}w$ which is one-third the way from v to $v + w$.
(c) Set $t = \frac{3}{4}$ to get the point $v + \frac{3}{4}w$ which is three-fourths the way from v to $v + w$.

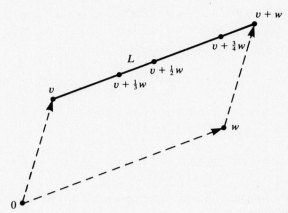

Fig. 10-1

Problems 10.123–10.125 refer to the line segment L between vectors v and u.

10.123 Show that L consists of the points $(1-t)v + tu$ for $0 \le t \le 1$.

▌ Let $w = u - v$. Then $u = v + w$. By Problem 10.122, L consists of the points $v + tw = v + t(u-v) = v + tu - tv = (1-t)v + tu$ for $0 \le t \le 1$.

10.124 Show that L consists of the points $sv + (1-s)u$ for $0 \le s \le 1$.

▌ Let $s = 1 - t$. Then $t = 1 - s$. Also, when $0 \le t \le 1$, we have $0 \le s \le 1$. Thus L consists of the points $(1-t)v + tu = sv = sv + (1-s)u$ for $0 \le s \le 1$.

10.125 Show that L consists of the points $t_1 v + t_2 u$ for $t_1 + t_2 = 1$, $t_1 \ge 0$, $t_2 \ge 0$.

▌ Suppose $t_1 + t_2 = 1$, $t_1 \ge 0$, $t_2 \ge 0$. Then $0 \le t_1 \le 1$ and $t_2 = 1 - t_1$. Thus $t_1 v + t_2 u = t_1 v + (1 - t_1)u$. Hence $t_1 v + t_2 u$ belongs to L (Problem 10.124). Conversely, consider any element $sv + (1-s)u$ in L. Set $t_1 = s$ and $t_2 = 1 - s$. Then $sv + (1-s)u = t_1 v + t_2 u$. Also $t_1 + t_2 = 1$ and $t_1 \le 0$, $t_2 \le 0$.

10.126 Let $F: V \to U$ be linear. Show that the image $F(L)$ of a line segment L in V is a line segment in U.

▌ Suppose L is a line segment between v and u. Then L consists of the points $t_1 v + t_2 u$ with t_1, t_2 nonnegative and $t_1 + t_2 = 1$. Then $F(L)$ consists of the points $F(t_1 v + t_2 u) = t_1 F(v) + t_2 F(u)$ which is the line segment between $F(v)$ and $F(u)$ in U.

10.127 Define a convex set.

▌ A subset X of a vector space V is said to be convex if the line segment L between any two points (vectors) $P, Q \in X$ is contained in X.

10.128 Is the rectangular area X in Fig. 10-2(a) convex?

▌ Yes, since the line segment between any two points $P, Q \in X$ is contained in X.

10.129 Is the elliptical area Y in Fig. 10-2(b) convex?

▌ Yes, since the line segment between any two points $P, Q \in Y$ is contained in Y.

10.130 Is the U-shaped area area Z in Fig. 10-2(c) convex?

▌ No, since, as shown, the line segment between two points $P, Q \in Z$ need not be contained in Z.

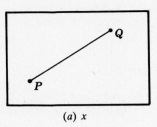
(a) x

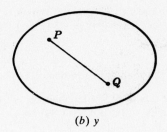
(b) y

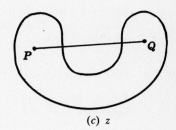
(c) z

Fig. 10-2

10.131 Prove that the intersection of any number of convex sets is convex.

▌ Let $\{X_i : i \in I\}$ be a collection of convex sets and let $Y = \cap_i X_i$. We need to show that Y is convex. Let $P, Q \in Y$. Then $P, Q \in X_i$, for every $i \in I$. Let L be the line segment between P and Q. Since each X_i is convex, $L \subset X_i$ for every $i \in I$. Thus $L \subset Y$, and so Y is convex.

10.132 Suppose W is a subspace of V. Show that W is convex.

▌ Let $u, v \in W$. Then $t_1 u + t_2 v \in W$ for any $t_1, t_2 \in \mathbf{R}$. Hence $t_1 u + t_2 v \in W$ for $t_1, t_2 \geq 0$ and $t_1 + t_2 = 1$. Thus W is convex.

10.133 Define the convex hull of a subset of a vector space V.

▌ The convex hull $H(X)$ of a subset X of V is the intersection of all convex sets which contain X. [By Problem 10.131, $H(X)$ is convex.] Alternatively, $H(X)$ is the smallest convex set containing X.

10.134 Describe the convex hull H of three vectors u, v, w in V.

▌ H consists of all vectors $t_1 u + t_2 v + t_3 w$ where $t_i \geq 0$ and $t_1 + t_2 + t_3 = 1$. [Geometrically, H is the triangle with vertices u, v, w as pictured in Fig. 10-3.]

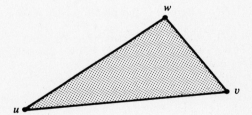

Fig. 10-3

10.135 Describe the convex hull H of vectors $v_1, v_2, \ldots, v_n$ in V.

▌ H consists of all vectors $t_1 v_1 + t_2 v_2 + \cdots + t_n v_n$ where each $t_i \geq 0$, and the sum $t_1 + \cdots + t_n = 1$.

10.136 Suppose $F: V \to U$ is a linear map and X is a convex subset of V. Show that the image $F(X)$ is a convex subset of U.

▌ Let $u_1, u_2 \in F(X)$. Then there exist vectors $v_1, v_2 \in X$ such that $F(v_1) = u_1$ and $F(v_2) = u_2$. Since X is convex, all the vectors $t_1 v_1 + t_2 v_2$, where $t_1, t_2 \geq 0$ and $t_1 + t_2 = 1$, belong to X. Hence the vectors $F(t_1 v_1 + t_2 v_2) = t_1 F(v_1) + t_2 F(v_2) = t_1 u_1 + t_2 u_2$ belong to $F(X)$. Thus $F(X)$ is convex.

10.137 Let $F: \mathbf{R}^2 \to \mathbf{R}^2$ be defined by $F(x, y) = (3x + 5y, 2x + 3y)$. Find the image $F(S)$ of the unit circle S in $\mathbf{R}^2$. [S consists of all points satisfying $x^2 + y^2 = 1$.]

▌ Set $F(x, y) = (s, t)$:

$$(3x + 5y, 2x + 3y) = (s, t) \quad \text{or} \quad \begin{cases} 3x + 5y = s \\ 2x + 3y = t \end{cases}$$

Solve for x and y in terms of s and t to get $x = -3s + 5t$, $y = 2s - 3t$. Substitute into $x^2 + y^2 = 1$ to obtain $13s^2 - 42st + 34t^2 = 1$ or $13x^2 - 42xy + 34y^2 = 1$ which is the image of S under F. [Note $F(S)$ is an ellipse.]

10.138 Find the preimage $F^{-1}(S)$ for the map F in Problem 10.137 and the unit circle S.

▌ Set $F(x, y) = (s, t)$ where $(s, t) \in S$, i.e., where $s^2 + t^2 = 1$. We get $3x + 5y = s$, $2x + 3y = t$. Substituting in $s^2 + t^2 = 1$ yields $13x^2 + 42xy + 24y^2 = 1$ which is $F^{-1}(S)$. [Note $F^{-1}(S)$ is an ellipse.]

Problems 10.139–10.141 refer to the linear map $G: \mathbf{R}^3 \to \mathbf{R}^3$ defined by

$$G(x, y, z) = (x + y + z, \ y - 2z, \ y - 3z)$$

and the unit sphere S_2 in $\mathbf{R}^3$ which consists of the points satisfying $x^2 + y^2 + z^2 = 1$.

10.139 Find the image $G(S_2)$ of the unit sphere S_2.

▌ Set $G(x, y, z) = (r, s, t)$:

$$(x + y + z, \ y - 2z, \ y - 3z) = (r, s, t) \qquad \text{or} \qquad \begin{aligned} x + y + z &= r \\ y - 2z &= s \\ y - 3z &= t \end{aligned}$$

Solve for x, y, z in terms of r, s, t to get $x = r - 4s + 3t$, $y = 3s - 2t$, $z = s - t$. Substitute in $x^2 + y^2 + z^2 = 1$ to get $(r - 4s + 3t)^2 + (3s - 2t)^2 + (s - t)^2 = 1$, $r^2 - 8rs + 16s^2 + 6rt - 38st + 14t^2 = 1$, or $x^2 - 8xy + 26y^2 + 6xz - 38yz + 14z^2 = 1$ which is $G(S_2)$.

10.140 Find the preimage $G^{-1}(S_2)$ of the unit sphere S_2.

▌ Set $G(x, y, z) = (r, s, t)$ where $(r, s, t) \in S_2$, i.e., where $r^2 + s^2 + t^2 = 1$. We get $x + y + z = r$, $y - 2z = s$, $y - 3z = t$. Substitute in $r^2 + s^2 + t^2 = 1$ to get $(x + y + z)^2 + (y - 2z)^2 + (y - 3z)^2 = 1$ or $x^2 + 2xy + 3y^2 + 2xz - 8yz + 14z^2 = 1$ which is the preimage $G^{-1}(S_2)$.

10.141 Find the preimage $G^{-1}(H)$ where H is the plane $x + 2y - 3z = 4$.

▌ Set $G(x, y, z) = (r, s, t)$ where $r + 2s - 3t = 4$. We get $x + y + z = r$, $y - 2z = s$, $y - 3z = t$. Substitute in $r + 2s - 3t = 4$ to get $(x + y + z) + 2(y - 2z) - 3(y - 3z) = 4$ or $x - 12z = 4$ which is $G^{-1}(H)$. [Note $G^{-1}(H)$ is also a plane.]

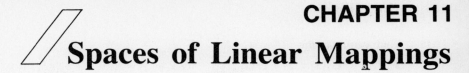

CHAPTER 11
Spaces of Linear Mappings

11.1 OPERATIONS WITH LINEAR MAPPINGS

11.1 Define the addition of linear mappings.

▮ Suppose $F: V \to U$ and $G: V \to U$ are linear mappings of vector spaces V and U over a field K. The sum $F + G$ is defined to be the mapping from V into U which assigns $F(v) + G(v)$ to $v \in V$; that is, $(F + G)(v) = F(v) + G(v)$.

11.2 Show that if $F: V \to U$ and $G: V \to U$ are linear, then $F + G$ is linear.

▮ For any vectors, $v, w \in V$ and any scalars $a, b \in K$,

$$\begin{aligned}
(F + G)(av + bw) &= F(av + bw) + G(av + bw) \\
&= aF(v) + bF(w) + aG(v) + bG(w) \\
&= a(F(v) + G(v)) + b(F(w) + G(w)) \\
&= a(F + G)(v) + b(F + G)(w)
\end{aligned}$$

Thus $F + G$ is linear.

11.3 Define the product of a scalar and a linear mapping.

▮ Suppose $F: V \to U$ is a linear mapping of vector spaces V and U over a field K. For any scalar $k \in K$, the product kF is defined to be the mapping from V into U which assigns $kF(v)$ to $v \in V$; that is, $(kF)(v) = kF(v)$.

11.4 Show that if $F: V \to U$ is linear, then kF is linear.

▮ For any vectors $v, w \in V$ and any scalars $a, b \in K$,

$$(kF)(av + bw) = kF(av + bw) = k(aF(v) + bF(w)) = akF(v) + bkF(w) = a(kF)(v) + b(kF)(w)$$

Thus kF is linear.

Problems 11.5–11.18 refer to the linear maps $F: \mathbf{R}^3 \to \mathbf{R}^2$, $G: \mathbf{R}^3 \to \mathbf{R}^2$, and $H: \mathbf{R}^2 \to \mathbf{R}^2$ defined by $F(x, y, z) = (2x, y + z)$, $G(x, y, z) = (x - z, y)$, and $H(x, y) = (y, x)$.

11.5 Find $(F + G)(v)$ where $v = (2, 3, 4)$.

▮ $(F + G)(v) = F(v) + G(v) = F(2, 3, 4) + G(2, 3, 4) = (4, 7) + (-2, 3) = (2, 10)$.

11.6 Find $(3F)(v)$ where $v = (2, 3, 4)$.

▮ $(3F)(v) = 3F(v) = 3F(2, 3, 4) = 3(4, 7) = (12, 21)$.

11.7 Find $(2F - 5G)(w)$ where $w = (5, 1, 3)$.

▮ $(2F - 5G)(w) = 2F(w) - 5G(w) = 2F(5, 1, 3) - 5G(5, 1, 3) = 2(10, 4) - 5(2, 1) = (20, 8) + (-10, -5)$
$= (10, 3)$.

11.8 Find a formula for $F + G$.

▮ $(F + G)(x, y, z) = F(x, y, z) + G(x, y, z) = (2x, y + z) + (x - z, y) = (3x - z, 2y + z)$.

11.9 Find a formula for $3F$.

▮ $(3F)(x, y, z) = 3F(x, y, z) = 3(2x, y + z) = (6x, 3y + 3z)$.

11.10   Find a formula for $2F - 5G$.

▌ $(2F - 5G)(x, y, z) = 2F(x, y, z) - 5G(x, y, z) = 2(2x, y + z) - 5(x - z, y) = (4x, 2y + 2z) + (-5x + 5z, -5y) = (-x + 5x, -3y + 2z)$.

11.11   Find $(H \circ F)(v)$ where $v = (2, 3, 4)$.

▌ $(H \circ F)(v) = H(F(v)) = H(F(2, 3, 4)) = H(4, 7) = (7, 4)$.

11.12   Find a formula for $H \circ F$.

▌ $(H \circ F)(x, y, z) = H(F(x, y, z)) = H(2x, y + z) = (y + z, 2x)$.

11.13   Find $(H \circ G)(v)$ where $v = (2, 3, 4)$.

▌ $(H \circ G)(v) = H(G(v)) = H(G(2, 3, 4)) = H(-2, 3) = (3, -2)$.

11.14   Find a formula for $H \circ G$.

▌ $(H \circ G)(x, y, z) = H(G(x, y, z)) = H(x - z, y) = (y, x - z)$.

11.15   Find a formula for $F + H$.

▌ $F + H$ is not defined since F and H have different domains.

11.16   Find a formula for $G \circ F$.

▌ $G \circ F$ is not defined since the co-domain of F is not the domain of G.

11.17   Find a formula for $5H$.

▌ $(5H)(x, y) = 5H(x, y) = 5(y, x) = (5y, 5x)$.

11.18   Find a formula for $H^2 = H \circ H$.

▌ $H^2(x, y) = H(H(x, y)) = H(y, x) = (x, y)$.   In other words, $H^2 = I$,   the identity mapping on $\mathbf{R}^2$.

Problems 11.19–11.28 refer to the linear maps $F: \mathbf{R}^3 \to \mathbf{R}^2$, $G: \mathbf{R}^3 \to \mathbf{R}^2$,   and $H: \mathbf{R}^2 \to \mathbf{R}^2$ defined by $F(x, y, z) = (y, x + z)$, $G(x, y, z) = (2z, x - y)$, $H(x, y) = (y, 2x)$,   and the vectors $v = (4, -1, 5)$ and $w = (3, 4, 1)$.

11.19   Find $(F + G)(v)$.

▌ $(F + G)(v) = F(v) + G(v) = F(4, -1, 5) + G(4, -1, 5) = (-1, 9) + (10, 5) = (9, 14)$.

11.20   Find $(F + G)(w)$.

▌ $(F + G)(w) = F(w) + G(w) = F(3, 4, 1) + G(3, 4, 1) = (4, 4) + (2, -1) = (6, 3)$.

11.21   Find a formula for $F + G$.

▌ $(F + G)(x, y, z) = F(x, y, z) + G(x, y, z) = (y, x + z) + (2z, x - y) = (y + 2z, 2x - y + z)$.

11.22   Find $(H \circ F)(v)$.

▌ $(H \circ F)(v) = H(F(v)) = H(F(4, -1, 5)) = H(-1, 9) = (9, -2)$.

11.23   Find a formula for $H \circ F$.

▌ $(H \circ F)(x, y, z) = H(F(x, y, z)) = H(y, x + z) = (x + z, 2y)$.

11.24 Find $(H \circ G)(w)$.

$\blacksquare$ $(H \circ G)(w) = H(G(w)) = H(G(3, 4, 1)) = H(2, 1) = (-1, 4)$.

11.25 Find a formula for $H \circ G$.

$\blacksquare$ $(H \circ G)(x, y, z) = H(G(x, y, z)) = H(2z, x - y) = (x - y, 4z)$.

11.26 Find a formula for $H \circ (F + G)$.

$\blacksquare$ Using Problem 11.21, $H \circ (F + G)(x, y, z) = H((F + G)(x, y, z)) = H(y + 2z, 2x - y + z) = (2x - y + z, 2y + 4z)$.

11.27 Find a formula for $H \circ F + H \circ G$. Compare with Problem 11.26.

$\blacksquare$ Using Problems 11.23 and 11.25, $(H \circ F + H \circ G)(x, y, z) = (H \circ F)(x, y, z) + (H \circ G)(x, y, z) = (x + z, 2y) + (x - y, 4z) = (2x - y + z, 2y + 4z)$. By Problem 11.26, $H \circ (F + G) = H \circ F + H \circ G$.

11.28 Find a formula for $H^2 = H \circ H$.

$\blacksquare$ $H^2(x, y) = H(H(x, y)) = H(y, 2x) = (2x, 2y)$.

Theorem 11.1: Let V, U, and W be vector spaces over K. Let F, F' be linear mappings from V into U and let G, G' be linear mappings from U into W; let $k \in K$. Then (i) $G \circ (F + F') = G \circ F + G \circ F'$; (ii) $(G + G') \circ F = G \circ F + G' \circ F$; (iii) $k(G \circ F) = (kG) \circ F = G \circ (kF)$.

11.29 Prove (i) of Theorem 11.1: $G \circ (F + F') = G \circ F + G \circ F'$.

$\blacksquare$ For every $v \in V$, $(G \circ (F + F'))(v) = G((F + F')(v)) = G(F(v) + F'(v)) = G(F(v)) + G(F'(v)) = (G \circ F)(v) + (G \circ F')(v) = (G \circ F + G \circ F')(v)$. Since $(G \circ (F + F'))(v) = (G \circ F + G \circ F')(v)$ for every $v \in V$, $G \circ (F + F') = G \circ F + G \circ F'$.

11.30 Prove (ii) of Theorem 11.1: $(G + G') \circ F = G \circ F + G' \circ F$.

$\blacksquare$ For every $v \in V$, $((G + G') \circ F)(v) = (G + G')(F(v)) = G(F(v)) + G'(F(v)) = (G \circ F)(v) + (G' \circ F)(v) = (G \circ F + G' \circ F)(v)$. Since $((G + G') \circ F)(v) = (G \circ F + G \circ F')(v)$ for every $v \in V$, $(G + G') \circ F = G \circ F + G' \circ F$.

11.31 Prove (iii) of Theorem 11.1: $k(G \circ F) = (kG \circ F) = G \circ (kF)$.

$\blacksquare$ For every $v \in V$, $(k(G \circ F))(v) = k(G \circ F)(v) = k(G(f(v))) = (kG)(F(v)) = (kG \circ F)(v)$ and $(k(G \circ F))(v) = k(G \circ F)(v) = k(G(F(v))) = G(kF(v)) = G((kF)(v)) = (G \circ kF)(v)$. Accordingly, $k(G \circ F) = (kG) \circ F = G \circ (kF)$. (We emphasize that two mappings are shown to be equal by showing that they assign the same image to each point in the domain.)

11.32 Suppose $F_1, F_2, \ldots, F_n$ are linear maps from V into U. Show that, for any scalars $a_1, a_2, \ldots, a_n$ and for any $v \in V$, $(a_1 F_1 + a_2 F_2 + \cdots + a_n F_n)(v) = a_1 F_1(v) + a_2 F_2(v) + \cdots + a_n F_n(v)$.

$\blacksquare$ Since $a_1 F_1$, $(a_1 F_1)(v) = a_1 F_1(v)$; the result holds for $n = 1$. Thus by induction,

$$(a_1 F_1 + a_2 F_2 + \cdots + a_n F_n)(v) = (a_1 F_1)(v) + (a_2 F_2 + \cdots + a_n F_n)(v) = a_1 F_1(v) + a_2 F_2(v) + \cdots + a_n F_n(v)$$

11.2 VECTOR SPACE OF LINEAR MAPPINGS

Theorem 11.2: Let V and U be vector spaces over a field K. Then the collection of all linear mappings from V into U with the above operations of addition and scalar multiplication form a vector space over k.

Remark: The space in Theorem 11.2 is usually denoted by $\text{Hom}(V, U)$. [Here Hom comes from the word homomorphism.] The proof of the theorem reduces to showing that $\text{Hom}(V, U)$ satisfies the eight axioms of a vector space [Section 7.1]. In the proof [Problems 11.33–11.40] F, G, H denote elements of $\text{Hom}(V, U)$ and k, a, b denote scalars in K.

11.33 Prove $\text{Hom}(V, U)$ satisfies $[A_1]$: $(F + G) + H = F + (G + H)$.

▌ For any vector $v \in V$, $((F + G) + H)(v) = (F + G)(v) + H(v) = (F(v) + G(v)) + H(v) = F(v) + (G(v) + H(v)) = F(v) + (G + H)(v) = (F + (G + H))(v)$. Thus $(F + G) + H = F + (G + H)$ and $[A_1]$ holds in $\text{Hom}(V, U)$.

11.34 Prove $\text{Hom}(V, U)$ satisfies $[A_2]$: There exists a zero element 0 such that $F + 0 = F$.

▌ Let 0 denote the zero mapping defined by $0(v) = 0$ for every $v \in V$. Then, for every $v \in V$, $(F + 0)(v) = F(v) + 0(v) = F(v) + 0 = F(v)$. Since $(F + 0)(v) = F(v)$ for every $v \in V$, $F + 0 = F$.

11.35 Prove $\text{Hom}(V, U)$ satisfies $[A_3]$: For each $F \in \text{Hom}(V, U)$, there exists an element $-F \in \text{Hom}(V, U)$ such that $F + (-F) = 0$.

▌ Let $-F$ be the map $(-1)F$. Then, for every $v \in V$,

$$(F + (-F))(v) = (F + (-1)F)(v) = F(v) + (-1)F(v) = F(v) - F(v) = 0 = 0(v)$$

11.36 Prove $\text{Hom}(V, U)$ satisfies $[A_4]$: $F + G = G + F$.

▌ For every $v \in V$, $(F + G)(v) = F(v) + G(v) = G(v) + F(v) = (G + F)(v)$. Hence $F + G = G + F$.

11.37 Prove $\text{Hom}(V, U)$ satisfies $[M_1]$: $k(F + G) = kF + kG$.

▌ For every $v \in V$, $(k(F + G))(v) = k[(F + G)(v)] = k[F(v) + G(v)] = kF(v) + kG(v) = (kF)(v) + (kG)(v) = (kF + kG)(v)$. Thus $k(F + G) = kF + kG$.

11.38 Prove $\text{Hom}(V, U)$ satisfies $[M_2]$: $(a + b)F = aF + bF$.

▌ For every $v \in V$, $((a + b)F)(v) = (a + b)[F(v)] = aF(v) + bF(v) = (aF)(v) + (bF)(v) = (aF + bF)(v)$. Thus $(a + b)F = aF + bF$.

11.39 Prove $\text{Hom}(V, U)$ satisfies $[M_3]$: $(ab)F = a(bF)$.

▌ For every $v \in V$, $((ab)F)(v) = (ab)[F(v)] = a(bF(v)) = a[(bF)(v)] = (a(bF))(v)$. Thus $(ab)F = a(bF)$.

11.40 Prove $\text{Hom}(V, U)$ satisfies $[M_4]$: $1F = F$.

▌ For every $v \in V$, $(1F)(v) = 1[F(v)] = F(v)$. Thus $1F = F$.

Problems 11.41–11.46 refer to the linear mappings $F: \mathbf{R}^3 \to \mathbf{R}^2$, $G: \mathbf{R}^3 \to \mathbf{R}^2$, $H: \mathbf{R}^3 \to \mathbf{R}^2$ defined by $F(x, y, z) = (x + y + z, x + y)$, $G(x, y, z) = (2x + z, x + y)$, $H(x, y, z) = (2y, x)$.

11.41 Which vector space, if any, do F, G, and H belong to?

▌ F, G, and H belong to $\text{Hom}(\mathbf{R}^3, \mathbf{R}^2)$ since they are linear maps from $\mathbf{R}^3$ to $\mathbf{R}^2$.

11.42 Find a formula for $F + G$.

▌ $(F + G)(x, y, z) = F(x, y, z) + G(x, y, z) = (x + y + z, x + y) + (2x + z, x + y) = (3x + y + 2z, 2x + 2y)$.

11.43 Find a formula for $F + H$.

▌ $(F + H)(x, y, z) = F(x, y, z) + H(x, y, z) = (x + y + z, x + y) + (2y, x) = (x + 3y + z, 2x + y)$.

11.44 Find a formula for $G \circ F$.

▮ $G \circ F$ is not defined since the co-domain of F is not the domain of G.

11.45 Find a formula for $3G + 2H$.

▮ $(3G + 2H)(x, y, z) = 3G(x, y, z) + 2H(x, y, z) = 3(2x + z, x + y) + 2(2y, x) = (6x + 4y + 3z, 5x + 3y)$.

11.46 Show that F, G, H are linearly independent [as elements in the vector space $\text{Hom}(V, U)$].

▮ Suppose, for scalars $a, b, c \in K$,

$$aF + bG + cH = \mathbf{0} \tag{1}$$

[Here $\mathbf{0}$ is the zero mapping.]   For $e_1 = (1, 0, 0) \in \mathbf{R}^3$, we have $(aF + bG + cH)(e_1) = aF(1, 0, 0) + bG(1, 0, 0) + cH(1, 0, 0) = a(1, 1) + b(2, 1) + c(0, 1) = (a + 2b, a + b + c)$ and $\mathbf{0}(e_1) = (0, 0)$.   Thus by *(1)*, $(a + 2b, a + b + c) = (0, 0)$ and so

$$a + 2b = 0 \quad \text{and} \quad a + b + c = 0 \tag{2}$$

Similarly for $e_2 = (0, 1, 0) \in \mathbf{R}^2$, we have $(aF + bG + cH)(e_2) = aF(0, 1, 0) + bG(0, 1, 0) + cH(0, 1, 0) = a(1, 1) + b(0, 1) + c(2, 0) = (a + 2c, a + b) = \mathbf{0}(e_2) = (0, 0)$.   Thus

$$a + 2c = 0 \quad \text{and} \quad a + b = 0 \tag{3}$$

Using *(2)* and *(3)*, we obtain

$$a = 0 \qquad b = 0 \qquad c = 0 \tag{4}$$

Since *(1)* implies *(4)*, the mappings F, G, and H are linearly independent.

Theorem 11.3: Suppose $\dim V = m$ and $\dim U = n$.   Then $\dim \text{Hom}(V, U) = mn$.

11.47 Prove Theorem 11.3.

▮ Suppose $\{v_1, \ldots, v_m\}$ is a basis of V and $\{u_1, \ldots, u_n\}$ is a basis of U.   By Theorem 10.1, a linear mapping in $\text{Hom}(V, U)$ is uniquely determined by arbitrarily assigning elements of U to the basis elements v_i of V.   We define

$$F_{ij} \in \text{Hom}(V, U) \qquad i = 1, \ldots, m, \; j = 1, \ldots, n$$

to be the linear mapping for which $F_{ij}(v_i) = u_j$ and $F_{ij}(v_k) = 0$ for $k \neq i$.   That is, F_{ij} maps v_i into u_j and the other v's into 0.   Observe that $\{F_{ij}\}$ contains exactly mn elements; hence the theorem is proved if we show that it is a basis of $\text{Hom}(V, U)$.

Proof that $\{F_{ij}\}$ generates $\text{Hom}(V, U)$. Let $F \in \text{Hom}(V, U)$.   Suppose $F(v_1) = w_1$, $F(v_2) = w_2, \ldots,$ $F(v_m) = w_m$.   Since $w_k \in U$, it is a linear combination of the u's; say,

$$w_k = a_{k1}u_1 + a_{k2}u_2 + \cdots + a_{kn}u_n \qquad k = 1, \ldots, m, \; a_{ij} \in K \tag{1}$$

Consider the linear mapping $G = \sum_{i=1}^{m} \sum_{j=1}^{n} a_{ij}F_{ij}$.   Since G is a linear combination of the F_{ij}, the proof that $\{F_{ij}\}$ generates $\text{Hom}(V, U)$ is complete if we show that $F = G$.

We now compute $G(v_k)$, $k = 1, \ldots, m$.   Since $F_{ij}(v_k) = 0$ for $k \neq i$ and $F_{ki}(v_k) = u_i$,

$$G(v_k) = \sum_{i=1}^{m} \sum_{j=1}^{n} a_{ij}F_{ij}(v_k) = \sum_{j=1}^{n} a_{kj}F_{kj}(v_k) = \sum_{j=1}^{n} a_{kj}u_j = a_{k1}u_1 + a_{k2}u_2 + \cdots + a_{kn}u_n$$

Thus by *(1)*, $G(v_k) = w_k$ for each k.   But $F(v_k) = w_k$ for each k.   Accordingly, by Theorem 10.1, $F = G$;   hence $\{F_{ij}\}$ generates $\text{Hom}(V, U)$.

Proof that $\{F_{ij}\}$ is linearly independent. Suppose, for scalars $a_{ij} \in K$,

$$\sum_{i=1}^{m} \sum_{j=1}^{n} a_{ij} F_{ij} = 0$$

For $v_k, k = 1, \ldots, m$,

$$0 = 0(v_k) = \sum_{i=1}^{m} \sum_{j=1}^{n} a_{ij} F_{ij}(v_k) = \sum_{j=1}^{n} a_{kj} F_{kj}(v_k) = \sum_{j=1}^{n} a_{kj} u_j = a_{k1} u_1 + a_{k2} u_2 + \cdots + a_{kn} u_n$$

But the u_i are linearly independent; hence for $k = 1, \ldots, m$, we have $a_{k1} = 0$, $a_{k2} = 0, \ldots$, $a_{kn} = 0$. In other words, all the $a_{ij} = 0$ and so $\{F_{ij}\}$ is linearly independent.

Thus $\{F_{ij}\}$ is a basis of $\text{Hom}(V, U)$; hence $\dim \text{Hom}(V, U) = mn$.

11.48 Find the dimension of $\text{Hom}(\mathbf{R}^3, \mathbf{R}^2)$.

I Since $\dim \mathbf{R}^3 = 3$ and $\dim \mathbf{R}^2 = 2$, we have [Theorem 11.3] $\dim(\text{Hom}(\mathbf{R}^3, \mathbf{R}^2)) = 3 \cdot 2 = 6$.

11.49 Find the dimension of $\text{Hom}(\mathbf{C}^3, \mathbf{R}^2)$.

I $\mathbf{C}^3$ is a vector space over $\mathbf{C}$, and $\mathbf{R}^2$ is a vector space over $\mathbf{R}$; hence $\text{Hom}(\mathbf{C}^3, \mathbf{R}^2)$ does not exist.

11.50 Let $V = \mathbf{C}^3$ be viewed as a vector space over $\mathbf{R}$. Find the dimension of $\text{Hom}(V, \mathbf{R}^2)$.

I As a vector space over $\mathbf{R}$, $V = \mathbf{C}^3$ has dimension 6. Hence [Theorem 11.3] $\dim(\text{Hom}(V, \mathbf{R}^2)) = 6 \cdot 2 = 12$.

11.3 ALGEBRA OF LINEAR MAPPINGS

This section considers the special case of linear mappings $T: V \rightarrow V$. They are also called *linear operators* or *linear transformations* on V. We will write $A(V)$, instead of $\text{Hom}(V, V)$, for the space of all such mappings.

11.51 Define an algebra and an associate algebra over a field K.

I An *algebra* $\mathbf{A}$ over a field K is a vector space over K in which an operation of multiplication is defined satisfying, for every $F, G, H \in \mathbf{A}$ and every $k \in K$. Distributive laws

$$F(G + H) = FG + FH \qquad \text{and} \qquad (G + H)F = GF + HF \tag{1}$$

$$K(GF) = (kG)F = G(kF) \tag{2}$$

If the associative law also holds for the multiplication, i.e., if for every $F, G, H \in \mathbf{A}$,

$$(FG)H = F(GH) \tag{3}$$

then the algebra $\mathbf{A}$ is said to be *associative*. [If the commutative law also holds for multiplication, i.e., if for every $F, G \in \mathbf{A}$

$$FG = GF \tag{4}$$

then the algebra is said to be *commutative*.]

11.52 Show how $A(V)$ may be viewed as an algebra over the base field K.

I The composition $G \circ F$ of two linear maps $F, G \in A(V)$ is defined and linear, and belongs to $A(V)$. By Theorem 11.1, the composition operation satisfies the properties in Problem 11.51. Thus $A(V)$ is an associative algebra over K with respect to composition of mappings; hence it is frequently called the *algebra of linear operators* on V. [We shall write GF for $G \circ F$ in the space $A(V)$.]

11.53 An algebra **A** is said to have an *identity element* 1 if $1 \cdot a = a \cdot 1 = a$ for every $a \in \mathbf{A}$. Show that $A(V)$ has an identity element.

    ❚ The identity mapping $I : V \to V$ belongs to $A(V)$. Also, for any $T \in A(V)$, we have $TI = IT = T$. Thus the identity mapping I is an identity element for the algebra $A(V)$.

11.54 Which of the following integers can be the dimension of an algebra $A(V)$ of linear maps: 5, 9, 18, 25, 31, 36, 44, 64, 88, 100?

    ❚ Suppose $\dim V = n$. Then $\dim(A(V)) = n^2$. Thus only the square integers can be the dimension of $A(V)$, that is, 9, 25, 36, 64, and 100.

11.55 Let T be an element of $A(V)$. Define the powers $T^2, T^3, \ldots$ of the linear map T.

    ❚ Since composition is the multiplication operation in $A(V)$, we have $T^2 = T \circ T$, $T^3 = T \circ T \circ T, \ldots$.

    Problems 11.56–11.61 refer to the linear operators $S, T \in A(\mathbf{R}^2)$ defined by $S(x, y) = (y, x)$ and $T(x, y) = (0, x)$.

11.56 Find a formula for $S + T$.

    ❚ $(S + T)(x, y) = S(x, y) + T(x, y) = (y, x) + (0, x) = (y, 2x)$.

11.57 Find a formula for $2S - 3T$.

    ❚ $(2S - 3T)(x, y) = 2S(x, y) - 3T(x, y) = 2(y, x) - 3(0, x) = (2y, -x)$.

11.58 Find a formula for ST.

    ❚ $(ST)(x, y) = S(T(x, y)) = S(0, x) = (x, 0)$.

11.59 Find a formula for TS.

    ❚ $(TS)(x, y) = T(S(x, y)) = T(y, x) = (0, y)$.

11.60 Find a formula for S^2.

    ❚ $S^2(x, y) = S(S(x, y)) = S(y, x) = (x, y)$. Note $S^2 = I$, the identity mapping.

11.61 Find a formula for T^2.

    ❚ $T^2(x, y) = T(T(x, y)) = T(0, x) = (0, 0)$. Note $T^2 = 0$, the zero mapping.

    Problems 11.62–11.64 refer to the linear operators $S, T \in A(\mathbf{R}^2)$ defined by $S(x, y) = (0, x)$ and $T(x, y) = (x, 0)$.

11.62 Show that $TS = 0$.

    ❚ $(TS)(x, y) = T(S(x, y)) = T(0, x) = (0, 0)$. Since TS assigns $0 = (0, 0)$ to every $(x, y) \in \mathbf{R}^2$, it is the zero mapping: $TS = 0$.

11.63 Show that $ST \neq 0$.

    ❚ $(ST)(x, y) = S(T(x, y)) = S(x, 0) = (0, x)$. For example, $(ST)(4, 2) = (0, 4)$. Thus $ST \neq 0$, since it does not assign $0 = (0, 0)$ to every element of $\mathbf{R}^2$.

11.64 Show that $T^2 = T$.

    ❚ $T^2(x, y) = T(T(x, y)) = T(x, 0) = (x, 0) = T(x, y)$. Hence $T^2 = T$.

Problems 11.65–11.70 refer to linear operators $S, T \in A(R^2)$ defined by $S(x, y) = (x + y, 0)$ and $T(x, y) = (-y, x)$.

11.65 Find a formula for $S + T$.

▮ $(S + T)(x, y) = S(x, y) + T(x, y) = (x + y, 0) + (-y, x) = (x, x)$.

11.66 Find a formula for $5S - 3T$.

▮ $(5S - 3T)(x, y) = 5S(x, y) - 3T(x, y) = 5(x + y, 0) - 3(-y, x) = (5x + 5y, 0) + (3y, -3x) = (5x + 8y, -3x)$.

11.67 Find a formula for ST.

▮ $(ST)(x, y) = S(T(x, y)) = S(-y, x) = (x - y, 0)$.

11.68 Find a formula for TS.

▮ $(TS)(x, y) = T(S(x, y)) = T(x + y, 0) = (0, x + y)$.

11.69 Show that $S^2 = S$.

▮ $S^2(x, y) = S(S(x, y)) = S(x + y, 0) = (x + y, 0) = S(x, y)$. Thus $S^2 = S$.

11.70 Show that $T^2 = -I$, where I is the identity operator.

▮ $T^2(x, y) = T(T(x, y)) = T(-y, x) = (-x, -y) = -(x, y) = -I(x, y)$. Thus $T^2 = -I$.

11.71 Consider a polynomial $p(x) = a_0 + a_1x + a_2x^2 + \cdots + a_nx^n$ over K, that is, $a_i \in K$. Define the operator $p(T)$ where $T \in A(V)$.

▮ The operator $p(T)$ is defined by $p(T) = a_0I + a_1T + a_2T^2 + \cdots + a_nT^n$. [For a scalar $k \in K$, the operator kI is frequently denoted by simply k.] In particular, if $p(T) = 0$, the zero mapping, then T is said to be a *zero* of the polynomial $p(x)$.

Problems 11.72–11.77 refer to the linear operator T on $\mathbf{R}^2$ defined by $T(x, y) = (x + 2y, 3x + 4y)$.

11.72 Find a formula for T^2.

▮ $T^2(x, y) = T(T(x, y)) = T(x + 2y, 3x + 4y)$
$= [(x + 2y) + 2(3x + 4y), 3(x + 2y) + 4(3x + 4y)] = (7x + 10y, 15x + 22y)$.

11.73 Find a formula for T^3.

▮ $T^3(x, y) = T(T^2(x, y)) = T(7x + 10y, 15x + 22y)$
$= [(7x + 10y) + 2(15x + 22y), 3(7x + 10y) + 4(15x + 22y)] = (37x + 54y, 81x + 118y)$.

11.74 Find $f(T)$ where $f(x) = x^2 - 3x + 4$.

▮ By definition, $f(T) = T^2 - 3T + 4I$. Thus

$$f(T)(x, y) = (T^2 - 3T + 4I)(x, y) = T^2(x, y) - 3T(x, y) + 4I(x, y)$$
$$= (7x + 10y, 15x + 22y) + (-3x - 6y, -9x - 12y) + (4x, 4y)$$
$$= (8x + 4y, 6x + 14y)$$

11.75 Is T a root of $f(x) = x^2 - 3x + 4$?

▮ No, since $f(T)$ is not the zero map.

11.76 Find $g(T)$ where $g(x) = x^2 - 5x - 2$.

▮ $g(T)(x, y) = (T^2 - 5T - 2I)(x, y) = T^2(x, y) - 5T(x, y) - 2I(x, y)$
$= (7x + 10y, 15x + 22y) + (-5x - 10y, -15x - 20y) + (-2x, -2y) = (0, 0)$

11.77 Is T a root of $g(x) = x^2 - 5x - 2$?

▮ Yes, since $g(T) = 0$, the zero map.

11.78 Let $T: \mathbf{R}^3 \to \mathbf{R}^3$ be defined by $T(x, y, z) = (0, x, y)$. Find $f(T)$ where $f(x) = x + 1$.

▮ $f(T)(x, y, z) = (T + I)(x, y, z) = (0, x, y) + (x, y, z) = (x, x + y, y + z)$.

11.79 Show that T in Problem 11.78 is a root of $p(x) = x^3$.

▮ $T^3(x, y, z) = T^2(0, x, y) = T(0, 0, x) = (0, 0, 0)$. Since $T^3 = 0$, T is a root of $p(x) = x^3$.

Problems 11.80–11.82 refer to a linear operator E in $A(V)$ for which $E^2 = E$. [Such an operator is termed a *projection*.]

11.80 Let U be the image of E. Show that if $u \in U$, then $E(u) = u$, that is, E is the identity map on U.

▮ If $u \in U$, the image of E, then $E(v) = u$ for some $v \in V$. Hence using $E^2 = E$, we have $u = E(v) = E^2(v) = E(E(v)) = E(u)$.

11.81 Show that if $E \neq I$, then E is singular, i.e., $E(v) = 0$ for some $v \neq 0$.

▮ If $E \neq I$ then, for some $v \in V$, $E(v) = u$ where $v \neq u$. Hence $u \in U$, the image of E, and so $E(u) = u$. Thus $E(v - u) = E(v) - E(u) = u - u = 0$ where $v - u \neq 0$.

11.82 Show that $V = U \oplus W$ where U is the image of E and W is the kernel of E.

▮ We first show that $V = U + W$. Let $v \in V$. Set $u = E(v)$ and $w = v - E(v)$. Then $v = E(v) + v - E(v) = u + w$. By definition, $u = E(v) \in U$, the image of E. We now show that $w \in W$, the kernel of E:

$$E(w) = E(v - E(v)) = E(v) - E^2(v) = E(v) - E(v) = 0$$

and thus $w \in W$. Hence $V = U + W$.
We next show that $U \cap W = \{0\}$. Let $v \in U \cap W$. Since $v \in U$, $E(v) = v$. Since $v \in W$, $E(v) = 0$. Thus $v = E(v) = 0$ and so $U \cap W = \{0\}$.
The above two properties imply that $V = U \oplus W$.

Problems 11.83–11.87 refer to linear operators E_1 and E_2 on a vector space $V = U \oplus W$ defined by $E_1(v) = u$, $E_2(v) = w$, where $v = u + w$, $u \in U$, $w \in W$.

11.83 Show that $E_1^2 = E_1$.

▮ For every $v = u + w$ in V, $E_1^2(v) = E_1(E_1(v)) = E_1(u) = E_1(u + 0) = u = E_1(v)$. Hence $E_1^2 = E_1$.

11.84 Show that $E_2^2 = E_2$.

▮ For every $v = u + w$ in V, $E_2^2(v) = E_2(E_2(v)) = E_2(w) = E_2(0 + w) + w = E_2(v)$. Hence $E_2^2 = E_2$.

11.85 Show that $E_1 E_2 = 0$.

▮ For every $v = u + w$ in V, $E_2 E_1(v) = E_1(w) = E_1(0 + w) = 0 = 0(v)$. Hence $E_1 E_2 = 0$.

11.86 Show that $E_2E_1 = 0$.

❚ For every $v = u + w$ in V, $E_2E_1(v) = E_2(u) = E_2(u + 0) = 0 = 0(v)$. Hence $E_2E_1 = 0$.

11.87 Show that $E_1 + E_2 = I$.

❚ For every $v = u + w$ in V, $(E_1 + E_2)(v) = E_1(v) + E_2(v) = u + w = v = I(v)$. Hence $E_1 + E_2 = I$.

Problems 11.88–11.91 refer to the vector space V of all real polynomials (in t).

11.88 Is V an algebra over **R**?

❚ Under ordinary multiplication of polynomials, V is an algebra.

11.89 Is V an associative algebra?

❚ Since multiplication of polynomials is associative, V is associative as an algebra.

11.90 Is V a commutative algebra?

❚ Since multiplication of polynomials is commutative, i.e., $f(t)g(t) = g(t)f(t)$, V is commutative.

11.91 Does V have an identity element?

❚ Yes; the constant polynomial $f(t) = 1$ is an identity element for V.

Problems 11.92–11.95 refer to the vector space M of all real n-square matrices.

11.92 Is M an algebra over **R**?

❚ Under multiplication of matrices, M is an algebra.

11.93 Is M an associative algebra?

❚ Yes, since multiplication of matrices is associative.

11.94 Is M a commutative algebra?

❚ For $n > 1$, M is not commutative since there are matrices $A, B \in M$ such that $AB \neq BA$.

11.95 Does M have an identity element?

❚ The identity matrix I is an identity element for M since $AI = IA = A$ for every $A \in M$.

11.96 Show how any vector space V may be made into an algebra.

❚ Define $uv = 0$ for any $u, v \in V$. Then V is an algebra.

11.4 INVERTIBLE OPERATORS

11.97 Define an invertible operator in $A(V)$, the collection of all linear mappings from V into V.

❚ A linear operator $T: V \rightarrow V$ is said to be *invertible* if it has an inverse, i.e., if there exists $T^{-1} \in A(V)$ such that $TT^{-1} = T^-T = I$.

Theorem 11.4: A linear operator $T: V \rightarrow V$ on a vector space of finite dimension is invertible if and only if it is nonsingular.

11.98 Prove Theorem 11.4.

▮ T is invertible iff T is one-one and onto. Thus in particular, if T is invertible then only $0 \in V$ can map into itself, i.e., T is nonsingular. On the other hand, suppose T is nonsingular, i.e., Ker $T = \{0\}$. By Problem 10.102, T is also one-to-one. Moreover, assuming V has finite dimension, we have, by Theorem 10.2, dim $V = \dim(\text{Im } T) + \dim(\text{Ker } T) = \dim(\text{Im } T) + \dim(\{0\}) = \dim(\text{Im } T) + 0 = \dim(\text{Im } T)$. Then Im $T = V$, i.e., the image of T is V; thus T is onto. Hence T is both one-one and onto and so is invertible.

11.99 Show that the finiteness of the dimensionality of V is necessary in Theorem 11.4.

▮ Let V be the vector space of polynomials over K, and let T be the operator on V defined by $T(a_0 + a_1 t + \cdots + a_n t^n) = a_0 t + a_1 t^2 + \cdots + a_n t^{n+1}$, that is, T increases the exponent of t in each term by 1. Now T is a linear mapping and is nonsingular. However, T is not onto and so is not invertible.

Problems 11.100–11.104 refer to the linear operator T on $\mathbf{R}^2$ defined by $T(x, y) = (y, 2x - y)$.

11.100 Show that T is invertible.

▮ By Theorem 11.4, we need only show that T is nonsingular, i.e., that $T^{-1}(0, 0) = (0, 0)$. Set $T(x, y) = (0, 0)$ and solve for x and y. We have $T(x, y) = (y, 2x - y) = (0, 0)$ or $2x - y = 0$, $y = 0$. The only solution is $x = 0$, $y = 0$. Thus Ker $T = \{0\}$ and so T is nonsingular; hence T is invertible.

11.101 Find a formula for T^{-1}.

▮ Set $T(x, y) = (s, t)$ [and hence $T^{-1}(s, t) = (x, y)$]. We have $T(x, y) = (y, 2x - y) = (s, t)$ and so $y = s$, $2x - y = t$. Solving for x and y in terms of s and t, we obtain $x = \frac{1}{2}s + \frac{1}{2}t$, $y = s$. Thus T^{-1} is given by the formula $T^{-1}(s, t) = (\frac{1}{2}s + \frac{1}{2}t, s)$.

11.102 Find $T(6, 2)$.

▮ Use the formula for T to get $T(6, 2) = (2, 12 - 2) = (2, 10)$.

11.103 Find $T^{-1}(6, 2)$.

▮ Use the formula for T^{-1} to get $T^{-1}(6, 2) = (3 + 1, 6) = (4, 6)$.

11.104 Find $T^{-1}(L)$ where L is the line $y = x$.

▮ Set $s = t$ in the formula for T^{-1} to get $T^{-1}(s, s) = (s, s)$. Thus $T^{-1}(L) = L$.

Problems 11.105–11.108 refer to the linear operator T on $\mathbf{R}^3$ defined by

$$T(x, y, z) = (2x, 4x - y, 2x + 3y - z)$$

11.105 Show that T is invertible.

▮ Let $W = \text{Ker } T$. We need only show that T is nonsingular, i.e., that $W = \{0\}$. Set $T(x, y, z) = (0, 0, 0)$, that is, $T(x, y, z) = (2x, 4x - y, 2x + 3y - z) = (0, 0, 0)$. Thus W is the solution space of the homogeneous system $2x = 0$, $4x - y = 0$, $2x + 3y - z = 0$ which has only the trivial solution $(0, 0, 0)$. Thus $W = \{0\}$; hence T is nonsingular and so T is invertible.

11.106 Find a formula for T^{-1}.

▮ Set $T(x, y, z) = (r, s, t)$ [and so $T^{-1}(r, s, t) = (x, y, z)$]. We have $(2x, 4x - y, 2x + 3y - z) = (r, s, t)$ or $2x = r$, $4x - y = s$, $2x + 3y - z = t$. Solve for x, y, z in terms of r, s, t to get $x = \frac{1}{2}r$, $y = 2r - s$, $z = 7r - 3s - t$. Thus $T^{-1}(r, s, t) = (\frac{1}{2}r, 2r - s, 7r - 3s - t)$.

11.107 Find $T^{-1}(2, 4, 6)$.

▮ Use the formula for T^{-1} to get $T^{-1}(2, 4, 6) = (1, 4 - 4, 14 - 12 - 6) = (1, 0, -4)$.

11.108 Find a formula for T^{-2}.

▌ Apply T^{-1} twice to get

$$\begin{aligned} T^{-2}(r, s, t) &= T^{-1}(\tfrac{1}{2}r, 2r - s, 7r - 3s - t) \\ &= (\tfrac{1}{4}r, r - (2r - s), \tfrac{7}{2}r - 3(2r - s) - (7r - 3s - t)) \\ &= (\tfrac{1}{4}r, -r + s, -\tfrac{19}{2}r + 6s + t) \end{aligned}$$

Problems 11.109–11.111 refer to the linear operator T on $\mathbf{R}^3$ defined by

$$T(x, y, z) = (x - 3y - 2z, y - 4z, z)$$

11.109 Show that T is invertible.

▌ Show that T is nonsingular. Set $T(x, y, z) = (0, 0, 0)$ to get the triangular homogeneous system $x - 3y - 2z = 0$, $y - 4z = 0$, $z = 0$. The only solution is $x = 0$, $y = 0$, $z = 0$. Hence T is nonsingular and so T is invertible.

11.110 Find a formula for T^{-1}.

▌ Set $T(x, y, z) = (r, s, t)$ to get $x - 3y - 2z = r$, $y - 4z = s$, $z = t$. Solve for x, y, z in terms of r, s, t to get $x = r + 3s + 14t$, $y = s + 4t$, $z = t$. Thus $T^{-1}(r, s, t) = (r + 3s + 14t, s + 4t, t)$ or $T^{-1}(x, y, z) = (x + 3y + 14z, y + 4z, z)$.

11.111 Find $T^{-1}(1, 2, 3)$.

▌ Use the formula for T^{-1} to get $T^{-1}(1, 2, 3) = (1 + 6 + 42, 2 + 12, 3) = (49, 14, 3)$.

Problems 11.112–11.114 refer to the linear operator T on $\mathbf{R}^3$ defined by $T(x, y, z) = (x + z, x - z, y)$.

11.112 Show that T is invertible.

▌ Set $T(x, y, z) = (0, 0, 0)$ to get the homogeneous system $x + z = 0$, $x - z = 0$, $y = 0$ whose only solution is $x = 0$, $y = 0$, $z = 0$. Thus T is nonsingular, and so T is invertible.

11.113 Find a formula for T^{-1}.

▌ Set $T(x, y, z) = (r, s, t)$ to get $x + z = r$, $x - z = s$, $y = t$. Solve for x, y, z to get $x = \tfrac{1}{2}r + \tfrac{1}{2}s$, $y = t$, $z = \tfrac{1}{2}r - \tfrac{1}{2}s$. Thus $T^{-1}(r, s, t) = (\tfrac{1}{2}r + \tfrac{1}{2}s, t, \tfrac{1}{2}r - \tfrac{1}{2}s)$ or $T^{-1}(x, y, z) = (\tfrac{1}{2}x + \tfrac{1}{2}y, z, \tfrac{1}{2}x - \tfrac{1}{2}y)$.

11.114 Find $T^{-1}(2, 4, 6)$.

▌ Use the formula for T^{-1} to get $T^{-1}(2, 4, 6) = (1 + 2, 6, 1 - 2) = (3, 6, -1)$.

Problems 11.115–11.118 refer to the linear operator T on $\mathbf{R}^2$ defined by $T(x, y) = (2x + 4y, 3x + 6y)$.

11.115 Find a formula for T^{-1}.

▌ T is singular, e.g., $T(2, -1) = (0, 0)$; hence the linear operator $T^{-1}: \mathbf{R}^2 \to \mathbf{R}^2$ does not exist.

11.116 Find $T^{-1}(8, 12)$.

▌ $T^{-1}(8, 12)$ means the preimage of $(8, 12)$ under T. Set $T(x, y) = (8, 12)$ to get the system

$$\begin{array}{ll} 2x + 4y = 8 & \\ 3x + 6y = 12 & \text{or} \quad x + 2y = 4 \end{array}$$

Here y is a free variable. Set $y = a$, where a is a parameter, to get the solution $x = -2a + 4$, $y = a$. Thus $T^{-1}(8, 12) = \{(-2a + 4, a): a \in \mathbf{R}\}$.

11.117   Find $T^{-1}(1, 2)$.

▌ Set $T(x, y) = (1, 2)$ to get the system $2x + 4y = 1$, $3x + 6y = 2$. The system has no solution. Thus $T^{-1}(1, 2) = \varnothing$, the empty set.

11.118   Is T an onto function?

▌ No, since, e.g., $(1, 2)$ has no preimage.

11.119   Let V be of finite dimension and let T be a linear operator on V. Recall that T is invertible if and only if T is nonsingular or one-to-one. Show that T is invertible if and only if T is onto.

▌ By Theorem 10.2, $\dim V = \dim(\operatorname{Im} T) + \dim(\operatorname{Ker} T)$. Hence the following statements are equivalent: (i) T is into, (ii) $\operatorname{Im} T = V$, (iii) $\dim(\operatorname{Im} T) = \dim V$, (iv) $\dim(\operatorname{Ker} T) = 0$, (v) $\operatorname{Ker} T = \{0\}$, (vi) T is nonsingular, (vii) T is invertible.

11.120   Let V be of finite dimension and T be a linear operator on V for which $TS = I$, for some operator S on V. [We call S a right inverse of T.] Show that T is invertible.

▌ Let $\dim V = n$. By the preceding problem, T is invertible if and only if T is onto; hence T is invertible if and only if $\operatorname{rank} T = n$. We have $n = \operatorname{rank} I = \operatorname{rank} TS \le \operatorname{rank} T \le n$. Hence $\operatorname{rank} T = n$ and T is invertible.

11.121   Show that $S = T^{-1}$ in Problem 11.120.

▌ $TT^{-1} = T^{-1}T = I$. Then $S = IS = (T^{-1}T)S = T^{-1}(TS) = T^{-1}I = T^{-1}$.

11.122   Give an example showing that the result in Problem 11.120 need not hold if V is of infinite dimension.

▌ Let V be the space of polynomials over K; say, $p(t) = a_0 + a_1 t + a_2 t^2 + \cdots + a_n t^n$. Let T and S be the linear maps defined by $T(p(t)) = 0 + a_1 + a_2 t + \cdots + a_n t^{n-1}$ and $S(p(t)) = a_0 t + a_1 t^2 + \cdots + a_n t^{n+1}$. We have $(TS)(p(t)) = T(S(p(t))) = T(a_0 t + a_1 t^2 + \cdots + a_n t^{n+1}) = a_0 + a_1 t + \cdots + a_n t^n = p(t)$ and so $TS = I$, the identity mapping. On the other hand, if $k \in K$ and $k \ne 0$, then $(ST)(k) = S(T(k)) = S(0) = 0 \ne k$. Accordingly, $ST \ne I$.

11.123   Show that a square matrix A is invertible if and only if it is nonsingular.

▌ Recall that A is invertible if and only if A is row equivalent to the identity matrix I. Thus the following statements are equivalent: (i) A is invertible. (ii) A and I are row equivalent. (iii) The equations $AX = 0$ and $IX = 0$ have the same solution space. (iv) $AX = 0$ has only the zero solution. (v) A is nonsingular.

11.124   Suppose S and T are invertible elements in $A(V)$. Show that ST is invertible and $(ST)^{-1} = T^{-1}S^{-1}$.

▌ $(ST)(T^{-1}S^{-1}) = S(TT^{-1})S^{-1} = SIS^{-1} = SS^{-1} = I$ and $(T^{-1}S^{-1})(ST) = T^{-1}(S^{-1}S)T = T^{-1}IT = T^{-1}T = I$. Hence ST is invertible with inverse $T^{-1}S^{-1}$.

11.125   Suppose $S \in A(V)$ is invertible. Show that S^{-1} is invertible and $(S^{-1})^{-1} = S$.

▌ We have $SS^{-1} = S^{-1}S = I$. Hence S^{-1} is invertible and $(S^{-1})^{-1} = S$.

11.126   Define similarity of operators in $A(V)$.

▌ Two operators $S, T \in A(V)$ are said to be *similar*, written $S \sim T$, if there exists an invertible operator $P \in A(V)$ for which $S = P^{-1}TP$.

Problems 11.127–11.129 show that the relation $S \sim T$ of similarity of linear operators in $A(V)$ is an equivalence relation, i.e., is reflexive, symmetric, and transitive.

11.127 Show that $\sim$ is reflexive, i.e., $S \sim S$ for every $S \in A(V)$.

❚ The identity operator $I \in A(V)$ is invertible and $S = I^{-1}SI$. Hence $S \sim S$.

11.128 Show that $\sim$ is symmetric, i.e., if $S \sim T$, then $T \sim S$.

❚ Suppose $S \sim T$, say $S = P^{-1}TP$ where P is invertible. Then P^{-1} is invertible and $T = PSP^{-1} = (P^{-1})^{-1}SP^{-1}$. Hence $T \sim S$.

11.129 Show that $\sim$ is transitive, i.e., if $F \sim G$ and $G \sim H$, then $F \sim H$.

❚ Suppose $F \sim G$ and $G \sim H$, say $F = P^{-1}GP$ and $G = Q^{-1}HQ$ where P and Q are invertible. Then QP is invertible and $F = P^{-1}GP = P^{-1}(Q^{-1}HQ)P = (P^{-1}Q^{-1})H(QP) = (QP)^{-1}H(QP)$. Hence $F \sim H$.

11.130 Let A be an associative algebra over a field K with an identity element $1 \in A$. Define an invertible element in A.

❚ An element $a \in A$ is invertible if there exists an inverse $a^{-1} \in A$ such that $aa^{-1} = a^{-1}a = 1$.

11.131 Let A be the algebra of polynomials (in t) over a field K. Which, if any, are the invertible elements in A?

❚ The nonzero constant polynomials $f(t) = k$, where $k \in K$, are the invertible elements in A.

11.132 Let A be the algebra of n-square matrices over a field K. Which, if any, are the invertible elements in A?

❚ The nonsingular matrices in A are the invertible elements in A.

11.5 LINEAR MAPPINGS AND SYSTEMS OF LINEAR EQUATIONS

11.133 Show how a system of linear equations is related to a linear mapping.

❚ Consider a system of m linear equations in n unknowns over a field K:

$$
\begin{aligned}
a_{11}x_1 + a_{12}x_2 + \cdots + a_{1n}x_n &= b_1 \\
a_{21}x_1 + a_{22}x_2 + \cdots + a_{2n}x_n &= b_2 \\
&\cdots\cdots\cdots\cdots\cdots\cdots\cdots \\
a_{m1}x_1 + a_{m2}x_2 + \cdots + a_{mn}x_n &= b_m
\end{aligned}
$$

The system is equivalent to the matrix equation $Ax = b$ where $A = (a_{ij})$ is the coefficient matrix and $x = (x_i)$ and $b = (b_i)$ are the column vectors of the unknowns and of the constants, respectively. Now the matrix A may also be viewed as the linear mapping $A: K^n \to K^m$. Thus the solution of the equation $Ax = b$ may be viewed as the preimage of $b \in K^m$ under the linear mapping $A: K^n \to K^m$. Furthermore, the solution of the associated homogeneous equation $Ax = 0$ may be viewed as the kernel of the linear mapping $A: K^n \to K^m$.

> **Theorem 11.5:** The dimension of the solution space W of the homogeneous system of linear equations $Ax = 0$ is $n - r$ where n is the number of unknowns and r is the rank of the coefficient matrix A.

11.134 Prove Theorem 11.5.

❚ Here A is viewed as a linear map, so $r = \text{rank}(A) = \dim(\text{Im } A)$ and $n = \dim K^n = \dim(\text{Domain of } A)$. Thus, by Theorem 10.2, $\dim W = \dim(\text{Ker } A) = \dim(\text{Domain of } A) - \dim(\text{Im } A) = n - r$.

11.135 Suppose the homogeneous system $Ax = 0$ has only the zero solution. Show that the nonhomogeneous system $Ax = b$ has at most one solution.

❚ If $Ax = 0$ has only the zero solution, then A is nonsingular. Hence A is one-to-one and so $Ax = b$ has at most one solution.

11.136   Suppose the homogeneous system $Ax = 0$ has nonzero solutions. Show that if $Ax = b$ has a solution, then it is not unique.

▮   Let $W = \text{Ker } A$, and suppose v is a solution of $Ax = b$. Then the preimage of b under the map A, that is, $A^{-1}(b)$, is the coset $v + W$. Since $Ax = 0$ has nonzero solutions, W contains more than one element. Hence $v + W$ contains more than one element.

Theorem 11.6:   Consider a system with the same number of equations as unknowns, i.e., consider the following system of linear equations:

$$a_{11}x_1 + a_{12}x_2 + \cdots + a_{1n}x_n = b_1$$
$$a_{21}x_1 + a_{22}x_2 + \cdots + a_{2n}x_n = b_2$$
$$\dots\dots\dots\dots\dots\dots\dots\dots\dots\dots\dots$$
$$a_{n1}x_1 + a_{n2}x_2 + \cdots a_{nn}x_n = b_n$$

(i)   Suppose the corresponding homogeneous system has only the zero solution. Then the above system has a unique solution for any values of the b_i.

(ii)   Suppose the corresponding homogeneous system has a nonzero solution. Then there are values for the b_i for which the above system does not have a solution.

Remark: Since the above system has the same number of equations as unknowns, we can represent this system by the matrix equation $Ax = b$ where A is an n-square matrix over K which we view as a linear operator on K^n.

11.137   Prove (i) of Theorem 11.6: If $Ax = 0$ has only the zero solution, then $Ax = b$ has a unique solution for any b.

▮   Here A is nonsingular since $Ax = 0$ has only the zero solution. Thus A is one-to-one and onto, i.e., A is invertible. Accordingly, $Ax = b$ has the unique solution $x = A^{-1}b$.

11.138   Prove (ii) of Theorem 11.6: If $Ax = 0$ has nonzero solutions, then there are vectors b such that $Ax = b$ has no solution.

▮   Here A is singular since $Ax = b$ has nonzero solutions. Hence $A : K^n \to K^n$ is not an onto mapping. Thus there exist $b \in K^n$ which do not belong to $\text{Im } A$, i.e., for which $Ax = b$ has no solution.

CHAPTER 12
Matrices and Linear Mappings

12.1 MATRIX REPRESENTATION OF A LINEAR OPERATOR

12.1 Define the matrix representation of a linear operator.

▮ Let T be a linear operator on a vector space V over a field K and suppose $B = \{e_1, \ldots, e_n\}$ is a basis of V. Now $T(e_1), \ldots, T(e_n)$ are vectors in V and so each is a linear combination of the elements of the basis $\{e_i\}$; say

$$
\begin{aligned}
T(e_1) &= a_{11}e_1 + a_{12}e_2 + \cdots + a_{1n}e_n \\
T(e_2) &= a_{21}e_1 + a_{22}e_2 + \cdots + a_{2n}e_n \\
&\cdots\cdots\cdots\cdots\cdots\cdots\cdots\cdots\cdots \\
T(e_n) &= a_{n1}e_1 + a_{n2}e_2 + \cdots + a_{nn}e_n
\end{aligned}
$$

The transpose of the above matrix of coefficients, denoted by $m_B(T)$ or $[T]_B$, is called the *matrix representation of T relative to the basis $\{e_i\}$* or simply the *matrix of T in the basis $\{e_i\}$*; that is,

$$
[T]_B = \begin{pmatrix} a_{11} & a_{21} & \cdots & a_{n1} \\ a_{12} & a_{22} & \cdots & a_{n2} \\ \cdots\cdots\cdots\cdots\cdots\cdots\cdots \\ a_{1n} & a_{2n} & \cdots & a_{nn} \end{pmatrix}
$$

[The subscript B may be omitted if the basis B is understood.]

Remark: Suppose $B = \{e_1, \ldots, e_n\}$ is a basis of a vector space V over a field K. Recall [Section 8.9] that any $v \in V$ can be written uniquely in the form $v = a_1e_1 + a_2e_2 + \cdots + a_ne_n$ and the coordinate vector of v relative to the basis B is denoted and defined by

$$
[v]_B = \begin{pmatrix} a_1 \\ a_2 \\ \vdots \\ a_n \end{pmatrix} = [a_1, a_2, \ldots, a_n]^T
$$

[where the subscript B may be omitted if the basis B is understood]. Using this notation, $m(T) = ([T(e_1)], [T(e_2)], \ldots, [T(e_n)])$. We emphasize that all coordinate vectors are assumed to be column vectors unless otherwise stated or implied.

Theorem 12.1, whose proof appears in Problem 12.50, is used throughout this section.

Theorem 12.1: Let $B = \{e_1, \ldots, e_n\}$ be a basis of V and T be any operator on V. Then, for any vector $v \in V$, we have $[T]_B[v]_B = [T(v)]_B$. [That is, if we multiply the coordinate vector of v by the matrix representation of T, then we obtain the coordinate vector of $T(v)$.]

Problems 12.2–12.6 find the matrix representation of the linear map $F: \mathbf{R}^2 \to \mathbf{R}^2$ defined by $F(x, y) = (2x - 5y, 3x + y)$ relative to the basis $\{u_1 = (2, 1), u_2 = (3, 2)\}$ of $\mathbf{R}^2$.

12.2 Find $F(u_1)$, the image of the first basis vector.

▮ $F(u_1) = F(2, 1) = (4 - 5, 6 + 1) = (-1, 7)$.

12.3 Write $F(u_1)$ as a linear combination of the basis vectors u_1 and u_2.

▮
$$
\begin{pmatrix} -1 \\ 7 \end{pmatrix} = x \begin{pmatrix} 2 \\ 1 \end{pmatrix} + y \begin{pmatrix} 3 \\ 2 \end{pmatrix} \quad \text{or} \quad \begin{aligned} 2x + 3y &= -1 \\ x + 2y &= 7 \end{aligned}
$$

The solution is $x = -23$, $y = 15$. Thus $F(u_1) = -23u_1 + 15u_2$.

12.4 Find $F(u_2)$.

▮ $F(u_2) = F(3, 2) = (6 - 10, 9 + 2) = (-4, 11)$.

12.5 Write $F(u_2)$ as a linear combination of u_1 and u_2.

▮
$$\binom{-4}{11} = x\binom{2}{1} + y\binom{3}{2} \quad \text{or} \quad \begin{matrix} 2x + 3y = -4 \\ x + 2y = 7 \end{matrix}$$

The solution is $x = -29$, $y = 18$. Thus $F(u_2) = -29u_1 + 18u_2$.

12.6 Find $[F]$, the matrix representation of F in the given basis.

▮ Write the coordinates of $F(u_1)$ and $F(u_2)$ as columns to get

$$[F] = \begin{pmatrix} -23 & -29 \\ 15 & 18 \end{pmatrix}$$

Remark: Observe that the matrix of coefficients of the systems of linear equations in Problems 12.3 and 12.5 are the same. Hence it is usually advantageous to first find the coordinates of an arbitrary vector $(a, b) \in \mathbf{R}^2$ in the given basis; i.e., to first solve $(a, b) = xu_1 + yu_2 = x(2, 1) + y(3, 2)$ to obtain x and y in terms of a and b. This will be done in subsequent problems.

Problems 12.7–12.12 find the matrix of the linear map $G: \mathbf{R}^2 \to \mathbf{R}^2$ defined by $G(x, y) = (2x - 3y, 4x + y)$ in the basis $\{u_1 = (1, -2), u_2 = (2, -5)\}$ of $\mathbf{R}^2$.

12.7 Find the coordinates of an arbitrary vector $(a, b) \in \mathbf{R}^2$ with respect to the given basis.

▮ We have

$$\binom{a}{b} = x\binom{1}{-2} + y\binom{2}{-5} \quad \text{or} \quad \begin{matrix} x + 2y = a \\ -2x - 5y = b \end{matrix}$$

Solve for x and y in terms of a and b to get $x = 5a + 2b$, $y = -2a - b$. Thus

$$(a, b) = (5a + 2b)u_1 + (-2a - b)u_2 \quad \text{or, equivalently,} \quad [(a, b)] = [5a + 2b, -2a - b]^T$$

[*Remark*: This formula for (a, b) will be used repeatedly below.]

12.8 Find $G(u_1)$, the image of the first basis vector.

▮ $G(u_1) = G(1, -2) = (2 + 6, 4 - 2) = (8, 2)$.

12.9 Write $G(u_1)$ as a linear combination of the basis vectors u_1 and u_2.

▮ By Problem 12.7, $G(u_1) = (8, 2) = (40 + 4)u_1 + (-16 - 2)u_2 = 44u_1 - 18u_2$.

12.10 Find $G(u_2)$.

▮ $G(u_2) = G(2, -5) = (4 + 15, 8 - 5) = (19, 3)$.

12.11 Write $G(u_2)$ as a linear combination of u_1 and u_2.

▮ By Problem 12.7, $G(u_2) = (19, 3) = (95 + 6)u_1 + (-38 - 3)u_2 = 101u_1 - 41u_2$.

12.12 Find $[G]$, the matrix representation of G in the given basis.

▮ Write the coordinates of $G(u_1)$ and $G(u_2)$ as columns:

$$[G] = \begin{pmatrix} 44 & 101 \\ -18 & -41 \end{pmatrix}$$

Problems 12.13–12.16 refer to the vector $v = (4, -3)$ and the above linear map G and the above basis $\{u_1 = (1, -2), u_2 = (2, -5)\}$.

12.13 Find $[v]$, the coordinate vector of v relative to the given basis.

▮ By Problem 12.7, $[v] = [(4, -3)] = [20 - 6, -8 + 3]^T = [14, -5]^T$.

12.14 Find $G(v)$.

▮ $G(v) = G(4, -3) = (8 + 9, 16 - 3) = (17, 13)$.

12.15 Find $[G(v)]$, the coordinate vector of $G(v)$ in the given basis.

▮ By Problem 12.7, $[G(v)] = [(17, 13)] = [85 + 26, -34 - 13]^T = [111, -47]^T$.

12.16 Verify Theorem 12.1 that $[G][v] = [G(v)]$.

▮
$$[G][v] = \begin{pmatrix} 44 & 101 \\ -18 & -41 \end{pmatrix}\begin{pmatrix} 14 \\ -5 \end{pmatrix} = \begin{pmatrix} 616 - 505 \\ -252 + 205 \end{pmatrix} = \begin{pmatrix} 111 \\ -47 \end{pmatrix} = [G(v)]$$

Problems 12.17–12.20 refer to the vector space P_3 of real polynomials $p(t)$ of degree ≤ 3.

12.17 Let $D: P_3 \to P_3$ be the differential operator defined by $D(p(t)) = dp/dt$. Find the matrix of D in the basis $\{1, t, t^2, t^3\}$ of P_3.

▮ Find $D(1)$, $D(t)$, $D(t^2)$, and $D(t^3)$ and write them as linear combinations of 1, t, t^2, and t^3:

$$
\begin{aligned}
D(1) &= 0 = 0 - 0t + 0t^2 + 0t^3 \\
D(t) &= 1 = 1 + 0t + 0t^2 + 0t^3 \\
D(t^2) &= 2t = 0 + 2t + 0t^2 + 0t^3 \\
D(t^3) &= 3t^2 = 0 + 0t + 3t^2 + 0t^3
\end{aligned}
\quad \text{and} \quad
[D] = \begin{pmatrix} 0 & 1 & 0 & 0 \\ 0 & 0 & 2 & 0 \\ 0 & 0 & 0 & 3 \\ 0 & 0 & 0 & 0 \end{pmatrix}
$$

[Note the coordinate vectors of $D(1)$, $D(t)$, $D(t^2)$, and $D(t^3)$ are the columns, not rows, in $[D]$.]

12.18 Let $p(t) = a + bt + ct^2 + dt^3$. Find $D(p(t))$.

▮ Take the derivative to get $D(p(t)) = b + 2ct + 3dt^2$.

12.19 Find $[p(t)]$ and $[D(p(t))]$, the coordinate vectors relative to the usual basis $\{1, t, t^2, t^3\}$ of P_3.

▮ Write down the coefficients of $p(t)$ and $D(p(t))$ to get

$$[p(t)] = [a, b, c, d]^T \quad \text{and} \quad [D(p(t))] = [b, 2c, 3d, 0]^T$$

12.20 Verify that Theorem 12.1 does hold here.

▮
$$[D][p(t)] = \begin{pmatrix} 0 & 1 & 0 & 0 \\ 0 & 0 & 2 & 0 \\ 0 & 0 & 0 & 3 \\ 0 & 0 & 0 & 0 \end{pmatrix}\begin{pmatrix} a \\ b \\ c \\ d \end{pmatrix} = \begin{pmatrix} b \\ 2c \\ 3d \\ 0 \end{pmatrix} = [D(p(t))]$$

Problems 12.21–12.26 find the matrix representation of the linear operator $S: \mathbf{R}^2 \to \mathbf{R}^2$ defined by $S(x, y) = (2y, 3x - y)$ relative to the following basis:

$$B = \{v_1 = (1, 3), v_2 = (2, 5)\}$$

12.21 Find the coordinates of an arbitrary vector $(a, b) \in \mathbf{R}^2$ with respect to the basis B.

▮ We have

$$\begin{pmatrix} a \\ b \end{pmatrix} = x\begin{pmatrix} 1 \\ 3 \end{pmatrix} + y\begin{pmatrix} 2 \\ 5 \end{pmatrix} \quad \text{or} \quad \begin{array}{l} x + 2y = a \\ 3x + 5y = b \end{array}$$

Solve for x and y in terms of a and b to get $x = 2b - 5a$ and $y = 3a - b$. Thus

$$(a, b) = (-5a + 2b)v_1 + (3a - b)v_2$$

[*Remark*: This formula for (a, b) will be used repeatedly below.]

12.22 Find $S(v_1)$.

▌ $S(v_1) = S(1, 3) = (6, 3 - 3) = (6, 0)$.

12.23 Write $S(v_1)$ as a linear combination of v_1 and v_2.

▌ Use the formula in Problem 12.21 to get $S(v_1) = (6, 0) = -30v_1 + 18v_2$.

12.24 Find $S(v_2)$.

▌ $S(v_2) = S(2, 5) = (10, 6 - 5) = (10, 1)$.

12.25 Write $S(v_2)$ as a linear combination of v_1 and v_2.

▌ By Problem 12.21, $S(v_2) = S(10, 1) = (-50 + 2)v_1 + (30 - 1)v_2 = -48v_1 + 29v_2$.

12.26 Find $[S]_B$, the matrix of S in the above basis B.

▌ Write the coordinates of $S(v_1)$ and $S(v_2)$ as columns to obtain

$$[S]_B = \begin{pmatrix} -30 & -48 \\ 18 & 29 \end{pmatrix}$$

12.27 Find $[S]_E$, the matrix representation of S relative to the usual basis $E = \{e_1 = (1, 0), e_2 = (0, 1)\}$.

▌ Recall that if $(a, b) \in \mathbf{R}^2$ then $(a, b) = ae_1 + be_2$. Thus

$$\begin{aligned} S(e_1) &= S(1, 0) = (0, 3) = 0e_1 + 3e_2 \\ S(e_2) &= S(0, 1) = (2, -1) = 2e_1 - e_2 \end{aligned} \quad \text{and} \quad [S]_E = \begin{pmatrix} 0 & 2 \\ 3 & -1 \end{pmatrix}$$

Problems 12.28–12.32 find the matrix representation of the linear map $T: \mathbf{R}^2 \to \mathbf{R}^2$ defined by $T(x, y) = (3x - 4y, x + 5y)$ relative to the above basis B.

12.28 Find $T(v_1)$.

▌ $T(v_1) = T(1, 3) = (3 - 12, 1 + 15) = (-9, 16)$.

12.29 Write $T(v_1)$ as a linear combination of v_1 and v_2.

▌ By Problem 12.21, $T(v_1) = (-9, 16) = (45 + 32)v_1 + (-27 - 16)v_2 = 77v_1 - 43v_2$.

12.30 Find $T(v_2)$.

▌ $T(v_2) = T(2, 5) = (6 - 20, 2 + 25) = (-14, 27)$.

12.31 Write $T(v_2)$ as a linear combination of v_1 and v_2.

▌ By Problem 12.21, $T(v_2) = (-14, 27) = (70 + 54)v_1 + (-42 - 27)v_2 = 124v_1 - 69v_2$.

12.32 Find $[T]_B$, the matrix representation of T in the basis B.

▮ Write the coordinates of $T(v_1)$ and $T(v_2)$ as columns:

$$[T]_B = \begin{pmatrix} -30 & -48 \\ 18 & 29 \end{pmatrix}$$

12.33 Find $[T]_E$, the matrix representation of T in the usual basis E.

▮ $\begin{aligned} T(e_1) &= T(1, 0) = (3, 1) = 3e_1 + e_2 \\ T(e_2) &= T(0, 1) = (-4, 5) = -4e_1 + 5e_2 \end{aligned}$ and thus $[T]_E = \begin{pmatrix} 3 & -4 \\ 1 & 5 \end{pmatrix}$

12.34 Let $A = \begin{pmatrix} 1 & 2 \\ 3 & 4 \end{pmatrix}$ and T be the linear operator on $\mathbf{R}^2$ defined by $T(v) = Av$ [where v is written as a column vector]. Find the matrix representation of T relative to the above basis B.

▮ By Problem 12.21, $(a, b) = (-5a + 2b)v_1 + (3a - b)v_2$; hence

$$T(v_1) = \begin{pmatrix} 1 & 2 \\ 3 & 4 \end{pmatrix}\begin{pmatrix} 1 \\ 3 \end{pmatrix} = \begin{pmatrix} 7 \\ 15 \end{pmatrix} = -5v_1 + 6v_2$$

$$T(v_2) = \begin{pmatrix} 1 & 2 \\ 3 & 4 \end{pmatrix}\begin{pmatrix} 2 \\ 5 \end{pmatrix} = \begin{pmatrix} 12 \\ 26 \end{pmatrix} = -8v_1 + 10v_2$$

and thus $[T]_B = \begin{pmatrix} -5 & -8 \\ 6 & 10 \end{pmatrix}$

12.35 Find the matrix representation of the linear operator T in Problem 12.34 relative to the usual basis E.

▮ $$T(e_1) = \begin{pmatrix} 1 & 2 \\ 3 & 4 \end{pmatrix}\begin{pmatrix} 1 \\ 0 \end{pmatrix} = \begin{pmatrix} 1 \\ 3 \end{pmatrix} = 1e_1 + 3e_2$$

$$T(e_2) = \begin{pmatrix} 1 & 2 \\ 3 & 4 \end{pmatrix}\begin{pmatrix} 0 \\ 1 \end{pmatrix} = \begin{pmatrix} 2 \\ 4 \end{pmatrix} = 2e_1 + 4e_2$$

and thus $[T]_E = \begin{pmatrix} 1 & 2 \\ 3 & 4 \end{pmatrix}$

Remark: Observe that the matrix of T in the usual basis is precisely the original matrix A which defined T. This is not unusual. In fact, we show in the next problem that this is true for any matrix A when using the usual basis.

12.36 Recall that any n-square matrix $A = (a_{ij})$ may be viewed as the linear operator T on K^n defined by $T(v) = Av$, where v is written as a column vector. Show that the matrix representation of T relative to the usual basis $\{e_i\}$ of K^n is the matrix A, that is, $[T]_e = A$.

▮ $$T(e_1) = Ae_1 = \begin{pmatrix} a_{11} & a_{12} & \cdots & a_{1n} \\ a_{21} & a_{22} & \cdots & a_{2n} \\ \cdots\cdots\cdots\cdots\cdots \\ a_{n1} & a_{n2} & \cdots & a_{nn} \end{pmatrix}\begin{pmatrix} 1 \\ 0 \\ .. \\ 0 \end{pmatrix} = \begin{pmatrix} a_{11} \\ a_{21} \\ .. \\ a_{n1} \end{pmatrix} = a_{11}e_1 + a_{21}e_2 + \cdots + a_{n1}e_n$$

$$T(e_2) = Ae_2 = \begin{pmatrix} a_{11} & a_{12} & \cdots & a_{1n} \\ a_{21} & a_{22} & \cdots & a_{2n} \\ \cdots\cdots\cdots\cdots\cdots \\ a_{n1} & a_{n2} & \cdots & a_{nn} \end{pmatrix}\begin{pmatrix} 0 \\ 1 \\ .. \\ 0 \end{pmatrix} = \begin{pmatrix} a_{12} \\ a_{22} \\ .. \\ a_{n2} \end{pmatrix} = a_{12}e_1 + a_{22}e_2 + \cdots + a_{n2}e_n$$

$$\cdots\cdots\cdots\cdots\cdots\cdots\cdots\cdots\cdots\cdots\cdots\cdots\cdots\cdots\cdots\cdots\cdots$$

$$T(e_n) = Ae_n = \begin{pmatrix} a_{11} & a_{12} & \cdots & a_{1n} \\ a_{21} & a_{22} & \cdots & a_{2n} \\ \cdots\cdots\cdots\cdots\cdots \\ a_{n1} & a_{n2} & \cdots & a_{nn} \end{pmatrix}\begin{pmatrix} 0 \\ 0 \\ .. \\ 1 \end{pmatrix} = \begin{pmatrix} a_{1n} \\ a_{2n} \\ .. \\ a_{nn} \end{pmatrix} = a_{1n}e_1 + a_{2n}e_2 + \cdots + a_{nn}e_n$$

[That is, $T(e_i) = Ae_i$ is the ith column of A.] Accordingly,

$$[T]_e = \begin{pmatrix} a_{11} & a_{12} & \cdots & a_{1n} \\ a_{21} & a_{22} & \cdots & a_{2n} \\ \cdots\cdots\cdots\cdots\cdots \\ a_{n1} & a_{n2} & \cdots & a_{nn} \end{pmatrix} = A$$

12.37 The set $\{1, t, e^t, te^t\}$ is a basis of a vector space V of functions $f: \mathbf{R} \to \mathbf{R}$. Let D be the differential operator on V, that is, $D(f) = df/dt$. Find the matrix of D in the given basis.

$$
\begin{aligned}
D(1) &= 0 &&= 0(1) + 0(t) + 0(e^t) + 0(te^t) \\
D(t) &= 1 &&= 1(1) + 0(t) + 0(e^t) + 0(te^t) \\
D(e^t) &= e^t &&= 0(1) + 0(t) + 1(e^t) + 0(te^t) \\
D(te^t) &= e^t + te^t = 0(1) + 0(t) + 1(e^t) + 1(te^t)
\end{aligned}
\quad \text{and thus} \quad [D] = \begin{pmatrix} 0 & 1 & 0 & 0 \\ 0 & 0 & 0 & 0 \\ 0 & 0 & 1 & 1 \\ 0 & 0 & 0 & 1 \end{pmatrix}
$$

12.38 The set $\{e^{3t}, te^{3t}, t^2 e^{3t}\}$ is a basis of a vector space V of functions $f: \mathbf{R} \to \mathbf{R}$. Let D be the differential operator on V, that is, $D(f) = df/dt$. Find the matrix of D in the given basis.

$$
\begin{aligned}
D(e^{3t}) &= 3e^{3t} &&= 3(e^{3t}) + 0(te^{3t}) + 0(t^2 e^{3t}) \\
D(te^{3t}) &= e^{3t} + 3te^{3t} &&= 1(e^{3t}) + 3(te^{3t}) + 0(t^2 e^{3t}) \\
D(t^2 e^{3t}) &= 2te^{3t} + 3t^2 e^{3t} &&= 0(e^{3t}) + 2(te^{3t}) + 3(t^2 e^{3t})
\end{aligned}
\quad \text{and thus} \quad [D] = \begin{pmatrix} 3 & 1 & 0 \\ 0 & 3 & 2 \\ 0 & 0 & 3 \end{pmatrix}
$$

12.39 Consider the basis $S = \{(1, 0), (1, 1)\}$ of $\mathbf{R}^2$. Let $L: \mathbf{R}^2 \to \mathbf{R}^2$ be defined by $L(1, 0) = (6, 4)$ and $L(1, 1) = (1, 5)$. [Recall that a linear map is completely defined by its action on a basis.] Find the matrix representation of L with respect to the basis S.

    Write $(6, 4)$ and then $(1, 5)$ each as a linear combination of the basis vectors to get

$$
\begin{aligned}
L(1, 0) &= (6, 4) = 2(1, 0) + 4(1, 1) \\
L(1, 1) &= (1, 5) = -4(1, 0) + 5(1, 1)
\end{aligned}
\quad \text{and} \quad [L] = \begin{pmatrix} 2 & -4 \\ 4 & 5 \end{pmatrix}
$$

12.40 Consider the usual basis $E = \{e_1, e_2, \ldots, e_n\}$ of K^n. Let $L: K^n \to K^n$ be defined by $L(e_i) = v_i$. Show that the matrix A representing L relative to the usual basis E is obtained by writing the image vectors $v_1, v_2, \ldots, v_n$ as columns.

    Suppose $v_i = (a_{i1}, a_{i2}, \ldots, a_{in})$. Then $L(e_i) = v_i = a_{i1}e_1 + a_{i2}e_2 + \cdots + a_{in}e_n$. Thus

$$
[L] = \begin{pmatrix} a_{11} & a_{21} & \cdots & a_{n1} \\ a_{12} & a_{22} & \cdots & a_{n2} \\ \cdots & \cdots & \cdots & \cdots \\ a_{1n} & a_{2n} & \cdots & a_{nn} \end{pmatrix}
$$

as claimed.

12.41 Let $F: \mathbf{R}^2 \to \mathbf{R}^2$ be defined by $F(1, 0) = (2, 4)$ and $F(0, 1) = (5, 8)$. Find the matrix A representing F with respect to the usual basis of $\mathbf{R}^2$.

    Since $(1, 0)$ and $(0, 1)$ do form the usual basis of $\mathbf{R}^2$, write their images under F as columns to get $A = \begin{pmatrix} 2 & 5 \\ 4 & 8 \end{pmatrix}$.

12.42 Let L be the rotation in $\mathbf{R}^2$ counterclockwise by $90°$. Find the matrix A which represents L with respect to the usual basis of $\mathbf{R}^2$.

    Under the rotation L, we have $L(1, 0) = (0, 1)$ and $L(0, 1) = (-1, 0)$. Thus $A = \begin{pmatrix} 0 & -1 \\ 1 & 0 \end{pmatrix}$.

12.43 Let T denote the relection in $\mathbf{R}^2$ about the line $y = -x$. Find the matrix of T with respect to the usual basis of $\mathbf{R}^2$.

    Under the reflection T, we have $T(1, 0) = (0, -1)$ and $T(0, 1) = (-1, 0)$. Thus $[T] = \begin{pmatrix} 0 & -1 \\ -1 & 0 \end{pmatrix}$.

    Problems 12.44–12.45 refer to the complex field $\mathbf{C}$ as a vector space over the real field $\mathbf{R}$ and the conjugation operator T on $\mathbf{C}$, i.e., defined by $T(z) = \bar{z}$.

12.44 Find the matrix of T relative to the usual basis $\{1, i\}$.

$$
\begin{aligned}
T(1) &= \bar{1} = 1 = 1(1) + 0(i) \\
T(i) &= \bar{i} = -i = 0(1) - 1(i)
\end{aligned}
\quad \text{and thus} \quad [T] = \begin{pmatrix} 1 & 0 \\ 0 & -1 \end{pmatrix}
$$

12.45 Find the matrix of T relative to the basis $\{1 + i, 1 + 2i\}$.

$$T(1 + i) = 1 - i = 3(1 + i) - 2(1 + 2i)$$
$$T(1 + 2i) = 1 - 2i = 4(1 + i) - 3(1 + 2i) \quad \text{and thus} \quad [T] = \begin{pmatrix} 3 & 4 \\ -2 & -3 \end{pmatrix}$$

12.46 Let 1_V denote the identity operator of a vector space V, that is, $1_V(v) = v$ for any $v \in V$. Show that, for any basis $B = \{v_i\}$ of V, $[1_V]_B = I$, the identity matrix.

■ For $i = 1, 2, \ldots, n$, $1_V(v_i) = v_i = 0 \cdot v_1 + \cdots + 1 \cdot v_i + \cdots + 0 \cdot v_n$. Thus $[1_V]_B = I$.

12.47 Let 0_V denote the zero operator on V, i.e., defined by $0_V(v) = 0$, for any $v \in V$. Show that, for any basis $B = \{v_i\}$ of V, $[0_V]_B = 0$, the zero matrix.

■ For any basis vector v_i, we have $0_V(v_i) = 0 = 0v_1 + 0v_2 + \cdots + 0v_n$. Hence $[0_V]_B = 0$.

Problems 12.48–12.49 refer to the vector space V of 2×2 matrices over **R** and the following usual basis E of V:

$$E = \left\{ E_1 = \begin{pmatrix} 1 & 0 \\ 0 & 0 \end{pmatrix}, E_2 = \begin{pmatrix} 0 & 1 \\ 0 & 0 \end{pmatrix}, E_3 = \begin{pmatrix} 0 & 0 \\ 1 & 0 \end{pmatrix}, E_4 = \begin{pmatrix} 0 & 0 \\ 0 & 1 \end{pmatrix} \right\}$$

12.48 Let $M = \begin{pmatrix} 1 & 2 \\ 3 & 4 \end{pmatrix}$ and T be the linear operator on V defined by $T(A) = MA$. Find the matrix representation of T relative to the above usual basis of V.

■ We have

$$T(E_1) = ME_1 = \begin{pmatrix} 1 & 2 \\ 3 & 4 \end{pmatrix} \begin{pmatrix} 1 & 0 \\ 0 & 0 \end{pmatrix} = \begin{pmatrix} 1 & 0 \\ 3 & 0 \end{pmatrix} = 1E_1 + 0E_2 + 3E_3 + 0E_4$$

$$T(E_2) = ME_2 = \begin{pmatrix} 1 & 2 \\ 3 & 4 \end{pmatrix} \begin{pmatrix} 0 & 1 \\ 0 & 0 \end{pmatrix} = \begin{pmatrix} 0 & 1 \\ 0 & 3 \end{pmatrix} = 0E_1 + 1E_2 + 0E_3 + 3E_4$$

$$T(E_3) = ME_3 = \begin{pmatrix} 1 & 2 \\ 3 & 4 \end{pmatrix} \begin{pmatrix} 0 & 0 \\ 1 & 0 \end{pmatrix} = \begin{pmatrix} 2 & 0 \\ 4 & 0 \end{pmatrix} = 2E_1 + 0E_2 + 4E_3 + 0E_4$$

$$T(E_4) = ME_4 = \begin{pmatrix} 1 & 2 \\ 3 & 4 \end{pmatrix} \begin{pmatrix} 0 & 0 \\ 0 & 1 \end{pmatrix} = \begin{pmatrix} 0 & 2 \\ 0 & 4 \end{pmatrix} = 0E_1 + 2E_2 + 0E_3 + 4E_4$$

Hence

$$[T]_E = \begin{pmatrix} 1 & 0 & 2 & 0 \\ 0 & 1 & 0 & 2 \\ 3 & 0 & 4 & 0 \\ 0 & 3 & 0 & 4 \end{pmatrix}$$

[Since $\dim V = 4$, any matrix representation of a linear operator on V must be a 4-square matrix.]

12.49 Let $M = \begin{pmatrix} a & b \\ c & d \end{pmatrix}$ and $S : V \to V$ be the linear map defined by $S(A) = AM$. Find the matrix representation of S in the above usual basis of V.

■ We have

$$S(E_1) = E_1M = \begin{pmatrix} 1 & 0 \\ 0 & 0 \end{pmatrix} \begin{pmatrix} a & b \\ c & d \end{pmatrix} = \begin{pmatrix} a & b \\ 0 & 0 \end{pmatrix} = aE_1 + bE_2 + 0E_3 + 0E_4$$

$$S(E_2) = E_2M = \begin{pmatrix} 0 & 1 \\ 0 & 0 \end{pmatrix} \begin{pmatrix} a & b \\ c & d \end{pmatrix} = \begin{pmatrix} c & d \\ 0 & 0 \end{pmatrix} = cE_1 + dE_2 + 0E_3 + 0E_4$$

$$S(E_3) = E_3M = \begin{pmatrix} 0 & 0 \\ 1 & 0 \end{pmatrix} \begin{pmatrix} a & b \\ c & d \end{pmatrix} = \begin{pmatrix} 0 & 0 \\ a & b \end{pmatrix} = 0E_1 + 0E_2 + aE_3 + bE_4$$

$$S(E_4) = E_4M = \begin{pmatrix} 0 & 0 \\ 0 & 1 \end{pmatrix} \begin{pmatrix} a & b \\ c & d \end{pmatrix} = \begin{pmatrix} 0 & 0 \\ c & d \end{pmatrix} = 0E_1 + 0E_2 + cE_3 + dE_4$$

Thus

$$[S]_E = \begin{pmatrix} a & b & 0 & 0 \\ c & d & 0 & 0 \\ 0 & 0 & a & b \\ 0 & 0 & c & d \end{pmatrix} = \begin{pmatrix} M & 0 \\ 0 & M \end{pmatrix}$$

12.50 Prove Theorem 12.1

▌ Suppose, for $i = 1, \ldots, n$,

$$T(e_i) = a_{i1}e_1 + a_{i2}e_2 + \cdots + a_{in}e_n = \sum_{j=1}^{n} a_{ij}e_j$$

Then $[T]_e$ is the n-square matrix whose jth row is

$$(a_{1j}, a_{2j}, \ldots, a_{nj}) \tag{1}$$

Now suppose

$$v = k_1 e_1 + k_2 e_2 + \cdots + k_n e_n = \sum_{i=1}^{n} k_i e_i$$

Writing a column vector as the transpose of a row vector,

$$[v]_e = (k_1, k_2, \ldots, k_n)^T \tag{2}$$

Furthermore, using the linearity of T,

$$T(v) = T\left(\sum_{i=1}^{n} k_i e_i\right) = \sum_{i=1}^{n} k_i T(e_i) = \sum_{i=1}^{n} k_i \left(\sum_{j=1}^{n} a_{ij}e_j\right)$$

$$= \sum_{j=1}^{n} \left(\sum_{i=1}^{n} a_{ij}k_i\right)e_j = \sum_{j=1}^{n} (a_{1j}k_1 + a_{2j}k_2 + \cdots + a_{nj}k_n)e_j$$

Thus $[T(v)]_e$ is the column vector whose jth entry is

$$a_{1j}k_1 + a_{2j}k_2 + \cdots + a_{nj}k_n \tag{3}$$

On the other hand, the jth entry of $[T]_e[v]_e$ is obtained by multiplying the jth row of $[T]_e$ by $[v]_e$, i.e., multiply *(1)* by *(2)*. But the product of *(1)* and *(2)* is *(3)*; hence $[T]_e[v]_e$ and $[T(v)]_e$ have the same entries. Thus $[T]_e[v]_e = [T(v)]_e$.

12.2 MATRICES AND LINEAR OPERATORS ON $\mathbf{R}^3$

This section restricts itself to linear operators on $\mathbf{R}^3$. By the usual basis of $\mathbf{R}^3$ we mean

$$E = \{e_1 = (1, 0, 0), e_2 = (0, 1, 0), e_3 = (0, 0, 1)\}$$

12.51 Suppose that T is the linear operator on $\mathbf{R}^3$ defined by $T(x, y, z) = (a_1 x + a_2 y + a_3 z, b_1 x + b_2 y + b_3 z, c_1 x + c_2 y + c_3 z)$. Show that the matrix of T in the usual basis $\{e_i\}$ is given by

$$[T]_e = \begin{pmatrix} a_1 & a_2 & a_3 \\ b_1 & b_2 & b_3 \\ c_1 & c_2 & c_3 \end{pmatrix}$$

That is, the rows of $[T]_e$ are obtained from the coefficients of x, y, and z in the components of $T(x, y, z)$.

▌ We have

$$T(e_1) = T(1, 0, 0) = (a_1, b_1, c_1) = a_1 e_1 + b_1 e_2 + c_1 e_3$$
$$T(e_2) = T(0, 1, 0) = (a_2, b_2, c_2) = a_2 e_1 + b_2 e_2 + c_2 e_3$$
$$T(e_3) = T(0, 0, 1) = (a_3, b_3, c_3) = a_3 e_1 + b_3 e_2 + c_3 e_3$$

Accordingly,

$$[T]_e = \begin{pmatrix} a_1 & a_2 & a_3 \\ b_1 & b_2 & b_3 \\ c_1 & c_2 & c_3 \end{pmatrix}$$

Remark: This property holds for any space K^n but only relative to the usual basis $\{e_1 = (1, 0, \ldots, 0),$ $e_2 = (0, 1, 0, \ldots, 0), \ldots, e_n = (0, \ldots, 0, 1)\}$.

12.52 Let $F: \mathbf{R}^3 \to \mathbf{R}^3$ be defined by $F(x, y, z) = (2x - 3y + 4z, 5x - y + 2z, 4x + 7y)$. Find the matrix of F relative to the usual basis of $\mathbf{R}^3$.

▌ By Problems 12.51, $\qquad\qquad [F] = \begin{pmatrix} 2 & -3 & 4 \\ 5 & -1 & 2 \\ 4 & 7 & 0 \end{pmatrix}$

12.53 Let $G: \mathbf{R}^3 \to \mathbf{R}^3$ be defined by $G(x, y, z) = (2y + z, x - 4y, 3x)$. Find the matrix representation of G in the usual basis of $\mathbf{R}^3$.

▌ By Problem 12.51, $\qquad\qquad [G] = \begin{pmatrix} 0 & 2 & 1 \\ 1 & -4 & 0 \\ 3 & 0 & 0 \end{pmatrix}$

Problems 12.54–12.61 find the matrix representation of $S: \mathbf{R}^3 \to \mathbf{R}^3$ defined by $S(x, y, z) = (x + 2y - 3z, 2x + y + z, 5x - y + z)$ relative to the following basis of $\mathbf{R}^3$:

$$B = \{u_1 = (1, 1, 0), \; u_2 = (1, 2, 3), \; u_3 = (1, 3, 5)\}$$

12.54 Find the coordinates of an arbitrary vector $(a, b, c) \in \mathbf{R}^3$ with respect to the above basis B.

▌ Write (a, b, c) as a linear combination of u_1, u_2, u_3 using unknowns x, y, z:

$$(a, b, c) = x(1, 1, 0) + y(1, 2, 3) + z(1, 3, 5) = (x + y + z, x + 2y + 3z, 3y + 5z)$$

$$\text{or} \quad \begin{array}{l} x + y + z = a \\ x + 2y + 3z = b \\ 3y + 5z = c \end{array} \quad \text{or} \quad \begin{array}{l} x + y + z = a \\ y + 2z = -a + b \\ 3y + 5z = c \end{array} \quad \text{or} \quad \begin{array}{l} x + y + z = a \\ y + 2z = -a + b \\ z = -3a + 3b - c \end{array}$$

Solving for x, y, z in terms of a, b, c yields $x = -a + 2b - c$, $y = 5a - 5b + 2c$, $z = -3a + 3b - c$. Thus

$$(a, b, c) = (-a + 2b - c)u_1 + (5a - 5b + 2c)u_2 + (-3a + 3b - c)u_3$$

or, equivalently,

$$[(a, b, c)] = [-a + 2b - c, 5a - 5b + 2c, -3a + 3b - c]^T$$

[*Remark*: This formula for (a, b, c) will be used repeatedly below.]

12.55 Find $S(u_1)$.

▌ $S(u_1) = S(1, 1, 0) = (1 + 2 - 0, 2 + 1 + 0, 5 - 1 + 0) = (3, 3, 4)$.

12.56 Write $S(u_1)$ as a linear combination of $u_1, u_2,$ and u_3.

▌ Use Problem 12.54, to get

$$S(u_1) = (3, 3, 4) = (-3 + 6 - 4)u_1 + (15 - 15 + 8)u_2 + (-9 + 9 - 4)u_3 = -u_1 + 8u_2 - 4u_3$$

12.57 Find $S(u_2)$.

▌ $S(u_2) = S(1, 2, 3) = (1 + 4 - 9, 2 + 2 + 3, 5 - 2 + 3) = (-4, 7, 6)$.

12.58 Write $S(u_2)$ as a linear combination of $u_1, u_2,$ and u_3.

▌ $S(u_2) = (-4, 7, 6) = (4 + 14 - 6)u_1 + (-20 - 35 + 12)u_2 + (12 + 21 - 6)u_3 = 12u_1 - 43u_2 + 27u_3$.

12.59 Find $S(u_3)$.

▌ $S(u_3) = S(1, 3, 5) = (1 + 6 - 15, 2 + 3 + 5, 5 - 3 + 5) = (-8, 10, 7)$.

12.60   Write $S(u_3)$ as a linear combination of u_1, u_2, and u_3.

∎ $S(u_3) = (-8, 10, 7) = (8 + 20 - 7)u_1 + (-40 - 50 + 14)u_2 + (24 + 30 - 7)u_3 = 21u_1 - 76u_2 + 47u_3$.

12.61   Find $[S]$, the matrix of S in the above basis B.

∎ Write the coordinates of $S(u_1)$, $S(u_2)$, and $S(u_3)$ as columns to get

$$[S] = \begin{pmatrix} -1 & 12 & 21 \\ 8 & -43 & -76 \\ -4 & 27 & 47 \end{pmatrix}$$

Problems 12.62–12.65 refer to the vector $v = (1, 1, 1)$ and the above linear map $S: \mathbf{R}^3 \to \mathbf{R}^3$ and basis B.

12.62   Find $[v]$.

∎ By Problem 12.54, $[v] = [(1, 1, 1)] = [-1 + 2 - 1, 5 - 5 + 2, -3 + 3 - 1]^T = [0, 2, -1]^T$.

12.63   Find $S(v)$.

∎ $S(1, 1, 1) = (1 + 2 - 3, 2 + 1 + 1, 5 - 1 + 1) = (0, 4, 5)$.

12.64   Find $[S(v)]$.

∎ By Problem 12.54, $[S(v)] = [(0, 4, 5)] = [0 + 8 - 5, 0 - 20 + 10, 0 + 12 - 5]^T = [3, -10, 7]^T$.

12.65   Verify Theorem 12.1 that $[S][v] = [S(v)]$.

∎
$$[S][v] = \begin{pmatrix} -1 & 12 & 21 \\ 8 & -43 & -76 \\ -4 & 27 & 47 \end{pmatrix} \begin{pmatrix} 0 \\ 2 \\ -1 \end{pmatrix} = \begin{pmatrix} 0 + 24 - 21 \\ 0 - 86 + 76 \\ 0 + 54 - 47 \end{pmatrix} = \begin{pmatrix} 3 \\ -10 \\ 7 \end{pmatrix} = [S(v)]$$

Problems 12.66–12.73 find the matrix representation of $T: \mathbf{R}^3 \to \mathbf{R}^3$ defined by $T(x, y, z) = (2y + z, x - 4y, 3x)$ relative to the basis $B = \{w_1 = (1, 1, 1), \ w_2 = (1, 1, 0), \ w_3 = (1, 0, 0).\}$

12.66   Find the coordinates of an arbitrary vector $(a, b, c) \in \mathbf{R}^3$ with respect to the above basis B.

∎ Write (a, b, c) as a linear combination of w_1, w_2, w_3 using unknown scalars x, y, and z:

$(a, b, c) = x(1, 1, 1) + y(1, 1, 0) + z(1, 0, 0) = (x + y + z, x + y, x)$ or $x + y + z = a, \ x + y = b, \ x = c$

Solve the system of x, y, and z in terms of a, b, and c to find $x = c, \ y = b - c, \ z = a - b$. Thus

$(a, b, c) = cw_1 + (b - c)w_2 + (a - b)w_3$ or, equivalently, $[(a, b, c)] = [c, b - c, a - b]^T$

[This formula for (a, b, c) will be used repeatedly below.]

12.67   Find $T(w_1)$.

∎ $T(w_1) = T(1, 1, 1) = (2 + 1, 1 - 4, 3) = (3, -3, 3)$.

12.68   Write $T(w_1)$ as a linear combination of w_1, w_2, w_3.

∎ By Problem 12.66, $T(w_1) = (3, -3, 3) = 3w_1 + (-3 - 3)w_2 + (3 + 3)w_3 = 3w_1 - 6w_2 + 6w_3$.

12.69   Find $T(w_2)$.

∎ $T(w_2) = T(1, 1, 0) = (2 + 0, 1 - 4, 3) = (2, -3, 3)$.

12.70   Write $T(w_2)$ as a linear combination of w_1, w_2, w_3.

∎ $T(w_2) = T(2, -3, 3) = 3w_1 + (-3 - 3)w_2 + (2 + 3)w_3 = 3w_1 - 6w_2 + 5w_3$.

12.71 Find $T(w_3)$.

▮ $T(w_3) = T(1, 0, 0) = (0 + 0, 1 - 0, 3) = (0, 1, 3)$.

12.72 Write $T(w_3)$ as a linear combination of w_1, w_2, w_3.

▮ $T(w_3) = (0, 1, 3) = 3w_1 + (1 - 3)w_2 + (0 - 1)w_3 = 3w_1 - 2w_2 - w_3$.

12.73 Find $[T]$, the matrix of T relative to the basis B.

▮ Write the coordinates of $T(w_1)$, $T(w_2)$, $T(w_3)$ as columns to obtain

$$[T] = \begin{pmatrix} 3 & 3 & 3 \\ -6 & -6 & -2 \\ 6 & 5 & -1 \end{pmatrix}$$

12.74 Write $T(v)$ as a linear combination of w_1, w_2, w_3 where $v = (a, b, c)$ is an arbitrary vector in $\mathbf{R}^3$.

▮ By Problem 12.66, $T(v) = T(a, b, c) = (2b + c, a - 4b, 3a) = 3aw_1 + (-2a - 4b)w_2 + (-a + 6b + c)w_3$ or, equivalently, $[T(v)] = [3a, -2a - 4b, -a + 6b + c]^T$.

12.75 Verify Theorem 12.1 that $[T][v] = [T(v)]$ where $v = (a, b, c)$.

▮ Using Problems 12.66 and 12.74,

$$[T][v] = \begin{pmatrix} 3 & 3 & 3 \\ -6 & -6 & -2 \\ 6 & 5 & -1 \end{pmatrix} \begin{pmatrix} c \\ b - c \\ a - b \end{pmatrix} = \begin{pmatrix} 3a \\ -2a - 4b \\ -a + 6b + c \end{pmatrix} = [T(v)]$$

12.3 MATRICES AND LINEAR MAPPING OPERATIONS

Theorem 12.2, whose proof appears in Problems 12.104–12.107, is used throughout this section.

> **Theorem 12.2:** Let $B = \{e_1, e_2, \ldots, e_n\}$ be a basis of V over K and $\mathcal{M}$ be the algebra of n-square matrices over K. Then the mapping $m: A(V) \to \mathcal{M}$ defined by $m(T) = [T]_B$ is an algebra isomorphism from $A(V)$ onto $\mathcal{M}$. That is, for any $S, T \in A(V)$ and any $k \in K$, we have
> (i) $m(T + S) = m(T) + m(S)$, that is, $[T + S] = [T] + [S]$.
> (ii) $m(kT) = km(T)$, that is, $[kT] = k[T]$.
> (iii) $m(S \circ T) = m(S)m(T)$, that is, $[S \circ T] = [S][T]$.
> (iv) The mapping m is one-to-one and onto.

Problems 12.76–12.90 illustrate Theorem 12.2 using the basis $B = \{u_1 = (1, 1), u_2 = (1, 2)\}$ of $\mathbf{R}^2$ and the linear maps S and T defined by $S(x, y) = (x + 2y, 4x)$ and $T(x, y) = (y, x + 3y)$.

12.76 Find the coordinates of an arbitrary vector $(a, b) \in \mathbf{R}^2$ with respect to the above basis B.

▮ Write (a, b) as a linear combination of u_1 and u_2 using unknowns x and y: $(a, b) = x(1, 1) + y(1, 2)$ or $x + y = a$, $x + 2y = b$. Solve for x and y to get $x = 2a - b$, $y = -a + b$. Thus

$$(a, b) = (2a - b)u_1 + (-a + b)u_2 \quad \text{or, equivalently,} \quad [(a, b)] = [2a - b, -a + b]^T$$

[This formula for (a, b) will be used repeatedly

12.77 Find $[S]$, the matrix representation of S in the basis B.

▮ $\begin{aligned} S(u_1) &= S(1, 1) = (3, 4) = 2u_1 + u_2 \\ S(u_2) &= S(1, 2) = (5, 4) = 6u_1 - u_2 \end{aligned}$ and thus $[S] = \begin{pmatrix} 2 & 6 \\ 1 & -1 \end{pmatrix}$

12.78 Find $[T]$, the matrix of T relative to the basis B.

▮ $\begin{aligned} T(u_1) &= T(1, 1) = (1, 4) = -2u_1 + 3u_2 \\ T(u_2) &= T(1, 2) = (2, 7) = -3u_1 + 5u_2 \end{aligned}$ and thus $[T] = \begin{pmatrix} -2 & -3 \\ 3 & 5 \end{pmatrix}$

12.79 Write $(S + T)(u_1)$ as a linear combination of u_1 and u_2.

▮ $(S + T)(u_1) = S(u_1) + T(u_1) = (2u_1 + u_2) + (-2u_1 + 3u_2) = 0u_1 + 4u_2$.

12.80 Write $(S + T)(u_2)$ as a linear combination of u_1 and u_2.

▮ $(S + T)(u_2) = S(u_2) + T(u_2) = (6u_1 - u_2) + (-3u_1 + 5u_2) = 3u_1 + 4u_2$.

12.81 Find $[S + T]$.

▮ Write the coordinates of $(S + T)(u_1)$ and $(S + T)(u_2)$ as columns:

$$[S + T] = \begin{pmatrix} 0 & 3 \\ 4 & 4 \end{pmatrix}$$

12.82 Verify Theorem 12.2(i): $[S] + [T] = [S + T]$.

▮ $$[S] + [T] = \begin{pmatrix} 2 & 6 \\ 1 & -1 \end{pmatrix} + \begin{pmatrix} -2 & -3 \\ 3 & 5 \end{pmatrix} = \begin{pmatrix} 0 & 3 \\ 4 & 4 \end{pmatrix} = [S + T]$$

12.83 Write $(3T)(u_1)$ as a linear combination of u_1 and u_2.

▮ $(3T)(u_1) = 3T(u_1) = 3(-2u_1 + 3u_2) = -6u_1 + 9u_2$.

12.84 Write $(3T)(u_2)$ as a linear combination of u_1 and u_2.

▮ $(3T)(u_2) = 3T(u_2) = 3(-3u_1 + 5u_2) = -9u_1 + 15u_2$.

12.85 Find $[3T]$.

▮ Write the coordinates of $(3T)(u_1)$ and $(3T)(u_2)$ as columns:

$$[3T] = \begin{pmatrix} -6 & -9 \\ 9 & 15 \end{pmatrix}$$

12.86 Verify that $3[T] = [3T]$.

▮ $$3[T] = 3\begin{pmatrix} -2 & -3 \\ 3 & 5 \end{pmatrix} = \begin{pmatrix} -6 & -9 \\ 9 & 15 \end{pmatrix} = [3T]$$

12.87 Write $(S \circ T)(u_1)$ as a linear combination of u_1 and u_2.

▮ $(S \circ T)(u_1) = S(T(u_1)) = S(T(1, 1)) = S(1, 4) = (9, 4) = (18 - 4)u_1 + (-9 + 4)u_2 = 14u_1 - 5u_2$.

12.88 Write $(S \circ T)(u_2)$ as a linear combination of u_1 and u_2.

▮ $(S \circ T)(u_2) = S(T(u_2)) = S(T(1, 2)) = S(2, 7) = (16, 8) = (32 - 8)u_1 + (-16 + 8)u_2 = 24u_1 - 8u_2$.

12.89 Find $[S \circ T]$.

▮ Write the coordinates of $(S \circ T)(u_1)$ and $(S \circ T)(u_2)$ as columns:

$$[S \circ T] = \begin{pmatrix} 14 & 24 \\ -5 & -8 \end{pmatrix}$$

12.90 Verify Theorem 12.2(iii): $[S][T] = [S \circ T]$.

▮ $$[S][T] = \begin{pmatrix} 2 & 6 \\ 1 & -1 \end{pmatrix}\begin{pmatrix} -2 & -3 \\ 3 & 5 \end{pmatrix} = \begin{pmatrix} -4 + 18 & -6 + 30 \\ -2 - 3 & -3 - 5 \end{pmatrix} = \begin{pmatrix} 14 & 24 \\ -5 & -8 \end{pmatrix} = [S \circ T]$$

Problems 12.91–12.103 illustrate Theorem 12.2 for $\dim V = 2$, i.e., for a basis $B = \{e_1, e_2\}$ of V and linear operators T and S on V defined by

$$T(e_1) = a_1 e_1 + a_2 e_2 \qquad S(e_1) = c_1 e_1 + c_2 e_2$$
$$T(e_2) = b_1 e_1 + b_2 e_2 \qquad S(e_2) = d_1 e_1 + d_2 e_2$$

12.91 Find $[T]$ and $[S]$.

▮ Write the coordinates of $T(e_1)$, $T(e_2)$ and of $S(e_1)$, $S(e_2)$ as columns to get

$$[T] = \begin{pmatrix} a_1 & b_1 \\ a_2 & b_2 \end{pmatrix} \quad \text{and} \quad [S] = \begin{pmatrix} c_1 & d_1 \\ c_2 & d_2 \end{pmatrix}$$

12.92 Write $(T + S)(e_1)$ as a linear combination of e_1 and e_2.

▮ $(T + S)(e_1) = T(e_1) + S(e_1) = a_1 e_1 + a_2 e_2 + c_1 e_1 + c_2 e_2 = (a_1 + c_1)e_1 + (a_2 + c_2)e_2$.

12.93 Write $(T + S)(e_2)$ as a linear combination of e_1 and e_2.

▮ $(T + S)(e_2) = T(e_2) + S(e_2) = b_1 e_1 + b_2 e_2 + d_1 e_1 + d_2 e_2 = (b_1 + d_1)e_1 + (b_2 + d_2)e_2$.

12.94 Find $[T + S]$.

▮ Write the coordinates of $(T + S)(e_1)$ and $(T + S)(e_2)$ as columns:

$$[T + S] = \begin{pmatrix} a_1 + c_1 & b_1 + d_1 \\ a_2 + c_2 & b_2 + c_2 \end{pmatrix}$$

12.95 Verify Theorem 12.2(i): $[T] + [S] = [T + S]$.

▮ $$[T] + [S] = \begin{pmatrix} a_1 & b_1 \\ a_2 & b_2 \end{pmatrix} + \begin{pmatrix} c_1 & d_1 \\ c_2 & d_2 \end{pmatrix} = \begin{pmatrix} a_1 + c_1 & b_1 + d_1 \\ a_2 + c_2 & b_2 + d_2 \end{pmatrix} = [T + S]$$

12.96 Write $(kT)(e_1)$ as a linear combination of e_1 and e_2, where $k \in K$.

▮ $(kT)(e_1) = kT(e_1) = k(a_1 e_1 + a_2 e_2) = ka_1 e_1 + ka_2 e_2$.

12.97 Write $(kT)(e_2)$ as a linear combination of e_1 and e_2.

▮ $(kT)(e_2) = kT(e_2) = k(b_1 e_1 + b_2 e_2) = kb_1 e_1 + kb_2 e_2$.

12.98 Find $[kT]$.

▮ Write the coordinates of $(kT)(e_1)$ and $(kT)(e_2)$ as columns:

$$[kT] = \begin{pmatrix} ka_1 & kb_1 \\ ka_2 & kb_2 \end{pmatrix}$$

12.99 Verify Theorem 12.2(ii): $k[T] = [kT]$.

▮ $$k[T] = k\begin{pmatrix} a_1 & b_1 \\ a_2 & b_2 \end{pmatrix} = \begin{pmatrix} ka_1 & kb_1 \\ ka_2 & kb_2 \end{pmatrix} = [kT]$$

12.100 Write $(S \circ T)(e_1)$ as a linear combination of e_1 and e_2.

▮ $(S \circ T)(e_1) = S(T(e_1)) = S(a_1 e_1 + a_2 e_2) = a_1 S(e_1) + a_2 S(e_2) = a_1(c_1 e_1 + c_2 e_2) + a_2(d_1 e_1 + d_2 e_2)$
$= (a_1 c_1 + a_2 d_1)e_1 + (a_1 c_2 + a_2 d_2)e_2$.

12.101 Write $(S \circ T)(e_2)$ as a linear combination of e_1 and e_2.

$\blacksquare$ $(S \circ T)(e_2) = S(T(e_2)) = S(b_1 e_1 + b_2 e_2) = b_1 S(e_1) + b_2 S(e_2) = b_1(c_1 e_1 + c_2 e_2) + b_2(d_1 e_1 + d_2 e_2)$
$= (b_1 c_1 + b_2 d_1)e_1 + (b_1 c_2 + b_2 d_2)e_2.$

12.102 Find $[S \circ T]$.

$\blacksquare$ Write the coordinates of $(S \circ T)(e_1)$ and $(S \circ T)(e_2)$ as columns:

$$[S \circ T] = \begin{pmatrix} a_1 c_1 + a_2 d_1 & b_1 c_1 + b_2 d_1 \\ a_1 c_2 + a_2 d_2 & b_1 c_2 + b_2 d_2 \end{pmatrix}$$

12.103 Verify Theorem 12.2(iii): $[S][T] = [S \circ T]$.

$\blacksquare$
$$[S][T] = \begin{pmatrix} c_1 & d_1 \\ c_2 & d_2 \end{pmatrix}\begin{pmatrix} a_1 & b_1 \\ a_2 & b_2 \end{pmatrix} = \begin{pmatrix} a_1 c_1 + a_2 d_1 & b_1 c_1 + b_2 d_1 \\ a_1 c_2 + a_2 d_2 & b_1 c_2 + b_2 d_2 \end{pmatrix} = [S \circ T]$$

12.104 Prove (i) of Theorem 12.2: $[T + S] = [T] + [S]$.

$\blacksquare$ Suppose, for $i = 1, \ldots, n$,

$$T(e_i) = \sum_{j=1}^{n} a_{ij} e_j \qquad \text{and} \qquad S(e_i) = \sum_{j=1}^{n} b_{ij} e_j$$

Let A and B be the matrices $A = (a_{ij})$ and $B = (b_{ij})$. Then $[T] = A^T$ and $[S] = B^T$. We have, for $i = 1, \ldots, n$,

$$(T + S)(e_i) = T(e_i) + S(e_i) = \sum_{j=1}^{n} (a_{ij} + b_{ij})e_j$$

Observe that $A + B$ is the matrix $(a_{ij} + b_{ij})$. Accordingly, $[T + S] = (A + B)^T = A^T + B^T = [T] + [S]$.

12.105 Prove (ii) of Theorem 12.2: $[kT] = k[T]$.

$\blacksquare$ For $i = 1, \ldots, n$,

$$(kT)(e_i) = kT(e_i) = k \sum_{j=1}^{n} a_{ij} e_j = \sum_{j=1}^{n} (ka_{ij})e_j$$

Observe that kA is the matrix (ka_{ij}). Accordingly, $[kT] = (kA)^T = kA^T = k[T]$.

12.106 Prove (iii) of Theorem 12.2: $[S \circ T] = [S][T]$.

$\blacksquare$ For $i = 1, \ldots, n$,

$$(S \circ T)(e_i) = S(T(e_i)) = S\left(\sum_{j=1}^{n} a_{ij} e_j\right) = \sum_{j=1}^{n} a_{ij} S(e_j) = \sum_{j=1}^{n} a_{ij}\left(\sum_{k=1}^{n} b_{jk} e_k\right) = \sum_{k=1}^{n}\left(\sum_{j=1}^{n} a_{ij} b_{jk}\right)e_k$$

Recall that AB is the matrix $AB = (c_{ik})$ where $c_{ik} = \sum_{j=1}^{n} a_{ij} b_{jk}$. Accordingly, $[S \circ T] = (AB)^T = B^T A^T = [S][T]$.

12.107 Prove (iv) of Theorem 12.2: The mapping $m: A(V) \rightarrow \mathcal{M}$ defined by $m(T) = [T]$ is one-to-one and onto $\mathcal{M}$.

$\blacksquare$ The mapping is one-to-one since a linear mapping is completely determined by its values on a basis. The mapping is onto since each matrix $M \in \mathcal{M}$ is the image of the linear operator

$$F(e_i) = \sum_{j=1}^{n} m_{ij} e_j \qquad i = 1, \ldots, n$$

where (m_{ij}) is the transpose of the matrix M.

12.4 MATRICES AND LINEAR MAPPINGS

The preceding section of the chapter only considered linear mappings from a vector space into itself. This section considers the general case of linear mappings from one space into another.

12.108 Let V and U be vector spaces over the same field K. Define the matrix representation of a linear mapping $F: V \to U$.

▮ Let $\{e_1, \ldots, e_m\}$ and $\{f_1, \ldots, f_n\}$ be arbitrary but fixed bases of V and U, respectively. Then the vectors $F(e_1), \ldots, F(e_m)$ belong to U and so each is a linear combination of the f_i, say

$$F(e_1) = a_{11}f_1 + a_{12}f_2 + \cdots + a_{1n}f_n$$
$$F(e_2) = a_{21}f_1 + a_{22}f_2 + \cdots + a_{2n}f_n$$
$$\cdots\cdots\cdots\cdots\cdots\cdots\cdots\cdots\cdots\cdots\cdots\cdots$$
$$F(e_m) = a_{m1}f_1 + a_{m2}f_2 + \cdots + a_{mn}f_n$$

The transpose of the above matrix of coefficients, denoted by $[F]_e^f$ or simply $[F]$, is called the *matrix representation* of F relative to the bases $\{e_i\}$ and $\{f_i\}$, or the matrix of F in the bases $\{e_i\}$ and $\{f_i\}$. That is,

$$[F] = \begin{pmatrix} a_{11} & a_{21} & \cdots & a_{m1} \\ a_{12} & a_{22} & \cdots & a_{m2} \\ \cdots\cdots\cdots\cdots\cdots\cdots \\ a_{1n} & a_{2n} & \cdots & a_{mn} \end{pmatrix}$$

[Observe that $[F]$ is an $n \times m$ matrix where $\dim V = m$ and $\dim U = n$.]

Theorem 12.3, whose proof appears in Problem 12.138, is used below.

Theorem 12.3: For any vector $v \in V$, $[F]_e^f [v]_e = [F(v)]_f$.
[That is, multiplying the coordinate vector of v in the basis $\{e_i\}$ by the matrix $[F]_e^f$ we obtain the coordinate vector of $F(v)$ in the basis $\{f_i\}$.]

Problems 12.109–12.113 find the matrix representation of the linear map $F: \mathbf{R}^3 \to \mathbf{R}^2$ defined by $F(x, y, z) = (2x + 3y - z, 4x - y + 2z)$ relative to the following basis of $\mathbf{R}^3$ and $\mathbf{R}^2$, respectively:

$$B_1 = \{u_1 = (1, 1, 0), u_2 = (1, 2, 3), u_3 = (1, 3, 5)\} \quad \text{and} \quad B_2 = \{v_1 = (1, 2), v_2 = (2, 3)\}$$

12.109 Find the coordinates of an arbitrary vector $(a, b) \in \mathbf{R}^2$ with respect to the basis vectors v_1 and v_2.

▮
$$\binom{a}{b} = x\binom{1}{2} + y\binom{2}{3} \quad \text{or} \quad \begin{array}{l} x + 2y = a \\ 2x + 3y = b \end{array}$$

The solution is $x = -3a + 2b$, $y = 2a - b$. Thus

$$(a, b) = (-3a + 2b)v_1 + (2a - b)v_2 \quad \text{or} \quad [(a, b)]_{B_2} = [-3a + 2b, 2a - b]^T$$

12.110 Write $F(u_1)$, the image of the first basis vector of $\mathbf{R}^3$, as a linear combination of the basis vectors v_1 and v_2 of $\mathbf{R}^2$.

▮ $F(u_1) = F(1, 1, 0) = (2 + 3 + 0, 4 - 1 + 0) = (5, 3) = (-15 + 6)v_1 + (10 - 3)v_2 = -9v_1 + 7v_2$.

12.111 Write $F(u_2)$ as a linear combination of v_1 and v_2.

▮ $F(u_2) = F(1, 2, 3) = (2 + 6 - 3, 4 - 2 + 6) = (5, 8) = (-15 + 16)v_1 + (10 - 8)v_2 = v_1 + 2v_2$.

12.112 Write $F(u_3)$ as a linear combination of v_1 and v_2.

▮ $F(u_3) = F(1, 3, 5) = (2 + 9 - 5, 4 - 3 + 10) = (6, 11) = (-18 + 22)v_1 + (12 - 11)v_2 = 4v_1 + v_2$.

12.113 Find $[F]$, the matrix representation of F relative to the bases B_1 and B_2.

▮ Write the coordinates of $F(u_1)$, $F(u_2)$, $F(u_3)$ in the basis $\{v_1, v_2\}$ as columns:

$$[F] = \begin{pmatrix} -9 & 1 & 4 \\ 7 & 2 & 1 \end{pmatrix}$$

Problems 12.114–12.117 refer to the vector $v = (2, 5, -3)$ and the above linear map $F: \mathbf{R}^3 \to \mathbf{R}^2$ and bases B_1 and B_2.

12.114 Find $[v]_{B_1}$, the coordinate vector of v in the basis B_1.

▮ By Problem 12.54, $[(a, b, c)]_{B_1} = [-a + 2b - c, 5a - 5b + 2c, -3a + 3b - c]^T$, hence
$$[v]_{B_1} = [-2 + 10 + 3, 10 - 25 - 6, -6 + 15 + 3]^T = [11, -21, 12]^T$$

12.115 Find $F(v)$.

▮ $F(v) = F(2, 5, -3) = (4 + 15 + 3, 8 - 5 - 6) = (22, -3)$.

12.116 Find $[F(v)]_{B_2}$, the coordinate vector of $F(v)$ in the basis B_2.

▮ Using Problem 12.109, $[F(v)]_{B_2} = [(22, -3)]_{B_2} = [-66 - 6, 44 + 3]^T = [-72, 47]^T$.

12.117 Verify Theorem 12.3: $[F][v]_{B_1} = [F(v)]_{B_2}$.

▮
$$[F][v]_{B_1} = \begin{pmatrix} -9 & 1 & 4 \\ 7 & 2 & 1 \end{pmatrix} \begin{pmatrix} 11 \\ -21 \\ 12 \end{pmatrix} = \begin{pmatrix} -99 - 21 + 48 \\ 77 - 42 + 12 \end{pmatrix} = \begin{pmatrix} -72 \\ 47 \end{pmatrix} = [F(v)]_{B_2}$$

Problems 12.118–12.121 find the matrix representation of the linear map $F: \mathbf{R}^3 \to \mathbf{R}^2$ defined by $F(x, y, z) = (3x + 2y - 4z, x - 5y + 3z)$ relative to the following bases of $\mathbf{R}^3$ and $\mathbf{R}^2$, respectively:

$$B_1 = \{u_1 = (1, 1, 1), u_2 = (1, 1, 0), u_3 = (1, 0, 0)\} \quad \text{and} \quad B_2 = \{v_1 = (1, 3), v_2 = (2, 5)\}$$

[Recall that $(a, b) = (-5a + 2b)v_1 + (3a - b)v_2$ by Problem 12.21.]

12.118 Write $F(u_1)$ as a linear combination of the basis vectors v_1 and v_2.

▮ $F(u_1) = F(1, 1, 1) = (3 + 2 - 4, 1 - 5 + 3) = (1, -1) = (-5 - 2)v_1 + (3 + 1)v_2 = -7v_1 + 4v_2$.

12.119 Write $F(u_2)$ as a linear combination of v_1 and v_2.

▮ $F(u_2) = F(1, 1, 0) = (5, -4) = (-25 - 8)v_1 + (15 + 4)v_2 = -33v_1 + 19v_2$.

12.120 Write $F(u_3)$ as a linear combination of v_1 and v_2.

▮ $F(u_3) = F(1, 0, 0) = (3, 1) = (-15 + 2)v_1 + (9 - 1)v_2 = -13v_1 + 8v_2$.

12.121 Find $[F]$, the matrix representation of F in the bases B_1 and B_2.

▮ Write the coordinates of $F(u_1)$, $F(u_2)$, $F(u_3)$ as columns:
$$[F] = \begin{pmatrix} -7 & -33 & -13 \\ 4 & 19 & 8 \end{pmatrix}$$

Problems 12.122–12.124 refer to an arbitrary vector $v = (x, y, z) \in \mathbf{R}^3$ and the above linear map F and bases B_1 and B_2.

12.122 Find $[v]_{B_1}$, the coordinate vector of v in the basis B_1.

▮ By Problem 12.66, $[(a, b, c)]_{B_1} = [c, b - c, a - b]^T$ and so $[v]_{B_1} = [x, y - z, x - y]^T$.

12.123 Find $[F(v)]_{B_2}$, the coordinate vector of $F(v)$ in the basis B_2.

❚ $F(v) = (3x + 2y - 4z, x - 5y + 3z) = (-13x - 20y + 26z)v_1 + (8x + 11y - 15z)$ and so $[F(v)]_{B_2} = [-13x - 20y + 26z, 8x + 11y - 15z]^T$.

12.124 Verify Theorem 12.3: $[F][v]_{B_1} = [F(v)]_{B_2}$.

❚
$$[F][v]_{B_1} = \begin{pmatrix} -7 & -33 & -13 \\ 4 & 19 & 8 \end{pmatrix} \begin{pmatrix} z \\ y - z \\ x - y \end{pmatrix} = \begin{pmatrix} -13x - 20y + 26z \\ 8x + 11y - 15z \end{pmatrix} = [F(v)]_{B_2}$$

12.125 Let $F: K^n \to K^m$ be the linear mapping defined by
$$F(x_1, x_2, \ldots, x_n) = (a_{11}x_{11} + \cdots + a_{1n}x_n, a_{21}x_1 + \cdots + a_{2n}x_n, \ldots, a_{m1}x_1 + \cdots + a_{mn}x_n)$$
Show that the matrix representation of F relative to the usual bases of K^n and of K^m is given by
$$[F] = \begin{pmatrix} a_{11} & a_{12} & \cdots & a_{1n} \\ a_{21} & a_{22} & \cdots & a_{2n} \\ \cdots\cdots\cdots\cdots\cdots\cdots \\ a_{m1} & a_{m2} & \cdots & a_{mn} \end{pmatrix}$$
That is, the rows of $[F]$ are obtained from the coefficients of the x_i in the components of $F(x_1, \ldots, x_n)$, respectively.

❚ We have
$$\begin{array}{l} F(1, 0, \ldots, 0) = (a_{11}, a_{21}, \ldots, a_{m1}) \\ F(0, 1, \ldots, 0) = (a_{12}, a_{22}, \ldots, a_{m2}) \\ \cdots\cdots\cdots\cdots\cdots\cdots\cdots\cdots\cdots\cdots\cdots \\ F(0, 0, \ldots, 1) = (a_{1n}, a_{2n}, \ldots, a_{mn}) \end{array} \quad \text{and} \quad [F] = \begin{pmatrix} a_{11} & a_{12} & \cdots & a_{1n} \\ a_{21} & a_{22} & \cdots & a_{2n} \\ \cdots\cdots\cdots\cdots\cdots\cdots \\ a_{m1} & a_{m2} & \cdots & a_{mn} \end{pmatrix}$$

12.126 Let $F: \mathbf{R}^2 \to \mathbf{R}^3$ be defined by $F(x, y) = (3x - y, 2x + 4y, 5x - 6y)$. Find the matrix representation $[F]$ of F relative to the usual bases of $\mathbf{R}^n$.

❚ By Problem 12.125, we need only look at the coefficients of the unknowns in $F(x, y, \ldots)$. Thus
$$[F] = \begin{pmatrix} 3 & -1 \\ 2 & 4 \\ 5 & -6 \end{pmatrix}$$

12.127 Let $G: \mathbf{R}^4 \to \mathbf{R}^2$ be defined by $G(x, y, s, t) = (3x - 4y + 2s - 5t, 5x + 7y - s - 2t)$. Find $[G]$ relative to the usual bases of $\mathbf{R}^n$.

❚ By Problem 12.125, $[G] = \begin{pmatrix} 3 & -4 & 2 & -5 \\ 5 & 7 & -1 & -2 \end{pmatrix}$

12.128 Let $H: \mathbf{R}^3 \to \mathbf{R}^4$ be defined by $H(x, y, z) = (2x + 3y - 8z, x + y + z, 4x - 5z, 6y)$. Find $[H]$ relative to the usual bases of $\mathbf{R}^n$.

❚ By Problem 12.125, $[H] = \begin{pmatrix} 2 & 3 & -8 \\ 1 & 1 & 1 \\ 4 & 0 & -5 \\ 0 & 6 & 0 \end{pmatrix}$

12.129 Let $A = \begin{pmatrix} 2 & 5 & -3 \\ 1 & -4 & 7 \end{pmatrix}$. Recall that A determines a linear mapping $F: \mathbf{R}^3 \to \mathbf{R}^2$ defined by $F(v) = Av$ where v is written as a column vector. Show that the matrix representation of F relative to the usual basis of $\mathbf{R}^3$ and of $\mathbf{R}^2$ is the matrix A itself: that is, $[F] = A$.

❚ We have
$$F(1, 0, 0) = \begin{pmatrix} 2 & 5 & -3 \\ 1 & -4 & 7 \end{pmatrix} \begin{pmatrix} 1 \\ 0 \\ 0 \end{pmatrix} = \begin{pmatrix} 2 \\ 1 \end{pmatrix} = 2e_1 + 1e_2$$

$$F(0, 1, 0) = \begin{pmatrix} 2 & 5 & -3 \\ 1 & -4 & 7 \end{pmatrix}\begin{pmatrix} 0 \\ 1 \\ 0 \end{pmatrix} = \begin{pmatrix} 5 \\ -4 \end{pmatrix} = 5e_1 - 4e_2$$

$$F(0, 0, 1) = \begin{pmatrix} 2 & 5 & -3 \\ 1 & -4 & 7 \end{pmatrix}\begin{pmatrix} 0 \\ 0 \\ 1 \end{pmatrix} = \begin{pmatrix} -3 \\ 7 \end{pmatrix} = -3e_1 + 7e_2$$

from which $[F] = \begin{pmatrix} 2 & 5 & -3 \\ 1 & -4 & 7 \end{pmatrix} = A$.

Problems 12.130–12.133 find the matrix representation of the linear mapping $F: \mathbf{R}^3 \to \mathbf{R}^2$ defined in Problem 12.129 relative to the bases $B_1 = \{u_1, u_2, u_3\}$ and $B_2 = \{v_1, v_2\}$ in Problems 12.118–12.121.

12.130 Write $F(u_1)$ as a linear combination of v_1 and v_2.

▌ By Problem 12.21, $(a, b) = (-5a + 2b)v_1 + (3a - b)v_2$; hence

$$F(u_1) = \begin{pmatrix} 2 & 5 & -3 \\ 1 & -4 & 7 \end{pmatrix}\begin{pmatrix} 1 \\ 1 \\ 1 \end{pmatrix} = \begin{pmatrix} 4 \\ 4 \end{pmatrix} = (-20 + 8)v_1 + (12 - 4)v_2 = -12v_1 + 8v_2$$

12.131 Write $F(u_2)$ as a linear combination of v_1 and v_2.

▌ $F(u_2) = \begin{pmatrix} 2 & 5 & -3 \\ 1 & -4 & 7 \end{pmatrix}\begin{pmatrix} 1 \\ 1 \\ 0 \end{pmatrix} = \begin{pmatrix} 7 \\ -3 \end{pmatrix} = (-35 - 6)v_1 + (21 + 3)v_2 = -41v_1 + 24v_2.$

12.132 Write $F(u_3)$ as a linear combination of v_1 and v_2.

▌ $F(u_3) = \begin{pmatrix} 2 & 5 & -3 \\ 1 & -4 & 7 \end{pmatrix}\begin{pmatrix} 1 \\ 0 \\ 0 \end{pmatrix} = \begin{pmatrix} 2 \\ 1 \end{pmatrix} = (-10 + 2)v_1 + (6 - 1)v_2 = -8v_1 + 5v_2.$

12.133 Find $[F]$ with respect to the bases B_1 and B_2.

▌ Write the coordinates of $F(u_1)$, $F(u_2)$, $F(u_3)$ as columns:

$$[F] = \begin{pmatrix} -12 & -41 & -8 \\ 8 & 24 & 5 \end{pmatrix}$$

Problems 12.134–12.137 refer to the linear operator $T: \mathbf{R}^2 \to \mathbf{R}^2$ defined by $T(x, y) = (2x - 3y, x + 4y)$ and the following bases of $\mathbf{R}^2$: $E = \{e_1 = (1, 0), e_2 = (0, 1)\}$ and $B = \{v_1 = (1, 3), v_2 = (2, 5)\}$. [We can view T as a linear mapping from one space into another, each having its own basis.]

12.134 Find $[T]_E^B$, the matrix representation of T relative to the bases E and B.

▌ By Problem 1.21, $(a, b) = (-5a + 2b)v_1 + (3a - b)v_2$; hence

$$\begin{aligned} T(e_1) = T(1, 0) = (2, 1) &= -8v_1 + 5v_2 \\ T(e_2) = T(0, 1) = (-3, 4) &= 23v_1 - 13v_2 \end{aligned} \quad \text{and so} \quad [T]_E^B = \begin{pmatrix} -8 & 23 \\ 5 & -13 \end{pmatrix}$$

12.135 Find $[T]_B^E$, the matrix representation of T relative to the bases B and E.

▌ $$\begin{aligned} T(v_1) = T(1, 3) = (-7, 13) &= -7e_1 + 13e_2 \\ T(v_2) = T(2, 5) = (-11, 22) &= -11e_1 + 22e_2 \end{aligned} \quad \text{and so} \quad [T]_B^E = \begin{pmatrix} -7 & -11 \\ 13 & 22 \end{pmatrix}$$

12.136 Find $[T]_E^E$.

▌ $$\begin{aligned} T(e_1) = T(1, 0) = (2, 1) &= 2e_1 + e_2 \\ T(e_2) = T(0, 1) = (-3, 4) &= -3e_1 + 4e_2 \end{aligned} \quad \text{and so} \quad [T]_E^E = \begin{pmatrix} 2 & -3 \\ 1 & 4 \end{pmatrix}$$

12.137 Find $[T]_B^B$.

▌
$$T(v_1) = T(1, 3) = (-7, 13) = 61v_1 - 34v_2$$
$$T(v_2) = T(2, 5) = (-11, 22) = 99v_1 - 55v_2 \quad \text{and so} \quad [T]_B^B = \begin{pmatrix} 61 & 99 \\ -34 & -55 \end{pmatrix}$$

Remark: $[T]_E^E$ and $[T]_B^B$ are simply the matrix representations of T as a linear operator as discussed in Section 12.1.

12.138 Prove Theorem 12.3: Let $F: V \to U$ be linear. Then $[F][v]_e = [F(v)]_f$.

▌ Suppose $\{e_1, \ldots, e_m\}$ is a basis of V and $\{f_1, \ldots, f_n\}$ is a basis of U; and suppose, for $i = 1, \ldots, m$,

$$F(e_i) = a_{i1}f_1 + a_{i2}f_2 + \cdots + a_{in}f_n = \sum_{j=1}^{n} a_{ij}f_j$$

Then $[F]$ is the $n \times m$ matrix whose jth row is

$$(a_{1j}, a_{2j}, \ldots, a_{mj}) \tag{1}$$

Now suppose $v = k_1 e_1 + \cdots + k_m e_m = \sum_{i=1}^{m} k_i e_i$. Writing a column vector as the transpose of a row vector,

$$[v]_e = [k_1, k_2, \ldots, k_m]^T \tag{2}$$

Furthermore, using the linearity of F,

$$F(v) = F\left(\sum_{i=1}^{m} k_i e_i\right) = \sum_{i=1}^{m} k_i F(e_i) = \sum_{i=1}^{m} k_i\left(\sum_{j=1}^{n} a_{ij}f_j\right) = \sum_{j=1}^{n}\left(\sum_{i=1}^{m} a_{ij}k_i\right)f_j = \sum_{j=1}^{n}(a_{1j}k_1 + a_{2j}k_2 + \cdots + a_{mj}k_m)f_j$$

Thus $[F(v)]_f$ is the column vector whose jth entry is

$$a_{1j}k_1 + a_{2j}k_2 + \cdots + a_{mj}k_m \tag{3}$$

On the other hand, the jth entry of $[F][v]_e$ is obtained by multiplying *(1)* by *(2)*. But the product of *(1)* and *(2)* is *(3)*; hence $[F][v]_e$ and $[F(v)]_f$ have the same entries. Thus $[F][v]_e = [F(v)]_f$. [*Remark:* Observe the similarity between the proof of Theorem 12.3 and the proof of Theorem 12.1 in Problem 11.50.]

Theorem 12.4: Let $F: V \to U$ be linear. Then there exists a basis of V and a basis of U such that the matrix representation A of F has the form $A = \begin{pmatrix} I & 0 \\ 0 & 0 \end{pmatrix}$ where I is the r-square identity matrix and r is the rank of F.

12.139 Prove Theorem 12.4.

▌ Suppose $\dim V = m$ and $\dim U = n$. Let W be the kernel of F and U' the image of F. We are given that rank $F = r$; hence the dimension of the kernel of F is $m - r$. Let $\{w_1, \ldots, w_{m-r}\}$ be a basis of the kernel of F and extend this to a basis of V: $\{v_1, \ldots, v_r, w_1, \ldots, w_{m-r}\}$.

Set $u_1 = F(v_1), u_2 = F(v_2), \ldots, u_r = F(v_r)$. We note that $\{u_1, \ldots, u_r\}$ is a basis of U', the image of F. Extend this to a basis $\{u_1, \ldots, u_r, u_{r+1}, \ldots, u_n\}$ *of* U. Observe that

$$
\begin{aligned}
F(v_1) &= u_1 = 1u_1 + 0u_2 + \cdots + 0u_r + 0u_{r+1} + \cdots + 0u_n \\
F(v_2) &= u_2 = 0u_1 + 1u_2 + \cdots + 0u_r + 0u_{r+1} + \cdots + 0u_n \\
&\cdots\cdots\cdots\cdots\cdots\cdots\cdots\cdots\cdots\cdots\cdots\cdots\cdots\cdots\cdots \\
F(v_r) &= u_r = 0u_1 + 0u_2 + \cdots + 1u_r + 0u_{r+1} + \cdots + 0u_n \\
F(w_1) &= 0 = 0u_1 + 0u_2 + \cdots + 0u_r + 0u_{r+1} + \cdots + 0u_n \\
&\cdots\cdots\cdots\cdots\cdots\cdots\cdots\cdots\cdots\cdots\cdots\cdots\cdots\cdots\cdots \\
F(w_{m-r}) &= 0 = 0u_1 + 0u_2 + \cdots + 0u_r + 0u_{r+1} + \cdots + 0u_n
\end{aligned}
$$

Thus the matrix of F in the above bases has the required form.

12.140 Suppose $\dim V = m$ and $\dim U = n$. Recall that the space $\mathrm{Hom}(V, U)$ of all linear maps from V into U is a vector space of dimension mn. Describe the connection between $\mathrm{Hom}(V, U)$ and the vector space $\mathcal{M}$ of all $n \times m$ matrices over the base field K.

❙ The correspondence between $\mathrm{Hom}(V, U)$ and $\mathcal{M}$ is given by the Theorem 12.5.

Theorem 12.5: Let $\{e_i\}$ be a basis of V and let $\{f_i\}$ be a basis of U. Then the map $m: \mathrm{Hom}(V, U) \rightarrow \mathcal{M}$ defined by $m(F) = [F]_e^f$ is a vector space isomorphism. That is, for any $F, G \in \mathrm{Hom}(V, U)$ and any $k \in K$, we have
  (i) $m(F + G) = m(F) + m(G)$, that is, $[F + G] = [F] + [G]$.
  (ii) $m(kF) = km(F)$, that is, $[kF] = k[F]$.
  (iii)   The mapping m is one-to-one and onto $\mathcal{M}$.

[The proof of this theorem is essentially the same as the proofs of Parts (i), (ii), and (iv) of Theorem 12.2 which appear in Problems 12.104, 12.105, and 12.107; and hence will be omitted.]

CHAPTER 13
Change of Basis, Similarity

13.1 CHANGE-OF-BASIS (TRANSITION) MATRIX

13.1 Define the change-of-basis matrix for a vector space V.

▌ Let $\{e_1, \ldots, e_n\}$ be a basis of V and $\{f_1, \ldots, f_n\}$ be another basis. Suppose

$$
\begin{aligned}
f_1 &= a_{11}e_1 + a_{12}e_2 + \cdots + a_{1n}e_n \\
f_2 &= a_{21}e_1 + a_{22}e_2 + \cdots + a_{2n}e_n \\
&\quad\cdots\cdots\cdots\cdots\cdots\cdots\cdots\cdots\cdots \\
f_n &= a_{n1}e_1 + a_{n2}e_2 + \cdots + a_{nn}e_n
\end{aligned}
$$

Then the transpose P of the above matrix of coefficients is called the *change-of-basis matrix* or the *transition matrix* from the "old" basis $\{e_i\}$ to the "new" basis $\{f_i\}$. In other words, the columns of P are, respectively, the coordinates of the vectors $f_1, f_2, \ldots, f_n$ with respect to the "old" basis $\{e_i\}$.

Theorems 13.1 and 13.2, whose proofs appear in Problems 13.43 and 13.44, will be used below.

Theorem 13.1: Let P be the change-of-basis matrix from a basis $\{e_i\}$ to a basis $\{f_i\}$ and Q be the change-of-basis matrix from the basis $\{f_i\}$ back to the basis $\{e_i\}$. Then P is invertible and $Q = P^{-1}$.

Theorem 13.2: Let P be the change-of-basis matrix from a basis $\{e_i\}$ to a basis $\{f_i\}$ in a vector space V. Then, for any vector $v \in V$: (i) $P[v]_f = [v]_e$ and (ii) $P^{-1}[v_e] = [v]_f$.

Remark: Although P is called the transition matrix from the old basis $\{e_i\}$ to the new basis $\{f_i\}$, its effect is to transform the coordinates of a vector in the new basis $\{f_i\}$ back to the coordinates in the old basis $\{e_i\}$.

Problems 13.2–13.12 refer to the following bases of $\mathbf{R}^2$: $S_1 = \{u_1 = (1, -2), u_2 = (3, -4)\}$ and $S_2 = \{v_1 = (1, 3), v_2 = (3, 8)\}$. In particular, Problems 13.2–13.5 find the change-of-basis matrix P from S_1 to S_2 and Problems 13.6–13.9 find the change-of-basis matrix Q from S_2 back to S_1.

13.2 Find the coordinates of an arbitrary vector (a, b) in $\mathbf{R}^2$ with respect to the basis $S_1 = \{u_1, u_2\}$.

▌ We have

$$
\binom{a}{b} = x\binom{1}{-2} + y\binom{3}{-4} \quad \text{or} \quad \begin{aligned} x + 3y &= a \\ -2x - 4y &= b \end{aligned} \quad \text{or} \quad \begin{aligned} x + 3y &= a \\ 2y &= 2a + b \end{aligned}
$$

Solve for x and y in terms of a and b to get $x = -2a - \frac{3}{2}b$, $y = a + \frac{1}{2}b$. Thus

$$
(a, b) = (-2a - \tfrac{3}{2}b)u_1 + (a + \tfrac{1}{2}b)u_2 \quad \text{or} \quad [(a, b)]_{S_1} = [-2a - \tfrac{3}{2}b, \, a + \tfrac{1}{2}b]^T
$$

13.3 Write v_1, the first basis vector of S_2, as a linear combination of the basis vectors u_1 and u_2 of S_1.

▌ Use Problem 13.2 to get $v_1 = (1, 3) = (-2 - \frac{9}{2})u_1 + (1 + \frac{3}{2})u_2 = (-\frac{13}{2})u_1 + (\frac{5}{2})u_2$.

13.4 Write v_2 as a linear combination of u_1 and u_2.

▌ $v_3 = (3, 8) = (-6 - 12)u_1 + (3 + 4)u_2 = -18u_1 + 7u_2$.

13.5 Find the change-of-basis matrix P from S_1 to S_2.

❚ Write the coordinates of v_1 and v_2 in the basis S_1 as columns:

$$P = \begin{pmatrix} -\frac{13}{2} & -18 \\ \frac{5}{2} & 7 \end{pmatrix}$$

13.6 Find the coordinates of an arbitrary vector $(a, b) \in \mathbf{R}^2$ with respect to the basis $S_2 = \{v_1, v_2\}$.

❚ We have

$$\begin{pmatrix} a \\ b \end{pmatrix} = x\begin{pmatrix} 1 \\ 3 \end{pmatrix} + y\begin{pmatrix} 3 \\ 8 \end{pmatrix} \qquad \text{or} \qquad \begin{matrix} x + 3y = a \\ 3x + 8y = b \end{matrix}$$

Solve for x and y to get $x = -8a + 3b$, $y = 3a - b$. Thus

$$(a, b) = (-8a + 3b)v_1 + (3a - b)v_2 \qquad \text{or} \qquad [(a, b)]S_2 = [-8a + 3b, 3a - b]^T$$

13.7 Write u_1, the first basis vector of S_1, as a linear combination of the basis vectors v_1 and v_2 of S_2.

❚ Use Problem 13.6 to get $u_1 = (1, -2) = (-8 - 6)v_1 + (3 + 2)v_2 = -14v_1 + 5v_2$.

13.8 Write u_2 as a linear combination of v_1 and v_2.

❚ $u_2 = (3, -4) = (-24 - 12)v_1 + (9 + 4)v_2 = -36v_1 + 13v_2$.

13.9 Find the change-of-basis matrix Q from S_2 back to S_1.

❚ Write the coordinates of u_1 and u_2 in the basis S_2 as columns:

$$Q = \begin{pmatrix} -14 & -36 \\ 5 & 13 \end{pmatrix}$$

13.10 Verify that $Q = P^{-1}$ [Theorem 13.1].

❚
$$QP = \begin{pmatrix} -14 & -36 \\ 5 & 13 \end{pmatrix}\begin{pmatrix} -\frac{13}{2} & -18 \\ \frac{5}{2} & 7 \end{pmatrix} = \begin{pmatrix} 1 & 0 \\ 0 & 1 \end{pmatrix} = I$$

13.11 Show that $P[v]_{S_2} = [v]_{S_1}$ for any vector $v = (a, b)$ [Theorem 13.2(i)].

❚ Using Problems 13.2, 13.5, and 13.6,

$$P[v]_{S_1} = \begin{pmatrix} -\frac{13}{2} & -18 \\ \frac{5}{2} & 7 \end{pmatrix}\begin{pmatrix} -8a + 3b \\ 3a - b \end{pmatrix} = \begin{pmatrix} -2a - \frac{3}{2}b \\ a + \frac{1}{2}b \end{pmatrix} = [v]_{S_1}$$

13.12 Show that $P^{-1}[v]_{S_1} = [v]_{S_2}$ for any vector $v = (a, b)$ [Theorem 13.2(ii)].

❚
$$P^{-1}[v]_{S_1} = Q[v]_{S_1} = \begin{pmatrix} -14 & -36 \\ 5 & 13 \end{pmatrix}\begin{pmatrix} -2a + \frac{3}{2}b \\ a + \frac{1}{2}b \end{pmatrix} = \begin{pmatrix} -8a + 3b \\ 3a - b \end{pmatrix} = [v]_{S_2}$$

Problems 13.13–13.25 refer to the following bases of $\mathbf{R}^3$:

$$S = \{u_1 = (1, 2, 0), \ u_2 = (1, 3, 2), \ u_3 = (0, 1, 3)\} \quad \text{and} \quad S' = \{v_1 = (1, 2, 1), \ v_2 = (0, 1, 2), \ v_3 = (1, 4, 6)\}$$

In particular, Problems 13.27–13.17 find the change-of-basis matrix P from S to S', and Problems 13.18–13.22 find the change-of-basis matrix Q from B' to S.

13.13 Find the coordinates of an arbitrary vector $(a, b, c) \in \mathbf{R}^3$ with respect to the basis $S = \{u_1, u_2, u_3\}$.

❚ We have

$$\begin{pmatrix} a \\ b \\ c \end{pmatrix} = x\begin{pmatrix} 1 \\ 2 \\ 0 \end{pmatrix} + y\begin{pmatrix} 1 \\ 3 \\ 2 \end{pmatrix} + z\begin{pmatrix} 0 \\ 1 \\ 3 \end{pmatrix} \qquad \text{or} \qquad \begin{matrix} x + \ y \quad\ \ = a \\ 2x + 3y + \ z = b \\ 2y + 3z = c \end{matrix}$$

Solve for x, y, z to get $x = 7a - 3b + c$, $y = -6a + 3b - c$, $z = 4a - 2b + c$. Thus

$$(a, b, c) = (7a - 3b + c)u_1 + (-6a + 3b - c)u_2 + (4a - 2b + c)u_3$$

or $[(a, b, c)]_S = [7a - 3b + c, -6a + 3b - c, 4a - 2b + c]^T$.

13.14 Write v_1, the first basis vector in S', as a linear combination of the basis vectors u_1, u_2, u_3 of S.

▌ Use Problem 13.13 to get $v_1 = (1, 2, 1) = (7 - 6 + 1)u_1 + (-6 + 6 - 1)u_2 + (4 - 4 + 1)u_3 = 2u_1 - u_2 + u_3$.

13.15 Write v_2 as a linear combination of u_1, u_2, and u_3.

▌ $v_2 = (0, 1, 2) = (0 - 3 + 2)u_1 + (0 + 3 - 2)u_2 + (0 - 2 + 2)u_3 = -u_1 + u_2 + 0u_3$.

13.16 Write v_3 as a linear combination of u_1, u_2, and u_3.

▌ $v_3 = (1, 4, 6) = (7 - 12 + 6)u_1 + (-6 + 12 - 6)u_2 + (4 - 8 + 6)u_3 = u_1 + 0u_2 + 2u_3$.

13.17 Find the change-of-basis matrix P from the basis S to the basis S'.

▌ Write the coordinates of v_1, v_2, and v_3 with respect to the basis S as columns:

$$P = \begin{pmatrix} 2 & -1 & 1 \\ -1 & 1 & 0 \\ 1 & 0 & 2 \end{pmatrix}$$

13.18 Find the coordinates of an arbitrary vector $v = (a, b, c) \in \mathbf{R}^3$ with respect to the basis $S' = \{v_1, v_2, v_3\}$.

▌ We have

$$\begin{pmatrix} a \\ b \\ c \end{pmatrix} = x\begin{pmatrix} 1 \\ 2 \\ 1 \end{pmatrix} + y\begin{pmatrix} 0 \\ 1 \\ 2 \end{pmatrix} + z\begin{pmatrix} 1 \\ 4 \\ 6 \end{pmatrix} \quad \text{or} \quad \begin{aligned} x + \quad z &= a \\ 2x + y + 4z &= b \\ x + 2y + 6z &= c \end{aligned}$$

Solve for x, y, z to get $x = -2a + 2b - c$, $y = -8a + 5b - 2c$, $z = 3a - 2b + c$. Thus

$$v = (a, b, c) = (-2a + 2b - c)v_1 + (8a + 5b - 2c)v_2 + (3a - 2b + c)v_3$$

or $[v]_{S'} = [(a, b, c)]_{S'} = [-2a + 2b - c, -8a + 5b - 2c, 3a - 2b + c]^T$.

13.19 Write u_1, the first basis vector of S, as a linear combination of the basis vectors v_1, v_2, v_3 of S'.

▌ By Problem 13.18, $u_1 = (1, 2, 0) = (-2 + 4 + 0)v_1 + (-8 + 10 + 0)v_2 + (3 - 4 + 0)v_3 = 2v_1 + 2v_2 - v_3$.

13.20 Write u_2 as a linear combination of v_1, v_2, and v_3.

▌ $u_2 = (1, 3, 2) = (-2 + 6 - 2)v_1 + (-8 + 15 - 4)v_2 + (3 - 6 + 2)v_3 = 2v_1 + 3v_2 - v_3$.

13.21 Write u_3 as a linear combination of v_1, v_2, and v_3.

▌ $u_3 = (0, 1, 3) = (0 + 2 - 3)v_1 + (0 + 5 - 6)v_2 + (0 - 2 + 3)v_3 = -v_1 - v_2 + v_3$.

13.22 Find the change-of-basis matrix Q from the basis S' back to the basis S.

▌ Write the coordinates of u_1, u_2, and u_3 with respect to the basis S' as columns:

$$Q = \begin{pmatrix} 2 & 2 & -1 \\ 2 & 3 & -1 \\ -1 & -1 & 1 \end{pmatrix}$$

13.23 Verify that $Q = P^{-1}$ [Theorem 13.1].

/ $$QP = \begin{pmatrix} 2 & 2 & -1 \\ 2 & 3 & -1 \\ -1 & -1 & 1 \end{pmatrix}\begin{pmatrix} 2 & -1 & 1 \\ -1 & 1 & 0 \\ 1 & 0 & 2 \end{pmatrix} = \begin{pmatrix} 1 & 0 & 0 \\ 0 & 1 & 0 \\ 0 & 0 & 1 \end{pmatrix} = I$$

13.24 Show that $P[v]_{S'} = [v]_S$ for any vector $v = (a, b, c)$ [Theorem 13.2(i)].

/ $$P[v]_{S'} = \begin{pmatrix} 2 & -1 & 1 \\ -1 & 1 & 0 \\ 1 & 0 & 2 \end{pmatrix}\begin{pmatrix} -2a + 2b - c \\ -8a + 5b - 2c \\ 3a - 2b + c \end{pmatrix} = \begin{pmatrix} 7a + 3b + c \\ -6a + 3b - c \\ 4a - 2b + c \end{pmatrix} = [v]_S$$

13.25 Show that $P^{-1}[v]_S = [v]_{S'}$ for any vector $v = (a, b, c)$ [Theorem 13.2(ii)].

/ $$P^{-1}[v]_S = Q[v]_S = \begin{pmatrix} 2 & 2 & -1 \\ 2 & 3 & -1 \\ -1 & -1 & 1 \end{pmatrix}\begin{pmatrix} 7a - 3b + c \\ -6a + 3b - c \\ 4a + 2b + c \end{pmatrix} = \begin{pmatrix} -2a + 2b - c \\ -8a + 5b - 2c \\ 3a - 2b + c \end{pmatrix} = [v]_{S'}$$

13.26 Suppose $v_1 = (a_1, a_2, \ldots, a_n)$, $v_2 = (b_1, b_2, \ldots, b_n) \ldots$, $v_n = (c_1, c_2, \ldots, c_n)$ form a basis S of K^n. Show that the change-of-basis matrix from the usual basis $E = \{e_i\}$ of K^n to the basis S is the matrix P whose columns are the vectors $v_1, v_2, \ldots, v_n$, respectively.

/ We have

$$v_1 = (a_1, a_2, \ldots, a_n) = a_1 e_1 + a_2 e_2 + \cdots + a_n e_n$$
$$v_2 = (b_1, b_2, \ldots, b_n) = b_1 e_1 + b_2 e_2 + \cdots + b_n e_n$$
$$\cdots\cdots\cdots\cdots\cdots\cdots\cdots\cdots\cdots\cdots\cdots$$
$$v_n = (c_1, c_2, \ldots, c_n) = c_1 e_1 + c_2 e_2 + \cdots + c_n e_n$$

Writing the coordinates as columns, we get

$$P = \begin{pmatrix} a_1 & b_1 & \cdots & c_1 \\ a_2 & b_2 & \cdots & c_2 \\ \cdots\cdots\cdots\cdots\cdots \\ a_m & b_n & \cdots & c_n \end{pmatrix}$$

as claimed.

13.27 Find the change-of-basis matrix P from the usual basis $E = \{e_1, e_2, e_3\}$ of $\mathbf{R}^3$ to the basis $S = \{w_1 = (1, 1, 1), w_2 = (1, 1, 0), w_3 = (1, 0, 0)\}$.

/ By Problem 13.26, write the basis vectors w_1, w_2, w_3 as columns:

$$P = \begin{pmatrix} 1 & 1 & 1 \\ 1 & 1 & 0 \\ 1 & 0 & 0 \end{pmatrix}$$

13.28 Find the change-of-basis matrix Q from the above basis S back to the usual basis E of $\mathbf{R}^3$.

/ Recall [Problem 12.66] that $(a, b, c) = cw_1 + (b - c)w_2 + (a - b)w_3$. Thus

$$\begin{array}{l} e_1 = (1, 0, 0) = 0w_1 + 0w_2 + 1w_3 \\ e_2 = (0, 1, 0) = 0w_1 + 1w_2 - 1w_3 \\ e_3 = (0, 0, 1) = 1w_1 - 1w_2 + 0w_3 \end{array} \quad \text{and} \quad Q = \begin{pmatrix} 0 & 0 & 1 \\ 0 & 1 & -1 \\ 1 & -1 & 0 \end{pmatrix}$$

13.29 Verify that $Q = P^{-1}$ for the above matrices P and Q [Theorem 13.1].

/ $$PQ = \begin{pmatrix} 1 & 1 & 1 \\ 1 & 1 & 0 \\ 1 & 0 & 0 \end{pmatrix}\begin{pmatrix} 0 & 0 & 1 \\ 0 & 1 & -1 \\ 1 & -1 & 0 \end{pmatrix} = \begin{pmatrix} 1 & 0 & 0 \\ 0 & 1 & 0 \\ 0 & 0 & 1 \end{pmatrix} = I$$

13.30 Show that $P^{-1}[v]_E = [v]_S$ for any vector $v = (a, b, c)$ in $\mathbf{R}^3$.

▮ We have $[v]_E = [a, b, c]^T$ and $[v]_S = [c, b - c, a - b]^T$. Thus

$$P^{-1}[v]_E = \begin{pmatrix} 0 & 0 & 1 \\ 0 & 1 & -1 \\ 1 & -1 & 0 \end{pmatrix}\begin{pmatrix} a \\ b \\ c \end{pmatrix} = \begin{pmatrix} c \\ b - c \\ a - b \end{pmatrix} = [v]_S$$

13.31 Let $v = (a, b)$ be an element of $\mathbf{R}^2$. Then $v = ae_1 + be_2$ where $E = \{e_1, e_2\}$ is the usual basis of $\mathbf{R}^2$. Suppose another basis, say $S = \{u_1 = (2, 1), u_2 = (-1, 1)\}$, is chosen. Give a geometrical interpretation of the coordinate vector $[v]_S = [a', b']$.

▮ The basis S determines a new coordinate system for the plane $\mathbf{R}^2$ with new axes x' and y' as pictured in Fig. 13-1. That is, the vectors u_1 and u_2 indicate, respectively, the positive directions of the new axes x' and y', and the lengths of u_1 and u_2 determine, respectively, the unit lengths on the new axes x' and y'. With this new coordinate system, a' is the intersection of the x' axis and a line through v parallel to y', and b' is the intersection of the y' axis and a line through v parallel to x'.

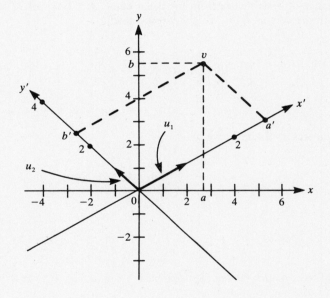

Fig. 13-1

13.32 In the preceding problem, find the change-of-basis matrix P from the E basis to the S basis, and find the change-of-basis Q from the S basis back to the E basis.

▮ Since E is the usual basis, write u_1 and u_2 as columns to obtain P, that is

$$P = \begin{pmatrix} 2 & -1 \\ 1 & 1 \end{pmatrix}$$

Also, using the formula for the inverse of a 2×2 matrix,

$$Q = P^{-1} = \begin{pmatrix} \frac{1}{3} & \frac{1}{3} \\ -\frac{1}{3} & \frac{2}{3} \end{pmatrix}$$

13.33 In Problem 13.31, express a' and b' in terms of a and b.

▮ By Theorem 13.2,

$$\begin{pmatrix} a' \\ b' \end{pmatrix} = [v]_S = P^{-1}[v]_E = \begin{pmatrix} \frac{1}{3} & \frac{1}{3} \\ -\frac{1}{3} & \frac{2}{3} \end{pmatrix}\begin{pmatrix} a \\ b \end{pmatrix} = \begin{pmatrix} \frac{1}{3}a + \frac{1}{3}b \\ -\frac{1}{3}a + \frac{2}{3}b \end{pmatrix}$$

That is, $a' = a/3 + b/3$ and $b' = -a/3 + 2b/3$.

13.34 Consider the bases $S = \{1, i\}$ and $S' = \{1 + i, 1 + 2i\}$ of the complex field $\mathbf{C}$ over the real field $\mathbf{R}$. Find the change-of-basis matrix P from the S basis to the S' basis.

   ▮ We have
   $$1 + i = 1(1) + 1(i) \qquad \text{and so} \qquad P = \begin{pmatrix} 1 & 1 \\ 1 & 2 \end{pmatrix}$$
   $$1 + 2i = 1(1) + 2(i)$$

13.35 In Problem 13.34, find the change-of-basis matrix from the S' basis to the S basis.

   ▮ Using the formula for the inverse of a 2×2 matrix [Problem 4.87],
   $$Q = P^{-1} = \begin{pmatrix} 2 & -1 \\ -1 & 1 \end{pmatrix}$$

13.36 Consider the bases $E = \{e_1 = (1, 0), e_2 = (0, 1)\}$ and $S = \{v_1 = (1, 3), v_2 = (2, 5)\}$ of $\mathbf{R}^2$. Find the transition matrix P from the E basis to the S basis.

   ▮ Since E is the usual basis of $\mathbf{R}^2$, write the vectors v_1 and v_2 as columns:
   $$P = \begin{pmatrix} 1 & 2 \\ 3 & 5 \end{pmatrix}$$

13.37 Find the change-of-basis matrix Q from the above basis S back to the usual basis E of $\mathbf{R}^2$.

   ▮ Recall [Problem 12.21] that $(a, b) = (-5a + 2b)v_1 + (3a - b)v_2$. Thus
   $$e_1 = (1, 0) = -5v_1 + 3v_2 \qquad \text{and} \qquad Q = \begin{pmatrix} -5 & 2 \\ 3 & -1 \end{pmatrix}$$
   $$e_2 = (0, 1) = 2v_1 - v_2$$

13.38 Verify that $Q = P^{-1}$ for the above matrices P and Q [Theorem 13.1].

   ▮
   $$QP = \begin{pmatrix} -5 & 2 \\ 3 & -1 \end{pmatrix} \begin{pmatrix} 1 & 2 \\ 3 & 5 \end{pmatrix} = \begin{pmatrix} 1 & 0 \\ 0 & 1 \end{pmatrix} = I$$

13.39 Show that $P^{-1}[v]_E = [v]_S$ for any vector $v = (a, b) \in \mathbf{R}^2$.

   ▮ We have $[v]_E = [a, b]^T$ and $[v]_S = [-5a + 2b, 3a - b]^T$. Hence
   $$P^{-1}[v]_E = \begin{pmatrix} -5 & 2 \\ 3 & -1 \end{pmatrix} \begin{pmatrix} a \\ b \end{pmatrix} = \begin{pmatrix} -5a + 2b \\ 3a - b \end{pmatrix}$$

13.40 Suppose the x and y axes in the plane $\mathbf{R}^2$ are rotated counterclockwise $45°$ so that the new x' axis is along the line $y = x$ and the new y' axis is along the line $y = -x$. Find the change-of-basis matrix P.

   ▮ Here $u_1 = (\sqrt{2}/2, \sqrt{2}/2)$ is the unit vector in the new x' axis and $u_2 = (-\sqrt{2}/2, \sqrt{2}/2)$ is the unit vector in the new y' axis. [Note the usual basis vectors are, respectively, the unit vectors of the original x and y axes.] Thus
   $$P = \begin{pmatrix} \sqrt{2}/2 & -\sqrt{2}/2 \\ \sqrt{2}/2 & \sqrt{2}/2 \end{pmatrix}$$

13.41 Find the new coordinates of a point $A(5, 6)$ in $\mathbf{R}^2$ under the rotation in Problem 13.40.

   ▮ Multiply the coordinates of the point by P^{-1}:
   $$\begin{pmatrix} \sqrt{2}/2 & \sqrt{2}/2 \\ -\sqrt{2}/2 & \sqrt{2}/2 \end{pmatrix} \begin{pmatrix} 5 \\ 6 \end{pmatrix} = \begin{pmatrix} 11\sqrt{2}/2 \\ \sqrt{2}/2 \end{pmatrix}$$

13.42 Illustrate Theorem 13.2 in the case $\dim V = 3$. Specifically, suppose P is the change-of-basis matrix from a basis $\{e_1, e_2, e_3\}$ of V to a basis $\{f_1, f_2, f_3\}$ of V; say,
   $$\begin{aligned} f_1 &= a_1 e_1 + a_2 e_2 + a_3 e_3 \\ f_2 &= b_1 e_1 + b_2 e_2 + b_3 e_3 \qquad \text{Hence} \qquad P = \begin{pmatrix} a_1 & b_1 & c_1 \\ a_2 & b_2 & c_2 \\ a_3 & b_3 & c_3 \end{pmatrix} \\ f_3 &= c_1 e_1 + c_2 e_2 + c_3 e_3 \end{aligned}$$

Also, suppose $v \in V$ and, say, $v = k_1 f_1 + k_2 f_2 + k_3 f_3$. Show that $P[v]_f = [v]_e$ and $P^{-1}[v]_e = [v]_f$.

▌ Substituting for the f_i in $v = k_1 f_1 + k_2 f_2 + k_3 f_3$, we obtain

$$v = k_1(a_1 e_1 + a_2 e_2 + a_3 e_3) + k_2(b_1 e_1 + b_2 e_2 + b_3 e_3) + k_3(c_1 e_1 + c_2 e_2 + c_3 e_3)$$
$$= (a_1 k_1 + b_1 k_2 + c_1 k_3)e_1 + (a_2 k_1 + b_2 k_2 + c_2 k_3)e_2 + (a_3 k_1 + b_3 k_2 + c_3 k_3)e_3$$

Thus $[v]_f = [k_1, k_2, k_3]^T$ and $[v]_e = [a_1 k_1 + b_1 k_2 + c_1 k_3, a_2 k_1 + b_2 k_2 + c_2 k_3, a_3 k_1 + b_3 k_2 + c_3 k_3]^T$, so

$$P[v]_f = \begin{pmatrix} a_1 & b_1 & c_1 \\ a_2 & b_2 & c_2 \\ a_3 & b_3 & c_3 \end{pmatrix} \begin{pmatrix} k_1 \\ k_2 \\ k_3 \end{pmatrix} = \begin{pmatrix} a_1 k_1 + b_1 k_2 + c_1 k_3 \\ a_2 k_1 + b_2 k_2 + c_2 k_3 \\ a_3 k_1 + b_3 k_2 + c_3 k_3 \end{pmatrix} = [v]_e$$

Also, multiplying the above equation by P^{-1}, we have $P^{-1}[v]_e = P^{-1}P[v]_f = I[v]_f = [v]_f$.

13.43 Prove Theorem 13.1: Let P be the change-of-basis matrix from a basis $\{e_i\}$ to a basis $\{f_i\}$, and let Q be the change-of-basis matrix from the basis $\{f_i\}$ back to the basis $\{e_i\}$. Then P is invertible and $Q = P^{-1}$.

▌ Suppose, for $i = 1, 2, \ldots, n$,

$$f_i = a_{i1} e_1 + q_{i2} e_2 + \cdots + a_{in} e_n = \sum_{j=1}^{n} a_{ij} e_j \qquad (1)$$

and, for $j = 1, 2, \ldots, n$,

$$e_j = b_{j1} f_1 + b_{j2} f_2 + \cdots + b_{jn} f_n = \sum_{k=1}^{n} b_{jk} f_k \qquad (2)$$

Let $A = (a_{ij})$ and $B = (b_{jk})$. Then $P = A^T$ and $Q = B^T$. Substituting (2) into (1) yields

$$f_i = \sum_{j=1}^{n} a_{ij} \left(\sum_{k=1}^{n} b_{jk} f_k \right) = \sum_{k=1}^{n} \left(\sum_{j=1}^{n} a_{ij} b_{jk} \right) f_k$$

Since the $\{f_i\}$ is a basis $\sum a_{ij} b_{jk} = \delta_{ik}$ where δ_{ik} is the Kronecker delta function, i.e., $\delta_{ik} = 1$ if $i = k$ but $\delta_{ik} = 0$ if $i \neq k$. Suppose $AB = (c_{ik})$. Then $c_{ik} = \delta_{ik}$. Accordingly, $AB = I$, and so $QP = B^T A^T = (AB)^T = I^T = I$. Thus $Q = P^{-1}$.

13.44 Prove Theorem 13.2: Let P be the change-of-basis matrix from a basis $\{e_i\}$ to a basis (f_i) in a vector space V. Then for any $v \in V$, (i) $P[v]_f = [v]_e$ and (ii) $P^{-1}[v]_e = [v]_f$.

▌ Suppose, for $i = 1, \ldots, n$, $f_i = a_{i1} e_1 + a_{i2} e_2 + \cdots + a_{in} e_n = \sum_{j=1}^{n} a_{ij} e_j$. Then P is the n-square matrix whose jth row is

$$(a_{1j}, a_{2j}, \ldots, a_{nj}) \qquad (1)$$

Also suppose $v = k_1 f_1 + k_2 f_2 + \cdots + k_n f_n = \sum_{i=1}^{n} k_i f_i$. Then writing a column vector as the transpose of a row vector,

$$[v]_f = (k_1, k_2, \ldots, k_n)^T \qquad (2)$$

Substituting for f_i in the equation for v,

$$v = \sum_{i=1}^{n} k_i f_i = \sum_{i=1}^{n} k_i \left(\sum_{j=1}^{n} a_{ij} e_j \right) = \sum_{j=1}^{n} \left(\sum_{i=1}^{n} a_{ij} k_i \right) e_j = \sum_{j=1}^{n} (a_{1j} k_1 + a_{2j} k_2 + \cdots + a_{nj} k_n) e_j$$

Accordingly, $[v]_e$ is the column vector whose jth entry is

$$a_{1j} k_1 + a_{2j} k_2 + \cdots + a_{nj} k_n \qquad (3)$$

On the other hand, the jth entry of $P[v]_f$ is obtained by multiplying the jth row of P by $[v]_f$, i.e., multiplying (1) by (2). But the product of (1) and (2) is (3); hence $P[v]_f$ and $[v]_e$ have the same entries and thus $P[v]_f = [v]_e$.

Furthermore, multiplying the above by P^{-1} gives $P^{-1}[v]_e = P^{-1}P[v]_f = [v]_f$.

13.2 CHANGE OF BASIS AND LINEAR OPERATORS

The preceding section discusses the effect of a change of basis on coordinate vectors. This section discusses the effect of a change of basis on the matrix representation of a linear operator. In particular, the following theorems, whose proofs appear in Problems 13.60 and 13.61 are used below.

Theorem 13.3: Let P be the change-of-basis matrix from a basis S_1 to a basis S_2 in a vector space V. Then for any linear operator T on V, $[T]_{S_2} = P^{-1}[T]_{S_1}P$.

Theorem 13.4: Let A be an n-square matrix over K (which may be viewed as a linear operator on K^n) and let $\{u_1, u_2, \ldots, u_n\}$ be a basis of K^n. Then the matrix representation of A relative to the given basis is the matrix $B = P^{-1}AP$, where P is the matrix whose columns are $u_1, u_2, \ldots, u_n$, respectively.

13.45 Let $T: \mathbf{R}^2 \to \mathbf{R}^2$ be the linear operator defined by $T(x, y) = (3x - 5y, 2x + 7y)$. Find the matrix representation $[T]_E$ of T relative to the usual basis E of $\mathbf{R}^2$.

❙ Since E is the usual basis of $\mathbf{R}^2$, simply write the coefficients of x and y as the rows to get

$$[T]_E = \begin{pmatrix} 3 & -5 \\ 2 & 7 \end{pmatrix}.$$

13.46 Find the matrix representation $[T]_S$ of the above linear map T relative to the basis $S = \{v_1 = (1, 3), v_2 = (2, 5)\}$.

❙ Method 1: Use the definition of $[T]_S$. Using $(a, b) = (-5a + 2b)v_1 + (3a - b)v_2$ [Problem 13.37], we have

$$T(v_1) = T(1, 3) = (3 - 15, 2 + 23) = (-12, 25) = (60 + 46)v_1 + (-36 - 23)v_2 = 106v_1 - 59v_2$$
$$T(v_2) = T(2, 5) = (6 - 25, 4 + 35) = (-19, 39) = (95 + 78)v_1 + (-57 - 39)v_2 = 173v_1 - 96v_2$$

Write the coordinates of $T(v_1)$ and $T(v_2)$ as columns to get $[T]_S = \begin{pmatrix} 106 & 173 \\ -59 & -96 \end{pmatrix}$.

Method 2: Use Theorem 13.3. By Problems 13.36 and 13.38, the change-of-basis matrix from the E basis to the S basis is $P = \begin{pmatrix} 1 & 2 \\ 3 & 5 \end{pmatrix}$ and $P^{-1} = \begin{pmatrix} -5 & 2 \\ 3 & -1 \end{pmatrix}$. Thus

$$[T]_S = P^{-1}[T]_E P = \begin{pmatrix} -5 & 2 \\ 3 & -1 \end{pmatrix} \begin{pmatrix} 3 & -5 \\ 2 & 7 \end{pmatrix} \begin{pmatrix} 1 & 2 \\ 3 & 5 \end{pmatrix} = \begin{pmatrix} 106 & 173 \\ -59 & -96 \end{pmatrix}$$

13.47 Let $L: \mathbf{R}^2 \to \mathbf{R}^2$ be defined by $L(x, y) = (2y, 3x - y)$. Find the matrix representation $[L]_S$ of L relative to the above basis S.

❙ We have $[L]_E = \begin{pmatrix} 0 & 2 \\ 3 & -1 \end{pmatrix}$ where E is the usual basis of $\mathbf{R}^2$. Thus

$$[L]_S = P^{-1}[L]_E P = \begin{pmatrix} -5 & 2 \\ 3 & -1 \end{pmatrix} \begin{pmatrix} 0 & 2 \\ 3 & -1 \end{pmatrix} \begin{pmatrix} 1 & 2 \\ 3 & 5 \end{pmatrix} = \begin{pmatrix} -30 & -48 \\ 18 & 29 \end{pmatrix}$$

13.48 Let $F: \mathbf{R}^3 \to \mathbf{R}^3$ be defined by $F(x, y, z) = (x + 3y + 2z, x - 4z, y + 3z)$. Find the matrix representation of F relative to the usual basis E of $\mathbf{R}^3$.

❙ Since E is the usual basis of $\mathbf{R}^3$, simply write the coefficients of x, y, z as the rows to get

$$[F]_E = \begin{pmatrix} 1 & 3 & 2 \\ 1 & 0 & -4 \\ 0 & 1 & 3 \end{pmatrix}$$

13.49 Find the matrix representation of the above linear map F relative to the following basis of R^3:

$$S = \{w_1 = (1, 1, 1),\ w_2 = (1, 1, 0),\ w_3 = (1, 0, 0)\}$$

▮ By Problems 13.27 and 13.29, the change-of-basis matrix from the E basis to the S basis is

$$P = \begin{pmatrix} 1 & 1 & 1 \\ 1 & 1 & 0 \\ 1 & 0 & 0 \end{pmatrix} \quad \text{and} \quad P^{-1} = \begin{pmatrix} 0 & 0 & 1 \\ 0 & 1 & -1 \\ 1 & -1 & 0 \end{pmatrix}$$

Thus, by Theorem 13.3,

$$[F]_S = P^{-1}[F]_E P = \begin{pmatrix} 0 & 0 & 1 \\ 0 & 1 & -1 \\ 1 & -1 & 0 \end{pmatrix}\begin{pmatrix} 1 & 3 & 2 \\ 1 & 0 & -4 \\ 0 & 1 & 3 \end{pmatrix}\begin{pmatrix} 1 & 1 & 1 \\ 1 & 1 & 0 \\ 1 & 0 & 0 \end{pmatrix} = \begin{pmatrix} 4 & 1 & 0 \\ -7 & 0 & 1 \\ 9 & 3 & 0 \end{pmatrix}$$

13.50 Let $G: \mathbf{R}^3 \to \mathbf{R}^3$ be defined by $G(x, y, z) = (2y + z, x - 4y, 3x)$. Find the matrix representation of G relative to the above basis S.

▮ By Theorem 13.3,

$$[G]_S = P^{-1}[G]_E P = \begin{pmatrix} 0 & 0 & 1 \\ 0 & 1 & -1 \\ 1 & -1 & 0 \end{pmatrix}\begin{pmatrix} 0 & 2 & 1 \\ 1 & -4 & 0 \\ 3 & 0 & 0 \end{pmatrix}\begin{pmatrix} 1 & 1 & 1 \\ 1 & 1 & 0 \\ 1 & 0 & 0 \end{pmatrix} = \begin{pmatrix} 3 & 3 & 3 \\ -6 & -6 & -2 \\ 6 & 5 & -1 \end{pmatrix}$$

13.51 Let $A: \mathbf{R}^3 \to \mathbf{R}^3$ be the linear operator defined by the matrix

$$A = \begin{pmatrix} 1 & 2 & 1 \\ 3 & -1 & 0 \\ 0 & 1 & 2 \end{pmatrix}$$

Find the matrix representation of A relative to the usual basis E of $\mathbf{R}^3$.

▮ Relative to the usual basis E, we get back the matrix A, i.e.,

$$[A]_E = A = \begin{pmatrix} 1 & 2 & 1 \\ 3 & -1 & 0 \\ 0 & 1 & 2 \end{pmatrix}$$

13.52 Find the matrix representation of the above linear map A relative to the basis S in Problem 13.49.

▮ By Theorem 13.3,

$$[A]_S = P^{-1}AP = \begin{pmatrix} 0 & 0 & 1 \\ 0 & 1 & -1 \\ 1 & -1 & 0 \end{pmatrix}\begin{pmatrix} 1 & 2 & 1 \\ 3 & -1 & 0 \\ 0 & 1 & 2 \end{pmatrix}\begin{pmatrix} 1 & 1 & 1 \\ 1 & 1 & 0 \\ 1 & 0 & 0 \end{pmatrix} = \begin{pmatrix} 3 & 1 & 0 \\ -1 & 1 & 3 \\ 2 & 1 & -2 \end{pmatrix}$$

13.53 Let $A: \mathbf{R}^2 \to \mathbf{R}^2$ be defined by the matrix $A = \begin{pmatrix} 5 & -7 \\ 2 & 3 \end{pmatrix}$. Let B be the matrix representation of A relative to the basis $\{(1, 4), (3, 10)\}$. Find B.

▮ Write the basis vectors as columns to get $P = \begin{pmatrix} 1 & 3 \\ 4 & 10 \end{pmatrix}$. Use the formula for the inverse of a 2-square matrix [Problem 4.87] to get $P^{-1} = \begin{pmatrix} -5 & \frac{3}{2} \\ 2 & -\frac{1}{2} \end{pmatrix}$. Thus

$$B = P^{-1}AP = \begin{pmatrix} -5 & \frac{3}{2} \\ 2 & -\frac{1}{2} \end{pmatrix}\begin{pmatrix} 5 & -7 \\ 2 & 3 \end{pmatrix}\begin{pmatrix} 1 & 3 \\ 4 & 10 \end{pmatrix} = \begin{pmatrix} 136 & 329 \\ -53 & -128 \end{pmatrix}$$

13.54 Let $A = \begin{pmatrix} 4 & 5 \\ 2 & -1 \end{pmatrix}$. Let B be the matrix representation of the linear map $A: \mathbf{R}^2 \to \mathbf{R}^2$ relative to the basis $\{(1, 4), (2, 9)\}$. Find B.

▮ By Theorem 13.4,

$$B = P^{-1}AP = \begin{pmatrix} 9 & -2 \\ -4 & 1 \end{pmatrix}\begin{pmatrix} 4 & 5 \\ 2 & -1 \end{pmatrix}\begin{pmatrix} 1 & 2 \\ 4 & 9 \end{pmatrix} = \begin{pmatrix} 220 & 487 \\ -98 & -217 \end{pmatrix}$$

13.55 Let $P = \begin{pmatrix} 1 & 2 & 1 \\ 1 & 3 & 4 \\ 2 & 5 & 6 \end{pmatrix}$. Find P^{-1}.

▮ Use the Gaussian elimination algorithm described in Problem 4.92:

$$(P, I) = \begin{pmatrix} 1 & 2 & 1 & \bigm| & 1 & 0 & 0 \\ 1 & 3 & 4 & \bigm| & 0 & 1 & 0 \\ 2 & 5 & 6 & \bigm| & 0 & 0 & 1 \end{pmatrix} \sim \begin{pmatrix} 1 & 2 & 1 & \bigm| & 1 & 1 & 0 \\ 0 & 1 & 3 & \bigm| & -1 & 1 & 0 \\ 0 & 1 & 4 & \bigm| & -2 & 0 & 1 \end{pmatrix}$$

$$\sim \begin{pmatrix} 1 & 2 & 1 & \bigm| & 1 & 0 & 0 \\ 0 & 1 & 3 & \bigm| & -1 & 1 & 0 \\ 0 & 0 & 1 & \bigm| & -1 & -1 & 1 \end{pmatrix} \sim \begin{pmatrix} 1 & 2 & 0 & \bigm| & 2 & 1 & -1 \\ 0 & 1 & 0 & \bigm| & 2 & 4 & -3 \\ 0 & 0 & 1 & \bigm| & -1 & -1 & 1 \end{pmatrix}$$

$$\sim \begin{pmatrix} 1 & 0 & 0 & \bigm| & -2 & -7 & 5 \\ 0 & 1 & 0 & \bigm| & 2 & 4 & -3 \\ 0 & 0 & 1 & \bigm| & -1 & -1 & 1 \end{pmatrix}$$

Hence

$$P^{-1} = \begin{pmatrix} -2 & -7 & 5 \\ 2 & 4 & -3 \\ -1 & -1 & 1 \end{pmatrix}$$

13.56 Let

$$A = \begin{pmatrix} 2 & 3 & -4 \\ 4 & -6 & 3 \\ 1 & 4 & -2 \end{pmatrix}$$

Let B be the matrix representation of the linear map $A: \mathbf{R}^3 \to \mathbf{R}^3$ relative to the basis $\{(1, 1, 2), (2, 3, 5), (1, 4, 6)\}$. Find B.

▮ By Theorem 13.3 and Problem 13.55,

$$B = P^{-1}AP = \begin{pmatrix} -2 & -7 & 5 \\ 2 & 4 & -3 \\ -1 & -1 & 1 \end{pmatrix}\begin{pmatrix} 2 & 3 & -4 \\ 4 & -6 & 3 \\ 1 & 4 & -2 \end{pmatrix}\begin{pmatrix} 1 & 2 & 1 \\ 1 & 3 & 4 \\ 2 & 5 & 6 \end{pmatrix} = \begin{pmatrix} -17 & -1 & 59 \\ 7 & -6 & -43 \\ 0 & 6 & 17 \end{pmatrix}$$

13.57 Find the inverse of $P = \begin{pmatrix} 1 & 1 & 1 \\ 0 & 1 & 1 \\ 0 & 0 & 1 \end{pmatrix}$.

▮ The inverse of P is of the form $P^{-1} = \begin{pmatrix} 1 & x & y \\ 0 & 1 & z \\ 0 & 0 & 1 \end{pmatrix}$. Set $PP^{-1} = I$, the identity matrix:

$$PP^{-1} = \begin{pmatrix} 1 & 1 & 1 \\ 0 & 1 & 1 \\ 0 & 0 & 1 \end{pmatrix}\begin{pmatrix} 1 & x & y \\ 0 & 1 & x \\ 0 & 0 & 1 \end{pmatrix} = \begin{pmatrix} 1 & x+1 & y+z+1 \\ 0 & 1 & z+1 \\ 0 & 0 & 1 \end{pmatrix} = \begin{pmatrix} 1 & 0 & 0 \\ 0 & 1 & 0 \\ 0 & 0 & 1 \end{pmatrix} = I$$

Set corresponding entries equal to each other to obtain the system $x + 1 = 0$, $y + z + 1 = 0$, $z + 1 = 0$. The solution is $x = -1$, $y = 0$, $z = -1$. Thus

$$P^{-1} = \begin{pmatrix} 1 & -1 & 0 \\ 0 & 1 & -1 \\ 0 & 0 & 1 \end{pmatrix}$$

13.58 Let

$$A = \begin{pmatrix} 1 & 3 & 5 \\ 2 & 4 & 6 \\ 7 & 8 & 9 \end{pmatrix}$$

Let B be the matrix representation of the linear map $A: \mathbf{R}^3 \to \mathbf{R}^3$ relative to the basis $\{(1, 0, 0), (1, 1, 0), (1, 1, 1)\}$. Find B.

❚ The above matrix P is the change-of-basis matrix from the usual basis of $\mathbf{R}^3$ to the given basis. Thus

$$B = P^{-1}AP = \begin{pmatrix} 1 & -1 & 0 \\ 0 & 1 & -1 \\ 0 & 0 & 1 \end{pmatrix}\begin{pmatrix} 1 & 3 & 5 \\ 2 & 4 & 6 \\ 7 & 8 & 9 \end{pmatrix}\begin{pmatrix} 1 & 1 & 1 \\ 0 & 1 & 1 \\ 0 & 0 & 1 \end{pmatrix} = \begin{pmatrix} -1 & -2 & -3 \\ -5 & -9 & -12 \\ 7 & 15 & 24 \end{pmatrix}$$

13.59 Let $T: \mathbf{C} \to \mathbf{C}$ be the conjugate operator where $\mathbf{C}$ is the complex field viewed as a vector space over the real field $\mathbf{R}$. Find the matrix representation of T relative to the basis $S = \{1 + 2i, 3 + 4i\}$ of $\mathbf{C}$.

❚ Consider the usual basis $E = \{1, i\}$ of $\mathbf{C}$. Since $T(1) = 1$ and $T(i) = -i$, $[T]_E = \begin{pmatrix} 1 & 0 \\ 0 & -1 \end{pmatrix}$.

Also, the change-of-basis matrix from the E basis to the S basis is $P = \begin{pmatrix} 1 & 3 \\ 2 & 4 \end{pmatrix}$. Furthermore,

$$P^{-1} = \begin{pmatrix} -2 & \frac{3}{2} \\ 1 & -\frac{1}{2} \end{pmatrix} \quad \text{and so} \quad [T]_S = P^{-1}[T]_E P = \begin{pmatrix} -2 & \frac{3}{2} \\ 1 & -\frac{1}{2} \end{pmatrix}\begin{pmatrix} 1 & 0 \\ 0 & -1 \end{pmatrix}\begin{pmatrix} 1 & 3 \\ 2 & 4 \end{pmatrix} = \begin{pmatrix} -5 & -12 \\ 2 & 5 \end{pmatrix}$$

13.60 Prove Theorem 13.3: Let P be the change-of-basis matrix from a basis $\{e_i\}$ to a basis $\{f_i\}$ in a vector space V. Then, for any linear operator T on V, $[T]_f = P^{-1}[T]_e P$.

❚ For any vector $v \in V$, $P^{-1}[T]_e P[v]_f = P^{-1}[T]_e[v]_e = P^{-1}[T(v)]_e = [T(v)]_f$. But $[T]_f[v]_f = [T(v)]_f$; hence $P^{-1}[T]_e P[v]_f = [T]_f[v]_f$. Since the mapping $v \mapsto [v]_f$ is onto K^n, $P^{-1}[T]_e PX = [T]_f X$ for every $X \in K^n$. Accordingly, $P^{-1}[T]_e P = [T]_f$.

13.61 Prove Theorem 13.4: Let A be an n-square matrix over K and let $\{u_1, u_2, \ldots, u_n\}$ be a basis of K^n. Then the matrix representation of A relative to the given basis is the matrix $B = P^{-1}AP$, where P is the matrix whose columns are $u_1, \ldots, u_n$, respectively.

❚ The matrix representation of A relative to the usual basis of K^n is the matrix A itself. Also, P is the change-of-basis matrix from the usual basis to the given basis. Thus, by Theorem 13.3, $B = P^{-1}AP$ is the matrix representation of A relative to the given basis.

13.3 SIMILARITY AND SIMILARITY TRANSFORMATIONS

13.62 Define similarity of matrices and a similarity transformation.

❚ Suppose A and B are square matrices for which there exists an invertible matrix P such that $B = P^{-1}AP$. Then B is said to be *similar* to A or is said to be obtained from A by a *similarity transformation*.

Problems 13.63–13.65 show that similarity of matrices is an equivalence relation.

13.63 Show that A is similar to A for any (square) matrix A.

❚ The identity matrix I is invertible and $I = I^{-1}$. Since $A = I^{-1}AI$, A is similar to A.

13.64 Show that if A is similar to B, then B is similar to A.

❚ Since A is similar to B there exists an invertible matrix P such that $A = P^{-1}BP$. Hence $B = PAP^{-1} = (P^{-1})^{-1}AP^{-1}$ and P^{-1} is invertible. Thus B is similar to A.

13.65 Show that if A is similar to B and B is similar to C then A is similar to C.

❚ Since A is similar to B there exists an invertible matrix P such that $A = P^{-1}BP$, and since B is similar to C there exists an invertible matrix Q such that $B = Q^{-1}CQ$. Hence $A = P^{-1}BP = P^{-1}(Q^{-1}CQ)P = (QP)^{-1}C(QP)$ and QP is invertible. Thus A is similar to C.

Remark: Since similarity of matrices is an equivalence relation, all n-square matrices are partitioned into equivalence classes of similar matrices.

Theorem 13.5: Suppose A is a matrix representation of a linear operator T. Then B is also a matrix representation of T if and only if B is similar to A. [Thus all the matrix representations of T form an equivalence class of similar matrices.]

13.66 Prove Theorem 13.5.

❚ Suppose A is the matrix representation of T relative to the basis $\{e_i\}$. Suppose B is similar to A, say $B = P^{-1}AP$, where $P = (p_{ij})$. Since P is invertible, the n vectors $f_i = p_{1i}e_1 + p_{2i}e_2 + \cdots + p_{ni}e_n$, $i = 1, 2, \ldots, n$, are linearly independent and so form another basis of V. Also, P is the change-of-basis matrix from the basis $\{e_i\}$ to the basis $\{f_i\}$. Thus $B = P^{-1}AP$ is the matrix representation of T relative to the basis $\{f_i\}$.

Conversely, suppose B is the matrix representation of T relative to a basis $\{f_i\}$. Let P be the change-of-basis matrix from $\{e_i\}$ to the basis $\{f_i\}$. By Theorem 13.3, $B = P^{-1}AP$ and so B is similar to A.

Remark: Suppose f is a function on square matrices which assigns the same value to similar matrices; i.e., $f(A) = f(B)$ whenever A is similar to B. Then f induces a function, also denoted by f, on linear operators T in the following natural way: $f(T) = f([T]_e)$, where $\{e_i\}$ is any basis. The function is well-defined by Theorem 13.5.

Remark: Recall that the vector space M_n of all n-square matrices over a field K and the vector space $A(V)$ of all linear operators on a vector space V over K are each algebras over K. The notion of similarity is also defined for an abstract algebra $\mathscr{A}$ over a field K; i.e., the elements $A, B \in \mathscr{A}$ are similar if there exists an invertible element $P \in \mathscr{A}$ such that $B = P^{-1}AP$. Theorem 13.6, proved in Problems 13.67–13.70, applies.

Theorem 13.6: Let $\mathscr{A}$ be an algebra over a field K and P be an invertible element in $\mathscr{A}$. Then the mapping $T_P: \mathscr{A} \to \mathscr{A}$ defined by $T_P(A) = P^{-1}AP$ is an algebra isomorphism. That is, for every $A, B \in \mathscr{A}$ and any $k \in K$:
(i) $T_P(A + B) = T_P(A) + T_P(B)$. (iii) $T_P(AB) = T_P(A)T_P(B)$.
(ii) $T_P(kA) = kT_P(A)$. (iv) T_P is one-to-one and onto.
[The map T_P is called a *similarity transformation*.]

13.67 Prove (i) of Theorem 13.6: $T_P(A + B) = T_P(A) + T_P(B)$.

❚ $T_P(A + B) = P^{-1}(A + B)P = P^{-1}AP + P^{-1}BP = T_P(A) + T_P(B)$.

13.68 Prove (ii) of Theorem 13.6: $T_P(kA) = kT_P(A)$.

❚ $T_P(kA) = P^{-1}(kA)P = k(P^{-1}AP) = kT_P(A)$.

13.69 Prove (iii) of Theorem 13.6: $T_P(AB) = T_P(A)T_P(B)$.

❚ $T_P(AB) = P^{-1}(AB)P = (P^{-1}AP)(P^{-1}BP) = T_P(A)T_P(B)$.

13.70 Prove (iv) of Theorem 13.6: T_P is one-to-one and onto $\mathscr{A}$.

❚ Suppose $T_P(A) = T_P(B)$. Then $P^{-1}AP = P^{-1}BP$. Multiplying by P on the left and P^{-1} on the right yields $A = B$. Hence T_P is one-to-one. Now suppose $B \in \mathscr{A}$. Let $A = PBP^{-1}$. Then $T_P(A) = P^{-1}(PBP^{-1}P) = B$. Thus T_P is onto V.

Problems 13.71–13.73 list additional properties of similarity in an algebra $\mathscr{A}$.

13.71 Suppose B is similar to A, say $B = P^{-1}AP$. Show that B^{-1} is similar to A^{-1}.

❚ $B^{-1} = (P^{-1}AP)^{-1} = P^{-1}A^{-1}(P^{-1})^{-1} = P^{-1}A^{-1}P$ and so B^{-1} is similar to A^{-1}.

13.72 Suppose B is similar to A, say $B = P^{-1}AP$. Show that $B^n = P^{-1}A^nP$ and so B^n is similar to A^n.

▐ The proof is by induction on n. The result holds for $n = 1$ by hypothesis. Suppose $n > 1$ and the result holds for $n - 1$. Then $B^n = BB^{n-1} = (P^{-1}AP)(P^{-1}A^{n-1}P) = P^{-1}A^nP$.

13.73 Suppose D in $\mathscr{A}$ is a diagonal element, i.e., $D = kI$ for some $k \in K$. Show that D is the only element similar to itself.

▐ Suppose B is similar to D, say $B = P^{-1}DP$. Then $B = P^{-1}DP = P^{-1}(kI)P = k(P^{-1}IP) = kI = D$.

13.4 TRACE AND DETERMINANT OF LINEAR OPERATORS

The *trace* of a square matrix $A = (a_{ij})$, written tr(A), is the sum of its diagonal elements, i.e., tr(A) = $a_{11} + a_{22} + \cdots + a_{nn}$. Theorem 13.7, whose proof appears in Problem 13.89, is used below.

Theorem 13.7: Suppose matrix B is similar to matrix A. Then tr(B) = tr(A).

13.74 Define the trace of a linear operator T, written tr(T). Why is the definition well-defined?

▐ By definition tr(T) = tr([T]), where [T] is any matrix representation of T. By Theorem 13.7, all similar matrices have the same trace, and so all matrix representations of T will have the same trace.

13.75 Define the determinant of a linear operator T, written det(T). Why is the definition well-defined?

▐ By definition det(T) = det([T]), where [T] is any matrix representation of T. Since similar matrices have the same determinant, any matrix representation of T will yield the same determinantal value.

Problems 13.76–13.79 refer to the linear operator on $\mathbf{R}^2$ defined by $F(x, y) = (3x - 7y, 4x + 8y)$.

13.76 Find the trace of F.

▐ We first must find a matrix representation of F. Choosing the usual basis, $[F] = \begin{pmatrix} 3 & 4 \\ -7 & 8 \end{pmatrix}$. Hence tr($F$) = tr([$F$]) = $3 + 8 = 11$.

13.77 Do we ever get another value for tr(F) by choosing another basis?

▐ No. All matrix representations of F are similar and hence all have the same trace value 11.

13.78 Find the determinant of F.

▐ Relative to the usual basis, $[F] = \begin{pmatrix} 3 & 4 \\ -7 & 8 \end{pmatrix}$. Hence det($F$) = $\begin{vmatrix} 3 & 4 \\ -7 & 8 \end{vmatrix} = 24 + 28 = 52$.

13.79 Do we ever get another value for det(F) by choosing another basis?

▐ No. All matrix representations of F are similar and hence all have the same determinantal value 52.

13.80 Find det(T) for the linear operator on $\mathbf{R}^3$ defined by $T(x, y, z) = (2x - z, x + 2y - 4z, 3x - 3y + z)$.

▐ Find the matrix representation of T relative to, say, the usual basis by writing down the coefficients of x, y, z as rows to get

$$[T] = \begin{pmatrix} 2 & 0 & -1 \\ 1 & 2 & -4 \\ 3 & -3 & 1 \end{pmatrix}$$

Then

$$\det(T) = \begin{vmatrix} 2 & 0 & -1 \\ 1 & 2 & -4 \\ 3 & -3 & 1 \end{vmatrix} = 4 + 0 + 3 + 6 - 24 - 0 = -11$$

13.81 Find the trace of the above linear operator T.

$$\text{tr}(T) = \text{tr}\begin{pmatrix} 2 & 0 & -1 \\ 1 & 2 & -4 \\ 3 & -3 & 1 \end{pmatrix} = 2 + 2 + 1 = 5$$

13.82 Find the trace of the following operator on $\mathbf{R}^3$:

$$T(x, y, z) = (a_1 x + a_2 y + a_3 z, \; b_1 x + b_2 y + b_3 z, \; c_1 x + c_2 y + c_3 z)$$

We first must find a matrix representation of T. Choosing the usual basis $\{e_i\}$,

$$[T] = \begin{pmatrix} a_1 & a_2 & a_3 \\ b_1 & b_2 & b_3 \\ c_1 & c_2 & c_3 \end{pmatrix}$$

and $\quad \text{tr}(T) = \text{tr}([T]) = a_1 + b_2 + c_3$.

13.83 Find the determinant of the above linear operator T.

$$\det(T) = \begin{vmatrix} a_1 & a_2 & a_3 \\ b_1 & b_2 & b_3 \\ c_1 & c_2 & c_3 \end{vmatrix} = a_1 b_2 c_3 + a_2 b_3 c_1 + a_3 b_1 c_2 - a_3 b_2 c_1 - a_1 b_3 c_2 - a_2 b_1 c_3$$

13.84 Consider the complex field $\mathbf{C}$ as a vector space over the real field $\mathbf{R}$. Let T be the conjugate operator on $\mathbf{C}$, that is, $T(z) = \bar{z}$. Find $\det(T)$.

Since $T(1) = 1$ and $T(i) = -i$, we have $[T] = \begin{pmatrix} 1 & 0 \\ 0 & -1 \end{pmatrix}$ relative to the usual basis $\{1, i\}$ of $\mathbf{C}$ over $\mathbf{R}$. Then $\det(T) = \begin{vmatrix} 1 & 0 \\ 0 & -1 \end{vmatrix} = -1$.

13.85 Find the trace of the above conjugate operator T on $\mathbf{C}$.

$$\text{tr}(T) = \text{tr}\begin{pmatrix} 1 & 0 \\ 0 & -1 \end{pmatrix} = 1 - 1 = 0.$$

13.86 Suppose T is the operator on the vector space V of 2-squares matrices over K defined by $T(A) = MA$ where $M = \begin{pmatrix} a & b \\ c & d \end{pmatrix}$. Find $\det(T)$.

Find a matrix representation of T in some basis of V, say,

$$\left\{ E_1 = \begin{pmatrix} 1 & 0 \\ 0 & 0 \end{pmatrix}, \; E_2 = \begin{pmatrix} 0 & 1 \\ 0 & 0 \end{pmatrix}, \; E_3 = \begin{pmatrix} 0 & 0 \\ 1 & 0 \end{pmatrix}, \; E_4 = \begin{pmatrix} 0 & 0 \\ 0 & 1 \end{pmatrix} \right\}$$

Then

$$T(E_1) = \begin{pmatrix} a & b \\ c & d \end{pmatrix}\begin{pmatrix} 1 & 0 \\ 0 & 0 \end{pmatrix} = \begin{pmatrix} a & 0 \\ c & 0 \end{pmatrix} = aE_1 + 0E_2 + cE_3 + 0E_4$$

$$T(E_2) = \begin{pmatrix} a & b \\ c & d \end{pmatrix}\begin{pmatrix} 0 & 1 \\ 0 & 0 \end{pmatrix} = \begin{pmatrix} 0 & a \\ 0 & c \end{pmatrix} = 0E_1 + aE_2 + 0E_3 + cE_4$$

$$T(E_3) = \begin{pmatrix} a & b \\ c & d \end{pmatrix}\begin{pmatrix} 0 & 0 \\ 1 & 0 \end{pmatrix} = \begin{pmatrix} b & 0 \\ d & 0 \end{pmatrix} = bE_1 + 0E_2 + dE_3 + 0E_4$$

$$T(E_4) = \begin{pmatrix} a & b \\ c & d \end{pmatrix}\begin{pmatrix} 0 & 0 \\ 0 & 1 \end{pmatrix} = \begin{pmatrix} 0 & b \\ 0 & d \end{pmatrix} = 0E_1 + bE_2 + 0E_3 + dE_4$$

Thus

$$[T]_E = \begin{pmatrix} a & 0 & c & 0 \\ 0 & a & 0 & c \\ b & 0 & d & 0 \\ 0 & b & 0 & d \end{pmatrix} \quad \text{and} \quad \det(T) = \begin{vmatrix} a & 0 & c & 0 \\ 0 & a & 0 & c \\ b & 0 & d & 0 \\ 0 & b & 0 & d \end{vmatrix} = a\begin{vmatrix} a & 0 & c \\ 0 & d & 0 \\ b & 0 & d \end{vmatrix} + c\begin{vmatrix} 0 & a & c \\ b & 0 & 0 \\ 0 & b & d \end{vmatrix} = a^2 d^2 + b^2 c^2 - 2abcd$$

13.87 Find the trace of the above linear operator T.

$$\text{tr}(T) = \text{tr}\begin{pmatrix} a & 0 & c & 0 \\ 0 & a & 0 & c \\ b & 0 & d & 0 \\ 0 & b & 0 & d \end{pmatrix} = 2a + 2d$$

13.88 Show that $\text{tr}(AB) = \text{tr}(BA)$ for any n-square matrices A and B.

∎ Suppose $A = (a_{ij})$ and $B = (b_{ij})$. Then $AB = (c_{ik})$ where $c_{ik} = \sum_{j=1}^{n} a_{ij} b_{jk}$. Thus

$$\text{tr}(AB) = \sum_{i=1}^{n} c_{ii} = \sum_{i=1}^{n} \sum_{j=1}^{n} a_{ij} b_{ji}$$

On the other hand, $BA = (d_{jk})$ where $d_{jk} = \sum_{i=1}^{n} b_{ji} a_{ik}$. Thus

$$\text{tr}(BA) = \sum_{j=1}^{n} d_{jj} = \sum_{j=1}^{n} \sum_{i=1}^{n} b_{ji} a_{ij} = \sum_{i=1}^{n} \sum_{j=1}^{n} a_{ij} b_{ji} = \text{tr}(AB)$$

13.89 Prove Theorem 13.7: If matrix B is similar to matrix A, then $\text{tr}(B) = \text{tr}(A)$.

∎ If A is similar to B, there exists an invertible matrix P such that $A = P^{-1}BP$. Using Problem 13.88, $\text{tr}(A) = \text{tr}(P^{-1}BP) = \text{tr}(BPP^{-1}) = \text{tr}(B)$.

13.5 CHANGE OF BASIS AND LINEAR MAPPINGS

This section discusses the effect of a change of basis on the matrix representation of a linear mapping from one vector space into another. Theorem 13.8 is used below.

Theorem 13.8: Suppose P is th change-of-basis matrix from a basis $\{e_i\}$ to a basis $\{e_i'\}$ in a vector space V, and suppose Q is the change-of-basis matrix from a basis $\{f_j\}$ to a basis $\{f_i'\}$ in a vector space U. Let A be a matrix representation of a linear mapping $F: V \to U$ relative to the bases $\{e_i\}$ and $\{f_j\}$. Then
  (i)   The matrix representation of F relative to the bases $\{e_i'\}$ and $\{f_j\}$ is $Q^{-1}AP$; that is, $[F]_{e'}^{f'} = Q^{-1}[F]_e^f P$.
  (ii) The matrix representation of F relative to the bases $\{e_i'\}$ and $\{f_j\}$, i.e., when a change of basis only takes place in V, is AP; that is, $[F]_{e'}^f = [F]_e^f P$.
  (iii) The matrix representation of F relative to the bases $\{e_i\}$ and $\{f_j'\}$, i.e., when a change of basis only takes place in U, is $Q^{-1}A$; that is, $[F]_e^{f'} = Q^{-1}[F]_e^f$.

Throughout this section, E_2, E_3, and E_4 will denote, respectively, the usual basis for $\mathbf{R}^2$, $\mathbf{R}^3$, and $\mathbf{R}^4$; and S_2, S_3, and S_4 will denote, respectively, the following basis for $\mathbf{R}^2$, $\mathbf{R}^3$, and $\mathbf{R}^4$:

$$S_2 = \{(1,3),(2,5)\} \qquad S_3 = \{(1,1,1),(1,1,0),(1,0,0)\} \qquad \text{and}$$
$$S_4 = \{(1,2,3,4),(1,2,4,7),(0,1,1,1),(0,1,1,2)\}$$

13.90 Let P_2, P_3, and P_4 denote, respectively, the change-of-basis matrix from E_2 to S_2, from E_2 to S_3, and from E_4 to S_4. Find P_2, P_3, and P_4.

∎ By Problem 13.26, we need only write the new basis vectors as columns since E_2, E_3, and E_4 are the usual bases:

$$P_2 = \begin{pmatrix} 1 & 2 \\ 3 & 5 \end{pmatrix} \qquad P_3 = \begin{pmatrix} 1 & 1 & 1 \\ 1 & 1 & 0 \\ 1 & 0 & 0 \end{pmatrix} \qquad P_4 = \begin{pmatrix} 1 & 1 & 0 & 0 \\ 2 & 2 & 1 & 1 \\ 3 & 4 & 1 & 1 \\ 4 & 7 & 1 & 2 \end{pmatrix}$$

Remark: These matrices will be used below.

Problems 13.91–13.94 refer to the linear mapping $F: \mathbf{R}^3 \to \mathbf{R}^2$ defined by

$$F(x, y, z) = (2x + y - z, 3x - 2y + 4z)$$

13.91 Let A be the matrix representation of F relative to the usual bases E_3 and E_2. Find A.

❚ Since E_2 and E_3 are the usual bases, simply write the coefficients of x, y, z as rows to get $A = [F]_{E_3}^{E_2} = \begin{pmatrix} 2 & 1 & -1 \\ 3 & -2 & 4 \end{pmatrix}$.

13.92 Suppose a change of basis from E_3 to S_3 only takes place in $\mathbf{R}^3$. Find the matrix representation of F relative to the bases S_3 and E_2.

❚ By Theorem 13.8(ii),

$$[F]_{S_3}^{E_2} = AP_3 = \begin{pmatrix} 2 & 1 & -1 \\ 3 & -2 & 4 \end{pmatrix}\begin{pmatrix} 1 & 1 & 1 \\ 1 & 1 & 0 \\ 1 & 0 & 0 \end{pmatrix} = \begin{pmatrix} 2 & 3 & 2 \\ 5 & 1 & 3 \end{pmatrix}$$

13.93 Suppose a change of basis from E_2 to S_2 takes place only in $\mathbf{R}^2$. Find the matrix representation of F relative to the bases E_3 and S_2.

❚ The inverse of $P_2 = \begin{pmatrix} 1 & 2 \\ 3 & 5 \end{pmatrix}$, is $P_2^{-1} = \begin{pmatrix} -5 & 2 \\ 3 & -1 \end{pmatrix}$. Thus, by Theorem 13.8(iii),

$$[F]_{E_3}^{S_2} = P_2^{-1}A = \begin{pmatrix} -5 & 2 \\ 3 & -1 \end{pmatrix}\begin{pmatrix} 2 & 1 & -1 \\ 3 & -2 & 4 \end{pmatrix} = \begin{pmatrix} -4 & -9 & 13 \\ 3 & 5 & -7 \end{pmatrix}$$

13.94 Find the matrix representation B of F in the bases S_3 and S_2.

❚ By Theorem 13.8(i), $B = [F]_{S_3}^{S_2} = P_2^{-1}AP_3 = \begin{pmatrix} -5 & 2 \\ 3 & -1 \end{pmatrix}\begin{pmatrix} 2 & 1 & -1 \\ 3 & -2 & 4 \end{pmatrix}\begin{pmatrix} 1 & 1 & 1 \\ 1 & 1 & 0 \\ 1 & 0 & 0 \end{pmatrix} = \begin{pmatrix} 0 & -13 & -4 \\ 1 & 8 & 3 \end{pmatrix}$

13.95 Let $L: \mathbf{R}^4 \to \mathbf{R}^2$ be defined by $L(v) = Av$ where $A = \begin{pmatrix} 1 & 3 & 1 & 4 \\ 2 & 3 & 4 & 5 \end{pmatrix}$. Find the matrix representation of L relative to the usual bases E_4 and E_2.

❚ Since E_4 and E_2 are the usual bases, the matrix representation of L is the matrix A itself, i.e., $[L]_{E_4}^{E_2} = A = \begin{pmatrix} 1 & 3 & 1 & 4 \\ 2 & 3 & 4 & 5 \end{pmatrix}$.

13.96 Find the matrix representation B of the above linear map L relative to the bases S_4 and S_2.

❚ By Theorem 13.8(iii),

$$B = [L]_{S_4}^{S_2} = P_2^{-1}AP_4 = \begin{pmatrix} -5 & 2 \\ 3 & -1 \end{pmatrix}\begin{pmatrix} 1 & 3 & 1 & 4 \\ 2 & 3 & 4 & 5 \end{pmatrix}\begin{pmatrix} 1 & 1 & 0 & 0 \\ 2 & 2 & 1 & 1 \\ 3 & 4 & 1 & 1 \\ 4 & 7 & 1 & 2 \end{pmatrix} = \begin{pmatrix} -50 & -77 & -16 & -26 \\ 38 & 58 & 12 & 19 \end{pmatrix}$$

13.97 Let $F: \mathbf{R}^4 \to \mathbf{R}^3$ be defined by $F(x, y, z, t) = (2x + 3y - z + 2t, x - 5y + 6t, 2y + z + t)$. Find the matrix A which represents F using the usual bases E_4 and E_3.

❚ Write the coordinates of x, y, z, t as rows to get $A = \begin{pmatrix} 2 & 3 & -1 & 2 \\ 1 & -5 & 0 & 6 \\ 0 & 2 & 1 & 1 \end{pmatrix}$.

13.98 Find the matrix B which represents the above linear map F relative to the bases S_4 and S_3.

❚ The inverse of P_3, the change-of-basis matrix from E_3 to S_3, is $P_3^{-1} = \begin{pmatrix} 0 & 0 & 1 \\ 0 & 1 & -1 \\ 1 & -1 & 0 \end{pmatrix}$ [Problem 13.28]. Hence

$$B = P_3^{-1}AP_4 = \begin{pmatrix} 0 & 0 & 1 \\ 0 & 1 & -1 \\ 1 & -1 & 0 \end{pmatrix}\begin{pmatrix} 2 & 3 & -1 & 2 \\ 1 & -5 & 0 & 6 \\ 0 & 2 & 1 & 1 \end{pmatrix}\begin{pmatrix} 1 & 1 & 0 & 0 \\ 2 & 2 & 1 & 1 \\ 3 & 4 & 1 & 1 \\ 4 & 7 & 1 & 2 \end{pmatrix} = \begin{pmatrix} 11 & 15 & 4 & 5 \\ 4 & 18 & -3 & 2 \\ -2 & -15 & 3 & -1 \end{pmatrix}$$

13.99 Let $A: \mathbf{R}^2 \to \mathbf{R}^3$ be defined by the matrix $A = \begin{pmatrix} 2 & 3 \\ 1 & 4 \\ 0 & 2 \end{pmatrix}$. Find the matrix B which represents A relative to the bases S_2 and S_3.

▮
$$B = P_3^{-1} A P_2 = \begin{pmatrix} 0 & 0 & 1 \\ 0 & 1 & -1 \\ 1 & -1 & 0 \end{pmatrix} \begin{pmatrix} 2 & 3 \\ 1 & 4 \\ 0 & 2 \end{pmatrix} \begin{pmatrix} 1 & 2 \\ 3 & 5 \end{pmatrix} = \begin{pmatrix} 6 & 10 \\ 7 & 12 \\ -2 & -3 \end{pmatrix}$$

13.100 Prove Theorem 13.8(i): $[F]_{e'}^{f'} = Q^{-1}[F]_e^f P$.

▮ For any $v \in V$, $Q^{-1}[F]_e^f P[v]_{e'} = (Q^{-1}[F]_e^f)(P[v]_{e'}) = (Q^{-1}[F]_e^f)[v]_e = Q^{-1}([F]_e^f[v]_e) = Q^{-1}[F(v)]_f = [F(v)]_{f'}$. But $[F]_{e'}^{f'}[v]_{e'} = [F(v)]_{f'}$. Hence $Q^{-1}[F]_e^f P[v]_{e'} = [F]_{e'}^{f'}[v]_{e'}$. Since the map $v \mapsto [v]_{e'}$ is onto K^m, $Q^{-1}[F]_e^f P X = [F]_{e'}^{f'} X$, for every $X \in K^m$. Thus $[F]_{e'}^{f'} = Q^{-1}[F]_e^f P$.

13.101 Prove Theorem 13.8(ii): $[F]_e^{f'} = Q^{-1}[F]_e^f$.

▮ Consider $e' = e$. Then $P = I$, the identity matrix. Then, by above, $[F]_e^{f'} = [F]_{e'}^{f'} = Q^{-1}[F]_e^f I = Q^{-1}[F]_e^f$.

13.102 Prove Theorem 13.8(iii): $[F]_{e'}^f = [F]_e^f P$.

▮ Consider $f' = f$. Then $Q = I$ and $Q^{-1} = I$. Then, by above, $[F]_{e'}^f = [F]_{e'}^{f'} = I[F]_e^f P = [F]_e^f P$.

13.103 Define equivalence of matrices.

▮ Suppose A and B are $m \times n$ matrices for which there exists a nonsingular n-square matrix P and a nonsingular n-square matrix Q such that $B = QAP$. Then B is said to be equivalent to A.

13.104 Suppose that A and B are matrix representations of a linear map $L: V \to U$. Show that B is equivalent to A.

▮ By Theorem 13.8, there exists change-of-basis matrices P and Q such that $B = Q^{-1}AP$. Since Q^{-1} and P are nonsingular, B is equivalent to A.

Problems 13.105–13.107 show that equivalence of matrices is an equivalence relation. [Thus all matrix representations of a linear mapping $L: V \to U$ belong to the same equivalence class of equivalent matrices.]

13.105 Show that A is equivalent to A for any $m \times n$ matrix A.

▮ The identity matrices I_m and I_n are nonsingular. Since $A = I_m A I_n$, A is equivalent to A.

13.106 Show that if A is equivalent to B, then B is equivalent to A.

▮ Since A is equivalent to B, there exist nonsingular matrices P and Q such that $A = QBP$. Then $B = Q^{-1}AP^{-1}$ and Q^{-1} and P^{-1} are nonsingular. Thus B is equivalent to A.

13.107 Show that if A is equivalent to B and B is equivalent to C, then A is equivalent to C.

▮ We have $A = QBP$ and $B = Q'CP'$ where P, Q, P', and Q' are nonsingular. Then $A = QBP = QQ'CP'P$ where QQ' and $P'P$ are nonsingular. Hence A is equivalent to C.

CHAPTER 14
Inner Product Spaces, Orthogonality

The definition of a vector space V involves an arbitrary field K. In this chapter we restrict K to be either the real field **R** or the complex field **C**. Specifically, we first assume, unless otherwise stated or implied, that $K = \mathbf{R}$, in which case V is called a *real vector space*, and in the last sections we extend our results to the case that $K = \mathbf{C}$, in which case V is called a *complex vector space*.

Recall that the concepts of "length" and "orthogonality" did not appear in the investigation of arbitrary vector spaces [although they did appear in Chapter 1 on the spaces $\mathbf{R}^n$ and $\mathbf{C}^n$]. In this chapter we place an additional structure on a vector space V to obtain an *inner product space*, and in this context these concepts are defined.

14.1 INNER PRODUCT SPACES

14.1 Define an inner product and an inner product space.

▌ Let V be a real vector space. Suppose to each pair of vectors $u, v \in V$ there is assigned a real number, denoted by $\langle u, v \rangle$. This function is called a (real) inner product on V if it satisfies the following axioms [where $u_1, u_2, u, v \in V$ and $a, b, k \in \mathbf{R}$]:

[RIP$_1$] (Linear Property) $\langle au_1 + bu_2, v \rangle = a\langle u_1, v \rangle + b\langle u_2, v \rangle$ or, equivalently,
    (a) $\langle u_1 + u_2, v \rangle = \langle u_1, v \rangle + \langle u_2, v \rangle$ and (b) $\langle ku, v \rangle = k\langle u, v \rangle$.

[RIP$_2$] (Symmetric Property) $\langle u, v \rangle = \langle v, u \rangle$.

[RIP$_3$] (Positive Definite Property) If $u \neq 0$, then $\langle u, u \rangle > 0$.

The vector space V with an inner product is called an *inner product space*. [Sometimes a real inner product space is called a *Euclidean space*.]

14.2 Show that $\langle 0, v \rangle = 0 = \langle v, 0 \rangle$ for every v in V. [Thus, in particular, $\langle 0, 0 \rangle = 0$.]

▌ $\langle 0, v \rangle = \langle 0v, v \rangle = 0\langle v, v \rangle = 0$. Also, $\langle v, 0 \rangle = \langle 0, v \rangle = 0$.

[RIP$_1$] says that an inner product is linear with respect to its first position. Problems 14.3–14.4 show that a real inner product is also linear with respect to its second position.

14.3 Show that $\langle u, v_1 + v_2 \rangle = \langle u, v_1 \rangle + \langle u, v_2 \rangle$.

▌ By [RIP$_1$] and [RIP$_2$], we have $\langle u, v_1 + v_2 \rangle = \langle v_1 + v_2, u \rangle = \langle v_1, u \rangle + \langle v_2, u \rangle = \langle u, v_1 \rangle + \langle u, v_2 \rangle$.

14.4 Show that $\langle u, kv \rangle = k\langle u, v \rangle$.

▌ $$\langle u, kv \rangle = \langle kv, u \rangle = k\langle v, u \rangle = k\langle u, v \rangle.$$

[*Remark*: We emphasize that this result is slightly different for complex inner product spaces as seen by Problem 14.219.]

14.5 Define the norm or length of a vector u in an inner product space V.

▌ By [RPI$_3$], $\langle u, u \rangle$ is nonnegative and hence its positive real square root exists. We use the notation $\|u\| = \sqrt{\langle u, u \rangle}$. This nonnegative real number $\|u\|$ is called the *norm* or *length* of u. [The relation $\|u\|^2 = \langle u, u \rangle$ will be frequently used.]

14.6 Expand $\langle 5u_1 + 8u_2, 6v_1 - 7v_2 \rangle$.

▌ Use the linearity in both positions to get $\langle 5u_1 + 8u_2, 6v_1 - 7v_2 \rangle = \langle 5u_1, 6v_1 \rangle + \langle 5u_1, -7v_2 \rangle + \langle 8u_2, 6v_1 \rangle + \langle 8u_2, -7v_2 \rangle = 30\langle u_1, v_1 \rangle - 35\langle u_1, v_2 \rangle + 48\langle u_2, v_1 \rangle - 56\langle u_2, v \rangle$. [*Remark*: Observe the similarity between the above expansion and the expansion of $(5a + 8b)(6c - 7d)$ in ordinary algebra.]

14.7 Expand $\langle 3u + 5v, 4u - 6v \rangle$.

▌ $\langle 3u + 5v, 4u - 6v \rangle = 12\langle u, u \rangle - 18\langle u, v \rangle + 20\langle v, u \rangle - 30\langle v, v \rangle = 12\langle u, u \rangle - 18\langle u, v \rangle + 20\langle u, v \rangle - 30\langle v, v \rangle = 12\langle u, u \rangle + 2\langle u, v \rangle - 30\langle v, v \rangle = 12\|u\|^2 + 2\langle u, v \rangle - 30\|v\|^2$.

14.8   Expand $\|2u - 3v\|^2$.

▮ $\|2u - 3v\|^2 = \langle 2u - 3v, 2u - 3v \rangle = 4\langle u, u \rangle - 6\langle u, v \rangle - 6\langle v, u \rangle + 9\langle v, v \rangle = 4\|u\|^2 - 12\langle u, v \rangle + 9\|v\|^2$.

14.9   Expand $\langle 3u_1 + 2u_2, 5v_1 - 6v_2 + 4v_3 \rangle$.

▮   Take the scalar product of each term on the left with each term on the right to get

$$\langle 3u_1 + 2u_2, 5v_1 - 6v_2 + 4v_3 \rangle = 15\langle u_1, v_1 \rangle - 18\langle u_1, v_2 \rangle + 12\langle u_1, v_3 \rangle + 10\langle u_2, v_1 \rangle - 12\langle u_2, v_2 \rangle + 8\langle u_2, v_3 \rangle.$$

14.10   Define the usual or standard inner product on $\mathbf{R}^n$.

▮   Let $u = (a_i)$ and $v = (b_i)$ be vectors in $\mathbf{R}^n$.   Then the dot product in $\mathbf{R}^n$ defined by $u \cdot v = a_1 b_1 + a_2 b_2 + \cdots + a_n b_n$ is an inner product on $\mathbf{R}^n$.   Although there are many different ways to define an inner product on $\mathbf{R}^n$ [see Problem 14.18] we shall assume this inner product on $\mathbf{R}^n$ unless otherwise stated or implied, and we denote this inner product by $u \cdot v$ rather than $\langle u, v \rangle$.

Remark: Assuming u and v are column vectors, then the above inner product may be defined by $\langle u, v \rangle = u^T v$ where $u^T v$ refers to the product of the row vector u^T and the column vector v under matrix multiplication, e.g.,

$$\left\langle \begin{pmatrix} a_1 \\ a_2 \\ a_3 \end{pmatrix}, \begin{pmatrix} b_1 \\ b_2 \\ b_3 \end{pmatrix} \right\rangle = (a_1, a_2, a_3) \begin{pmatrix} b_1 \\ b_2 \\ b_3 \end{pmatrix} = a_1 b_1 + a_2 b_2 + a_3 b_3$$

Problems 14.11–14.17 refer to the following vectors in $\mathbf{R}^3$: $u = (1, 2, 4)$, $v = (2, -3, 5)$, $w = (4, 2, -3)$.

14.11   Find $u \cdot v$.

▮   Multiply corresponding components and add to get $u \cdot v = 2 - 6 + 20 = 16$.

14.12   Find $u \cdot w$.

▮ $u \cdot w = 4 + 4 - 12 = -4$.

14.13   Find $v \cdot w$.

▮ $v \cdot w = 8 - 6 - 15 = -13$.

14.14   Find $(u + v) \cdot w$.

▮   First find $u + v = (3, -1, 9)$.   Then $(u \cdot v) \cdot w = 12 - 2 - 27 = -17$.   Alternatively, using [RIP₁], $(u + v) \cdot w = u \cdot w + v \cdot w = -4 - 13 = -17$.

14.15   Find $\|u\|$.

▮   First find $\|u\|^2$ by squaring the components of u and adding: $\|u\|^2 = 1^2 + 2^2 + 4^2 = 1 + 4 + 16 = 21$. Then $\|u\| = \sqrt{21}$.

14.16   Find $\|v\|$.

▮ $\|v\|^2 = 4 + 9 + 25 = 38$ and so $\|v\| = \sqrt{38}$.

14.17   Find $\|u + v\|$.

▮   First find $u + v = (3, -1, 9)$.   Hence $\|u + v\|^2 = 9 + 1 + 81 = 91$.   Thus $\|u + v\| = \sqrt{91}$.

14.18   Verify that the following is an inner product in $\mathbf{R}^2$. $\langle u, v \rangle = x_1 y_1 - x_1 y_2 - x_2 y_1 + 3x_2 y_2$,   where $u = (x_1, x_2)$, $v = (y_1, y_2)$.

▮   We verify the three axioms of an inner product.   Letting $w = (z_1, z_2)$,   we find $au + bw = a(x_1, x_2) + b(z_1, z_2) = (ax_1 + bz_1, ax_2 + bz_2)$.

Thus

$$\langle au + bw, v \rangle = \langle (ax_1 + bz_1, ax_2 + bz_2), (y_1, y_2) \rangle$$
$$= (ax_1 + bz_1)y_1 - (ax_1 + bz_1)y_2 - (ax_2 + bz_2)y_1 + 3(ax_2 + bz_2)y_2$$
$$= a(x_1y_1 - x_1y_2 - x_2y_1 + 3x_2y_2) + b(z_1y_1 - z_1y_2 - z_2y_1 + 3z_2y_2)$$
$$= a\langle u, v \rangle + b\langle w, v \rangle$$

Accordingly, [RIP$_1$] is satisfied. Also $\langle v, u \rangle = y_1x_1 - y_1x_2 - y_2x_1 + 3y_2x_2 = x_1y_1 - x_1y_2 - x_2y_1 + 3x_2y_2 = \langle u, v \rangle$ and so axiom [RIP$_2$] is satisfied. Finally when $x \neq 0$ $\langle u, u \rangle = x_1^2 - 2x_1x_2 + 3x_2^2 = x_1^2 - 2x_1x_2 + x_2^2 + 2x_2^2 = (x_1 - x_2)^2 + 2x_2^2 > 0$. Hence the last axiom [RIP$_3$] is satisfied.

14.19 Show that $\langle u, v \rangle = x_1y_1x_2y_2$ is not an inner product on $\mathbf{R}^2$ where $u = (x_1, x_2)$, $v = (y_1, y_2)$.

▌ Let $k = 2$ and $u = (1, 3)$, $v = (1, 1)$. Then $ku = (2, 6)$ and we have $\langle u, v \rangle = 1 \cdot 3 \cdot 1 \cdot 1 = 3$ and $\langle ku, v \rangle = 2 \cdot 6 \cdot 1 \cdot 1 = 12$. Thus $k\langle u, v \rangle = 2 \cdot 3 = 6$ is not equal to $\langle ku, v \rangle$; and so axiom [RIP$_1$] is not satisfied.

14.20 Show that $\langle u, v \rangle = x_1y_1 + x_2y_2 - x_3y_3$ is not an inner product on $\mathbf{R}^3$ where $u = (x_1, x_2, x_3)$ and $v = (y_1, y_2, y_3)$.

▌ Let $u = (3, 4, 5)$. Then $\langle u, u \rangle = 3 \cdot 3 + 4 \cdot 4 - 5 \cdot 5 = 9 + 16 - 25 = 0$; and so axiom [RIP$_3$] is not satisfied.

Problems 14.21–14.28 refer to the following vectors in $\mathbf{R}^2$: $u = (1, 5)$, $v = (3, 4)$, $w = (7, -2)$.

14.21 Find $\langle u, v \rangle$ with respect to the usual inner product in $\mathbf{R}^2$.

▌ $\langle u, v \rangle = 3 + 20 = 23$.

14.22 Find $\langle u, v \rangle$ with respect to the inner product in $\mathbf{R}^2$ in Problem 14.18.

▌ $\langle u, v \rangle = 1 \cdot 3 - 1 \cdot 4 - 5 \cdot 3 + 3 \cdot 5 \cdot 4 = 3 - 4 - 15 + 60 = 44$.

14.23 Find $\langle u, w \rangle$ with respect to the usual inner product in $\mathbf{R}^2$.

▌ $\langle u, w \rangle = 7 - 10 = -3$.

14.24 Find $\langle u, w \rangle$ using the inner product in $\mathbf{R}^2$ in Problem 14.18.

▌ $\langle u, w \rangle = 1 \cdot 7 - 1 \cdot (-2) - 5 \cdot 7 + 3 \cdot 5 \cdot (-2) = 7 + 2 - 35 - 30 = -56$.

14.25 Find $\|v\|$ using the usual inner product in $\mathbf{R}^2$.

▌ $\|v\|^2 = \langle v, v \rangle = \langle (3, 4), (3, 4) \rangle = 9 + 16 = 25$; hence $\|v\| = 5$.

14.26 Find $\|v\|$ using the inner product in $\mathbf{R}^2$ in Problem 14.18.

▌ $\|v\|^2 = \langle v, v \rangle = \langle (3, 4), (3, 4) \rangle = 9 - 12 - 12 + 48 = 33$; hence $\|v\| = \sqrt{33}$.

14.27 Find $\|w\|$ using the usual inner product in $\mathbf{R}^2$.

▌ $\|w\|^2 = \langle w, w \rangle = 49 + 4 = 53$; hence $\|w\| = \sqrt{53}$.

14.28 Find $\|w\|$ using the inner product in $\mathbf{R}^2$ in Problem 14.18.

▌ $\|w\|^2 = \langle w, w \rangle = 49 + 14 + 14 + 12 = 89$; hence $\|w\| = \sqrt{89}$.

14.29 Define a unit vector.

▌ If $\|u\| = 1$ or, equivalently, if $\langle u, u \rangle = 1$, then u is called a unit vector and is said to be normalized.

14.30 Show that $\|v\| > 0$ for any vector $v \neq 0$.

▌ By [RIP$_3$], $\langle v, v \rangle$ is positive. Hence $\|v\| = \sqrt{\langle v, v \rangle}$ is also positive.

14.31 Show that if $v \neq 0$, then $\|\hat{v}\| = \dfrac{1}{\|v\|}v$ is the unique unit vector that is a positive multiple of v. [The process of obtaining $\hat{v}$ from v is called normalizing v.]

▮ Suppose $\hat{v} = kv$ where $k > 0$ and $\|\hat{v}\| = 1$. Then $1 = \|v\|^2 = \langle kv, kv \rangle = k^2 \langle v, v \rangle = k^2 \|v\|^2$. Since k is positive, we get $k = 1/\|v\|$.

14.32 Normalize $u = (2, 1, -1)$ in Euclidean 3-space $\mathbf{R}^3$.

▮ Note $\langle u, u \rangle$ is the sum of the squares of the entries of u; that is, $\langle u, u \rangle = 2^2 + 1^2 + (-1)^2 = 6$. Hence divide u by $\|u\| = \sqrt{\langle u, u \rangle} = \sqrt{6}$ to obtain the required unit vector: $\hat{u} = u/\|u\| = (2/\sqrt{6}, 1/\sqrt{6}, -1/\sqrt{6})$.

14.33 Normalize $v = (\frac{1}{2}, \frac{2}{3}, -\frac{1}{4})$ in Euclidean 3-space $\mathbf{R}^3$.

▮ First multiply v by 12 to "clear" of fractions obtaining $12v = (6, 8, -3)$. We have $\langle 12v, 12v \rangle = 6^2 + 8^2 + (-3)^2 = 109$. Then the required unit vector is $\hat{v} = 12v/\|12v\| = (6/\sqrt{109}, 8/\sqrt{109}, -3/\sqrt{109})$.

14.34 Normalize $v = (3, 4)$ in $\mathbf{R}^2$: (a) using the usual inner product in $\mathbf{R}^2$, (b) using the inner product in $\mathbf{R}^2$ defined in Problem 14.18.
▮ (a) By Problem 14.25, $\|v\| = 5$; hence $\hat{v} = v/\|v\| = (\frac{3}{5}, \frac{4}{5})$.
(b) By Problem 14.26, $\|v\| = \sqrt{33}$. Thus $\hat{v} = v/\|v\| = (3/\sqrt{33}, 4/\sqrt{33})$.

14.35 Define the distance between vectors u and v, denoted by $d(u, v)$, in an inner product space V.

▮ The distance $d(u, v)$ is defined in terms of the norm as follows: $d(u, v) = \|u - v\|$.

14.36 Show how the above definition of distance in an inner product space V agrees with the usual notion of (Euclidean) distance in $\mathbf{R}^3$.

▮ Let $P(a_1, a_2, a_3)$ and $Q(b_1, b_2, b_3)$ be points in $\mathbf{R}^3$ [with corresponding vectors u and v from the origin 0 to P and Q, respectively] as pictured in Fig. 14-1. Then the distance d between P and Q is as follows: $d = \sqrt{(a_1 - b_1)^2 + (a_2 - b_2)^2 + (a_3 - b_3)^2}$ which agrees with $d(u, v) = \|u - v\| = \|(a_1 - b_1, a_2 - b_2, a_3 - b_3)\| = (a_1 - b_1)^2 + (a_2 - b_2)^2 + (a_3 - b_3)^2$.

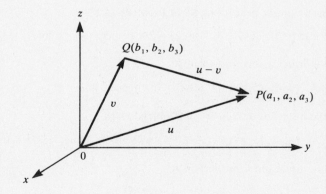

Fig. 14-1

Problems 14.37–14.39 refer to the following vectors in Euclidean space $\mathbf{R}^4$: $u = (5, 5, 8, 8)$, $v = (1, 2, 3, 4)$, $w = (4, -3, 2, -1)$.

14.37 Find $d(u, v)$.

▮ First find $u - v = (5 - 1, 5 - 2, 8 - 3, 8 - 4) = (4, 3, 5, 4)$. Then find $\|u - v\|^2 = 4^2 + 3^2 + 5^2 + 4^2 = 16 + 9 + 25 + 16 = 66$. Hence $d(u, v) = \sqrt{66}$.

14.38 Find $d(u, w)$.

▮ $u - w = (1, 8, 6, 9)$ and $\|u - w\|^2 = 1 + 64 + 36 + 81 = 182$. Thus $d(u, w) = \sqrt{182}$.

14.39 Find $d(v, w)$.

$\blacksquare$ $v - w = (-3, 5, 1, 5)$ and $\|v - w\|^2 = 9 + 25 + 1 + 25 = 60$. Thus $d(v, w) = \sqrt{60} = 2\sqrt{15}$.

14.40 Find $d(u, v)$ where $u = (5, 4)$, $v = (2, -6)$ in Euclidean space $\mathbf{R}^2$.

$\blacksquare$ $u - v = (3, 10)$ and $\|u - v\|^2 = 9 + 100 = 109$. Hence $d(u, v) = \sqrt{109}$.

14.41 Find $d(u, v)$ for u, v in Problem 14.40 using the inner product in $\mathbf{R}^2$ in Problem 14.18.

$\blacksquare$ We have $u - v = (3, 10)$. Hence $\|u - v\|^2 = \langle (3, 10), (3, 10) \rangle = 3 \cdot 3 - 3 \cdot 10 - 10 \cdot 3 + 3 \cdot 10 \cdot 10 = 9 - 30 - 30 + 300 = 249$. Thus $d(u, v) = \sqrt{249}$.

Remark: The above examples show that the distance between vectors depends on the way the inner product is defined.

14.42 Let V be a vector space of real continuous functions on the interval $a \le t \le b$. Show that the following is an inner product on V:

$$\langle f, g \rangle = \int_a^b f(t)g(t)\, dt$$

$\blacksquare$ Let f, g, h be functions in V. Then, using results from calculus,

$$\langle f + g, h \rangle = \int_a^b (f(t) + g(t))h(t)\, dt = \int_a^b f(t)h(t)\, dt + \int_a^b g(t)h(t)\, dt = \langle f, h \rangle + \langle g, h \rangle$$

and

$$\langle kf, g \rangle = \int_a^b (kf(t))g(t)\, dt = k \int_a^b f(t)g(t)\, dt = k\langle f, g \rangle$$

Thus [RIP$_1$] is satisfied. Also $\langle f, g \rangle = \int_a^b f(t)g(t)\, dt = \int_a^b g(t)f(t)\, dt = \langle g, f \rangle$ and so [RIP$_2$] is satisfied.

Finally, if $f \ne 0$, then $\langle f, f \rangle = \int_a^b (f(t))^2\, dt > 0$. Thus [RIP$_3$] is satisfied. Consequently, this product is an inner product on V.

Problems 14.43–14.49 refer to the vector space V of polynomials with inner product defined by $\int_0^1 f(t)g(t)\, dt$ and the polynomials $f(t) = t + 2$, $g(t) = 3t - 2$, and $h(t) = t^2 - 2t - 3$.

14.43 Find $\langle f, g \rangle$.

$\blacksquare$
$$\langle f, g \rangle = \int_0^1 (t + 2)(3t - 2)\, dt = \int_0^1 (3t^2 + 4t - 4)\, dt = [t^3 + 2t^2 - 4t]_0^1 = -1$$

14.44 Find $\langle f, h \rangle$.

$\blacksquare$
$$\langle f, h \rangle = \int_0^1 (t + 2)(t^2 - 2t - 3)\, dt = \left[\frac{t^4}{4} - \frac{7t^2}{2} - 6t \right]_0^1 = -\frac{37}{4}$$

14.45 Find $\|f\|$.

$\blacksquare$
$$\langle f, f \rangle = \int_0^1 (t + 2)(t + 2)\, dt = \tfrac{19}{3} \quad \text{and} \quad \|f\| = \sqrt{\langle f, f \rangle} = \sqrt{\tfrac{19}{3}} = \tfrac{1}{3}\sqrt{57}.$$

14.46 Find $\|g\|$.

$\blacksquare$
$$\langle g, g \rangle = \int_0^1 (3t - 2)(3t - 2)\, dt = 1; \text{ hence } \|g\| = \sqrt{1} = 1.$$

14.47 Normalize f.

$\blacksquare$ Since $\|f\| = \tfrac{1}{3}\sqrt{57}$,

$$\hat{f} = \frac{1}{\|f\|}\, f = \frac{3}{\sqrt{57}}\, (t + 2)$$

14.48 Normalize g.

▮ Note g is already a unit vector since $\|g\| = 1$; hence $\hat{g} = g = 3t - 2$.

14.49 Find $d(f, g)$.

▮ We have $f(t) - g(t) = -2t + 4$. Then

$$\|f - g\|^2 = \langle f - g, f - g \rangle = \int_0^1 (-2t + 4)(-2t + 4)\,dt = \int_0^1 (4t^2 - 16t + 16) = [\tfrac{4}{3}t^3 - 8t^2 + 16t]_0^1 = \tfrac{28}{3}$$

Hence $d(f, g) = \sqrt{\tfrac{28}{3}} = \tfrac{2}{3}\sqrt{21}$.

Problems 14.50–14.65 refer to the vector space V of $m \times n$ matrices over $\mathbf{R}$ and the inner product $\langle\,,\,\rangle$ on V defined by $\langle A, B \rangle = \text{tr}(B^T A)$ where tr stands for trace, the sum of the diagonal elements. In particular, Problems 14.50–14.52 show that $\langle\,,\,\rangle$ does satisfy the three axioms of an inner product.

14.50 Show that $\langle\,,\,\rangle$ satisfies [RIP$_1$].

▮ Using properties of the trace function, $\langle A_1 + A_2, B \rangle = \text{tr}[B^T(A_1 + A_2)] = \text{tr}[B^T A_1 + B^T A_2] = \text{tr}(B^T A_1) + \text{tr}(B^T A_2) = \langle A_1, B \rangle + \langle A_2, B \rangle$ and $\langle kA, B \rangle = \text{tr}[B^T(kA)] = \text{tr}[k(B^T A)] = k\,\text{tr}(B^T A) = k\langle A, B \rangle$.

14.51 Show that $\langle\,,\,\rangle$ satisfies [RIP$_2$].

▮ Using the fact that $\text{tr}(M) = \text{tr}(M^T)$, we have $\langle A, B \rangle = \text{tr}(B^T A) = \text{tr}[(B^T A)^T] = \text{tr}[A^T B^{TT})] = \text{tr}(A^T B) = \langle B, A \rangle$.

14.52 Let $A = (a_{ij})$. Show that $\langle A, A \rangle = \text{tr}(A^T A) = \sum_{i=1}^{m}\sum_{j=1}^{n} a_{ij}^2$, the sum of squares of all the elements of A. Thus $\langle\,,\,\rangle$ satisfies [RIP$_3$].

▮ Let $A^T = (b_{ij})$ and so $b_{ij} = a_{ji}$, and let $A^T A = (c_{ij})$. Then

$$c_{ii} = \sum_{j=1}^{n} b_{ij} a_{ji} = \sum_{j=1}^{n} a_{ji}^2$$

and so

$$\text{tr}(A^T A) = \sum_{i=1}^{m} c_{ii} = \sum_{i=1}^{m}\sum_{j=1}^{n} a_{ji}^2 = \sum_{i=1}^{m}\sum_{j=1}^{n} a_{ij}^2$$

14.53 Let $A = [C_1, C_2, \ldots, C_n]$ and $B = [D_1, D_2, \ldots, D_n]$ here the C_i and D_i are, respectively, the columns of the matrices A and B. Show that $\langle A, B \rangle = D_1^T C_1 + D_2^T C_2 + \cdots + D_n^T C_n$.

▮ Let $B^T A = (c_{ij})$. Then $c_{ii} = D_i^T C_i$ and so $\langle A, B \rangle = \text{tr}(B^T A) = D_1^T C_1 + \cdots + D_n^T C_n$.

Problems 14.54–14.65 refer to the following matrices:

$$A = \begin{pmatrix} 9 & 8 & 7 \\ 6 & 5 & 4 \end{pmatrix} \qquad B = \begin{pmatrix} 1 & 2 & 3 \\ 4 & 5 & 6 \end{pmatrix} \qquad C = \begin{pmatrix} 3 & -5 & 2 \\ 1 & 0 & -4 \end{pmatrix}$$

14.54 Find $\langle A, B \rangle$.

▮ $\langle A, B \rangle = (1, 4)\begin{pmatrix} 9 \\ 6 \end{pmatrix} + (2, 5)\begin{pmatrix} 8 \\ 5 \end{pmatrix} + (3, 6)\begin{pmatrix} 7 \\ 4 \end{pmatrix} = (9 + 24) + (16 + 25) + (21 + 24) = 119$

14.55 Find $\langle A, C \rangle$.

▮ $\langle A, C \rangle = (27 + 6) + (-40 + 0) + (14 - 16) = -9$.

14.56 Find $\langle B, C \rangle$.

▮ $\langle B, C \rangle = (3 + 4) + (-10 + 0) + (6 - 24) = -21$.

14.57 Find $\langle A, B + C \rangle$.

▌ First find $B + C = \begin{pmatrix} 4 & -3 & 5 \\ 5 & 5 & 2 \end{pmatrix}$. Then $\langle A, B + C \rangle = (36 + 30) + (-24 + 25) + (35 + 8) = 110$.

Alternatively, $\langle A, B + C \rangle = \langle A, B \rangle + \langle A, C \rangle = 119 + (-9) = 110$.

14.58 Find $\langle 2A + 3B, 4C \rangle$.

▌ $\langle 2A + 3B, 4C \rangle = 8\langle A, C \rangle + 12\langle B, C \rangle = 8(-9) + 12(-21) = -324$.

14.59 Find $\|A\|$.

▌ First find $\langle A, A \rangle = \|A\|^2$ by squaring the components of A and adding: $\langle A, A \rangle = 9^2 + 8^2 + 7^2 + 6^2 + 5^2 + 4^2 = 271$. Hence $\|A\| = \sqrt{271}$.

14.60 Find $\|B\|$.

▌ $\langle B, B \rangle = \|B\|^2 = 1^2 + 2^2 + 3^2 + 4^2 + 5^2 + 6^2 = 91$ and so $\|B\| = \sqrt{91}$.

14.61 Normalize B.

▌
$$\hat{B} = \frac{1}{\|B\|} B = \frac{1}{\sqrt{91}} B = \begin{pmatrix} 1/\sqrt{91} & 2/\sqrt{91} & 3/\sqrt{91} \\ 4/\sqrt{91} & 5/\sqrt{91} & 6/\sqrt{91} \end{pmatrix}$$

14.62 Find $\|C\|$.

▌ $\langle C, C \rangle = \|C\|^2 = 9 + 25 + 4 + 1 + 16 = 55$ and so $\|C\| = \sqrt{55}$.

14.63 Normalize C.

▌ Divide each entry of C by $\|C\|$ to get

$$\hat{C} = \frac{1}{\|C\|} C = \begin{pmatrix} 3/\sqrt{55} & -5/\sqrt{55} & 2/\sqrt{55} \\ 1/\sqrt{55} & 0 & -4/\sqrt{55} \end{pmatrix}$$

14.64 Find $d(A, B)$.

▌ First find $A - B = \begin{pmatrix} 8 & 6 & 4 \\ 2 & 0 & -2 \end{pmatrix}$. Then $\|A - B\|^2 = 64 + 36 + 16 + 4 + 0 + 4 = 124$. Thus $d(A, B) = \|A - B\| = \sqrt{124} = 2\sqrt{31}$.

14.65 Find $d(A, C)$.

▌ $A - C = \begin{pmatrix} 6 & 13 & 5 \\ 5 & 5 & 8 \end{pmatrix}$, so $\|A - C\|^2 = 36 + 169 + 25 + 25 + 25 + 64 = 344$. Hence $d(A, C) = \sqrt{344} = 2\sqrt{86}$.

14.2 PROPERTIES OF INNER PRODUCTS AND NORMS

14.66 Show that an inner product $\langle \, , \, \rangle$ satisfies the following nondegeneracy axiom:
[ND] $\langle u, v \rangle = 0$ for every $v \in V$ if and only if $u = 0$.

▌ If $u = 0$, then $\langle u, v \rangle = \langle 0, v \rangle = \langle 0v, v \rangle = 0\langle v, v \rangle = 0$. On the other hand, if $u \neq 0$, then, for $v = u$, we have $\langle u, v \rangle = \langle u, u \rangle \neq 0$.

14.67 Show that $\langle a_1 u_1 + \cdots + a_r u_r, v \rangle = a_1 \langle u_1, v \rangle + a_2 \langle u_2, v \rangle + \cdots + a_r \langle u_r, v \rangle$.

▌ The proof is by induction on r. By [RIP$_1$], the result holds for $r = 1$. Suppose $r > 1$. Then $\langle a_1 u_1 + \cdots + a_r u_r, v \rangle = \langle a_1 u_1 + \cdots + a_{r-1} u_{r-1}, v \rangle + \langle a_r u_r, v \rangle = a_1 \langle u_1, v \rangle + a_2 \langle u_2, v \rangle + \cdots + a_{r-1} \langle u_{r-1}, v \rangle + a_r \langle u_r, v \rangle$.

14.68 Show that $\left(\sum_{i=1}^{r} a_i u_i, \sum_{j=1}^{s} b_j v_j \right) = \sum_{i=1}^{r} \sum_{j=1}^{s} a_i b_j (u_i, v_j)$.

$\blacksquare$ By Problem 14.7 and [RIP$_2$],

$$\left\langle \sum_{i=1}^{r} a_i u_i, \sum_{j=1}^{s} b_j v_j \right\rangle = \sum_{i=1}^{r} a_i \left\langle u_i, \sum_{j=1}^{s} b_j v_j \right\rangle = \sum_{i=1}^{r} a_i \left\langle \sum_{j=1}^{s} b_j v_j, u_i \right\rangle$$

$$= \sum_{i=1}^{r} \sum_{j=1}^{s} a_i b_j \langle v_j, u_i \rangle = \sum_{i=1}^{r} \sum_{j=1}^{s} a_i b_j \langle u_i, v_j \rangle$$

14.69 Show that $\|u+v\|^2 = \|u\|^2 + 2\langle u, v \rangle + \|v\|^2$.

$\blacksquare$ $\|u+v\|^2 = \langle u+v, u+v \rangle = \langle u, u \rangle + \langle u, v \rangle + \langle v, u \rangle + \langle v, v \rangle = \langle u, u \rangle + \langle u, v \rangle + \langle u, v \rangle + \langle v, v \rangle = \|u\|^2 + 2\langle u, v \rangle + \|v\|^2$.

14.70 Show that $\|u-v\|^2 = \|u\|^2 - 2\langle u, v \rangle + \|v\|^2$.

$\blacksquare$ $\|u-v\|^2 = \langle u-v, u-v \rangle = \langle u, u \rangle - \langle u, v \rangle - \langle v, u \rangle + \langle v, v \rangle = \langle u, u \rangle - \langle u, v \rangle - \langle u, v \rangle + \langle v, v \rangle = \|u\|^2 - 2\langle u, v \rangle + \|v\|^2$.

14.71 Show that $\langle u+v, u-v \rangle = \|u\|^2 - \|v\|^2$.

$\blacksquare$ $\langle u+v, u-v \rangle = \langle u, u \rangle - \langle u, v \rangle + \langle v, u \rangle - \langle v, v \rangle = \|u\|^2 - \langle u, v \rangle + \langle u, v \rangle - \|v\|^2 = \|u\|^2 - \|v\|^2$.

14.72 Verify the Parallelogram Law: $\|u+v\|^2 + \|u-v\|^2 = 2\|u\| + 2\|v\|$. [See Fig. 14-2.]

$\blacksquare$ Add the equations in Problems 14.69 and 14.70 to get $\|u+v\|^2 + \|u-v\|^2 = 2\|u\|^2 + 2\|v\|^2$.

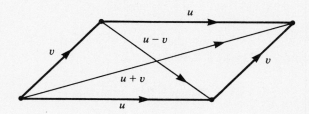

Fig. 14-2

14.73 Verify the following polar form for $\langle u, v \rangle$ [which shows that the inner product can be obtained from the norm function]: $\langle u, v \rangle = \frac{1}{4}(\|u+v\|^2, \|u-v\|^2)$.

$\blacksquare$ Subtract the equation in Problem 14.70 from the equation in Problem 14.69 to get $\|u+v\|^2 - \|u-v\|^2 = 4\langle u, v \rangle$. Dividing by 4 gives us the result.

Problems 14.74–14.80 refer to two inner products f and g on the same vector space V. [Here we use the functional notation $f(u, v)$ and $g(u, v)$ to denote the inner products of u and v under f and g, respectively.]

14.74 Show that if f and g have equal associated norm functions, i.e., $\|v\|_f = \|v\|_g$ for every $v \in V$, then $f = g$.

$\blacksquare$ Use the polar form of the inner product in Problem 14.73 to get $f(u, v) = \frac{1}{4}(\|u+v\|_f^2 - \|u-v\|_f^2) = \frac{1}{4}(\|u+v\|_g^2 - \|u-v\|_g^2) = g(u, v)$. Thus $f = g$.

Problems 14.75–14.77 show that the sum $f + g$, defined by $(f + g)(u, v) = f(u, v) + g(u, v)$, is also an inner product on V.

14.75 Show that $f + g$ satisfies axiom [RIP$_1$].

$\blacksquare$
$$(f+g)(au_1 + bu_2, v) = f(au_1 + bu_2, v) + g(au_1 + bu_2, v)$$
$$= af(u_1, v) + bf(u_2, v) + ag(u_1, v) + bg(u_2, v)$$
$$= a[f(u_1, v) + g(u_1, v)] + b[f(u_2, v) + g(u_2, v)]$$
$$= a[(f+g)(u_1, v)] + b[(f+g)(u_2, v)]$$

Thus $f + g$ satisfies [RIP$_1$].

14.76   Show that $f + g$ satisfies axiom [RIP$_2$].

▮ $(f + g)(u, v) = f(u, v) + g(u, v) = f(v, u) + g(v, u) = (f + g)(v, u)$.   Thus $f + g$ satisfies [RIP$_2$].

14.77   Show that $f + g$ satisfies axiom [RIP$_3$].

▮ Suppose $u \neq 0$.   Then $f(u, u)$ and $g(u, u)$ are both positive.   Hence $(f + g)(u, u) = f(u, u) + g(u, u)$ is positive.   Thus $f + g$ satisfies [RIP$_3$].

Problems 14.78–14.80 show that the scalar multiple kf, defined by $(kf)(u, v) = kg(u, v)$,   is also an inner product on V when $k > 0$.

14.78   Show that kf satisfies [RIP$_1$].

▮ $(kf)(au_1 + bu_2, v) = k[f(au_1 + bu_2, v)] = k[af(u_1, v) + bf(u_2, v)] = a[kf(u_1, v)] + b[kf(u_2, v)] = a(kf)(u_1, v) + b(kf)(u_2, v)$.   Thus fk satisfies [RIP$_1$].

14.79   Show that kf satisfies [RIP$_2$].

▮ $(kf)(u, v) = kf(u, v) = kf(v, u) = (kf)(u, v)$.   Thus kf satisfies [RIP$_2$].

14.80   Show that kf satisfies [RIP$_3$].

▮ If $u \neq 0$,   then $f(u, u)$ is positive.   But k is positive.   Thus $(kf)(u, u) = kf(u, u)$ is also positive. Thus kf satisfies [RIP$_3$].

14.3   CAUCHY-SCHWARZ INEQUALITY AND APPLICATIONS

This section uses the following important theorem, proved in Problem 14.92.

Theorem 14.1 (Cauchy-Schwarz Inequality):   For any vectors $u, v \in V$, $\langle u, v \rangle^2 \leq \|u\|^2 \|v\|^2$ [or, equivalently, $|\langle u, v \rangle| \leq \|u\| \|v\|$].

14.81   Let $x_1, x_2, \ldots, x_n$ and $y_1, y_2, \ldots, y_n$ be real numbers.   Show that $(x_1 y_1 + x_2 y_2 + \cdots + x_n y_n) \leq (x_1^2 + \cdots + x_n^2)(y_1^2 + \cdots + y_n^2)$.

▮ Let $u = (x_i)$ and $v = (y_i)$ be the corresponding vectors in $\mathbf{R}^n$.   By the Cauchy-Schwarz inequality $\langle u, v \rangle^2 \leq \|u\|^2 \cdot \|v\|^2$.   This gives the desired inequality.

14.82   Let f and g be any real continuous functions on a closed interval $D = [a, b]$.   Show that

$$\left(\int_a^b f(t)g(t) \, dt \right)^2 \leq \int_a^b f^2(t) \, dt \int_a^b g^2(t) \, dt$$

▮ Let V be the vector space of continuous functions on D.   By Problem 14.42, the following is an inner product on V, $\langle f, g \rangle = \int_a^b f(t)g(t) \, dt$.   Thus, by the Cauchy-Schwarz inequality, $\langle f, g \rangle^2 \leq \|f\|^2 \cdot \|g\|^2$. This, however, is our desired result.

14.83   Define the inner product space V known as l_2-space (or Hilbert space).

▮ V is the vector space of infinite sequences of real numbers $(a_1, a_2, \ldots)$ satisfying $\sum_{i=1}^{\infty} a_i^2 = a_1^2 + a_2^2 + \cdots < \infty$,   i.e., the sum converges.   Addition and scalar multiplication are defined componentwise:

$$(a_1, a_2, \ldots) + (b_1, b_2, \ldots) = (a_1 + b_1, a_2 + b_2, \ldots)$$
$$k(a_1, a_2, \ldots) = (ka_1, ka_2, \ldots)$$

An inner product is defined in V by $\langle (a_1, a_2, \ldots), (b_1, b_2, \ldots) \rangle = a_1 b_1 + a_2 b_2 + \cdots$.

14.84   Show that the inner product in the above l_2-space is well-defined, i.e., show that the sum $\sum_{i=1}^{\infty} a_i b_i = a_1 b_1 + a_2 b_2 + \cdots$ converges absolutely.

▮ By the Cauchy-Schwarz inequality,

$$|a_1 b_1| + \cdots + |a_n b_n| \le \sqrt{\sum_{i=1}^{n} a_i^2} \; \sqrt{\sum_{i=1}^{n} b_i^2} \le \sqrt{\sum_{i=1}^{\infty} a_i^2} \; \sqrt{\sum_{i=1}^{\infty} b_i^2}$$

which holds for every n. Thus the (monotonic) sequence of sums $S_n = |a_1 b_1| + \cdots + |a_n b_n|$ is bounded, and therefore converges. Hence the infinite sum converges absolutely.

14.85 Define angles in a (real) inner product space V.

▮ The angle θ between nonzero vectors $u, v \in V$ is defined to be the unique angle θ such that $0 \le \theta \le \pi$ and $\cos \theta = \langle u, v \rangle / \|u\| \, \|v\|$.

14.86 Why does the above angle θ always exist?

▮ By the Cauchy-Schwarz inequality, $-1 \le \cos \theta \le 1$ and so an angle θ can always be obtained. By the restriction $0 \le \theta \le \pi$, the angle θ is unique.

14.87 Find $\cos \theta$ for the angle θ between $u = (1, -3, 2)$, $v = (2, 1, 5)$ in $\mathbf{R}^3$.

▮ Compute $\langle u, v \rangle = 2 - 3 + 10 = 9$, $\|u\|^2 = 1 + 9 + 4 = 14$, $\|v\|^2 = 4 + 1 + 25 = 30$. Thus,

$$\cos \theta = \frac{9}{\sqrt{14}\sqrt{30}} = \frac{9}{\sqrt{105}}$$

14.88 Find $\cos \theta$ for the angle θ between $u = (5, 1)$ and $v = (-2, 3)$ is Euclidean 2-space $\mathbf{R}^2$. In which quadrant does θ lie?

▮ Compute $\langle u, v \rangle = -10 + 3 = -7$, $\|u\|^2 = 25 + 1 = 26$, $\|v\|^2 = 4 + 9 = 13$. Thus

$$\cos \theta = \frac{-7}{\sqrt{13}\sqrt{26}} = -\frac{7}{13\sqrt{2}}$$

Since $\cos \theta$ is negative, θ lies in the second quadrant.

14.89 Find $\cos \theta$ for the angle θ between $u = (5, 1)$ and $v = (-2, 3)$ in $\mathbf{R}^2$ and the inner product defined in Problem 14.18. [Compare with Problem 14.88.]

▮ Compute $\langle u, v \rangle = -10 - 15 + 2 + 9 = -14$, $\|u\|^2 = \langle u, u \rangle = 25 - 5 - 5 + 3 = 18$, $\|v\|^2 = \langle v, v \rangle = 4 + 6 + 6 + 27 = 43$. Thus

$$\cos \theta = \frac{-14}{\sqrt{18}\sqrt{43}} = -\frac{14}{3\sqrt{86}}$$

14.90 Find $\cos \theta$ for the angle θ between $f(t) = 2t - 1$ and $g(t) = t^2$ in the vector space V of polynomials with inner product $\langle f, g \rangle = \int_0^1 f(t) g(t) \, dt$.

▮ Compute

$$\langle f, g \rangle = \int_0^1 (2t^3 - t^2) \, dt = \left[\frac{t^4}{2} - \frac{t^3}{3} \right]_0^1 = \tfrac{1}{2} - \tfrac{1}{3} = \tfrac{1}{6}$$

$$\|f\|^2 = \langle f, f \rangle = \int_0^1 (4t^2 - 4t + 1) \, dt = \tfrac{1}{3}$$

$$\|g\|^2 = \langle g, g \rangle = \int_0^1 t^4 \, dt = \tfrac{1}{5}$$

Thus

$$\cos \theta = \frac{\tfrac{1}{6}}{(1/\sqrt{3})(1/\sqrt{5})} = \frac{\sqrt{15}}{6}$$

14.91 Find $\cos \theta$ for the angle θ between $A = \begin{pmatrix} 2 & 1 \\ 3 & -1 \end{pmatrix}$, $B = \begin{pmatrix} 0 & -1 \\ 2 & 3 \end{pmatrix}$ in the vector space of 2×2 real matrices with inner product defined by $\langle A, B \rangle = \operatorname{tr}(B^T A)$. [See Problems 14.50–14.53.]

▮ Compute $\langle A, B \rangle = (0 + 6) + (-1 - 3) = 2$, $\|A\|^2 = 4 + 1 + 9 + 1 = 15$, $\|B\|^2 = 0 + 1 + 4 + 9 = 14$. Thus

$$\cos \theta = \frac{2}{\sqrt{15}\sqrt{14}} = \frac{2}{\sqrt{210}}$$

14.92 Prove Theorem 14.1 (Cauchy-Schwarz): $\langle u, v \rangle^2 \le \|u\|^2 \cdot \|v\|^2$.

▮ For any real number t, $\langle tu + v, tu + v \rangle = t^2 \langle u, u \rangle + 2t \langle u, v \rangle + \langle v, v \rangle = t^2 \|u\|^2 + 2t \langle u, v \rangle + \|v\|^2$. Let $a = \|u\|^2$, $b = 2 \langle u, v \rangle$, and $c = \|v\|^2$. Since $\|tu + v\|^2 \ge 0$, we have $at^2 + bt + c \ge 0$ for every value of t. This means that the quadratic polynomial cannot have two real roots. This implies that $b^2 - 4ac \le 0$ or $b^2 \le 4ac$. Thus $4 \langle u, v \rangle^2 \le 4 \|u\|^2 \|v\|^2$. Dividing by 4 gives us our result.

Remark: The Cauchy-Schwarz inequality for complex inner product spaces appears in Problem 14.233.

Problems 14.93–14.95 show that the norm coming from an inner product does satisfy the three axioms of a norm. [See Section 14.10.]

14.93 Prove [N_1]: $\|v\| \ge 0$; and $\|v\| = 0$ if and only if $v = 0$.

▮ If $v \ne 0$, then $\langle v, v \rangle > 0$ and hence $\|v\| = \sqrt{\langle v, v \rangle} > 0$. If $v = 0$ then $\langle 0, 0 \rangle = 0$ and so $\|0\| = \sqrt{0} = 0$.

14.94 Prove [N_2]: $\|kv\| = |k| \, \|v\|$.

▮ We have $\|kv\|^2 = \langle kv, kv \rangle = k^2 \langle v, v \rangle = k^2 \|v\|^2$. Taking the square root of both sides gives [N_2].

14.95 Prove [N_3]: $\|u + v\| \le \|u\| + \|v\|$.

▮ Using the Cauchy-Schwarz inequality, we obtain $\|u + v\|^2 = \langle u + v, u + v \rangle = \langle u, u \rangle + \langle u, v \rangle + \langle u, v \rangle + \langle v, v \rangle \le \|u\|^2 + 2\|u\| \, \|v\| + \|v\|^2 = (\|u\| + \|v\|)^2$. Taking the square root of both sides yields [N_3].

14.96 Axiom [N_3] is frequently called the *triangle inequality*. Why?

▮ If we view $u + v$ as the side of the triangle formed with u and v [as pictured in Fig. 14-3], then [N_3] states that the length of one side of a triangle is less than or equal to the sum of the lengths of the other two sides.

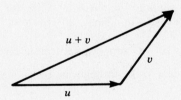

Fig. 14-3

14.4 ORTHOGONALITY, ORTHOGONAL COMPLEMENTS

14.97 Define orthogonality in an inner product space V.

▮ The vectors $u, v \in V$ are said to be *orthogonal* or, equivalently, u is said to be orthogonal to v, written $u \perp v$, if $\langle u, v \rangle = 0$.

14.98 Show that the relation of orthogonality is symmetric, i.e., if $u \perp v$, then $v \perp u$.

▮ If $u \perp v$, then $\langle u, v \rangle = 0$. Hence $\langle v, u \rangle = \langle u, v \rangle = 0$ and so $v \perp u$.

14.99 Show that $0 \in V$ is orthogonal to every $v \in V$.

▮ We have $\langle 0, v \rangle = \langle 0v, v \rangle = 0 \langle v, v \rangle = 0$; hence 0 is orthogonal to every $v \in V$.

14.100 Show that if u is orthogonal to every $v \in V$ then $u = 0$.

▮ If $\langle u, v \rangle = 0$ for every $v \in V$, then $\langle u, u \rangle = 0$ and hence $u = 0$.

Remark: Observe that Problems 14.99 and 14.100 are a restatement that an inner product satisfies the nondegeneracy axiom [ND] in Problem 14.66.

14.101 Suppose u and v are nonzero in V. Show that u and v are orthogonal if and only if they are "perpendicular," i.e., $\theta = \pi/2$ (or $\theta = 90°$) where θ is the angle between u and v.

▮ We have u and v are orthogonal iff $\langle u, v \rangle = 0$ iff $\cos \theta = 0$ iff $\theta = \pi/2$.

14.102 Show that if u is orthogonal to v, then every scalar multiple of u is also orthogonal to v.

▮ If $\langle u, v \rangle = 0$ then $\langle ku, v \rangle = k \langle u, v \rangle = k \cdot 0 = 0$, as required.

14.103 Find a unit vector orthogonal to $v_1 = (1, 1, 2)$ and $v_2 = (0, 1, 3)$ in $\mathbf{R}^3$.

▮ Let $w = (x, y, z)$. We want $0 = \langle w, v_1 \rangle = x + y + 2z$ and $0 = \langle w, v_2 \rangle = y + 3z$. Thus we obtain the homogeneous system $x + y + 2z = 0$, $y + 3z = 0$. Set $z = 1$ to find $y = -3$ and $x = 1$; then $w = (1, -3, 1)$. Normalize w to obtain the required unit vector w' orthogonal to v_1 and v_2: $w' = w/\|w\| = (1/\sqrt{11}, -3/\sqrt{11}, 1/\sqrt{11})$.

14.104 Use cross products (Section 1.12) to find a unit vector orthogonal to $u = (1, 2, 3)$ and $v = (3, -1, 4)$.

▮ Find $w = u \times v$ from the array $\begin{pmatrix} 1 & 2 & 3 \\ 3 & -1 & 4 \end{pmatrix}$ to get $w = (11, 5, -7)$. Normalize w to obtain the required unit vector $\hat{w} = w/\|w\| = (11/\sqrt{195}, 5/\sqrt{195}, -7/\sqrt{195})$.

Remark: We emphasize that cross products only exist in $\mathbf{R}^3$ and hence cross products can only be used in $\mathbf{R}^3$ in problems involving orthogonality.

14.105 Suppose W is a subset of V. Define the orthogonal complement of W, denoted by $W^\perp$ (read "W perp").

▮ $W^\perp$ consists of those vectors in V which are orthogonal to every $w \in W$; that is, $W^\perp = \{v \in V : \langle v, w \rangle = 0$ for every $w \in W\}$.

14.106 Show that $W^\perp$ is a subspace of V.

▮ Clearly, $0 \in W^\perp$. Now suppose $u, v \in W^\perp$. Then for any $a, b \in K$ and any $w \in W$, $\langle au + bv, w \rangle = a \langle u, w \rangle + b \langle v, w \rangle = a \cdot 0 + b \cdot 0 = 0$. Thus $au + bv \in W^\perp$ and therefore W is a subspace of V.

14.107 Let u be a nonzero vector in $\mathbf{R}^3$. Give a geometrical description of $u^\perp$.

▮ The subspace $u^\perp$ is the plane in $\mathbf{R}^3$ through the origin 0 and perpendicular to the vector u, as pictured in Fig. 14-4.

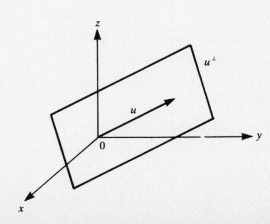

Fig. 14-4

14.108 Let $u = (1, 3, -4)$ in $\mathbf{R}^3$. Find a basis for $u^\perp$.

$\blacksquare$ Note $u^\perp$ consists of all vectors (x, y, z) such that $\langle (x, y, z), (1, 3, -4) \rangle = 0$ or $x + 3y - 4z = 0$. The free variables are y and z. Set $y = -1$, $z = 0$ to obtain the solution $w_1 = (3, -1, 0)$ and set $y = 0$, $z = 1$ to obtain the solution $w_2 = (4, 0, 1)$. The vectors w_1 and w_2 form a basis for the solution space of the equation and hence a basis for $u^\perp$.

14.109 Let W consist of the vectors $u = (1, 2, 3, -1, 2)$ and $v = (2, 4, 7, 2, -1)$ in $\mathbf{R}^5$. Find a basis of the orthogonal complement $W^\perp$ of W.

$\blacksquare$ We seek all vectors $w = (x, y, z, s, t)$ such that

$$\langle w, u \rangle = x + 2y + 3z - s + 2t = 0$$
$$\langle w, v \rangle = 2x + 4y + 7z + 2s - t = 0$$

Eliminating x from the second equation, we find the equivalent system

$$x + 2y + 3z - s + 2t = 0$$
$$z + 4s - 5t = 0$$

The free variables are y, s, and t. Set $y = -1$, $s = 0$, $t = 0$ to obtain the solution $w_1 = (2, -1, 0, 0, 0)$. Set $y = 0$, $s = 1$, $t = 0$ to find the solution $w_2 = (13, 0, -4, 1, 0)$. Set $y = 0$, $s = 0$, $t = 1$ to obtain the solution $w_3 = (-17, 0, 5, 0, 1)$. The set $\{w_1, w_2, w_3\}$ is a basis of $W^\perp$.

14.110 Consider $u = (0, 1, -2, 5)$ in $\mathbf{R}^4$. Find a basis for the orthogonal complement $u^\perp$ of u.

$\blacksquare$ We seek all vectors (x, y, z, t) in $\mathbf{R}^4$ such that $\langle (x, y, z, t), (0, 1, -2, 5) \rangle = 0$ or $0x + y - 2z + 5t = 0$. The free variables are x, z, and t. Set $x = 1$, $z = 0$, $t = 0$ to obtain the solution $w_1 = (1, 0, 0, 0)$. Set $x = 0$, $z = 1$, $t = 0$ to obtain the solution $w_2 = (0, 2, 1, 0)$. Set $x = 0$, $z = 0$, $t = 1$ to obtain the solution $w_3 = (0, -5, 0, 1)$. The vectors w_1, w_2, w_3 form a basis of the solution space of the equation and hence a basis for $u^\perp$.

14.111 Consider a homogeneous system of linear equations over $\mathbf{R}$:

$$a_{11}x_1 + a_{12}x_2 + \cdots + a_{1n}x_n = 0$$
$$a_{21}x_1 + a_{22}x_2 + \cdots + a_{2n}x_n = 0$$
$$\cdots\cdots\cdots\cdots\cdots\cdots\cdots\cdots\cdots$$
$$a_{m1}x_1 + a_{m2}x_2 + \cdots + a_{mn}x_n = 0$$

or in matrix notation $AX = 0$. Recall that the solution space W may be viewed as the kernel of the linear mapping A. Give another interpretation of W using the notion of orthogonality.

$\blacksquare$ Each solution vector $v = (x_1, x_2, \ldots, x_n)$ is orthogonal to each row of A. Thus W is the orthogonal complement of the row space of A.

14.112 Show that $0^\perp = V$.

$\blacksquare$ Each $v \in V$ is orthogonal to 0, hence $0^\perp = V$.

14.113 Show that $V^\perp = 0$.

$\blacksquare$ Since $\langle 0, v \rangle = 0$ for every $v \in V$, $0 \in V^\perp$. If $u \neq 0$, then $\langle u, u \rangle \neq 0$; hence $u \notin V^\perp$. Thus $V^\perp = 0$.

14.114 Suppose $W_1 \subseteq W_2$. Show that $W_2^\perp \subseteq W_1^\perp$.

$\blacksquare$ Let $w \in W_2^\perp$. Then $\langle w, v \rangle = 0$ for every $v \in W_2$. Since $W_1 \subseteq W_2$, $\langle w, v \rangle = 0$ for every $v \in W_1$. Thus $w \in W_1^\perp$, and hence $W_2^\perp \subseteq W_1^\perp$.

14.115 Show that $W^\perp = \text{span}(W)^\perp$.

$\blacksquare$ Since $W \subseteq \text{span}(W)$, we have $\text{span}(W)^\perp \subseteq W^\perp$. Suppose $u \in W^\perp$ and suppose $v \in \text{span}(W)$. Then there exist $w_1, w_2, \ldots, w_k$ in W such that $v = a_1 w_1 + a_2 w_2 + \cdots + a_k w_k$. Then, using $u \in W^\perp$, we have $\langle u, v \rangle = \langle u, a_1 w_1 + a_2 w_2 + \cdots + a_k w_k \rangle = a_1 \langle u, w_1 \rangle + a_2 \langle u, w_2 \rangle + \cdots + a_k \langle u, w_k \rangle = a_1 \cdot 0 + a_2 \cdot 0 + \cdots + a_k \cdot 0 = 0$. Thus $u \in \text{span}(W)^\perp$. Accordingly, $W^\perp \subseteq \text{span}(W)^\perp$. Both inclusions give $W^\perp = \text{span}(W)^\perp$.

14.116 Show that $W \subseteq W^{\perp\perp}$.

❚ Let $w \in W$. Then $\langle w, v \rangle = 0$ for every $v \in W^{\perp}$; hence $w \in W^{\perp\perp}$. Accordingly, $W \subseteq W^{\perp\perp}$.

14.117 Suppose W is a subspace of a finite-dimensional space V. Space that $W = W^{\perp\perp}$.

❚ By Theorem 14.11, $V = W \oplus W^{\perp}$ and, also, $V = W^{\perp} \oplus W^{\perp\perp}$. Hence $\dim W = \dim V - \dim W^{\perp}$ and $\dim W^{\perp\perp} = \dim V - \dim W^{\perp}$. This yields $\dim W = \dim W^{\perp\perp}$. But $W \subseteq W^{\perp\perp}$ by the above; hence $W = W^{\perp\perp}$, as required.

14.5 ORTHOGONAL SETS AND BASES

The following definitions are used throughout the section.

Definitions: Consider a set $S = \{u_1, u_2, \ldots, u_k\}$ of vectors in an inner product space V. S is said to be *orthogonal* if each of its vectors are nonzero and if its vectors are mutually orthogonal, i.e., if $\langle u_i, u_i \rangle \neq 0$ but $\langle u_i, u_j \rangle = 0$ for $i \neq j$. S is said to be orthonormal if S is orthogonal and if each of its vectors have unit length or, in other words, if

$$\langle u_i, u_j \rangle = \delta_{ij} = \begin{cases} 1 & \text{if } i = j \\ 0 & \text{if } i \neq j \end{cases}$$

Normalizing refers to the process of dividing each vector in an orthogonal set S by its length so S is transformed into an orthonormal set. An orthogonal (orthonormal) basis refers to a basis S which is also orthogonal (orthonormal).

14.118 Show that the following set S of vectors in $\mathbf{R}^4$ is orthogonal: $S = \{u = (1, 2, -3, 4), \quad v = (3, 4, 1, -2), \quad w = (3, -2, 1, 1)\}$.

❚
$$\langle u, v \rangle = 3 + 8 - 3 - 8 = 0$$
$$\langle u, w \rangle = 3 - 4 - 3 + 4 = 0$$
$$\langle v, w \rangle = 9 - 8 + 1 - 2 = 0$$

Each pair of vectors is orthogonal; hence S is orthogonal.

14.119 Normalize the orthogonal set S in Problem 14.118 to obtain an orthonormal set.

❚ Divide each vector in S by its length. First find $\|u\|^2 = 1 + 4 + 9 + 16 = 30$, $\|v\|^2 = 9 + 16 + 1 + 4 = 30$, $\|w\|^2 = 9 + 4 + 1 + 1 = 15$. Then $\hat{u} = (1/\sqrt{30}, 2/\sqrt{30}, -3/\sqrt{30}, 4/\sqrt{30})$, $\hat{v} = (3/\sqrt{30}, 4/\sqrt{30}, 1/\sqrt{30}, -2/\sqrt{30})$, $\hat{w} = (3/\sqrt{15}, -2/\sqrt{30}, 1/\sqrt{15}, 1/\sqrt{15})$ form the desired orthonormal set of vectors.

14.120 Consider the usual basis of Euclidean 3-space $\mathbf{R}^3$: $E = \{e_1 = (1, 0, 0), \quad e_2 = (0, 1, 0), \quad e_3 = (0, 0, 1)\}$. Is E orthogonal? Is E orthonormal?

❚ We have $\langle e_1, e_2 \rangle = 0$, $\langle e_1, e_3 \rangle = 0$, and $\langle e_2, e_3 \rangle = 0$. Thus E is orthogonal. Furthermore, $\langle e_1, e_1 \rangle = 1$, $\langle e_2, e_2 \rangle = 1$, and $\langle e_3, e_3 \rangle = 1$. Thus E is an orthonormal basis of $\mathbf{R}^3$.

Remark: The above is true in general; i.e., the usual basis of $\mathbf{R}^n$ is orthonormal for every n.

14.121 Let V be the vector space of real continuous functions on the interval $-\pi \leq t \leq \pi$ with inner product defined by $\langle f, g \rangle = \int_{-\pi}^{\pi} f(t)g(t)\, dt$. The following set S of functions plays a fundamental role in the theory of Fourier series: $S = \{1, \sin t, \cos t, \sin 2t, \cos 2t, \ldots\}$. Is S orthogonal? Is S orthonormal?

❚ S is orthogonal since, for any function $f, g \in S$, we have $\int_{-\pi}^{\pi} f(t)g(t)\, dt = 0$. On the other hand, S is not orthonormal since, for example, $\langle \cos t, \cos t \rangle = \int_{-\pi}^{\pi} \cos^2 t\, dt = \pi$.

14.122 Show that an orthogonal set S of vectors is linearly independent.

❚ Suppose $S = \{u_1, u_2, \ldots, u_r\}$ and suppose

$$a_1 u_1 + a_2 u_2 + \cdots + a_r u_r = 0 \tag{1}$$

Taking the inner product of *(1)* with u_1 we get $0 = \langle 0, u_1 \rangle = \langle a_1 u_1 + a_2 u_2 + \cdots + a_r u_r, u_1 \rangle = a_1 \langle u_1, u_1 \rangle + a_2 \langle u_2, u_1 \rangle + \cdots + a_r \langle u_r, u_1 \rangle = a_1 \langle u_1, u_1 \rangle + a_2 \cdot 0 + \cdots + a_r \cdot 0 = a_1 \langle u_1, u_1 \rangle$. Since S is orthogonal, $\langle u_1, u_1 \rangle \neq 0$; hence $a_1 = 0$. Similarly, for $i = 2, \ldots, r$, taking the inner product of *(1)* with u_i, $0 = \langle 0, u_i \rangle = \langle a_1 u_1 + \cdots + a_r u_r, u_i \rangle = a_1 \langle u_1, u_i \rangle + \cdots + a_i \langle u_i, u_i \rangle + \cdots + a_r \langle u_r, u_i \rangle = a_i \langle u_i, u_i \rangle$. But $\langle u_i, u_i \rangle \neq 0$ and hence $a_i = 0$. Thus S is linearly independent.

Problems 14.123–14.127 refer to the following set S of vectors in $\mathbf{R}^3$: $S = \{u_1 = (1, 2, 1), \quad u_2 = (2, 1, -4), \quad u_3 = (3, -2, 1)\}$.

14.123 Show that S is orthogonal.

▮ $u_1 \cdot u_2 = 2 + 2 - 4 = 0$, $\quad u_1 \cdot u_3 = 3 - 4 + 1 = 0$, $\quad u_2 \cdot u_3 = 6 - 2 - 4 = 0$. Thus S is orthogonal.

14.124 Is S a basis of $\mathbf{R}^3$?

▮ Since S is orthogonal it is linearly independent, and any three linearly independent vectors form a basis for $\mathbf{R}^3$.

14.125 Write $v = (4, 1, 18)$ as a linear combination of u_1, u_2, u_3.

▮ Set v as a linear combination of u_1, u_2, u_3 using unknowns x, y, z as follows:

$$(4, 1, 18) = x(1, 2, 1) + y(2, 1, -4) + z(3, -2, 1) \tag{1}$$

Method 1: Expand *(1)* obtaining the system

$$x + 2y + 3z = 4 \qquad 2x + y - 2z = 1 \qquad x - 4y + z = 18$$

Solve the system to obtain $x = 4$, $y = -3$, $z = 2$. Thus $v = 4u_1 - 3u_2 + 2u_3$.

Method 2: [This method uses the fact that the basis vectors are orthogonal, and the arithmetic is much simpler.] Take the inner product of *(1)* with u_1 to get $(4, 1, 18) \cdot (1, 2, 1) = x(1, 2, 1) \cdot (1, 2, 1)$ or $24 = 6x$ or $x = 4$. [The two last terms drop out since u_1 is orthogonal to u_2 and to u_3.] Take the inner product of *(1)* with u_2 to get $(4, 1, 18) \cdot (2, 1, -4) = y(2, 1, -4) \cdot (2, 1, -4)$ or $-63 = 21y$ or $y = -3$. Take the inner product of *(1)* with u_3 to get $(4, 1, 18) \cdot (3, -2, 1) = z(3, -2, 1) \cdot (3, -2, 1)$ or $28 = 14z$ or $z = 2$. Thus $v = 4u_1 - 3u_2 + 2u_3$.

14.126 Write $w = (3, 4, 5)$ as a linear combination of u_1, u_2, u_3.

▮ First form the equation

$$(3, 4, 5) = x(1, 2, 1) + y(2, 1, -4) + z(3, -2, 1) \tag{1}$$

Take the inner product of *(1)* with respect to u_1 to get $(3, 4, 5) \cdot (1, 2, 1) = x(1, 2, 1) \cdot (1, 2, 1)$ or $16 = 6x$ or $x = \frac{8}{3}$. Take the inner product of *(1)* with respect to u_2 to get $(3, 4, 5) \cdot (2, 1, -4) = y(2, 1, -4) \cdot (2, 1, -4)$ or $-10 = 21y$ or $y = -\frac{10}{21}$. Take the inner product of *(1)* with respect to u_3 to get $(3, 4, 5) \cdot (3, -2, 1) = z(3, -2, 1) \cdot (3, -2, 1)$ or $6 = 14z$ or $z = \frac{3}{7}$. Thus $w = \frac{8}{3}u_1 - \frac{10}{21}u_2 + \frac{3}{7}u_3$.

14.127 Normalize S to obtain an orthonormal basis of $\mathbf{R}^3$.

▮ $\qquad \|u_1\|^2 = 1 + 4 + 1 = 6 \qquad \|u_2\|^2 = 4 + 1 + 16 = 21 \qquad \|u_3\|^2 = 9 + 4 + 1 = 14$

Thus $\hat{u}_1 = (1/\sqrt{6}, 2/\sqrt{6}, 1/\sqrt{6})$, $\hat{u}_2 = (2/\sqrt{21}, 1/\sqrt{21}, -4/\sqrt{21})$, $\hat{u}_3 = (3/\sqrt{14}, -2/\sqrt{14}, 1/\sqrt{14})$ form the desired orthonormal basis of $\mathbf{R}^3$.

Theorem 14.2: Suppose $\{u_1, u_2, \ldots, u_n\}$ is an orthogonal basis for V. Then, for any $v \in V$,

$$v = \frac{\langle v, u_1 \rangle}{\langle u_1, u_1 \rangle} u_1 + \frac{\langle v, u_2 \rangle}{\langle u_2, u_2 \rangle} u_2 + \cdots + \frac{\langle v, u_n \rangle}{\langle u_n, u_n \rangle} u_n$$

14.128 Prove Theorem 14.2.

▌ Suppose $v = k_1 u_1 + k_2 u_2 + \cdots + k_n u_n$. Taking the inner product of both sides with u_1 yields $\langle v, u_1 \rangle = \langle k_1 u_1 + k_2 u_2 + \cdots + k_n u_n, u_1 \rangle = k_1 \langle u_1, u_1 \rangle + k_2 \langle u_2, u_1 \rangle + \cdots + k_n \langle u_n, u_1 \rangle = k_1 \langle u_1, u_1 \rangle + k_2 \cdot 0 + \cdots + k_n \cdot 0 = k_1 \langle u_1, u_1 \rangle$. Thus $k_1 = \langle v, u_1 \rangle / \langle u_1, u_1 \rangle$. Similarly, for $i = 2, \ldots, n$, $\langle v, u_i \rangle = \langle k_1 u_1 + k_2 u_2 + \cdots + k_n u_n, u_i \rangle = k_1 \langle u_1, u_i \rangle + k_2 \langle u_2, u_i \rangle + \cdots + k_n \langle u_n, u_i \rangle = k_1 \cdot 0 + \cdots + k_i \langle u_i, u_i \rangle + \cdots + k_n \cdot 0 = k_i \langle u_i, u_i \rangle$. Thus $k_i = \langle v, u_i \rangle / \langle u_i, u_i \rangle$. Substituting for k_i in the equation $u = k_1 u_1 + \cdots + k_n u_n$, we obtain the desired result.

Remark: The above scalar

$$k_i = \frac{\langle v, u_i \rangle}{\langle u_i, u_i \rangle} = \frac{\langle v, u_i \rangle}{\|u_i\|^2}$$

is called the *component* of v along u_i or the Fourier coefficient of v with respect to u_i.

Problems 14.129–14.132 refer to the following set S of vectors in $\mathbf{R}^4$: $S = \{u_1 = (1, 1, 0, -1), \quad u_2 = (1, 2, 1, 3), \quad u_3 = (1, 1, -9, 2), \quad u_4 = (16, -13, 1, 3)\}$.

14.129 Show that S is orthogonal.

▌ $u_1 \cdot u_2 = 1 + 2 + 0 - 3 = 0 \qquad u_1 \cdot u_3 = 1 + 1 + 0 - 2 = 0 \qquad u_1 \cdot u_4 = 16 - 13 + 0 - 3 = 0$
$u_2 \cdot u_3 = 1 + 2 - 9 + 6 = 0 \qquad u_2 \cdot u_4 = 16 - 26 + 1 + 9 = 0 \qquad u_3 \cdot u_4 = 16 - 13 - 9 + 6 = 0$

Thus S is orthogonal.

14.130 Is S a basis of $\mathbf{R}^4$?

▌ Yes, since S is orthogonal it is linearly independent, and any four linearly independent vectors form a basis of $\mathbf{R}^4$.

14.131 Find the coordinates of an arbitrary vector $v = (a, b, c, d)$ in $\mathbf{R}^4$ relative to the basis S.

▌ Since S is orthogonal, we need only find the Fourier coefficients of v with respect to the basis vectors as in Theorem 14.2. Thus

$$k_1 = \frac{\langle v, u_1 \rangle}{\langle u_1, u_1 \rangle} = \frac{a + b - d}{3}$$

$$k_2 = \frac{\langle v, u_2 \rangle}{\langle u_2, u_2 \rangle} = \frac{a + 2b + c + 3d}{15}$$

$$k_3 = \frac{\langle v, u_3 \rangle}{\langle u_3, u_3 \rangle} = \frac{a + b - 9c + 2d}{87}$$

$$k_4 = \frac{\langle v, u_4 \rangle}{\langle u_4, u_4 \rangle} = \frac{16a - 13b + c + 3d}{435}$$

are the coordinates of v with respect to the basis S.

14.132 Normalize S to obtain an orthonormal basis of $\mathbf{R}^4$.

▌ We have $\|u_1\|^2 = 3$, $\|u_2\|^2 = 15$, $\|u_3\|^2 = 87$, and $\|u_4\|^2 = 435$. Thus $\hat{u}_1 = (1/\sqrt{3}, 1/\sqrt{3}, 0, -1/\sqrt{3})$, $\hat{u}_2 = (1/\sqrt{15}, 2/\sqrt{15}, 1/\sqrt{15}, 3/\sqrt{15})$, $\hat{u}_3 = (1/\sqrt{87}, 1/\sqrt{87}, -9/\sqrt{87}, 2/\sqrt{87})$, $\hat{u}_4 = (16/\sqrt{435}, -13/\sqrt{435}, 1/\sqrt{435}, 3/\sqrt{435})$ is the desired orthonormal basis of $\mathbf{R}^4$.

14.133 Let $u = (1, 1, 1, 1)$ be a vector in $\mathbf{R}^4$. Find an orthogonal basis of $u^\perp$.

▌ Note that $u^\perp$ is the solution space of the linear equation

$$x + y + z + t = 0 \qquad\qquad (1)$$

Find a nonzero solution v_1 of *(1)*, say, $v_1 = (0, 0, 1, -1)$. We want our second basis vector v_2 to be a solution to *(1)* and also orthogonal to v_1, i.e., to be a solution of the system

$$x + y + z + t = 0 \qquad z - t = 0 \qquad (2)$$

Find a nonzero solution v_2 of *(2)*, say, $v_2 = (0, 2, -1, -1)$. We want our third basis vector to be a solution of *(1)* and also orthogonal to v_1 and v_2, i.e., to be a solution of the system

$$x + y + z + t = 0 \qquad 2y - z - t = 0 \qquad z - t = 0 \qquad (3)$$

Find a nonzero solution of *(3)*, say $v_3 = (-1, 1, 1, 1)$. Then $\{v_1, v_2, v_3\}$ is an orthogonal basis of $u^\perp$. [*Remark*: Observe that we chose the intermediate solutions v_1 and v_2 in such a way that each new system is already in echelon form. This makes the calculations simpler.]

14.134 Find an orthonormal basis of the orthogonal complement $u^\perp$ of the vector $u = (1, 1, 1, 1)$ in $\mathbf{R}^4$.

▮ Normalize the orthogonal basis of $u^\perp$ obtained above.

$$\|v_1\|^2 = 0 + 0 + 1 + 1 = 2 \qquad \|v_2\|^2 = 0 + 4 + 1 + 1 = 6 \qquad \|v_3\|^2 = 9 + 1 + 1 + 1 = 12$$

Thus the following is an orthonormal basis for $u^\perp$.

$$v_1 = (0, 0, 1/\sqrt{2}, -1/\sqrt{2}) \qquad v_2 = (0, 2/\sqrt{6}, -1/\sqrt{6}, -1/\sqrt{6}) \qquad v_3 = (-3/\sqrt{12}, 1/\sqrt{12}, 1/\sqrt{12}, 1/\sqrt{12})$$

14.135 Let $w = (1, 2, 3)$ be a vector in Euclidean space $\mathbf{R}^3$. Find an orthogonal basis for $w^\perp$.

▮ Find a nonzero solution of $x + 2y + 3z = 0$, say $v_1 = (1, 1, -1)$. Now find a nonzero solution to the system $x + 2y + 3z = 0$, $x + y - z = 0$ to obtain $v_2 = (5, -4, 1)$. [Alternatively, v_2 can be obtained by taking the cross product $w \times v_1$.] Then $\{v_1, v_2\}$ is an orthogonal basis for $w^\perp$.

14.136 Find an orthonormal basis for $w^\perp$ where $w = (1, 2, 3)$.

▮ Normalize the orthogonal basis obtained above: $\|v_1\|^2 = 1 + 1 + 1 = 3$, $\|v_2\|^2 = 25 + 16 + 1 = 42$. Thus $\hat{v}_1 = (1/\sqrt{3}, 1/\sqrt{3}, -1/\sqrt{3})$ and $v_2 = (5/\sqrt{42}, -4/\sqrt{42}, 1/\sqrt{42})$ form an orthonormal basis of $w^\perp$.

Theorem 14.3 (Generalized Phythagorean Theorem): Suppose $\{u_1, u_2, \ldots, u_r\}$ is an orthogonal set of vectors. Then $\|u_1 + u_2 + \cdots + u_r\|^2 = \|u_1\|^2 + \|u_2\|^2 + \cdots + \|u_r\|^2$.

14.137 Prove Theorem 14.3.

▮ We have $\|u_1 + u_2 + \cdots + u_r\|^2 = \langle u_1 + u_2 + \cdots + u_r, u_1 + u_2 + \cdots + u_r \rangle = \langle u_1, u_1 \rangle + \langle u_2, u_2 \rangle + \cdots + \langle u_r, u_r \rangle + \sum_{i \neq j} \langle u_i, u_j \rangle$. The theorem follows from the fact that $\langle u_i, u_i \rangle = \|u_i\|^2$ and $\langle u_i, u_j \rangle = 0$ for $i \neq j$.

14.138 Verify the Pythagorean Theorem for the following orthogonal set in $\mathbf{R}^4$ [see Problem 14.118]: $\{u = (1, 2, -3, 4), v = (3, 4, 1, -2), w = (3, -2, 1, 1)\}$.

▮ We have $u + v + w = (7, 4, -1, 3)$ and $\|u + v + w\|^2 = 49 + 16 + 1 + 9 = 75$. By Problem 14.119, $\|u\|^2 = 30$, $\|v\|^2 = 30$, and $\|w\|^2 = 15$. Thus $\|u\|^2 + \|v\|^2 + \|w\|^2 = 30 + 30 + 15 = 75 = \|u + v + w\|^2$.

Problems 14.139–14.143 refer to the vector space $V = \mathbf{R}^2$ with inner product defined as follows [see Problem 14.18]: $\langle u, v \rangle = x_1y_1 - x_1y_2 - x_2y_1 + 3x_2y_2$, where $u = (x_1, x_2)$, $v = (y_1, y_2)$.

14.139 Is $E = \{(1, 0), (0, 1)\}$ a basis of V?

▮ Yes, the question of a basis of V is not affected by the inner product of V.

14.140 Is E an orthogonal or orthonormal basis of V?

▮ $$\langle (1, 0), (0, 1) \rangle = 0 - 1 + 0 + 0 = -1$$

Thus E is not an orthogonal basis of V, and so E is not an orthonormal basis of V.

14.141 Find an orthogonal basis S of V which includes the vector $u_1 = (1, 2)$.

❚ Since $\dim V = 2$, we need only find a nonzero vector $u_2 = (x, y)$ such that $\langle u_1, u_2 \rangle = 0$. We have $\langle (1, 2), (x, y) \rangle = x - y - 2x + 6y = -x + 5y = 0$. A nonzero solution to the equation is $x = 5$, $y = 1$. Thus $S = \{u_1 = (1, 2), u_2 = (5, 1)\}$ is an orthogonal basis of V.

14.142 Find an orthonormal basis of v.

❚ Normalize the above orthogonal basis of V.

$$\|u_1\|^2 = 1 - 2 - 2 + 12 = 9 \qquad \|u_2\|^2 = 25 - 5 - 5 + 3 = 18$$

Thus $\hat{u}_1 = (\tfrac{1}{3}, \tfrac{2}{3})$ and $\hat{u}_2 = (5/\sqrt{18}, 1/\sqrt{18})$ form an orthonormal basis of V.

14.143 Verify the Phytagorean Theorem for u_1 and u_2.

❚ We have $u_1 + u_2 = (1, 2) + (5, 1) = (6, 3)$ and $\|u_1 + u_2\|^2 = \langle (6, 3), (6, 3) \rangle = 36 - 18 - 18 + 27 = 27$. Thus $\|u_1\|^2 + \|u_2\|^2 = 9 + 18 = 27 = \|u_1 + u_2\|^2$.

14.144 Show that if $\{u_1, u_2, \ldots, u_r\}$ is orthogonal, then $\{a_1 u_1, a_2 u_2, \ldots, a_r u_r\}$ is orthogonal for any choice of nonzero scalars $a_1, \ldots, a_r$ in **R**.

❚ Since $u_i \neq 0$ and $a_i \neq 0$, we have $a_i u_i \neq 0$. Also, for $i \neq j$, $\langle u_i, u_j \rangle = 0$ and hence $\langle a_i u_i, a_j u_j \rangle = a_i a_j \langle u_i, u_j \rangle = a_i a_j \cdot 0 = 0$. Thus $\{a_i u_i\}$ is orthogonal.

Problems 14.145–14.148 refer to an orthonormal basis $E = \{e_1, \ldots, e_n\}$ of an inner product space V.

14.145 Show that for any $u \in V$, we have $u = \langle u, e_1 \rangle e_1 + \langle u, e_2 \rangle e_2 + \cdots + \langle u, e_n \rangle e_n$. [Compare with Theorem 14.2.]

❚ Suppose $u = k_1 e_1 + k_2 e_2 + \cdots + k_e e_n$. Taking the inner product of u with e_1, $\langle u, e_1 \rangle = \langle k_1 e_1 + k_2 e_2 + \cdots + k_n e_n, e_1 \rangle = k_1 \langle e_1, e_1 \rangle + k_2 \langle e_2, e_1 \rangle + \cdots + k_n \langle e_n, e_1 \rangle = k_1 \cdot 1 + k_2 \cdot 0 + \cdots + k_n \cdot 0 = k_1$. Similarly, for $i = 2, \ldots, n$, $\langle u, e_i \rangle = \langle k_1 e_1 + \cdots + k_i e_i + \cdots + k_n e_n, e_i \rangle = k_1 \langle e_1, e_i \rangle + \cdots + k_i \langle e_i, e_i \rangle + \cdots + k_n \langle e_n, e_i \rangle = k_1 \cdot 0 + \cdots + k_i \cdot 1 + \cdots + k_n \cdot 0 = k_i$. Substituting $\langle u, e_i \rangle$ for k_i in the equation $u = k_1 e_1 + \cdots + k_n e_n$, we obtain the desired result.

14.146 Show that $\langle a_1 e_1 + \cdots + a_n e_n, b_1 e_1 + \cdots + b_n e_n \rangle = a_1 b_1 + a_2 b_2 + \cdots + a_n b_n$.

❚ We have

$$\left\langle \sum_{i=1}^n a_i e_i, \sum_{j=1}^n b_j e_j \right\rangle = \sum_{i,j=1}^n a_i b_j \langle e_i, e_j \rangle = \sum_{i=1}^n a_i b_i \langle e_i, e_i \rangle + \sum_{i \neq j} a_i b_j \langle e_i, e_j \rangle$$

But $\langle e_i, e_j \rangle = 0$ for $i \neq j$, and $\langle e_i, e_j \rangle = 1$ for $i = j$; hence, as required,

$$\left\langle \sum_{i=1}^n a_i e_i, \sum_{j=1}^n b_j e_j \right\rangle = \sum_{i=1}^n a_i b_i = a_1 b_1 + a_2 b_2 + \cdots + a_n b_n$$

14.147 Show that, for any $u, v \in V$, $\langle u, v \rangle = \langle u, e_1 \rangle \langle v, e_1 \rangle + \cdots + \langle u, e_n \rangle \langle v, e_n \rangle$.

❚ By Problem 14.145, $u = \langle u, e_1 \rangle e_1 + \cdots + \langle u, e_n \rangle e_n$ and $v = \langle v, e_1 \rangle e_1 + \cdots + \langle v, e_n \rangle e_n$. Thus, by Problem 14.146, $\langle u, v \rangle = \langle u, e_1 \rangle \langle v, e_1 \rangle + \langle u, e_2 \rangle \langle v, e_2 \rangle + \cdots + \langle u, e_n \rangle \langle v, e_n \rangle$.

14.148 Suppose $T : V \to V$ is linear and suppose A is the matrix representing T in the given basis $\{e_i\}$. Show that $\langle T(e_j), e_i \rangle$ is the ij entry of A.

❚ By Problem 14.145,

$$T(e_1) = \langle T(e_1), e_1 \rangle e_1 + \langle T(e_1), e_2 \rangle e_2 + \cdots + \langle T(e_1), e_n \rangle e_n$$
$$T(e_2) = \langle T(e_2), e_1 \rangle e_1 + \langle T(e_2), e_2 \rangle e_2 + \cdots + \langle T(e_2), e_n \rangle e_n$$
$$\cdots\cdots\cdots\cdots\cdots\cdots\cdots\cdots\cdots\cdots\cdots\cdots\cdots\cdots\cdots\cdots$$
$$T(e_n) = \langle T(e_n), e_1 \rangle e_1 + \langle T(e_n), e_2 \rangle e_2 + \cdots + \langle T(e_n), e_n \rangle e_n$$

The matrix A representing T in the basis $\{e_i\}$ is the transpose of the above matrix of coefficients; hence the ij entry of A is $\langle T(e_j), e_i \rangle$.

Problems 14.149–14.154 refer to the vector space V of real 2×2 matrices with inner product defined by $\langle A, B \rangle = \operatorname{tr}(B^T A)$.

14.149 Show that the following usual basis S of V is orthogonal:

$$S = \left\{ E_1 = \begin{pmatrix} 1 & 0 \\ 0 & 0 \end{pmatrix}, E_2 = \begin{pmatrix} 0 & 1 \\ 0 & 0 \end{pmatrix}, E_3 = \begin{pmatrix} 0 & 0 \\ 1 & 0 \end{pmatrix}, E_4 = \begin{pmatrix} 0 & 0 \\ 0 & 1 \end{pmatrix} \right\}$$

▐ We have $\langle E_1, E_2 \rangle = \operatorname{tr}(E_2^T E_1) = \operatorname{tr}\left[\begin{pmatrix} 0 & 0 \\ 1 & 0 \end{pmatrix} \begin{pmatrix} 1 & 0 \\ 0 & 0 \end{pmatrix} \right] = \operatorname{tr}\begin{pmatrix} 0 & 0 \\ 1 & 0 \end{pmatrix} = 0 + 0 = 0.$ Similarly, $\langle E_i, E_j \rangle = \operatorname{tr}(E_j^T E_i) = 0$ for $i \neq j$. Thus S is orthogonal.

14.150 Show that S is, in fact, orthonormal.

▐ We have $\langle E_1, E_1 \rangle = \operatorname{tr}(E_1^T E_1) = \operatorname{tr}\left[\begin{pmatrix} 1 & 0 \\ 0 & 0 \end{pmatrix} \begin{pmatrix} 1 & 0 \\ 0 & 0 \end{pmatrix} \right] = \operatorname{tr}\begin{pmatrix} 1 & 0 \\ 0 & 0 \end{pmatrix} = 1 + 0 = 1,$ $\langle E_2, E_2 \rangle = \operatorname{tr}(E_2^T E_2) = \operatorname{tr}\left[\begin{pmatrix} 0 & 0 \\ 1 & 0 \end{pmatrix} \begin{pmatrix} 0 & 1 \\ 0 & 0 \end{pmatrix} \right] = \operatorname{tr}\begin{pmatrix} 0 & 0 \\ 0 & 1 \end{pmatrix} = 0 + 1 = 1.$ Similarly, $\langle E_3, E_3 \rangle = 1$ and $\langle E_4, E_4 \rangle = 1.$ Thus S is orthonormal.

14.151 Let W be the subspace of V consisting of the diagonal matrices. Find an orthogonal basis of $W^\perp$, the orthogonal complement of W.

▐ W is spanned by the matrices $\begin{pmatrix} 1 & 0 \\ 0 & 0 \end{pmatrix}$ and $\begin{pmatrix} 0 & 0 \\ 0 & 1 \end{pmatrix}$ which are part of the above usual basis S of V. Thus the other matrices $\begin{pmatrix} 0 & 1 \\ 0 & 0 \end{pmatrix}$ and $\begin{pmatrix} 0 & 0 \\ 1 & 0 \end{pmatrix}$ in S form an orthogonal basis for $W^\perp$.

14.152 Find an orthonormal basis of $W^\perp$.

▐ Since $\begin{pmatrix} 0 & 1 \\ 0 & 0 \end{pmatrix}$ and $\begin{pmatrix} 0 & 0 \\ 1 & 0 \end{pmatrix}$ are already unit vectors, they form an orthonormal basis of $W^\perp$.

14.153 Let U be the subspace of V consisting of the symmetric matrices. Find an orthogonal basis for $U^\perp$.

▐ U is spanned by the matrices $A = \begin{pmatrix} 1 & 0 \\ 0 & 0 \end{pmatrix}$, $B = \begin{pmatrix} 0 & 1 \\ 1 & 0 \end{pmatrix}$, $C = \begin{pmatrix} 0 & 0 \\ 0 & 1 \end{pmatrix}$. $U^\perp$ consists of all matrices $M = \begin{pmatrix} x & y \\ z & t \end{pmatrix}$ orthogonal to A, B, and C. Thus

$$\langle M, A \rangle = \operatorname{tr}\left[\begin{pmatrix} 1 & 0 \\ 0 & 0 \end{pmatrix} \begin{pmatrix} x & y \\ z & t \end{pmatrix} \right] = \operatorname{tr}\begin{pmatrix} x & y \\ 0 & 0 \end{pmatrix} = x = 0$$

$$\langle M, B \rangle = \operatorname{tr}\left[\begin{pmatrix} 0 & 1 \\ 1 & 0 \end{pmatrix} \begin{pmatrix} x & y \\ z & t \end{pmatrix} \right] = \operatorname{tr}\begin{pmatrix} z & t \\ x & y \end{pmatrix} = y + z = 0$$

$$\langle M, C \rangle = \operatorname{tr}\left[\begin{pmatrix} 0 & 0 \\ 0 & 1 \end{pmatrix} \begin{pmatrix} x & y \\ z & t \end{pmatrix} \right] = \operatorname{tr}\begin{pmatrix} 0 & 0 \\ z & t \end{pmatrix} = t = 0$$

The system $x = 0$, $y + z = 0$, $t = 0$ has only one free variable which is z. Thus $x = 0$, $y = 1$, $z = -1$, $t = 0$ is a basis for the solution space of the system. Accordingly, $M = \begin{pmatrix} 0 & 1 \\ -1 & 0 \end{pmatrix}$ forms a basis for $U^\perp$. In particular it is an orthogonal basis for $U^\perp$.

14.154 Find an orthonormal basis for $U^\perp$.

▐ Simply normalize M. We have $\langle M, M \rangle = 1 + 1 = 2$. Thus $\hat{M} = \begin{pmatrix} 0 & 1/\sqrt{2} \\ -1/\sqrt{2} & 0 \end{pmatrix}$ forms an orthonormal basis of $U^\perp$.

14.6 ORTHOGONAL MATRICES

Recall [Section 4.13] that a real matrix P is said to be orthogonal if P is invertible and if $P^T = P^{-1}$, i.e., if $PP^T = P^T P = I$. This section further investigates these matrices. Furthermore, since our vectors in

$\mathbf{R}^n$ may be multiplied by matrices, we assume that such vectors are column vectors unless otherwise stated or implied. Thus, for vectors $u, v \in \mathbf{R}^n$, $\langle u, v \rangle = u^T v$ with respect to the usual inner product on $\mathbf{R}^n$.

Theorem 14.4: Let P be a real n-square matrix. Then the following three properties are equivalent:
(i) P is orthogonal, i.e., $P^T = P^{-1}$.
(ii) The rows of P form an orthonormal set of vectors.
(iii) The columns of P form an orthonormal set of vectors.

14.155 Discuss the validity of Theorem 14.4 [See Theorem 4.9].

❚ This theorem [proved in Problems 4.235–4.236] is true only with respect to the usual inner product on $\mathbf{R}_n$. It is not true if $\mathbf{R}^n$ is given any other inner product.

14.156 Show that $\begin{pmatrix} \cos\theta & -\sin\theta \\ \sin\theta & \cos\theta \end{pmatrix}$ is orthogonal for any real number θ.

❚ We have $(\cos\theta, -\sin\theta) \cdot (\sin\theta, \cos\theta) = \sin\theta\cos\theta - \sin\theta\cos\theta = 0$ and so the rows are orthogonal. Also $\|(\cos\theta, -\sin\theta)\|^2 = \cos^2\theta + \sin^2\theta = 1$, $\|(\sin\theta, \cos\theta)\|^2 = \sin^2\theta + \cos^2\theta = 1$ and so the rows are unit vectors. Thus the matrix is orthogonal.

Remark: Actually, we have the following stronger result proved in Problems 4.241–4.242:

Theorem 14.5: Any 2×2 orthogonal matrix P has the form

$$\begin{pmatrix} \cos\theta & \sin\theta \\ -\sin\theta & \cos\theta \end{pmatrix} \quad \text{or} \quad \begin{pmatrix} \cos\theta & \sin\theta \\ \sin\theta & -\cos\theta \end{pmatrix}$$

for some real number θ.

14.157 Find an orthogonal matrix P with first row $(1/\sqrt{10}, 3/\sqrt{10})$.

❚ By Theorem 14.5, $P = \begin{pmatrix} 1/\sqrt{10} & 3/\sqrt{10} \\ -3/\sqrt{10} & 1/\sqrt{10} \end{pmatrix}$ or $\begin{pmatrix} 1/\sqrt{10} & 3/\sqrt{10} \\ 3/\sqrt{10} & -1/\sqrt{10} \end{pmatrix}$.

14.158 Find an orthogonal matrix P whose first row is $u_1 = (\frac{1}{3}, \frac{2}{3}, \frac{2}{3})$.

❚ First find a nonzero vector $w_2 = (x, y, z)$ which is orthogonal to u_1, or, equivalently, to $w_1 = 3u_1 = (1, 2, 2)$. We have:

$$\langle w_1, w_2 \rangle = (1, 2, 2) \cdot (x, y, z) = 0 \quad \text{or} \quad x + 2y + 2z = 0$$

One such solution is $w_2 = (0, 1, -1)$. Next find a nonzero vector $w_3 = (x, y, z)$ which is orthogonal to both w_1 and w_2. We have

$$\langle w_1, w_3 \rangle = (1, 2, 2) \cdot (x, y, z) = x + 2y + 2z = 0$$
$$\langle w_2, w_3 \rangle = (0, 1, -1) \cdot (x, y, z) = \qquad y - z = 0$$

Set $z = -1$ and find the solution $w_3 = (4, -1, -1)$. Normalize w_2 and w_3 to obtain, respectively,

$$u_2 = (0, 1/\sqrt{2}, -1/\sqrt{2}) \quad \text{and} \quad u_3 = (4/\sqrt{18}, -1/\sqrt{18}, -1/\sqrt{18})$$

Thus

$$P = \begin{pmatrix} \frac{1}{3} & \frac{2}{3} & \frac{2}{3} \\ 0 & 1/\sqrt{2} & -1/\sqrt{2} \\ 4/3\sqrt{2} & -1/3\sqrt{2} & -1/3\sqrt{2} \end{pmatrix}$$

We emphasize that the above matrix P is not unique.

Problems 14.159–14.164 refer to the matrix $A = \begin{pmatrix} 1 & 1 & -1 \\ 1 & 3 & 4 \\ 7 & -5 & 2 \end{pmatrix}$.

14.159 Are the rows of A orthogonal?

▮ Yes, since

$$(1, 1, -1) \cdot (1, 3, 4) = 1 + 3 - 4 = 0$$
$$(1, 1, -1) \cdot (7, -5, 2) = 7 - 5 - 2 = 0$$
$$(1, 3, 4) \cdot (7, -5, 2) = 7 - 15 + 8 = 0$$

14.160 Is A an orthogonal matrix?

▮ No, since the rows of A are not unit vectors, e.g., $(1, 1, -1)^2 = 1 + 1 + 1 = 3$.

14.161 Are the columns of A orthogonal?

▮ No, e.g., $(1, 1, 7) \cdot (1, 3, -5) = 1 + 3 - 35 = -31 \neq 0$.

14.162 Let B be the matrix obtained by normalizing each row of A. Find B.

▮ We have $\|(1, 1, -1)\|^2 = 1 + 1 + 1 = 3$, $\|(1, 3, 4)\|^2 = 1 + 9 + 16 = 26$, $\|(7, -5, 2)\|^2 = 49 + 25 + 4 = 78$. Thus

$$B = \begin{pmatrix} 1/\sqrt{3} & 1/\sqrt{3} & -1/\sqrt{3} \\ 1/\sqrt{26} & 3/\sqrt{26} & 4/\sqrt{26} \\ 7/(6\sqrt{2}) & -5/(6\sqrt{2}) & 2/(6\sqrt{2}) \end{pmatrix}$$

14.163 Is B an orthogonal matrix?

▮ Yes, since the rows of B are still orthogonal and are now unit vectors.

14.164 Are the columns of B orthogonal?

▮ Yes, since the rows of B form an orthonormal set of vectors then, by Theorem 14.4, the columns of B must automatically form an orthonormal set.

14.165 Find a symmetric orthogonal matrix P whose first row is $(\frac{1}{3}, \frac{2}{3}, \frac{2}{3})$. [Compare with Problem 14.158.]

▮ Since P is symmetric, P must have the form

$$P = \begin{pmatrix} \frac{1}{3} & \frac{2}{3} & \frac{2}{3} \\ \frac{2}{3} & x & y \\ \frac{2}{3} & y & z \end{pmatrix}$$

Since the first and second rows are orthogonal, we get $\frac{2}{9} + \frac{2}{3}x + \frac{2}{3}y = 0$ or $1 + 3x + 3y = 0$. Since the second row is a unit vector, we get $\frac{4}{9} + x^2 + y^2 = 1$ or $9x^2 + 9y^2 = 5$. Substitute $y = -(1 + 3x)/3$ into $9x^2 + 9y^2 = 5$ to get $9x^2 + 3x - 2 = 0$ or $(3x - 1)(3x + 2) = 0$. There are two cases.
Case (i): $x = \frac{1}{3}$. Then $y = -\frac{2}{3}$. Since the first and third rows are orthogonal, we get $\frac{2}{9} - \frac{4}{9} + \frac{2}{3}z = 0$ or $z = \frac{1}{3}$. Thus

$$P = \begin{pmatrix} \frac{1}{3} & \frac{2}{3} & \frac{2}{3} \\ \frac{2}{3} & \frac{1}{3} & -\frac{2}{3} \\ \frac{2}{3} & -\frac{2}{3} & \frac{1}{3} \end{pmatrix}$$

Case (ii): $x = -\frac{2}{3}$. Then $y = \frac{1}{3}$. Since the first and third rows are orthogonal, we get $\frac{2}{9} + \frac{2}{9} + \frac{2}{3}z = 0$ or $z = -\frac{2}{3}$. Thus

$$P = \begin{pmatrix} \frac{1}{3} & \frac{2}{3} & \frac{2}{3} \\ \frac{2}{3} & -\frac{2}{3} & \frac{1}{3} \\ \frac{2}{3} & \frac{1}{3} & -\frac{2}{3} \end{pmatrix}$$

14.166 Prove:
(a) P is orthogonal if and only if P^T is orthogonal.
(b) If P is orthogonal, then P^{-1} is orthogonal.
(c) If P is orthogonal, then PQ is orthogonal.
(d) If P is orthogonal, then $\det(P) = 1$ or $\det(P) = -1$.

▌ (a) We have $(P^T)^T = P$. Thus P is orthogonal iff $PP^T = I$ iff $P^{TT}P^T = I$ iff P^T is orthogonal. [Here "iff" is short for "if and only if."]
(b) We have $P^T = P^{-1}$ since P is orthogonal. Thus, by (a), P^{-1} is orthogonal.
(c) We have $P^T = P^{-1}$ and $Q^T = Q^{-1}$. Thus $(PQ)(PQ)^T = PQQ^T P^T = PQQ^{-1}P^{-1} = I$. Thus $(PQ)^T = (PQ)^{-1}$, and so PQ is orthogonal.
(d) We have $PP^T = I$. Using $|P| = |P^T|$, $1 = |I| = |PP^T| = |P||P^T| = |P|^2$. Thus $|P| = 1$ or -1.

14.167 Let $\mathcal{O}_n$ denote the collection of all n-square orthogonal matrices. Show that $\mathcal{O}_n$ is a group under multiplication [called the *orthogonal group*].

▌ The identity matrix $I \in \mathcal{O}_n$ since I is orthogonal. By Problem 14.166, $\mathcal{O}_n$ is closed under multiplication and inverses. Thus $\mathcal{O}_n$ is a group.

14.168 Let P be a matrix with rows R_i^T and columns C_j. Show that (a) the ij entry of PP^T is $\langle R_i, R_j \rangle$ and (b) the ij entry of P^TP is $\langle C_i, C_j \rangle$.

▌ (a) The columns of P^T are $R_1, R_2, \ldots$. Thus the ij entry of PP^T is $R_i^T R_j = \langle R_i, R_j \rangle$. (b) The rows of P^T are $C_1^T, C_2^T, \ldots$. Thus the ij entry of P^TP is $C_i^T C_j = \langle C_i, C_j \rangle$.

> **Theorem 14.6:** Suppose $E = \{e_i\}$ and $F = \{f_i\}$ are orthonormal bases of V. Let P be the change-of-basis matrix from the E basis to the F basis. Then P is orthogonal.

14.169 Prove Theorem 14.6.

▌ Suppose

$$f_i = b_{i1}e_1 + b_{i2}e_2 + \cdots + b_{in}e_n \qquad i = 1, \ldots, n \tag{1}$$

Using Problem 14.146 and the fact that $\{f_i\}$ is orthogonal, we get

$$\delta_{ij} = \langle f_i, f_j \rangle = b_{i1}b_{j1} + b_{i2}b_{j2} + \cdots + b_{in}b_{jn} \tag{2}$$

Let $B = (b_{ij})$ be the matrix of coefficients in (1). [Then $P = B^T$.] Suppose $BB^T = (c_{ij})$. Then, by Problem 14.168 and (2), $c_{ij} = b_{i1}b_{j1} + b_{i2}b_{j2} + \cdots + b_{in}b_{jn} = \delta_{ij}$. Thus $BB^T = I$. Accordingly, B is orthogonal, and hence $P = B^T$ is orthogonal.

> **Theorem 14.7:** Let $(e_1, \ldots, e_n\}$ be an orthonormal basis of an inner product space V. Let $P = (a_{ij})$ be an orthogonal matrix. Then the following is an orthonormal basis: $\{e_i' = a_{1i}e_1 + a_{2i}e_2 + \cdots + a_{ni}e_n : i = 1, \ldots, n\}$.

14.170 Prove Theorem 14.7.

▌ Since $\{e_i\}$ is orthonormal, we get, by Problem 14.146, $\langle e_i', e_i' \rangle = a_{1i}a_{1j} + a_{2i}a_{2j} + \cdots + a_{ni}a_{nj} = \langle C_i, C_j \rangle$ where C_i denotes the ith column of the orthogonal matrix $P = (a_{ij})$. Since P is orthogonal, its columns form an orthonormal set. This implies $\langle e_i', e_j' \rangle = \langle C_i, C_j \rangle = \delta_{ij}$. Thus $\{e_i'\}$ is an orthonormal basis.

14.171 Suppose P is an orthogonal matrix. Show that $\langle Pu, Pv \rangle = \langle u, v \rangle$ for any $u, v \in V$.

▌ Using $P^TP = I$, we have $\langle Pu, Pv \rangle = (Pu)^T(Pv) = u^T P^T Pv = u^T v = \langle u, v \rangle$. [*Remark*: This states that P, viewed as a linear map, preserves inner products.]

14.172 Suppose P is orthogonal. Show that $\|Pu\| = \|u\|$ for every $u \in V$.

▌ Using $P^TP = I$, we have $\|Pu\|^2 = \langle Pu, Pu \rangle = u^T P^T Pu = u^T u = \langle u, u \rangle = \|u\|^2$. Taking the square root of both sides gives us our result. [*Remark*: This states that P, viewed as a linear map, preserves lengths.]

14.173 Define orthogonally equivalent matrices.

▌ Real matrices A and B are orthogonally equivalent if there exists an orthogonal matrix P such that $B = P^T AP = P^{-1}AP$.

Problems 14.174–14.176 show that orthogonally equivalent is an equivalence relation.

14.174 Show that any matrix A is orthogonally equivalent to A.

▌ The identity matrix I is orthogonal, and $I^T = I$. Since $A = I^T A I$, we have A is orthogonally equivalent to A.

14.175 Suppose A is orthogonally equivalent to B. Show that B is orthogonally equivalent to A.

▌ There exists an orthogonal matrix P such that $A = P^T B P = P^{-1} B P$. Then $B = P A P^{-1} = P A P^T = (P^T)^T A P^T$. Thus B is orthogonally equivalent to A.

14.176 Suppose A is orthogonally equivalent to B and B is orthogonally equivalent to C. Show that A is orthogonally equivalent to C.

▌ There exist orthogonal matrices P and Q such that $A = P^T B P$ and $B = Q^T C Q$. Then $A = P^T B P = P^T (Q^T C Q) P = (QP)^T C (QP)$. However, QP is also orthogonal. Thus A is orthogonally equivalent to C.

14.7 PROJECTIONS, GRAM-SCHMIDT ALGORITHM, APPLICATIONS

14.177 Suppose $w \neq 0$. Let v be any vector in V. Show that

$$c = \frac{\langle v, w \rangle}{\langle w, w \rangle} = \frac{\langle v, w \rangle}{\|w\|^2}$$

is the unique scalar such that $v' = v - cw$ is orthogonal to w.

▌ In order for v' to be orthogonal to w we must have $\langle v - cw, w \rangle = 0$ or $\langle v, w \rangle - c\langle w, w \rangle = 0$ or $\langle v, w \rangle = c\langle w, w \rangle$. Thus $c = \langle v, w \rangle / \langle w, w \rangle$. Conversely, suppose $c = \langle v, w \rangle / \langle w, w \rangle$. Then

$$\langle v - cw, w \rangle = \langle v, w \rangle - c\langle w, w \rangle = \langle v, w \rangle - \frac{\langle v, w \rangle}{\langle w, w \rangle} \langle w, w \rangle = 0$$

Remark: The above scalar c is called the Fourier coefficient of v with respect to w or the component of v along w. *Note*: cw is called the projection of v along w as indicated by Fig. 14-5.

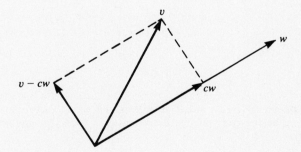

Fig. 14-5

14.178 Find the Fourier coefficient c and the projection cw of $v = (1, -1, 2)$ along $w = (0, 1, 1)$ in $\mathbf{R}^3$.

▌ Compute $\langle v, w \rangle = 0 - 1 + 2 = 1$ and $\|w\|^2 = 0 + 1 + 1 = 2$. Hence $c = \frac{1}{2}$ and $cw = (0, \frac{1}{2}, \frac{1}{2})$ is the projection of v along w.

14.179 Find the component c and the projection cw of $v = (1, 2, 3, 4)$ along $w = (1, -3, 4, -2)$ in $\mathbf{R}^4$.

▌ Compute $\langle v, w \rangle = 1 - 6 + 12 - 8 = -1$ and $\|w\|^2 = 1 + 9 + 16 + 4 = 30$. Thus $c = -\frac{1}{30}$ and $cw = (-\frac{1}{30}, \frac{1}{10}, -\frac{2}{15}, \frac{1}{15})$ is the projection of v along w.

14.180 Let V be the vector space of polynomials with inner product $\langle f, g \rangle = \int_0^1 f(t)g(t)\, dt$. Find the Fourier coefficient c and the projection cg of $f(t) = 2t - 1$ along $g(t) = t^2$.

❚ Compute

$$\langle f, g \rangle = \int_0^1 (2t^3 - t^2) \, dt = \left[\frac{t^4}{2} - \frac{t^3}{3} \right]_0^1 = \frac{1}{6}$$

$$\langle g, g \rangle = \int_0^1 t^4 \, dt = \left[\frac{t^5}{5} \right]_0^1 = \frac{1}{5}$$

Thus $c = \frac{5}{6}$ and $cg(t) = \frac{5}{6} t^2$ is the projection of f along g.

14.181 Let V be the vector space of 2×2 real matrices with inner product $\langle A, B \rangle = \text{tr}(B^T A)$. Find the component c and the projection cB of $A = \begin{pmatrix} 1 & 2 \\ 1 & -3 \end{pmatrix}$ along $B = \begin{pmatrix} 0 & -1 \\ 1 & 2 \end{pmatrix}$.

❚ Compute $\langle A, B \rangle = \text{tr}(B^T A) = 1 - 8 = -7$ and $\|B\|^2 = 0 + 1 + 1 + 4 = 6$. Thus $c = -\frac{7}{6}$ and $cB = \begin{pmatrix} 0 & \frac{7}{6} \\ -\frac{7}{6} & -\frac{7}{3} \end{pmatrix}$ is the projection of A along B.

14.182 Suppose $\{w_1, w_2\}$ is an orthogonal set of vectors. Suppose v is any vector in V. Find c_1 and c_2 such that $v' = v - c_1 w_1 - c_2 w_2$ is orthogonal to w_1 and w_2.

❚ If v' is orthogonal to w_1, then $0 = \langle v - c_1 w_1 - c_2 w_2, w_1 \rangle = \langle v, w_1 \rangle - c_1 \langle w_1, w_1 \rangle - c_2 \langle w_2, w_1 \rangle = \langle v, w_1 \rangle - c_1 \langle w_1, w_1 \rangle - c_2 \cdot 0 = \langle v, w_1 \rangle - c_1 \langle w_1, w_1 \rangle$. Thus $c_1 = \langle v, w_1 \rangle / \langle w_1, w_1 \rangle$. [That is, c_1 is the component of v along w_1.] Similarly, if v' is orthogonal to w_2, then $0 = \langle v - c_1 w_1 - c_2 w_2, w_2 \rangle = \langle v, w_2 \rangle - c_2 \langle w_2, w_2 \rangle$. Thus $c_2 = \langle v, w_2 \rangle / \langle w_2, w_2 \rangle$. [That is, c_2 is the component of v along w_2.]

Lemma 14.8: Suppose $\{w_1, w_2, \ldots, w_r\}$ is an orthogonal set of vectors in V. Let v be any vector in V. Define $v' = v - c_1 w_1 - c_2 w_2 - \cdots - c_r w_r$ where

$$c_i = \frac{\langle v, w_i \rangle}{\langle w_i, w_i \rangle} = \frac{\langle v, w_i \rangle}{\|w_i\|^2}$$

Then v' is orthogonal to $w_1, w_2, \ldots, w_r$. [Note that the c's are, respectively, the components of v along the w's.]

14.183 Prove Lemma 14.8.

❚ For $i = 1, 2, \ldots, r$ and using $\langle w_i, w_j \rangle = 0$ for $i \neq j$,

$$\begin{aligned}
\langle v - c_1 w_1 - c_2 w_2 - \cdots - c_r w_r, w_i \rangle &= \langle v, w_i \rangle - c_1 \langle w_1, w_i \rangle - \cdots - c_i \langle w_i w_i \rangle - \cdots - c_r \langle w_r, w_i \rangle \\
&= \langle v, w_i \rangle - c_1 \cdot 0 - \cdots - c_i \langle w_i, w_i \rangle - \cdots - c_r \cdot 0 \\
&= \langle v, w_i \rangle = c_i \langle w_i, w_i \rangle = \langle v, w_i \rangle - \frac{\langle v, w_i \rangle}{\langle w_i, w_i \rangle} \langle w_i, w_i \rangle \\
&= 0
\end{aligned}$$

Theorem 14.9: Suppose $\{w_1, \ldots, w_r\}$ is an orthogonal set of vectors in V. Let v be any vector in V and let c_i be the component of v along w_i. Then, for any scalars $a_1, \ldots, a_r$,

$$\left\| v - \sum_{k=1}^r c_k w_k \right\| \leq \left\| v - \sum_{k=1}^r a_k w_k \right\|$$

14.184 Prove Theorem 14.9 which shows that $c_1 w_1 + \cdots + c_r w_r$ is the closest approximation to v as a linear combination of $w_1, \ldots, w_r$.

❚ By Lemma 14.8, $v - \sum c_k w_k$ is orthogonal to every w_i and hence orthogonal to any linear combination of $w_1, \ldots, w_r$. Therefore, using the Pythagorean Theorem, and summing from $k = 1$ to r,

$$\begin{aligned}
\left\| v - \sum a_k w_k \right\|^2 &= \left\| v - \sum c_k w_k + \sum (c_k - a_k) w_k \right\|^2 = \left\| v - \sum c_k w_k \right\|^2 + \left\| \sum (c_k - a_k) w_k \right\|^2 \\
&\geq \left\| v - \sum c_k w_k \right\|^2
\end{aligned}$$

The square root of both sides gives us our theorem.

14.185 Suppose $v_1, v_2, \ldots, v_r$ form a basis for a subspace U of an inner product space V. Describe the Gram-Schmidt algorithm which yields an orthogonal basis [and by normalization an orthonormal basis] of U.

$\blacksquare$ Set

$$w_1 = v_1$$

$$w_2 = v_2 - c_{21}w_1 = v_2 - \frac{\langle v_2, w_1 \rangle}{\|w_1\|^2} w_1$$

$$w_3 = v_3 - c_{31}w_1 - c_{32}w_2 = v_3 - \frac{\langle v_3, w_1 \rangle}{\|w_1\|^2} w_1 - \frac{\langle v_3, w_2 \rangle}{\|w_2\|^2} w_2$$

$$\cdots\cdots\cdots\cdots\cdots\cdots\cdots\cdots\cdots\cdots\cdots\cdots\cdots\cdots\cdots\cdots\cdots\cdots\cdots\cdots$$

$$w_r = v_r - c_{r1}w_1 - c_{r2}w_2 - \cdots - c_{r-1}w_{r-1}$$

where $c_{ri} = \langle v_r, w_i \rangle / \|w_i\|^2$. The set $\{w_1, w_2, \ldots, w_r\}$ is the required orthogonal basis of U.

Remark: In hand calculations, it may be simpler to clear the fraction in any new w_k by multiplying w_k by an appropriate scalar as this does not affect the orthogonality.

14.186 Show that in the above Gram-Schmidt algorithm, $\operatorname{span}(v_1, \ldots, v_k) = \operatorname{span}(w_1, \ldots, w_k)$ for $k = 1, \ldots, r$.

$\blacksquare$ The proof is by induction on k. For $k = 1$, $w_1 = v_1$ and so $\operatorname{span}(v_1) = \operatorname{span}(w_1)$. Suppose $k > 1$. Since v_k is a linear combination of $w_1, \ldots, w_k$, we have $\operatorname{span}(v_1, \ldots, v_k) \subseteq \operatorname{span}(w_1, \ldots, w_k)$. On the other hand, w_k is a linear combination of v_k and $w_1, \ldots, w_{k-1}$. By induction, $\operatorname{span}(w_1, \ldots, w_{k-1}) = \operatorname{span}(v_1, \ldots, v_{k-1})$. Thus w_k is a linear combination of $v_1, \ldots, v_k$ and hence $\operatorname{span}(w_1, \ldots, w_k) \subseteq \operatorname{span}(v_1, \ldots, v_k)$. Both inclusions give us our result.

14.187 Show that in the above Gram-Schmidt algorithm, the vectors $w_1, w_2, \ldots, w_r$ form an orthogonal set.

$\blacksquare$ First we have $w_1 = v_1 \neq 0$. For $k > 1$, we have $v_k \notin \operatorname{span}(v_1, \ldots, v_{k-1})$ since $v_1, \ldots, v_r$ are linearly independent. Hence $w_k \neq 0$.

By Lemma 14.8, each w_k is orthogonal to the preceding $w_1, \ldots, w_{k-1}$. Thus $\{w_1, \ldots, w_r\}$ is an orthogonal set.

Theorem 14.10: Let $\{v_1, v_2, \ldots, v_n\}$ be any basis of an inner product space V. Then there exists an orthonormal basis $\{u_1, u_2, \ldots, u_n\}$ of V such that the change-of-basis matrix from $\{v_i\}$ to $\{u_i\}$ is triangular; i.e., for $k = 1, \ldots, n$, $u_k = a_{k1}u_1 + a_{k2}u_2 + \cdots + a_{kk}u_k$.

14.188 Theorem 14.10.

$\blacksquare$ The proof follows from the Gram-Schmidt algorithm and Problems 14.186 and 14.187. Specifically, apply the algorithm to $\{v_i\}$ to obtain an orthogonal basis $\{w_1, \ldots, w_n\}$ and then normalize $\{w_i\}$ to obtain an orthonormal basis $\{u_i\}$ of V. The specific algorithm guarantees than each w_k is a linear combination of $v_1, \ldots, v_k$ and hence each u_k is a linear combination of $v_1, \ldots, v_k$.

14.189 Find an orthonormal basis for the subspace U of $\mathbf{R}^4$ spanned by $v_1 = (1, 1, 1, 1)$, $v_2 = (1, 2, 4, 5)$, $v_3 = (1, -3, -4, -2)$.

$\blacksquare$ First find an orthogonal basis of U using the Gram-Schmidt algorithm. First set $w_1 = u_1 = (1, 1, 1, 1)$. Next find

$$v_2 - \frac{\langle v_2, w_1 \rangle}{\|w_1\|^2} w_1 = (1, 2, 4, 5) - \tfrac{12}{4}(1, 1, 1, 1) = (-2, -1, 1, 2)$$

Set $w_2 = (-2, -1, 1, 2)$. Then find

$$v_3 - \frac{\langle v_3, w_1 \rangle}{\|w_1\|^2} w_1 - \frac{\langle v_3, w_2 \rangle}{\|w_2\|^2} w_2 = (1, -3, -4, -2) - \frac{-8}{4}(1, 1, 1, 1) - \frac{-7}{10}(-2, -1, 1, 2) = (\tfrac{8}{5}, -\tfrac{17}{10}, -\tfrac{13}{10}, \tfrac{7}{5})$$

Clear fractions to obtain $w_3 = (16, -17, -13, 14)$. [In hand calculations, it is usually simpler to clear fractions as this does not affect the orthogonality.] Last, normalize the orthogonal basis $w_1 = (1, 1, 1, 1)$,

$w_2 = (-2, -1, 1, 2)$, $w_3 = (16, -17, -13, 14)$. Since $\|w_1\|^2 = 4$, $\|w_2\|^2 = 10$, $\|w_3\|^2 = 910$, the following vectors form an orthonormal basis of U:

$$u_1 = \frac{1}{2}(1, 1, 1, 1) \qquad u_2 = \frac{1}{\sqrt{10}}(-2, -1, 1, 2) \qquad u_3 = \frac{1}{\sqrt{910}}(16, -17, -13, 14)$$

14.190 Consider the following basis of Euclidean space $\mathbf{R}^3$: $\{v_1 = (1, 1, 1), v_2 = (0, 1, 1), v_3 = (0, 0, 1)\}$. Use the Gram-Schmidt algorithm to transform $\{v_i\}$ into an orthonormal basis $\{u_i\}$ of $\mathbf{R}^3$.

▮ First set $w_1 = v_1 = (1, 1, 1)$. Then find

$$v_2 - \frac{\langle v_2, w_1\rangle}{\|w_1\|^2}\, w_1 = (0, 1, 1) - \tfrac{2}{3}(1, 1, 1) = (-\tfrac{2}{3}, \tfrac{1}{3}, \tfrac{1}{3})$$

Clear fractions to obtain $w_2 = (-2, 1, 1)$. Next find

$$v_3 - \frac{\langle v_3, w_1\rangle}{\|w_1\|^2} - \frac{\langle v_3, w_2\rangle}{\|w_2\|^2} = (0, 0, 1) - \tfrac{1}{3}(1, 1, 1) - \tfrac{1}{6}(-2, 1, 1) = (0, -\tfrac{1}{2}, \tfrac{1}{2})$$

Clear fractions to obtain $w_3 = (0, -1, 1)$. Normalize $\{w_1, w_2, w_3\}$ to obtain the following required orthonormal basis of $\mathbf{R}^3$:

$$\left\{u_1 = \left(\frac{1}{\sqrt{3}}, \frac{1}{\sqrt{3}}, \frac{1}{\sqrt{3}}\right), u_2 = \left(-\frac{2}{\sqrt{6}}, \frac{1}{\sqrt{6}}, \frac{1}{\sqrt{6}}\right), u_3 = \left(0, -\frac{1}{\sqrt{2}}, \frac{1}{\sqrt{2}}\right)\right\}$$

14.191 Find an orthonormal basis of the subspace W of $\mathbf{R}^5$ spanned by $v_1 = (1, 1, 1, 0, 1)$, $v_2 = (1, 0, 0, -1, 1)$, $v_3 = (3, 1, 1, -2, 3)$, $v_4 = (0, 2, 1, 1, -1)$.

▮ First set $w_1 = u_1 = (1, 1, 1, 0, 1)$. Then find

$$v_2 - \frac{\langle v_2, w_1\rangle}{w_1^2}\, w_1 = (1, 0, 0, -1, 1) - \tfrac{2}{4}(1, 1, 1, 0, 1) = (\tfrac{1}{2}, -\tfrac{1}{2}, -\tfrac{1}{2}, -1, \tfrac{1}{2})$$

Clear fractions to obtain $w_2 = (1, -1, -1, -2, 1)$. Next find

$$v_3 - \frac{\langle v_3, w_1\rangle}{\|w_1\|^2} - \frac{\langle v_3, w_2\rangle}{\|w_2\|^2} = (3, 1, 1, -2, 3) - \tfrac{8}{4}(1, 1, 1, 0, 1) - \tfrac{8}{8}(1, -1, -1, -2, 1) = (0, 0, 0, 0, 0, 0)$$

This shows that v_3 is a linear combination of v_1 and v_2 and hence v_3 is omitted. Next form

$$v_4 - \frac{\langle v_4, w_1\rangle}{\|w_1\|^2}\, w_1 - \frac{\langle v_4, w_2\rangle}{\|w_2\|^2}\, w_2 = (0, 2, 1, 1, -1) - \tfrac{2}{4}(1, 1, 1, 0, 1)\frac{-6}{8}(1, -1, -1, -2, 1)$$

$$= (\tfrac{1}{4}, \tfrac{3}{4}, -\tfrac{1}{4}, -\tfrac{1}{2}, -\tfrac{3}{4})$$

Clear fractions to obtain $w_3 = (1, 3, -1, -2, -3)$. Normalize $\{w_1, w_2, w_3\}$ to obtain the required orthonormal basis of W:

$$u_1 = \frac{1}{2}(1, 1, 1, 0, 1) \qquad u_2 = \frac{1}{2\sqrt{2}}(1, -1, -1, -2, 1) \qquad u_3 = \frac{1}{2\sqrt{6}}(1, 3, -1, -2, -3)$$

14.192 Let V be the vector space of polynomials $f(t)$ with inner product $\langle f, g\rangle = \int_{-1}^{1} f(t)g(t)\, dt$. Apply the Gram-Schmidt algorithm to the set $\{1, t, t^2, t^3\}$ to obtain an orthonormal set $\{f_0, f_1, f_2, f_3\}$.

▮ Here we use the fact that if $r + s = n$ then

$$\langle t^r, t^s\rangle = \int_{-1}^{1} t^n\, dt = \left[\frac{t^{n+1}}{n+1}\right]_{-1}^{1} = \begin{cases} 2/(n+1) & \text{if } n \text{ is even} \\ 0 & \text{if } n \text{ is odd} \end{cases}$$

First set $f_0 = 1$. Then find

$$f_1 = t - \frac{\langle t, 1\rangle}{\langle 1, 1\rangle} \cdot 1 = t - \frac{0}{2} \cdot 1 = t - \frac{0}{2} \cdot 1 = t$$

Next find

$$f_2 = t^2 - \frac{\langle t^2, 1 \rangle}{\langle 1, 1 \rangle} \cdot 1 - \frac{\langle t^2, t \rangle}{\langle t, t \rangle} \cdot t = t^2 - \frac{\frac{2}{3}}{2} \cdot 1 - \frac{0}{\frac{2}{3}} \cdot t = t^2 - \frac{1}{3}$$

Last, find

$$f_3 = t^3 - \frac{\langle t^3, 1 \rangle}{\langle 1, 1 \rangle} \cdot 1 - \frac{\langle t^3, t \rangle}{\langle t, t \rangle} t - \frac{\langle t^3, t^2 - \frac{1}{3} \rangle}{\langle t^2 - \frac{1}{3}, t^2 - \frac{1}{3} \rangle} (t^2 - \frac{1}{3}) = t^3 - 0 \cdot 1 - \frac{\frac{2}{5}}{\frac{2}{3}} t - 0(t^2 - \frac{1}{3}) = t^3 - \frac{3}{5} t$$

That is, $\{1, t, t^2 - \frac{1}{3}, t^3 - \frac{3}{5} t\}$ is the required orthonormal set of polynomials.

14.193 Find the first four Legendre polynomials.

▎ Take multiples of the orthogonal polynomials obtained in Problem 14.192 so that $p(1) = 1$ for any polynomial $p(t)$ in the set. This gives $\{1, t, \frac{1}{2}(3t^2 - 1), \frac{1}{2}(5t^3 - 3t)\}$. These are the first four Legendre polynomials [which are important in the study of differential equations].

14.194 Let W be a subspace of an inner product space V. Show that there is an orthonormal basis of W which is part of an orthonormal basis of V.

▎ We choose a basis $\{v_1, \ldots, v_r\}$ of W and extend it to a basis $\{v_1, \ldots, v_n\}$ of V. We then apply the Gram-Schmidt orthogonalization process to $\{v_1, \ldots, v_n\}$ to obtain an orthonormal basis $\{u_1, \ldots, u_n\}$ of V where, for $i = 1, \ldots, n$, $u_i = a_{i1} v_1 + \cdots + a_{ii} v_i$. Thus $u_1, \ldots, u_r \in W$ and therefore $\{u_1, \ldots, u_r\}$ is an orthonormal basis of W.

Theorem 14.11: Let W be a subspace of V; then $V = W \oplus W^\perp$.

14.195 Prove Theorem 14.11.

▎ By Problem 14.194, there exists an orthonormal basis $\{u_1, \ldots, u_r\}$ of W which is part of an orthonormal basis $\{u_1, \ldots, u_n\}$ of V. Since $\{u_1, \ldots, u_n\}$ is orthonormal, $u_{r+1}, \ldots, u_n \in W^\perp$. If $v \in V$, $v = a_1 u_1 + \cdots + a_n u_n$ where $a_1 u_1 + \cdots + a_r u_r \in W$, $a_{r+1} u_{r+1} + \cdots + a_n u_n \in W^\perp$. Accordingly, $V = W + W^\perp$. On the other hand, if $w \in W \cap W^\perp$, then $\langle w, w \rangle = 0$. This yields $w = 0$; hence $W \cap W^\perp = \{0\}$. The two conditions, $V = W + W^\perp$ and $W \cap W^\perp = \{0\}$, give the desired result $V = W \oplus W^\perp$.

Note that we have proved the theorem only for the case that V has finite dimension; we remark that the theorem also holds for spaces of arbitrary dimension.

14.196 Let W be a subspace of an inner product space V. Define the orthogonal projection mapping of V onto W, denoted by E_W. What is the image and kernel of E_W?

▎ Let $v \in V$. Since $V = W \oplus W^\perp$, there exists unique $w \in W$ and $w' \in W^\perp$ such that $v = w + w'$. Define $E_W : V \to V$ by $E_W(v) = w$. Then this mapping E_W is called the *orthogonal projection* of V onto W. It is linear and $\text{Im}(E_W) = W$ and $\text{Ker}(E_W) = W^\perp$.

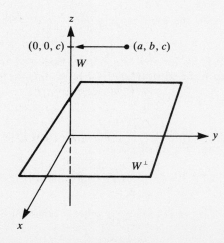

Fig. 14-6

14.197 Let W be the z axis in $\mathbf{R}^3$, that is, $W = \{(0, 0, c): c \in \mathbf{R}\}$. What is $W^\perp$? Find the projection mapping E_W.

❚ $W^\perp$ is the xy plane, i.e., $W^\perp = \{(a, b, 0): a, b \in \mathbf{R}\}$. [See Fig. 14-6.] The orthogonal projection E_W of $\mathbf{R}^3$ onto W is given by $E_W(x, y, z) = (0, 0, z)$.

Theorem 14.12: Suppose $\{u_1, u_2, \ldots, u_r\}$ is an orthonormal set of vectors in V. Let v be any vector in V and let c_i be the Fourier coefficient of v with respect to u_i. Then $\sum_{k=1}^{r} c_k{}^2 \leq \|v\|^2$.

14.198 Prove Theorem 14.12, known as the *Bessel inequality*.

❚ Note that $c_i = \langle v, u_i \rangle$ since $\|u_i\| = 1$. Using $\langle u_i, u_j \rangle = 0$ for $i \neq j$, we get $0 \leq \langle v - \sum c_k u_k, v - \sum c_k, u_k \rangle = \langle v, v \rangle - 2\langle v, \sum c_k u_k \rangle + \sum c_k{}^2 = \langle v, v \rangle - \sum 2c_k \langle v, u_k \rangle + \sum c_k{}^2 = \langle v, v \rangle - \sum 2c_k{}^2 + \sum c_k{}^2 = \langle v, v \rangle - \sum c_k{}^2$. This gives us our inequality.

14.8 INNER PRODUCTS AND POSITIVE DEFINITE MATRICES

14.199 Let V be an inner product space and let $B = \{e_1, \ldots, e_n\}$ be a basis of V. Define the matrix A which represents the inner product on V with respect to the basis B.

❚ The matrix $A = (a_{ij})$ is defined by $a_{ij} = \langle e_i, e_j \rangle$; that is,

$$A = \begin{pmatrix} \langle e_1, e_1 \rangle \langle e_1, e_2 \rangle \cdots \langle e_1, e_n \rangle \\ \langle e_2, e_1 \rangle \langle e_2, e_2 \rangle \cdots \langle e_2, e_n \rangle \\ \cdots\cdots\cdots\cdots\cdots\cdots\cdots \\ \langle e_n, e_1 \rangle \langle e_n, e_2 \rangle \cdots \langle e_n, e_n \rangle \end{pmatrix}$$

[*Remark*: Observe that A is symmetric since $\langle e_i, e_j \rangle = \langle e_j, e_i \rangle$ for any basis vectors e_i and e_j and that A depends on the inner product on V and also the basis of V.]

14.200 Consider the basis $B = \{u_1 = (1, 1, 0), u_2 = (1, 2, 3), u_3 = (1, 3, 5)\}$ of $\mathbf{R}^3$. Find the matrix A which represents the usual inner product on $\mathbf{R}^3$ with respect to the basis B.

❚ Compute: $\langle u_1, u_1 \rangle = 1 + 1 + 0 = 2$, $\langle u_1, u_2 \rangle = 1 + 2 + 0 = 3$, $\langle u_1, u_3 \rangle = 1 + 3 + 0 = 4$, $\langle u_2, u_2 \rangle = 1 + 4 + 9 = 14$, $\langle u_2, u_3 \rangle = 1 + 6 + 15 = 22$, $\langle u_3, u_3 \rangle = 1 + 9 + 25 = 35$.

Thus

$$A = \begin{pmatrix} 2 & 3 & 4 \\ 3 & 14 & 22 \\ 4 & 22 & 35 \end{pmatrix}$$

14.201 Consider the usual basis $E = \{e_1 = (1, 0, 0), e_2 = (0, 1, 0), e_3 = (0, 0, 1)\}$ of $\mathbf{R}^3$. Find the matrix which represents the usual inner product on $\mathbf{R}^3$ with respect to the usual basis E.

❚ We have $\langle e_1, e_1 \rangle = 1$, $\langle e_1, e_2 \rangle = 0$, $\langle e_1, e_3 \rangle = 0$, $\langle e_2, e_2 \rangle = 1$, $\langle e_2, e_3 \rangle = 0$, $\langle e_3, e_3 \rangle = 1$. Thus the identity matrix I represents the usual inner product on $\mathbf{R}^3$ with respect to the usual basis E of $\mathbf{R}^3$.

Remark: The above result holds for any orthonormal basis $\{e_i\}$ of an inner product space V. That is, if $\langle e_i, e_j \rangle = \delta_{ij}$, then the identity matrix I represents the inner product on V with respect to the basis $\{e_i\}$.

14.202 Consider the basis $B = \{v_1 = (1, 3), v_2 = (2, 5)\}$ of $\mathbf{R}^2$. Find the matrix A_1 which represents the usual inner product on $\mathbf{R}^2$ with respect to the basis B.

❚ Compute $\langle v_1, v_1 \rangle = 1 + 9 = 10$, $\langle v_1, v_2 \rangle = 2 + 15 = 17$, $\langle v_2, v_2 \rangle = 4 + 25 = 29$. Thus $A_1 = \begin{pmatrix} 10 & 17 \\ 17 & 29 \end{pmatrix}$.

Problems 14.203–14.204 refer to the following inner product on $\mathbf{R}^2$ (see Problem 14.18): $\langle u, v \rangle = x_1 y_1 - x_1 y_2 - x_2 y_1 + 3x_2 y_2$, where $u = (x_1, x_2)$, $v = (y_1, y_2)$.

14.203 Find the matrix A which represents the given inner product on $\mathbf{R}^2$ with respect to the usual basis $\{(1, 0), (0, 1)\}$ of $\mathbf{R}^2$.

▮ Compute $\langle (1, 0), (1, 0) \rangle = 1 - 0 - 0 + 1 = 1$, $\langle (1, 0), (0, 1) \rangle = 0 - 1 - 0 + 0 = -1$, $\langle (0, 1), (0, 1) \rangle = 0 - 0 - 0 + 3 = 3$. Thus $A = \begin{pmatrix} 1 & -1 \\ -1 & 3 \end{pmatrix}$.

Remark: Assuming $u = \begin{pmatrix} x_1 \\ x_2 \end{pmatrix}$ and $v = \begin{pmatrix} y_1 \\ y_2 \end{pmatrix}$ are column vectors, observe that

$$u^T A v = (x_1, x_2) \begin{pmatrix} 1 & -1 \\ -1 & 3 \end{pmatrix} \begin{pmatrix} y_1 \\ y_2 \end{pmatrix} = x_1 y_1 - x_1 y_2 - x_2 y_1 + 3 x_2 y_2 = \langle u, v \rangle$$

[See Theorem 14.3.]

14.204 Find the matrix A_2 which represents the given inner product on $\mathbf{R}^2$ with respect to the basis $B = \{v_1 = (1, 3), v_2 = (2, 5)\}$ of $\mathbf{R}^2$. [Compare with Problem 14.202.]

▮ Compute $\langle (1, 3), (1, 3) \rangle = 1 - 3 - 3 + 27 = 22$, $\langle (1, 3), (2, 5) \rangle = 2 - 5 - 6 + 45 = 36$, $\langle (2, 5), (2, 5) \rangle = 4 - 10 - 10 + 75 = 59$. Thus $A_2 = \begin{pmatrix} 22 & 36 \\ 36 & 59 \end{pmatrix}$.

Remark: Problems 14.202–14.204 indicate that the matrix representing an inner product depends on both the basis and the inner product on V.

Theorem 14.13: Let A be the matrix representing an inner product on V with respect to a basis $B = \{e_1, \ldots, e_n\}$. Then, for any vectors $u, v \in V$, $\langle u, v \rangle = [u]^T A [v]$ where $[u]$ and $[v]$ denote, respectively, the (column) coordinate vectors of u and v relative to the basis B.

14.205 Prove Theorem 14.13.

▮ Suppose $A = (k_{ij})$, so $k_{ij} = \langle e_i, e_j \rangle$. Suppose $u = a_1 e_1 + a_2 e_2 + \cdots + a_n e_n$ and $v = b_1 e_1 + b_2 e_2 + \cdots + b_n e_n$. Then

$$\langle u, v \rangle = \sum_{i=1}^{n} \sum_{j=1}^{n} a_i b_j \langle e_i, e_j \rangle \tag{1}$$

On the other hand,

$$[u]^T A [v] = (a_1, a_2, \ldots, a_n) \begin{pmatrix} k_{11} & k_{12} & \cdots & k_{1n} \\ k_{21} & k_{22} & \cdots & k_{2n} \\ \cdots\cdots\cdots\cdots\cdots\cdots \\ k_{n1} & k_{ne} & \cdots & k_{nn} \end{pmatrix} \begin{pmatrix} b_1 \\ b_2 \\ \vdots \\ b_n \end{pmatrix}$$

$$= \left(\sum_{i=1}^{n} a_i k_{i1}, \sum_{i=1}^{n} a_i k_{i2}, \ldots, \sum_{i=1}^{n} a_i k_{in} \right) \begin{pmatrix} b_1 \\ b_2 \\ \vdots \\ b_n \end{pmatrix} = \sum_{j=1}^{n} \sum_{i=1}^{n} a_i b_j k_{ij} \tag{2}$$

Since $k_{ij} = \langle e_i, e_j \rangle$, the final sums in *(1)* and *(2)* are equal. Thus $\langle u, v \rangle = [u]^T A [v]$.

Problems 14.206–14.208 refer to the vector space V of polynomials $f(t)$ of degree ≤ 2 with inner product defined by $\int_{-1}^{1} f(t) g(t) \, dt$.

14.206 Find $\langle f, g \rangle$ where $f(t) = t + 2$ and $g(t) = t^2 - 3t + 4$.

▮ $\langle f, g \rangle = \int_{-1}^{1} (t + 2)(t^2 - 3t + 4) \, dt = \int_{-1}^{1} (t^3 - t^2 - 2t + 8) \, dt = \left[\frac{t^4}{4} - \frac{t^3}{3} - t^2 + 8t \right]_{-1}^{1} = \frac{46}{3}$

14.207 Find the matrix A of the inner product with respect to the basis $\{1, t, t^2\}$ of V.

∎ Here we use the fact that if $r + s = n$ then

$$\langle t^r, t^s \rangle = \int_{-1}^{1} t^n \, dt = \left[\frac{t^{n+1}}{n+1} \right]_{-1}^{1} = \begin{cases} 2/(n+1) & \text{if } n \text{ is even} \\ 0 & \text{if } n \text{ is odd} \end{cases}$$

Then $\langle 1, 1 \rangle = 2$, $\langle 1, t \rangle = 0$, $\langle 1, t^2 \rangle = \frac{2}{3}$, $\langle t, t \rangle = \frac{2}{3}$, $\langle t, t^2 \rangle = 0$, $\langle t^2, t^2 \rangle = \frac{2}{5}$. Thus

$$A = \begin{pmatrix} 2 & 0 & \frac{2}{3} \\ 0 & \frac{2}{3} & 0 \\ \frac{2}{3} & 0 & \frac{2}{5} \end{pmatrix}$$

14.208 Verify Theorem 14.13 that $\langle f, g \rangle = [f]^T A[g]$ with respect to the basis $\{1, t, t^2\}$.

∎ We have $[f]^T = (2, 1, 0)$ and $[g]^T = (4, -3, 1)$ relative to the given basis. Then

$$[f]^T A[g] = (2, 1, 0) \begin{pmatrix} 2 & 0 & \frac{2}{3} \\ 0 & \frac{2}{3} & 0 \\ \frac{2}{3} & 0 & \frac{2}{5} \end{pmatrix} \begin{pmatrix} 4 \\ -3 \\ 1 \end{pmatrix} = (4, \tfrac{2}{3}, \tfrac{4}{3}) \begin{pmatrix} 4 \\ -3 \\ 1 \end{pmatrix} = \tfrac{46}{3} = \langle f, g \rangle$$

14.209 Define a positive definite matrix.

∎ A square matrix A is positive definite if A is symmetric and if $X^T A X > 0$ for any nonzero vector X.

Theorem 14.14: Let A be a matrix which represents an inner product on V with respect to any basis $B = \{e_i\}$. Then A is positive definite.

14.210 Prove Theorem 14.14.

∎ A is symmetric since $\langle e_i, e_j \rangle = \langle e_j, e_i \rangle$. Let X be any nonzero vector in $\mathbf{R}^n$. Then $[u] = X$ for some nonzero vector $u \in V$. Using Theorem 14.13, we have $X^T A X = [u]^T A[u] = \langle u, u \rangle > 0$. Thus A is positive definite.

Theorem 14.15: Let A be a positive definite n-square matrix. Define $\langle u, v \rangle_A = u^T A v$ for any vectors $u, v \in \mathbf{R}^n$. Then $\langle \, , \, \rangle_A$ is an inner product on $\mathbf{R}^n$, that is, $\langle \, , \, \rangle_A$ satisfies axioms $[\text{RIP}_1]$, $[\text{RIP}_2]$, and $[\text{RIP}_3]$. [For notational convenience, we will omit the subscript A on $\langle \, , \, \rangle_A$.]

14.211 Show that $\langle \, , \, \rangle$ satisfies $[\text{RIP}_1]$.

∎ For any vectors u_1, u_2, and v, $\langle v_1 + u_2, v \rangle = (u_1 + u_2)^T A v = (u_1{}^T + u_2{}^T) A v = u_1{}^T A v + u_2{}^T A v = \langle u_1, v \rangle + \langle u_2, v \rangle$ and, for any scalar k and vectors u, v, $\langle ku, v \rangle = (ku)^T A v = k u^T A v = k \langle u, v \rangle$. Thus $\langle \, , \, \rangle$ satisfies $[\text{RIP}_1]$.

14.212 Show that $\langle \, , \, \rangle$ satisfies $[\text{RIP}_2]$.

∎ Since $u^T A v$ is a scalar, $(u^T A v)^T = u^T A v$. Also, $A^T = A$ since A is symmetric. Therefore, $\langle u, v \rangle = u^T A v = (u^T A v)^T = v^T A^T u^{TT} = v^T A u = \langle v, u \rangle$. Thus $\langle \, , \, \rangle$ satisfies $[\text{RIP}_2]$.

14.213 Show that $\langle \, , \, \rangle$ satisfies $[\text{RIP}_3]$.

∎ Since A is positive definite, $X^T A X > 0$ for any nonzero $X \in \mathbf{R}^n$. Thus, for any nonzero vector v, $\langle v, v \rangle = v^T A v > 0$. Thus $\langle \, , \, \rangle$ satisfies $[\text{RIP}_3]$.

14.214 Suppose A and B are positive definite matrices. Show that the sum $A + B$ is also positive definite.

∎ Since A and B are symmetric, $(A + B)^T = A^T + B^T = A + B$, so $A + B$ is symmetric. Also, for any nonzero vector X, we have $X^T A X > 0$ and $X^T B X > 0$. Thus $X^T(A + B)X = X^T A X + X^T B X > 0$ Accordingly, $A + B$ is positive definite.

14.215 Suppose A is positive definite, and $k > 0$. Show that kA is also positive definite.

∎ We have $(kA)^T = kA^T = kA$, so kA is symmetric. Also, for any nonzero vector X, we have $X^T A X > 0$; hence $X^T(kA)X = k(X^T A X) > 0$. Thus kA is positive definite.

14.216 Suppose B is a real nonsingular matrix. Show that B^TB is positive definite.

$\blacksquare$ We have $(B^TB)^T = B^TB^{TT} = B^TB$, so B^TB is symmetric. Suppose X is a nonzero vector in $\mathbf{R}^n$. Since B is nonsingular, BX is also nonzero. Therefore, $\langle BX, BX \rangle > 0$ [for the usual inner product in $\mathbf{R}^n$]. Hence $X^T(B^TB)X = (BX)^T(BX) = \langle BX, BX \rangle > 0$. Thus B^TB is positive definite.

14.9 COMPLEX INNER PRODUCT SPACES

This section considers vector spaces V over the complex field $\mathbf{C}$. First we recall some properties of complex numbers. Suppose $z \in \mathbf{C}$, say, $z = a + bi$ where $a, b \in \mathbf{R}$. Then $\bar{z} = a - bi$, $z\bar{z} = a^2 + b^2$, and $|z| = \sqrt{a^2 + b^2}$. Also, for any $z, z_1, z_2 \in \mathbf{C}$, $\overline{z_1 + z_2} = \bar{z}_1 + \bar{z}_2$, $\overline{z_1 z_2} = \bar{z}_1 \cdot \bar{z}_2$, $\bar{\bar{z}} = z$, and z is real if and only if $\bar{z} = z$.

14.217 Define a complex inner product and a complex inner product space V.

$\blacksquare$ Suppose to each pair of vectors $u, v \in V$ there is assigned a complex number, denoted by $\langle u, v \rangle$. Then this function $\langle \, , \, \rangle$ is called a *complex inner product* on V if it satisfies the following axioms [where $u_1, u_2, u, v \in V$ and $a, b, k \in \mathbf{C}$]:
[CIP$_1$] (Linear Property) $\langle au_1 + bu_2, v \rangle = a\langle u_1, v \rangle + b\langle u_2, v \rangle$ or, equivalently, (a) $\langle u_1 + u_2, v \rangle = \langle u_1, v \rangle + \langle u_2, v \rangle$ and (b) $\langle ku, v \rangle = k\langle u, v \rangle$.
[CIP$_2$] (Conjugate Symmetric Property) $\langle u, v \rangle = \overline{\langle v, u \rangle}$.
[CIP$_3$] (Positive Definite Property) If $u \neq 0$, then $\langle u, u \rangle > 0$.
The complex vector space V with an inner product is called a complex inner product space.

Remark: Observe that a complex inner product differs only slightly from a real inner product space [only [CIP$_2$] differs from [RIP$_2$]]. In fact, many of the definitions and properties of a complex inner product space are the same as that of a real inner product space. However, some of the proofs must be adapted to the complex case.

14.218 Show that $\langle 0, v \rangle = 0 = \langle v, 0 \rangle$ for every v in V. [Thus, in particular, $\langle 0, 0 \rangle = 0$.] [Compare with Problem 14.2.]

$\blacksquare$ $\langle 0, v \rangle = \langle 0v, v \rangle = 0\langle v, v \rangle = 0$. Also, since 0 is real and $\bar{0} = 0$, we have $\langle v, 0 \rangle = \overline{\langle 0, v \rangle} = \bar{0} = 0$.

14.219 Show that $\langle u, kv \rangle = \bar{k}\langle u, v \rangle$. [In other words, we must take the conjugate of a complex scalar when it is taken out of the second position of the inner product.]

$\blacksquare$ $\langle u, kv \rangle = \overline{\langle kv, u \rangle} = \overline{k\langle v, u \rangle} = \bar{k}\overline{\langle v, u \rangle} = \bar{k}\overline{\overline{\langle u, v \rangle}} = \bar{k}\langle u, v \rangle$.

14.220 Verify the relation $\langle u, av_1 + bv_2 \rangle = \bar{a}\langle u, v_1 \rangle = \bar{b}\langle u, v_2 \rangle$.

$\blacksquare$ $\langle u, av_1 + bv_2 \rangle = \overline{\langle av_1 + bv_2, u \rangle} = \overline{a\langle v_1, u \rangle + b\langle v_2, u \rangle} = \bar{a}\overline{\langle v_1, u \rangle} + \bar{b}\overline{\langle v_2, u \rangle} = \bar{a}\langle u, v_1 \rangle + \bar{b}\langle u, v_2 \rangle$.

Remark: One can analogously prove $\langle a_1u_1 + a_2u_2, b_1v_1 + b_2v_2 \rangle = a_1\bar{b}_1\langle u_1, v_1 \rangle + a_1\bar{b}_2\langle u_1, v_2 \rangle + a_2\bar{b}_1\langle u_2, v_1 \rangle + a_2\bar{b}_2\langle u_2, v_2 \rangle$ and, by induction, one can prove

$$\left\langle \sum_{i=1}^m a_iu_i, \sum_{j=1}^n b_jv_j \right\rangle = \sum_{i,j} a_i\bar{b}_j\langle u_i, v_j \rangle$$

[Compare with Problem 14.68.]

In Problems 14.221–14.223, we are given $\langle u, v \rangle = 3 + 2i$.

14.221 Find $\langle (2 - 4i)u, v \rangle$.

$\blacksquare$ $\langle (2 - 4i)u, v \rangle = (2 - 4i)\langle u, v \rangle = (2 - 4i)(3 + 2i) = 14 - 4i$.

14.222 Find $\langle u, (4 + 3i)v \rangle$.

$\blacksquare$ $\langle u, (4 + 3i)v \rangle = \overline{(4 + 3i)}\langle u, v \rangle = (4 - 3i)(3 + 2i) = 18 - i$.

14.223 Find $\langle(3-6i)u,(5-2i)v\rangle$.

▮ $\langle(3-6i)u,(5-2i)v\rangle=(3-6i)\overline{(5-2i)}\langle u,v\rangle=(3-6i)(5+2i)(3+2i)=137-30i$.

14.224 Axiom [CIP_3] assumes that $\langle u,u\rangle$ is real. Show that this fact follows from [CIP_2]. Also, define the length or norm of a vector u in complex inner product space V.

▮ By [CIP_2], $\langle u,u\rangle=\overline{\langle u,u\rangle}$. Thus $\langle u,u\rangle$ must be real. By [CIP_3], $\langle u,u\rangle$ must be nonnegative, and hence its positive real square root exists. As with real inner product spaces, we define $\|u\|=\sqrt{\langle u,u\rangle}$ to be the norm or length of u.

Remark: Besides the norm, we define the notions of orthogonality, orthogonal complement, orthogonal and orthonormal sets as before. In fact, the definitions of distance and Fourier coefficient and projection are the same as with the real case.

14.225 Define the usual or standard inner product in $\mathbf{C}^n$ and show that this definition reduces to the analogous one in $\mathbf{C}^n$ when all entries are real.

▮ Let $u=(z_i)$ and $v=(w_i)$ be vectors in $\mathbf{C}^n$. Then $\langle u,v\rangle=\sum_{k=1}^{n} z_k\bar{w}_k=z_1\bar{w}_1+z_2\bar{w}_2+\cdots+z_n\bar{w}_n$ is the usual or standard inner product on $\mathbf{C}^n$. [We assume this inner product on $\mathbf{C}^n$ unless otherwise stated or implied.] If the entries in u and v are real, then $\bar{w}_k=w_k$; hence $\langle u,v\rangle=z_1\bar{w}_1+z_2\bar{w}_2+\cdots+z_n\bar{w}_n=z_1w_1+z_2w_2+\cdots+z_nw_n$ which is the definition for $\mathbf{R}^n$.

Remark: Assuming u and v are column vectors, then the above inner product may be defined by $\langle u,v\rangle=u^T\bar{v}$ where $u^T\bar{v}$ refers to the product of the transpose u^T of u by the conjugate $\bar{v}$ of v under matrix multiplication, e.g.,

$$\left\langle\begin{pmatrix}z_1\\z_2\\z_3\end{pmatrix},\begin{pmatrix}w_1\\w_2\\w_3\end{pmatrix}\right\rangle=(z_1,z_2,z_3)\begin{pmatrix}\bar{w}_1\\\bar{w}_2\\\bar{w}_3\end{pmatrix}=z_1\bar{w}_1+z_2\bar{w}_2+z_3\bar{w}_3$$

14.226 Define the usual inner product on each of the following complex vector spaces: (*a*) U is the vector space of $m\times n$ matrices over $\mathbf{C}$. (*b*) V is the vector space of complex continuous functions on the (real) interval $a\le t\le b$.

▮ (*a*) The following is the usual inner product on U: $\langle A,B\rangle=\operatorname{tr}(B^*A)$. As usual, B^* denotes the conjugate transpose of the matrix B. (*b*) The following is the usual inner product on V: $\langle f,g\rangle=\int_a^b f(t)\overline{g(t)}\,dt$.

Problems 14.227–14.231 refer to the vectors $u=(1-i,2+3i)$ and $v=(2-5i,3-i)$ in $\mathbf{C}^2$.

14.227 Find $\langle u,v\rangle$.

▮ Recall that the conjugate of the second vector appears in the inner product: $\langle u,v\rangle=(1-i)\overline{(2-5i)}+(2+3i)\overline{(3-i)}=(1-i)(2+5i)+(2+3i)(3+i)=7+3i+3+11i=10+14i$.

14.228 Find $\langle v,u\rangle$.

▮ $\langle v,u\rangle=(2-5i)\overline{(1-i)}+(3-i)\overline{(2+3i)}=(2-5i)(1+i)+(3-i)(2-3i)=7-3i+3-11i=10-14i$. [As expected from [$\mathrm{CIP}_2$], $\langle v,u\rangle=\overline{\langle u,v\rangle}$.]

14.229 Find $\|u\|$.

▮ Recall that $z\bar{z}=a^2+b^2$ when $z=a+bi$. Use $\|u\|^2=\langle u,u\rangle=z_1\bar{z}_1+z_2\bar{z}_2+\cdots+z_n\bar{z}_n$ where $u=(z_1,z_2,\ldots,z_n)$. Compute $\|u\|^2=1^2+(-1)^2+2^2+3^2=1+1+2+9=13$ or $\|u\|=\sqrt{13}$.

14.230 Find $\|v\|$.

▮ $\|v\|^2=4+25+9+1=39$ and so $\|v\|=\sqrt{39}$.

14.231 Find $d(u, v)$, the distance between u and v.

▮ Recall $d(u, v) = \|u - v\|$. First find $u - v = (-1 + 4i, -1 + 4i)$. Then $\|u - v\|^2 = 1 + 16 + 1 + 16 = 34$; hence $d(u, v) = \|u - v\| = \sqrt{34}$.

14.232 Find the Fourier coefficient (component) c and the projection cw of $v = (3 + 4i, 2 - 3i)$ along $w = (5 + i, 2i)$ in $\mathbf{C}^2$.

▮ Recall $c = \langle v, w \rangle / \langle w, w \rangle$. Compute

$$\langle v, w \rangle = (3 + 4i)(\overline{5 + i}) + (2 - 3i)(\overline{2i}) = (3 + 4i)(5 - i) + (2 - 3i)(-2i) = 19 + 17i - 6 - 4i = 13 + 13i$$

$$\langle w, w \rangle = 25 + 1 + 4 = 30$$

Thus $c = (13 + 13i)/30 = 13/30 + 13i/30$. Accordingly, $cw = (26/15 + 39i/15, -13/15 + i/15)$.

Theorem 14.16 (Cauchy-Schwarz): $|\langle u, v \rangle| \le \|u\| \, \|v\|$.

14.233 Prove Theorem 14.16 for complex inner product spaces V.

▮ If $v = 0$, the inequality reduces to $0 \le 0$ and hence is valid. Now suppose $v \ne 0$. Using $z\bar{z} = |z|^2$ [for any complex number z] and $\langle v, u \rangle = \langle \overline{u, v} \rangle$, we expand $\|u - \langle u, v \rangle tv\|^2 \ge 0$ where t is any real value:

$$\begin{aligned}
0 \le \|u - \langle u, v \rangle tv\|^2 &= \langle u - \langle u, v \rangle tv, \, u - \langle u, v \rangle tv \rangle \\
&= \langle u, u \rangle - \langle \overline{u, v} \rangle t \langle u, v \rangle - \langle u, v \rangle t \langle v, u \rangle + \langle u, v \rangle \langle \overline{u, v} \rangle t^2 \langle v, v \rangle \\
&= \|u\|^2 - 2t |\langle u, v \rangle|^2 + |\langle u, v \rangle|^2 t^2 \|v\|^2
\end{aligned}$$

Set $t = 1/\|v\|^2$ to find $0 \le \|u\|^2 - (|\langle u, v \rangle|^2 / \|v\|^2)$, from which $|\langle u, v \rangle|^2 \le \|u\|^2 \|v\|^2$. Taking the square root of both sides, we obtain the required inequality.

14.234 Find an orthonormal basis of the subspace W of $\mathbf{C}^3$ spanned by $v_1 = (1, i, 0)$ and $v_2 = (1, 2, 1 - i)$.

▮ Apply the Gram-Schmidt algorithm. Set $w_1 = v_1 = (1, i, 0)$. Compute

$$v_2 - \frac{\langle v_2, w_1 \rangle}{\|w_1\|} w_1 = (1, 2, 1 - i) - \frac{1 - 2i}{2}(1, i, 0) = (\tfrac{1}{2} + i, \, 1 - \tfrac{1}{2}i, \, 1 - i)$$

Multiply by 2 to clear fractions obtaining $w_2 = (1 + 2i, 2 - i, 2 - 2i)$. Next find $\|w_1\| = \sqrt{2}$ and $\|w_2\| = \sqrt{18}$. Normalizing $\{w_1, w_2\}$ we obtain the following required orthonormal basis of W:

$$\left\{ u_1 = \left(\frac{1}{\sqrt{2}}, \frac{i}{\sqrt{2}}, 0 \right), \, u_2 = \left(\frac{1 + 2i}{\sqrt{18}}, \frac{2 - i}{\sqrt{18}}, \frac{2 - 2i}{\sqrt{18}} \right) \right\}$$

Following are a list of properties of a complex inner product space V which are analogous to properties of real inner product spaces and whose proof are analogous to the real case and hence are omitted.

Theorem 14.17: Let W be a subspace of a complex inner product space V. Then $V = W \oplus W$.

Lemma 14.18: Let $\{e_1, \ldots, e_n\}$ be an orthonormal basis of V. Then
  (a) For any $u \in V$, $u = \langle u, e_1 \rangle e_1 + \langle u, e_2 \rangle e_2 + \cdots + \langle u, e_n \rangle e_n$.
  (b) $\langle a_1 e_1 + \cdots + a_n e_n, \, b_1 e_1 + \cdots + b_n e_n \rangle = a_1 \bar{b} + a_2 \bar{b}_2 + \cdots + a_n \bar{b}_n$.
  (c) For any $u, v \in V$, $\langle u, v \rangle = \langle u, e_1 \rangle \langle \overline{v, e_1} \rangle + \cdots + \langle u, e_n \rangle \langle \overline{v, e_n} \rangle$.
  (d) If $T: V \to V$ is linear, then $\langle T(e_j), e_i \rangle$ is the ij entry of the matrix A representing T in the given basis $\{e_i\}$.

Theorem 14.19: Let $\{u_1, \ldots, u_n\}$ be a basis of V. Let $A = (a_{ij})$ be the complex matrix defined by $a_{ij} = \langle u_i, u_j \rangle$. Then, for any $u, v \in V$, $\langle u, v \rangle = [u]^T A [\bar{v}]$ where $[u]$ and $[v]$ are the coordinate column vectors in the given basis $\{u_i\}$. [Remark: This matrix A is said to represent the inner product on V.]

Theorem 14.20: Let A be a Hermitian matrix [i.e., $A^* = \bar{A}^T = A$] such that $X^T A \bar{X}$ is real and positive for every nonzero vector $X \in \mathbf{C}^n$. Then $\langle u, v \rangle = u^T A \bar{v}$ is an inner product on $\mathbf{C}^n$.

Theorem 14.21: Let A be the matrix which represents an inner product on V. Then A is Hermitian, and $X^T A X$ is real and positive for any nonzero vector in $\mathbf{C}^n$.

14.10 NORMED VECTOR SPACES

14.235 Define a normed vector space.

▌ Let V be a real or complex vector space. Suppose to each $v \in V$ there is assigned a real number, denoted by $\|v\|$. This function $\|\cdot\|$ is called a *norm* on V if it satisfies the following axioms [where $u, v \in V$ and $k \in K$]:
[N_1] $\|u + v\| \le \|u\| + \|v\|$.
[N_2] $\|kv\| = |k|\,\|v\|$.
[N_3] If $v \ne 0$, then $\|v\| > 0$.
The vector space V with a norm is called a *normed vector space*.

14.236 Show that $\|0\| = 0$.

▌ $\|0\| = \|0v\| = 0\|v\| = 0$.

14.237 Show that every inner product space V is a normed vector space.

▌ The norm on V defined by $\|v\| = \sqrt{\langle v, v \rangle}$ does satisfy [N_1], [N_2], and [N_3]. [See Problems 14.93–14.95.] Thus V is a normed vector space.

14.238 Define distance in a normed vector space V.

▌ The distance between vectors $u, v \in V$ is denoted and defined by $d(u, v) = \|u - v\|$.

Problems 14.239–14.241 show that $d(u, v)$ satisfies the following three axioms of a metric space:
[M_1] If $u \ne v$ then $d(u, v) > 0$ and $d(u, u) = 0$.
[M_2] $d(u, v) = d(v, u)$.
[M_3] $d(u, v) \le d(u, w) + d(w, v)$.

14.239 Show that if $u \ne v$ then $d(u, v) > 0$ and $d(u, u) = 0$.

▌ If $u \ne v$ then $u - v \ne 0$, and hence $d(u, v) = |(u - v)| > 0$. Also, $d(u, u) = \|u - u\| = \|0\| = 0$.

14.240 Show that $d(u, v) = d(v, u)$.

▌ $d(u, v) = \|u - v\| = \|-1(v - u)\| = |-1|\,\|v - u\| = \|v - u\| = d(v, u)$.

14.241 Show that $d(u, v) \le d(u, w) + d(w, v)$.

▌ $d(u, v) = \|u - v\| = \|(u - w) + (w - v)\| \le \|u - w\| + \|w - v\| = d(u, w) + d(w, v)$.

The following three norms on $\mathbf{R}^n$ and $\mathbf{C}^n$ will be used throughout this section:

$$\|(a_1, \ldots, a_n)\|_\infty = \max(|a_i|)$$
$$\|(a_1, \ldots, a_n)\|_1 = |a_1| + |a_2| + \cdots + |a_n|$$
$$\|(a_1, \ldots, a_n)\|_2 = \sqrt{|a_1|^2 + |a_2|^2 + \cdots + |a_n|^2}$$

The norms $\|\cdot\|_\infty$, $\|\cdot\|_1$, and $\|\cdot\|_2$ are called the *infinity-norm*, *one-norm*, and *two-norm*, respectively. Observe that $\|\cdot\|_2$ is the norm on $\mathbf{R}^n$ ($\mathbf{C}^n$) induced by the usual inner product on $\mathbf{R}^n$ ($\mathbf{C}^n$). [We will let d_∞, d_1, and d_2 denote, respectively, the corresponding distance functions.]

Problems 14.242–14.245 refer to the vectors $u = (1, 3, -6, 4)$ and $v = (3, -5, 1, -2)$ in $\mathbf{R}^4$.

14.242 Find $\|u\|_\infty$ and $\|v\|_\infty$.

▮ The infinity-norm chooses the maximum of the absolute values of the vectors. Hence $\|u\|_\infty = 6$ and $\|v\|_\infty = 5$.

14.243 Find $\|u\|_1$ and $\|v\|_1$.

▮ The one-norm adds the absolute values of the components. Thus $\|u\|_1 = 1 + 3 + 6 + 4 = 14$, $\|v\|_1 = 3 + 5 + 1 + 2 = 11$.

14.244 Find $\|u\|_2$ and $\|v\|_2$.

▮ The two-norm is equal to the square root of the sum of the square of the components [i.e., the norm induced by the usual inner product on $\mathbf{R}^4$]. Thus $\|u\|_2 = \sqrt{1 + 9 + 36 + 16} = \sqrt{62}$ and $\|v\|_2 = \sqrt{9 + 25 + 1 + 4} = \sqrt{39}$.

14.245 Find $d_\infty(u, v)$, $d_1(u, v)$, and $d_2(u, v)$.

▮ First find $u - v = (-2, 8, -7, 6)$. Then $d_\infty(u, v) = \|u - v\|_\infty = 8$, $d_1(u, v) = \|u - v\|_1 = 2 + 8 + 7 + 6 = 23$, $d_2(u, v) = \|u - v\|_2 = \sqrt{4 + 64 + 49 + 36} = \sqrt{153}$.

14.246 Let D_1 be the set of points $u = (x, y)$ in $\mathbf{R}^2$ such that $\|u\|_2 = 1$. Plot D_1 in the coordinate plane $\mathbf{R}^2$.

▮ Plot the points (x, y) such that $\|u\|_2^2 = x^2 + y^2 = 1$. Thus D_1 is the unit circle as pictured in Fig. 14-7.

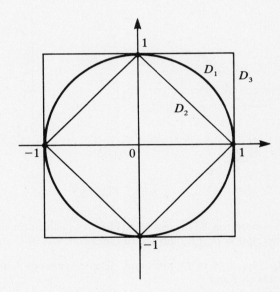

Fig. 14-7

14.247 Let D_2 be the set of points $u = (x, y)$ in $\mathbf{R}^2$ such that $\|u\|_1 = 1$. Plot D_2 in the coordinate plane $\mathbf{R}^2$.

▮ Plot the points (x, y) such that $\|u\|_1 = |x| + |y| = 1$. Thus D_2 is the diamond inside the unit circle as pictured in Fig. 14-7.

14.248 Let D_3 be the set of points $u = (x, y)$ in $\mathbf{R}^2$ such that $\|u\|_\infty = 1$. Plot D_3 in the coordinate plane $\mathbf{R}^2$.

▮ Plot the points (x, y) such that $\|u\|_\infty = \max(|x|, |y|) = 1$. Thus D_3 is the square circumscribing the unit circle as pictured in Fig. 14-7.

Problems 14.249–14.252 refer to the vectors $u = (5 - 2i, 3 + 4i)$ and $v = (2 + i, 2 - 3i)$ in $\mathbf{C}^2$.

14.249 Find $\|u\|_1$ and $\|v\|_1$.

▮ $\qquad \|u\|_1 = |5 - 2i| + |3 + 4i| = \sqrt{29} + 5 \qquad \|v\|_1 = |2 + i| + |2 - 3i| = \sqrt{5} + \sqrt{13}$.

14.250 Find $\|u\|_\infty$ and $\|v\|_\infty$.

❚ $\|u\|_\infty = \max(|5 - 2i|, |3 + 4i|) = \max(\sqrt{29}, 5) = \sqrt{29}, \quad \|v\|_\infty = \max(|2 + i|, |2 - 3i|) = \max(\sqrt{5}, \sqrt{13}) = \sqrt{13}.$

14.251 Find $\|u\|_2$ and $\|v\|_2$.

❚ $\|u\|_2^{\,2} = |5 - 2i|^2 + |3 + 4i|^2 = 29 + 25 = 54;$ so $\|u\|_2 = \sqrt{54} = 3\sqrt{6}.$ $\|v\|_2^{\,2} = |2 + i|^2 + |2 - 3i|^2 = 5 + 13 = 18;$ so $\|v\|_2 = \sqrt{18} = 3\sqrt{2}.$

14.252 Find $d_1(u, v)$, $d_\infty(u, v)$, and $d_2(u, v)$.

❚ First find $u - v = (3 - 3i, 1 + 7i)$. Then $d_1(u, v) = |3 - 3i| + |1 + 7i| = \sqrt{18} + \sqrt{50} = 3\sqrt{2} + 5\sqrt{2} = 8\sqrt{2},$ $d_\infty(u, v) = \max(|3 - 3i|, |1 + 7i|) = \max(3\sqrt{2}, 5\sqrt{2}) = 5\sqrt{2}.$ Also, $\|u - v\|^2 = 9 + 9 + 1 + 49 = 68,$ so $d_2(u, v) = \sqrt{68}.$

Problems 14.253–14.254 refer to the vector space V of continuous functions on the interval $a \leq t \leq b$.

14.253 The following defines a norm on V: $\|f\| = \int_a^b |f(t)|\, dt.$ [This norm is analogous to the $\|\cdot\|_1$ norm on $\mathbf{R}^n$.] Give a geometrical description of $\|f\|$ and the distance $d(f, g)$.

❚ As pictured in Fig. 14-8, $\|f\|$ is the area between the function $|f|$ and the t axis, and $d(f, g)$ is the area between the functions f and g.

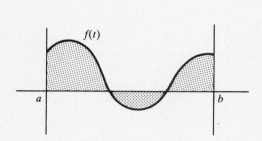

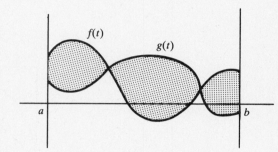

(a) $\|f\|$ is shaded.

(b) $d(f, g)$ is shaded.

Fig. 14-8

14.254 The following defines a norm on V: $\|f\| = \max(|f(t)|).$ [This norm is analogous to the $\|\cdot\|_\infty$ on $\mathbf{R}^n$.] Give a geometrical description of $\|f\|$ and the distance function $d(f, g)$.

❚ As pictured in Fig. 14-9, $\|f\|$ is the maximum distance between f and the x axis, and $d(f, g)$ is the maximum distance between f and g.

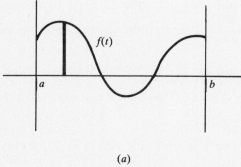

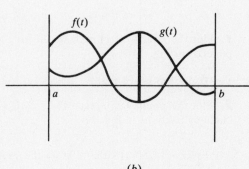

(a)

(b)

Fig. 14-9

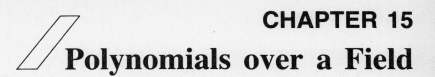

CHAPTER 15
Polynomials over a Field

This chapter investigates the ring $K[t]$ of polynomials over a field K and shows that $K[t]$ has many properties which are analogous to properties of the integers. These results play an important role in obtaining canonical forms for a linear operator T on a vector space V over K.

15.1 RING OF POLYNOMIALS

15.1   Define a polynomial over a field K and its degree.

▌ Let K be a field. Formally, a polynomial f over K is an infinite sequence of elements from K in which all except a finite number of them are 0: $f = (\ldots, 0, a_n, \ldots, a_1, a_0)$. Alternatively, using a symbol, say t, as an indeterminant, the above polynomial is usually denoted in the form $f(t) = a_n t^n + \cdots + a_1 t + a_0$. The entry a_k is called the kth coefficient of f. If n is the largest integer for which $a_n \neq 0$, then we say that the *degree* of f is n, written $\deg f = n$. We also call a_n the *leading coefficient* of f, and if $a_n = 1$ we call f a *monic polynomial*. On the other hand, if every coefficient of f is 0 then f is called the *zero polynomial*, written $f = 0$. The degree of the zero polynomial is not defined.

15.2   Define the ring of polynomials over the field K.

▌ Let $K[t]$ be the collection of all polynomials $f(t)$. Addition and multiplication is defined in $K[t]$ as follows. Suppose f is the polynomial in Problem 15.1 and suppose g is another polynomial over K, say $g(t) = b_m t^m + \cdots + b_1 t + b_0$. Then the *sum* $f + g$ is the polynomial obtained by adding corresponding coefficients. That is, if $m \leq n$ then $f(t) + g(t) = a_n t^n + \cdots + (a_m + b_m)t^m + \cdots + (a_1 + b_1)t + (a_0 + b_0)$ [or $f + g = (\ldots, 0, a_n, \ldots, a_m + b_m, \ldots, a_1 + b_1, a_0 + b_0)$]. Furthermore, the product of f and g is the polynomial $f(t)g(t) = a_n b_m t^{n+m} + \cdots + (a_1 b_0 + a_0 b_1)t + a_0 b_0$ [or $fg = (\ldots, 0, a_n b_m, \ldots, a_1 b_0 + a_0 b_1, a_0 b_0)$], i.e., the kth coefficient c_k of fg is $c_k = \sum_{i=0}^{k} a_i b_{k-1} = a_0 b_k + a_1 b_{k-1} + \cdots + a_k b_0$. Theorem 15.1 applies.

> **Theorem 15.1:** $K[t]$ under the above operations of addition and multiplication is a commutative ring with a unity element and with no zero divisors. [That is, $K[t]$ is an integral domain.]

15.3   Show how K may be viewed as a subset of $K[t]$.

▌ We identify the scalar $a_0 \in K$ with the polynomial $f(t) = a_0$ or $a_0 = (\ldots, 0, a_0)$. Then the operations of addition and multiplication of elements of K are preserved under this identification:

$$(\ldots, 0, a_0) + (\ldots, 0, b_0) = (\ldots, 0, a_0 + b_0)$$
$$(\ldots, 0, a_0) \cdot (\ldots, 0, b_0) = (\ldots, 0, a_0 b_0)$$

> **Theorem 15.2:**   Suppose f and g are polynomials in $K[t]$. Then $\deg(fg) = \deg f + \deg g$.

15.4   Prove Theorem 15.2.

▌ Suppose $f(t) = a_n t^n + \cdots + a_0$ and $g(t) = b_m t^m + \cdots + b_0$ and $a_n \neq 0$ and $b_m \neq 0$. Then $f(t)g(t) = a_n b_m t^{n+m} +$ terms of lower degree. Also, since the field K has no zero divisors, $a_n b_m \neq 0$. Thus $\deg(fg) = n + m = \deg f + \deg g$.

15.5   Show that the nonzero elements of K are the units of $K[t]$.

▌ Suppose $f(t)g(t) = 1$. Then $0 = \deg(1) = \deg(fg) = \deg f + \deg g$. Hence $\deg f = 0$ and $\deg g = 0$ and f and g are scalars in K. On the other hand, if $a \in K$ and $a \neq 0$, then $a \cdot a^{-1} = 1$ and a is a unit of $K[t]$.

> **Remark:**   A polynomial g is said to divide a polynomial f if there exists a polynomial h such that
> $f(t) = g(t)h(t)$.

15.6 Suppose $g(t)$ divides $f(t)$. Show that $\deg g \le \deg f$.

▮ If g divides f, then there exists h such that $f(t) = g(t)h(t)$. Then, by Theorem 15.2, $\deg f = \deg g + \deg h \ge \deg g$.

15.7 Suppose f and g are polynomials such that f divides g and g divides f. Show that (a) $\deg f = \deg g$ and (b) f and g are associates, i.e., $f(t) = kg(t)$ where $k \in K$.

▮ (a) By Theorem 15.2, $\deg f \le \deg g$ and $\deg g \le \deg f$. Hence $\deg f = \deg g$. (b) Since g divides f, there exist h such that $f(t) = h(t)g(t)$. Since $\deg f = \deg g$, we have $\deg h = 0$. In other words, $h(t) = k$, an element of K.

15.8 Suppose d and d' are monic polynomials such that d divides d' and d' divides d. Then $d = d'$.

▮ By Problem 15.7, $d(t) = kd'(t)$ where $k \in K$. The leading coefficient of d is 1 since d is monic and the leading coefficient of kd' is k since d' is monic. Hence $k = 1$ and $d = d'$.

15.2 EUCLIDEAN ALGORITHM, ROOTS OF POLYNOMIALS

This section uses the following theorems whose proofs appear in Problems 15.13–15.15.

> **Theorem 15.3 (Euclidean Division Algorithm):** Let $f(t)$ and $g(t)$ be polynomials over a field K with $g(t) \ne 0$. Then there exist polynomials $q(t)$ and $r(t)$ such that $f(t) = q(t)g(t) + r(t)$ where either $r(t) \equiv 0$ or $\deg r < \deg g$.

[The above theorem formalizes the process known as "long division."]

> **Theorem 15.4:** Suppose $a \in K$ is a root of a polynomial $f(t)$ over K which $\deg f = n$. Then there exists a polynomial $q(t)$ with $\deg q = n - 1$ such that $f(t) = (t - a)q(t)$. [That is, $t - a$ divides $f(t)$.]

> **Theorem 15.5:** Suppose a rational number p/q [reduced to lowest terms] is a root of the polynomial $f(t) = a_n t^n + \cdots + a_1 t + a_0$ where $a_n, \ldots, a_1, a_0 \in Z$. Then p divides the constant term a_0 and q divides the leading coefficient a_n. [In particular, if $c = p/q$ is an integer, then c divides a_0.]

15.9 Suppose $f(t) = t^3 + t^2 - 8t + 4$. Assuming $f(t)$ has a rational root, find all the roots of $f(t)$.

▮ Since the leading coefficient is 1, the only rational roots of $f(t)$ must be integers. Also, these integers are among ± 1, ± 2, ± 4. Note $f(1) \ne 0$ and $f(-1) \ne 0$. By synthetic division [or dividing by $t - 2$], we get

$$
2 \begin{array}{|l}
1 + 1 - 8 + 4 \\
2 + 6 - 4 \\
\hline
1 + 3 - 2 + 0
\end{array}
$$

Therefore, $t = 2$ is a root and $f(t) = (t - 2)(t^2 + 3t - 2)$. Using the quadratic formula for $t^2 + 3t - 2 = 0$, we obtain the following roots of $f(t)$: $t = 2, t = (-3 + \sqrt{17})/2, t = (-3 - \sqrt{17})/2$.

15.10 Suppose $g(t) = t^3 - 2t^2 - 6t - 3$. Find the roots of $g(t)$ assuming $g(t)$ has an integer root.

▮ The only integer roots of $g(t)$ must be among ± 1, ± 3. Note $f(1) \ne 0$. Using synthetic division [or dividing by $t + 1$], we get

$$
-1 \begin{array}{|l}
1 - 2 - 6 - 3 \\
{-1} + 3 + 3 \\
\hline
1 - 3 - 3 + 0
\end{array}
$$

Therefore, $t = -1$ is a root and $g(t) = (t + 1)(t^2 - 3t - 3)$. We can now use the quadratic formula on $t^2 - 3t - 3$ to obtain the following three roots of $g(t)$: $t = -1$, $t = (3 + \sqrt{21})/2$, $t = (3 - \sqrt{21})/2$.

15.11 Suppose $h(t) = t^4 - 2t^3 + 11t - 10$. Find all the real roots of $h(t)$ assuming there are two integer roots.

$\blacksquare$ The integer roots must be among ± 1, ± 2, ± 5, ± 10. By synthetic division [or dividing by $t - 1$ and then $t + 2$] we get

$$
\begin{array}{r|rrrrr}
1 & 1 - 2 + 0 + 11 - 10 \\
& \quad\ 1 - 1 - \ \ 1 + 10 \\
\hline
-2 & 1 - 1 - 1 + 10 + \ \ 0 \\
& \quad\quad\ -2 + 6 - 10 \\
\hline
& 1 - 3 + 5 + \ \ 0
\end{array}
$$

Thus $t = 1$ and $t = -2$ are roots and $h(t) = (t - 1)(t + 2)(t^2 - 3t + 5)$. The quadratic formula with $t^2 - 3t + 5$ tells us that there are no other real roots. That is, $t = 1$ and $t = -2$ are the only real roots of $h(t)$.

15.12 Suppose $f(t) = 2t^3 - 3t^2 - 6t - 2$. Find all the roots of $f(t)$ knowing that there is a rational root.

$\blacksquare$ The rational roots must be among ± 1, ± 2, $\pm \frac{1}{2}$. Testing each possible rational root, we get, by synthetic division (or dividing by $2t + 1$),

$$
\begin{array}{r|rrrr}
-\frac{1}{2} & 2 - 3 - 6 - 2 \\
& \quad\ -1 + 2 + 2 \\
\hline
& 2 - 4 - 4 + 0
\end{array}
$$

Thus $t = -\frac{1}{2}$ is a root and $f(t) = (t + \frac{1}{2})(2t^2 - 4t - 4) = (2t + 1)(t^2 - 2t - 2)$. We can now use the quadratic formula on $t^2 - 2t - 2$ to obtain the following three roots of $f(t)$: $t = -\frac{1}{2}, t = 1 + \sqrt{3}, t = 1 - \sqrt{3}$.

15.13 Prove Theorem 15.3.

$\blacksquare$ If $f(t) \equiv 0$ or if $\deg f < \deg g$, then we have the required representation $f(t) = 0g(t) + f(t)$. Now suppose $\deg f \geq \deg g$, say $f(t) = a_n t^n + \cdots + a_1 t + a_0$ and $g(t) = b_m t^m + \cdots + b_1 t + b_0$ where $a_n, b_m \neq 0$ and $n \geq m$. We form the polynomial

$$f_1(t) = f(t) - \frac{a_n}{b_m} t^{n-m} g(t) \tag{1}$$

[This is the first step in "long division."] Then $\deg f_1 < \deg f$. By induction, there exist polynomials $q_1(t)$ and $r(t)$ such that $f_1(t) = q_1(t)g(t) + r(t)$ where either $r(t) \equiv 0$ or $\deg r < \deg g$. Substituting this into *(1)* and solving for $f(t)$, we get

$$f(t) = \left[q_1(t) + \frac{a_n}{b_m} t^{n-m} \right] g(t) + r(t)$$

which is the desired representation.

15.14 Prove Theorem 15.4.

$\blacksquare$ By Theorem 15.3, there exist $q(t)$ and $r(t)$ such that

$$f(t) = (t - a)(t) + r(t) \tag{1}$$

with $r(t) \equiv 0$ or $\deg r < \deg(t - a) = 1$. Thus $r(t) = k$, a constant. Substituting $t = a$ and $r(t) = k$ into *(1)* yields $f(a) = (a - a)q(t) + k$. Since $f(a) = 0$ and $a - a = 0$, we get $k = r(t) = 0$. Thus $f(t) = (t - a)q(t)$. Also $n = \deg f = \deg(t - a) + \deg q = 1 + \deg q$. Hence $\deg q = n - 1$.

15.15 Prove Theorem 15.5.

$\blacksquare$ Substitute $t = p/q$ into $f(t) = 0$ to obtain $a_n(p/q)^n + \cdots + a_1(p/q) + a_0 = 0$. Multiply both sides of the equation by q^n to obtain

$$a_n p^n + a_{n-1} p^{n-1} q + a_{n-2} p^{n-2} q^2 + \cdots + a_1 p q^{n-1} + a_0 q^n = 0 \tag{1}$$

Since p divides all of the first n terms of *(1)*, p must divide the last term $a_0 q^n$. Assuming p and q are relatively prime, p divides a_0. Similarly, q divides the last n terms of *(1)*, hence q divides the first term $a_n p^n$. Since p and q are relatively prime, q divides a_n.

Theorem 15.6: Suppose $f(t)$ is a polynomial over a field K and $\deg f = n$. Then $f(t)$ has at most n roots in K.

15.16 Prove Theorem 15.6.

▌ The proof is by induction on n. If $n = 1$, then $f(t) = at + b$ and $f(t)$ has the unique root $t = -b/a$. Suppose $n > 1$. If $f(t)$ has no roots, then the theorem is true. Suppose $a \in K$ is a root of $f(t)$. Then

$$f(t) = (t - a)g(t) \tag{1}$$

where $\deg g = n - 1$. We claim that any other root of $f(t)$ must also be a root of $g(t)$. Suppose $b \neq a$ is another root of $f(t)$. Substituting $t = b$ in *(1)* yields $0 = f(b) = (b - a)g(b)$. Since K has no zero divisors, and $b - a \neq 0$, we must have $g(b) = 0$. By induction, $g(t)$ has at most $n - 1$ roots. Thus $f(t)$ has at most $n - 1$ roots other than a. Thus $f(t)$ has at most n roots.

Theorem 15.7: Suppose $f(t)$ is a polynomial over the real field **R**, and suppose the complex number $z = a + bi, b \neq 0$, is a root of $f(t)$. Then the complex conjugate $\bar{z} = a - bi$ is also a root of $f(t)$ and hence $c(t) = (t - z)(t - z) = t^2 - 2at + a^2 + b^2$ is a factor of $f(t)$.

15.17 Prove Theorem 15.7.

▌ Since $\deg c = 2$, there exist $q(t)$ and $M, N \in \mathbf{R}$ such that

$$f(t) = c(t)q(t) + Mt + N \tag{1}$$

Since $z = a + bi$ is a root of $f(t)$ and $c(t)$, we have by substituting $t = a + bi$ in *(1)*

$$f(z) = c(z)q(z) + M(z) + N \quad \text{or} \quad 0 = 0q(z) + M(z) + N \quad \text{or} \quad M(a + bi) + N = 0$$

Thus $Ma + N = 0$ and $Mb = 0$. Since $b \neq 0$, we must have $M = 0$. Then $0 + N = 0$ or $N = 0$. Accordingly, $f(t) = c(t)q(t)$ and $\bar{z} = a - bi$ is a root of $f(t)$.

15.18 Suppose $f(t) = t^4 - 3t^3 + 6t^2 + 25t - 39$. Find all the roots of $f(t)$ given that $t = 2 + 3i$ is a root.

▌ Since $2 + 3i$ is a root, then $2 - 3i$ is a root and $c(t) = t^2 - 4t + 13$ is a factor of $f(t)$. Dividing $f(t)$ by $c(t)$ we get $f(t) = (t^2 - 4t + 13)(t^2 + t - 3)$. The quadratic formula with $t^2 + t - 3$ gives us the other roots of $f(t)$. That is, the four roots of $f(t)$ follow: $2 + 3i$, $2 - 3i$, $(-1 + \sqrt{13})/2$, $(-1 - \sqrt{13})/2$.

15.19 Suppose $f(t)$ is a real polynomial with odd degree. Show that $f(t)$ must have a real root.

▌ The complex roots of $f(t)$ come in pairs by Theorem 15.7. Hence one root of $f(t)$ must be real.

15.20 Give a geometrical proof of the fact that a real polynomial $f(t)$ of odd degree has a real root.

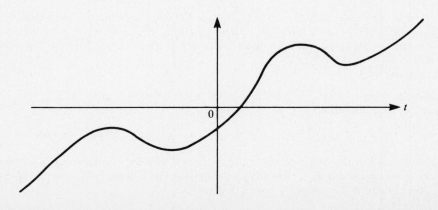

Fig. 15-1

▮ Suppose the leading coefficient of $f(t)$ is positive [otherwise, multiply $f(t)$ by -1]. Since deg $f = n$ where n is odd, we have

$$\lim_{t \to \infty} f(t) = +\infty \qquad \text{and} \qquad \lim_{t \to -\infty} f(t) = -\infty$$

Thus the graph of $f(t)$ must cross the t axis in at least one point as pictured in Fig. 15-1.

15.3   PRINCIPAL IDEAL DOMAIN, UNIQUE FACTORIZATION DOMAIN

This section proves that the ring $K[t]$ of polynomials over a field K is a principal ideal domain and a unique factorization domain. [The reader is referred to Section 6.8 for relevant definition.]

Theorem 15.8:   The ring $K[t]$ of polynomials over a field K is a principal ideal domain. If J is an ideal in $K[t]$, then there exists a unique monic polynomial d which generates J, i.e., such that d divides every polynomial $f \in J$.

15.21   Prove Theorem 15.8.

▮ Let d be a polynomial of lowest degree in J. Since we can multiply d by a nonzero scalar and still remain in J, we can assume without loss in generality that d is a monic polynomial. Now suppose $f \in J$. By the division algorithm there exist polynomials q and r such that $f = qd + r$ where either $r = 0$ or deg $r <$ deg d. Now $f, d \in J$ implies $qd \in J$ and hence $r = f - qd \in J$. But d is a polynomial of lowest degree in J. Accordingly, $r = 0$ and $f = qd$, that is, d divides f. It remains to show that d is unique. If d' is another monic polynomial which generates J, then d divides d' and d' divides d. This implies that $d = d'$, because d and d' are monic. Thus the theorem is proved.

Theorem 15.9:   Let f and g be nonzero polynomials in $K[t]$. Then there exists a unique monic polynomial d such that (i) d divides f and g and (ii) if d' divides f and g, then d' divides d.

15.22   Prove Theorem 15.9.

▮ The set $I = \{mf + ng : m, n \in K[t]\}$ is an ideal. Let d be the monic polynomial which generates I. Note $f, g \in I$; hence d divides f and g. Now suppose d' divides f and g. Let J be the ideal generated by d'. Then $f, g \in J$ and hence $I \subset J$. Accordingly, $d \in J$ and so d' divides d as claimed. It remains to show that d is unique. If d_1 is another [monic] greatest common divisor of f and g, then d divides d_1 and d_1 divides d. This implies that $d = d_1$ because d and d_1 are monic. Thus the theorem is proved.

Remark: The above polynomial d is called the *greatest common divisor* of f and g. If $d = 1$, then f and g are said to be *relatively prime*.

Corollary 15.10:   Let d be the greatest common divisor of the polynomials f and g. Then there exist polynomials m and n such that $d = mf + ng$. In particular, if f and g are relatively prime then there exist polynomials m and n such that $mf + ng = 1$.

15.23   Prove Corollary 15.10.

▮ From the proof of Theorem 15.9, d generates the ideal $I = \{mf + ng : m, n \in K[t]\}$. Thus there exists $m, n \in K[t]$ such that $d = mf + ng$.

15.24   Define an irreducible polynomial.

▮ A polynomial $p \in K[t]$ is said to be irreducible if p has positive degree [i.e., p is not a constant] and if $p = fg$ implies f or g is a scalar.

Lemma 15.11:   Suppose $p \in K[t]$ is irreducible. If p divides the product fg of polynomials $f, g \in K[t]$, then p divides f or p divides g. More generally, if p divides the product of n polynomials $f_1 f_2 \ldots f_n$, then p divides one of them.

15.25   Prove Lemma 15.11.

❚   Suppose p divides fg but not f.   Since p is irreducible, the polynomials f and p must then be relatively prime.   Thus there exist polynomials $m, n \in K[t]$ such that $mf + np = 1$.   Multiplying this equation by g,   we obtain $mfg + npg = g$.   But p divides fg and so mfg, and p divides npg; hence p divides the sum $g = mfg + npg$.

Now suppose p divides $f_1 f_2 \ldots f_n$.   If p divides f_1, then we are through.   If not, then by the above result p divides the product $f_2 \ldots f_n$.   By induction on n, p divides one of the polynomials $f_2, \ldots, f_n$.   Thus the lemma is proved.

> **Theorem 15.12 (*Unique Factorization Theorem*):**   Let f be a nonzero polynomial in $K[t]$.   Then f can be written uniquely [except for order] as a product $f = kp_1 p_2 \ldots p_n$ where $k \in K$ and the p_i are monic irreducible polynomials in $K[t]$.

15.26   Prove Theorem 15.12.

❚   We prove the existence of such a product first.   If f is irreducible or if $f \in K$,   then such a product clearly exists.   On the other hand, suppose $f = gh$ where f and g are nonscalars.   Then g and h have degrees less than that of f.   By induction, we can assume $g = k_1 g_1 g_2 \ldots g_r$ and $h = k_2 h_1 h_2 \ldots h_s$ where $k_1, k_2 \in K$ and the g_i and h_j are monic irreducible polynomials.   Accordingly, $f = (k_1 k_2) g_1 g_2 \ldots g_r h_1 h_2 \ldots h_s$ is our desired representation.

We next prove uniqueness [except for order] of such a product for f.   Suppose $f = kp_1 p_1 \ldots p_n = k' q_1 q_2 \ldots q_m$ where $k, k' \in K$ and the $p_1, \ldots, p_n, q_1, \ldots, q_m$ are monic irreducible polynomials.   Now p_1 divides $k' q_1 \ldots q_m$.   Since p_1 is irreducible it must divide one of the q_i by the above lemma.   Say p_1 divides q_1.   Since p_1 and q_1 are both irreducible and monic, $p_1 = q_1$.   Accordingly, $kp_2 \cdots p_n = k' q_2 \ldots q_m$.   By induction, we have that $n = m$ and $p_2 = q_2, \ldots, p_n = q_m$ for some rearrangement of the q_i.   We also have that $k = k'$.   Thus the theorem is proved.

15.27   State the Fundamental Theorem of Algegra.   [The proof of this theorem lies beyond the scope of this text.]

❚   **Fundamental Theorem of Algebra:**   The complex field **C** is closed.   That is, any nonzero polynomial $f(t)$ over **C** has a root in **C** and therefore $f(t)$ can be written uniquely [except for order] as a product $f(t) = k(t - r_1)(t - r_2) \cdots (t - r_n)$ where $k, r_i \in \mathbf{C}$,   i.e., as a product of linear polynomials.

> **Theorem 15.13:**   Let $f(t)$ be a nonzero polynomial over the real field **R**.   Then $f(t)$ can be written uniquely [except for order] as a product $f(t) = kp_1(t) p_2(t) \cdots p_m(t)$ where $k \in \mathbf{R}$ and the $p_i(t)$ are monic irreducible polynomials of degree one or two.

15.28   Prove Theorem 15.13.

❚   By the fundamental theorem of algebra, $f(t) = k(t - r_1)(t - r_2) \cdots (t - r_n)$ where $k, r_i \in \mathbf{C}$.   Since k is the leading coefficient of $f(t)$ we have $k \in \mathbf{R}$.   Also, if $r_i = a + bi$ is a nonreal root, then there exists a root $r_j = a - bi$.   Furthermore, $p(t) = (t - r_i)(t - r_j) = t^2 - 2at + a^2 + b^2$ is a polynomial over **R**, $p(t)$ is monic, and $p(t)$ is irreducible over **R** since its roots are nonreal. The theorem follows.

15.4 POLYNOMIALS WITH MATRICES AND LINEAR OPERATORS

This section considers polynomials of matrices.   Specifically, if $f(t) = a_n t^n + \cdots + a_1 t + a_0$ is a polynomial over a field K and A is an n-square matrix over K, then we define $f(A) = a_n A^n + \cdots + a_1 A + a_0 I$ where I is the identity matrix.   In particular, we say that A is a *root* or *zero* of the polynomial $f(t)$ if $f(A) = 0$.

Problems 15.29–15.32 refer to the matrices $A = \begin{pmatrix} 1 & 2 \\ 3 & 4 \end{pmatrix}$ and $B = \begin{pmatrix} 1 & -2 \\ 4 & 5 \end{pmatrix}$.

15.29   Find $f(A)$ where $f(t) = 2t^2 - 3t + 7$.   Is A a root of $f(t)$?

❚
$$f(A) = 2A^2 - 3A + 7I = 2\begin{pmatrix} 1 & 2 \\ 3 & 4 \end{pmatrix}^2 - 3\begin{pmatrix} 1 & 2 \\ 3 & 4 \end{pmatrix} + 7\begin{pmatrix} 1 & 0 \\ 0 & 1 \end{pmatrix}$$
$$= \begin{pmatrix} 14 & 20 \\ 30 & 44 \end{pmatrix} + \begin{pmatrix} -3 & -6 \\ -9 & -12 \end{pmatrix} + \begin{pmatrix} 7 & 0 \\ 0 & 7 \end{pmatrix} = \begin{pmatrix} 18 & 14 \\ 21 & 39 \end{pmatrix}$$

A is not a root of $f(t)$ since $f(A)$ is not the zero matrix.

15.30 Find $g(A)$ where $g(t) = t^2 - 5t - 2$. Is A a root of $g(t)$?

$$g(A) = A^2 - 5A - 2I = \begin{pmatrix} 1 & 2 \\ 3 & 4 \end{pmatrix}^2 - 5\begin{pmatrix} 1 & 2 \\ 3 & 4 \end{pmatrix} - 2\begin{pmatrix} 1 & 0 \\ 0 & 1 \end{pmatrix}$$

$$= \begin{pmatrix} 7 & 10 \\ 15 & 22 \end{pmatrix} + \begin{pmatrix} -5 & -10 \\ -15 & -20 \end{pmatrix} + \begin{pmatrix} -2 & 0 \\ 0 & -2 \end{pmatrix} = \begin{pmatrix} 0 & 0 \\ 0 & 0 \end{pmatrix}$$

A is a zero of $g(t)$ since $g(t)$ is the zero matrix.

15.31 Find $f(B)$ where $f(t) = 2t^2 - 3t + 7$.

$$f(B) = 2B^2 - 3B + 7I = 2\begin{pmatrix} 1 & -2 \\ 4 & 5 \end{pmatrix}^2 - 3\begin{pmatrix} 1 & -2 \\ 4 & 5 \end{pmatrix} + 7\begin{pmatrix} 1 & 0 \\ 0 & 1 \end{pmatrix}$$

$$= \begin{pmatrix} -14 & -24 \\ 48 & 34 \end{pmatrix} + \begin{pmatrix} -3 & 6 \\ -12 & -15 \end{pmatrix} + \begin{pmatrix} 7 & 0 \\ 0 & 7 \end{pmatrix} = \begin{pmatrix} -10 & -18 \\ 36 & 26 \end{pmatrix}$$

15.32 Find $h(B)$ where $h(t) = t^2 - 6t + 13$.

$$h(B) = B^2 - 6B + 13I = \begin{pmatrix} -7 & -12 \\ 24 & 17 \end{pmatrix} + \begin{pmatrix} -6 & 12 \\ -24 & -30 \end{pmatrix} + \begin{pmatrix} 13 & 0 \\ 0 & 13 \end{pmatrix} = \begin{pmatrix} 0 & 0 \\ 0 & 0 \end{pmatrix}$$

[Thus B is a root of $h(t)$.]

15.33 Show that $A = \begin{pmatrix} 1 & 4 \\ 2 & 3 \end{pmatrix}$ is a zero of $f(t) = t^2 - 4t - 5$.

$$f(A) = A^2 - 4A - 5I = \begin{pmatrix} 9 & 16 \\ 8 & 17 \end{pmatrix} + \begin{pmatrix} -4 & -16 \\ -8 & -12 \end{pmatrix} + \begin{pmatrix} -5 & 0 \\ 0 & -5 \end{pmatrix} = \begin{pmatrix} 0 & 0 \\ 0 & 0 \end{pmatrix}$$

15.34 Let $A = \begin{pmatrix} 1 & 1 \\ 0 & 1 \end{pmatrix}$. Find A^2, A^3, A^n.

$$A^2 = \begin{pmatrix} 1 & 1 \\ 0 & 1 \end{pmatrix}\begin{pmatrix} 1 & 1 \\ 0 & 1 \end{pmatrix} = \begin{pmatrix} 1 & 2 \\ 0 & 1 \end{pmatrix} \qquad A^3 = AA^2 = \begin{pmatrix} 1 & 1 \\ 0 & 1 \end{pmatrix}\begin{pmatrix} 1 & 2 \\ 0 & 1 \end{pmatrix} = \begin{pmatrix} 1 & 3 \\ 0 & 1 \end{pmatrix}$$

We claim that $A^n = \begin{pmatrix} 1 & n \\ 0 & 1 \end{pmatrix}$. It is true for $n = 1, 2$, and 3. Assuming it is true for $n - 1$, we have

$$A^n = AA^{n-1} = \begin{pmatrix} 1 & 1 \\ 0 & 1 \end{pmatrix}\begin{pmatrix} 1 & n-1 \\ 0 & 1 \end{pmatrix} = \begin{pmatrix} 1 & n \\ 0 & 1 \end{pmatrix}$$

Theorem 15.14: Let f and g be polynomials over K and let A be an n-square matrix over K. Then
    (i) $(f + g)(A) = f(A) + g(A)$.
    (ii) $(fg)(A) = f(A)g(A)$.
    (iii) $(kf)(A) = kf(A)$ where $k \in K$.

15.35 Prove (i) of Theorem 15.14.

Suppose $f = a_n t^n + \cdots + a_1 t + a_0$ and $g = b_m t^m + \cdots + b_1 t + b_0$. Then by definition, $f(A) = a_n A^n + \cdots + a_1 A + a_0 I$ and $g(A) = b_m A^m + \cdots + b_1 A + b_0 I$. Suppose $m \leq n$ and let $b_i = 0$ if $i > m$. Then $f + g = (a_n + b_n)t^n + \cdots + (a_1 + b_1)t + (a_0 + b_0)$. Hence $(f + g)(A) = (a_n + b_n)A^n + \cdots + (a_1 + b_1)A + (a_0 + b_0)I = a_n A^n + b_n A^n + \cdots + a_1 A + b_1 A + a_0 I + b_0 I = f(A) + g(A)$.

15.36 Prove (ii) of Theorem 15.14.

By definition, $fg = c_{n+m}t^{n+m} + \cdots + c_1 t + c_0 = \displaystyle\sum_{k=0}^{n+m} c_k t^k$ where $c_k = \displaystyle\sum_{i=0}^{k} a_i b_{k-i}$. Hence $(fg)(A) = \displaystyle\sum_{k=0}^{n+m} c_k A^k$ and

$$f(A)g(A) = \left(\sum_{i=0}^{n} a_i A^i\right)\left(\sum_{j=0}^{m} b_j A^j\right) = \sum_{i=0}^{n}\sum_{j=0}^{m} a_i b_j A^{i+j} = \sum_{k=0}^{n+m} c_k A^k = (fg)(A).$$

15.37 Prove (iii) of Theorem 15.14.

▮ By definition, $kf = ka_nt^n + \cdots + ka_1t + ka_0$, and so $(kf)(A) = ka_nA^n + \cdots + ka_1A + ka_0I = k(a_nA^n + \cdots + a_1A + a_0I) = kf(A)$.

15.38 Show that any two polynomials in a matrix A commute, i.e., $f(A)g(A) = g(A)f(A)$ for any polynomials $f(t)$ and $g(t)$.

▮ Since $f(t)g(t) = g(t)f(t)$, Theorem 15.14 tells us that $f(A)g(A) = g(A)f(A)$.

Remark: Suppose $T: V \to V$ is a linear operator on a vector space V over K and suppose $f(t) = a_nt^n + \cdots + a_1t + a_0$. Then we define $f(A)$ in the same way as we did for matrices: $f(T) = a_nT^n + \cdots + a_1T + a_0I$ where I is now the identity mapping. We also say that T is a *zero* or *root* of $f(t)$ if $f(T) = 0$. Furthermore, Theorem 15.14 holds for operators as it does for matrices. Thus, in particular, any two polynomials in T commute.

15.39 Let V be the vector space of functions which has $\{\sin\theta, \cos\theta\}$ as a basis and let D be the differential operator on V. Show that D is a zero of $f(t) = t^2 + 1$.

▮ Apply $f(D)$ to each basis vector:

$$f(D)(\sin\theta) = (D^2 + I)(\sin\theta) = D^2(\sin\theta) + I(\sin\theta) = -\sin\theta + \sin\theta = 0$$
$$f(D)(\cos\theta) = (D^2 + I)(\cos\theta) = D^2(\cos\theta) + I(\cos\theta) = -\cos\theta + \cos\theta = 0$$

Since each basis vector is mapped into 0, every vector $v \in V$ is also mapped into 0 by $f(D)$. Thus $f(D) = 0$.

15.40 Let A be a matrix representation of an operator T. Show that $f(A)$ is the matrix representation of $f(T)$, for any polynomial $f(t)$.

▮ Let ϕ be the mapping $T \mapsto A$, i.e., which sends the operator T into its matrix representation A. We need to prove that $\phi(f(T)) = f(A)$. Suppose $f(t) = a_nt^n + \cdots + a_1t + a_0$. The proof is by induction on n, the degree of $f(t)$.

Suppose $n = 0$. Recall that $\theta(I') = I$ where I' is the identity mapping and I is the identity matrix. Thus $\phi(f(T)) = \phi(a_0I') = a_0\phi(I') = a_0I = f(A)$ and so the theorem holds for $n = 0$.

Now assume the theorem holds for polynomials of degree less than n. Then since ϕ is an algebra isomorphism, $\phi(f(T)) = \phi(a_nT^n + a_{n-1}T^{n-1} + \cdots + a_1T + a_0I') = a_n\phi(T)\phi(T^{n-1}) + \phi(a_{n-1}T^{n-1} + \cdots + a_1T + a_0I') = a_nAA^{n-1} + (a_{n-1}A^{n-1} + \cdots + a_1A + a_0I) = f(A)$ and the theorem is proved.

15.41 Let A be any square matrix and let P be a nonsingular matrix of the same order. Show that (a) $(P^{-1}AP)^n = P^{-1}A^nP$, for every positive n, and (b) $f(P^{-1}AP) = P^{-1}f(A)P$, for any polynomial $f(t)$.

▮ (a) The condition trivially holds for $n = 1$. Then, by induction and $n > 1$,

$$(P^{-1}AP)^n = (P^{-1}AP)(P^{-1}AP)^{n-1} = (P^{-1}AP)(P^{-1}A^{n-1}P) = p^{-1}A^nP^{-1}.$$

(b) Suppose $f(t) = a_nt^n + \cdots + a_1t + a_0$. Then

$$f(P^{-1}AP) = a_n(P^{-1}AP)^n + a_{n-1}(P^{-1}AP)^{n-1} + \cdots + a_1(P^{-1}AP) + a_0I$$
$$= a_n(P^{-1}A^nP) + a_{n-1}(P^{-1}A^{n-1}P) + \cdots + a_1(P^{-1}AP) + a_0(P^{-1}IP)$$
$$= P^{-1}(a_nA^n + a_{n-1}A^{n-1} + \cdots + a_1A + a_0I)P = P^{-1}f(A)P.$$

15.42 Suppose a matrix B is similar to a matrix A. Show that $f(B)$ is similar to $f(A)$ for any polynomial $f(t)$.

▮ Since B is similar to A, there exists a nonsingular matrix P such that $B = P^{-1}AP$. Then, by Problem 15.41, $f(B) = f(P^{-1}AP) = P^{-1}f(A)P$. Thus $f(B)$ is similar to $f(A)$.

15.43 Let A be any square matrix. Show that (a) $(A^T)^n = (A^n)^T$ for any positive n, and (b) $f(A^T) = [f(A)]^T$ for any polynomial $f(t)$.

❚ (a)   The condition holds trivially for $n = 1$.   Then, by induction and $n > 1$,

$$(A^T)^n = A^T(A^T)^{n-1} = A^T(A^{n-1})^T = (A^{n-1}A)^T = (A^n)^T.$$

(b)   Suppose $f(t) = a_n t^n + a_{n-1} t^{n-1} + \cdots + a_1 t + a_0$.   Then, using the fact that $(P + Q)^T = P^T + Q^T$ and $(kP)^T = kP^T$,   we have

$$\begin{aligned}
f(A^T) &= a_n(A^T)^n + a_{n-1}(A^T)^{n-1} + \cdots + a_1 A^T + a_0 I \\
&= a_n(A^n)^T + a_{n-1}(A^{n-1})^T + \cdots + a_1 A^T + a_0 I^T \\
&= [a_n A^n + a_{n-1} a^{n-1} + \cdots + a_1 A + a_0 I]^T = [f(A)]^T.
\end{aligned}$$

15.44   Suppose A is symmetric.   Show that $f(A)$ is symmetric for any polynomial $f(t)$.

❚ Since A is symmetric, $A^T = A$.   Then, by Problem 15.43, $[f(A)]^T = f(A^T) = f(A)$.   Hence $f(A)$ is symmetric.

15.45   Let A be an n-square matrix.   Show that A is a zero of some nonzero polynomial.

❚ Let $N = n^2$.   Consider the following $N + 1$ matrices: $I, A, A^2, \ldots, A^N$.   Recall that the vector space of V of n-square matrices has dimension $N = n^2$.   Thus the above $N + 1$ matrices are linearly dependent.   Hence there exist scalars $a_0, a_1, \ldots, a_N$ such that $a_N A^N + \cdots + a_1 A + a_0 I = 0$.   Accordingly, A is a zero of the polynomial $f(t) = a_n t^n + \cdots + a_1 t + a_0$.

Remark: The above result is an existence proof; it does not tell us how to find a polynomial in which A is a root.   The next chapter gives us such a polynomial, the characteristic polynomial of A.

15.46   Consider a block diagonal matrix

$$M = \begin{pmatrix} A_1 & 0 & \cdots & 0 \\ 0 & A_2 & \cdots & 0 \\ \multicolumn{4}{c}{\dotfill} \\ 0 & 0 & \cdots & A_n \end{pmatrix}$$

Describe $f(M)$ for any polynomial $f(t)$.

❚ Since the diagonal blocks add and multiply independently, $f(M)$ has the following form where the diagonal blocks are $f(A_1), \ldots, f(A_n)$:

$$f(M) = \begin{pmatrix} f(A_1) & 0 & \cdots & 0 \\ 0 & f(A_2) & \cdots & 0 \\ \multicolumn{4}{c}{\dotfill} \\ 0 & 0 & \cdots & f(A_n) \end{pmatrix}$$

15.47   Consider a block triangular matrix

$$N = \begin{pmatrix} A_1 & B & \cdots & C \\ 0 & A_2 & \cdots & D \\ \multicolumn{4}{c}{\dotfill} \\ 0 & 0 & \cdots & A_n \end{pmatrix}$$

where the A_i are square matrices.   Describe $f(N)$ for any polynomial $f(t)$.

❚ Since the sum and product of block diagonal matrices are again block triangular matrices, and since the diagonal elements add and multiply independently, $f(N)$ has the following form where $f(A_1), \ldots, f(A_n)$ are the diagonal blocks:

$$f(N) = \begin{pmatrix} f(A_1) & X & \cdots & Y \\ 0 & f(A_2) & \cdots & Z \\ \multicolumn{4}{c}{\dotfill} \\ 0 & 0 & \cdots & f(A_n) \end{pmatrix}$$

CHAPTER 16

Eigenvalues and Eigenvectors, Diagonalization

This chapter studies the conditions for a matrix A to be similar to a diagonal matrix and, equivalently, for a linear operator T to be represented by a diagonal matrix. This question is closely related to the roots of certain polynomials associated with A (and T). The particular underlying field K also plays an important part in this theory since the existence of roots of a polynomial depends on K.

16.1 CHARACTERISTIC POLYNOMIAL, CAYLEY-HAMILTON THEOREM

16.1 Consider an n-square matrix A over a field K:

$$A = \begin{pmatrix} a_{11} & a_{22} & \cdots & a_{1n} \\ a_{21} & a_{22} & \cdots & a_{2n} \\ \cdots\cdots\cdots\cdots\cdots\cdots \\ a_{n1} & a_{n2} & \cdots & a_{nn} \end{pmatrix}$$

Define the characteristic polynomial of A.

▌ The matrix $tI_n - A$, where I_n is the n-square identity matrix and t is an indeterminant, is called the *characteristic matrix* of A:

$$tI_n - A = \begin{pmatrix} t - a_{11} & -a_{12} & \cdots & -a_{2n} \\ -a_{21} & t - a_{22} & \cdots & -a_{2n} \\ \cdots\cdots\cdots\cdots\cdots\cdots\cdots\cdots \\ -a_{n1} & -a_{n2} & \cdots & t - a_{nn} \end{pmatrix}$$

Its determinant $\Delta_A(t) = \det(tI_n - A)$, which is a polynomial in t, is called the *characteristic polynomial* of A. We also call $\Delta_A(t) = \det(tI_n - A) = 0$ the *characteristic equation* of A.

Theorem 16.1, which is proved in Problem 16.20 and used below, is one of the most important theorems in linear algebra.

Theorem 16.1 (Cayley-Hamilton): Every square matrix A is a zero of its characteristic polynomial.

16.2 Find the characteristic polynomial of $A = \begin{pmatrix} 2 & -3 \\ 5 & 1 \end{pmatrix}$.

▌ Form the characteristic matrix $tI - A$:

$$tI - A = \begin{pmatrix} t & 0 \\ 0 & t \end{pmatrix} + \begin{pmatrix} -2 & 3 \\ -5 & -1 \end{pmatrix} = \begin{pmatrix} t - 2 & 3 \\ -5 & t - 1 \end{pmatrix}$$

The characteristic polynomial $\Delta(t)$ of A is its determinant:

$$\Delta(t) = |tI - A| = \begin{vmatrix} t - 2 & 3 \\ -5 & t - 1 \end{vmatrix} = (t - 2)(t - 1) + 15 = t^2 - 3t + 17$$

16.3 Find the characteristic polynomial $\Delta(t)$ of $B = \begin{pmatrix} 1 & 3 & 0 \\ -2 & 2 & -1 \\ 4 & 0 & -2 \end{pmatrix}$.

▌ $\quad \Delta(t) = |tI - B| = \begin{vmatrix} t - 1 & -3 & 0 \\ 2 & t - 2 & 1 \\ -4 & 0 & t + 2 \end{vmatrix} = (t - 1)(t - 2)(t + 2) + 12 + 6(t + 2) = t^3 - t^2 + 2t + 28$

16.4 Find the characteristic polynomial $\Delta(t)$ of $A = \begin{pmatrix} 1 & 2 \\ 3 & 2 \end{pmatrix}$.

▌ $\quad \Delta(t) = |tI - A| = \begin{vmatrix} t - 1 & -2 \\ -3 & t - 2 \end{vmatrix} = (t - 1)(t - 2) - 6 = t^2 - 3t - 4$

16.5 Verify the Cayley-Hamilton theorem for the matrix A in Problem 16.4, i.e., verify that A is a root of its characteristic polynomial.

▮ Given $\Delta(t) = t^2 - 3t - 4$, we have

$$\Delta(A) = A^2 - 3A - 4I = \begin{pmatrix} 7 & 6 \\ 9 & 10 \end{pmatrix} + \begin{pmatrix} -3 & -6 \\ -9 & -6 \end{pmatrix} + \begin{pmatrix} -4 & 0 \\ 0 & -4 \end{pmatrix} = \begin{pmatrix} 0 & 0 \\ 0 & 0 \end{pmatrix}$$

16.6 Consider an n-square matrix $A = (a_{ij})$ [as in Problem 16.1]. Determine the first and second terms and the constant term of the characteristic polynomial $\Delta_A(t)$ of A.

▮ Each term in the determinant contains one and only one entry from each row and from each column; hence the above characteristic polynomial is of the form

$$\Delta_A(t) = (t - a_{11})(t - a_{22})\cdots(t - a_{nn}) + \text{terms with at most } n - 2 \text{ factors of the form } t - a_{ii}$$

Accordingly,

$$\Delta_A(t) = t^n - (a_{11} + a_{22} + \cdots + a_{nn})t^{n-1} + \text{terms of lower degree}$$

Recall that the trace of A is the sum of its diagonal elements. Thus the characteristic polynomial $\Delta_A(t) = \det(tI_n - A)$ of A is a monic polynomial of degree n, and the coefficient of t^{n-1} is the negative of the trace of A. [A polynomial is *monic* if its leading coefficient is 1.]

Furthermore, if we set $t = 0$ in $\Delta_A(t)$, we obtain $\Delta_A(0) = |-A| = (-1)^n|A|$. But $\Delta_A(0)$ is the constant term of the polynomial $\Delta_A(t)$. Thus the constant term of the characteristic polynomial of the matrix A is $(-1)^n|A|$ where n is the order of A.

16.7 Find the characteristic polynomial $\Delta(t)$ of $A = \begin{pmatrix} -2 & -6 \\ 4 & 9 \end{pmatrix}$.

▮ Here $\operatorname{tr}(A) = -2 + 9 = 7$ and $\det(A) = -18 + 24 = 6$. Hence $\Delta(t) = t^2 - 7t + 6$.

16.8 Find the characteristic polynomial $\Delta(t)$ of $B = \begin{pmatrix} 4 & 5 \\ -3 & -7 \end{pmatrix}$.

▮ Here $\operatorname{tr}(B) = 4 + (-7) = -3$ and $\det(B) = -28 + 15 = -13$. Hence $\Delta(t) = t^2 + 3t - 13$. [We emphasize that it is the negative of the trace which is the coefficient of t^{n-1}.]

16.9 Suppose $A = \begin{pmatrix} a_{11} & a_{12} & a_{13} \\ a_{21} & a_{22} & a_{23} \\ a_{31} & a_{32} & a_{33} \end{pmatrix}$. The following is a formula for the characteristic polynomial $\Delta(t)$ of A:

$$\Delta(t) = t^3 - (a_{11} + a_{22} + a_{33})t^2 + \left(\begin{vmatrix} a_{22} & a_{23} \\ a_{32} & a_{33} \end{vmatrix} + \begin{vmatrix} a_{11} & a_{13} \\ a_{31} & a_{33} \end{vmatrix} + \begin{vmatrix} a_{11} & a_{12} \\ a_{21} & a_{22} \end{vmatrix} \right)t - \begin{vmatrix} a_{11} & a_{12} & a_{13} \\ a_{21} & a_{22} & a_{23} \\ a_{31} & a_{32} & a_{33} \end{vmatrix}$$

$$= t^3 - \operatorname{tr}(A)t^2 + (A_{11} + A_{22} + A_{33})t - \det(A)$$

[Here A_{11}, A_{22}, A_{33} denote, respectively, the cofactors of the diagonal elements a_{11}, a_{22}, a_{33}.] Find $\Delta(t)$ when

$$A = \begin{pmatrix} 1 & 2 & 3 \\ 5 & 4 & 1 \\ 2 & 7 & 2 \end{pmatrix}$$

▮ Here $\operatorname{tr}(A) = 1 + 4 + 2 = 7$, $A_{11} = \begin{vmatrix} 4 & 1 \\ 7 & 2 \end{vmatrix} = 1$, $A_{22} = \begin{vmatrix} 1 & 3 \\ 2 & 2 \end{vmatrix} = -4$, $A_{33} = \begin{vmatrix} 1 & 2 \\ 5 & 4 \end{vmatrix} = -6$, $A_{11} + A_{22} + A_{33} = -9$, and

$$|A| = \begin{vmatrix} 1 & 2 & 3 \\ 5 & 4 & 1 \\ 2 & 7 & 2 \end{vmatrix} = 8 + 4 + 105 - 24 - 7 - 20 = 66$$

Thus $\Delta(t) = t^3 - 7t^2 - 9t - 66$.

Problems 16.10–16.12 refer to the following matrices:

$$A = \begin{pmatrix} 2 & 5 \\ 1 & -3 \end{pmatrix} \qquad B = \begin{pmatrix} 2 & -3 \\ 7 & -4 \end{pmatrix} \qquad C = \begin{pmatrix} 1 & 4 & -3 \\ 0 & 3 & 1 \\ 0 & 2 & -1 \end{pmatrix}$$

16.10 Find a polynomial $f(t)$ for which A is a root.

▌ By the Cayley-Hamiltonian theorem every matrix is a root of its characteristic polynomial. Therefore, let $f(t)$ be the characteristic polynomial of A:

$$f(t) = |tI - A| = \begin{vmatrix} t-2 & -5 \\ -1 & t+3 \end{vmatrix} = t^2 + t - 11$$

16.11 Find a polynomial $g(t)$ for which B is a zero.

▌ Let $g(t)$ be the characteristic polynomial of B:

$$g(t) = |tI - B| = \begin{vmatrix} t-2 & 3 \\ -7 & t+4 \end{vmatrix} = t^2 + 2t + 13$$

16.12 Find a polynomial $h(t)$ for which C is a root.

▌

$$h(t) = |tI - C| = \begin{vmatrix} t-1 & -4 & 3 \\ 0 & t-3 & -1 \\ 0 & -2 & t+1 \end{vmatrix} = (t-1)(t^2 - 2t - 5)$$

16.13 Find the characteristic polynomial $\Delta(t)$ of a triangular matrix

$$A = \begin{pmatrix} a_{11} & a_{12} & \cdots & a_{1n} \\ 0 & a_{22} & \cdots & a_{2n} \\ \cdots\cdots\cdots\cdots\cdots\cdots \\ 0 & 0 & \cdots & a_{nn} \end{pmatrix}$$

▌ Since A is triangular and tI is diagonal, $tI - A$ is also triangular with diagonal elements $t - a_{ii}$:

$$tI - A = \begin{pmatrix} t - a_{11} & -a_{12} & \cdots & -a_{1n} \\ 0 & t - a_{22} & \cdots & -a_{2n} \\ \cdots\cdots\cdots\cdots\cdots\cdots\cdots \\ 0 & 0 & \cdots & t - a_{nn} \end{pmatrix}$$

Then $\Delta(t) = |tI - A|$ is the product of the diagonal elements $t - a_{ii}$: $\Delta(t) = (t - a_{11})(t - a_{22})\cdots(t - a_{nn})$.

16.14 Show that a matrix A and its transpose A^T have the same characteristic polynomial.

▌ By the transpose operation, $(tI - A)^T = tI^T - A^T$. Since a matrix and its transpose have the same determinant, $tI - A = (tI - A)^T = tI - A^T$. Thus A and A^T have the same characteristic polynomial.

16.15 Suppose $M = \begin{pmatrix} A_1 & B \\ 0 & A_2 \end{pmatrix}$ where A_1 and A_2 are square matrices. Show that the characteristic polynomial of M is the product of the characteristic polynomials of A_1 and A_2.

▌ Here $tI - M = \begin{pmatrix} tI - A_1 & -B \\ 0 & tI - A_2 \end{pmatrix}$. However, the determinant of a block triangular matrix is the product of the determinants of the diagonal blocks. Thus $|tI - M| = \begin{vmatrix} tI - A_1 & -B \\ 0 & tI - A_2 \end{vmatrix} = |tI - A||tI - B|$, as required.

16.16 Generalize the statement in Problem 16.15.

▌ The characteristic polynomial $\Delta_M(t)$ of the triangular block matrix

$$M = \begin{pmatrix} A_1 & B & \cdots & C \\ 0 & A_2 & \cdots & D \\ \cdots\cdots\cdots\cdots\cdots\cdots \\ 0 & 0 & \cdots & A_n \end{pmatrix}$$

is the product of the characteristic polynomials of the diagonal blocks A_i, that is, $\Delta_M(t) = \Delta_{A_1}(t)\Delta_{A_2}(t) \cdots \Delta_{A_n}(t)$.

Problems 16.17–16.19 refer to the following matrices:

$$R = \begin{pmatrix} 1 & 2 & 3 & 4 \\ 0 & 2 & 8 & -6 \\ 0 & 0 & 3 & -5 \\ 0 & 0 & 0 & 4 \end{pmatrix} \qquad S = \begin{pmatrix} 2 & 5 & 7 & -9 \\ 1 & 4 & -6 & 4 \\ 0 & 0 & 6 & -5 \\ 0 & 0 & 2 & 3 \end{pmatrix} \qquad T = \begin{pmatrix} 5 & 8 & -1 & 0 \\ 0 & 3 & 6 & 7 \\ 0 & -3 & 5 & -4 \\ 0 & 0 & 0 & 7 \end{pmatrix}$$

16.17 Find the characteristic polynomial $\Delta(t)$ of R.

▌ Since R is triangular, $\Delta(t) = (t-1)(t-2)(t-3)(t-4)$.

16.18 Find the characteristic polynomial $\Delta(t)$ of S.

▌ Note S is block triangular with diagonal blocks $A_1 = \begin{pmatrix} 2 & 5 \\ 1 & 4 \end{pmatrix}$ and $A_2 = \begin{pmatrix} 6 & -5 \\ 2 & 3 \end{pmatrix}$. Thus $\Delta(t) = \Delta_{A_1}(t)\Delta_{A_2}(t) = (t^2 - 6t + 3)(t^2 - 9t + 28)$.

16.19 Find the characteristic polynomial $\Delta(t)$ of T.

▌ Note T is block triangular with diagonal blocks (5), $\begin{pmatrix} 3 & 6 \\ -3 & 5 \end{pmatrix}$, and (7). Thus $\Delta(t) = (t-5)(t^2 - 8t + 33)(t-7)$.

16.20 Prove the Cayley-Hamilton theorem 16.1.

▌ Let A be an arbitrary n-square matrix and let $\Delta(t)$ be its characteristic polynomial; say, $\Delta(t) = |tI - A| = t^n + a_{n-1}t^{n-1} + \cdots + a_1 t + a_0$. Now let $B(t)$ denote the classical adjoint of the matrix $tI - A$. The elements of $B(t)$ are cofactors of the matrix $tI - A$ and hence are polynomials in t of degree not exceeding $n - 1$. Thus $B(t) = B_{n-1}t^{n-1} + \cdots + B_1 t + B_0$, where the B_i are n-square matrices over K which are independent on t. By the fundamental property of the classical adjoint [Theorem 8.11],

$$(tI - A)B(t) = |tI - A|I$$

or $$(tI - A)(B_{n-1}t^{n-1} + \cdots + B_1 t + B_0) = (t^n + a_{n-1}t^{n-1} + \cdots + a_1 t + a_0)I$$

Removing parentheses and equating the coefficients of corresponding powers of t,

$$\begin{aligned} B_{n-1} &= I \\ B_{n-2} - AB_{n-1} &= a_{n-1}I \\ B_{n-3} - AB_{n-2} &= a_{n-2}I \\ &\cdots\cdots\cdots\cdots \\ B_0 - AB_1 &= a_1 I \\ -AB_0 &= a_0 I \end{aligned}$$

Multiplying the above matrix equations by $A^n, A^{n-1}, \ldots, A, I$, respectively,

$$\begin{aligned} A^n B_{n-1} &= A^n \\ A^{n-1}B_{n-2} - A^n B_{n-1} &= a_{n-1}A^{n-1} \\ A^{n-2}B_{n-3} - A^{n-1}B_{n-2} &= a_{n-2}A^{n-2} \\ &\cdots\cdots\cdots\cdots\cdots\cdots\cdots \\ AB_0 - A^2 B_1 &= a_1 A \\ -AB_0 &= a_0 I \end{aligned}$$

Adding the above matrix equations, $0 = A^n + a_{n-1}A^{n-1} + \cdots + a_1 A + a_0 I$. In other words, $\Delta(A) = 0$. That is, A is a zero of its characteristic polynomial.

Theorem 16.2: Similar matrices have the same characteristic polynomial.

16.21 Prove Theorem 16.2.

▌ Suppose A and B are similar matrices, say $B = P^{-1}AP$ where P is invertible. Using $tI = P^{-1}tIP$, $|tI - B| = |tI - P^{-1}AP| = |P^{-1}tIP - P^{-1}AP| = |P^{-1}(tI - A)P| = |P^{-1}||tI - A||P|$. Since determinants are scalars and commute, and since $|P^{-1}||P| = 1$, we finally obtain $|tI - B| = |tI - A|$. That is, A and B have the same characteristic polynomial.

16.22 Suppose $L: V \rightarrow V$ is a linear operator on a vector space V with finite dimension. Define the characteristic polynomial $\Delta(t)$ of L.

▎ Let A be the matrix representation of L relative to some basis of V. Then $\Delta(t)$ is defined to be the characteristic polynomial of A.

16.23 Since a linear map $L: V \rightarrow V$ may have many matrix representations, is it possible for L to have many characteristic polynomials?

▎ No. All matrix representations of L are similar matrices and similar matrices have the same characteristic polynomial by Theorem 16.2. In other words, the characteristic polynomial $\Delta(t)$ of L is unique.

16.24 Let $L: \mathbf{R}^3 \rightarrow \mathbf{R}^3$ be defined by $L(x, y, z) = (2x + 3y - 2z, 5y + 4z, x - z)$. Find the characteristic polynomial $\Delta(t)$ of L.

▎ Find a matrix representation of L. Using the usual basis of $\mathbf{R}$, we have

$$[L] = \begin{pmatrix} 2 & 3 & -2 \\ 0 & 5 & 4 \\ 1 & 0 & -1 \end{pmatrix}$$

Thus

$$\Delta(t) = \begin{vmatrix} t-2 & -3 & 2 \\ 0 & t-5 & -4 \\ -1 & 0 & t+1 \end{vmatrix} = t^3 - 6t^2 + 5t - 12$$

16.25 Let V be the vector space of functions which have $B = \{\sin \theta, \cos \theta\}$ as a basis, and let D be the differential operator on V. Find the characteristic polynomial $\Delta(t)$ of D.

▎ First find the matrix A which represents D in the basis B:

$$D(\sin \theta) = \cos \theta = 0(\sin \theta) + 1(\cos \theta)$$
$$D(\cos \theta) = -\sin \theta = -1(\sin \theta) + 0(\cos \theta)$$

Thus $[D] = A = \begin{pmatrix} 0 & 1 \\ -1 & 0 \end{pmatrix}$ and $\det(tI - A) = \begin{vmatrix} t & -1 \\ 1 & t \end{vmatrix} = t^2 + 1$. Accordingly, $\Delta(t) = t^2 + 1$ is the characteristic polynomial of D.

16.2 EIGENVALUES AND EIGENVECTORS

The following definitions are used throughout this section. Each definition has two forms, one for matrices and one for linear operators.

Definitions A: Let A be an n-square matrix over a field K. A scalar $\lambda \in K$ is called an eigenvalue of A if there exists a nonzero (column) vector $v \in K^n$ for which $Av = \lambda v$. Every vector satisfying this relation is then called an eigenvector of A belonging to the eigenvalue λ. The set E_λ of all eigenvectors belonging to λ is a subspace of K^n called the *eigenspace* of λ.

The terms *characteristic value* and *characteristic vector* [or *proper value* and *proper vector*] are frequently used instead of eigenvalue and eigenvector.

Definitions B: Let $T: V \rightarrow V$ be a linear operator on a vector space V over a field K. A scalar $\lambda \in K$ is called an *eigenvalue* of T if there exists a nonzero vector $v \in V$ for which $T(v) = \lambda v$. Every vector satisfying this relation is then called an *eigenvector* of T belonging to the eigenvalue λ. The set E_λ of all such vectors is a subspace of V [Problem 16.30] called the *eigenspace* of λ.

Theorem 16.3, proved in Problem 16.37, is used below.

Theorem 16.3: Nonzero eigenvectors belonging to distinct eigenvalues are linearly independent.

16.26 Let $I: V \rightarrow V$ be the identity mapping on any nonzero vector space V. Show that $\lambda = 1$ is an eigenvalue of I. What is the eigenspace E_1 of $\lambda = 1$?

▮ For every $v \in V$, $I(v) = v = 1v$. This $\lambda = 1$ is an eigenvalue of I, and $E_1 = V$ since every vector in V is an eigenvector belonging to 1.

16.27 Let $L: \mathbf{R}^2 \to \mathbf{R}^2$ be the linear operator which rotates each vector $v \in \mathbf{R}^2$ by an angle $\theta = \pi/2 = 90°$. Show geometrically that L has no eigenvalues and hence no eigenvectors.

▮ Observe that no nonzero vector is a multiple of itself, which is the defining condition for an eigenvalue. Thus L has no eigenvalues and hence no eigenvectors.

16.28 Let $A = \begin{pmatrix} 1 & 2 \\ 3 & 2 \end{pmatrix}$. Show that (a) $v_1 = (2, 3)^T$ is an eigenvector of A belonging to the eigenvalue $\lambda_1 = 4$ of A; (b) $v_2 = (1, 1)^T$ is an eigenvector of A belonging to the eigenvalue $\lambda_2 = -1$ of A.

(a)
$$Av_1 = \begin{pmatrix} 1 & 2 \\ 3 & 2 \end{pmatrix} \begin{pmatrix} 2 \\ 3 \end{pmatrix} = \begin{pmatrix} 8 \\ 12 \end{pmatrix} = 4 \begin{pmatrix} 2 \\ 3 \end{pmatrix} = 4v_1$$

Thus v_1 is an eigenvector of A belonging to $\lambda_1 = 4$.

(b)
$$Av_2 = \begin{pmatrix} 1 & 2 \\ 3 & 2 \end{pmatrix} \begin{pmatrix} 1 \\ -1 \end{pmatrix} = \begin{pmatrix} -1 \\ 1 \end{pmatrix} = (-1)v_2$$

Thus v_2 is an eigenvector of A belonging to $\lambda_2 = -1$.

16.29 Let V be the vector space of differentiable functions on $\mathbf{R}$ and $D: V \to V$ be the differential operator. Show that the functions $e^{a_1 t}, e^{a_2 t}, \ldots, e^{a_n t}$, where $a_1, \ldots, a_n$ are distinct nonzero scalars, are eigenvectors of D. To which eigenvalue λ_i does $e^{a_i t}$ belong? Show that the functions are linearly independent.

▮ $D(e^{a_i t}) = a_i e^{a_i t}$; hence $e^{a_i t}$ is an eigenvector of D belonging to the eigenvalue $\lambda_i = a_i$. By Theorem 16.3 the functions are linearly independent since they are nonzero eigenvectors of D belonging to distinct eigenvalues.

16.30 Let λ be an eigenvalue of a linear operator $T: V \to V$. Let E_λ be the eigenspace of λ, i.e., the set of all eigenvectors of T belonging to λ. Show that E_λ is a subspace of V, i.e., show that
(a) If $v \in E_\lambda$, then $kv \in E_\lambda$ for any scalar $k \in K$.
(b) If $u, v \in E_\lambda$, then $u + v \in E_\lambda$.

▮ (a) Since $v \in E_\lambda$, we have $T(v) = \lambda v$. Then $T(kv) = kT(v) = k(\lambda v) = \lambda(kv)$. Thus $kv \in E_\lambda$.
(b) Since $u, v \in E_\lambda$, we have $T(u) = \lambda v$ and $T(v) = \lambda v$. Then $T(u + v) = T(u) + T(v) = \lambda u + \lambda v = \lambda(u + v)$. Thus $u + v \in E_\lambda$.

Theorem 16.4: Let $T: V \to V$ be a linear operator on a vector space over K. Then $\lambda \in K$ is an eigenvalue of T if and only if the operator $\lambda I - T$ is singular. The eigenspace of λ is then the kernel of $\lambda I - T$.

16.31 Prove Theorem 16.4 which gives an important characterization of eigenvalues which is frequently used as its definition.

▮ The scalar λ is an eigenvalue of T if and only if there exists a nonzero vector v such that $T(v) = \lambda v$ or $(\lambda I)(v) - T(v) = 0$ or $(\lambda I - T)(v) = 0$, that is, $\lambda I - T$ is singular. We also have that v is in the eigenspace of λ if and only if the above relations hold; hence v is in the kernel of $\lambda I - T$.

Theorem 16.5: A linear operator $T: V \to V$ can be represented by a diagonal matrix B if and only if V has a basis consisting of eigenvectors of T. In this case the diagonal elements of B are the corresponding eigenvalues.

16.32 Prove Theorem 16.5.

▮ T can be represented by a diagonal matrix

$$\begin{pmatrix} k_1 & 0 & \cdots & 0 \\ 0 & k_2 & \cdots & 0 \\ \cdots\cdots\cdots\cdots\cdots\cdots \\ 0 & 0 & \cdots & k_n \end{pmatrix}$$

if and only if there exists a basis $\{v_1, \ldots, v_n\}$ of V for which

$$
\begin{aligned}
T(v_1) &= k_1 v_1 \\
T(v_2) &= k_2 v_2 \\
&\cdots\cdots\cdots\cdots\cdots\cdots \\
T(v_n) &= k_n v_n
\end{aligned}
$$

i.e., such that the vectors $v_1, \ldots, v_n$ are eigenvectors of T belonging respectively to eigenvalues $k_1, \ldots, k_n$.

> **Theorem 16.6:** An n-square matrix A is similar to a diagonal matrix B if and only if A has n linearly independent eigenvectors. In this case, the diagonal elements of B are the corresponding eigenvalues and $B = P^{-1}AP$ where P is the matrix whose columns are the eigenvectors.

16.33 Prove Theorem 16.6.

❚ The first part of the theorem is simply a restatement of Theorem 16.5 for matrices. We need only show that the columns of P are the eigenvectors. Now A may be viewed as the matrix of a linear map T on K^n relative to the usual basis $\{e_i\}$ of K^n, B may be viewed as the matrix of T relative to the basis $\{v_i\}$ of eigenvectors, and P is the change-of-basis matrix from $\{e_i\}$ to $\{v_i\}$. However, since $\{e_i\}$ is the usual basis, P is the matrix whose columns are the vectors v_i. Thus the theorem is proved.

16.34 Verify Theorem 16.6 for the matrix $A = \begin{pmatrix} 1 & 2 \\ 3 & 2 \end{pmatrix}$ in Problem 16.28.

❚ By Problem 16.28, A has two linearly independent eigenvectors $\begin{pmatrix} 2 \\ 3 \end{pmatrix}$ and $\begin{pmatrix} 1 \\ -1 \end{pmatrix}$. Set $P = \begin{pmatrix} 2 & 1 \\ 3 & -1 \end{pmatrix}$,

and so $P^{-1} = \begin{pmatrix} \frac{1}{5} & \frac{1}{5} \\ \frac{3}{5} & -\frac{2}{5} \end{pmatrix}$. Then A is similar to the diagonal matrix

$$
B = P^{-1}AP = \begin{pmatrix} \frac{1}{5} & \frac{1}{5} \\ \frac{3}{5} & -\frac{2}{5} \end{pmatrix}\begin{pmatrix} 1 & 2 \\ 3 & 2 \end{pmatrix}\begin{pmatrix} 2 & 1 \\ 3 & -1 \end{pmatrix} = \begin{pmatrix} 4 & 0 \\ 0 & -1 \end{pmatrix}
$$

As expected, the diagonal elements 4 and -1 of the diagonal matrix B are the eigenvalues corresponding to the given eigenvectors.

> **Theorem 16.7:** Let A be an n-square matrix over a field K. A scalar $\lambda \in K$ is an eigenvalue of A if and only if λ is a root of the characteristic polynomial $\Delta(t)$ of A.

16.35 Prove Theorem 16.7 which is used in the diagonalization algorithm in Section 16.3.

❚ Now λ is an eigenvalue of A if and only if the matrix $\lambda I - A$ is singular. However, $\lambda I - A$ is singular if and only if $\det(\lambda I - A) = 0$. But $\det(\lambda I - A) = 0$ if and only if λ is a root of $\Delta(t)$. Thus the theorem is proved.

> **Corollary 16.8:** Suppose the characteristic polynomial $\Delta(t)$ of an n-square matrix A is a product of n distinct factors, say, $\Delta(t) = (t - a_1)(t - a_2)\cdots(t - a_n)$. Then A is similar to a diagonal matrix whose diagonal elements are the a_i.

16.36 Prove Corollary 16.8 which gives us a sufficient condition for a matrix to be diagonalizable.

❚ By Theorem 16.7, the a_i are eigenvalues of A. Let v_i be corresponding eigenvectors. By Theorem 16.3, the vectors v_i are linearly independent and hence form a basis of K^n. Thus A is diagonalizable by Theorem 16.6.

16.37 Prove Theorem 16.3: Let $v_1, \ldots, v_n$ be nonzero eigenvectors of an operator $T: V \to V$ belonging to distinct eigenvalues $\lambda_1, \ldots, \lambda_n$. Then $v_1, \ldots, v_n$ are linearly independent.

❚ The proof is by induction on n. If $n = 1$, then v_1 is linearly independent since $v_1 \neq 0$. Assume $n > 1$. Suppose

$$
a_1 v_1 + a_2 v_2 + \cdots + a_n v_n = 0 \tag{1}
$$

where the a_i are scalars. Applying T to the above relation, we obtain by linearity $a_1 T(v_1) + a_2 T(v_2) + \cdots + a_n T(v_n) = T(0) = 0$. But by hypothesis $T(v_i) = \lambda_i v_i$; hence

$$a_1 \lambda_1 v_1 + a_2 \lambda_2 v_2 + \cdots + a_n \lambda_n v_n = 0 \tag{2}$$

On the other hand, multiplying (1) by λ_n,

$$a_1 \lambda_n v_1 + a_2 \lambda_n v_2 + \cdots + a_n \lambda_n v_n = 0 \tag{3}$$

Now subtracting (3) from (2), $a_1(\lambda_1 - \lambda_n)v_1 + a_2(\lambda_2 - \lambda_n)v_2 + \cdots + a_{n-1}(\lambda_{n-1} - \lambda_n)v_{n-1} = 0$. By induction, $v_1, v_2, \ldots, v_{n-1}$ are linearly independent; hence each of the above coefficients is 0. Since the λ_i are distinct, $\lambda_i - \lambda_n \neq 0$ for $i \neq n$. Hence $a_1 = \cdots = a_{n-1} = 0$. Substituting this into (1) we get $a_n v_n = 0$, and hence $a_n = 0$. Thus the v_i are linearly independent.

16.38 Let λ be an eigenvalue of a linear operator $T: V \to V$. Define the algebraic multiplicity and the geometric multiplicity of λ.

❚ The algebraic multiplicity of λ is defined to be the multiplicity of λ as a root of the characteristic polynomial of T. The *geometric multiplicity* of λ is defined to be the dimension of its eigenspace.

Theorem 16.9: Let λ be an eigenvalue of a linear operator $T: V \to V$. Then the geometric multiplicity of λ does not exceed its algebraic multiplicity.

16.39 Prove Theorem 16.9.

❚ Suppose the geometric multiplicity of λ is r. Then λ contains r linearly independent eigenvectors $v_1, \ldots, v_r$. Extend the set $\{v_i\}$ to a basis of V: $\{v_1, \ldots, v_r, w_1, \ldots, w_s\}$. We have

$$
\begin{aligned}
T(v_1) &= \lambda v_1 \\
T(v_2) &= \lambda v_2 \\
&\cdots\cdots\cdots\cdots\cdots\cdots\cdots\cdots\cdots\cdots \\
T(v_r) &= \lambda v_r \\
T(w_1) &= a_{11}v_1 + \cdots + a_{1r}v_r + b_{11}w_1 + \cdots + b_{1s}w_s \\
T(w_2) &= a_{21}v_1 + \cdots + a_{2r}v_r + b_{21}w_1 + \cdots + b_{2s}w_s \\
&\cdots\cdots\cdots\cdots\cdots\cdots\cdots\cdots\cdots\cdots \\
T(w_s) &= a_{s1}v_1 + \cdots + a_{sr}v_r + b_{s1}w_1 + \cdots + b_{ss}w_s
\end{aligned}
$$

The matrix of T in the above basis is

$$
M = \left(
\begin{array}{cccc|cccc}
\lambda & 0 & \cdots & 0 & a_{11} & a_{21} & \cdots & a_{s1} \\
0 & \lambda & \cdots & 0 & a_{12} & a_{22} & \cdots & a_{s2} \\
\multicolumn{8}{c}{\cdots\cdots\cdots\cdots\cdots\cdots\cdots\cdots} \\
0 & 0 & \cdots & \lambda & a_{1r} & a_{2r} & \cdots & a_{sr} \\
\hline
0 & 0 & \cdots & 0 & b_{11} & b_{21} & \cdots & b_{r1} \\
0 & 0 & \cdots & 0 & b_{12} & b_{22} & \cdots & b_{r2} \\
\multicolumn{8}{c}{\cdots\cdots\cdots\cdots\cdots\cdots\cdots\cdots} \\
0 & 0 & \cdots & 0 & b_1 & b_{2s} & \cdots & b_{ss}
\end{array}
\right) = \left(
\begin{array}{c|c}
\lambda I_r & A \\
\hline
0 & B
\end{array}
\right)
$$

where $A = (a_{ij})^T$ and $B = (b_{ij})^T$.

Since M is a block triangular matrix, the characteristic polynomial of λI_r, which is $(t - \lambda)^r$, must divide the characteristic polynomial of M and hence T. Thus the algebraic multiplicity of λ for the operator T is at least r, as required.

16.40 Show that 0 is an eigenvalue of T if and only if T is singular.

❚ We have that 0 is an eigenvalue of T if and only if there exists a nonzero vector v such that $T(v) = 0v = 0$, i.e., that T is singular.

16.41 Let A and B be n-square matrices. Show that AB and BA have the same eigenvalues.

❚ Since the product of nonsingular matrices is nonsingular, the following statements are equivalent: (i) 0 is an eigenvalue of AB, (ii) AB is singular, (iii) A or B is singular, (iv) BA is singular, (v) 0 is an eigenvalue of BA.

Now suppose λ is a nonzero eigenvalue of AB. Then there exists a nonzero vector v such that $ABv = \lambda v$. Set $w = Bv$. Since $\lambda \neq 0$ and $v \neq 0$, $Aw = ABv = \lambda v \neq 0$ and so $w \neq 0$. But w is an eigenvector of BA belonging to the eigenvalue λ since $BAw = BABv = B\lambda v = \lambda Bv = \lambda w$. Hence λ is an eigenvalue of BA. Similarly, any nonzero eigenvalue of BA is also an eigenvalue of AB.

Thus AB and BA have the same eigenvalues.

16.3 COMPUTING EIGENVALUES AND EIGENVECTORS, DIAGONALIZING MATRICES

This section computes the eigenvalues and eivenvectors for a given square matrix A and determines whether or not a nonsingular matrix P exists such that $P^{-1}AP$ is diagonal. Specifically, the following algorithm will be applied to the matrix A.

Diagonalization Algorithm
Step 1. Find the characteristic polynomial $\Delta(t)$ of A.
Step 2. Find the roots of $\Delta(t)$ to obtain the eigenvalues of A.
Step 3. Repeat (a) and (b) for each eigenvalue λ of A:
   (a) Form $M = A - \lambda I$ by subtracting λ down the diagonal of A, or form $M' = \lambda I - A$ by substituting $t = \lambda$ in $tI - A$.
   (b) Find a basis for the solution space of the homogeneous system $MX = 0$. [These basis vectors are linearly independent eigenvectors of A belonging to λ.]
Step 4. Consider the collection $S = \{v_1, v_2, \dots, v_m\}$ of all eigenvectors obtained in Step 3:
   (a) If $m \neq n$, then A is not diagonalizable.
   (b) If $m = n$, let P be the matrix whose columns are the eigenvectors $v_1, v_2, \dots, v_n$. Then

$$B = P^{-1}AP = \begin{pmatrix} \lambda_1 & & & \\ & \lambda_2 & & \\ & & \dots & \\ & & & \lambda_n \end{pmatrix}$$

where λ_i is the eigenvalue corresponding to the eigenvector v_i.

Problems 16.42–16.46 apply the Diagonalization Algorithm to $A = \begin{pmatrix} 1 & 4 \\ 2 & 3 \end{pmatrix}$.

16.42 Find the characteristic polynomial $\Delta(t)$ of A.

▮ Form the characteristic matrix $tI - A$ of A:

$$tI - A = \begin{pmatrix} t & 0 \\ 0 & t \end{pmatrix} - \begin{pmatrix} 1 & 4 \\ 2 & 3 \end{pmatrix} = \begin{pmatrix} t-1 & -4 \\ -2 & t-3 \end{pmatrix}$$

The characteristic polynomial $\Delta(t)$ of A is its determinant:

$$\Delta(t) = |tI - A| = \begin{vmatrix} t-1 & -4 \\ -2 & t-3 \end{vmatrix} = t^2 - 4t - 5 = (t-5)(t+1)$$

Alternatively, $\operatorname{tr}(A) = 1 + 3 = 4$ and $A = 3 - 8 = -5$, so $\Delta(t) = t^2 - 4t - 5$.

16.43 Find the eigenvalues of A.

▮ The roots $\lambda_1 = 5$ and $\lambda_2 = -1$ of the characteristic polynomial $\Delta(t)$ are the eigenvalues of A.

16.44 Find an eigenvector v_1 of A belonging to the eigenvalue $\lambda_1 = 5$.

▮ Substitute $t = 5$ in the matrix $tI - A$ to obtain the matrix $M = \begin{pmatrix} 4 & -4 \\ -2 & 2 \end{pmatrix}$. The eigenvectors belonging to $\lambda_1 = 5$ form the solution of the homogeneous system $MX = 0$, that is,

$$\begin{pmatrix} 4 & -4 \\ -2 & 2 \end{pmatrix}\begin{pmatrix} x \\ y \end{pmatrix} = \begin{pmatrix} 0 \\ 0 \end{pmatrix} \quad \text{or} \quad \begin{cases} 4x - 4y = 0 \\ -2x - 2y = 0 \end{cases} \quad \text{or} \quad x - y = 0$$

The system has only one independent solution; for example, $x = 1$, $y = 1$. Thus $v_1 = (1, 1)$ is an eigenvector which spans the eigenspace of $\lambda_1 = 5$.

16.45 Find an eigenvector v_2 of A belonging to the eigenvalue $\lambda_2 = -1$.

▮ Substitute $t = -1$ into $tI - A$ to obtain $M = \begin{pmatrix} -2 & -4 \\ -2 & -4 \end{pmatrix}$ which yields the homogeneous system

$$\begin{pmatrix} -2 & -4 \\ -2 & -4 \end{pmatrix} \begin{pmatrix} x \\ y \end{pmatrix} = \begin{pmatrix} 0 \\ 0 \end{pmatrix} \quad \text{or} \quad \begin{cases} -2x - 4y = 0 \\ -2x - 4y = 0 \end{cases} \quad \text{or} \quad x + 2y = 0$$

The system has only one independent solution; for example, $x = 2$, $y = -1$. Thus $v_2 = (2, -1)$ is an eigenvector which spans the eigenspace of $\lambda_2 = -1$.

16.46 Find an invertible matrix P such that $P^{-1}AP$ is diagonal.

▮ Let P be the matrix whose columns are the above eigenvectors: $P = \begin{pmatrix} 1 & 2 \\ 1 & -1 \end{pmatrix}$. Then $B = P^{-1}AP$ is the diagonal matrix whose diagonal entries are the respective eigenvalues:

$$B = P^{-1}AP = \begin{pmatrix} \frac{1}{3} & \frac{2}{3} \\ \frac{1}{3} & -\frac{1}{3} \end{pmatrix} \begin{pmatrix} 1 & 4 \\ 2 & 3 \end{pmatrix} \begin{pmatrix} 1 & 2 \\ 1 & -1 \end{pmatrix} = \begin{pmatrix} 5 & 0 \\ 0 & -1 \end{pmatrix}$$

[*Remark*: Here P is the transition matrix from the usual basis of $\mathbf{R}^2$ to the basis $\{v_1, v_2\}$. Hence B is the matrix representation of the operator A in this new basis.]

16.47 Find all eigenvalues and a maximum set of linearly independent eigenvectors of the matrix $B = \begin{pmatrix} 1 & 3 \\ 2 & -4 \end{pmatrix}$.

▮ Find the characteristic polynomial $\Delta(t) = t^2 + 3t - 10 = (t + 5)(t - 2)$. Then eigenvalues are $\lambda_1 = -5$ and $\lambda_2 = 2$.

(i) Subtract $\lambda_1 = -5$ (or add 5) down the diagonal of B to obtain $M = \begin{pmatrix} 6 & 3 \\ 2 & 1 \end{pmatrix}$ which corresponds to the homogeneous system

$$\begin{pmatrix} 6 & 3 \\ 2 & 1 \end{pmatrix} \begin{pmatrix} x \\ y \end{pmatrix} = 0 \quad \text{or} \quad \begin{matrix} 6x + 3y = 0 \\ 2x + y = 0 \end{matrix} \quad \text{or} \quad 2x + y = 0$$

Here $v_1 = (1, -2)$ is a nonzero solution of the system and hence v_2 is an eigenvector of B belonging to $\lambda_1 = -5$.

(ii) Subtract $\lambda_2 = 2$ down the diagonal of B to obtain $M = \begin{pmatrix} -1 & 3 \\ 2 & -6 \end{pmatrix}$ which corresponds to the homogeneous system $x - 3y = 0$. Here $v_2 = (3, 1)$ is a nonzero of the system and hence v_2 is an eigenvector belonging to $\lambda_2 = 2$.

The set $\{v_1 = (1, -2), v_2 = (3, 1)\}$ is a maximum set of independent eigenvectors of the matrix B.

16.48 Is the above matrix B diagonalizable? If yes, find P such that $P^{-1}BP$ is diagonal.

▮ Since the eigenvectors $v_1 = (1, -2)$ and $v_2 = (3, 1)$ form a basis of $\mathbf{R}^2$, B is diagonalizable. Let P be the matrix whose columns are v_1 and v_2, that is, $P = \begin{pmatrix} 1 & 3 \\ -2 & 1 \end{pmatrix}$. Then $P^{-1}BP = \begin{pmatrix} -5 & 0 \\ 0 & 2 \end{pmatrix}$.

16.49 Find all eigenvalues and a maximum set of linearly independent eigenvectors of the matrix $C = \begin{pmatrix} 5 & -1 \\ 1 & 3 \end{pmatrix}$.

▮ Find $\Delta(t) = t^2 - 8t + 16 = (t - 4)^2$. Thus $\lambda = 4$ is the only eigenvalue. Subtract $\lambda = 4$ down the diagonal of C to obtain $M = \begin{pmatrix} 1 & -1 \\ 1 & -1 \end{pmatrix}$ which corresponds to the homogeneous system $x + y = 0$. Here $v = (1, 1)$ is a nonzero solution of the system and hence v is an eigenvector of C belonging to $\lambda = 4$. Since there are no other eigenvalues, $\{v = (1, 1)\}$ is a maximum set of linearly independent eigenvectors.

16.50 Is the above matrix C diagonalizable? If yes, find P such that $P^{-1}CP$ is diagonal.

▮ C is not diagonalizable since the number of independent eigenvector is not equal to the dimension of $V = \mathbf{R}^2$. Thus no such matrix P exists.

16.51 Find all eigenvalues and a maximum set of linearly independent eigenvectors of the matrix $D = \begin{pmatrix} 5 & 6 \\ 3 & -2 \end{pmatrix}$.

▮ Here $\Delta(t) = |tI - D| = t^2 - 3t - 28 = (t - 7)(t + 4)$. Thus the eigenvalues of D are $\lambda_1 = 7$ and $\lambda_2 = -4$.

(i) Subtract $\lambda_1 = 7$ down the diagonal of D to obtain $M = \begin{pmatrix} -2 & 6 \\ 3 & -9 \end{pmatrix}$ which corresponds to the system

$$-2x + 6y = 0$$
$$3x - 9y = 0 \quad \text{or} \quad x - 3y = 0$$

Here $v_1 = (3, 1)$ is the eigenvector of $\lambda_1 = 7$.

(ii) Subtract $\lambda_2 = -4$ [or add 4] down the diagonal of D to obtain $= \begin{pmatrix} 9 & 6 \\ 3 & 2 \end{pmatrix}$ which corresponds to the system $3x + 2y = 0$.

Here $v_2 = (2, -3)$ is a solution and hence an eigenvector of $\lambda_2 = -4$. Thus $\{v_1 = (3, 1), v_2 = (2, -3)\}$ is a maximum set of linearly independent eigenvectors of D.

Problems 16.52–16.55 refer to the matrix $A = \begin{pmatrix} 1 & -1 \\ 2 & -1 \end{pmatrix}$.

16.52 Find all eigenvalues and corresponding eigenvectors of A assuming A is a real matrix.

▌ Here $\Delta(t) = |tI - A| = t^2 + 1$. Since $t^2 + 1$ has no solution in **R**, A has no eigenvalues and hence no eigenvectors.

16.53 Is A diagonalizable? If yes, find P such that $P^{-1}AP$ is diagonal.

▌ Viewed as a real matrix A has no eigenvectors and hence A is not diagonalizable.

16.54 Find all eigenvalues and corresponding eigenvectors of A assuming A is a complex matrix.

▌ Again $\Delta(t) = |tI - A| = t^2 + 1$. Now, however, $\lambda_1 = i$ and $\lambda_2 = -i$ are eigenvalues of A.

(i) Substitute $t = i$ in $tI - B$ to obtain the homogeneous system

$$\begin{pmatrix} i-1 & 1 \\ -2 & i+1 \end{pmatrix}\begin{pmatrix} x \\ y \end{pmatrix} = \begin{pmatrix} 0 \\ 0 \end{pmatrix} \quad \text{or} \quad \begin{cases} (i-1)x + \qquad\ y = 0 \\ -2x + (i+1)y = 0 \end{cases} \quad \text{or} \quad (i-1)x + y = 0$$

The system has only one independent solution, i.e., $x = 1$, $y = 1 - i$. Thus $v_1 = (1, 1 - i)$ is an eigenvector which spans the eigenspace of $\lambda_1 = i$.

(ii) Substitute $t = -i$ into $tI - B$ to obtain the homogeneous system

$$\begin{pmatrix} -i-1 & 1 \\ -2 & -i-1 \end{pmatrix}\begin{pmatrix} x \\ y \end{pmatrix} = \begin{pmatrix} 0 \\ 0 \end{pmatrix} \quad \text{or} \quad \begin{cases} (-i-1)x + \qquad\ y = 0 \\ -2x + (-i-1)y = 0 \end{cases} \quad \text{or} \quad (-i-1)x + y = 0$$

The system has only one independent solution, i.e., $x = 1$, $y = 1 + i$. Thus $v_2 = (1, 1 + i)$ is an eigenvector of A which spans the eigenspace of $\lambda_2 = -i$.

16.55 Is A diagonalizable? If yes, find P such that $P^{-1}AP$ is diagonal.

▌ As a complex matrix, A is diagonalizable. Let P be the matrix whose columns are v_1 and v_2, that is, $P = \begin{pmatrix} 1 & 1 \\ 1-i & 1+i \end{pmatrix}$. Then $P^{-1}AP = \begin{pmatrix} i & 0 \\ 0 & -i \end{pmatrix}$.

***Remark*:** Problems 16.52–16.55 indicate that the question of eigenvalues and eigenvectors and diagonalizability of a matrix A depends on the underlying field K since the roots of the characteristic polynomial $\Delta(t)$ depend on the underlying field K.

Problems 16.56–16.60 refer to the matrix $A = \begin{pmatrix} 1 & -3 & 3 \\ 3 & -5 & 3 \\ 6 & -6 & 4 \end{pmatrix}$.

16.56 Find the characteristic polynomial $\Delta(t)$ of A.

▌
$$\Delta(t) = |tI - A| = \begin{vmatrix} t-1 & 3 & -3 \\ -3 & t+5 & -3 \\ -6 & 6 & t-4 \end{vmatrix} = t^3 - 12t - 16$$

Alternatively, $\Delta(t) = t^3 - \text{tr}(A)t^2 + (A_{11} + A_{22} + A_{33})t - |A| = t^3 - 12t - 16$. [Here A_{ii} is the cofactor of a_{ii} in the matrix A.]

16.57 Find the eigenvalues of A.

▌ Assuming $\Delta(t)$ has a rational root, it must be among ± 1, ± 2, ± 4, ± 16. Testing, we get

$$
\begin{array}{r|l}
-2 & 1 + 0 - 12 - 16 \\
& -2 + 4 + 16 \\
\hline
& 1 - 2 - 8 + 0
\end{array}
$$

Thus $t = -2$ is a root of $\Delta(t)$ and hence $\Delta(t) = (t + 2)(t^2 - 2t - 8) = (t + 2)(t - 4)(t + 2) = (t + 2)^2(t - 4)$. Accordingly, $\lambda_1 = -2$ and $\lambda_2 = 4$ are the eigenvalues of A.

16.58 Find a basis for the eigenspace of $\lambda_1 = -2$.

▌ Substitute $t = -2$ into $tI - A$ to obtain the homogeneous system

$$
\begin{pmatrix} -3 & 3 & -3 \\ -3 & 3 & -3 \\ -6 & 6 & -6 \end{pmatrix}\begin{pmatrix} x \\ y \\ z \end{pmatrix} = \begin{pmatrix} 0 \\ 0 \\ 0 \end{pmatrix} \quad \text{or} \quad \begin{cases} -3x + 3y - 3z = 0 \\ -3x + 3y - 3z = 0 \\ -6x + 6y - 6z = 0 \end{cases} \quad \text{or} \quad x - y + z = 0
$$

The system has two independent solutions, i.e., $x = 1$, $y = 1$, $z = 0$ and $x = 1$, $y = 0$, $z = -1$. Thus $u = (1, 1, 0)$ and $v = (1, 0, -1)$ are independent eigenvectors which span the eigenspace of $\lambda_1 = -2$. That is, u and v form a basis for the eigenspace of $\lambda_1 = -2$. This means that every other eigenvector belonging to $\lambda_1 = -2$ is a linear combination of u and v.

16.59 What is the algebraic multiplicity and geometric multiplicity of the eigenvalue $\lambda_1 = -2$?

▌ Since $t + 2$ occurs twice in the characteristic polynomial $\Delta(t) = (t + 2)^2(t - 4)$, the algebraic multiplicity of λ_1 is two. Also, the geometric multiplicity of λ_1 is two since $\dim E_{\lambda_1} = 2$ where E_{λ_1} is the eigenspace of λ_1. [Compare with Problem 16.64.]

16.60 Find a basis for the eigenspace of $\lambda_2 = 4$.

▌ Substitute $t = 4$ into $tI - A$ to obtain the homogeneous system

$$
\begin{pmatrix} 3 & 3 & -3 \\ -3 & 9 & -3 \\ -6 & 6 & 0 \end{pmatrix}\begin{pmatrix} x \\ y \\ z \end{pmatrix} = \begin{pmatrix} 0 \\ 0 \\ 0 \end{pmatrix} \quad \text{or} \quad \begin{cases} 3x + 3y - 3z = 0 \\ -3x + 9y - 3z = 0 \\ -6x + 6y = 0 \end{cases} \quad \text{or} \quad \begin{cases} x + y - z = 0 \\ 2y - z = 0 \end{cases}
$$

The system has only one free variable; hence any particular nonzero solution, e.g., $x = 1$, $y = 1$, $z = 2$, spans the solution space. Thus $w = (1, 1, 2)$ is an eigenvector of A which spans, and so forms a basis, of the eigenspace of $\lambda_2 = 4$.

16.61 Is A diagonalizable? If yes, find P such that $P^{-1}AP$ is diagonal.

▌ Since A has three linearly independent eigenvectors, A is diagonalizable. Let P be the matrix whose columns are the three independent eigenvectors:

$$
P = \begin{pmatrix} 1 & 1 & 1 \\ 1 & 0 & 1 \\ 0 & -1 & 2 \end{pmatrix} \quad \text{Then} \quad P^{-1}AP = \begin{pmatrix} -2 & 0 & 0 \\ 0 & -2 & 0 \\ 0 & 0 & 4 \end{pmatrix}
$$

As expected, the diagonal elements of $P^{-1}AP$ are the eigenvalues of A corresponding to the columns of P.

Problems 16.62–16.66 refer to the matrix $B = \begin{pmatrix} -3 & 1 & -1 \\ -7 & 5 & -1 \\ -6 & 6 & -2 \end{pmatrix}$.

16.62 Find the characteristic polynomial $\Delta(t)$ and eigenvalues of B.

▌ $$\Delta(t) = |tI - B| = \begin{vmatrix} t + 3 & -1 & 1 \\ 7 & t - 5 & 1 \\ 6 & -6 & t + 2 \end{vmatrix} = t^3 - 12t - 16$$

By Problem 16.57, $\Delta(t) = (t + 2)^2(t - 4)$. Thus $\lambda_1 = -2$ and $\lambda_2 = 4$ are the eigenvalues of B.

16.63 Find a basis for the eigenspace of $\lambda_1 = -2$.

▌ Substitute $t = -2$ into $tI - B$ to obtain the homogeneous system

$$\begin{pmatrix} 1 & -1 & 1 \\ 7 & -7 & 1 \\ 6 & -6 & 0 \end{pmatrix}\begin{pmatrix} x \\ y \\ z \end{pmatrix} = \begin{pmatrix} 0 \\ 0 \\ 0 \end{pmatrix} \quad \text{or} \quad \begin{cases} x - y + z = 0 \\ 7x - 7y + z = 0 \\ 6x - 6y = 0 \end{cases} \quad \text{or} \quad \begin{cases} x - y + z = 0 \\ x - y = 0 \end{cases}$$

The system has only one independent solution, i.e., $x = 1$, $y = 1$, $z = 0$. Thus $u = (1, 1, 0)$ forms a basis for the eigenspace of $\lambda_1 = -2$.

16.64 What is the algebraic multiplicity and geometric multiplicity of $\lambda_1 = -2$.

▌ The algebraic multiplicity of λ_1 is two since $t + 2$ occurs twice in the characteristic polynomial $\Delta(t) = (t + 2)^2(t - 4)$. However, the geometric multiplicity of λ_1 is one since $\dim E_{\lambda_1} = 1$ where E_{λ_1} is the eigenspace of λ_1. [Compare with Problem 16.59.]

16.65 Find a basis for the eigenspace of $\lambda_2 = 4$.

▌ Substitute $t = 4$ into $tI - B$ to obtain the homogeneous system

$$\begin{pmatrix} 7 & -1 & 1 \\ 7 & -1 & 1 \\ 6 & -6 & 6 \end{pmatrix}\begin{pmatrix} x \\ y \\ z \end{pmatrix} = \begin{pmatrix} 0 \\ 0 \\ 0 \end{pmatrix} \quad \text{or} \quad \begin{cases} 7x - y + z = 0 \\ 7x - y + z = 0 \\ 6x - 6y + 6z = 0 \end{cases} \quad \text{or} \quad \begin{cases} 7x - y + z = 0 \\ x = 0 \end{cases}$$

The system has only one independent solution, i.e., $x = 0$, $y = 1$, $z = 1$. Thus $v = (0, 1, 1)$ forms a basis of the eigenspace of $\lambda_2 = 4$.

16.66 Is B diagonalizable? If yes, find P such that $P^{-1}BP$ is diagonal.

▌ Since B has a maximum of two independent eigenvectors, B is not similar to a diagonal matrix, i.e., B is not diagonalizable.

16.67 Are the above matrices A and B similar?

▌ Since A can be diagonalized but B cannot, A and B are not similar matrices, even though they have the same characteristic polynomial.

16.68 Show that $A = \begin{pmatrix} 1 & 1 \\ 0 & 1 \end{pmatrix}$ is not diagonalizable.

▌ The characteristic polynomial of A is $\Delta(t) = (t - 1)^2$; hence 1 is the only eigenvalue of A. We find a basis of the eigenspace of the eigenvalue 1. Substitute $t = 1$ into the matrix $tI - A$ to obtain the homogeneous system.

16.69 Let $A = \begin{pmatrix} 2 & 2 \\ 1 & 3 \end{pmatrix}$. Is A diagonalizable? If yes, find P such that $P^{-1}AP$ is diagonal.

▌ Here $\Delta(t) = t^2 - \text{tr}(A)t + |A| = t^2 - 5t + 4 = (t - 1)(t - 4)$. Hence $\lambda_1 = 1$ and $\lambda_2 = 4$ are eigenvalues of A. We find corresponding eigenvectors:

(i) Subtract $\lambda_1 = 1$ down the diagonal of A to obtain $M = \begin{pmatrix} 1 & 2 \\ 1 & 2 \end{pmatrix}$ which corresponds to the homogeneous system $x + 2y = 0$. Here $v_1 = (2, -1)$ is a nonzero solution of the system and so is an eigenvector of A belonging to $\lambda_1 = 1$.

(ii) Subtract $\lambda_2 = 4$ down the diagonal of A to obtain $M = \begin{pmatrix} -2 & 2 \\ 1 & -1 \end{pmatrix}$ which corresponds to the homogeneous system $x - y = 0$. Here $v_2 = (1, 1)$ is a nonzero solution and so is an eigenvector of A belonging to $\lambda_2 = 4$.

Let P be the matrix whose columns are v_1 and v_2, that is, $P = \begin{pmatrix} 2 & 1 \\ -1 & -1 \end{pmatrix}$. Then $P^{-1}AP = \begin{pmatrix} 1 & 0 \\ 0 & 4 \end{pmatrix}$.

16.70 Suppose in Problem 16.69 we let $P = \begin{pmatrix} 1 & 2 \\ -1 & -1 \end{pmatrix}$ [by interchanging the eigenvectors]. Does P still diagonalize A?

▌ Yes. However, now $P^{-1}AP = \begin{pmatrix} 4 & 0 \\ 0 & 1 \end{pmatrix}$. In other words, the order of the eigenvalues in $P^{-1}AP$ corresponds to the corresponding eigenvectors in P.

16.71 Let $B = \begin{pmatrix} 2 & 4 \\ 3 & 1 \end{pmatrix}$. Is B diagonalizable? If yes, find P such that $P^{-1}BP$ is diagonal.

▋ Here $\Delta(t) = t^2 - \text{tr}(B)t + |B| = t^2 - 3t - 10 = (t - 5)(t + 2)$. Thus $\lambda_1 = 5$ and $\lambda_2 = -2$ are the eigenvalues of B.

(i) Subtract $\lambda_1 = 5$ down the diagonal of B to obtain $M = \begin{pmatrix} -3 & 4 \\ 3 & -4 \end{pmatrix}$ which corresponds to the homogeneous system $3x - 4y = 0$. Here $v_1 = (4, 3)$ is a nonzero solution.

(ii) Subtract $\lambda_2 = -2$ [or add 2] down the diagonal of B to obtain $M = \begin{pmatrix} 4 & 4 \\ 3 & 3 \end{pmatrix}$ which corresponds to the system $x + y = 0$ which has a nonzero solution $v_2 = (1, -1)$.

Since B has two independent eigenvectors, B is diagonalizable. Let $P = \begin{pmatrix} 4 & 1 \\ 3 & -1 \end{pmatrix}$. Then $P^{-1}BP = \begin{pmatrix} 5 & 0 \\ 0 & -2 \end{pmatrix}$.

Problems 16.72–16.76 refer to the matrix $C = \begin{pmatrix} 4 & 1 & -1 \\ 2 & 5 & -2 \\ 1 & 1 & 2 \end{pmatrix}$.

16.72 Find the characteristic polynomial $\Delta(t)$ of C.

▋
$$\Delta(t) = |tI - C| = \begin{vmatrix} t-4 & -1 & 1 \\ -2 & t-5 & 2 \\ -1 & -1 & t-2 \end{vmatrix} = t^3 - 11t^2 + 39t - 45$$

Alternatively, $\Delta(t) = t^3 - \text{tr}(C)t^2 + (C_{11} + C_{22} + C_{33})t - |C| = t^3 - 11t^2 - 39t - 45$. [Here C_{ii} is the cofactor of c_{ii} in C. See Problem 16.9.]

16.73 Find the eigenvalues of C.

▋ Assuming $\Delta(t)$ has a rational root, it must be among $\pm 1, \pm 3, \pm 5, \pm 9, \pm 45$. Testing, we get

$$3 \,\big|\, \begin{array}{rrrr} 1 & -11 & +39 & -45 \\ & 3 & -24 & +45 \\ \hline 1 & -8 & +15 & +0 \end{array}$$

Thus $t = 3$ is a root of $\Delta(t)$ and $\Delta(t) = (t - 3)(t^2 - 8t + 15) = (t - 3)^2(t - 5)$. Accordingly, $\lambda_1 = 3$ and $\lambda_2 = 5$ are the eigenvalues of C.

16.74 Find a maximum set of linearly independent eigenvectors of C.

▋ Find independent eigenvectors for each eigenvalue of C.

(i) Subtract $\lambda_1 = 3$ down the diagonal of C to obtain the matrix

$$M = \begin{pmatrix} 1 & 1 & -1 \\ 2 & 2 & -2 \\ 1 & 1 & -1 \end{pmatrix}$$

which corresponds to the homogeneous $x + y - z = 0$. Here $u = (1, -1, 0)$ and $v = (1, 0, 1)$ are two independent solutions.

(ii) Subtract $\lambda_2 = 5$ down the diagonal of C to obtain

$$M = \begin{pmatrix} -1 & 1 & -1 \\ 2 & 0 & -2 \\ 1 & 1 & -3 \end{pmatrix}$$

which corresponds to the homogeneous system

$$\begin{array}{r} -x + y - z = 0 \\ 2x \quad - 2z = 0 \\ x + y - 3z = 0 \end{array} \quad \text{or} \quad \begin{array}{r} x \quad - z = 0 \\ y - 2z = 0 \end{array}$$

Only z is a free variable. Here $w = (1, 2, 1)$ is a solution. Thus $\{u = (1, -1, 0), v = (1, 0, 1), w = (1, 2, 1)\}$ is a maximum set of linearly independent eigenvectors of C.

16.75 How does one know that u, v, and w are linearly independent?

▮ The vectors u and v were chosen so they were independent solutions of the homogeneous system $x + y - z = 0$. On the other hand, w is automatically independent of u and v since w belongs to a different eigenvalue of C.

16.76 Is C diagonalizable? If yes, find P such that $P^{-1}CP$ is diagonal.

▮ C is diagonalizable since C has three linearly independent eigenvectors. Let P be the matrix whose columns are u, v, w, respectively; that is, let

$$P = \begin{pmatrix} 1 & 1 & 1 \\ -1 & 0 & 2 \\ 0 & 1 & 1 \end{pmatrix} \quad \text{Then} \quad P^{-1}CP = \begin{pmatrix} 3 & & \\ & 3 & \\ & & 5 \end{pmatrix}$$

Problems 16.77–16.81 refer to the linear operator $T: \mathbf{R}^3 \to \mathbf{R}^3$ defined by $T(x, y, z) = (2x + y, y - z, 2y + 4z)$.

16.77 Find the characteristic polynomial $\Delta(t)$ of T.

▮ First find a matrix representation of T, say relative to the usual basis of $\mathbf{R}^3$:

$$A = [T] = \begin{pmatrix} 2 & 1 & 0 \\ 0 & 1 & -1 \\ 0 & 2 & 4 \end{pmatrix}$$

The characteristic polynomial $\Delta(t)$ of T is then

$$\Delta(t) = |tI - A| = \begin{vmatrix} t-2 & -1 & 0 \\ 0 & t-1 & 1 \\ 0 & -2 & t-4 \end{vmatrix} = t^3 - 7t^2 + 16t - 12$$

16.78 Find the eigenvalues of T.

▮ Assuming $\Delta(t)$ has a rational root, it must be among ± 1, ± 2, ± 3, ± 4, ± 6, ± 12. Testing, we get

$$
\begin{array}{r|rrrr}
2 & 1 - 7 + 16 - 12 \\
  & 2 - 10 + 12 \\
\hline
  & 1 - 5 + 6 + 0
\end{array}
$$

Thus $t = 2$ is a root of $\Delta(t)$ and $\Delta(t) = (t - 2)(t^2 - 5t + 6) = (t - 2)^2(t - 3)$. Accordingly, $\lambda_1 = 2$ and $\lambda_2 = 3$ are the eigenvalues of T.

16.79 Find a basis of the eigenspace of $\lambda_1 = 2$.

▮ Substitute $t = 2$ into $tI - A$ to obtain the homogeneous system

$$\begin{pmatrix} 0 & -1 & 0 \\ 0 & 1 & 1 \\ 0 & -2 & -2 \end{pmatrix}\begin{pmatrix} x \\ y \\ z \end{pmatrix} = \begin{pmatrix} 0 \\ 0 \\ 0 \end{pmatrix} \quad \text{or} \quad \begin{cases} -y & = 0 \\ y + z = 0 \\ -2y - 2z = 0 \end{cases} \quad \text{or} \quad \begin{cases} y & = 0 \\ y + z = 0 \end{cases}$$

The system has only one independent solution, i.e., $x = 1$, $y = 0$, $z = 0$. Thus $u = (1, 0, 0)$ forms a basis for the eigenspace of $\lambda_1 = 2$.

16.80 Find a basis of the eigenspace of $\lambda_2 = 3$.

▮ Substitute $t = 3$ into $tI - A$ to obtain the homogeneous system

$$\begin{pmatrix} 1 & -1 & 0 \\ 0 & 2 & 1 \\ 0 & -2 & -1 \end{pmatrix}\begin{pmatrix} x \\ y \\ z \end{pmatrix} = \begin{pmatrix} 0 \\ 0 \\ 0 \end{pmatrix} \quad \text{or} \quad \begin{cases} x - y & = 0 \\ 2y + z = 0 \\ -2y - z = 0 \end{cases} \quad \text{or} \quad \begin{cases} x - y & = 0 \\ 2y + z = 0 \end{cases}$$

The system has only one independent solution, i.e., $x = 1$, $y = 1$, $z = -2$. Thus $v = (1, 1, -2)$ forms a basis of the eigenspace of $\lambda_2 = 3$.

16.81   Is T diagonalizable, i.e., can T be represented by a diagonal matrix with respect to some basis of $\mathbf{R}^3$?

▮ T is not diagonalizable since T has only two linearly independent eigenvectors but $\dim \mathbf{R}^3 = 3$.

Problems 16.82–16.84 refer to the matrix $A = \begin{pmatrix} 3 & 0 & 0 \\ 0 & 2 & -5 \\ 0 & 1 & -2 \end{pmatrix}$.

16.82   Find the characteristic polynomial of A.

▮
$$\Delta(t) = \begin{vmatrix} t-3 & 0 & 0 \\ 0 & t-2 & 5 \\ 9 & -1 & t+2 \end{vmatrix} = (t-3)(t^2+1)$$

16.83   Assuming A is a matrix over the real field $\mathbf{R}$, is A diagonalizable?

▮   As a real matrix, A has only one eigenvalue $\lambda_1 = 3$,   with algebraic multiplicity one.   Thus $\lambda_1 = 3$ has only one independent eigenvector and hence A is not diagonalizable over the real field $\mathbf{R}$.

16.84   Assuming A is a matrix over the complex field $\mathbf{C}$, is A diagonalizable?

▮   Now A has three distinct eigenvalues, 3, i, and $-i$, which have corresponding eigenvectors which are linearly independent.   Thus there exists an invertible matrix P over the complex field $\mathbf{C}$ for which

$$P^{-1}AP = \begin{pmatrix} 3 & 0 & 0 \\ 0 & i & 0 \\ 0 & 0 & -i \end{pmatrix}$$

i.e., A is diagonalizable.

16.85   Problem 16.52 shows that the real matrix $A = \begin{pmatrix} 1 & -1 \\ 2 & -1 \end{pmatrix}$ has no eigenvalues and no eigenvectors.   Prove that every real 3×3 matrix M has at least one eigenvalue and one eigenvector.   Generalize.

▮   The characteristic polynomial $\Delta(t)$ of M has degree 3, and every real polynomial of degree 3 has a real root since complex roots come in pairs.   Thus M has an eigenvalue λ which, by definition, has an eigenvector.   Similarly, every real matrix of odd order must have an eigenvalue and hence an eigenvector.

16.86   Is there an analogous result for complex matrices?

▮   By the Fundamental Theorem of Algebra [every polynomial over $\mathbf{C}$ has a root], the characteristic polynomial $\Delta(t)$ must have a root.   [See Theorem 16.10.]

Theorem 16.10:   Let A be an n-square matrix over the complex field $\mathbf{C}$.   Then A has at least one eigenvalue.

16.4   MINIMUM POLYNOMIAL

Definition: Let A be an n-square matrix over a field K and let $J(A)$ denote the collection of all polynomials $f(t)$ for which $f(A) = 0$.   [Note $J(A)$ is not empty since the characteristic polynomial $\Delta_A(t)$ of A belongs to $J(A)$.]   Let $m(t)$ be the monic polynomial of minimal degree in $J(A)$.   Then $m(t)$ is called the *minimal polynomial* of A.

The following theorems, proved below, are used throughout this section.

Theorem 16.11:   The minimum polynomial $m(t)$ of A divides every polynomial which has A as a zero.   In particular, $m(t)$ divides the characteristic polynomial $\Delta(t)$ of A.

Theorem 16.12:   The characteristic and minimum polynomials of a matrix A have the same irreducible factors.

This theorem does not say that $m(t) = \Delta(t)$; only that any irreducible factor of one must divide the other. In particular, since a linear factor is irreducible, $m(t)$ and $\Delta(t)$ have the same linear factors; hence they have the same roots.

Theorem 16.13: A scalar λ is an eigenvalue for a matrix A if and only if λ is a root of the minimum polynomial of A.

Theorem 16.14: Consider a block diagonal matrix

$$M = \begin{pmatrix} A_1 & 0 & \cdots & 0 \\ 0 & A_2 & \cdots & 0 \\ \cdots\cdots\cdots\cdots\cdots \\ 0 & 0 & \cdots & A_n \end{pmatrix}$$

Then the minimum polynomial $m(t)$ of M is the least common multiple of the minimum polynomials of the A_i.

Problems 16.87–16.88 refer to the matrix $A = \begin{pmatrix} 4 & -2 & 2 \\ 6 & -3 & 4 \\ 3 & -2 & 3 \end{pmatrix}$.

16.87 Find the characteristic polynomial $\Delta(t)$ of A.

∎ $$\Delta(t) = |tI - A| = \begin{pmatrix} t-4 & 2 & -2 \\ -6 & t+3 & -4 \\ -3 & 2 & t-3 \end{pmatrix} = t^3 - 4t^2 + 5t - 2 = (t-2)(t-1)^2$$

Alternatively, $\Delta(t) = t^3 - \text{tr}(A)t^2 + (A_{11} + A_{22} + A_{33})t - |A| = t^3 - 4t^2 + 5t - 2 = (t-2)(t-1)^2$. [Here A_{ii} is the cofactor of a_{ii} in A. See Problem 16.9.]

16.88 Find the minimum polynomial $m(t)$ of A.

∎ The minimum polynomial $m(t)$ must divide $\Delta(t)$. Also, each irreducible factor of $\Delta(t)$, that is, $t-2$ and $t-1$ must also be a factor of $m(t)$. Thus $m(t)$ is exactly only one of the following: $f(t) = (t-2)(t-1)$ or $g(t) = (t-2)(t-1)^2$. Testing $f(t)$ we have

$$f(A) = (A - 2I)(A - I) = \begin{pmatrix} 2 & -2 & 2 \\ 6 & -5 & 4 \\ 3 & -2 & 1 \end{pmatrix}\begin{pmatrix} 3 & -2 & 2 \\ 6 & -4 & 4 \\ 3 & 2 & 2 \end{pmatrix} = \begin{pmatrix} 0 & 0 & 0 \\ 0 & 0 & 0 \\ 0 & 0 & 0 \end{pmatrix}$$

Thus $f(t) = m(t) = (t-2)(t-1) = t^2 - 3t + 2$ is the minimal polynomial of A.

Problems 16.89–16.90 refer to the matrix $B = \begin{pmatrix} 3 & -2 & 2 \\ 4 & -4 & 6 \\ 2 & -3 & 5 \end{pmatrix}$.

16.89 Find the characteristic polynomial $\Delta(t)$ of A.

∎ $$\Delta(t) = |tI - B| = \begin{vmatrix} t-3 & 2 & -2 \\ -4 & t+4 & -6 \\ -2 & 3 & t-5 \end{vmatrix} = t^3 - 4t^2 + 5t - 2 = (t-2)(t-1)^2$$

16.90 Find the minimum polynomial $m(t)$ of B.

∎ The minimum polynomial $m(t)$ is exactly one of the following: $f(t) = (t-2)(t-1)$ or $g(t) = (t-2)(t-1)^2$. Testing $f(t)$ we have

$$f(B) = (B - 2I)(B - I) = \begin{pmatrix} 1 & -2 & 2 \\ 4 & -6 & 6 \\ 2 & -3 & 3 \end{pmatrix}\begin{pmatrix} 2 & -2 & 2 \\ 4 & -5 & 6 \\ 2 & -3 & 4 \end{pmatrix} = \begin{pmatrix} -2 & 2 & -2 \\ -4 & 4 & -4 \\ -2 & 2 & -2 \end{pmatrix} \neq 0$$

Thus $f(t) \neq m(t)$. Accordingly $m(t) = g(t) = (t-2)(t-1)^2$ is the minimum polynomial of B. [We do not need to compute $g(B)$; we know $g(B) = 0$ by the Cayley-Hamilton theorem.]

Problems 16.91–16.93 refer to the following polynomials [where $a \neq 0$]:

$$A = \begin{pmatrix} \lambda & a \\ 0 & \lambda \end{pmatrix} \qquad B = \begin{pmatrix} \lambda & a & 0 \\ 0 & \lambda & a \\ 0 & 0 & \lambda \end{pmatrix} \qquad C = \begin{pmatrix} \lambda & a & 0 & 0 \\ 0 & \lambda & a & 0 \\ 0 & 0 & \lambda & a \\ 0 & 0 & 0 & \lambda \end{pmatrix}$$

16.91   Find the minimum polynomial $m(t)$ of A.

$\blacksquare$ The characteristic polynomial of A is $\Delta(t) = (t - \lambda)^2$. We find $A - \lambda I \neq 0$; hence $m(t) = \Delta(t) = (t - \lambda)^2$.

16.92   Find the minimum polynomial $m(t)$ of B.

$\blacksquare$ The characteristic polynomial of B is $\Delta(t) = (t - \lambda)^3$. [Note $m(t)$ is exactly one of $t - \lambda$, $(t - \lambda)^2$, or $(t - \lambda)^3$.] We find $(B - \lambda I)^2 \neq 0$; thus $m(t) = \Delta(t) = (t - \lambda)^3$.

16.93   Find the minimum polynomial $m(t)$ of C.

$\blacksquare$ The characteristic polynomial of C is $\Delta(t) = (t - \lambda)^4$. We find $(C - \lambda I)^3 \neq 0$; hence $m(t) = \Delta(t) = (t - \lambda)^3$.

16.94   Generalize the result of Problems 16.91–16.93.

$\blacksquare$ Consider the following n-square matrix M with λ's on the diagonal and a's on the superdiagonal $(a \neq 0)$:

$$M = \begin{pmatrix} \lambda & a & 0 & \cdots & 0 & 0 \\ 0 & \lambda & a & \cdots & 0 & 0 \\ \multicolumn{6}{c}{\dotfill} \\ 0 & 0 & 0 & \cdots & \lambda & a \\ 0 & 0 & 0 & \cdots & 0 & \lambda \end{pmatrix}$$

Then $f(t) = (t - \lambda)^n$ is both the characteristic and minimum polynomial of M.

16.95   Let $M = \begin{pmatrix} A & 0 \\ 0 & B \end{pmatrix}$ where A and B are square matrices. Show that the minimum polynomial $m(t)$ of M is the least common multiple of the minimum polynomials $g(t)$ and $h(t)$ of A and B, respectively. [Theorem 16.14, which generalizes this result, follows directly from this result using induction.]

$\blacksquare$ Since $m(t)$ is the minimum polynomial of M, $m(M) = \begin{pmatrix} m(A) & 0 \\ 0 & m(B) \end{pmatrix} = 0$ and hence $m(A) = 0$ and $m(B) = 0$. Since $g(t)$ is the minimum polynomial of A, $g(t)$ divides $m(t)$. Similarly, $h(t)$ divides $m(t)$. Thus $m(t)$ is a multiple of $g(t)$ and $h(t)$.

Now let $f(t)$ be another multiple of $g(t)$ and $h(t)$; then $f(M) = \begin{pmatrix} f(A) & 0 \\ 0 & f(B) \end{pmatrix} = \begin{pmatrix} 0 & 0 \\ 0 & 0 \end{pmatrix} = 0$. But $m(t)$ is the minimum polynomial of M; hence $m(t)$ divides $f(t)$. Thus $m(t)$ is the least common multiple of $g(t)$ and $h(t)$.

16.96   Find the minimum polynomial $m(t)$ of

$$M = \begin{pmatrix} 2 & 8 & 0 & 0 & 0 & 0 & 0 \\ 0 & 2 & 0 & 0 & 0 & 0 & 0 \\ 0 & 0 & 4 & 2 & 0 & 0 & 0 \\ 0 & 0 & 1 & 3 & 0 & 0 & 0 \\ 0 & 0 & 0 & 0 & 0 & 3 & 0 \\ 0 & 0 & 0 & 0 & 0 & 0 & 0 \\ 0 & 0 & 0 & 0 & 0 & 0 & 5 \end{pmatrix}$$

$\blacksquare$ Observe that $M = \begin{pmatrix} A & & & \\ & B & & \\ & & C & \\ & & & D \end{pmatrix}$ where $A = \begin{pmatrix} 2 & 8 \\ 0 & 2 \end{pmatrix}$, $B = \begin{pmatrix} 4 & 2 \\ 1 & 3 \end{pmatrix}$, $C = \begin{pmatrix} 0 & 3 \\ 0 & 0 \end{pmatrix}$, $D = (5)$. Thus $m(t)$ is the least common multiple of the minimum polynomials of A, B, C, and D. Using Problem 16.94, the minimum polynomials of A, C, and D are $(t - 2)^2$, t^2, and $t - 5$, respectively. The characteristic polynomial of B is

$$|tI - B| = \begin{vmatrix} t-4 & -2 \\ -1 & t-3 \end{vmatrix} = t^2 - 7t + 10 = (t-2)(t-5)$$

Since the factors are distinct, it is also the minimum polynomial of B. Thus $m(t)$ is the least common multiple of $(t-2)^2$, t^2, $t-5$, and $(t-2)(t-5)$. Accordingly, $m(t) = t^2(t-2)^2(t-5)$.

Problems 16.97–16.100 refer to the following matrices:

$$A = \begin{pmatrix} 2 & 5 & 0 & 0 & 0 \\ 0 & 2 & 0 & 0 & 0 \\ 0 & 0 & 4 & 2 & 0 \\ 0 & 0 & 3 & 5 & 0 \\ 0 & 0 & 0 & 0 & 7 \end{pmatrix} \qquad B = \begin{pmatrix} 3 & 1 & 0 & 0 & 0 \\ 0 & 3 & 0 & 0 & 0 \\ 0 & 0 & 3 & 1 & 0 \\ 0 & 0 & 0 & 3 & 1 \\ 0 & 0 & 0 & 0 & 3 \end{pmatrix} \qquad C = \begin{pmatrix} \lambda & 0 & 0 & 0 & 0 \\ 0 & \lambda & 0 & 0 & 0 \\ 0 & 0 & \lambda & 0 & 0 \\ 0 & 0 & 0 & \lambda & 0 \\ 0 & 0 & 0 & 0 & \lambda \end{pmatrix}$$

16.97 Find the characteristic polynomial $\Delta(t)$ of A.

▌ Note A is a block diagonal matrix with diagonal blocks

$$A_1 = \begin{pmatrix} 2 & 5 \\ 0 & 2 \end{pmatrix} \qquad A_2 = \begin{pmatrix} 4 & 2 \\ 3 & 5 \end{pmatrix} \qquad A_3 = (7)$$

Then $\Delta(t)$ is the product of the characteristic polynomials $\Delta_1(t)$, $\Delta_2(t)$, and $\Delta_3(t)$ of A_1, A_2, and A_3, respectively. Since A_1 and A_3 are triangular, $\Delta_1(t) = (t-2)^2$ and $\Delta_3(t) = (t-7)$. Also, $\Delta_2(t) = t^2 - 9t + 14 = (t-2)(t-7)$. Thus $\Delta(t) = (t-2)^3(t-7)^2$. [As expected, $\deg m(t) = 5$.]

16.98 Find the minimum polynomial $m(t)$ of A.

▌ Note that the minimum polynomials $m_1(t)$, $m_2(t)$, and $m_3(t)$ of the diagonal blocks A_1, A_2, and A_3, respectively, are equal to the characteristic polynomials; i.e., $m_1(t) = (t-2)^2$, $m_2(t) = (t-2)(t-7)$, $m_3(t) = t-7$. But $m(t)$ is equal to the least common multiple of $m_1(t)$, $m_2(t)$, $m_3(t)$. Thus $m(t) = (t-2)^2(t-7)$.

16.99 Find the characteristic polynomial $\Delta(t)$ and minimum polynomial $m(t)$ of B.

▌ B is a block diagonal matrix with diagonal blocks

$$B_1 = \begin{pmatrix} 3 & 1 \\ 0 & 3 \end{pmatrix} \qquad \text{and} \qquad B_2 = \begin{pmatrix} 3 & 1 & 0 \\ 0 & 3 & 1 \\ 0 & 0 & 3 \end{pmatrix}$$

By Problem 16.94, the characteristic and minimum polynomial of B_1 is $f(t) = (t-3)^2$ and the characteristic and minimum polynomial of B_2 is $g(t) = (t-3)^3$. Thus $\Delta(t) = f(t)g(t) = (t-3)^5$ but $m(t) = gcd(f(t), g(t)) = (t-3)^3$ [which is the size of the largest block].

16.100 Find the characteristic polynomial $\Delta(t)$ and the minimum polynomial $m(t)$ of C. [Note C is a scalar matrix, i.e., $C = \lambda I$.]

▌ Since C is triangular, $\Delta(t) = (t-\lambda)^5$. On the other hand, $m(t) = t - \lambda$ since $C - \lambda I = 0$.

16.101 Let A be any square matrix. Suppose $m_1(t)$ and $m_2(t)$ are monic polynomials of minimal degree for which A is a root. Show that $m_1(t) = m_2(t)$.

▌ Let $n = \deg m_1 = \deg m_2$. Suppose $m_1(t) \neq m_2(t)$. Then the difference $f(t) = m_1(t) - m_2(t)$ is a polynomial which has A as a root and $\deg f < n$. Dividing f by its leading coefficient gives a monic polynomial f' which has A as a root and $\deg f' = \det f < n$. This contradicts the minimality of m_1 and m_2. Thus $m_1(t) = m_2(t)$.

16.102 Prove Theorem 16.11.

▌ Suppose $f(t)$ is a polynomial for which $f(A) = 0$. By the division algorithm there exist polynomials $q(t)$ and $r(t)$ for which $f(t) = m(t)q(t) + r(t)$ and $r(t) = 0$ or $\deg r(t) < \deg m(t)$. Substituting $t = A$ in this equation, and using $f(A) = 0$ and $m(A) = 0$, we obtain $r(A) = 0$. If $r(t) \neq 0$, then $r(t)$ is a polynomial of degree less than $m(t)$ which has A as a zero; this contradicts the definition of the minimum polynomial. Thus $r(t) = 0$ and so $f(t) = m(t)q(t)$, that is, $m(t)$ divides $f(t)$.

16.103 Let $m(t)$ be the minimum polynomial of an n-square matrix A. Show that the characteristic polynomial of A divides $(m(t))^n$.

▮ Suppose $m(t) = t^r + c_1 t^{r-1} + \cdots + c_{r-1} t + c_r$. Consider the following matrices:

$$
\begin{aligned}
B_0 &= I \\
B_1 &= A + c_1 I \\
B_2 &= A^2 + c_1 A + c_2 I \\
&\cdots\cdots\cdots\cdots\cdots\cdots\cdots \\
B_{r-1} &= A^{r-1} + c_1 A^{r-2} + \cdots + c_{r-1} I
\end{aligned}
$$

Then

$$
\begin{aligned}
B_0 &= I \\
B_1 - AB_0 &= c_1 I \\
B_2 - AB_1 &= c_2 I \\
&\cdots\cdots\cdots\cdots\cdots \\
B_{r-1} - AB_{r-2} &= c_{r-1} I
\end{aligned}
$$

Also,

$$
\begin{aligned}
-AB_{r-1} &= c_r I - (A^r + c_1 A^{r-1} + \cdots + c_{r-1} A + c_r I) \\
&= c_r I - m(A) \\
&= c_r I
\end{aligned}
$$

Set

$$
B(t) = t^{r-1} B_0 + t^{r-2} B_1 + \cdots + t B_{r-2} + B_{r-1}
$$

Then

$$
\begin{aligned}
(tI - A) \cdot B(t) &= (t^r B_0 + t^{r-1} B_1 + \cdots + t B_{r-1}) = (t^{r-1} A B_0 + t^{r-2} A B_1 + \cdots + A B_{r-1}) \\
&= t^r B_0 + t^{r-1}(B_1 - A B_0) + t^{r-2}(B_2 - A B_1) + \cdots + t(B_{r-1} - A B_{r-2}) - A B_{r-1} \\
&= t^r I + c_1 t^{r-1} I + c_2 t^{r-2} I + \cdots + c_{r-1} t I + c_r I \\
&= m(t)_I
\end{aligned}
$$

The determinant of both sides gives $|tI - A| |B(t)| = |m(t)I| = (m(t))^n$. Since $|B(t)|$ is a polynomial, $|tI - A|$ divides $(m(t))^n$; i.e., the characteristic polynomial of A divides $(m(t))^n$.

16.104 Prove Theorem 16.12.

▮ Suppose $f(t)$ is an irreducible polynomial. If $f(t)$ divides $m(t)$ then, since $m(t)$ divides $\Delta(t)$, $f(t)$ divides $\Delta(t)$. On the other hand, if $f(t)$ divides $\Delta(t)$ then, by the preceding problem, $f(t)$ divides $(m(t))^n$. But $f(t)$ is irreducible; hence $f(t)$ also divides $m(t)$. Thus $m(t)$ and $\Delta(t)$ have the same irreducible factors.

16.105 Let T be a linear operator on a vector space V of finite dimension. Show that T is invertible if and only if the constant term of the minimal (characteristic) polynomial of T is not zero.

▮ Suppose the minimal (characteristic) polynomial of T is $f(t) = t^r + a_{n-1} t^{r-1} + \cdots + a_1 t + a_0$. Each of the following statements is equivalent to the succeeding one by previous results: (i) T is invertible; (ii) T is nonsingular; (iii) 0 is not an eigenvalue of T; (iv) 0 is not a root of $m(t)$; (v) the constant term a_0 is not zero. Thus the result is proved.

16.106 Suppose $\dim V = n$. Let $T : V \to V$ be an invertible operator. Show that T^{-1} is equal to a polynomial in T of degree not exceeding n.

▮ Let $m(t)$ be the minimal polynomial of T. Then $m(t) = t^r + a_{r-1} t^{r-1} + \cdots + a_1 t + a_0$, where $r \le n$. Since T is invertible, $a_0 \ne 0$. We have $m(T) = T^r + a_{r-1} T^{r-1} + \cdots + a_1 T + a_0 I = 0$. Hence

$$
-\frac{1}{a_0}(T^{r-1} + a_{r-1} T^{r-2} + \cdots + a_1 I)T = I \quad \text{and} \quad T^{-1} = -\frac{1}{a_0}(T^{r-1} + a_{r-1} T^{r-2} + \cdots + a_1 I)
$$

16.107 Let F be an extension of a field K. Let A be an n-square matrix over K. Note that A may also be viewed as a matrix $\hat{A}$ over F. Clearly $|tI - A| = |tI - \hat{A}|$, that is, A and $\hat{A}$ have the same characteristic polynomial. Show that A and $\hat{A}$ also have the same minimum polynomial.

▮ Let $m(t)$ and $m'(t)$ be the minimum polynomials of A and $\hat{A}$, respectively. Now $m'(t)$ divides every polynomial over F which has A as a zero. Since $m(t)$ has A as a zero and since $m(t)$ may be viewed as a polynomial over F, $m'(t)$ divides $m(t)$. We show now that $m(t)$ divides $m'(t)$.

Since $m'(t)$ is a polynomial over F which is an extension of K, we may write $m'(t) = f_1(t)b_1 + f_2(t)b_2 + \cdots + f_n(t)b_n$ where $f_i(t)$ are polynomials over K, and $b_1, \ldots, b_n$ belong to F and are linearly independent over K. We have

$$m'(A) = f_1(A)b_1 + f_2(A)b_2 + \cdots + f_n(A)b_n = 0 \qquad (1)$$

Let $a_{ij}^{(k)}$ denote the ij entry of $f_k(A)$. The above matrix equation implies that, for each pair (i, j), $a_{ij}^{(1)}b_1 + a_{ij}^{(2)}b_2 + \cdots + a_{ij}^{(n)}b_n = 0$. Since the b_i are linearly independent over K and since the $a_{ij}^{(k)} \in K$, every $a_{ij}^{(k)} = 0$. Then $f_1(A) = 0$, $f_2(A) = 0, \ldots, f_n(A) = 0$. Since the $f_i(t)$ are polynomials over K which have A as a zero and since $m(t)$ is the minimum polynomial of A as a matrix over K, $m(t)$ divides each of the $f_i(t)$. Accordingly, by (1), $m(t)$ must also divide $m'(t)$. But monic polynomials which divide each other are necessarily equal. That is, $m(t) = m'(t)$, as required.

16.108 Let $\{v_1, \ldots, v_n\}$ be a basis of V. Let $T: V \to V$ be an operator for which $T(v_1) = 0$, $T(v_2) = a_{21}v_1$, $T(v_3) = a_{31}v_1 + a_{32}v_2, \ldots, T(v_n) = a_{n1}v_1 + \cdots + a_{n,n-1}v_{n-1}$. Show that $T^n = 0$. Thus the minimum polynomial of T has the form $m(t) = t^r$ where $r \leq n$.

▌ It suffices to show that

$$T^j(v_j) = 0 \qquad (1)$$

for $j = 1, \ldots, n$. For then it follows that $T^n(v_j) = T^{n-j}(T^j(v_j)) = T^{n-j}(0) = 0$, for $j = 1, \ldots, n$, and, since $\{v_1, \ldots, v_n\}$ is a basis, $T^n = 0$.

We prove (1) by induction on j. The case $j = 1$ is true by hypothesis. The inductive step follows [for $j = 2, \ldots, n$] from

$$\begin{aligned} T^j(v_j) &= T^{j-1}(T(v_j)) = T^{j-1}(a_{j1}v_1 + \cdots + a_{j,j-1}v_{j-1}) \\ &= a_{j1}T^{j-1}(v_1) + \cdots + a_{j,j-1}T^{j-1}(v_{j-1}) \\ &= a_{j1}0 + \cdots + a_{j,j-1}0 = 0 \end{aligned}$$

Remark: Observe that the matrix representation of T in the above basis is triangular with diagonal elements 0:

$$\begin{pmatrix} 0 & a_{21} & a_{31} & \cdots & a_{n1} \\ 0 & 0 & a_{32} & \cdots & a_{n2} \\ \cdots\cdots\cdots\cdots\cdots\cdots\cdots \\ 0 & 0 & 0 & \cdots & a_{n,n-1} \\ 0 & 0 & 0 & \cdots & 0 \end{pmatrix}$$

16.5 FURTHER PROPERTIES OF EIGENVALUES AND EIGENVECTORS

16.109 Suppose λ is an eigenvalue of an invertible operator T. Show that λ^{-1} is an eigenvalue of T^{-1}.

▌ Since T is invertible, it is also nonsingular; hence $\lambda \neq 0$.

By definition of an eigenvalue, there exists a nonzero vector v for which $T(v) = \lambda v$. Applying T^{-1} to both sides, we obtain $v = T^{-1}(\lambda v) = \lambda T^{-1}(v)$. Hence $T^{-1}(v) = \lambda^{-1}v$; that is, λ^{-1} is an eigenvalue of T^{-1}.

16.110 Suppose v is a nonzero eigenvector of linear maps S and T. Show that v is an eigenvector of $S + T$.

▌ Suppose $S(v) = \lambda_1(v)$ and $T(v) = \lambda_2(v)$. Then $(S + T)(v) = S(v) + T(v) = \lambda_1 v + \lambda_2 v = (\lambda_1 + \lambda_2)v$. Thus v is an eigenvector of $S + T$ belonging to the eigenvalue $\lambda_1 + \lambda_2$.

16.111 Suppose v is a nonzero eigenvector of T. Show that, for any $k \in K$, v is an eigenvector of kT.

▌ Suppose $T(v) = \lambda v$. Then $(kT)(v) = kT(v) = k(\lambda v) = (k\lambda)v$. Thus v is an eigenvector of kT belonging to the eigenvalue $k\lambda$.

16.112 Suppose λ is an eigenvalue of a linear operator T. (a) Show that λ^2 is an eigenvalue of T^2. (b) More generally, show that λ^n is an eigenvalue of T^n for $n \geq 1$.

❚ Since λ is an eigenvalue of T, there exists a nonzero vector v such that $T(v) = \lambda v$.

(a) We have $T^2(v) = T(T(v)) = T(\lambda v) = \lambda(T(v)) = \lambda(\lambda v) = \lambda^2 v$. Thus λ^2 is an eigenvalue of T^2.

(b) Suppose $n > 1$ and the result holds for $n - 1$. Then $T^n(v) = T(T^{n-1}(v)) = T(\lambda^{n-1}v) = \lambda^{n-1}(T(v)) = \lambda^{n-1}(\lambda v) = \lambda^n v$. Thus λ^n is an eigenvalue of T^n.

16.113 Suppose λ is an eigenvalue of a linear operator T. Show that $f(\lambda)$ is an eigenvalue of $f(T)$ for any polynomial $f(t)$.

❚ There exists a nonzero vector v such that $T(v) = \lambda v$. Suppose $f(t) = a_n t^n + \cdots + a_1 t + a_0$. Then

$$f(T)(v) = (a_n T^n + \cdots + a_1 T + a_0 I)(v) = a_n T^n(v) + \cdots + a_1 T(v) + a_0 I(v)$$
$$= a_n \lambda^n v + \cdots + a_1 \lambda v + a_0 v = (a_n \lambda^n + \cdots + a_1 \lambda + a_0)(v)$$
$$= f(\lambda)v$$

Thus $f(\lambda)$ is an eigenvalue of $f(T)$.

16.114 Let A be an n-square matrix for which $A^k = 0$ for some $k > n$. Show that $A^n = 0$.

❚ Here A is a root of $f(t) = t^k$. Since the minimum polynomial $m(t)$ of A must divide $f(t)$, we have $m(t) = t^r$ for $r \leq k$. However, the degree of $m(t)$ cannot exceed the degree of the characteristic polynomial $\Delta(t)$ of A which has degree n. Hence A is a root of $m(t) = t^r$ for $r \leq n$. Thus A is a root of t^n.

16.115 Let $E: V \rightarrow V$ be a projection operator, i.e., $E^2 = E$. Show that E is diagonalizable and, in fact, can be represented by the diagonal matrix $A = \begin{pmatrix} I_r & 0 \\ 0 & 0 \end{pmatrix}$.

❚ Since $E^2 = E$, the projection operator E is a root of $f(t) = t^2 - t = t(t - 1)$. The minimum polynomial $m(t)$ of E divides $m(t)$. Thus $m(t)$ has distinct roots and hence E is diagonalizable. The eigenvalues must be 0 or 1 or both 0 and 1. Thus the diagonal matrix A representing E has 1s and/or 0s on the diagonal. By putting the eigenvectors with eigenvalue 1 first, A will have the given form.

16.116 Consider an arbitrary monic polynomial $f(t) = t^n + a_{n-1} t^{n-1} + \cdots + a_1 t + a_0$. Define the companion matrix A of $f(t)$. What is the minimum polynomial $m(t)$ and the characteristic polynomial $\Delta(t)$ of A?

❚ A is an n-square matrix with 1s on the subdiagonal, the negatives of the coefficient in the last column, and 0s elsewhere as follows:

$$A = \begin{pmatrix} 0 & 0 & \cdots & 0 & -a_0 \\ 1 & 0 & \cdots & 0 & -a_1 \\ 0 & 1 & \cdots & 0 & -a_2 \\ \cdots\cdots\cdots\cdots\cdots\cdots \\ 0 & 0 & \cdots & 1 & -a_{n-1} \end{pmatrix}$$

The minimum polynomial $m(t)$ and the characteristic polynomial $\Delta(t)$ are both equal to $f(t)$.

16.117 Find a matrix A whose minimum polynomial is $t^3 - 5t^2 + 6t + 8$.

❚ Let A be the companion matrix, i.e., $A = \begin{pmatrix} 0 & 0 & -8 \\ 1 & 0 & -6 \\ 0 & 1 & 5 \end{pmatrix}$.

16.118 Find a matrix B whose minimum polynomial is $t^4 - 5t^3 - 2t^2 + 7t + 4$.

❚ Let B be the companion matrix, i.e.,

$$B = \begin{pmatrix} 0 & 0 & 0 & -4 \\ 1 & 0 & 0 & -7 \\ 0 & 1 & 0 & 2 \\ 0 & 0 & 1 & 5 \end{pmatrix}$$

CHAPTER 17
Canonical Forms

Let T be a linear operator on a vector space of finite dimension. As seen in the preceding chapter, T may not have a diagonal matrix representation. However, it is still possible to "simplify" the matrix representation of T in a number of ways. This is the main topic of this chapter. In particular, we obtain the *primary decomposition theorem* and the *triangular*, *Jordan*, and *rational* canonical forms.

17.1 INVARIANT SUBSPACES

17.1 Define an invariant subspace of a linear operator.

▌ Let $T: V \to V$ be linear. A subspace W of V is said to be *invariant under T* or *T-invariant* if T maps W into itself, i.e., if $v \in W$ implies $T(v) \in W$. In this case T restricted to W defines a linear operator on W; that is, T induces a linear operator $\hat{T}: W \to W$ defined by $\hat{T}(w) = T(w)$ for every $w \in W$.

Problems 17.2–17.5 refer to the linear operator $T: \mathbf{R}^3 \to \mathbf{R}^3$ which rotates each vector about the z axis by an angle θ as pictured in Fig. 17-1; i.e., $T(x, y, z) = (x \cos \theta - y \sin \theta, x \sin \theta + y \cos \theta, z)$.

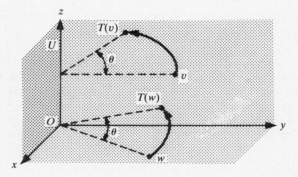

Fig. 17-1

17.2 Let W be the xy plane in $\mathbf{R}^3$. Is W invariant under T?

▌ Each vector $w = (a, b, 0)$ in the xy plane W remains in W under the mapping W as indicated in Fig. 17-1. Thus W is invariant under T. The restriction of T to W rotates each vector in W about the origin O.

17.3 Let W' be the yz plane in $\mathbf{R}^3$. Is W' invariant under T?

▌ A nonzero vector $w' = (0, b, c)$ in W' does not remain in W' under T [unless $\theta = \pi$ or a multiple of π]. Thus W' is not T-invariant.

17.4 Let U be the z axis in $\mathbf{R}^3$. Is U invariant under T?

▌ For any vector $u = (0, 0, z)$ in U, we have $T(u) = u$. Thus U is invariant under T. In fact, the restriction of T to U is the identity mapping on U.

17.5 Let U' be the x axis in $\mathbf{R}^3$. Is U' invariant under T?

▌ A nonzero vector $u' = (a, 0, 0)$ in U' does not remain in U' under T [unless $\theta = \pi$ or a multiple of π]. Thus U' is not invariant under T.

17.6 What, if any, is the relationship between eigenvectors of a linear operator T and invariant subspaces of T?

▌ If v is any nonzero eigenvector of T, then span(v) is a one-dimensional invariant subspace of T. Conversely, if W is a one-dimensional invariant subspace of T, then any nonzero vector in W is an eigenvector of T.

Problems 17.7–17.10 refer to any linear operator $T: V \to V$.

17.7 Show that $\{0\}$ is invariant under T.

▮ We have $T(0) = 0 \in \{0\}$; hence $\{0\}$ is invariant under T.

17.8 Show that V is invariant under T.

▮ For every $v \in V$, $T(v) \in V$; hence V is invariant under T.

17.9 Show that the kernel of T is invariant under T.

▮ Let $u \in \text{Ker } T$. Then $T(u) = 0 \in \text{Ker } T$ since the kernel of T is a subspace of V. Thus Ker T is invariant under T.

17.10 Show that the image of T is invariant under T.

▮ Since $T(v) \in \text{Im } T$ for every $v \in V$, it is certainly true if $v \in \text{Im } T$. Hence the image of T is invariant under T.

17.11 Find all invariant subspaces of $A = \begin{pmatrix} 2 & -5 \\ 1 & -2 \end{pmatrix}$ viewed as an operator on $\mathbf{R}^2$.

▮ First of all, we have that $\mathbf{R}^2$ and $\{0\}$ are invariant under A. Now if A has any other invariant subspaces, then it must be one-dimensional. However, the characteristic polynomial of A is

$$\Delta(t) = |tI - A| = \begin{vmatrix} t-2 & 5 \\ -1 & t+2 \end{vmatrix} = t^2 + 1$$

Hence A has no eigenvalues (in $\mathbf{R}$) and so A has no eigenvectors. But the one-dimensional invariant subspaces correspond to the eigenvectors; thus $\mathbf{R}^2$ and $\{0\}$ are the only subspaces invariant under A.

17.12 Suppose $\{W_i\}$ is a collection of T-invariant subspaces of a vector space V. Show that the intersection $W = \cap_i W_i$ is also T-invariant.

▮ Suppose $v \in W$; then $v \in W_i$ for every i. Since W_i is T-invariant, $T(v) \in W_i$ for every i. Thus $T(v) \in W = \cap_i W_i$ and so W is T-invariant.

Theorem 17.1: Let $T: V \to V$ be linear, and let $f(t)$ be any polynomial. Then the kernel of $f(T)$ is invariant under T.

17.13 Prove Theorem 17.1

▮ Suppose $v \in \text{Ker } f(T)$, that is, $f(T)(v) = 0$. We need to show that $T(v)$ also belongs to the kernel of $f(T)$, that is, $f(T)(T(v)) = 0$. Since $f(t)t = tf(t)$, we have $f(T)T = Tf(T)$. Thus $f(T)T(v) = Tf(T)(v) = T(0) = 0$ as required.

Theorem 17.2: Suppose W is an invariant subspace of $T: V \to V$. Then T has a block matrix representation $\begin{pmatrix} A & B \\ 0 & C \end{pmatrix}$ where A is a matrix representation of the restriction $\hat{T}$ of T to W.

17.14 Prove Theorem 17.2.

▮ We choose a basis $\{w_1, \ldots, w_r\}$ of W and extend it to a basis $\{w_1, \ldots, w_r, v_1, \ldots, v_s\}$ of V. Then

$$\begin{aligned}
\hat{T}(w_1) &= T(w_1) = a_{11}w_1 + \cdots + a_{1r}w_r \\
\hat{T}(w_2) &= T(w_2) = a_{21}w_1 + \cdots + a_{2wr} \\
&\cdots\cdots\cdots\cdots\cdots\cdots\cdots\cdots\cdots\cdots \\
\hat{T}(w_r) &= T(w_r) = a_{r1}w_1 + \cdots + a_{rr}w_r \\
T(v_1) &= b_{11}w_1 + \cdots + b_{1r}w_r + c_{11}v_1 + \cdots + c_{1s}v_s \\
T(v_2) &= b_{21}w_1 + \cdots + b_{2r}w_r + c_{21}v_1 + \cdots + c_{2s}v_s \\
&\cdots\cdots\cdots\cdots\cdots\cdots\cdots\cdots\cdots\cdots\cdots\cdots \\
T(v_s) &= b_{s1}w_1 + \cdots + b_{sr}w_r + c_{s1}v_1 + \cdots + c_{ss}v_s
\end{aligned}$$

But the matrix of T in this basis is the transpose of the matrix of coefficient in the above system of equations. Therefore it has the form $\begin{pmatrix} A & B \\ 0 & C \end{pmatrix}$ where A is the transpose of the matrix of coefficients for the obvious subsystem. By the same argument, A is the matrix of $\hat{T}$ relative to the basis $\{w_i\}$ of W.

Problems 17.15–17.16 refer to the restriction $\hat{T}$ of a linear operator T to an invariant subspace W, that is, $\hat{T}(w) = T(w)$ for every $w \in W$.

17.15 Prove: For any polynomial $f(t)$, $f(\hat{T})(w) = f(T)(w)$.

▌ If $f(t) = 0$ or if $f(t)$ is a constant, i.e., of degree 1, then the result clearly holds. Assume $\deg f = n > 1$ and that the result holds for polynomials of degree less than n. Suppose that $f(t) = a_n t^n + a_{n-1} t^{n-1} + \cdots + a_1 t + a_0$. Then

$$\begin{aligned}
f(\hat{T})(w) &= (a_n \hat{T}^n + a_{n-1} \hat{T}^{n-1} + \cdots + a_0 I)(w) \\
&= (a_n \hat{T}^{n-1})(\hat{T}(w)) + (a_{n-1} \hat{T}^{n-1} + \cdots + a_0 I)(w) \\
&= (a_n T^{n-1})(T(w)) + (a_{n-1} T^{n-1} + \cdots + a_0 I)(w) \\
&= f(T)(w)
\end{aligned}$$

17.16 Prove: The minimum polynomial of $\hat{T}$ divides the minimum polynomial of T.

▌ Let $m(t)$ denote the minimum polynomial of T. Then by Problem 17.15, $m(\hat{T})(w) = m(T)(w) = 0(w) = 0$ for every $w \in W$; that is, $\hat{T}$ is a zero of the polynomial $m(t)$. Hence the minimum polynomial of $\hat{T}$ divides $m(t)$.

17.17 Show that every subspace of V is invariant under I and 0, the identity and zero operators.

▌ Suppose W is a subspace of V and $w \in W$. Then $I(w) = w \in W$ and $0(w) = 0 \in W$. Thus W is invariant under I and 0.

17.18 Determine the invariant subspaces of $A = \begin{pmatrix} 2 & -4 \\ 5 & -2 \end{pmatrix}$ viewed as a linear operator $\mathbf{R}^2$.

▌ Here $\Delta(t) = t^2 + 16$ is the characteristic polynomial of A. There are no eigenvalues (in $\mathbf{R}$) and hence there are no eigenvectors. Thus there are no one-dimensional invariant subspaces. Accordingly, $\{0\}$ and $\mathbf{R}^2$ are the only A-invariant subspaces.

17.19 Determine the invariant subspace of the above matrix A viewed as a linear operator on $\mathbf{C}^2$.

▌ Since $\Delta(t) = t^2 + 16 = (t + 4i)(t - 4i)$, there are two eigenvalues, $\lambda_i = 4i$ and $\lambda_2 = -4i$. Setting $\lambda_1 I - A = 0$ yields a nonzero solution $v_1 = (2, 1 - 2i)$ and setting $\lambda_2 I - A = 0$ yields a nonzero solution $v_2 = (2, 1 + 2i)$. Thus the only invariant subspaces are the following: $\{0\}, \mathbf{C}^2, W_1 = \text{span}(2, 1 - 2i)$, $W_2 = \text{span}(2, 1 + 2i)$.

Problems 17.20–17.23 refer to a subspace W which is invariant under $S: V \to V$ and $T: V \to V$.

17.20 Show that W is invariant under $S + T$.

▌ Let $w \in W$. Then $S(w) \in W$ and $T(w) \in W$. Since W is a subspace $S(w) + T(w) \in W$. Therefore, $(S + T)(w) = S(w) + T(w)$ belongs to W. Thus W is invariant under $S + T$.

17.21 Show that W is invariant under the composition $S \circ T$.

▌ Let $w \in W$. Then $T(w) \in W$ and hence $(S \circ T)(w) = S(T(w)) \in W$. Thus W is invariant under $S \circ T$.

17.22 Show that W is invariant under kT for any scalar $k \in K$.

▌ Let $w \in W$. Then $T(w) \in W$. Since W is a subspace $kT(w) \in W$. Thus $(kT)(w) = kT(w)$ belongs to W. Hence W is invariant under kT.

17.23 Show that W is invariant under $f(T)$ for any polynomial $f(t)$.

$\boldsymbol{I}$ By Problem 17.21, W is invariant under T^2 and, by induction, W is invariant under T^k for any $k \geq 1$. By Problem 17.22, W is invariant under $a_k T^k$ for any scalar a_k. Also, W is invariant under I by Problem 17.17 [where I is the identity map]. Last, by Problem 17.20, W is invariant under $a_n T^n + \cdots + a_1 T + a_0 I$. In other words, W is invariant under $f(T)$ for any polynomial $f(t)$.

17.2 DIRECT SUMS, PROJECTIONS

17.24 Define a direct sum of subspaces and the corresponding projections.

$\boldsymbol{I}$ A vector space V is termed the *direct sum* of its subspaces $W_1, \ldots, W_r$, written $V = W_1 \oplus W_2 \oplus \cdots \oplus W_r$, if every vector $v \in V$ can be written uniquely in the form $v = w_1 + w_2 + \cdots + w_r$ with $w_i \in W_i$. In such a case, the projection of V into its subspace W_i is the mapping $E: V \to V$ defined by $E(v) = w_i$. [The projection E is well-defined since the sum for v is unique and there is a projection mapping for each subspace W_i.]

Problems 17.25–17.28 refer to the following subspaces of $\mathbf{R}^3$: $U = xy$ plane, $W = yz$ plane, $Z = z$ axis, $L = \{(k, k, k): k \in \mathbf{R}\}$.

17.25 Is $\mathbf{R}^3 = U \oplus W$?

$\boldsymbol{I}$ $\mathbf{R}^3 = U + W$ since every vector in $\mathbf{R}^3$ is the sum of a vector in U and a vector in W. However, $\mathbf{R}^3$ is not the direct sum of U and W since such sums are not unique; e.g., $(1, 2, 3) = (1, 1, 0) + (0, 1, 1) = (1, 3, 0) + (0, -1, 3)$.

17.26 Is $\mathbf{R}^3 = U \oplus Z$?

$\boldsymbol{I}$ Any vector $(a, b, c) \in \mathbf{R}^3$ can be written as the sum of a vector in U and a vector in Z in one and only one way: $(a, b, c) = (a, b, 0) = (0, 0, c)$. Thus $\mathbf{R}^3 = U \oplus Z$.

17.27 Given $\mathbf{R}^3 = U \oplus L$, find the projections E_U and E_L of V into U and L, respectively.

$\boldsymbol{I}$ For any vector $(a, b, c) \in \mathbf{R}^3$, the unique representation is as follows: $(a, b, c) = (a - c, b - c, 0) + (c, c, c)$. Thus E_U and E_L are defined by $E_U(a, b, c) = (a - c, b - c, 0)$ and $E_L(a, b, c) = (c, c, c)$.

17.28 Given $\mathbf{R}^3 = W \oplus L$, find the projections E_W and E_L onto W and L, respectively.

$\boldsymbol{I}$ We have $(a, b, c) = (0, b - a, c - a) + (a, a, a)$ is the unique representation; hence $E_W(a, b, c) = (0, b - a, c - a)$ and $E_L(a, b, c) = (a, a, a)$.

Theorem 17.3: Suppose $W_1, \ldots, W_r$ are subspaces of V and suppose $B_i = \{w_{i1}, \ldots, w_{in_i}\}$ is a basis for W_i for $i = 1, \ldots, r$. Let B be the union of all the basis vectors, i.e.,
(i) If B is a basis of V, then $V = W_1 \oplus \cdots \oplus W_r$.
(ii) If $V = W_1 \oplus \cdots \oplus W_r$, then B is a basis of V.

17.29 Prove (i) of Theorem 17.3.

$\boldsymbol{I}$ Let $v \in V$. Since B is a basis for V $v = a_{11}w_{11} + \cdots + a_{1n_1}w_{1n_1} + \cdots + a_{r1}w_{r1} + \cdots + a_{rn_r}w_{rn_r} = w_1 + w_2 + \cdots + w_r$ where $w_i = a_{i1}w_{i1} + \cdots + a_{in_i}w_{in_i} \in W_i$. We next show that such a sum is unique. Suppose $v = w'_1 + w'_2 + \cdots + w'_r$ where $w'_i \in W_i$. Since $\{w_{i1}, \ldots, w_{in_i}\}$ is a basis of W_i, $w'_i = b_{1n_1}w_{1n_1} + \cdots + b_{in_i}w_{in_i}$ and so $v = b_{11}w_{11} + \cdots + b_{1n_1}w_{1n_1} + \cdots + b_{r1}w_{r1} + \cdots + b_{rn_r}w_{rn_r}$. Since B is a basis of V, $a_{ij} = b_{ij}$ for each i and each j. Hence $w_i = w'_i$ and so the sum for v is unique. Accordingly, V is the direct sum of the W_i.

17.30 Prove (ii) of Theorem 17.3.

$\boldsymbol{I}$ Let $v \in V$. Since V is the direct sum of the W_i, we have $v = w_1 + \cdots + w_r$ where $w_i \in W_i$. Since $\{w_{ij}\}$ is a basis of W_i, each w_i is a linear combination of the w_{ij} and so v is a linear combination of the elements of B. Thus B spans V. We now show that B is linearly independent. Suppose $a_{11}w_{11} + \cdots + a_{1n_1}w_{1n_1} + \cdots + a_{r1}w_{r1} + \cdots + a_{rn_r}w_{rn_r} = 0$. Note that $a_{i1}w_{i1} + \cdots + a_{in_i}w_{in_i} \in W_i$. We also have that $0 = $

$0 + 0 + \cdots + 0$ where $0 \in W_i$.   Since such a sum for 0 is unique, $a_{i1}w_{i1} + \cdots + a_{in_i}w_{in_i} = 0$ for $i = 1, \ldots, r$.   The independence of the bases $\{w_{ij_i}\}$ implies that all the a's are 0.   Thus B is linearly independent and hence is a basis of V.

17.31   Let $V = W_1 \oplus \cdots \oplus W_r$ and let $E: V \to V$ be the projection mapping $E: V \to V$ defined by $E(v) = w_k$ where $v = w_1 + \cdots + w_r$, $w_i \in W_i$.   Show that E is linear.

▌   Suppose, for $u \in V$, $u = w_1' + \cdots + w_r'$, $w_i' \in W_i$.   Then $v + u = (w_1 + w_1') + \cdots + (w_r + w_r')$ and $kv = kw_1 + \cdots + kw_r$, $kw_i, w_i + w_i' \in W_i$,   are the unique sums corresponding to $v + u$ and kv.   Hence $E(v + u) = w_k + w_k' = E(v) + E(u)$ and $E(kv) = kw_k = kE(v)$ and therefore E is linear.

17.32   Show that $E^2 = E$ for the above projection map E.

▌   First we have that $w_k = 0 + \cdots + 0 + w_k + 0 + + \cdots + 0$ is the unique sum corresponding to $w_k \in W_k$; hence $E(w_k) = w_k$.   Then for any $v \in V$, $E^2(v) = E(E(v)) = E(w_k) = w_k = E(v)$.   Thus $E^2 = E$,   as required.

Theorem 17.4:   Suppose $E: V \to V$ is linear and $E^2 = E$.   Then
    (i) $E(u) = u$ for any $u \in \operatorname{Im} E$.
    (ii) $V = \operatorname{Im} E \oplus \operatorname{Ker} E$.
    (iii) E is the projection of V into $\operatorname{Im} E$.

Remark: By this theorem and Problems 17.31 and 17.32, a linear mapping $T: V \to V$ is a projection if and only if $T^2 = T$;   This characterization of a projection is frequently used as its definition.

17.33   Prove (i) of Theorem 17.4.

▌   If $u \in \operatorname{Im} E$,   then there exists $v \in V$ for which $E(v) = u$;   hence $E(u) = E(E(v)) = E^2(v) = E(v) = u$ as required.

17.34   Prove (ii) of Theorem 17.4.

▌   Let $v \in V$.   We can write v in the form $v = E(v) + v - E(v)$.   Now $E(v) \in \operatorname{Im} E$ and, since $E(v - E(v)) = E(v) - E^2(v) = E(v) - E(v) = 0$, $v - E(v) \in \operatorname{Ker} E$.   Accordingly, $V = \operatorname{Im} E + \operatorname{Ker} E$.
    Now suppose $w \in \operatorname{Im} E \cap \operatorname{Ker} E$.   By (i) of Theorem 17.4, $E(w) = w$ because $w \in \operatorname{Im} E$.   On the other hand, $E(w) = 0$ because $w \in \operatorname{Ker} E$.   Thus $w = 0$ and so $\operatorname{Im} E \cap \operatorname{Ker} E = \{0\}$.   These two conditions imply that V is the direct sum of the image and kernel of E.

17.35   Prove (iii) of Theorem 17.4.

▌   Let $v \in V$ and suppose $v = u + w$ where $u \in \operatorname{Im} E$ and $w \in \operatorname{Ker} E$.   Note $E(u) = u$ by (i) of Theorem 17.4, and $E(w) = 0$ because $w \in \operatorname{Ker} E$.   Hence $E(v) = E(u + w) = E(u) + E(w) = u + 0 = u$. That is, E is the projection of V into its image.

17.36   Suppose $E: V \to V$ is a projection, that is, $E^2 = E$.   Show that $I - E$ is a projection.

▌ $(I - E)^2 = (I - 2E + E^2) = (I - 2E + E) = I - E$.   Thus $I - E$ is a projection.

17.3   INVARIANT DIRECT-SUM DECOMPOSITIONS

17.37   Define an invariant direct-sum decomposition of a vector space with respect to a linear operator.

▌   Let $T: V \to V$ be linear.   Suppose V is the direct sum of [nonzero] T-invariant subspaces $W_1, \ldots, W_r$;   i.e., suppose $V = W_1 \oplus \cdots \oplus W_r$ and $T(W_i) \subseteq W_i$, $i = 1, \ldots, r$.   Then the subspaces $W_1, \ldots, W_r$ are said to reduce T or to form a T-invariant direct-sum decomposition of V.   Furthermore, if T_i is the restriction of T to W_i, then T is said to be decomposable into the operators T_i, or T is said to be the direct sum of the T_i, written $T = T_1 \oplus \cdots \oplus T_r$.

17.38   Let $T: \mathbf{R}^3 \to \mathbf{R}^3$ be the linear operator which rotates each vector about the z axis by an angle θ [as pictured in Fig. 17-1]; i.e., $T(x, y, z) = (x \cos \theta - y \sin \theta, x \sin \theta + y \cos \theta, z)$.   Show that the xy plane W and the z axis U form a T-invariant direct-sum decomposition of $\mathbf{R}^3$.

❚ Note first that $\mathbf{R}^3 = W \oplus U$ since the only way that $v = (a, b, c)$ in $\mathbf{R}^3$ can be written as the sum of a vector in W and a vector in U is as follows: $(a, b, c) = (a, b, 0) + (0, 0, c)$. Furthermore, W and U are invariant under T. Thus W and U form a T-invariant direct-sum decomposition of $\mathbf{R}^3$.

The following three theorems [proved in Problems 17.39, 17.44, and 17.45] indicate the main content of this section.

Theorem 17.5: Suppose $T: V \rightarrow V$ is linear and V is the direct sum of T-invariant subspaces $W_1, \ldots, W_r$. If A_i is a matrix representation of the restriction of T to W_i, then T can be represented by the block diagonal matrix

$$M = \begin{pmatrix} A_1 & 0 & \cdots & 0 \\ 0 & A_2 & \cdots & 0 \\ \cdots\cdots\cdots\cdots\cdots \\ 0 & 0 & \cdots & A_r \end{pmatrix}$$

Theorem 17.6 [Primary Decomposition Theorem]: Let $T: V \rightarrow V$ be a linear operator with minimal polynomial $m(t) = f_1(t)^{n_1} f_2(t)^{n_2} \cdots f_r(t)^{n_r}$, where the $f_i(t)$ are distinct monic irreducible polynomials. Then V is the direct sum of T-invariant subspaces $W_1, \ldots, W_r$ where W_i is the kernel of $f_i(T)^{n_i}$. Moreover, $f_i(t)^{n_i}$ is the minimal polynomial of the restriction of T to W_i.

Theorem 17.7: A linear operator $T: V \rightarrow V$ has a diagonal matrix representation if and only if its minimal polynomial $m(t)$ is a product of distinct linear polynomials.

Theorem 17.8 [Alternate Form of Theorem 17.7]: A matrix A is similar to a diagonal matrix if and only if its minimal polynomial is a product of distinct linear polynomials.

Remark: Theorem 17.8 is a useful characterization of diagonalizable operators, e.g., see Problem 17.46.

17.39 Suppose $T: V \rightarrow V$ is linear and $V = U \oplus W$ is a T-invariant direct-sum decomposition of V. Prove Theorem 17.5 in the case that $\dim U = 2$ and $\dim W = 3$.

❚ Suppose $\{u_1, u_2\}$ and $\{w_1, w_2, w_3\}$ are bases of U and W, respectively. If T_1 and T_2 denote the restrictions of T to U and W, respectively, then

$$\begin{array}{ll} T_1(u_1) = a_{11} u_1 + q_{12} u_2 & \begin{array}{l} T_2(w_1) = b_{11} w_1 + b_{12} w_2 + b_{13} w_3 \\ T_1(u_2) = a_{21} u_1 + a_{22} u_2 & T_2(w_2) = b_{21} w_1 + b_{22} w_2 + b_{23} w_3 \\ & T_2(w_3) = b_{31} w_1 + b_{32} w_2 + b_{33} w_3 \end{array} \end{array}$$

Hence

$$A = \begin{pmatrix} a_{11} & a_{21} \\ a_{12} & a_{22} \end{pmatrix} \quad \text{and} \quad B = \begin{pmatrix} b_{11} & b_{21} & b_{31} \\ b_{12} & b_{22} & b_{32} \\ b_{13} & b_{23} & b_{33} \end{pmatrix}$$

are matrix representations of T_1 and T, respectively. By Theorem 17.3, $\{u_1, u_2, w_1, w_2, w_3\}$ is a basis of V. Since $T(u_i) = T_1(u_i)$ and $T(w_j) = T_2(w_j)$, the matrix of T in this basis is the block diagonal matrix $\begin{pmatrix} A & 0 \\ 0 & B \end{pmatrix}$.

Remark: The proof of Theorem 17.5 is exactly the same as the above proof and will be omitted.

17.40 Suppose $T: V \rightarrow V$ is linear and suppose $T = T_1 \oplus T_2$ with respect to a T-invariant direct-sum decomposition $V = U \oplus W$. Let $m(t)$, $m_1(t)$, and $m_2(t)$ denote, respectively, the minimum polynomials of T, T_1, and T_2. Show that $m(t)$ is the least common multiple of $m_1(t)$ and $m_2(t)$.

❚ By Problem 17.16, each of $m_1(t)$ and $m_2(t)$ divides $m(t)$. Now suppose $f(t)$ is a multiple of both $m_1(t)$ and $m_2(t)$; then $f(T_1)(U) = 0$ and $f(T_2)(W) = 0$. Let $v \in V$; then $v = u + w$ with $u \in U$ and $w \in W$. Now $f(T)v = f(T)u + f(T)w = f(T_1)u + f(T_2)w = 0 + 0 = 0$. That is, T is a zero of $f(t)$. Hence $m(t)$ divides $f(t)$, and so $m(t)$ is the least common multiple of $m_1(t)$ and $m_2(t)$.

17.41 In the above problem, let $\Delta(t)$, $\Delta_1(t)$, and $\Delta_2(t)$ denote, respectively, the characteristic polynomials of T, T_1, and T_2. Show that $\Delta(t) = \Delta_1(t)\Delta_2(t)$.

▌ By Theorem 17.5, T has a matrix representation $M = \begin{pmatrix} A & 0 \\ 0 & B \end{pmatrix}$ where A and B are matrix representations of T_1 and T_2, respectively. Then,

$$\Delta(t) = |tI - M| = \begin{vmatrix} tI - A & 0 \\ 0 & tI - B \end{vmatrix} = |tI - A|\,|tI - B| = \Delta_1(t)\Delta_2(t)$$

as required.

Theorem 17.9: Suppose $T: V \to V$ is linear, and suppose $f(t) = g(t)h(t)$ are polynomials such that $f(T) = 0$ and $g(t)$ and $h(t)$ are relatively prime. Then V is the direct sum of the T-invariant subspaces U and W, where $U = \operatorname{Ker} g(T)$ and $W = \operatorname{Ker} H(T)$.

17.42 Prove Theorem 17.9.

▌ Note first that U and W are T-invariant by Theorem 17.1. Since $g(T)$ and $h(T)$ are relatively prime, there exist polynomials $r(t)$ and $s(t)$ such that $r(t)g(t) + s(t)h(t) = 1$. Hence for the operator T,

$$r(T)g(T) + s(T)h(T) = I \tag{1}$$

Let $v \in V$; then by *(1)*, $v = r(T)g(T)v + s(T)h(T)v$. But the first term in this sum belongs to $W = \operatorname{Ker} h(T)$ since $h(T)r(T)g(T)v = r(T)g(T)h(T)v = r(T)f(T)v = r(T)0v = 0$. Similarly, the second term belongs to U. Hence V is the sum of U and W.

To prove that $V = U \oplus W$, we must show that a sum $v = u + w$, with $u \in U$, $w \in W$, is uniquely determined by v. Applying the operator $r(T)g(T)$ to $v = u + w$ and using $g(T)u = 0$, we obtain $r(T)g(T)v = r(T)g(T)u + r(T)g(T)w = r(T)g(T)w$. Also, applying *(1)* to w alone and using $h(T)w = 0$, we obtain $w = r(T)g(T)w + s(T)h(T)w = r(T)g(T)w$. Both of the above formulas give us $w = r(T)g(T)v$ and so w is uniquely determined by v. Similarly u is uniquely determined by v. Hence $V = U \oplus W$, as required.

Theorem 17.10: Suppose in Theorem 17.9 that $f(t)$ is the minimum polynomial of T [and $g(t)$ and $h(t)$ are monic]. Then $g(t)$ and $h(t)$ are the minimum polynomials of T_1 and T_2, respectively [where T_1 is the restriction of T to U and T_2 is the restriction of T to W].

17.43 Prove Theorem 17.10.

▌ Let $m_1(t)$ and $m_2(t)$ be the minimal polynomials of T_1 and T_2, respectively. Note that $g(T_1) = 0$ and $h(T_2) = 0$ because $U = \operatorname{Ker} g(t)$ and $W = \operatorname{Ker} h(t)$. Thus

$$m_1(t) \text{ divides } g(t) \qquad \text{and} \qquad m_2(t) \text{ divides } h(t) \tag{1}$$

By Problem 17.40, $f(t)$ is the least common multiple of $m_1(t)$ and $m_2(t)$. But $m_1(t)$ and $m_2(t)$ are relatively prime since $g(t)$ and $h(t)$ are relatively prime. Accordingly, $f(t) = m_1(t)m_2(t)$. We also have that $f(t) = g(t)h(t)$. These two equations together with *(1)* and the fact that all the polynomials are monic imply that $g(t) = m_1(t)$ and $h(t) = m_2(t)$, as required.

17.44 Prove the Primary Decomposition Theorem 17.6.

▌ The proof is by induction on r. The case $r = 1$ is trivial. Suppose that the theorem has been proved for $r - 1$. By Theorem 17.9 we can write V as the direct sum of T-invariant subspaces W_1 and V_1 where W_1 is the kernel of $f_1(T)^{n_1}$ and where V_1 is the kernel of $f_2(T)^{n_2} \cdots f_r(T)^{n_r}$. By Theorem 17.10, the minimal polynomial of the restrictions of T to W_1 and V_1 are, respectively, $f_1(t)^{n_1}$ and $f_2(t)^{n_2} \cdots f_r(t)^{n_r}$.

Denote the restriction of T to V_1 by T_1. By the inductive hypothesis, V_1 is the direct sum of subspaces $W_2, \ldots, W_r$ such that W_i is the kernel of $f_i(T_1)^{n_i}$ and such that $f_i(T)^{n_i}$ is the minimal polynomial for the restriction of T_1 to W_i. But the kernel of $f_i(T)^{n_i}$, for $i = 2, \ldots, r$, is necessarily contained in V_1 since $f_i(t)^{n_i}$ divides $f_2(t)^{n_2} \cdots f_r(t)^{n_r}$. Thus the kernel of $f_i(T)^{n_i}$ is the same as the kernel of $f_i(T_1)^{n_i}$, which is W_i. Also, the restriction of T to W_i is the same as the restriction of T_1 to W_i (for $i = 2, \ldots, r$); hence $f_i(t)^{n_i}$ is also the minimal polynomial for the restriction of T to W_i. Thus $V = W_1 \oplus W_2 \oplus \cdots \oplus W_r$ is the desired decomposition of T.

17.45 Prove Theorem 17.7.

❚ Suppose $m(t)$ is a product of distinct linear polynomials; say, $m(t) = (t - \lambda_1)(t - \lambda_2) \cdots (t - \lambda_r)$ where the λ_i are distinct scalars. By the Primary Decomposition theorem, V is the direct sum of subspaces $W_1, \ldots, W_r$ where $W_i = \text{Ker}(T - \lambda_i I)$. Thus if $v \in W_i$, then $(T - \lambda_i I)(v) = 0$ or $T(v) = \lambda_i v$. In other words, every vector in W_i is an eigenvector belonging to the eigenvalue λ_i. By Theorem 10.4, the union of bases for $W_1, \ldots, W_r$ is a basis of V. This basis consists of eigenvectors and so T is diagonalizable.

Conversely, suppose T is diagonalizable, i.e., V has a basis consisting of eigenvectors of T. Let $\lambda_1, \ldots, \lambda_s$ be the distinct eigenvalues of T. Then the operator $f(T) = (T - \lambda_1 I)(T - \lambda_2 I) \cdots (T - \lambda_s I)$ maps each basis vector into 0. Thus $f(T) = 0$ and hence the minimum polynomial $m(t)$ of T divides the polynomial $f(t) = (t - \lambda_1)(t - \lambda_2) \cdots (t - \lambda_s)$. Accordingly, $m(t)$ is a product of distinct linear polynomials.

17.46 Suppose $A \neq I$ is a square matrix for which $A^3 = I$. Determine whether or not A is similar to a diagonal matrix if A is a matrix over (i) the real field **R**, (ii) the complex field **C**.

❚ Since $A^3 = I$, A is a zero of the polynomial $f(t) = t^3 - 1 = (t - 1)(t^2 + t + 1)$. The minimal polynomial $m(t)$ of A cannot be $t - 1$, since $A \neq I$. Hence $m(t) = t^2 + t + 1$ or $m(t) = t^3 - 1$. Since neither polynomial is a product of linear polynomials over **R**, A is not diagonalizable over **R**. On the other hand, each of the polynomials is a product of distinct linear polynomials over **C**. Hence A is diagonalizable over **C**.

17.4 NILPOTENT OPERATORS AND MATRICES

17.47 Define a nilpotent operator and a nilpotent matrix.

❚ A linear operator $T: V \to V$ is termed *nilpotent* if $T^n = 0$ for some positive integer n; we call k the *index of nilpotency* of T if $T^k = 0$ but $T^{k-1} \neq 0$. Analogously, a square matrix A is termed *nilpotent* if $A^n = 0$ for some positive integer n, and of *index* k if $A^k = 0$ but $A^{k-1} \neq 0$.

Problems 17.48–17.51 refer to an n-square nilpotent matrix A of index k.

17.48 What is the minimum polynomial $m(t)$ of A?

❚ Since $A^k = 0$, but $A^{k-1} \neq 0$, we have $m(t) = t^k$.

17.49 Find the eigenvalues of A.

❚ Since $m(t) = t^k$ is the minimum polynomial of A, only 0 is an eigenvalue of A.

17.50 Show that $k \leq n$, i.e., that the index of A does not exceed its order.

❚ Since the degree of the characteristic polynomial $\Delta(t)$ of A is n, $k = \deg m(t) \leq \deg \Delta(t) = n$. Thus $k \leq n$.

17.51 Show that A is singular.

❚ Since $A^k = 0$, we have A^k is singular. Recall that the product of nonsingular matrices is nonsingular; hence A must be singular.

Problems 17.52–17.5 refer to the following matrices:

$$A = \begin{pmatrix} -2 & 1 & 1 \\ -3 & 1 & 2 \\ -2 & 1 & 1 \end{pmatrix} \qquad B = \begin{pmatrix} 1 & 3 & -2 \\ 1 & 3 & -2 \\ 1 & 3 & -2 \end{pmatrix} \qquad C = \begin{pmatrix} 1 & -3 & 2 \\ 1 & -3 & 2 \\ 1 & -3 & 2 \end{pmatrix}$$

17.52 Is A nilpotent? If yes, what is its index?

❚ Compute $$A^2 = \begin{pmatrix} -1 & 0 & 1 \\ -1 & 0 & 1 \\ -1 & 0 & 1 \end{pmatrix} \qquad A^3 = 0$$

Thus A is nilpotent of index 3.

17.53 Is B nilpotent? If yes, what is its index?

❚ Compute $$B^2 = \begin{pmatrix} 2 & 6 & -4 \\ 2 & 6 & -4 \\ 2 & 6 & -4 \end{pmatrix} \qquad B^3 = \begin{pmatrix} 4 & 12 & -8 \\ 4 & 12 & -8 \\ 4 & 12 & -8 \end{pmatrix}$$

Thus B is not nilpotent. [We do not need to test higher than the order of B.]

17.54 Is C nilpotent? If yes, what is its index.

❚ Compute $C^2 = 0$. Thus C is nilpotent of index 2.

17.55 Define a basic nilpotent block N of index k.

❚ N is the k-square matrix with 1s on the superdiagonal and 0s elsewhere, i.e.,

$$N = \begin{pmatrix} 0 & 1 & 0 & \cdots & 0 & 0 \\ 0 & 0 & 1 & \cdots & 0 & 0 \\ \cdots & \cdots & \cdots & \cdots & \cdots & \cdots \\ 0 & 0 & 0 & \cdots & 0 & 1 \\ 0 & 0 & 0 & \cdots & 0 & 0 \end{pmatrix}$$

[*Remark*: The fact that N is nilpotent of index k is proved in Problem 17.67.]

17.56 Write down the basic nilpotent blocks of orders 1, 2, 3, and 4.

❚ The matrices follow:

$$(0) \qquad \begin{pmatrix} 0 & 1 \\ 0 & 0 \end{pmatrix} \qquad \begin{pmatrix} 0 & 1 & 0 \\ 0 & 0 & 1 \\ 0 & 0 & 0 \end{pmatrix} \qquad \begin{pmatrix} 0 & 1 & 0 & 0 \\ 0 & 0 & 1 & 0 \\ 0 & 0 & 0 & 1 \\ 0 & 0 & 0 & 0 \end{pmatrix}$$

Note that the basic nilpotent block N of order 1 is simply the 1×1 zero matrix.

The main content of this section is the following fundamental theorem [proved in Problem 17.70] on nilpotent operators.

Theorem 17.11: Let $T: V \to V$ be a nilpotent operator of index k. Then T has a block diagonal matrix representation, say,

$$M = \begin{pmatrix} N_1 & & & \\ & N_2 & & \\ & & \cdots & \\ & & & N_m \end{pmatrix}$$

such that each diagonal entry N_i is a basic nilpotent block. Also
(i) There is at least one block N of order k and all other N are of order $\le k$.
(ii) The number of N of each possible order is uniquely determined by T.
(iii) The number m of blocks N is equal to the nullity of T.

Theorem 17.12 [*Alternate Form of Theorem 17.11*]: Every nilpotent matrix A is similar to a unique nilpotent matrix M in the above form.

Remark: The above matrix M is called a *canonical nilpotent matrix*, and M is called the *canonical form* of T and of A. Two such canonical matrices M are assumed to be equal if they have the same set of diagonal blocks [i.e., the orders of the blocks may differ].

17.57 Describe all canonical nilpotent matrices of order 3.

❚ Such canonical matrices of index 1, 2, and 3 follow:

$$\text{Index 1:} \begin{pmatrix} 0 & & \\ & 0 & \\ & & 0 \end{pmatrix} \qquad \text{Index 2:} \begin{pmatrix} 0 & 1 & \\ 0 & 0 & \\ & & 0 \end{pmatrix} \qquad \text{Index 3:} \begin{pmatrix} 0 & 1 & 0 \\ 0 & 0 & 1 \\ 0 & 0 & 0 \end{pmatrix}$$

17.58 Show that there are two nonsimilar canonical nilpotent matrices of order 4 and index 2.

▌ The two such matrices follow:

$$\left(\begin{array}{cc|cc} 0 & 1 & & \\ 0 & 0 & & \\ \hline & & 0 & \\ & & & 0 \end{array}\right) \qquad \left(\begin{array}{cc|cc} 0 & 1 & & \\ 0 & 0 & & \\ \hline & & 0 & 1 \\ & & 0 & 0 \end{array}\right)$$

17.59 Find the canonical nilpotent form of matrix A in Problem 17.52.

▌ Since the index of A is 3, its canonical form follows:

$$\begin{pmatrix} 0 & 1 & 0 \\ 0 & 0 & 1 \\ 0 & 0 & 0 \end{pmatrix}$$

17.60 Find the canonical nilpotent form of matrix B in Problem 17.53.

▌ B is not nilpotent; hence it is not similar to any canonical nilpotent matrix.

17.61 Find the canonical nilpotent form of matrix C in Problem 17.54.

▌ Since the index of C is 2, its canonical form follows:

$$\left(\begin{array}{cc|c} 0 & 1 & \\ 0 & 0 & \\ \hline & & 0 \end{array}\right)$$

17.62 Let $A = \begin{pmatrix} 0 & 1 & 1 & 0 & 1 \\ 0 & 0 & 1 & 1 & 1 \\ 0 & 0 & 0 & 0 & 0 \\ 0 & 0 & 0 & 0 & 0 \\ 0 & 0 & 0 & 0 & 0 \end{pmatrix}$. Is A nilpotent? If yes, what is its index?

▌ Compute $\qquad A^2 = \begin{pmatrix} 0 & 0 & 1 & 1 & 1 \\ 0 & 0 & 0 & 0 & 0 \\ 0 & 0 & 0 & 0 & 0 \\ 0 & 0 & 0 & 0 & 0 \\ 0 & 0 & 0 & 0 & 0 \end{pmatrix}$

and $A^3 = 0$. Thus A is nilpotent of index 3.

17.63 Find the canonical form M of the above matrix A.

▌ Since A is nilpotent of index 3, M contains a diagonal block of order 3 and none greater than 3. There are two possibilities for the other diagonal blocks, one 2×2 block, or two 1×1 blocks. Since rank $A = 2$; the nullity of $A = 5 - 2 = 3$. Thus M must contain three diagonal blocks. Thus M must contain one diagonal block of order 3 and two of order 1; i.e.,

$$M = \left(\begin{array}{ccc|c|c} 0 & 1 & 0 & & \\ 0 & 0 & 1 & & \\ 0 & 0 & 0 & & \\ \hline & & & 0 & \\ \hline & & & & 0 \end{array}\right)$$

Lemma 17.13: Let $T: V \to V$ be linear. Suppose, for $v \in V$, $T^k(v) = 0$ but $T^{k-1}(v) \neq 0$.
 (i) The set $S = \{v, T(v), \ldots, T^{k-1}(v)\}$ is linearly independent.
 (ii) The subspace W generated by S is T-invariant.
 (iii) The restriction $\hat{T}$ of T to W is nilpotent of index k.
 (iv) Relative to the basis $\{T^{k-1}(v), \ldots, T(v), v\}$ of W, the matrix of $\hat{T}$ is the following k-square canonical matrix N:

$$N = \begin{pmatrix} 0 & 1 & 0 & \cdots & 0 & 0 \\ 0 & 0 & 1 & \cdots & 0 & 0 \\ \cdots & \cdots & \cdots & \cdots & \cdots & \cdots \\ 0 & 0 & 0 & \cdots & 0 & 1 \\ 0 & 0 & 0 & \cdots & 0 & 0 \end{pmatrix}$$

Thus the k-square matrix N is nilpotent of index k.

17.64 Prove (i) of Lemma 17.13.

▎ Suppose

$$av + a_1 T(v) + a_2 T^2(v) + \cdots + a_{k-1} T^{k-1}(v) = 0 \qquad (1)$$

Applying T^{k-1} to *(1)* and using $T^k(v) = 0$, we obtain $aT^{k-1}(v) = 0$; since $T^{k-1}(v) \neq 0$, $a = 0$. Now applying T^{k-2} to *(1)* and using $T^k(v) = 0$ and $a = 0$, we find $a_1 T^{k-1}(v) = 0$; hence $a_1 = 0$. Next applying T^{k-3} to *(1)* and using $T^k(v) = 0$ and $a = a_1 = 0$, we obtain $a_2 T^{k-1}(v) = 0$; hence $a_2 = 0$. Continuing this process, we find that all the a's are 0; hence S is independent.

17.65 Prove (ii) of Lemma 17.13.

▎ Let $v \in W$. Then $v = bv + b_1 T(v) + b_2 T^2(v) + \cdots + b_{k-1} T^{k-1}(v)$. Using $T^k(v) = 0$, we have that $T(v) = bT(v) + b_1 T^2(v) + \cdots + b_{k-2} T^{k-1}(v) \in W$. Thus W is T-invariant.

17.66 Prove (iii) of Lemma 17.13.

▎ By hypothesis $T^k(v) = 0$. Hence, for $i = 0, \ldots, k-1$, $\hat{T}^k(T^i(v)) = T^{k+i}(v) = 0$. That is, applying $\hat{T}^k$ to each generator of W, we obtain 0; hence $\hat{T}^k = 0$ and so $\hat{T}$ is nilpotent of index at most k. On the other hand, $\hat{T}^{k-1}(v) = T^{k-1}(v) \neq 0$; hence T is nilpotent of index exactly k.

17.67 Prove (iv) of Lemma 17.13.

▎ For the basis $\{T^{k-1}(v), T^{k-2}(v), \ldots, T(v), v\}$ of W,

$$\hat{T}(T^{k-1}(v)) = T^k(v) = 0$$
$$\hat{T}(T^{k-2}(v)) = \qquad\qquad T^{k-1}(v)$$
$$\hat{T}(T^{k-3}(v)) = \qquad\qquad\qquad T^{k-2}(v)$$
$$\cdots\cdots\cdots\cdots\cdots\cdots\cdots\cdots\cdots\cdots\cdots\cdots$$
$$\hat{T}(T(v)) = \qquad\qquad\qquad T^2(v)$$
$$\hat{T}(v) = \qquad\qquad\qquad\qquad T(v)$$

Hence the matrix of $\hat{T}$ in this basis is N.

17.68 Let $T : V \to V$ be linear. Let $U = \text{Ker } T^i$ and $W = \text{Ker } T^{i+1}$. Show that (i) $U \subset W$, (ii) $T(W) \subset U$.

▎ (i) Suppose $u \in U = \text{Ker } T^i$. Then $T^i(u) = 0$ and so $T^{i+1}(u) = T(T^i(u)) = T(0) = 0$. Thus $u \in \text{Ker } T^{i+1} = W$. But this is true for every $u \in U$; hence $U \subset W$.

(ii) Similarly, if $w \in W = \text{Ker } T^{i+1}$, then $T^{i+1}(w) = 0$. Thus $T^{i+1}(W) = T^i(T(w)) = T^i(0) = 0$ and so $T(W) \subset U$.

17.69 Let $T : V \to V$ be linear. Let $X = \text{Ker } T^{i-2}$, $Y = \text{Ker } T^{i-1}$, and $Z = \text{Ker } T^i$. By the preceding problem, $X \subset Y \subset Z$. Suppose $\{u_1, \ldots, u_r\}$, $\{u_1, \ldots, u_r, v_1, \ldots, v_s\}$, $\{u_1, \ldots, u_r, v_1, \ldots, v_s, w_1, \ldots, w_t\}$ are bases of X, Y, and Z, respectively. Show that $S = \{u_1, \ldots, u_r, T(w_1), \ldots, T(w_t)\}$ is contained in Y and is linearly independent.

▎ By the preceding problem, $T(Z) \subset Y$ and hence $S \subset Y$. Now suppose S is linearly dependent. Then there exists a relation $a_1 u_1 + \cdots + a_r u_r + b_1 T(w_1) + \cdots + b_t T(w_t) = 0$ where at least one coefficient is not zero. Furthermore, since $\{u_i\}$ is independent, at least one of the b_k must be nonzero. Transposing, we find $b_1 T(w_1) + \cdots + b_t T(w_t) = -a_1 u_1 - \cdots - a_r u_r \in X = \text{Ker } T^{i-2}$. Hence $T^{i-2}(b_1 T(w_1) + \cdots + b_t T(w_t)) = 0$. Thus $T^{i-1}(b_1 w_1 + \cdots + b_t w_t) = 0$ and so $b_1 w_1 + \cdots + b_t w_t \in Y = \text{Ker } T^{i-1}$. Since

$\{u_i, v_j\}$ generates Y, we obtain a relation among the u_i, v_j, and w_k where one of the coefficients, i.e., one of the b_k, is not zero. This contradicts the fact that $\{u_i, v_j, w_k\}$ is independent. Hence S must also be independent.

17.70 Prove Theorem 17.11. Let $T: V \to V$ be a nilpotent operator of index k. Then T has a block diagonal matrix representation whose diagonal entries are of the form

$$
N = \begin{pmatrix}
0 & 1 & 0 & \cdots & 0 & 0 \\
0 & 0 & 1 & \cdots & 0 & 0 \\
\cdots & \cdots & \cdots & \cdots & \cdots & \cdots \\
0 & 0 & 0 & \cdots & 0 & 1 \\
0 & 0 & 0 & \cdots & 0 & 0
\end{pmatrix}
$$

There is at least one N of order k and all other N are of orders $\leq k$. The number of N of each possible order is uniquely determined by T. Moreover, the total number of N of all orders is the nullity of T.

■ Suppose $\dim V = n$. Let $W_1 = \operatorname{Ker} T$, $W_2 = \operatorname{Ker} T^2, \ldots, W_k = \operatorname{Ker} T^k$. Set $m_i = \dim W_i$, for $i = 1, \ldots, k$. Since T is of index k, $W_k = V$ and $W_{k-1} \neq V$ and so $m_{k-1} < m_k = n$. By Problem 10.17, $W_1 \subset W_2 \subset \cdots \subset W_k = V$. Thus, by induction, we can choose a basis $\{u_1, \ldots, u_n\}$ of V such that $\{u_1, \ldots, u_{m_i}\}$ is a basis of W_i.

We now choose a new basis for V with respect to which T has the desired form. It will be convenient to label the members of this new basis by pairs of indices. We begin by setting $v(1, k) = u_{m_{k-1}+1}$, $v(2, k) = u_{m_{k-1}+2}, \ldots, v(m_k - m_{k-1}, k) = u_{m_k}$ and setting $v(1, k-1) = Tv(1, k)$, $v(2, k-1) = Tv(2, k), \ldots, v(m_k - m_{k-1}, k-1) = Tv(m_k - m_{k-1}, k)$. By the preceding problem, $S_1 = \{u_1, \ldots, u_{m_{k-2}}, v(1, k-1), \ldots, v(m_k - m_{k-1}, k-1)\}$ is a linearly independent subset of W_{k-1}. We extend S_1 to a basis of W_{k-1} by adjoining new elements [if necessary] which we denote by $v(m_k - m_{k-1} + 1, k-1), v(m_k - m_{k-1} + 2, k-1), \ldots, v(m_{k-1} - m_{k-2}, k-1)$. Next we set $v(1, k-2) = Tv(1, k-1), v(2, k-2) = Tv(2, k-1), \ldots, v(m_{k-1} - m_{k-2}, k-2) = Tv(m_{k-1} - m_{k-2}, k-1)$. Again by the preceding problem, $S_2 = \{u_1, \ldots, u_{m_{k-3}}, v(1, k-2), \ldots, v(m_{k-1} - m_{k-2}, k-2)\}$ is a linearly independent subset of W_{k-2} which we can extend to a basis of W_{k-2} by adjoining elements $v(m_{k-1} - m_{k-2} + 1, k-2), v(m_{k-1} - m_{k-2} + 2, k-2), \ldots, v(m_{k-2} - m_{k-3}, k-2)$. Continuing in this manner we get a new basis for V which, for convenient reference, we arrange as follows:

$$
\begin{aligned}
&v(1, k), && \ldots, v(m_k - m_{k-1}, k) \\
&v(1, k-1), & \ldots, & v(m_k - m_{k-1}, k-1), \ldots, v(m_{k-1} - m_{k-2}, k-1) \\
&\cdots \cdots \cdots \\
&v(1, 2), && \ldots, v(m_k - m_{k-1}, 2), \quad \ldots, v(m_{k-1} - m_{k-2}, 2), \ldots, v(m_2 - m_1, 2) \\
&v(1, 1), && \ldots, v(m_k - m_{k-1}, 1), \quad \ldots, v(m_{k-1} - m_{k-2}, 1), \ldots, v(m_2 - m_1, 1), \ldots, v(m_1, 1)
\end{aligned}
$$

The bottom row forms a basis of W_1, the bottom two rows form a basis of W_2, etc. But what is important for us is that T maps each vector into the vector immediately below it in the table or into 0 if the vector is in the bottom row. That is,

$$
Tv(i, j) = \begin{cases} v(i, j-1) & \text{for } j > 1 \\ 0 & \text{for } j = 1 \end{cases}
$$

Now it is clear that T will have the desired form if the $v(i, j)$ are ordered lexicographically: beginning with $v(1, 1)$ and moving up the first column to $v(1, k)$, then jumping to $v(2, 1)$ and moving up the second column as far as possible, etc.

Moreover, there will be exactly

$$
\begin{array}{ll}
m_k - m_{k-1} & \text{diagonal entries of order } k \\
(m_{k-1} - m_{k-2}) - (m_k - m_{k-1}) = 2m_{k-1} - m_k - m_{k-2} & \text{diagonal entries of order } k-1 \\
\cdots \cdots \cdots \\
2m_2 - m_1 - m_3 & \text{diagonal entries of order } 2 \\
2m_1 - m_2 & \text{diagonal entries of order } 1
\end{array}
$$

as can be read off directly from the table. In particular, since the numbers $m_1, \ldots, m_k$ are uniquely determined by T, the number of diagonal entries of each order is uniquely determined by T. Finally, the identity $m_1 = (m_k - m_{k-1}) + (2m_{k-1} - m_k - m_{k-2}) + \cdots + (2m_2 - m_1 - m_3) + (2m_1 - m_2)$ shows that the nullity m_1 of T is the total number of diagonal entries of T.

17.71 Suppose A is nilpotent of index k. Show that A^T and cA, $c^k \neq 0$, are nilpotent of index k.

❚ We have $A^k = 0$ if and only if $(A^T)^k = (a^k)^T = 0^T = 0$. Thus A^T is also nilpotent of index k. Also, $A^k = 0$ if and only if $(cA)^k = c^kA^k = 0$. Thus cA is also nilpotent of index k.

17.72 Suppose nilpotent matrices A and B commute, i.e., $AB = BA$. Show that AB is nilpotent.

❚ Suppose $A^m = 0$ and $B^n = 0$ and, say, $m \le n$. Then $(AB)^n = A^nB^n = A^n0 = 0$. Thus AB is nilpotent.

17.73 Suppose nilpotent matrices A and B commute, i.e., show that $A + B$ is nilpotent.

❚ Suppose $A^m = 0$ and $B^n = 0$. Since A and B commute,

$$(A + B)^{m+n} = \sum_{i+1}^{m+n} \binom{m+n}{i} A^iB^{m+n-i}$$

If $i \ge m$, then $A^i = 0$. If $i < m$, then $m + n - i \ge n$ and hence $B^{m+n-i} = 0$. Thus each term in the expansion of $(A + B)^{m+n}$ is equal to 0. Accordingly, $(A + B)^{m+n} = 0$ and $A + B$ is nilpotent.

17.74 Suppose A is nilpotent of index k. Show that $A^n, n > 1$, is nilpotent of index $\le k$.

❚ Since $A^k = 0$, we have $(A^n)^k = (A^k)^n = 0^n = 0$. Thus A^n is nilpotent of index $\le k$.

17.75 Suppose A and B are similar. Show that A is nilpotent of index k if and only if B is nilpotent of index k.

❚ Suppose $B = P^{-1}AP$. If $A^r = 0$, then $B^r = (P^{-1}AP)^r = P^{-1}A^rP = P^{-1}0P = 0$. Similarly, if $B^r = 0$, then $A^r = 0$. Thus A is nilpotent if and only if B is nilpotent, and, in such a case, they have the same index.

17.5 JORDAN CANONICAL FORM

17.76 Define a Jordan block J or order k belonging to the eigenvalue λ.

❚ J is the k-square matrix with λs on the diagonal, 1s on the superdiagonal, and 0s elsewhere; i.e.,

$$J = \begin{pmatrix} \lambda & 1 & 0 & \cdots & 0 & 0 \\ 0 & \lambda & 1 & \cdots & 0 & 0 \\ \cdots & \cdots & \cdots & \cdots & \cdots & \cdots \\ 0 & 0 & 0 & \cdots & \lambda & 1 \\ 0 & 0 & 0 & \cdots & 0 & \lambda \end{pmatrix}$$

17.77 Write down the Jordan blocks of orders $1, 2, 3$, and 4 belonging to the eigenvalue $\lambda = 7$.

❚ The matrices follow:

$$(7) \qquad \begin{pmatrix} 7 & 1 \\ 0 & 7 \end{pmatrix} \qquad \begin{pmatrix} 7 & 1 & 0 \\ 0 & 7 & 1 \\ 0 & 0 & 7 \end{pmatrix} \qquad \begin{pmatrix} 7 & 1 & 0 & 0 \\ 0 & 7 & 1 & 0 \\ 0 & 0 & 7 & 1 \\ 0 & 0 & 0 & 7 \end{pmatrix}$$

17.78 Show how a Jordan block J may be written as the sum of a scalar matrix and a canonical nilpotent block N.

❚ $J = \lambda I + N$ as follows:

$$\begin{pmatrix} \lambda & 1 & 0 & \cdots & 0 & 0 \\ 0 & \lambda & 1 & \cdots & 0 & 0 \\ \cdots & \cdots & \cdots & \cdots & \cdots & \cdots \\ 0 & 0 & 0 & \cdots & \lambda & 1 \\ 0 & 0 & 0 & \cdots & 0 & \lambda \end{pmatrix} = \begin{pmatrix} \lambda & 0 & \cdots & 0 & 0 \\ 0 & \lambda & \cdots & 0 & 0 \\ \cdots & \cdots & \cdots & \cdots & \cdots \\ 0 & 0 & \cdots & \lambda & 0 \\ 0 & 0 & \cdots & 0 & \lambda \end{pmatrix} + \begin{pmatrix} 0 & 1 & 0 & \cdots & 0 & 0 \\ 0 & 0 & 1 & \cdots & 0 & 0 \\ \cdots & \cdots & \cdots & \cdots & \cdots & \cdots \\ 0 & 0 & 0 & \cdots & 0 & 1 \\ 0 & 0 & 0 & \cdots & 0 & 0 \end{pmatrix}$$

Problems 17.79–17.81 refer to the following Jordan block A of order 4:

$$A = \begin{pmatrix} 7 & 1 & 0 & 0 \\ 0 & 7 & 1 & 0 \\ 0 & 0 & 7 & 1 \\ 0 & 0 & 0 & 7 \end{pmatrix}$$

17.79 What is the characteristic polynomial $\Delta(t)$ and minimum polynomial $m(t)$ of A? What are the eigenvalues of A?

▮ Both $\Delta(t)$ and $m(t)$ are equal to $(t-7)^4$; that is, $\Delta(t) = m(t) = (t-7)^4$. Thus $\lambda = 7$ is the only eigenvalue.

17.80 Find a basis for the eigenspace of the eigenvalue $\lambda = 7$.

▮ Substituting $t = 7$ in the matrix equation $tI - A = 0$ yields the following homogeneous system:

$$\begin{pmatrix} 0 & -1 & 0 & 0 \\ 0 & 0 & -1 & 0 \\ 0 & 0 & 0 & -1 \\ 0 & 0 & 0 & 0 \end{pmatrix} \begin{pmatrix} x \\ y \\ z \\ t \end{pmatrix} = \begin{pmatrix} 0 \\ 0 \\ 0 \\ 0 \end{pmatrix} \quad \text{or} \quad \begin{array}{rcl} -y & & = 0 \\ -z & & = 0 \\ -t & = 0 \\ 0 & = 0 \end{array}$$

There is only one free variable x; hence $v = (1, 0, 0, 0)$ forms a basis for the eigenspace of $\lambda = 7$.

17.81 What is the algebraic multiplicity and geometric multiplicity of the eigenvalue $\lambda = 7$?

▮ Since $\Delta(t) = (t-7)^4$, the algebraic multiplicity is 4. Since the eigenspace of $\lambda = 7$ has dimension one, the geometric multiplicity of A is 1.

17.82 Define a Jordan matrix M.

▮ A matrix M is a Jordan matrix if M is a block diagonal matrix whose diagonal blocks, say, $J_1, J_2, \ldots, J_r$, are Jordan blocks. [We emphasize that more than one diagonal block may belong to the same eigenvalue.]

17.83 Define equivalent Jordan matrices.

▮ A Jordan matrix M_2 is equivalent to a Jordan matrix M_1 if M_2 can be obtained from M_1 by rearranging the diagonal blocks.

Remark: We usually do not distinguish between equivalent Jordan matrices. In particular, the term "unique Jordan form" means unique up to equivalence.

Problems 17.84–17.87 refer to the following Jordan matrix:

$$M = \begin{pmatrix} -3 & 1 & 0 & & & & & \\ 0 & -3 & 1 & & & & & \\ 0 & 0 & -3 & & & & & \\ & & & 5 & 1 & & & \\ & & & 0 & 5 & & & \\ & & & & & 5 & 1 \\ & & & & & 0 & 5 \end{pmatrix}$$

17.84 Find all Jordan matrices equivalent to M.

▮ There are exactly two other ways of arranging the blocks on the diagonal as follows:

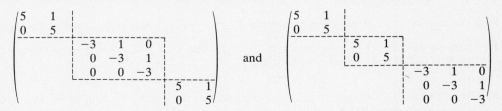

17.85 Find the characteristic polynomial $\Delta(t)$ and eigenvalues of M.

▮ Here $\Delta(t) = (t+3)^3(t-5)^4$. The exponent 3 comes from the fact that there are three -3s on the diagonal and the exponent 4 comes from the fact that there are four 5s on the diagonal. In particular, $\lambda_1 = -3$ and $\lambda_2 = 5$ are the eigenvalues.

17.86   Find the minimum polynomial $m(t)$ of M.

▌   Here $m(t) = (t + 3)^3(t - 5)^2$.   The exponent 3 comes from the fact that 3 is the order of the largest block belonging to $\lambda_1 = -3$ and the exponent 2 comes from the fact that 2 is the order of the largest block belonging to $\lambda_2 = 5$.   [Alternatively, $m(t)$ is the least common multiple of the minimal polynomials of the blocks.]

17.87   Find a maximum set S of linearly independent eigenvectors of M.

▌   Each block contributes one eigenvector to S.   Three such eigenvectors are $v_1 = (1, 0, 0, 0, 0, 0, 0)$, $v_2 = (0, 0, 0, 1, 0, 0, 0)$, $v_3 = (0, 0, 0, 0, 0, 1, 0)$ which correspond to the first, second, and third blocks, respectively.   The entry 1 in each vector is the position of the first entry in the corresponding block.

Problems 17.88–17.96 refer to the following Jordan matrices:

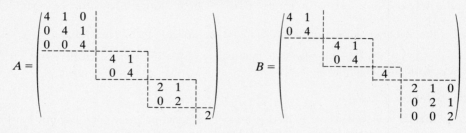

17.88   Find the characteristic polynomial $\Delta(t)$ and the eigenvalues of A.

▌   Here $\Delta(t) = (t - 4)^5(t - 2)^3$ since there are five 4s on the diagonal and three 2s on the diagonal.   Thus $\lambda_1 = 4$ and $\lambda_2 = 2$ are the eigenvalues of A.

17.89   Find the characteristic polynomial $\Delta(t)$ and eigenvalues of B.

▌   Here $\Delta(t) = (t - 4)^5(t - 2)^3$ since there are five 4s and three 2s on the diagonal.   Thus $\lambda_1 = 4$ and $\lambda_2 = 2$ are the eigenvalues of B.

17.90   Are A and B equivalent Jordan matrices?

▌   Although A and B have the same characteristic polynomial and the same eigenvalues, A and B are not equivalent since the diagonal blocks are different.

17.91   Find the minimum polynomial $m(t)$ of A.

▌   Here $m(t) = (t - 4)^3(t - 2)^2$ since 3 is the order of the largest block in A belonging to $\lambda_1 = 4$ and 2 is the order of the largest block in A belonging to $\lambda_2 = 2$.

17.92   Find the dimension d_1 of the eigenspace E_1 of $\lambda_1 = 4$ in A.   [In other words, find the geometric multiplicity of $\lambda_1 = 4$ in A.]   Also find a basis of the eigenspace E_1.

▌   Here $d_1 = 2$ since there are two blocks belonging to $\lambda_1 = 4$.   Also $v_1 = (1, 0, 0, 0, 0, 0, 0, 0)$ and $v_2 = (0, 0, 0, 1, 0, 0, 0, 0)$ form a basis for E_1.

17.93   Find the dimension d_2 of the eigenspace E_2 of $\lambda_2 = 2$ in A.   Also find a basis of the eigenspace E_2.

▌   There are two blocks in A belonging to $\lambda_2 = 2$;   hence $d_2 = 2$.   Also, $w_1 = (0, 0, 0, 0, 0, 1, 0, 0)$ and $w_2 = (0, 0, 0, 0, 0, 0, 0, 1)$ form a basis of E_2.

Remark: The entry 1 in each of the above eigenvectors of A is the position of the first entry in the corresponding block.

17.94   Find the minimum polynomial $m(t)$ of B.

▌   Note 2 is the order of the largest block in B belonging to $\lambda_1 = 4$ and 3 is the order of the largest block in B belonging to $\lambda_2 = 2$;   hence $m(t) = (t - 4)^2(t - 2)^3$.

17.95 Find the dimension d_1 of the eigenspace E_1 of $\lambda_1 = 4$ in B. [Note d_1 is the geometric multiplicity of $\lambda_1 = 4$.] Also, find a basis of the eigenspace E_1.

▮ There are three blocks in B belonging to $\lambda_1 = 4$; hence $d_1 = 3$. Also $v_1 = (1, 0, 0, 0, 0, 0, 0, 0)$, $v_2 = (0, 0, 1, 0, 0, 0, 0, 0)$, $v_3 = (0, 0, 0, 0, 1, 0, 0, 0)$ form a basis of E_1.

17.96 Find the dimension d_2 of the eigenspace E_2 of $\lambda_2 = 2$ in B and find a basis of E_2.

▮ There is only one block in B belonging to $\lambda_2 = 2$; hence $d_2 = 1$. Also, $w = (0, 0, 0, 0, 0, 1, 0, 0)$ forms a basis of E_2.

17.97 Find all (nonequivalent) Jordan matrices with characteristic polynomial $\Delta(t) = (t - 7)^4$.

▮ There are five such matrices which follow:

$$A_1 = \begin{pmatrix} 7 & 1 & 0 & 0 \\ 0 & 7 & 1 & 0 \\ 0 & 0 & 7 & 1 \\ 0 & 0 & 0 & 7 \end{pmatrix} \qquad A_2 = \begin{pmatrix} 7 & 1 & 0 & \\ 0 & 7 & 1 & \\ 0 & 0 & 7 & \\ \hline & & & 7 \end{pmatrix} \qquad A_3 = \begin{pmatrix} 7 & 1 & & \\ 0 & 7 & & \\ \hline & & 7 & 1 \\ & & 0 & 7 \end{pmatrix}$$

$$A_4 = \begin{pmatrix} 7 & 1 & & \\ 0 & 7 & & \\ \hline & & 7 & \\ & & & 7 \end{pmatrix} \qquad A_5 = \begin{pmatrix} 7 & & & \\ & 7 & & \\ & & 7 & \\ & & & 7 \end{pmatrix}$$

Since $\deg \Delta(t) = 4$, all the matrices are of order 4. Also, only 7 appears on the diagonal since $\lambda = 7$ is the only eigenvalue.

17.98 Find the minimum polynomial of each of the matrices in Problem 17.97.

▮ Let $m_i(t)$ denote the minimum polynomial of A_i. Then $m_i(t) = (t - 7)^k$ where k is the order of the largest block. Thus $m_1(t) = (t - 7)^4$, $m_2(t) = (t - 7)^3$, $m_3(t) = m_4(t) = (t - 7)^2$, $m_5(t) = t - 7$.

17.99 Find the geometric multiplicity of the eigenvalue $\lambda = 7$ in each of the matrices in Problem 17.97.

▮ Let d_i denote the geometric multiplicity of $\lambda = 7$ in A_i. Then d_i is equal to the number of blocks in A_i (belonging to $\lambda = 7$). Thus $d_i = 1$, $d_2 = d_3 = 2$, $d_4 = 3$, $d_5 = 4$.

Theorem 17.14: Let $T: V \to V$ be a linear operator whose characteristic and minimum polynomials are, respectively, $\Delta(t) = (t - \lambda_1)^{n_1} \cdots (t - \lambda_r)^{n_r}$ and $m(t) = (t - \lambda_1)^{m_1} \cdots (t - \lambda_r)^{m_r}$ where the λ_i are distinct scalars. Then T has a unique Jordan matrix representation M [called the Jordan canonical form of T]. Furthermore, the blocks J_{ij} of M belonging to the eigenvalue λ_i have the following properties:
(i) There is at least one J_{ij} of order m_i; all other J_{ij} are of order $\le m_i$.
(ii) The sum of the orders of the J_{ij} is n_i.
(iii) The number of J_{ij} equals the geometric multiplicity of λ_i.
(iv) The number of J_{ij} of each possible order is uniquely determined by T.

Theorem 17.15 [Alternate Form of Theorem 17.14]: Let A be a matrix whose characteristic polynomial $\Delta(t)$ is a product of linear factors. Then A is a similar to a unique Jordan matrix M with the above properties. [The matrix M is called the Jordan canonical form of A.]

17.100 Prove Theorem 17.14 which represents the main content of this section.

▮ By the Primary Decomposition theorem, T is decomposable into operators $T_1, \ldots, T_r$, that is, $T = T_1 \oplus \cdots \oplus T_r$, where $(t - \lambda_i)^{m_i}$ is the minimal polynomial of T_i. Thus in particular, $(T_1 - \lambda_1 I)^{m_1} = 0, \ldots, (T_r - \lambda_r I)^{m_r} = 0$. Set $N_i = T_i - \lambda_i I$. Then for $i = 1, \ldots, r$, $T_i = N_i + \lambda_i I$, where $N_i^{m_i} = 0$. That is, T_i is the sum of the scalar operator $\lambda_i I$ and a nilpotent operator N_i, which is of index m_i since $(t - \lambda_i)^{m_i}$ is the minimal polynomial of T_i.

Now by Theorem 17.11 on nilpotent operators, we can choose a basis so that N_i is in canonical form. In this basis, $T_i = N_i + \lambda_i I$ is represented by a block diagonal matrix M_i whose diagonal entries are the matrices J_{ij}. The direct sum J of the matrices M_i is in Jordan canonical form and, by Theorem 17.5, is a matrix representation of T.

Lastly, we must show that the blocks J_{ij} satisfy the required properties. Property (i) follows from the fact that N_i is of index m_i. Property (ii) is true since T and J have the same characteristic polynomial. Property (iii) is true since the nullity of $N_i = T_i - \lambda_i I$ is equal to the geometric multiplicity of the eigenvalue λ_i. Property (iv) follows from the fact that the T_i and hence the N_i are uniquely determined by T.

17.101 Suppose the characteristic and minimum polynomials of an operator T are, respectively, $\Delta(t) = (t-2)^4(t-3)^3$ and $m(t) = (t-2)^2(t-3)^2$. Find all possible Jordan canonical forms with above conditions.

   ❚ Since $\Delta(t) = (t-2)^4(t-3)^3$, there must be four 2s on the diagonal and three 3s on the diagonal. Also, since $m(t) = (t-2)^2(t-3)^2$, there must be a block of order 2, and none larger, belonging to the eigenvalue 2; and there must be a block of order 2, and none larger, belonging to the eigenvalue 3. There are two possiblities which follow:

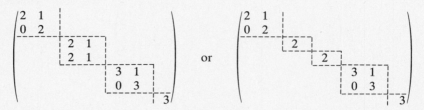

The first matrix occurs if T has two independent eigenvectors belonging to its eigenvalue 2; and the second matrix occurs if T has three independent eigenvectors belonging to 2.

17.102 Find all possible Jordan canonical forms for a linear map $T: V \to V$ whose characteristic polynomial is $\Delta(t) = (t-7)^5$ and whose minimum polynomial is $m(t) = (t-7)^2$.

   ❚ Since $\Delta(t) = (t-7)^5$ has degree 5, the matrix must have order 5 and have five 7s on the diagonal. Also, since $m(t) = (t-7)^2$, there must be a block of order 2, and none higher. There are two possibilities which follow:

$$\begin{pmatrix} 7 & 1 & & & \\ & 7 & & & \\ & & 7 & 1 & \\ & & & 7 & \\ & & & & 7 \end{pmatrix} \qquad \begin{pmatrix} 7 & 1 & & & \\ & 7 & & & \\ & & 7 & & \\ & & & 7 & \\ & & & & 7 \end{pmatrix}$$

The first occurs if $\lambda = 7$ has geometric multiplicity 3 and the second occurs if $\lambda = 7$ has geometric multiplicity 4.

17.103 Suppose $T: V \to V$ has characteristic polynomial $\Delta(t) = (t+8)^4(t-1)^3$ and minimum polynomial $m(t) = (t-8)^3(t-1)^2$. Find the Jordan canonical form M of T.

   ❚ Since $\deg \Delta(t) = 7$, the order of M is 7. Since $\Delta(t) = (t+8)^4(t-1)^3$, M has four -8s and three 1s on the diagonal. Also, since $m(t) = (t-8)^3(t-1)^2$ there must be a block of order 3 belonging to -8 and a block of order 2 belonging to 1. There is only one possibility which follows:

$$M = \begin{pmatrix} -8 & 1 & 0 & & & & \\ 0 & -8 & 1 & & & & \\ 0 & 0 & -8 & & & & \\ & & & 8 & & & \\ & & & & 1 & 1 & \\ & & & & 0 & 1 & \\ & & & & & & 1 \end{pmatrix}$$

17.104 Determine all possible Jordan canonical forms for a linear operator $T: V \to V$ whose characteristic polynomial is $\Delta(t) = (t-2)^3(t-5)^2$.

   ❚ Since $t-2$ has exponent 3 in $\Delta(t)$, 2 must appear three times on the main diagonal. Similarly 5 must appear twice. Thus the possible Jordan canonical forms are as follows:

$$B_1 = \begin{pmatrix} 2 & 1 & & & \\ & 2 & 1 & & \\ & & 2 & & \\ \hline & & & 5 & 1 \\ & & & & 5 \end{pmatrix} \qquad B_2 = \begin{pmatrix} 2 & 1 & & & \\ & 2 & & & \\ \hline & & 2 & & \\ \hline & & & 5 & 1 \\ & & & & 5 \end{pmatrix} \qquad B_3 = \begin{pmatrix} 2 & & & & \\ \hline & 2 & & & \\ \hline & & 2 & & \\ \hline & & & 5 & 1 \\ & & & & 5 \end{pmatrix}$$

$$B_4 = \begin{pmatrix} 2 & 1 & & & \\ & 2 & 1 & & \\ & & 2 & & \\ \hline & & & 5 & \\ & & & & 5 \end{pmatrix} \qquad B_5 = \begin{pmatrix} 2 & 1 & & & \\ & 2 & & & \\ \hline & & 2 & & \\ \hline & & & 5 & \\ & & & & 5 \end{pmatrix} \qquad B_6 = \begin{pmatrix} 2 & & & & \\ \hline & 2 & & & \\ \hline & & 2 & & \\ \hline & & & 5 & \\ & & & & 5 \end{pmatrix}$$

Problems 17.105–17.110 refer to the above matrices $B_1, B_2, \ldots, B_6$. Also, $m_i(t)$ denotes the minimum polynomial of B_i, and E_i and F_i denote, respectively, the eigenspaces of the eigenvalues 2 and 5 in B_i.

17.105 Find $m_1(t)$ and a basis for E_1 and F_1 in the matrix B_1.

▮ Here $m_1(t) = (t-2)^3(t-5)^2$. Also, $u = (1, 0, 0, 0, 0)$ forms a basis for E_1 and $v = (0, 0, 0, 1, 0)$ forms a basis for F_1.

17.106 Find $m_2(t)$ and the dimension of E_2 and of F_2 in the matrix B_2.

▮ Here $m_2(t) = (t-2)^2(t-5)^2$. Also, $\dim(E_2) = 2$ and $\dim(F_2) = 2$.

17.107 Find $m_3(t)$ and a basis for E_3 and F_3 in the matrix B_3.

▮ We have $m_3(t) = (t-2)(t-5)^2$. Also $u_1 = (1, 0, 0, 0, 0)$, $u_2 = (0, 1, 0, 0, 0)$, $u_3 = (0, 0, 1, 0, 0)$ form a basis of E_3 and $v = (0, 0, 0, 1, 0)$ forms a basis for F_3.

17.108 Find $m_4(t)$ and a basis for E_4 and F_4 in the matrix B_4.

▮ Here $m_4(t) = (t-2)^3(t-5)$. Also $u = (1, 0, 0, 0, 0)$ forms a basis of E_4, and $v_1 = (0, 0, 0, 1, 0)$ and $v_2 = (0, 0, 0, 0, 1)$ form a basis of F_4.

17.109 Find $m_5(t)$ and the dimension of E_5 and of F_5 in the matrix B_5.

▮ Here $m_5(t) = (t-2)^2(t-5)$ and $\dim(E_5) = 2$ and $\dim(F_5) = 2$.

17.110 Find $m_6(t)$ and the dimension of E_6 and of F_6 in the matrix B_6.

▮ $m_6(t) = (t-2)(t-5)$, and $\dim(E_6) = 3$ and $\dim(F_6) = 2$.

17.111 Suppose A is a 5-square matrix with minimum polynomial $m(t) = (t-2)^2$. Determine all possible Jordan canonical forms M for A.

▮ M must have one Jordan block of order 2 and the others must be of order 2 or 1. Thus there are only two possibilities:

$$M = \begin{pmatrix} 2 & 1 & & & \\ & 2 & & & \\ \hline & & 2 & 1 & \\ & & & 2 & \\ \hline & & & & 2 \end{pmatrix} \qquad \text{or} \qquad M = \begin{pmatrix} 2 & 1 & & & \\ & 2 & & & \\ \hline & & 2 & & \\ \hline & & & 2 & \\ \hline & & & & 2 \end{pmatrix}$$

Note that all the diagonal entries must be 2 since 2 is the only eigenvalue. The first matrix occurs when A has three independent eigenvectors, and the second matrix occurs when A has four independent eigenvectors.

17.112 Let A be a real [square] matrix. Is A similar to a Jordan matrix? If not, give a counterexample.

▮ A is similar to a Jordan matrix only if the characteristic polynomial $\Delta(t)$ of A is a product of linear factors. This is not always true. For example, the characteristic polynomial of $A = \begin{pmatrix} 0 & -1 \\ 1 & 0 \end{pmatrix}$ is $\Delta(t) = t^2 + 1$. Hence this matrix A is not similar to a Jordan matrix.

17.113   Let B be a complex [square] matrix.   Is B similar to a Jordan matrix?   If not, give a counterexample.

▌ Let $\Delta(t)$ be the characteristic polynomial of B. By the Fundamental Theorem of Algebra, $\Delta(t)$ factors into linear polynomials over the complex field **C**. Thus every complex matrix B is similar to a Jordan matrix.

17.6   QUOTIENT SPACES AND TRIANGULAR FORM

17.114   Let W be a subspace of a vector space V.   Define the cosets of W.

▌ For any vector $v \in V$, we write $u + W$ for the set of sums $u + w$ with $w \in W$; that is, $v + W = \{v + w : w \in W\}$. These sets are called the *cosets* of W in V. The dimension of $v + W$ is defined to be the dimension of W.

17.115   Let W be the subspace of $\mathbf{R}^2$ defined by $W = \{(a, b): a = b\}$.   Describe the cosets of W.

▌ W is the line in the plane $\mathbf{R}^2$ given by the equation $x - y = 0$. We can view $v + W$ as a translation of the line, obtained by adding the vector v to each point in W as pictured in Fig. 17-2. Note that $v + W$ is also a line and is parallel to W. Thus the cosets of W in $\mathbf{R}^2$ are precisely all lines parallel to W.

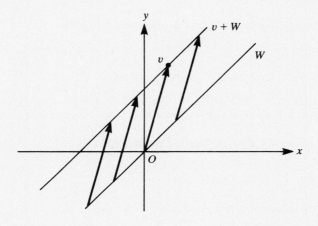

Fig. 17-2

17.116   Let W be the solution space of the homogeneous equation $2x + 3y + 4z = 0$.   Describe the cosets of W in $\mathbf{R}^3$.

▌ W is a plane through the origin $O = (0, 0, 0)$, and the cosets of W are the planes parallel to W. [See Fig. 17-3.] Equivalently, the cosets of W are the solution sets of the family of equations $2x + 3y + 4z = k$, $k \in \mathbf{R}$. In particular the coset $v + W$, where $v = (a, b, c)$, is the solution set of the linear equation $2x + 3y + 4z = 2a + 3b + 4c$ or $2(x - a) + 3(y - b) + 4(z - c) = 0$.

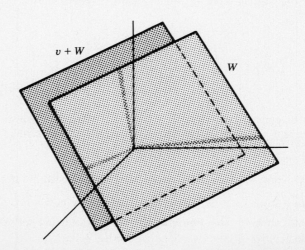

Fig. 17-3

17.117 Let $V = C[0, 2]$, the vector space of continuous functions on the interval $0 \le t \le 2$. Let W be the subset of V consisting of all functions $F(t)$ such that $f(1) = 0$. Show that W is a subspace of V.

❚ We have $\mathbf{0}(1) = 0$; hence the zero function $\mathbf{0}$ belongs to W. Suppose $f, g \in W$; then $f(1) = 0$ and $g(1) = 0$. Thus $(f + g)(1) = f(1) + g(1) = 0 + 0 = 0$ and $(kf)(1) = kf(1) = k \cdot 0 = 0$ for any scalar $k \in K$. Hence $f + g \in W$ and $kf \in W$. Accordingly, W is a subspace of V.

17.118 Describe geometrically cosets of W in V.

❚ W consists of all continuous functions passing through the point $A(1, 0)$ in the plane $\mathbf{R}^2$ as pictured in Fig. 17-4(a). A coset of W consists of all continuous functions passing through a point $B(1, k)$ for some fixed scalar k as pictured in Fig. 17-4(b).

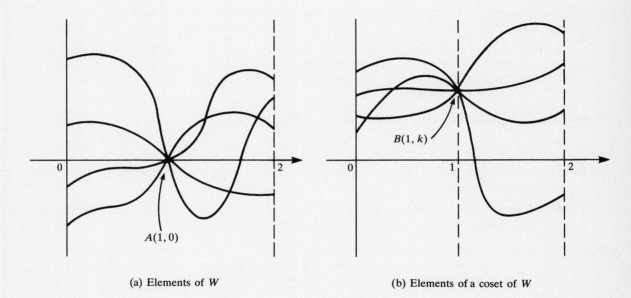

    (a) Elements of W (b) Elements of a coset of W

Fig. 17-4

Theorem 17.16:   Let W be a subspace of a vector space over a field K. Then the cosets of W in V form a vector space over K with the following operations of addition and scalar multiplication:
                (i) $(u + W) + (v + W) = (u + v) + W$.
                (ii) $k(u + W) = ku + W$,   where $k \in K$.

Remark: The above vector space consisting of the cosets of W in V is called the *quotient space* of V by W and is denoted by V/W.

Theorem 17.17:   Suppose W is a subspace invariant under a linear operator $T: V \to V$. Then T induces a linear operator $\bar{T}$ on V/W defined by $\bar{T}(v + W) = T(v) + W$. Moreover, if T is a zero of any polynomial, then so is $\bar{T}$. Thus the minimum polynomial of $\bar{T}$ divides the minimum polynomial of T.

17.119 Let W be a subspace of a vector space V. Show that the following are equivalent: (i) $u \in v + W$, (ii) $u - v \in W$, (iii) $v \in u + W$.

❚ Suppose $u \in v + W$. Then there exists $w_0 \in W$ such that $u = v + w_0$. Hence $u - v = w_0 \in W$. Conversely, suppose $u - v \in W$. Then $u - v = w_0$ where $w_0 \in W$. Hence $u = v + w_0 \in v + W$. Thus (i) and (ii) are equivalent.
    We also have $u - v \in W$ iff $-(u - v) = v - u \in W$ iff $v \in u + W$. Thus (ii) and (iii) are also equivalent.

17.120 Prove: The cosets of W in V partition V into mutually disjoint sets. That is, (i) any two cosets $u + W$ and $v + w$ are either identical or disjoint; and (ii) each $v \in V$ belongs to a coset; in fact, $v \in v + W$. Furthermore, $u + W = v + W$ if and only if $u - v \in W$, and so $(v + w) + W = v + W$ for any $w \in W$.

▮ Let $v \in V$. Since $0 \in W$, we have $v = v + 0 \in v + W$ which proves (ii).

Now suppose the cosets $u + W$ and $v + W$ are not disjoint; say, the vector x belongs to both $u + W$ and $v + W$. Then $u - x \in W$ and $x - v \in W$. The proof of (i) is complete if we show that $u + W = v + W$. Let $u + w_0$ be any element in the coset $u + W$. Since $u - x$, $x - v$, and w_0 belong to W, $(u + w_0) - v = (u - x) + (x - v) + w_0 \in W$. Thus $u + w_0 \in v + W$ and hence the coset $u + W$ is contained in the coset $v + W$. Similarly $v + W$ is contained in $u + W$ and so $u + W = v + W$.

The last statement folows from the fact that $u + W = v + W$ if and only if $u \in v + W$, and by the preceding problem this is equivalent to $u - v \in W$.

17.121 Show that the operations in Theorem 17.16 are well-defined; namely show that if $u + W = u' + W$ and $v + W = v' + W$, then (i) $(u + v) + W = (u' + v') + W$ and (ii) $ku + W = ku' + W$, for any $k \in K$.

▮ (i) Since $u + W = u' + W$ and $v + W = v' + W$, both $u - u'$ and $v - v'$ belong to W. But then $(u + v) - (u' + v') = (u - u') + (v - v') \in W$. Hence $(u + v) + W = (u' + v') + W$.

(ii) Also, since $u - u' \in W$ implies $k(u - u') \in W$, then $ku - ku' = k(u - u') \in W$; hence $ku + W = ku' + W$.

17.122 What is the zero element in the quotient space V/W?

▮ For every $v \in V$, we have $(v + W) + W = v + W$. Hence W, itself, is the zero element in V/W.

17.123 Let V be a vector space and W a subspace of V. Show that the natural map $\eta : V \to V/W$, defined by $\eta(v) = v + W$, is linear.

▮ For any $u, v \in V$ and any $k \in K$, we have $\eta(u + v) = u + v + W = u + W + v + W = \eta(u) + \eta(v)$ and $\eta(kv) = kv + W = k(v + W) = k\eta(v)$. Accordingly, η is linear.

17.124 Let W be a subspace of a vector space V. Suppose $\{w_1, \ldots, w_r\}$ is a basis of W and the set of cosets $\{\bar{v}_1, \ldots, \bar{v}_s\}$, where $\bar{v}_j = v_j + W$, is a basis of the quotient space. Show that $B = \{v_1, \ldots, v_s, w_1, \ldots, w_r\}$ is a basis of V. Thus $\dim V = \dim W + \dim(V/W)$.

▮ Suppose $u \in V$. Since $\{\bar{v}_j\}$ is a basis of V/W, $\bar{u} = u + W = a_1 \bar{v}_1 + a_2 \bar{v}_2 + \cdots + a_s \bar{v}_s$. Hence $u = a_1 v_1 + \cdots + a_s v_s + w$ where $w \in W$. Since $\{w_i\}$ is a basis of W, $u = a_1 v_1 + \cdots + a_s v_s + b_1 w_1 + \cdots + b_r w_r$. Accordingly, B generates V.

We now show that B is linearly independent. Suppose

$$c_1 v_1 + \cdots + c_s v_s + d_1 w_1 + \cdots + d_r w_r = 0 \qquad (1)$$

Then $c_1 \bar{v}_1 + \cdots + c_s \bar{v}_s = \bar{0} = W$. Since $\{\bar{v}_j\}$ is independent, the c's are all 0. Substituting into (1), we find $d_1 w_1 + \cdots + d_r w_r = 0$. Since $\{w_i\}$ is independent, the d's are all 0. Thus B is linearly independent and therefore a basis of V.

17.125 Prove Theorem 17.17.

▮ We first show that $\bar{T}$ is well-defined, i.e., if $u + W = v + W$ then $\bar{T}(u + W) = \bar{T}(v + W)$. If $u + W = v + W$ then $u - v \in W$ and, since W is T-invariant, $T(u - v) = T(u) - T(v) \in W$. Accordingly, $\bar{T}(u + W) = T(u) + W = T(v) + W = \bar{T}(v + W)$ as required.

We next show that $\bar{T}$ is linear. We have $\bar{T}((u + W) + (v + W)) = \bar{T}(u + v + W) = T(u + v) + W = T(u) + T(v) + W = T(u) + W + T(v) + W = \bar{T}(u + W) + \bar{T}(v + W)$ and $\bar{T}(k(u + W)) = \bar{T}(ku + W) = T(ku) + W = kT(u) + W = k(T(u) + W) = k\bar{T}(u + W)$. Thus $\bar{T}$ is linear.

Now, for any coset $u + W$ in V/W, $\overline{T^2}(u + W) = T^2(u) + W = T(T(u)) + W = \bar{T}(T(u) + W) = \bar{T}(\bar{T}(u + W)) = \bar{T}^2(u + W)$. Hence $\overline{T^2} = \bar{T}^2$. Similarly $\overline{T^n} = \bar{T}^n$ for any n. Thus for any polynomial

$$f(t) = a_n t^n + \cdots + a_0 ,$$

$$\overline{f(T)}(u + W) = f(T)(u) + W = \sum a_i T^i(u) + W = \sum a_i (T^i(u) + W)$$

$$= \sum a_i \overline{T^i}(u + W) = \sum a_i \bar{T}^i(u + W) = \left(\sum a_i \bar{T}^i \right)(u + W) = f(\bar{T})(u + W)$$

and so $\overline{f(T)} = f(\bar{T})$. Accordingly, if T is a root of $f(t)$ then $\overline{f(T)} = \bar{0} = W = f(\bar{T})$, that is, $\bar{T}$ is also a root of $f(t)$. Thus the theorem is proved.

Theorem 17.18: Let $T:V \to V$ be a linear operator whose characteristic polynomial factors into linear polynomials. Then V has a basis in which T is represented by a triangular matrix [called a triangular form of T].

Theorem 17.19 [Alternate Form of Theorem 17.18]: Let A be a [square] matrix whose characteristic polynomial factors into linear polynomials. Then A is similar to a triangular matrix.

17.126 Prove Theorem 17.18 which represents the main content of this section.

▌ The proof is by induction on the dimension of V. If $\dim V = 1$, then every matrix representation of T is a 1 by 1 matrix which is triangular.

Now suppose $\dim V = n > 1$ and that the theorem holds for spaces of dimension less than n. Since the characteristic polynomial of T factors into linear polynomials, T has at least one eigenvalue and so at least one nonzero eigenvector v, say $T(v) = a_{11}v$. Let W be the one-dimensional subspace spanned by v. Set $\bar{V} = V/W$. Then [Problem 17.124] $\dim \bar{V} = \dim V - \dim W = n - 1$. Note also that W is invariant under T. By Theorem 17.17, T induces a linear operator $\bar{T}$ on $\bar{V}$ whose minimum polynomial divides the minimum polynomial of T. Since the characteristic polynomial of T is a product of linear polynomials, so is its minimum polynomial; hence so are the minimum and characteristic polynomials of $\bar{T}$. Thus $\bar{V}$ and $\bar{T}$ satisfy the hypothesis of the theorem. Hence, by induction, there exists a basis $\{\bar{v}_2, \ldots, \bar{v}_n\}$ of $\bar{V}$ such that

$$\bar{T}(\bar{v}_2) = a_{22}\bar{v}_2$$
$$\bar{T}(\bar{v}_3) = a_{32}\bar{v}_2 + a_{33}\bar{v}_3$$
$$\cdots\cdots\cdots\cdots\cdots\cdots\cdots\cdots\cdots\cdots$$
$$\bar{T}(\bar{v}_n) = a_{n2}\bar{v}_2 + a_{n3}\bar{v}_3 + \cdots + a_{nn}\bar{v}_n$$

Now let $v_2, \ldots, v_n$ be elements of V which belong to the cosets $\bar{v}_2, \ldots, \bar{v}_n$ respectively. Then $\{v, v_2, \ldots, v_n\}$ is a basis of V [Problem 17.124]. Since $\bar{T}(\bar{v}_2) = a_{22}\bar{v}_2$, we have $\bar{T}(\bar{v}_2) - a_{22}\bar{v}_2 = 0$ and so $T(v_2) - a_{22}v_2 \in W$. But W is spanned by v; hence $T(v_2) - a_{22}v_2$ is a multiple of v, say $T(v_2) - a_{22}v_2 = a_{21}v$ and so $T(v_2) = a_{21}v + a_{22}v_2$. Similarly, for $i = 3, \ldots, n$, $T(v_i) - a_{i2}v_2 - a_{i3}v_3 - \cdots - a_{ii}v_i \in W$ and so $T(v_i) = a_{i1}v + a_{i2}v_2 + \cdots + a_{ii}v_i$. Thus

$$T(v) = a_{11}v$$
$$T(v_2) = a_{21}v + a_{22}v_2$$
$$\cdots\cdots\cdots\cdots\cdots\cdots\cdots\cdots\cdots\cdots$$
$$T(v_n) = a_{n1}v + a_{n2}v_2 + \cdots + a_{nn}v_n$$

and hence the matrix of T in this basis is triangular.

17.127 Let W be a subspace of V. Suppose the set of cosets $\{v_1 + W, v_2 + W, \ldots, v_n + W\}$ in V/W is linearly independent. Show that the set of vectors $\{v_1, v_2, \ldots, v_n\}$ in V is also linearly independent.

▌ Suppose $a_1v_1 + a_2v_2 + \cdots + a_nv_n = 0$. Then $(a_1v_1 + \cdots + a_nv_n) + W = 0 + W = W$. Hence $a_1(v_1 + W) + a_2(v_2 + w) + \cdots + a_n(v_n + w) = W$. Since the $v_i + W$ are linearly independent, $a_1 = 0, \ldots, a_n = 0$. Thus $v_1, v_2, \ldots, v_n$ are linearly independent.

17.128 Let W be a subspace of V. Suppose the set of vectors $\{u_1, u_2, \ldots, u_n\}$ in V is linearly independent, and that $\text{span}(u_i) \cap W = \{0\}$. Show that the set of cosets $\{u_1 + W, \ldots, u_n + W\}$ in V/W is also linearly independent.

▌ Suppose $a_1(u_1 + W) + a_2(u_2 + W) + \cdots + a_n(u_n + W) = W$. Then $(a_1u_1 + a_2u_2 + \cdots + a_nu_n) + W = W$ and $a_1u_1 + \cdots + a_nu_n \in W$. Since $\text{span}(u_i) \cap W = \{0\}$, we have $a_1u_1 + \cdots + a_nu_n = 0$. By hypothesis, $u_1, \ldots, u_n$ are linarly independent; hence $a_1 = 0, \ldots, a_n = 0$. Thus the cosets $u_i + W$ are linearly independent.

17.129 Let V be the vector space of polynomials over $\mathbf{R}$ and let W be the subspace of polynomials $h(t)$ which are divisible by t^4, that is, $h(t) = b_0t^4 + b_1t^5 + \cdots + b_{m-4}t^m$. Show that the quotient space V/W is of dimension 4.

∎ Let $f(t)$ be any polynomial in V, say $f(t) = a_0 + a_1 t + \cdots + a_n t^n$. Since $a_4 t^4 + \cdots + a_n t^n \in W$, we have $f(t) + W = a_0 + a_1 t + a_2 t^2 + a_3 t^3 + W = a_0(1 + W) + a_1(t + W) + a_2(t^2 + W) + a_3(t^3 + W)$. Thus $1 + W$, $t + W$, $t^2 + W$, $t^3 + W$ span V/W. They are also linearly independent since $1, t, t^2, t^3$ are linearly independent in V and $\text{span}(1, t, t^2, t^3) \cap W = 0$ [Problem 17.128]. Thus $\dim(V/W) = 4$.

17.7 CYCLIC SUBSPACES, $Z(v, T)$

This section assumes $T: V \to V$ is a linear operator, where V has finite dimension over K and $v \in V$ with $v \neq 0$.

17.130 Define $Z(v, T)$, called the T-cyclic subspace of V generated by v.

∎ $Z(v, T)$ is the set of all vectors of the form $f(T)(v)$ where $f(t)$ ranges over all polynomials over K.

17.131 Show that $Z(v, T)$ is a subspace of V.

∎ We have $0(v) = 0$ for the zero polynomial 0; hence $0 \in Z(v, T)$. Suppose $u, w \in Z(v, T)$. Then $u = f(T)(v)$ and $w = g(T)(v)$. Hence $u + w = f(T)(v) + g(T)(v) = [f(T) + g(T)](v)$ and therefore $u + w \in Z(v, T)$. Also, for any scalar $k \in K$, $ku = kf(T)(v) = [kf(T)](v)$. Thus $ku \in Z(v, T)$. Accordingly, $Z(v, T)$ is a subspace of V.

17.132 Show that $Z(v, T)$ is invariant under T. [The restriction of T to $Z(v, T)$ is denoted by T_v.]

∎ Let $u \in Z(v, T)$, say $u = f(T)(v)$. Then $T(u) = T[f(T)(v)] = [Tf(T)](v)$ and so $T(u) \in Z(v, T)$. Therefore, $Z(v, T)$ is T-invariant.

17.133 Define the T-annihilator of v in $Z(v, T)$, denoted by $m_v(T)$.

∎ Consider the sequence $v, T(v), T^2(v), T^3(v), \ldots$ of powers of T acting on v. Let k be the lowest integer such that $T^k(v)$ is a linear combination of those vectors which precede it in the sequence; say, $T^k(v) = -a_{k-1} T^{k-1}(v) - \cdots - a_1 T(v) - a_0 v$. Then the polynomial $m_v(t) = t^k + a_{k-1} t^{k-1} + \cdots + a_1 t + a_0$ is called the T-annihilator of v in $Z(v, T)$.

17.134 Show that $m_v(t)$ is the unique monic polynomial of lowest degree for which $m_v(T)(v) = 0$.

∎ Let $k = \deg m_v(T)$ and suppose $\deg f(t) = m$ where $m < k$, say $f(t) = b_0 + b_1 t + \cdots + b_m t^m$. If $f(T)(v) = 0$, then $b_0 v + b_1 T(v) + \cdots + b_{m-1} T^{m-1}(v) + b_m T^m(v) = 0$. Hence $T^m(v) = -b_m^{-1} b_{m-1} T^{m-1}(v) - \cdots - b_m^{-1} b_1 T(v) - b_m^{-1} b_0 v$. This contradicts the fact that k is the lowest integer such that $T^k(v)$ is a linear combination of $v, T(v), \ldots, T^{k-1}(v)$. Thus $m_v(t)$ is a polynomial of smallest degree such that $m_v(T)(v) = 0$. Suppose $m_v'(t)$ is another monic polynomial of degree k such that $m_v'(T)(v) = 0$. Then $m_v(t) - m_v'(t)$ has degree less than k and $[m_v(T) - m_v'(T)](v) = 0$. This also is a contradiction. Thus $m_v(t)$ is unique.

Theorem 17.19: Let $Z(v, T)$, T_v, and $m_v(t)$ be defined as above. Then

(i) $B = \{v, T(v), \ldots, T^{k-1}(v)\}$ is a basis of $Z(v, T)$; hence $\dim Z(v, T) = k$.

(ii) The minimal polynomial of T_v is $m_v(t)$.

(iii) The matrix representation of T_v in the above basis B is

$$C = \begin{pmatrix} 0 & 0 & 0 & \cdots & 0 & -a_0 \\ 1 & 0 & 0 & \cdots & 0 & -a_1 \\ 0 & 1 & 0 & \cdots & 0 & -a_2 \\ \cdots\cdots\cdots\cdots\cdots\cdots\cdots\cdots\cdots \\ 0 & 0 & 0 & \cdots & 0 & -a_{k-2} \\ 0 & 0 & 0 & \cdots & 1 & -a_{k-1} \end{pmatrix}$$

Remark: The above matrix C is called the *companion matrix* of the polynomial $m_v(t)$.

17.135 Prove (i) of Theorem 17.19.

∎ By definition of $m_v(t)$, $T^k(v)$ is the first vector in the sequence $v, T(v), T^2(v), \ldots$ which is a linear combination of those vectors which precede it in the sequence; hence $B = \{v, T(v), \ldots, T^{k-1}(v)\}$ is linearly independent. We now only have to show that $Z(v, T) = L(B)$, the linear span of B. By the above, $T^k(v) \in L(B)$. We prove by induction that $T^n(v) \in L(B)$ for every n. Suppose $n > k$ and $T^{n-1}(v) \in L(B)$, that is, $T^{n-1}(v)$ is a linear combination of $v, \ldots, T^{k-1}(v)$. Then $T^n(v) = T(T^{n-1}(v))$ is a linear combination of $T(v), \ldots, T^k(v)$. But $T^k(v) \in L(B)$; hence $T^n(v) \in L(B)$ for every n. Consequently $f(T)(v) \in L(B)$ for any polynomial $f(t)$. Thus $Z(v, T) = L(B)$ and so B is a basis as claimed.

17.136 Prove (ii) of Theorem 17.19.

▌ Suppose $m(t) = t^s + b_{s-1} t^{s-1} + \cdots + b_0$ is the minimal polynomial of T_v. Then, since $v \in Z(v, T)$, $0 = m(T_v)(v) = m(T)(v) = T^s(v) + b_{s-1} T^{s-1}(v) + \cdots + b_0 v$. Thus $T^s(v)$ is a linear combination of $v, T(v), \ldots, T^{s-1}(v)$, and therefore $k \le s$. However, $m_v(T) = 0$ and so $m_v(T_v) = 0$. Then $m(t)$ divides $m_v(t)$ and so $s \le k$. Accordingly $k = s$ and hence $m_v(t) = m(t)$.

17.137 Prove (iii) of Theorem 17.19.

▌ We have

$$
\begin{aligned}
T_v(v) &= && T(v) \\
T_v(T(v)) &= && T^2(v) \\
&\cdots\cdots\cdots\cdots\cdots\cdots\cdots\cdots\cdots\cdots\cdots\cdots\cdots\cdots\cdots\cdots \\
T_v(T^{k-2}(v)) &= && T^{k-1}(v) \\
T_v(T^{k-1}(v)) &= T^k(v) = -a_0 v - a_1 T(v) - a_2 T^2(v) - \cdots - a_{k-1} T^{k-1}(v)
\end{aligned}
$$

By definition, the matrix of T_v in this basis is the transpose of the matrix of coefficients of the above system of equations; hence it is C, as required.

17.138 Let $T: V \to V$ be linear. Let W be a T-invariant subspace of V and let $\bar{T}$ be the induced operator on V/W. Prove: (i) The T-annihilator of $v \in V$ divides the minimal polynomial of T. (ii) The $\bar{T}$-annihilator of $\bar{v} \in V/W$ divides the minimal polynomial of T.

▌ (i) The T-annihilator of $v \in V$ is the minimal polynomial of the restriction of T to $Z(v, T)$ and therefore, by Problem 17.16, it divides the minimal polynomial of T.

(ii) The $\bar{T}$-annihilator of $\bar{v} \in V/W$ divides the minimal polynomial of $\bar{T}$, which divides the minimal polynomial of T.

Remark: In case the minimal polynomial of T is $f(t)^n$ where $f(t)$ is a monic irreducible polynomial, then the T-annihilator of $v \in V$ and the $\bar{T}$-annihilator of $\bar{v} \in V/W$ are of the form $f(t)^m$ where $m \le n$.

17.139 Let W be the intersection of all T-invariant subspaces of V containing v. Show that $W = Z(v, T)$.

▌ Since $Z(v, T)$ is a T-invariant subspace containing v we have $W \subseteq Z(v, T)$. Since W is T-invariant and $v \in W$, we have $T^k(v) \in W$ for every k. Since W is a subspace, we have $f(T)(v) \in W$ for every polynomial $f(t)$. Thus $Z(v, T) \subset W$. Both inclusions give $W = Z(v, T)$.

17.8 RATIONAL CANONICAL FORM

This section presents the classical rational canonical form for a linear operator $T: V \to V$. This form exists even when the minimal polynomial and hence the characteristic polynomial cannot be factored into linear polynomials. [Recall that this is not the case for the Jordan canonical form.]

First we state a special case of the rational canonical form where the minimal polynomial and characteristic polynomial are powers of a single irreducible polynomial. The general case, which follows from the Primary Decomposition theorem, is stated afterwards.

Lemma 17.20: Left $T: V \to V$ be a linear operator whose minimal polynomial is $f(t)^n$ where $f(t)$ is a monic irreducible polynomial and whose characteristic polynomial is $f(t)^d$. Then V is the direct sum $V = Z(v_1, T) \oplus \cdots \oplus Z(v_r, T)$ of T-cyclic subspaces $Z(v_i, T)$ with corresponding T-annihilators $f(t)^{n_1}, f(t)^{n_2}, \ldots, f(t)^{n_r}$, where $n = n_1 \ge n_2 \ge \cdots \ge n_r$ and $d = n_1 + n_2 + \cdots + n_r$. Any other decomposition of V into T-cyclic subspaces has the same number of components and the same set of T-annihilators.

Remark: The above lemma does not say that the vectors v_i or the T-cyclic subspaces $Z(v_i, T)$ are uniquely determined by T; but it does say that the set of T-annihilators are uniquely determined by T. Thus T has a unique matrix representation

$$\begin{pmatrix} C_1 & & & \\ & C_2 & \cdots & \\ & & \ddots & \\ & & & C_r \end{pmatrix}$$

where the C_i are the companion matrices to the polynomials $f(t)^{n_i}$. Also, the polynomials $f(t)^{n_i}$ are called the elementary divisors of T.

The Primary Decomposition theorem and the above lemma give us the following basic result.

Theorem 17.21: Let $T: V \to V$ be a linear operator with minimal polynomial $m(t) = f_1(t)^{m_1} f_2(t)^{m_2} \cdots f_s(t)^{m_s}$, where the $f_i(t)$ are distinct monic irreducible polynomials, and with characteristic polynomial $\Delta(t) = f_1(t)^{d_1} f_2(t)^{d_2} \cdots f_s(t)^{d_s}$. Then T has a unique block diagonal matrix representation

$$M = \begin{pmatrix} C_{11} & & & & & & \\ & \ddots & & & & & \\ & & C_{1r_1} & & & & \\ & & & \ddots & & & \\ & & & & C_{s1} & & \\ & & & & & \ddots & \\ & & & & & & C_{sr_s} \end{pmatrix}$$

where the C_{ij} are companion matrices. In particular, the C_{ij} are the companion matrices of polynomials $f_i(t)^{n_{ij}}$ where $m_1 = n_{11} \geq n_{12} \geq \cdots \geq n_{1r_1}, \ldots, m_s = n_{s1} \geq n_{s2} \geq \cdots \geq n_{sr_s}$ and $d_1 = n_{11} + n_{12} + \cdots + n_{1r_1}, \ldots, d_s = n_{s1} + n_{s2} + \cdots + n_{sr_s}$.

Theorem 17.22 [Alternate Form of Theorem 17.21]: Every [square] matrix A is similar to a unique matrix M in rational canonical form as above.

Remark: The above matrix M is called the rational canonical form of T, and the polynomials $f_i(t)^{n_{ij}}$ are called the elementary divisors of T.

17.140 Find all rational canonical forms with minimal polynomial $m(t) = (t-1)^3$ and characteristic polynomial $\Delta(t) = (t-1)^7$.

❚ The elementary divisors consists of powers of $f(t) = t - 1$ with the exponents satisfying three conditions: (1) One exponent must equal 3, the exponent in $m(t)$. (2) No exponent can exceed 3. (3) The sum of the exponents must equal 7, the exponent in $\Delta(t)$. There are four possibilities:
(a) $7 = 3 + 3 + 1$, that is, $(t-1)^3$, $(t-1)^3$, $t-1$
(b) $7 = 3 + 2 + 2$, that is, $(t-1)^3$, $(t-1)^2$, $(t-1)^2$
(c) $7 = 3 + 2 + 1 + 1$, that is, $(t-1)^3$, $(t-1)^2$, $t-1$, $t-1$
(d) $7 = 3 + 1 + 1 + 1 + 1$, that is, $(t-1)^3$, $t-1$, $t-1$, $t-1$, $t-1$
Since $(t-1)^3 = t^3 - 3t^2 + 3t - 1$ and $(t-1)^2 = t^2 - 2t + 1$, the corresponding matrices follow:

$$\begin{pmatrix} 0 & 0 & 1 & & & & \\ 1 & 0 & -3 & & & & \\ 0 & 1 & 3 & & & & \\ & & & 0 & 0 & 1 & \\ & & & 1 & 0 & -3 & \\ & & & 0 & 1 & 3 & \\ & & & & & & 1 \end{pmatrix} \qquad \begin{pmatrix} 0 & 0 & 1 & & & & \\ 1 & 0 & -3 & & & & \\ 0 & 1 & 3 & & & & \\ & & & 0 & -1 & & \\ & & & 1 & 2 & & \\ & & & & & 0 & -1 \\ & & & & & 1 & 2 \end{pmatrix}$$

$$(a) \qquad\qquad\qquad\qquad\qquad\qquad (b)$$

$$
\begin{pmatrix}
0 & 0 & 1 \\
1 & 0 & -3 \\
0 & 1 & 3 \\
& & & 0 & -1 \\
& & & 1 & 2 \\
& & & & & 1 \\
& & & & & & 1
\end{pmatrix}
\qquad
\begin{pmatrix}
0 & 0 & 1 \\
1 & 0 & -3 \\
0 & 1 & 3 \\
& & & 1 \\
& & & & 1 \\
& & & & & 1 \\
& & & & & & 1
\end{pmatrix}
$$

(c) $\qquad\qquad\qquad\qquad\qquad\qquad$ (d)

17.141 Find all possible rational canonical forms with $m(t) = (t-2)^3$ and $\Delta(t) = (t-2)^5$.

▮ There are only two possible sets of elementary divisors: (a) $(t-2)^3$, $(t-2)^2$; (b) $(t-2)^3$, $t-2$, $t-2$. Using $(t-2)^3 = t^3 - 6t^2 + 12t - 8$ and $(t-2)^2 = t^2 - 4t + 4$, the corresponding rational canonical forms follow:

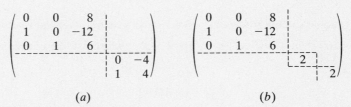

(a) $\qquad\qquad\qquad\qquad\qquad\qquad$ (b)

Problems 17.142–17.144 refer to a real matrix A of order 6 with minimum polynomial $m(t) = (t^2 - t + 3)(t-2)^2$.

17.142 Find the number of possible characteristic polynomials $\Delta(t)$ of A.

▮ The degree of $\Delta(t)$ must be 6 since A is a 6×6 matrix. Also, $\Delta(t)$ must be divisible by $m(t)$. Thus there are two possibilities: $\Delta_1(t) = (t^2 - t + 3)^2(t-2)^2$ or $\Delta_2(t) = (t^2 - t + 3)(t-2)^4$.

17.143 Suppose $\Delta_1(t)$ is the characteristic polynomial of A. Find the rational canonical form M of A.

▮ Here the elementary divisors of A are $t^2 - t + 3$, $t^2 - t + 3$, $(t-2)^2$. Thus

$$
M = \begin{pmatrix}
0 & -3 \\
1 & 1 \\
& & 0 & -3 \\
& & 1 & 1 \\
& & & & 0 & -4 \\
& & & & 1 & 4
\end{pmatrix}
$$

17.144 Suppose $\Delta_2(t)$ is the characteristic polynomial of A. Find the two possible rational canonical forms of A.

▮ There are two possible sets of elementary divisors as follows: (a) $t^2 - t + 3$, $(t-2)^2$, $(t-2)^2$; (b) $t^2 - t + 3$, $(t-2)^2$, $t-2$, $t-2$. The corresponding rational canonical forms follow:

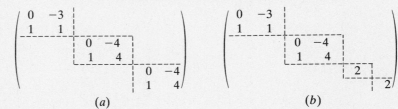

(a) $\qquad\qquad\qquad\qquad\qquad\qquad$ (b)

17.145 Find the rational canonical form M of the Jordan block $\begin{pmatrix} \lambda & 1 & 0 & 0 \\ 0 & \lambda & 1 & 0 \\ 0 & 0 & \lambda & 1 \\ 0 & 0 & 0 & \lambda \end{pmatrix}$.

▮ Here $\Delta(t) = m(t) = (t - \lambda)^4 = t^4 - 4\lambda t^3 + 6\lambda^2 t^2 - 4\lambda^3 t + \lambda^4$. Thus

$$M = \begin{pmatrix} 0 & 0 & 0 & -\lambda^4 \\ 1 & 0 & 0 & 4\lambda^3 \\ 0 & 1 & 0 & -6\lambda^2 \\ 0 & 0 & 1 & 4\lambda \end{pmatrix}$$

Problems 17.146–17.148 refer to a matrix A with $\Delta(t) = m(t) = (t^2 + 1)(t^2 - 3)$.

17.146 Find the rational canonical form M of A if A is a matrix over the rational field $\mathbf{Q}$.

▮ Since $t^2 + 1$ and $t^2 - 3$ are irreducible over $\mathbf{Q}$, the elementary divisors are $t^2 + 1$, $t^2 - 3$. Thus

$$M = \begin{pmatrix} 0 & -1 & & \\ 1 & 0 & & \\ & & 0 & 3 \\ & & 1 & 0 \end{pmatrix}$$

17.147 Find the rational canonical form M' of A if A is a matrix over the real field $\mathbf{R}$.

▮ Here $\Delta(t) = m(t) = (t^2 + 1)(t - \sqrt{3})(t + \sqrt{3})$, which are the elementary divisors of A. Thus

$$M' = \begin{pmatrix} 0 & -1 & & \\ 1 & 0 & & \\ & & \sqrt{3} & \\ & & & -\sqrt{3} \end{pmatrix}$$

17.148 Find the rational canonical form M'' for A if A is a matrix over the complex field $\mathbf{C}$.

▮ Here $\Delta(t) = m(t) = (t + i)(t - 1)(t + \sqrt{3})(t - \sqrt{3})$. Thus

$$M'' = \begin{pmatrix} i & & & \\ & -i & & \\ & & \sqrt{3} & \\ & & & -\sqrt{3} \end{pmatrix}$$

17.149 Let V be a vector space of dimension 7 over $\mathbf{R}$, and let $T : V \to V$ be a linear operator with minimal polynomial $m(t) = (t^2 + 2)(t + 3)^3$. Find all the possible rational canonical forms for T.

▮ The sum of the degrees of the elementary divisors must add up to 7 since $\dim V = 7$. Also, one elementary divisor must be $t^2 + 2$ and one must be $(t + 3)^3$. There are three possibilities: (a) $t^2 + 2$, $t^2 + 2$, $(t + 3)^3$; (b) $t^2 + 2$, $(t + 3)^3$, $(t + 3)^2$; (c) $t^2 + 2$, $(t + 3)^3$, $t + 3$, $t + 3$. The corresponding rational canonical forms follow:

$$\begin{pmatrix} 0 & -2 & & & & & \\ 1 & 0 & & & & & \\ & & 0 & -2 & & & \\ & & 1 & 0 & & & \\ & & & & 0 & 0 & -27 \\ & & & & 1 & 0 & -27 \\ & & & & 0 & 1 & -9 \end{pmatrix}$$

$$(a)$$

$$\begin{pmatrix} 0 & -2 & & & & & \\ 1 & 0 & & & & & \\ & & 0 & 0 & -27 & & \\ & & 1 & 0 & -27 & & \\ & & 0 & 1 & -9 & & \\ & & & & & 0 & -9 \\ & & & & & 1 & -6 \end{pmatrix}$$

$$(b)$$

$$\begin{pmatrix} 0 & -2 & & & & & \\ 1 & 0 & & & & & \\ & & 0 & 0 & -27 & & \\ & & 1 & 0 & -27 & & \\ & & 0 & 1 & -9 & & \\ & & & & & -3 & \\ & & & & & & -3 \end{pmatrix}$$

$$(c)$$

17.150 Find the characteristic polynomial in each of the cases in the preceding problem.

▌ The characteristic polynomial is equal to the product of the elementary divisors. Thus $\Delta_a(t) = (t^2+2)^2(t+3)^3$, $\Delta_b(t) = (t^2+2)(t+3)^6$, $\Delta_c(t) = (t^2+2)(t+3)^6$. [Observe that $\Delta_b(t) = \Delta_c(t)$.]

17.151 Prove Lemma 17.20.

▌ The proof is by induction of the dimension of V. If $\dim V = 1$, then V is itself T-cyclic and the lemma holds. Now suppose $\dim V > 1$ and that the lemma holds for those vector spaces of dimension less than that of V.

Since the minimal polynomial of T is $f(t)^n$, there exists $v_1 \in V$ such that $f(T)^{n-1}(v_1) \neq 0$; hence the T-annihilator of v_1 is $f(t)^n$. Let $Z_1 = Z(v_1, T)$ and recall that Z_1 is T-invariant. Let $\bar{V} = V/Z_1$ and let $\bar{T}$ be the linear operator on $\bar{V}$ induced by T. Then the minimal polynomial of $\bar{T}$ divides $f(t)^n$; hence the hypothesis holds for $\bar{V}$ and $\bar{T}$. Consequently, by induction, $\bar{V}$ is the direct sum of $\bar{T}$-cyclic subspaces; say, $\bar{V} = Z(\bar{v}_2, \bar{T}) \oplus \cdots \oplus Z(\bar{v}_r, \bar{T})$, where the corresponding $\bar{T}$-annihilators are $f(t)^{n_2}, \ldots, f(t)^{n_r}$, $n \geq n_2 \geq \cdots \geq n_r$.

We claim that there is a vector v_2 in the coset $\bar{v}_2$ whose T-annihilator is $f(t)^{n_2}$, the $\bar{T}$-annihilator of $\bar{v}_2$. Let w be any vector in $\bar{v}_2$. Then $f(T)^{n_2}(w) \in Z_1$. Hence there exists a polynomial $g(t)$ for which

$$f(T)^{n_2}(w) = g(T)(v_1) \tag{1}$$

Since $f(t)^n$ is the minimal polynomial of T, we have, by (1), $0 = f(T)^n(w) = f(T)^{n-n_2}g(T)(v_1)$. But $f(t)^n$ is the T-annihilator of v_1; hence $f(t)^n$ divides $f(t)^{n-n_2}g(t)$ and so $g(t) = f(t)^{n_2}h(t)$ for some polynomial $h(t)$. We set $v_2 = w - h(T)(v_1)$. Since $w - v_2 = h(T)(v_1) \in Z_1$, v_2 also belongs to the coset $\bar{v}_2$. Thus the T-annihilator of v_2 is a multiple of the $\bar{T}$-annihilator of $\bar{v}_2$. On the other hand, by (1), $f(T)^{n_2}(v_2) = f(T)^{n_2}(w - h(T)(v_1)) = f(T)^{n_2}(w) - g(T)(v_1) = 0$. Consequently the T-annihilator of v_2 is $f(t)^{n_2}$ as claimed.

Similarly, there exist vectors $v_3, \ldots, v_r \in V$ such that $v_i \in \bar{v}_i$ and that the T-annihilator of v_i is $f(t)^{n_i}$, the $\bar{T}$-annihilator of $\bar{v}i$. We set $Z_2 = Z(v_2, T), \ldots, Z_r = Z(v_r, T)$. Let d denote the degree of $f(t)$ so that $f(t)^{n_i}$ has degree dn_i. Then since $f(t)^{n_i}$ is both the T-annihilator of v_i and the $\bar{T}$-annihilator of $\bar{v}_i$, we know that $\{v_i, T(v_i), \ldots, T^{dn_i-1}(v_i)\}$ and $\{\bar{v}_i, \bar{T}(\bar{v}_i), \ldots, \bar{T}^{dn_i-1}(\bar{v}_i)\}$ are bases for $Z(v_i, T)$ and $Z(\bar{v}_i, \bar{T})$, respectively, for $i = 2, \ldots, r$. But $\bar{V} = Z(\bar{v}_2, \bar{T}) \oplus \cdots \oplus Z(\bar{v}_r, \bar{T})$; hence $\{\bar{v}_2, \ldots, \bar{T}^{dn_2-1}(\bar{v}_2), \ldots, \bar{v}_r, \ldots, \bar{T}^{dn_r-1}(\bar{v}_r)\}$ is a basis for $\bar{V}$. Therefore by the relation $\bar{T}^i(\bar{v}) = \overline{T^i(v)}$, $\{v_1, \ldots, T^{dn_1-1}(v_1), v_2, \ldots, T^{dn_2-1}(v_2), \ldots, v_r, \ldots, T^{dn_r-1}(v_r)\}$ is a basis for V. Thus $V = Z(v_1, T) \oplus \cdots \oplus Z(v_r, T)$, as required.

It remains to show that the exponents $n_1, \ldots, n_r$ are uniquely determined by T. Since d denotes the degree of $f(t)$, $\dim V = d(n_1 + \cdots + n_r)$ and $\dim Z_i = dn_i$, $i = 1, \ldots, r$. Also, if s is any positive integer then $f(T)^s(Z_i)$ is a cyclic subspace generated by $f(T)^s(v_i)$ and it has dimension $d(n_i - s)$ if $n_i > s$ and dimension 0 if $n_i \leq s$.

Now any vector $v \in V$ can be written uniquely in the form $v = w_1 + \cdots + w_r$ where $w_i \in Z_i$. Hence any vector in $f(T)^s(v) = f(T)^s(w_1) + \cdots + f(T)^s(w_r)$ where $f(T)^s(w_i) \in f(T)^s(Z_i)$. Let t be the integer, dependent on s, for which $n_1 > s, \ldots, n_t > s$, $n_{t+1} \leq s$. Then $f(T)^s(V) = f(T)^s(Z_1) \oplus \cdots \oplus f(T)^s(Z_t)$ and so

$$\dim(f(T)^s(V)) = d[(n_1 - s) + \cdots + (n_t - s)] \tag{2}$$

The numbers on the left of (2) are uniquely determined by T. Set $s = n - 1$ and (2) determines the number of n_i equal to n. Next set $s = n - 2$ and (2) determines the number of n_i (if any) equal to $n - 1$. We repeat the process until we set $s = 0$ and determine the number of n_i equal to 1. Thus the n_i are uniquely determined by T and V, and the lemma is proved.

17.152 Let T be a linear operator with minimum polynomial $f(t)^3$ and characteristic polynomial $f(t)^6$ where $f(t)$ is irreducible over the base field K. Find all possible sets of elementary divisors.

▌ The elementary divisors are powers of $f(t)$ with the exponents satisfying three conditions: One exponent must equal 3, the exponent in $m(t)$. The other exponents cannot exceed 3. The sum of the exponents must equal 6, the exponent in $\Delta(t)$. There are three possibilities:
(a) $6 = 3 + 3$ corresponding to $f(t)^3$, $f(t)^3$
(b) $6 = 3 + 2 + 1$ corresponding to $f(t)^3$, $f(t)^2$, $f(t)$
(c) $6 = 3 + 1 + 1 + 1$ corresponding to $f(t)^3$, $f(t)$, $f(t)$, $f(t)$

17.153 Find the maximum number of nonsimilar matrices with $m(t) = f(t)^3$ and $\Delta(t) = f(t)^6$ where $f(t)$ is irreducible.

▮ The nonsimilar matrices correspond to the number of distinct rational canonical forms which corresponds to the number of sets of elementary divisors. By Problem 17.152, there are three different sets of elementary divisors; hence there are three nonsimilar matrices with the given properties.

Problems 17.154–17.158 refer to the polynomial $f(t) = t^3 - 3t^2 + 3t + 2$ and a matrix A with minimum polynomial $m(t) = f(t)$ and characteristic polynomial $\Delta(t) = f(t)^2$.

17.154 Show that $f(t)$ is irreducible over the rational field **Q**.

▮ The rational roots of $f(t)$ must be among $\pm 1, \pm 2$. Testing each of the four integers we get $f(1) \neq 0$, $f(-1) \neq 0$, $f(2) \neq 0$, and $f(-2) \neq 0$. Thus $f(t)$ is irreducible over **Q**.

17.155 Suppose A is a matrix over the rational field **Q**. Find the rational canonical form M of A over **Q**.

▮ Since $f(t)$ is irreducible over **Q**, the elementary divisors of A can only be $f(t), f(t)$. Thus $M = C(t^3 - 3t^2 + 3t + 2) \oplus C(t^3 - 3t^2 + 3t + 2)$ where $C(f(T))$ denotes the companion polynomial of $f(t)$. That is,

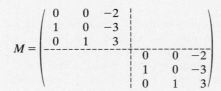

17.156 Find the number of real roots of $f(t)$.

▮ Plotting $f(t)$ as in Fig. 17-5 we see that $f(t)$ crosses the x axis at only one point; hence $f(t)$ has only one real root.

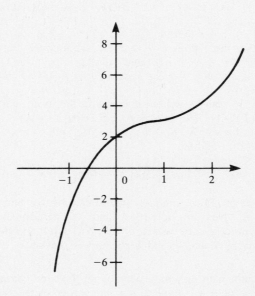

Fig. 17-5

17.157 Suppose A is a matrix over the real field **R**. Find the number of diagonal blocks in the rational canonical form M' of A over **R**.

▮ Since $f(t)$ has only one real root, $f(t)$ factors into $f(t) = g_1(t)g_2(t)$ where $g_1(t)$ and $g_2(t)$ are irreducible over **R**. Here the elementary divisors of A are as follows: $g_1(t), g_1(t), g_2(t), g_2(t)$. Thus M' has four diagonal blocks, two of order 1 and two of order 2.

17.158 Suppose A is a matrix over the complex field **C**. Show that the rational canonical form M'' of A over **C** is diagonal.

▮ Here $f(t)$ has one real root and two distinct complex roots. Thus $f(t)$ factors into $f(t) = h_1(t)h_2(t)h_3(t)$ over **C** where each $h_i(t)$ is linear. Thus the elementary divisors of A are as follows: $h_1(t)$, $h_1(t)$, $h_2(t)$, $h_2(t)$, $h_3(t)$, $h_3(t)$. Hence M'' has six diagonal blocks, each of order 1. In other words, M'' is diagonal.

17.159 Find all possible rational canonical forms for 6×6 matrices with minimum polynomial $m(t) = (t^2 + 3)(t + 1)^2$.

▮ Since the sum of the degrees of the elementary divisors must add up to 6, there are three possibilities: (a) $t^2 + 3$, $t^2 + 3$, $(t + 1)^2$; (b) $t^2 + 3$, $(t + 1)^2$, $(t + 1)^2$; (c) $t^2 + 3$, $(t + 1)^2$, $t + 1$, $t + 1$. The corresponding rational canonical forms follow:

$$
\begin{pmatrix}
0 & -3 & & & & \\
1 & 0 & & & & \\
& & 0 & -3 & & \\
& & 1 & 0 & & \\
& & & & 0 & -1 \\
& & & & 1 & -2
\end{pmatrix}
\quad
\begin{pmatrix}
0 & -3 & & & & \\
1 & 0 & & & & \\
& & 0 & -1 & & \\
& & 1 & -2 & & \\
& & & & 0 & -1 \\
& & & & 1 & -2
\end{pmatrix}
\quad
\begin{pmatrix}
0 & -3 & & & & \\
1 & 0 & & & & \\
& & 0 & -1 & & \\
& & 1 & -2 & & \\
& & & & -1 & \\
& & & & & -1
\end{pmatrix}
$$

17.160 Find the characteristic polynomial in each case in the preceding problem.

▮ The characteristic polynomial is equal to the product of the elementary divisors; hence $\Delta_a(t) = (t^2 + 3)^2(t + 1)^2$, $\Delta_b(t) = \Delta_c(t) = (t^2 + 3)(t + 1)^4$.

CHAPTER 18
Linear Functionals and the Dual Space

This chapter covers linear mappings from a vector space V into its field K of scalars [where K is viewed as a vector space over itself]. Naturally, all the theorems and results for arbitrary linear mappings on V hold for this special case. However, these mappings are also treated separately because of their fundamental importance and because the special relationship of V to K gives rise to new notions and results which do not apply in the general case.

18.1 LINEAR FUNCTIONALS AND THE DUAL SPACE

18.1 Define a linear functional.

▮ Let V be a vector space over a field K. A mapping $\phi: V \to K$ is termed a *linear functional* [or *linear form*] if, for every $u, v \in V$ and every $a, b \in K$, $\phi(au + bv) = a\phi(u) + b\phi(v)$. In other words, a linear functional on V is a linear mapping from V into K.

18.2 Let $V = K^n$. The mapping $\pi_i: V \to K$ defined by $\pi_i(a_1, a_2, \ldots, a_n) = a_i$ is called the *i*th *projection mapping*. Show that π_i is a linear functional on $V = K^n$.

▮ We need to show that π_i is linear. Suppose $u, v \in V$, say $u = (a_1, \ldots, a_n)$ and $v = (b_1, \ldots, b_n)$. Then $u + v = (a_1 + b, \ldots, a_n + b_n)$, $ku = (ka_i, \ldots, ka_n)$. Therefore, $\pi_i(u + v) = a_i + b_i = \pi_i(u) + \pi_i(v)$ and $\pi_i(ku) = ka_i = k\pi_i(u)$. Thus π_i is a linear functional.

18.3 Let V be a vector space of real continuous function on the interval $a \leq t \leq b$. Let $J: V \to \mathbf{R}$ be the integral operator defined by

$$J[f] = \int_a^b f(t)\, dt$$

Show that J is a linear functional on V.

▮ By calculus,

$$J[f + g] = \int_a^b [f(t) + g(t)]\, dt = \int_a^b f(t)\, dt + \int_a^b g(t)\, dt = J[f] + J[g]$$

and
$$J[kf] = \int_a^b kf(t)\, dt = k \int_a^b f(t)\, dt = kJ[f]$$

Thus J is linear and hence J is a linear functional on V.

18.4 Let V be the vector space of n-square matrices over K. Let $T: V \to K$ be the *trace mapping*, that is, $T(A) = a_{11} + a_{22} + \cdots + a_{nn}$, where $A = (a_{ij})$. That is, T assigns to a matrix A the sum of its diagonal elements. Is T a linear functional on V?

▮ T is a linear map, i.e., $T(A + B) = T(A) + T(B)$ and $T(kA) = kT(A)$. Thus the trace map is a linear functional on V.

18.5 Let V be the vector space of n-square matrices over K. Let $D: V \to K$ be the determinant function, i.e., $D(A) = \det(A)$. Is D a linear functional on V?

▮ D is not linear. For example, let $A = \begin{pmatrix} 1 & 0 \\ 0 & 0 \end{pmatrix}$ and $B = \begin{pmatrix} 0 & 0 \\ 0 & 1 \end{pmatrix}$. Then $D(A) = 0$, $D(B) = 0$ but $D(A + B) = D(I) = 1 \neq D(A) + D(B)$. Thus D is not a linear functional on V.

18.6 Let V be the vector space of all real polynomials. Let D be the derivative operator on V, that is, $D[f(t)] = df/dt$. Is D a linear functional on V?

▮ The derivative operator is linear. However, D does not map V into the scalar field $\mathbf{R}$. Hence D is not a linear functional on V.

18.7 Define the dual space of a vector space V.

▮ The set of linear functionals on a vector space V over a field K is also a vector space over K with addition and scalar multiplication defined by $(\phi + \sigma)(v) = \phi(v) + \sigma(v)$ and $(k\phi)(v) = k\phi(v)$ where ϕ and σ are linear functionals on V and $k \in K$. This space is called the *dual space* of V and is denoted by V^*.

Problems 18.8–18.10 refer to linear functionals $\phi: \mathbf{R}^2 \to \mathbf{R}$ and $\sigma: \mathbf{R}^2 \to \mathbf{R}$ defined by $\phi(x, y) = x + 2y$ and $\sigma(x, y) = 3x - y$.

18.8 Find $\phi + \sigma$.

▮ $(\phi + \sigma)(x, y) = \phi(x, y) + \sigma(x, y) = x + 2y + 3x - y = 4x + y$.

18.9 Find 4ϕ.

▮ $(4\phi)(x, y) = 4\phi(x, y) = 4(x + 2y) = 4x + 8y$.

18.10 Find $2\phi - 5\sigma$.

▮ $(2\phi - 5\sigma)(x, y) = 2\phi(x, y) - 5\sigma(x, y) = 2(x + 2y) - 5(3x - y) - 5(3x - y) = -13x + 9y$.

Problems 18.11–18.13 refer to linear functionals $\phi: \mathbf{R}^3 \to \mathbf{R}$ and $\sigma: \mathbf{R}^3 \to \mathbf{R}$ defined by $\phi(x, y, z) = 2x - 3y + z$ and $\sigma(x, y, z) = 4x - 2y + 3z$.

18.11 Find $\phi + \sigma$.

▮ $(\phi + \sigma)(x, y, z) = \phi(x, y, z) + \sigma(x, y, z) = (2x - 3y + z) + (4x - 2y + 3z) = 6x - 5y + 4z$.

18.12 Find 3ϕ.

▮ $(3\phi)(x, y, z) = 3\phi(x, y, z) = 3(2x - 3y + z) = 6x - 9y + 3z$.

18.13 Find $2\phi - 5\sigma$.

▮ $(2\phi - 5\sigma)(x, y, z) = 2\phi(x, y, z) - 5\sigma(x, y, z) = 2(2x - 3y + z) - 5(4x - 2y + 3z) = -16x + 4y - 13z$.

18.14 Let $V = K^n$. Show how the dual space V^* may be identified with the space of row vectors [where the elements of V are column vectors].

▮ Let σ be an element of the dual space V^*, i.e., a linear map $\sigma: V \to K$. Choosing a basis for V, say the usual basis, then σ is represented by a matrix $[\sigma]$. However, such a matrix $[\sigma]$ is a row vector. Also, the map $\sigma \mapsto [\sigma]$ is a vector space isomorphism.

On the other hand, any row vector $\phi = (a_1, \ldots, a_n)$ defines a linear functional $\phi: V \to K$ by

$$\phi(x_1, \ldots, x_n) = (a_1, a_2, \ldots, a_n)\begin{pmatrix} x_1 \\ x_2 \\ \vdots \\ x_n \end{pmatrix}$$

or simply $\phi(x_1, \ldots, x_n) = a_1 x_1 + a_2 x_2 + \cdots + a_n x_n$. Historically, the above formal expression was termed a *linear form*.

18.15 Let ϕ be the linear functional on $\mathbf{R}^2$ defined by $\phi(2, 1) = 15$ and $\phi(1, -2) = -10$. Find ϕ.

▮ Let $\phi = (a, b)$, a row vector. We are given

$$(a, b)\begin{pmatrix} 2 \\ 1 \end{pmatrix} = 15 \quad \text{or} \quad 2a + b = 15$$

$$(a, b)\begin{pmatrix} 1 \\ -2 \end{pmatrix} = -10 \quad \text{or} \quad a - 2b = -10$$

Both equations yield $a = 4$, $b = 7$. Thus $\phi = (4, 7)$ and $\phi(x, y) = 4x + 7y$.

18.16 Suppose dim $V = n$. What is the dimension of the dual space V^*?

▌ Note dim $K = 1$ where K is a vector space over itself. Since $V^* = \text{Hom}(V, K)$, the space of linear maps from V into K, dim $V^* = \dim[\text{Hom}(V, K)] = (\dim V) \cdot (\dim K) = n \cdot 1 = n$. [This is expected since V^* may be identified with the row vectors when V is identified with column vectors.]

18.2 DUAL BASIS

The main content of this section is the following theorem which is proved in Problem 18.23.

> **Theorem 18.1:** Suppose $\{v_1, \ldots, v_n\}$ is a basis of V over K. Let $\phi_1, \ldots, \phi_n \in V^*$ be the linear functionals defined by
>
> $$\phi_i(v_j) = \delta_{ij} = \begin{cases} 1 & \text{if } i = j \\ 0 & \text{if } i \neq j \end{cases}$$
>
> Then $\{\phi_1, \ldots, \phi_n\}$ is a basis of V^*.

The above basis $\{\phi_i\}$ is termed the basis *dual* to $\{v_i\}$ or the *dual basis*. The above formula which uses the Kronecker delta δ_{ij} is a short way of writing

$$\phi_1(v_1) = 1, \; \phi_1(v_2) = 0, \; \phi_1(v_3) = 0, \ldots, \phi_1(v_n) = 0$$
$$\phi_2(v_1) = 0, \; \phi_2(v_2) = 1, \; \phi_2(v_3) = 0, \ldots, \phi_2(v_n) = 0$$
$$\cdots\cdots\cdots\cdots\cdots\cdots\cdots\cdots\cdots\cdots\cdots\cdots\cdots\cdots\cdots$$
$$\phi_n(v_1) = 0, \; \phi_n(v_2) = 0, \ldots, \phi_n(v_{n-1}) = 0, \; \phi_n(v_n) = 1$$

These linear functions ϕ_i are unique and well-defined since they are defined on a basis of V.

18.17 Consider the following basis of $\mathbf{R}^2$: $u_1 = \begin{pmatrix} 1 \\ 2 \end{pmatrix}$ and $u_2 = \begin{pmatrix} 2 \\ 3 \end{pmatrix}$. Find row vectors w_1 and w_2 which form the dual basis.

▌ Let $w_1 = (a, b)$. Since $w_1 u_1 = 1$ and $w_1 u_2 = 0$, we get

$$w_1 u_1 = (a, b) \begin{pmatrix} 1 \\ 2 \end{pmatrix} = a + 2b = 1 \quad \text{and} \quad w_1 u_2 = (a, b) \begin{pmatrix} 2 \\ 3 \end{pmatrix} = 2a + 3b = 0$$

Thus $a = -3$, $b = 2$, and so $w_1 = (-3, 2)$.
Let $w_2 = (c, d)$. Since $w_2 u_1 = 0$ and $w_2 u_2 = 1$, we get

$$w_2 u_1 = (c, d) \begin{pmatrix} 1 \\ 2 \end{pmatrix} = c + 2d = 0 \quad \text{and} \quad w_2 u_2 = (c, d) \begin{pmatrix} 2 \\ 3 \end{pmatrix} = 2c + 3d = 1$$

Thus $c = 2$, $d = -1$, and so $w_2 = (2, -1)$.

18.18 Consider the following basis of $\mathbf{R}^2$: $\{v_1 = (2, 1), v_2 = (3, 1)\}$. Find the dual basis $\{\phi_1, \phi_2\}$.

▌ We seek linear functionals $\phi_1(x, y) = ax + by$ and $\phi_2(x, y) = cx + dy$ such that $\phi_1(v_1) = 1$, $\phi_1(v_2) = 0$, $\phi_2(v_1) = 0$, $\phi_2(v_2) = 1$. Thus

$$\left. \begin{array}{l} \phi_1(v_1) = \phi_1(2, 1) = 2a + b = 1 \\ \phi_1(v_2) = \phi_1(3, 1) = 3a + b = 0 \end{array} \right\} \quad \text{or} \quad a = -1, b = 3$$

$$\left. \begin{array}{l} \phi_2(v_1) = \phi_2(2, 1) = 2c + d = 0 \\ \phi_2(v_2) = \phi_2(3, 1) = 3c + d = 1 \end{array} \right\} \quad \text{or} \quad c = 1, d = -2$$

Hence the dual basis is $\{\phi_1(x, y) = -x + 3y, \phi_2(x, y) = x - 2y\}$.

18.19 Find the dual basis of the following basis of $\mathbf{R}^3$: $\{u_1 = (1, -2, 3), u_2 = (1, -1, 1), u_3 = (2, -4, 7)\}$.

▌ Let w_1, w_2, w_3 be the dual basis represented by row vectors. Suppose $w_1 = (a_1, a_2, a_3)$. Since $w_1 u_1 = 1$, $w_2 u_2 = 0$, $w_3 u_3 = 0$, we get

$$a_1 - 2a_2 + 3a_3 = 1 \qquad a_1 - a_2 + a_3 = 0 \qquad 2a_1 - 4a_2 + 7a_3 = 0 \qquad (1)$$

Solve to get $a_1 = -3$, $a_2 = -5$, $a_3 = -2$. Thus $w_1 = (-3, -5, -2)$ or $w_1(x, y, z) = -3x - 5y - 2z$.

Suppose $w_2 = (b_1, b_2, b_3)$. Since $w_2 u_1 = 0$, $w_2 u_2 = 1$, $w_2 u_3 = 0$, we get

$$b_1 - 2b_2 + 3b_3 = 0 \qquad b_1 - b_2 + b_3 = 1 \qquad 2b_1 - 4b_2 + 7b_3 = 0 \qquad (2)$$

Solve to get $b_1 = 2$, $b_2 = 1$, $b_3 = 0$. Thus $w_2 = (2, 1, 0)$ or $w_2(x, y, z) = 2x + y$.
Suppose $w_3 = (c_1, c_2, c_3)$. Since $w_3 u_1 = 0$, $w_3 u_2 = 0$, $w_3 u_3 = 1$, we get

$$c_1 - 2c_2 + 3c_3 = 0 \qquad c_1 - c_2 + c_3 = 0 \qquad 2c_1 - 4c_2 + 7c_3 = 1 \qquad (3)$$

Solve to get $c_1 = 1$, $c_2 = 2$, $c_3 = 1$. Thus $w_3 = (1, 2, 1)$ or $w_3(x, y, z) = x + 2y + z$.
   Remark: Observe the similarities in Equations *(1)*, *(2)*, and *(3)*.

18.20   Consider the following basis of $\mathbf{R}^3$: $\{v_1 = (1, -1, 3), v_2 = (0, 1, -1), v_3 = (0, 3, -2)\}$. Find the dual basis $\{\phi_1, \phi_2, \phi_3\}$.

❚   We seek linear functionals $\phi_1(x, y, z) = a_1 x + a_2 y + a_3 z$, $\phi_2(x, y, z) = b_1 x + b_2 y + b_3 z$, $\phi_3(x, y, z) = c_1 x + c_2 y + c_3 z$ such that

$$
\begin{array}{lll}
\phi_1(v_1) = 1 & \phi_1(v_2) = 0 & \phi_1(v_3) = 0 \\
\phi_2(v_1) = 0 & \phi_2(v_2) = 1 & \phi_2(v_3) = 0 \\
\phi_3(v_1) = 0 & \phi_3(v_2) = 0 & \phi_3(v_3) = 1
\end{array}
$$

We find ϕ_1 as follows:

$$
\begin{array}{l}
\phi_1(v_1) = \phi_1(1, -1, 3) = a_1 - a_2 + 3a_3 = 1 \\
\phi_1(v_2) = \phi_1(0, 1, -1) = \quad\;\; a_2 - a_3 = 0 \\
\phi_1(v_3) = \phi_1(0, 3, -2) = \quad\; 3a_2 - 2a_3 = 0
\end{array}
$$

Solving the system of equations, we obtain $a_1 = 1$, $a_2 = 0$, $a_3 = 0$. Thus $\phi_1(x, y, z) = x$.
   We next find ϕ_2:

$$
\begin{array}{l}
\phi_2(v_1) = \phi_2(1, -1, 3) = b_1 - b_2 + 3b_3 = 0 \\
\phi_2(v_2) = \phi_2(0, 1, -1) = \quad\;\; b_2 - b_3 = 1 \\
\phi_2(v_3) = \phi_2(0, 3, -2) = \quad\; 3b_2 - 2b_3 = 0
\end{array}
$$

Solving the system, we obtain $b_1 = 7$, $b_2 = -2$, $b_3 = -3$. Hence $\phi_2(x, y, z) = 7x - 2y - 3z$.
   Finally, we find ϕ_3:

$$
\begin{array}{l}
\phi_3(v_1) = \phi_3(1, -1, 3) = c_1 - c_2 + 3c_3 = 0 \\
\phi_3(v_2) = \phi_3(0, 1, -1) = \quad\;\; c_2 - c_3 = 0 \\
\phi_3(v_3) = \phi_3(0, 3, -2) = \quad\; 3c_2 - 2c_3 = 1
\end{array}
$$

Solving the system, we obtain $c_1 = -2$, $c_2 = 1$, $c_3 = 1$. Thus $\phi_3(x, y, z) = -2x + y + z$.

18.21   Find the basis $\{f_1, f_2, f_3\}$ which is dual to the usual basis $\{e_1, e_2, e_3\}$ of $\mathbf{R}^3$.

❚   Suppose $f_1 = (a, b, c)$. Since $f_1 e_1 = 1$, $f_1 e_2 = 0$, $f_1 e_3 = 0$, we get

$$1 = (a, b, c)\begin{pmatrix} 1 \\ 0 \\ 0 \end{pmatrix} = a \qquad 0 = (a, b, c)\begin{pmatrix} 0 \\ 1 \\ 0 \end{pmatrix} = b \qquad 0 = (a, b, c)\begin{pmatrix} 0 \\ 0 \\ 1 \end{pmatrix} = c$$

Thus $f_1 = (1, 0, 0)$. Similarly, $f_2 = (0, 1, 0)$ and $f_3 = (0, 0, 1)$. In other words, the dual of the usual basis of $\mathbf{R}^3$ is the usual basis of $\mathbf{R}^{3*}$.

18.22   Let V be the vector space of real polynomials $f(t) = a + bt$ [i.e., $\deg f \le 1$]. Let $\phi: V \to \mathbf{R}$ and $\phi_2: V \to \mathbf{R}$ be defined by

$$\phi_1(f(t)) = \int_0^1 f(t)\, dt \qquad \text{and} \qquad \phi_2(f(t)) = \int_0^2 f(t)\, dt$$

[We remark that ϕ_1 and ϕ_2 are linear and so belong to the dual space V^*.] Find the basis $\{v_1, v_2\}$ of V which is dual to $\{\phi_1, \phi_2\}$.

▮ Let $v_1 = a + bt$ and $v_2 = c + dt$. By definition of the dual basis, $\phi_1(v_1) = 1$, $\phi_2(v_1) = 0$ and $\phi_1(v_2) = 0$, $\phi_2(v_2) = 1$. Thus

$$\left.\begin{array}{l} \phi_1(v_1) = \displaystyle\int_0^1 (a + bt)\,dt = a + \tfrac{1}{2}b = 1 \\[2mm] \phi_2(v_1) = \displaystyle\int_0^2 (a + bt)\,dt = 2a + 2b = 0 \end{array}\right\} \quad \text{or} \quad a = 2,\, b = -2$$

$$\left.\begin{array}{l} \phi_1(v_2) = \displaystyle\int_0^1 (c + dt)\,dt = c + \tfrac{1}{2}d = 0 \\[2mm] \phi_2(v_2) = \displaystyle\int_0^2 (c + dt)\,dt = 2c - 2d = 1 \end{array}\right\} \quad \text{or} \quad c = -\tfrac{1}{2},\, d = 1$$

In other words, $\{2 - 2t, -\tfrac{1}{2} + t\}$ is the basis of V which is dual to $\{\phi_1, \phi_2\}$.

18.23 Prove Theorem 18.1.

▮ We first show that $\{\phi_1, \ldots, \phi_n\}$ spans V^*. Let ϕ be an arbitrary element of V^*, and suppose $\phi(v_1) = k_1, \phi(v_2) = k_2, \ldots, \phi(v_n) = k_n$. Set $\sigma = k_1\phi_1 + \cdots + k_n\phi_n$. Then

$$\begin{aligned} \sigma(v_1) &= (k_1\phi_1 + \cdots + k_n\phi_n)(v_1) \\ &= k_1\phi_1(v_1) + k_2\phi_2(v_1) + \cdots + k_n\phi_n(v_1) \\ &= k_1 \cdot 1 + k_2 \cdot 0 + \cdots + k_n \cdot 0 = k_1 \end{aligned}$$

Similarly, for $i = 2, \ldots, n$, $\sigma(v_i) = (k_1\phi_1 + \cdots + k_n\phi_n)(v_i) = k_1\phi_1(v_i) + \cdots + k_i\phi_i(v_i) + \cdots + k_n\phi_n(v_i) = k_i$. Thus $\phi(v_i) = \sigma(v_i)$ for $i = 1, \ldots, n$. Since ϕ and σ agree on the basis vectors, $\phi = \sigma = k_1\phi_1 + \cdots + k_n\phi_n$. Accordingly, $\{\phi_1, \ldots, \phi_n\}$ spans V^*.

It remains to be shown that $\{\phi_1, \ldots, \phi_n\}$ is linearly independent. Suppose $a_1\phi_1 + a_2\phi_2 + \cdots + a_n\phi_n = 0$. Applying both sides to v_1, we obtain

$$\begin{aligned} 0 = 0(v_1) &= (a_1\phi_1 + \cdots + a_n\phi_n)(v_1) \\ &= a_1\phi_1(v_1) + a_2\phi_2(v_1) + \cdots + a_n\phi_n(v_1) \\ &= a_1 \cdot 1 + a_2 \cdot 0 + \cdots + a_n \cdot 0 = a_1 \end{aligned}$$

Similarly, for $i = 2, \ldots, n$, $0 = 0(v_i) = (a_1\phi_1 + \cdots + a_n\phi_n)(v_i) = a_1\phi_1(v_i) + \cdots + a_i\phi_i(v_i) + \cdots + a_n\phi_n(v_i) = a_i$. That is, $a_1 = 0, \ldots, a_n = 0$. Hence $\{\phi_1, \ldots, \phi_n\}$ is linearly independent and so it is a basis of V^*.

Theorem 18.2: Let $\{v_1, \ldots, v_n\}$ be a basis of V and let $\{\phi_1, \ldots, \phi_n\}$ be the dual basis of V^*. Then for any vector $u \in V$,

$$u = \phi_1(u)v_1 + \phi_2(u)v_2 + \cdots + \phi_n(u)v_n \tag{1}$$

and, for any linear functional $\sigma \in V^*$,

$$\sigma = \sigma(v_1)\phi_1 + \sigma(v_2)\phi_2 + \cdots + \sigma(v_n)\phi_n \tag{2}$$

18.24 Prove Theorem 18.2.

▮ Suppose

$$u = a_1v_1 + a_2v_2 + \cdots + a_nv_n \tag{3}$$

Then $\phi_1(u) = a_1\phi_1(v_1) + a_2\phi_1(v_2) + \cdots + a_n\phi_1(v_n) = a_1 \cdot 1 + a_2 \cdot 0 + \cdots + a_n \cdot 0 = a_1$. Similarly, for $i = 2, \ldots, n$, $\phi_i(u) = a_1\phi_i(v_1) + \cdots + a_i\phi_i(v_i) + \cdots + a_n\phi_i(v_n) = a_i$. That is, $\phi_1(u) = a_1$, $\phi_2(u) = a_2, \ldots$, $\phi_n(u) = a_n$. Substituting these results into (3), we obtain (1).

Next we prove (2). Applying the linear functional σ to both sides of (1),

$$\begin{aligned} \sigma(u) &= \phi_1(u)\sigma(v_1) + \phi_2(u)\sigma(v_2) + \cdots + \phi_n(u)\sigma(v_n) \\ &= \sigma(v_1)\phi_1(u) + \sigma(v_2)\phi_2(u) + \cdots + \sigma(v_n)\phi_n(u) \\ &= (\sigma(v_1)\phi_1 + \sigma(v_2)\phi_2 + \cdots + \sigma(v_n)\phi_n)(u) \end{aligned}$$

Since the above holds for every $u \in V$, $\sigma = \sigma(v_1)\phi_1 + \sigma(v_2)\phi_2 + \cdots + \sigma(v_n)\phi_n$ as claimed.

Theorem 18.3: Let $\{v_1, \ldots, v_n\}$ and $\{w_1, \ldots, w_n\}$ be bases of V and let $\{\phi_1, \ldots, \phi_n\}$ and $\{\sigma_1, \ldots, \sigma_n\}$ be the bases of V^* dual to $\{v_i\}$ and $\{w_i\}$, respectively. Suppose P is the transition matrix from $\{v_i\}$ to $\{w_i\}$. Then $(P^{-1})^t$ is the transition matrix from $\{\phi_i\}$ to $\{\sigma_i\}$.

18.25 Prove Theorem 18.3.

▮ Suppose

$$
\begin{array}{ll}
w_1 = a_{11}v_1 + a_{12}v_2 + \cdots + a_{1n}v_n & \sigma_1 = b_{11}\phi_1 + b_{12}\phi_2 + \cdots + b_{1n}\phi_n \\
w_2 = a_{21}v_1 + a_{22}v_2 + \cdots + a_{2n}v_n & \sigma_2 = b_{21}\phi_1 + b_{22}\phi_2 + \cdots + b_{2n}\phi_n \\
\cdots\cdots\cdots\cdots\cdots\cdots\cdots\cdots & \cdots\cdots\cdots\cdots\cdots\cdots\cdots\cdots \\
w_n = a_{n1}v_1 + a_{n2}v_2 + \cdots + a_{nn}v_n & \sigma_n = b_{n1}\phi_1 + b_{n2}\phi_2 + \cdots + b_{nn}\phi_n
\end{array}
$$

where $P = (a_{ij})$ and $Q = (b_{ij})$. We seek to prove that $Q = (P^{-1})^T$.

Let R_i denote the ith row of Q and let C_j denote the jth column of P^T. Then $R_i = (b_{i1}, b_{i2}, \ldots, b_{in})$ and $C_j = (a_{j1}, a_{j2}, \ldots, a_{jn})^T$. By definition of the dual basis, $\sigma_i(w_j) = (b_{i1}\phi_1 + b_{i2}\phi_2 + \cdots + b_{in}\phi_n)(a_{j1}v_1 + a_{j2}v_2 + \cdots + a_{jn}v_n) = b_{i1}a_{j1} + b_{i2}a_{j2} + \cdots + b_{in}a_{jn} = R_iC_j = \delta_{ij}$ where δ_{ij} is the Kronecker delta.

Thus
$$
QP^T = \begin{pmatrix} R_1C_1 & R_1C_2 & \cdots & R_1C_n \\ R_2C_1 & R_2C_2 & \cdots & R_2C_n \\ \cdots\cdots\cdots\cdots\cdots\cdots \\ R_nC_1 & R_nC_2 & \cdots & R_nC_n \end{pmatrix} = \begin{pmatrix} 1 & 0 & \cdots & 0 \\ 0 & 1 & \cdots & 0 \\ \cdots\cdots\cdots\cdots \\ 0 & 0 & \cdots & 1 \end{pmatrix} = I
$$

and hence $Q = (P^T)^{-1} = (P^{-1})^T$ as claimed.

Problems 18.26–18.30 refer to the following basis of $\mathbf{R}^2$: $S_1 = \{v_1 = (1, 1), v_2 = (1, 0)\}$ and $S_2 = \{w_1 = (4, 3), w_2 = (3, 2)\}$.

18.26 Find the change-of-basis matrix P from S_1 to S_2.

▮ Write w_1 as a linear combination of v_1 and v_2: $(4, 3) = x(1, 1) + y(1, 0) = (x + y, x)$ or $x + y = 4$, $x = 3$. Thus $x = 3$, $y = 1$ is a solution; hence $w_1 = 3v_1 + v_2$.

Write w_2 as a linear combination of v_1 and v_2: $(3, 2) = x(1, 1) + y(1, 0) = (x + y, x)$ or $x + y = 3$, $x = 2$. Thus $x = 2$, $y = 1$ is a solution; hence $w_2 = 2v_1 + v_2$.

Write the coordinates of w_1 and w_2 as columns to get

$$P = \begin{pmatrix} 3 & 2 \\ 1 & 1 \end{pmatrix}$$

18.27 Find the basis $S_1' = \{\phi_1, \phi_2\}$ which is dual to S_1.

▮ Let $\phi_1 = (a, b)$. Then $\phi_1 v_1 = 1$ and $\phi_1 v_2 = 0$. Thus $a + b = 1$, $a = 0$, so $a = 0$, $b = 1$, and $\phi = (0, 1)$. Let $\phi_2 = (c, d)$. Then $\phi_2 v_1 = 0$, $\phi_2 v_2 = 1$. Thus $c + d = 0$, $c = 1$ so $c = 1$, $d = -1$, and $\phi_2 = (1, -1)$.

18.28 Find the basis $S_2' = \{\sigma_1, \sigma_2\}$ which is dual to S_2.

▮ Let $\sigma_1 = (a, b)$. Then $\sigma_1 w_1 = 1$ and $\sigma_1 w_2 = 0$. Thus $4a + 3b = 1$, $3a + 2b = 0$ or $a = -2$, $b = 3$, and $\sigma_1 = (-2, 3)$. Let $\sigma_2 = (c, d)$. Then $\sigma_2 w_1 = 0$ and $\sigma_2 w_2 = 1$. Thus $4c + 3d = 0$, $3c + 2d = 1$ and $c = 3$, $d = -4$, and $\sigma_2 = (3, -4)$.

18.29 Find the change-of-basis matrix Q from S_1' to S_2'.

▮ Write σ_1 as a linear combination of ϕ_1 and ϕ_2: $(-2, 3) = x(0, 1) + y(1, -1)$, so $y = -2$, $x - y = 3$. Thus $x = 1$, $y = -2$ is a solution; hence $\sigma_1 = \phi_1 - 2\phi_2$.

Write σ_2 as a linear combination of ϕ_1 and ϕ_2: $(3, -4) = x(0, 1) + y(1, -1)$ or $y = 3$, $x - y = -4$. Thus $x = -1$, $y = 3$ is a solution; hence $\sigma_2 = -\phi_1 + 3\phi_2$.

Write the coordinates of σ_1 and σ_2 as columns to get

$$Q = \begin{pmatrix} 1 & -1 \\ -2 & 3 \end{pmatrix}$$

18.30 Verify Theorem 18.3, i.e., that $Q = (P^{-1})^T$.

▮ $P^{-1} = \begin{pmatrix} 1 & -2 \\ -1 & 3 \end{pmatrix}$ and hence $(P^{-1})^T = \begin{pmatrix} 1 & -1 \\ -2 & 3 \end{pmatrix} = Q$, as expected.

18.3 SECOND DUAL SPACE, NATURAL MAPPING

18.31 Define the second dual space.

▮ Let V be any vector space over K. The dual space V^* is the vector space of linear maps from V to K. The second dual space V^{**} is the vector space of linear maps from V^* into K, that is, V^{**} is the dual of the dual of V.

18.32 Let $v \in V$. Then v determines a map $\hat{v} : V^* \to K$ as follows: For any $\phi \in V^*$, define $\hat{v}(\phi) = \phi(v)$. Show that $\hat{v}$ is linear.

▮ For any $\phi, \sigma \in V^*$ and for any $a, b \in K$, $\hat{v}(a\phi + b\sigma) = (a\phi + b\sigma)(v) = a\phi(v) + b\sigma(v) = a\hat{v}(\phi) + b\hat{v}(\sigma)$. Thus $\hat{v}$ is linear.

18.33 Show that $\hat{v} \in V^{**}$.

▮ By the preceding problem, $\hat{v}$ is a linear map from V^* into K; hence $\hat{v} \in V^{**}$.

18.34 Define the natural mapping of V into V^{**}.

▮ The mapping $v \mapsto \hat{v}$, where $\hat{v}$ is defined above, is called the natural mapping from V into V^{**}.

18.35 Show that the natural mapping from V into V^{**} is linear, i.e., show that for any vectors $v, w \in V$ and any scalars $a, b \in K$, we have $\widehat{av + bw} = a\hat{v} + b\hat{w}$. [Here $\hat{v} : V^* \to K$ is defined by $\hat{v}(\phi) = \phi(v)$.]

▮ For any linear functional $\phi \in V^*$, $\widehat{av + bw}(\phi) = \phi(av + bw) = a\phi(v) + b\phi(w) = a\hat{v}(\phi) + b\hat{w}(\phi) = (a\hat{v} + b\hat{w})(\phi)$. Since $\widehat{av + bw}(\phi) = (a\hat{v} + b\hat{w})(\phi)$ for every $\phi \in V^*$, we have $\widehat{av + bw} = a\hat{v} + b\hat{w}$. Thus the map $v \mapsto \hat{v}$ is linear.

18.36 Consider any nonzero vector $v \in V$. Show that there exists $\phi \in V^*$ such that $\phi(v) \neq 0$.

▮ Extend $\{v\}$ to a basis $\{v_i\}$ of V. Then there exists a unique linear map $\phi : V \to K$ such that $\phi(v) = 1$ but $\phi(v_i) = 0$ for $v_i \neq v$. Then ϕ has the required property.

18.37 Show that the natural map from V to V^{**} is one-to-one.

▮ Suppose $v \in V$, $v \neq 0$. Then, by the preceding problem, there exists $\phi \in V^*$ for which $\phi(v) \neq 0$. Hence $\hat{v}(\phi) = \phi(v) \neq 0$ and thus $\hat{v} \neq 0$. Since $v \neq 0$ implies $\hat{v} \neq 0$, the map $v \mapsto \hat{v}$ is nonsingular. Thus the natural map is one-to-one.

Theorem 18.4: If V has finite dimension, then the mapping $v \mapsto \hat{v}$ is an isomorphism of V onto V^{**}.

18.38 Prove Theorem 18.4.

▮ Now $\dim V = \dim V^* = \dim V^{**}$ because V has finite dimension. Accordingly, the nonsingular mapping $v \mapsto \hat{v}$ is an isomorphism of V onto V^{**}.

Remark: Suppose V does have finite dimension. By the above theorem, the natural mapping determines an isomorphism between V and V^{**}. Unless otherwise stated, V is identified with V^{**} by this mapping, and we write $V = V^{**}$. Also, if $\{\phi_i\}$ is the basis of V^* dual to a basis $\{v_i\}$ of V, then $\{v_i\}$ is the basis of $V = V^{**}$ which is dual to $\{\phi_i\}$.

18.4 ANNIHILATORS

18.39 Let W be a subset of a vector space V. Define the annihilator of W, denoted by W^0.

▮ A linear functional $\phi \in V^*$ is called an annihilator of W if $\phi(w) = 0$ for every $w \in W$ or, in other words, if $\phi(W) = 0$. The set of all such mappings, denoted by W^0, is called the annihilator of W.

18.40 Show that W^0 is a subspace of V^*.

$\blacksquare$ Clearly $0 \in W^0$. Now suppose $\phi, \sigma \in W^0$. Then, for any scalars $a, b \in K$ and for any $w \in W$, $(a\phi + b\sigma)(w) = a\phi(w) + b\sigma(w) = a0 + b0 = 0$. Thus $a\phi + b\sigma \in W^0$ and so W^0 is a subspace of V^*.

18.41 Show that if $\phi \in V^*$ annihilates a subset S of V, then ϕ annihilates the linear span(S) of S. Hence $S^0 = [\text{span}(S)]^0$.

$\blacksquare$ Suppose $v \in \text{span}(S)$. Then there exist $w_1, \ldots, w_r \in S$ for which $v = a_1 w_1 + a_2 w_2 + \cdots + a_r w_r$. Then $\phi(v) = a_1\phi(w_1) + a_2\phi(w_2) + \cdots + a_r\phi(w_r) = a_10 + a_20 + \cdots + a_r0 = 0$. Since v was an arbitrary element of span(S), ϕ annihilates span(S) as claimed.

18.42 Let W be the subspace of $\mathbf{R}^4$ spanned by $v_1 = (1, 2, -3, 4)$ and $v_2 = (0, 1, 4, -1)$. Find a basis of the annihilator of W.

$\blacksquare$ By the preceding problem, it suffices to find a basis of the set of linear functionals $\phi(x, y, z, w) = ax + by + cz + dw$ for which $\phi(v_1) = 0$ and $\phi(v_2) = 0$:

$$\phi(1, 2, -3, 4) = a + 2b - 3c - 4d = 0$$
$$\phi(0, 1, 4, -1) = b + 4c - d = 0$$

The system of equations in unknowns a, b, c, d is in echelon form with free variables c and d.

Set $c = 1$, $d = 0$ to obtain the solution $a = 11$, $b = -4$, $c = 1$, $d = 0$ and hence the linear functional $\phi_1(x, y, z, w) = 11x - 4y + z$.

Set $c = 0$, $d = -1$ to obtain the solution $a = 6$, $b = -1$, $c = 0$, $d = -1$ and hence the linear functional $\phi_2(x, y, z, w) = 6x - y - w$.

The set of linear functionals $\{\phi_1, \phi_2\}$ is a basis of W^0, the annihilator of W.

18.43 Let S be a subset of V. Show that $S \subseteq S^{00}$.

$\blacksquare$ Let $v \in S$. Then for every linear functional $\phi \in S^0$, $\hat{v}(\phi) = \phi(v) = 0$. Hence $\hat{v} \in (S^0)^0$. Therefore, under the identification of V and V^{**}, $v \in S^{00}$. Accordingly, $S \subseteq S^{00}$.

18.44 Suppose $S_1 \subseteq S_2$. Show that $S_2^0 \subseteq S_1^0$.

$\blacksquare$ Let $\phi \in S_2^0$. Then $\phi(v) = 0$ for every $v \in S_2$. But $S_1 \subset S_2$; hence ϕ annihilates every element of S_1, that is, $\phi \in S_1$. Therefore $S_2^0 \subseteq S_1^0$.

Theorem 18.5: Suppose V has finite dimension and W is a subspace of V. Then (i) $\dim W + \dim W^0 = \dim V$ and (ii) $W^{00} = W$.

18.45 Prove (i) of Theorem 18.5.

$\blacksquare$ Suppose $\dim V = n$ and $\dim W = r \le n$. We want to show that $\dim W^0 = n - r$. We choose a basis $\{w_1, \ldots, w_r\}$ of W and extend it to the following basis of V: $\{w_1, \ldots, w_r, v_1, \ldots, v_{n-r}\}$. Consider the dual basis $\{\phi_1, \ldots, \phi_r, \sigma_1, \ldots, \sigma_{n-r}\}$. By definition of the dual basis, each of the above σ's annihilates each w_i; hence $\sigma_1, \ldots, \sigma_{n-r} \in W^0$. We claim that $\{\sigma_j\}$ is a basis of W^0. Now $\{\sigma_j\}$ is part of a basis of V^* and so it is linearly independent.

We next show that $\{\sigma_j\}$ spans W^0. Let $\sigma \in W^0$. Then

$$\sigma = \sigma(w_1)\phi_1 + \cdots + \sigma(w_r)\phi_r + \sigma(v_1)\sigma_1 + \cdots + \sigma(v_{n-r})\sigma_{n-r}$$
$$= 0\phi_1 + \cdots + 0\phi_r + \sigma(v_1)\sigma_1 + \cdots + \sigma(v_{n-r})\sigma_{n-r}$$
$$= \sigma(v_1)\sigma_1 + \cdots + \sigma(v_{n-r})\sigma_{n-r}$$

Thus $\{\sigma_1, \ldots, \sigma_{n-r}\}$ spans W^0 and so it is a basis of W^0. Accordingly, $\dim W^0 = n - r = \dim V - \dim W$ as required.

18.46 Prove (ii) of Theorem 18.5.

$\blacksquare$ Suppose $\dim V = n$ and $\dim W = r$. Then $\dim V^* = n$ and, by (i), $\dim W^0 = n - r$. Thus by (i), $\dim W^{00} = n - (n - r) = r$; therefore $\dim W = \dim W^{00}$. Since $W \subseteq W^{00}$, we have $W = W^{00}$.

18.47 Using the concept of an annihilator, give another interpretation of a homogeneous system of linear equations, say

$$a_{11}x_1 + a_{12}x_2 + \cdots + a_{1n}x_n = 0$$
$$a_{21}x_1 + a_{22}x_2 + \cdots + a_{2n}x_n = 0$$
$$\cdots\cdots\cdots\cdots\cdots\cdots\cdots\cdots\cdots$$
$$a_{m1}x_1 + a_{m2}x_2 + \cdots + a_{mn}x_n = 0$$

(1)

▎ Each row $(a_{i1}, a_{i2}, \ldots, a_{in})$ of the coefficient matrix $A = (a_{ij})$ is viewed as an element of K^n and each solution vector $\phi = (x_1, x_2, \ldots, x_n)$ is viewed as an element of the dual space. In this context, the solution space S of *(1)* is the annihilator of the rows of A and hence the row space of A. Consequently, we again obtain the following fundamental result on the dimension of the solution space of a homogeneous system of linear equations: $\dim S = \dim K^n - \dim(\text{row space of } A) = n - \text{rank}(A)$.

18.48   Let U and W be subspaces of V. Prove: $(U + W)^0 = U^0 \cap W^0$.

▎ Let $\phi \in (U + W)^0$. Then ϕ annihilates $U + W$ and so, in particular, ϕ annihilates U and V. That is, $\phi \in U^0$ and $\phi \in W^0$; hence $\phi \in U^0 \cap W^0$. Thus $(U + W)^0 \subset U^0 \cap W^0$.

On the other hand, suppose $\sigma \in U^0 \cap W^0$. Then σ annihilates U and also W. If $v \in U + W$, then $v = u + w$ where $u \in U$ and $w \in W$. Hence $\sigma(v) = \sigma(u) + \sigma(w) = 0 + 0 = 0$. Thus σ annihilates $U + W$, that is, $\sigma \in (U + W)^0$. Accordingly, $U^0 + W^0 \subset (U + W)^0$.

Both inclusion relations give us the desired equality.

Remark: Observe that no dimension argument is employed in the proof; hence the result holds for spaces of finite or infinite dimension.

18.5   TRANSPOSE OF A LINEAR MAPPING

18.49   Let $T: V \to U$ be an arbitrary linear mapping from a vector space V into a vector space U. Define the transpose of T, denoted by T^t.

▎ For any linear functional $\phi \in U^*$, the composition $\phi \circ T$ is a linear mapping from V into K as pictured in Fig. 18-1. That is, $\phi \circ T \in V^*$. Thus the correspondence $\phi \mapsto \phi \circ T$ is a mapping from U^* into V^*; we denote it by T^t and call it the *transpose* of T. In other words, $T^t: U^* \to V^*$ is defined by $T^t(\phi) = \phi \circ T$. Thus $[T^t(\phi)](v) = \phi(T(v))$ for every $v \in V$.

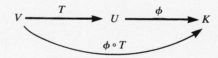

Fig. 18-1

Theorem 18.6:   The transpose mapping T^t defined above is linear.

18.50   Prove Theorem 18.6.

▎ For any scalars $a, b \in K$ and any linear functionals $\phi, \sigma \in U^*$, $T^t(a\phi + b\sigma) = (a\phi + b\sigma) \circ T = a(\phi \circ T) + b(\sigma \circ T) = aT^t(\phi) + bT^t(\sigma)$. That is, T^t is linear as claimed.

Problems 18.51–18.53 refer to a linear functional $\phi: \mathbf{R}^2 \to \mathbf{R}$ defined by $\phi(x, y) = x - 2y$ and a linear operator $T: \mathbf{R}^2 \to \mathbf{R}^2$ as given.

18.51   Find $[T^t(\phi)](x, y)$ when $T(x, y) = (x, 0)$.

▎ By definition of the transpose mapping, $T^t(\phi) = \phi \circ T$, that is, $[T^t(\phi)](v) = \phi(T(v))$ for every vector v. Hence $[T^t(\phi)](x, y) = \phi(T(x, y)) = \phi(x, 0) = x$.

18.52   Find $[T^t(\phi)](x, y)$ when $T(x, y) = (y, x + y)$.

▎ $[T^t(\phi)](x, y) = \phi(T(x, y)) = \phi(y, x + y) = y - 2(x + y) = -2x - y$.

18.53   Find $[T^t(\phi)](x, y)$ when $T(x, y) = (2x - 3y, 5x + 2y)$.

▎ $[T^t(\phi)](x, y) = \phi(T(x, y)) = \phi(2x - 3u, 5x + 2y) = (2x - 3y) - 2(5x + 2y) = -8x - 7y$.

Problems 18.54–18.56 refer to a linear functional $\phi: \mathbf{R}^2 \to \mathbf{R}$ defined by $\phi(x, y) = 3x - 2y$ and a linear operator $T: \mathbf{R}^3 \to \mathbf{R}^2$ as given.

18.54 Find $[T'(\phi)](x, y, z)$ when $T(x, y, z) = (x + y, y + z)$.

▌ Given $[T'(\phi)](v) = \phi(T(v))$, we have $[T'(\phi)](x, y, z) = \phi(T(x, y, z)) = \phi(x + y, y + z) = 3(x + y) - 2(y + z) = 3x + y - 2z$.

18.55 Find $[T'(\phi)](x, y, z)$ when $T(x, y, z) = (3z, x + y)$.

▌ $[T'(\phi)](x, y, z) = \phi(T(x, y, z)) = \phi(3z, x + y) = 3(3z) - 2(x + y) = -2x - 2y + 9z$.

18.56 Find $[T'(\phi)](x, y, z)$ when $T(x, y, z) = (x + y + z, 2x - y)$.

▌ $[T'(\phi)](x, y, z) = \phi(T(x, y, z)) = \phi(x + y + z, 2x - y) = 3(x - y + z) - 2(2x - y) = -x + 5y + 3z$.

18.57 Let $T: V \to U$ be linear and let $T': U^* \to V^*$ be its transpose. Show that the kernel of T' is the annihilator of the image of T, that is, $\operatorname{Ker} T' = (\operatorname{Im} T)^0$.

▌ Suppose $\phi \in \operatorname{Ker} T'$; that is, $T'(\phi) = \phi \circ T = 0$. If $u \in \operatorname{Im} T$, then $u = T(v)$ for some $v \in V$; hence $\phi(u) = \phi(T(v)) = (\phi \circ T)(v) = 0(v) = 0$. We have that $\phi(u) = 0$ for every $u \in \operatorname{Im} T$; hence $\phi \in (\operatorname{Im} T)^0$. Thus $\operatorname{Ker} T' \subset (\operatorname{Im} T)^0$.

On the other hand, suppose $\sigma \in (\operatorname{Im} T)^0$; that is, $\sigma(\operatorname{Im} T) = \{0\}$. Then, for every $v \in V$, $[T'(\sigma)](v) = (\sigma \circ T)(v) = \sigma(T(v)) = 0 = 0(v)$. We have that $[T'(\sigma)](v) = 0(v)$ for every $v \in V$; hence $T'(\sigma) = 0$. Therefore $\sigma \in \operatorname{Ker} T'$ and so $(\operatorname{Im} T)^0 \subset \operatorname{Ker} T'$.

Both inclusion relations give us the required equality.

18.58 Suppose V and U have finite dimension and suppose $T: V \to U$ is linear. Prove $\operatorname{rank}(T) = \operatorname{rank}(T')$.

▌ Suppose $\dim V = n$ and $\dim U = m$. Also suppose $\operatorname{rank}(T) = r$. Then, $\dim((\operatorname{Im} T)^0 = \dim U - \dim(\operatorname{Im} T) = m - \operatorname{rank}(T) = m - r$. By the preceding problem, $\operatorname{Ker} T' = (\operatorname{Im} T)^0$. Hence nullity $(T') = m - r$. It then follows that, as claimed, $\operatorname{rank}(T') = \dim U^* - \operatorname{nullity}(T') = m - (m - r) = r = \operatorname{rank}(T)$.

Theorem 18.7: Let $T: V \to U$ be linear, and let A be the matrix representation of T relative to bases $\{v_i\}$ of V and $\{u_i\}$ of U. Then the transpose matrix A^T is the matrix representation of $T': U^* \to V^*$ relative to the bases dual to $\{u_i\}$ and $\{v_i\}$.

18.59 Prove Theorem 18.7 which indicates the reason the name "transpose" is used for the mapping T'.

▌ Suppose

$$\begin{aligned} T(v_1) &= a_{11}u_1 + a_{12}u_2 + \cdots + a_{1n}u_n \\ T(v_2) &= a_{21}u_1 + a_{22}u_2 + \cdots + a_{2n}u_n \\ &\cdots\cdots\cdots\cdots\cdots\cdots\cdots\cdots\cdots\cdots \\ T(v_m) &= a_{m1}u_1 + a_{m2}u_2 + \cdots + a_{mn}u_n \end{aligned} \tag{1}$$

We want to prove that

$$\begin{aligned} T'(\sigma_1) &= a_{11}\phi_1 + a_{21}\phi_2 + \cdots + a_{m1}\phi_m \\ T'(\sigma_2) &= a_{12}\phi_1 + a_{22}\phi_2 + \cdots + a_{m2}\phi_m \\ &\cdots\cdots\cdots\cdots\cdots\cdots\cdots\cdots\cdots\cdots \\ T'(\sigma_n) &= a_{1n}\phi_1 + a_{2n}\phi_2 + \cdots + a_{mn}\phi_m \end{aligned} \tag{2}$$

where $\{\sigma_i\}$ and $\{\phi_j\}$ are the bases dual to $\{u_i\}$ and $\{v_j\}$, respectively.

Let $v \in V$ and suppose $v = k_1v_1 + k_2v_2 + \cdots + k_mv_m$. Then, by (1),

$$\begin{aligned} T(v) &= k_1 T(v_1) + k_2 T(v_2) + \cdots + k_m T(v_m) \\ &= k_1(a_{11}u_1 + \cdots + a_{1n}u_n) + k_2(a_{21}u_1 + \cdots + a_{2n}u_n) + \cdots + k_m(a_{m1}u_1 + \cdots + a_{mn}u_n) \\ &= (k_1 a_{11} + k_2 a_{21} + \cdots + k_m a_{m1})u_1 + \cdots + (k_1 a_{1n} + k_2 a_{2n} + \cdots + k_m a_{mn})u_n \\ &= \sum_{i=1}^{n} (k_1 a_{1i} + k_2 a_{2i} + \cdots + k_m a_{mi})u_i \end{aligned}$$

Hence for $j = 1, \ldots, n$,

$$(T'(\sigma_j)(v)) = \sigma_j(T(v)) + \sigma_j\left(\sum_{i=1}^{n}(k_1 a_{1i} + k_2 a_i + \cdots + k_m a_{mi})u_i\right) = k_1 a_{1j} + k_2 a_{2j} + \cdots + k_m a_{mj} \tag{3}$$

On the other hand, for $j = 1, \ldots, n$,

$$\begin{aligned} (a_{1j}\phi_1 + a_{2j}\phi_2 + \cdots + a_{mj}\phi_m)(v) &= (a_{1j}\phi_1 + a_{2j}\phi_2 + \cdots + a_{mj}\phi_m)(k_1v_1 + k_2v_2 + \cdots + k_mv_m) \\ &= k_1 a_{1j} + k_2 a_{2j} + \cdots + k_m a_{mj} \end{aligned} \tag{4}$$

Since $v \in V$ was arbitrary, (3) and (4) imply that $T'(\sigma_j) = a_{1j}\phi_1 + a_{2j}\phi_2 + \cdots + a_{mj}\phi_m$, $j = 1, \ldots, n$, which is (2). Thus the theorem is proved.

CHAPTER 19
Bilinear, Quadratic, and Hermitian Forms

19.1 BILINEAR FORMS

19.1 Define a bilinear form on a vector space V over a field K.

▮ A bilinear form on V is a mapping $f: V \times V \to K$ which satisfies
(i) $f(au_1 + bu_2, v) = af(u_1, v) + bf(u_2, v)$
(ii) $f(u, av_1 + bv_2) = af(u, v_1) + bf(u, v_2)$
for all $a, b \in K$ and all $u_i, v_i \in V$. We express condition (i) by saying f is linear in its first position or first variable and condition (ii) by saying f is linear in its second position or second variable.

19.2 Let f be the dot product on $\mathbf{R}^n$; that is, $f(u, v) = u \cdot v = a_1 b_1 + a_2 b_2 + \cdots + a_n b_n$ where $u = (a_i)$ and $v = (b_i)$. Is f a bilinear form on $\mathbf{R}^n$?

▮ Yes, since f is linear in both positions.

19.3 Let g be the dot product on $\mathbf{C}^n$; that is, $g(u, v) = z_1 \bar{w}_1 + z_2 \bar{w}_2 + \cdots + z_n \bar{w}_n$ where $u = (z_i)$ and $v = (w_i)$. Is g a bilinear form on $\mathbf{C}^n$?

▮ The complex dot product is linear in its first position. However, $g(u, kv) = \bar{k}g(u, v)$. Thus g is not linear in its second position and hence g is not a bilinear form.

19.4 Let A be any $n \times n$ matrix over K. Show that the map $f(X, Y) = X^T A Y$ is bilinear form on K^n.

▮ For any $a, b \in K$ and any $X_i, Y_i \in K^n$, $f(aX_1 + bX_2, Y) = (aX_1 + bX_2)^T A Y = (aX_1^T + bX_2^T)AY = aX_1^T A Y + bX_2^T A Y = af(X_1, Y) + bf(X_2, Y)$. Hence f is linear in the first variable. Also, $f(X, aY_1 + bY_2) = X^T A(aY_1 + bY_2) = aX^T A Y_1 + bX^T A Y_2 = af(X, Y_1) + bf(X, Y_2)$. Hence f is linear in the second variable, and so f is a bilinear form on K^n.

19.5 Let ϕ and σ be any linear functionals on a vector space V. Let $f: V \times V \to K$ be defined by $f(u, v) = \phi(u)\sigma(v)$. Show that f is a bilinear form.

▮ For any $a, b \in K$ and any $u_i, v_i \in V$,

$$f(au_1 + bu_2, v) = \phi(au_1 + bu_2)\sigma(v) = [a\phi(u_1) + b\phi(u_2)]\sigma(v)$$
$$= a\phi(u_1)\sigma(v) + b\phi(u_2)\sigma(v) = af(u_1, v) + bf(u_2, v)$$

Thus f is linear in its first position. Similarly,

$$f(u, av_1 + bv_2) = \phi(u)\sigma(av_1 + bv_2) = \phi(u)[a\sigma(v_1) + b\sigma(v_2)]$$
$$= a\phi(u)\sigma(v_1) + b\phi(u)\sigma(v_2) = af(u, v_1) + bf(u, v_2)$$

Thus f is linear in its second position. Accordingly, f is bilinear.

19.6 Define a polynomial bilinear form.

▮ A polynomial $f(x_i, y_i)$ in variables $x_1, \ldots, x_n$ and variables $y_1, \ldots, y_n$ is called a bilinear polynomial if

$$f(x_i, y_i) = \sum_{i=1}^{n} \sum_{j=1}^{n} a_{ij} x_j = a_{11} x_1 y_1 + a_{12} x_1 y_2 + \cdots + a_{nn} x_n y_n$$

The polynomial may be written in the matrix form

$$f(x_i, y_i) = (x_1, x_2, \ldots, x_n) \begin{pmatrix} a_{11} & a_{12} & \cdots & a_{1n} \\ a_{21} & a_{22} & \cdots & a_{2n} \\ \cdots\cdots\cdots\cdots\cdots\cdots \\ a_{n1} & a_{n2} & \cdots & a_{nn} \end{pmatrix} \begin{pmatrix} y_1 \\ y_2 \\ \vdots \\ y_n \end{pmatrix}$$

or simply $f(X, Y) = X^T A X$ where $X^T = (x_1, \ldots, x_n)$, $Y^T = (y_1, \ldots, y_n)$, and $A = (a_{ij})$. [Compare with Problem 19.4 where we are initially given the matrix A.]

19.7 Let $u = (x_1, x_2, x_3)$ and $v = (y_1, y_2, y_3)$, and let $f(u, v) = 3x_1 y_1 - 2x_1 y_2 + 5x_2 y_1 + 7x_2 y_2 - 8x_2 y_3 + 4x_3 y_2 - x_3 y_3$. Express f in matrix notation.

▮ Let A be the 3×3 matrix whose ij entry is the coefficient of $x_i y_i$. Then

$$f(u, v) = X^T A Y = (x_1, x_2, x_3) \begin{pmatrix} 3 & -2 & 0 \\ 5 & 7 & -8 \\ 0 & 4 & -1 \end{pmatrix} \begin{pmatrix} y_1 \\ y_2 \\ y_3 \end{pmatrix}$$

Problems 19.8–19.14 refer to a function $f(u, v)$ where $u = (x_1, x_2)$ and $v = (y_1, y_2)$. In each case, determine whether or not the given function is a bilinear form on $\mathbf{R}^2$. If yes, rewrite the function in matrix notation.

19.8 $f(u, v) = 2x_1 y_2 - 3x_2 y_1$.

▮ Yes, since each term is of the form $a_{ij} x_i y_k$. Also,

$$f(u, v) = (x_1, x_2) \begin{pmatrix} 0 & 2 \\ -3 & 0 \end{pmatrix} \begin{pmatrix} y_1 \\ y_2 \end{pmatrix}$$

19.9 $f(u, v) = x_1 + y_2$.

▮ No, the terms are not of the form $a_{ij} x_i y_j$.

19.10 $f(u, v) = 3x_2 y_2$.

▮ Yes, $f(u, v) = (x_1, x_2) \begin{pmatrix} 0 & 0 \\ 0 & 3 \end{pmatrix} \begin{pmatrix} y_1 \\ y_2 \end{pmatrix}$.

19.11 $f(u, v) = x_1 x_2 + y_1 y_2$.

▮ No, each term must contain one x_i and one y_j, not an x_i and an x_j.

19.12 $f(u, v) = x_1 y_1 + 2x_2 y_2 + 3$.

▮ No, a bilinear form cannot have a nonzero constant term.

19.13 $f(u, v) = 0$.

▮ Yes, $f(u, v) = (x_1, x_2) \begin{pmatrix} 0 & 0 \\ 0 & 0 \end{pmatrix} \begin{pmatrix} y_1 \\ y_2 \end{pmatrix}$.

19.14 $f(u, v) = 1$.

▮ No, a nonzero scalar function is not bilinear.

19.15 Let V be any vector space over K, and let $f: V \times V \to K$ be the zero function, i.e., $f(u, v) = 0$ for every $u, v \in V$. Show that f is bilinear.

▮ For every $a, b \in K$ and any $u_i, v_i \in V$,

$$af(u_1, v) + bf(u_2, v) = a \cdot 0 + b \cdot 0 = 0 = f(au_1 + bu_2, v)$$
$$af(u, v_1) + bf(u, v_2) = a \cdot 0 + b \cdot 0 = 0 = f(u, av_1 + bv_2)$$

Thus f is bilinear.

19.16 Let f and g be bilinear forms on V. Show that the sum $f + g$, defined by $(f + g)(u, v) = f(u, v) + g(u, v)$, is bilinear.

▮ For every $a, b \in K$ and any $u, v_i \in V$,

$$(f + g)(au_1 + bu_2, v) = f(au_1 + bu_2, v) + g(au_1 + bu_2, v) = af(u_1, v) + bf(u_2, v) + ag(u_1, v) + bg(u_2, v)$$
$$= a[f(u_1, v) + g(u_1, v)] + b[f(u_2, v) + g(u_2, v)] = a(f + g)(u_1, v) + b(f + g)(u_2, v)$$

Similarly, $(f + g)(u, av_1 + bv_2) = a(f + g)(u, v_1) + b(f + g)(u, v_2)$. Thus $f + g$ is bilinear.

19.17 Let f be a bilinear form on V and let $k \in K$. Show that the map kf, defined by $(kf)(u, v) = kf(u, v)$, is bilinear.

▮ For any $a, b \in K$ and any $u_i, v_i \in V$,

$$(kf)(au_1 + bu_2, v) = kf(au_1 + bu_2, v) = k[af(u_1, v) + bf(u_2, v)] = akf(u_1, v) + bkf(u_2, v)$$
$$= a(kf)(u_1, v) + b(kf)(u_2, v)$$

Thus kf is linear in its first position. Similarly,

$$(kf)(u, av_1 + bv_2) = kf(u, av_1 + bv_2) = k[af(u, v_1) + bf(u, v_2)] = akf(u, v_1) + bkf(u, v_2)$$
$$= a(kf)(u, v_1) + b(kf)(u, v_2)$$

Thus kf is linear in its second position. Accordingly, kf is bilinear.

19.18 Let $B(V)$ denote the collection of all bilinear forms on V. Show that $B(V)$ is a vector space with respect to the above operations of addition $f + g$ and scalar multiplication kf.

▮ **Method 1.** Show that $B(V)$ satisfies all eight axioms of a vector space.

Method 2. $B(V)$ is a subset of the vector space $\mathscr{J}$ of all functions from $V \times V$ into K. By Problems 19.15–19.17, $0 \in B(V)$ and, for any $f, g \in B(V)$ and any $k \in K$, we have $f + g \in B(V)$ and $kf \in B(V)$. Thus $B(V)$ is a subspace of $\mathscr{J}$.

> **Theorem 19.1:** Let V be a vector space of dimension n over K. Let $\{\phi_1, \ldots, \phi_n\}$ be a basis of the dual space V^*. Then $\{f_{ij}: i, j = 1, \ldots, n\}$ is a basis of $B(V)$ where f_{ij} is defined by $f_{ij}(u, v) = \phi_i(u)\phi_j(v)$. Thus, in particular, $\dim B(V) = n^2$.

19.19 Prove Theorem 19.1.

▮ Let $\{e_1, \ldots, e_n\}$ be the basis of V dual to $\{\phi_i\}$. We first show that $\{f_{ij}\}$ spans $B(V)$. Let $f \in B(V)$ and suppose $f(e_i, e_j) = a_{ij}$. We claim that $f = \sum a_{ij} f_{ij}$. It suffices to show that $f(e_s, e_t) = \left(\sum a_{ij} f_{ij}\right)(e_s, e_t)$ for $s, t = 1, \ldots, n$. We have

$$\left(\sum a_{ij} f_{ij}\right)(e_s, e_t) = \sum a_{ij} f_{ij}(e_s, e_t) = \sum a_{ij} \phi_i(e_s)\phi_j(e_t) = \sum a_{ij} \delta_{is}\delta_{jt} = a_{st} = f(e_s, e_t)$$

as required. Hence $\{f_{ij}\}$ spans $B(V)$.

It remains to show that $\{f_{ij}\}$ is linearly independent. Suppose $\sum a_{ij} f_{ij} = 0$. Then for $s, t = 1, \ldots, n$,

$$0 = 0(e_s, e_t) = \left(\sum a_{ij} f_{ij}\right)(e_s, e_t) = a_{rs}$$

The last step follows as above. Thus $\{f_{ij}\}$ is independent and hence is a basis of $B(V)$.

Problems 19.20–19.22 refer to a bilinear form f on V over K.

19.20 Show that $f(0, v) = 0$ and $f(v, 0) = 0$ for every $v \in V$.

▮ We have $f(0, v) = f(0v, v) = 0f(v, v) = 0$ and $f(v, 0) = f(v, 0v) = 0f(v, v) = 0$.

19.21 Let S be a subset of V. We write

$$S^{\perp} = \{v \in V: f(w, v) = 0 \text{ for every } w \in S\}$$
$$S^{\top} = \{v \in V: f(v, w) = 0 \text{ for every } w \in S\}$$

Show that $S^{\perp}$ and $S^{\top}$ are subspaces of V.

▮ Let $w \in S$. Since $f(w, 0) = 0$ we have $0 \in S^{\perp}$. Suppose $u, v \in S^{\perp}$ and $k \in K$. Then $f(w, u) = 0$ and $f(w, v) = 0$. Thus

$$f(w, u + v) = f(w, u) + f(w, v) = 0 + 0 = 0$$
$$f(w, ku) = kf(w, u) = k \cdot 0 = 0$$

Thus $u + v$, $ku \in S^{\perp}$. Therefore $S^{\perp}$ is a subspace of V.
    Similarly, $S^{\top}$ is a subset of V.

19.22   Suppose $S_1 \subseteq S_2$. Show that $S_2^{\perp} \subseteq S_1^{\perp}$ and $S_2^{\top} \subseteq S_1^{\top}$.

▮ Let $v \in S_2^{\perp}$. Then for every $w \in S_2$ we have $f(w, v) = 0$. Since $S_1 \subseteq S_2$, for every $w \in S_1$ we have $f(w, v) = 0$. Thus $v \in S_1^{\perp}$; and hence $S_2^{\perp} \subseteq S_1^{\perp}$. Similarly $S_2^{\top} \subseteq S_1^{\top}$.

19.2   BILINEAR FORMS AND MATRICES

19.23   Let f be a bilinear form on V and let $S = \{e_1, \ldots, e_n\}$ be a basis of V. Define the matrix representation of f relative to the basis S.

▮ Let A be the matrix whose ij entry is $f(e_i, e_j)$, that is,

$$A = \begin{pmatrix} f(e_1, e_1) & f(e_1, e_2) & \cdots & f(e_1, e_n) \\ f(e_2, e_1) & f(e_2, e_2) & \cdots & f(e_2, e_n) \\ \hdotsfor{4} \\ f(e_n, e_1) & f(e_n, e_2) & \cdots & f(e_n, e_n) \end{pmatrix}$$

Then A is called the matrix representation of f with respect to the basis S or, simply, the matrix of f in the basis S.

19.24   Show that the above matrix A represents f in the following way: If $u, v \in V$, then $f(u, v) = [u]^T A [v]$ {where $[u]$ denotes the coordinate (column) vector of u in the given basis S}.

▮ Suppose $u = a_1 e_1 + \cdots + a_n e_n$, $v = b_1 e_1 + \cdots + b_n e_n$. Then

$$f(u, v) = f(a_1 e_1 + \cdots + a_n e_n, b_1 e_1 + \cdots + b_n e_n) = a_1 b_1 f(e_1, e_1) + a_1 b_2 f(e_1, e_2) + \cdots + a_n b_n f(e_n, e_n)$$

$$= \sum_{i=1}^{n} \sum_{j=1}^{n} a_i b_j f(e_i, e_j) = (a_1, \ldots, a_n) A \begin{pmatrix} b_1 \\ b_2 \\ \vdots \\ b_n \end{pmatrix} = [u]^T A [v]$$

as required.

19.25   Define congruent matrices.

▮ A matrix B is said to be *congruent* to a matrix A if there exists an invertible [or nonsingular] matrix P such that $B = P^T A P$.

19.26   Do congruent matrices have the same rank?

▮ Multiplying a matrix A by a nonsingular matrix does not change its rank. If P is nonsingular, then P^T is nonsingular and $\operatorname{rank}(A) = \operatorname{rank}(P^T A P) = \operatorname{rank}(B)$. Thus congruent matrices have the same rank.

> **Theorem 19.2:**   Let P be the transition matrix from one basis of V to another. Let A be the matrix of a bilinear form f in the original basis. Then $B = P^T A P$ is the matrix of f in the new basis.

19.27   Prove Theorem 19.2.

▮ Let S be the original basis and let S' be the new basis. Then, for any $u, v \in V$, we have $P[u]_{S'} = [u]_S$ and $P[v]_{S'} = [v]_S$; hence $[u]_S^T = [u]_{S'}^T P^T$. Thus $f(u, v) = [u]_S^T A [v]_S = [u]_{S'}^T P^T A P [v]_{S'}$. Since u and v are arbitrary elements of V, $P^T A P$ is the matrix of f in the new basis S'.

Remark: The above theorem indicates one main difference between bilinear forms and linear operators, both of which can be represented by square matrices. Namely, if B and A represent the same linear operator, then B is similar to A, that is, $B = P^{-1}AP$ where P is the change-of-basis matrix; but if B and A represent the same bilinear form, then B is congruent to A, that is, $B = P^{T}AP$ where P is the change-of-basis matrix.

19.28 Define the rank of a bilinear form.

❚ The *rank* of a bilinear form f on V, written rank(f), is defined to be the rank of any matrix representation. [By Problem 19.26, the rank does not depend on the particular matrix representation.]

19.29 What is meant by a degenerate bilinear form?

❚ A bilinear form f on V is degenerate or nondegenerate according as to whether rank(f) < dim V or rank(f) = dim V.

Problems 19.30–19.33 refer to the bilinear form f on $\mathbf{R}^2$ defined by $f((x_1, x_2), (y_1, y_2)) = 2x_1 y_1 - 3x_1 y_2 + x_2 y_2$.

19.30 Find the matrix A of f in the basis $S = \{u_1 = (1, 0), u_2 = (1, 1)\}$.

❚ Set $A = (a_{ij})$ where $a_{ij} = f(u_i, u_j)$:

$$\begin{aligned}
a_{11} &= f(u_1, u_1) = f((1, 0), (1, 0)) = 2 - 0 + 0 = 2 \\
a_{12} &= f(u_1, u_2) = f((1, 0), (1, 1)) = 2 - 3 + 0 = -1 \\
a_{21} &= f(u_2, u_1) = f((1, 1), (1, 0)) = 2 - 0 + 0 = 2 \\
a_{22} &= f(u_2, u_2) = f((1, 1), (1, 1)) = 2 - 3 + 1 = 0
\end{aligned}$$

Thus $A = \begin{pmatrix} 2 & -1 \\ 2 & 0 \end{pmatrix}$ is the matrix of f in the basis $\{u_1, u_2\}$.

19.31 Find the matrix B of f in the basis $S' = \{v_1 = (2, 1), v_2 = (1, -1)\}$.

❚ Set $B = (b_{ij})$ where $b_{ij} = f(v_i, v_j)$:

$$\begin{aligned}
b_{11} &= f(v_1, v_1) = f((2, 1), (2, 1)) &&= 8 - 6 + 1 = 3 \\
b_{12} &= f(v_1, v_2) = f((2, 1), (1, -1)) &&= 4 + 6 - 1 = 9 \\
b_{21} &= f(v_2, v_1) = f((1, -1), (2, 1)) &&= 4 - 3 - 1 = 0 \\
b_{22} &= f(v_2, v_2) = f((1, -1), (1, -1)) &&= 2 + 3 + 1 = 6
\end{aligned}$$

Thus $B = \begin{pmatrix} 3 & 9 \\ 0 & 6 \end{pmatrix}$ is the matrix of f in the basis $\{v_1, v_2\}$.

19.32 Find the change-of-basis matrix P from the basis S to the basis S'.

❚ Write v_1 in terms of u_1 and u_2: $(2, 1) = x(1, 0) + y(1, 1)$; or $x + y = 2$, $y = 1$; or $x = 1$, $y = 1$. Hence, $v_1 = u_1 + u_2$. Next write v_2 in terms of u_1 and u_2: $(1, -1) = x(1, 0) + y(1, 1)$; or $x + y = 1$, $y = -1$; or $x = 2$, $y = -1$. Hence, $v_2 = 2u_1 - u_2$. Finally write the coordinates of v_1 and v_2 as columns to get

$$P = \begin{pmatrix} 1 & 2 \\ 1 & -1 \end{pmatrix}$$

19.33 Verify Theorem 19.2 that $B = P^{T}AP$.

❚ We have $P = \begin{pmatrix} 1 & 2 \\ 1 & -1 \end{pmatrix}$, and so $P^{T} = \begin{pmatrix} 1 & 1 \\ 2 & -1 \end{pmatrix}$. Thus

$$P^{T}AP = \begin{pmatrix} 1 & 1 \\ 2 & -1 \end{pmatrix}\begin{pmatrix} 2 & -1 \\ 2 & 0 \end{pmatrix}\begin{pmatrix} 1 & 2 \\ 1 & -1 \end{pmatrix} = \begin{pmatrix} 3 & 9 \\ 0 & 6 \end{pmatrix} = B$$

19.34 Let $[f]$ denote the matrix representation of a bilinear form f on V relative to a basis $\{e_1, \ldots, e_n\}$ of V. Show that the mapping $f \mapsto [f]$ is an isomorphism of $B(V)$ onto the vector space of n-square matrices.

❚ Since f is completely determined by the scalars $f(e_i, e_j)$, the mapping $f \mapsto [f]$ is one-to-one and onto. It suffices to show that the mapping $f \mapsto [f]$ is a homomorphism; i.e., that

$$[af + bg] = a[f] + b[g] \tag{1}$$

However, for $i, j = 1, \ldots, n$, $(af + bg)(e_i, e_j) = af(e_i, e_j) + bg(e_i, e_j)$ which is a restatement of (1). Thus the result is proved.

Problems 19.35–19.38 refer to the bilinear form f on $\mathbf{R}^2$ defined by $f(u, v) = 3x_1 y_1 - 2x_1 y_2 + 4x_2 y_1 - x_2 y_2$ where $u = (x_1, x_2)$ and $v = (y_1, y_2)$.

19.35 Express f in matrix notation.

❚ Let A be the 2×2 matrix whose ij entry is the coefficient of $x_i y_j$. Then

$$f(u, v) = X^T A Y = (x_1, x_2) \begin{pmatrix} 3 & -2 \\ 4 & -1 \end{pmatrix} \begin{pmatrix} y_1 \\ y_2 \end{pmatrix}$$

19.36 Find the matrix representation of f in the usual basis $E = \{e_1 = (1, 0), e_2 = (0, 1)\}$ of $\mathbf{R}^2$.

❚ Here $f(e_1, e_1) = 3$, $f(e_1, e_2) = -2$, $f(e_2, e_1) = 4$, $f(e_2, e_2) = -1$. In other words, $f(e_i, e_j)$ is the coefficient of $x_i y_j$. Thus we obtain the above matrix $A = \begin{pmatrix} 3 & -2 \\ 4 & -1 \end{pmatrix}$ as the matrix of f in the usual basis E of $\mathbf{R}^2$.

19.37 Find the matrix B which represents f in the basis $S = \{u_1 = (1, 1), u_2 = (1, 2)\}$.

❚ **Method 1.** Set $B = (b_{ij})$ where $b_{ij} = f(u_i, u_j)$:

$$\begin{aligned} b_{11} &= f(u_1, u_1) = 4 & b_{12} &= f(u_1, u_2) = 1 \\ b_{21} &= f(u_2, u_1) = 7 & b_{22} &= f(u_2, u_2) = 3 \end{aligned} \quad \text{so} \quad B = \begin{pmatrix} 4 & 1 \\ 7 & 3 \end{pmatrix}$$

Method 2. [Use Theorem 19.2.] Let P be the matrix whose columns are the basis vectors in S: that is, $P = \begin{pmatrix} 1 & 1 \\ 1 & 2 \end{pmatrix}$. Then

$$B = P^T A P = \begin{pmatrix} 1 & 1 \\ 1 & 2 \end{pmatrix} \begin{pmatrix} 3 & -2 \\ 4 & -1 \end{pmatrix} \begin{pmatrix} 1 & 1 \\ 1 & 2 \end{pmatrix} = \begin{pmatrix} 4 & 1 \\ 7 & 3 \end{pmatrix}$$

[Here P is the change-of-basis matrix from the usual basis E to the given basis S.]

19.38 Find the matrix C which represents f in the basis $S' = \{v_1 = (1, -1), v_2 = (3, 1)\}$.

❚ Let Q be the matrix whose columns are the basis vectors in S': $Q = \begin{pmatrix} 1 & 3 \\ -1 & 1 \end{pmatrix}$. Then

$$C = Q^T A Q = \begin{pmatrix} 1 & -1 \\ 3 & 1 \end{pmatrix} \begin{pmatrix} 3 & -2 \\ 4 & -1 \end{pmatrix} \begin{pmatrix} 1 & 3 \\ -1 & 1 \end{pmatrix} = \begin{pmatrix} 0 & -4 \\ 20 & 32 \end{pmatrix}$$

Problems 19.39–19.40 refer to the bilinear form f on $\mathbf{R}^3$ defined by the matrix

$$A = \begin{pmatrix} 1 & 2 & 3 \\ 4 & -3 & 2 \\ 1 & 5 & -2 \end{pmatrix}$$

i.e., where $f(u, v) = u^T A v$.

19.39 Find the matrix representation of f in the usual basis $E = \{e_1 = (1, 0, 0), e_2 = (0, 1, 0), e_3 = (0, 0, 1)\}$ of $\mathbf{R}^3$.

❚ Since $f(e_i, e_j) = e_i^T A e_j$ is the ij entry in A, the given matrix A is the matrix representation of f with respect to the usual basis E of $\mathbf{R}^3$.

19.40 Find the matrix B which represents f in the basis $S = \{u_1 = (1, 1, 1), u_2 = (1, 0, 1), u_3 = (2, -1, 0)\}$.

▮ Let P be the matrix whose columns are the basis vectors in S, i.e.,

$$P = \begin{pmatrix} 1 & 1 & 2 \\ 1 & 0 & -1 \\ 1 & 1 & 0 \end{pmatrix}$$

Then

$$B = P^T A P = \begin{pmatrix} 1 & 1 & 1 \\ 1 & 0 & 1 \\ 2 & -1 & 0 \end{pmatrix} \begin{pmatrix} 1 & 2 & 3 \\ 4 & -3 & 2 \\ 1 & 5 & -2 \end{pmatrix} \begin{pmatrix} 1 & 1 & 2 \\ 1 & 0 & -1 \\ 1 & 1 & 0 \end{pmatrix} = \begin{pmatrix} 13 & 9 & 8 \\ 10 & 3 & -3 \\ 9 & 2 & -11 \end{pmatrix}$$

[Here P is the change-of-basis matrix from the usual basis E of $\mathbf{R}^3$ to the given basis S.]

Problems 19.41–19.43 show that congruence of matrices is an equivalence relation.

19.41 Show that every matrix A is congruent to itself.

▮ The identity matrix I is nonsingular and $I^T = I$. Since $A = I^T A I$, we have A is congruent to A.

19.42 Show that if A is congruent to B, then B is congruent to A.

▮ Since A is congruent to B, there is a nonsingular matrix P such that $A = P^T B P$. Using $(P^T)^{-1} = (P^{-1})^T$, we have $B = (P^T)^{-1} A P^{-1} = (P^{-1})^T A P^{-1}$ and P^{-1} is nonsingular. Thus B is congruent to A.

19.43 Show that if A is congruent to B and B is congruent to C, then A is congruent to C.

▮ We have $A = P^T B P$ and $B = Q^T C Q$ where P and Q are nonsingular. Using $(QP)^T = P^T Q^T$, we have $A = P^T B P = P^T (Q^T C Q) P = (QP)^T C (QP)$ where QP is nonsingular. Thus A is congruent to C.

19.3 ALTERNATING BILINEAR FORMS

19.44 Define an alternating bilinear form.

▮ A bilinear form f on V is alternating if the following condition holds:

[ABF] $f(v, v) = 0$ for every $v \in V$

19.45 Define a skew symmetric (or antisymmetric) bilinear form.

▮ A bilinear form f on V is skew symmetric if the following condition holds:

[SSBF] $f(u, v) = -f(v, u)$ for every $u, v \in V$

19.46 Show that an alternating bilinear form f is skew symmetric.

▮ Since f is alternating, we have $0 = f(u + v, u + v) = f(u, u) + f(u, v) + f(v, u) + f(v, v)$. Using $f(u, u) = 0$ and $f(v, v) = 0$, we get $0 = f(u, v) + f(v, u)$. Thus $f(u, v) = -f(v, u)$ as required.

19.47 Suppose f is a skew symmetric bilinear form. Is f alternating?

▮ If $1 + 1 \neq 0$ in K, then condition [ABF] implies $f(v, v) = -f(v, v)$ which implies condition [SSBF]. However, if $1 + 1 = 0$ in K, then the conditions are not equivalent.

Remark: The condition $1 + 1 \neq 0$ in K plays an important role in the theory of bilinear and quadratic forms. This condition will be part of our hypothesis in many results in this chapter. Of course, this condition holds when K is the real field $\mathbf{R}$ or the complex field $\mathbf{C}$.

Theorem 19.3: Let f be an alternating bilinear form on V. Then there exists a basis of V in which f is represented by a matrix of the form

$$\begin{pmatrix} \begin{array}{cc} 0 & 1 \\ -1 & 0 \end{array} & & & & & \\ & \begin{array}{cc} 0 & 1 \\ -1 & 0 \end{array} & & & & \\ & & \ddots & & & \\ & & & \begin{array}{cc} 0 & 1 \\ -1 & 0 \end{array} & & \\ & & & & 0 & \\ & & & & & 0 \\ & & & & & & \ddots \\ & & & & & & & 0 \end{array} \end{pmatrix}$$

Moreover, the number of $\begin{pmatrix} 0 & 1 \\ -1 & 0 \end{pmatrix}$ is uniquely determined by f [because it is equal to $\frac{1}{2}\,\text{rank}(f)$].

19.48   Prove Theorem 19.3 which is the main structure theorem for alternating bilinear forms.

$\blacksquare$ If $f = 0$, then the theorem is obviously true. Also, if $\dim V = 1$, then $f(k_1 u, k_2 u) = k_1 k_2 f(u, u) = 0$ and so $f = 0$. Accordingly we can assume that $\dim V > 1$ and $f \neq 0$.

Since $f \neq 0$, there exist [nonzero] $u_1, u_2 \in V$ such that $f(u_1, u_2) \neq 0$. In fact, multiplying u_1 by an appropriate factor, we can assume that $f(u_1, u_2) = 1$ and so $f(u_2, u_1) = -1$. Now u_1 and u_2 are linearly independent; because if, say, $u_2 = k u_1$, then $f(u_1, u_2) = f(u_1, k u_1) = k f(u_1, u_1) = 0$. Let U be the subspace spanned by u_1 and u_2, that is, $U = \text{span}(u_1, u_2)$. Note:

(i) The matrix representation of the restriction of f to U in the basis $\{u_1, u_2\}$ is $\begin{pmatrix} 0 & 1 \\ -1 & 0 \end{pmatrix}$.

(ii) If $u \in U$, say $u = a u_1 + b u_2$, then

$$f(u, u_1) = f(a u_1 + b u_2, u_1) = -b$$
$$f(u, u_2) = f(a u_1 + b u_2, u_2) = a$$

Let W consist of those vectors $w \in V$ such that $f(w, u_1) = 0$ and $f(w, u_2) = 0$. Equivalently, $W = \{w \in V: f(w, u) = 0 \text{ for every } u \in U\}$. We claim that $V = U \oplus W$. It is clear that $U \cap W = \{0\}$, and so it remains to show that $V = U + W$. Let $v \in V$. Set

$$u = f(v, u_2) u_1 - f(v, u_1) u_2 \qquad \text{and} \qquad w = v - u \tag{1}$$

Since u is a linear combination of u_1 and u_2, $u \in U$. We show that $w \in W$. By *(1)* and (ii), $f(u, u_1) = f(v, u_1)$; hence $f(w, u_1) = f(v - u, u_1) = f(v, u_1) - f(u, u_1) = 0$. Similarly, $f(u, u_2) = f(v, u_2)$ and so $f(w, u_2) = f(v - u, u_2) = f(v, u_2) - f(u, u_2) = 0$. Then $w \in W$ and so, by *(1)*, $v = u + w$ where $u \in U$ and $w \in W$. This shows that $V = U + W$; and therefore $V = U \oplus W$.

Now the restriction of f to W is an alternating bilinear form on W. By induction, there exists a basis $u_3, \ldots, u_n$ of W in which the matrix representing f restricted to W has the desired form. Thus $u_1, u_2, u_3, \ldots, u_n$ is a basis of V in which the matrix representing f has the desired form.

19.49   Suppose f is an alternating bilinear form on V. Show that the rank of f is even.

$\blacksquare$ Here f can be represented by a matrix in the form in Theorem 19.3. However, each of the blocks $\begin{pmatrix} 0 & 1 \\ -1 & 0 \end{pmatrix}$ has rank 2. Thus if there are m such blocks then $\text{rank}(f) = 2m$. Thus the rank of f is even.

19.4 SYMMETRIC BILINEAR FORMS

19.50   Define a symmetric bilinear form.

$\blacksquare$ A bilinear form f on V is said to be symmetric if the following condition holds:

[SBF] $\qquad\qquad\qquad\qquad f(u, v) = f(v, u) \qquad \text{for every } u, v \in V$

19.51   Show that a bilinear form f is symmetric if and only if any matrix A representing f is symmetric.

$\blacksquare$ Suppose f is symmetric and A represents f. Then $f(X, Y) = X^T A Y = (X^T A Y)^T = Y^T A^T X$. [We use the fact that $X^T A Y$ is a scalar and therefore equals its transpose.] Since f is symmetric, $f(X, Y) = f(Y, X)$ and hence $Y^T A X = f(Y, X) = f(X, Y) = Y^T A^T X$. Since this condition is true for all vectors X, Y, it follows that $A = A^T$ or A is symmetric.

Conversely, suppose A is symmetric. Then $f(X, Y) = X^T A Y = (X^T A Y)^T = Y^T A^T X = Y^T A X = f(Y, X)$ and hence f is symmetric.

Theorem 19.4 is the main structure theorem for symmetric bilinear forms, which is proved in Problem 19.57.

Theorem 19.4: Let f be a symmetric bilinear form on V over K [in which $1 + 1 \neq 0$]. Then V has a basis $\{v_1, \ldots, v_n\}$ in which f is represented by a diagonal matrix, i.e., $f(v_i, v_j) = 0$ for $i \neq j$.

Theorem 19.5: [*Alternate Form of Theorem 19.4*]: Let A be a symmetric matrix over K [in which $1 + 1 \neq 0$]. Then there exists an invertible [or nonsingular] matrix P such that $P^T A P$ is diagonal. That is, A is congruent to a diagonal matrix.

19.52 Suppose $1 + 1 \neq 0$ in K. Give a formal algorithm to diagonalize [under congruence] a symmetric matrix $A = (a_{ij})$ over K.

▌ *Diagonalization Algorithm*

Case I: $a_{11} \neq 0$. Apply the row operations $R_i \rightarrow -a_{i1} R_1 + a_{11} R_i$, $i = 2, \ldots, n$, and then the corresponding column operations $C_i \rightarrow -a_{i1} C_1 + a_{11} C_i$ to reduce A to the form $\begin{pmatrix} a_{11} & 0 \\ 0 & B \end{pmatrix}$.

Case II: $a_{11} = 0$ but $a_{ii} \neq 0$, for some $i > 1$. Apply the row operation $R_1 \leftrightarrow R_i$ and then the corresponding column operation $C_1 \leftrightarrow C_i$ to bring a_{ii} into the first diagonal position. This reduces the matrix to Case I.

Case III: All diagonal entries $a_{ii} = 0$. Choose i, j such that $a_{ij} \neq 0$ and apply the row operation $R_i \rightarrow R_j + R_i$ and the corresponding column operation $C_i \rightarrow C_j + C_i$ to bring $2a_{ij} \neq 0$ into the ith diagonal position. This reduces the matrix to Case II.

In each of the cases, we can finally reduce A to the form $\begin{pmatrix} a_{11} & 0 \\ 0 & B \end{pmatrix}$ where B is a symmetric matrix of order less than A.

Remark: The hypothesis that $1 + 1 \neq 0$ in K is used in Case III where we state that $2a_{ij} \neq 0$. Repeat the above process with each new submatrix B until A is diagonalized.

19.53 Modify the algorithm in Problem 19.52 so it also finds the matrix P such that $P^T A P$ is diagonal.

▌ First form the matrix $M = (A, I)$. Then apply the row and column operations to M instead of to A alone. [Note that the row operation will change both halves of M, but the column operations will only change the left half of M.] The algorithm will finally transform M into the form $M' = (D, Q)$ where D is diagonal. Then $P = Q^T$ and $P^T A P = D$.

19.54 Let $A = \begin{pmatrix} 1 & 2 & -3 \\ 2 & 5 & -4 \\ -3 & -4 & 8 \end{pmatrix}$, a symmetric matrix. Find the nonsingular matrix P such that $P^T A P$ is diagonal and find $P^T A P$.

▌ First form the block matrix (A, I):

$$(A, I) = \begin{pmatrix} 1 & 2 & -3 & \vdots & 1 & 0 & 0 \\ 2 & 5 & -4 & \vdots & 0 & 1 & 0 \\ -3 & -4 & 8 & \vdots & 0 & 0 & 1 \end{pmatrix}$$

Apply the operations $R_2 \rightarrow -2R_1 + R_2$ and $R_3 \rightarrow 3R_1 + R_3$ to (A, I) and then the corresponding operations $C_2 \rightarrow -2C_1 + C_2$ and $C_3 \rightarrow 3C_1 + C_3$ to A to obtain

$$\begin{pmatrix} 1 & 2 & -3 & \vdots & 1 & 0 & 0 \\ 0 & 1 & 2 & \vdots & -2 & 1 & 0 \\ 0 & 2 & -1 & \vdots & 3 & 0 & 1 \end{pmatrix} \quad \text{and then} \quad \begin{pmatrix} 1 & 0 & 0 & \vdots & 1 & 0 & 0 \\ 0 & 1 & 2 & \vdots & -2 & 1 & 0 \\ 0 & 2 & -1 & \vdots & 3 & 0 & 1 \end{pmatrix}$$

Next apply the operation $R_3 \rightarrow -2R_2 + R_3$ and then the corresponding operation $C_3 \rightarrow -2C_2 + C_3$ to obtain

$$\begin{pmatrix} 1 & 0 & 0 & \vdots & 1 & 0 & 0 \\ 0 & 1 & 2 & \vdots & -2 & 1 & 0 \\ 0 & 0 & -5 & \vdots & 7 & -2 & 1 \end{pmatrix} \quad \text{and then} \quad \begin{pmatrix} 1 & 0 & 0 & \vdots & 1 & 0 & 0 \\ 0 & 1 & 0 & \vdots & -2 & 1 & 0 \\ 0 & 0 & -5 & \vdots & 7 & -2 & 1 \end{pmatrix}$$

Now A has been diagonalized.   Set

$$P = \begin{pmatrix} 1 & -2 & 7 \\ 0 & 1 & -2 \\ 0 & 0 & 1 \end{pmatrix} \quad \text{and then} \quad P^t A P = \begin{pmatrix} 1 & 0 & 0 \\ 0 & 1 & 0 \\ 0 & 0 & -5 \end{pmatrix}$$

19.55   Let $B = \begin{pmatrix} 1 & -3 & 2 \\ -3 & 7 & -5 \\ 2 & -5 & b \end{pmatrix}$,   a symmetric matrix.   Find a nonsingular matrix P such that $P^T B P$ is diagonal and find the diagonal matrix $P^T B P$.

▌   First form the block matrix (B, I):

$$(B, I) = \begin{pmatrix} 1 & -3 & 2 & \vdots & 1 & 0 & 0 \\ -3 & 7 & -5 & \vdots & 0 & 1 & 0 \\ 2 & -5 & 8 & \vdots & 0 & 0 & 1 \end{pmatrix}$$

Apply the row operations $R_2 \to 3R_1 + R_2$ and $R_3 \to -2R_3$ to (B, I) and then the corresponding column operations $C_2 \to 3C_1 + C_2$ and $C_3 \to -2C_1 + C_3$ to B to obtain

$$\begin{pmatrix} 1 & -3 & 2 & \vdots & 1 & 0 & 0 \\ 0 & -2 & 1 & \vdots & 3 & 1 & 0 \\ 0 & 1 & 4 & \vdots & -2 & 0 & 1 \end{pmatrix} \quad \text{and then} \quad \begin{pmatrix} 1 & 0 & 0 & \vdots & 1 & 0 & 0 \\ 0 & -2 & 1 & \vdots & 3 & 1 & 0 \\ 0 & 1 & 4 & \vdots & -2 & 0 & 1 \end{pmatrix}$$

Next apply the row operation $R_3 \to R_2 + 2R_3$ and then the corresponding column operation $C_3 \to C_2 + 2C_3$ to obtain

$$\begin{pmatrix} 1 & 0 & 0 & \vdots & 1 & 0 & 0 \\ 0 & -2 & 1 & \vdots & 3 & 1 & 0 \\ 0 & 0 & 9 & \vdots & -1 & 1 & 2 \end{pmatrix} \quad \text{and then} \quad \begin{pmatrix} 1 & 0 & 0 & \vdots & 1 & 0 & 0 \\ 0 & -2 & 0 & \vdots & 3 & 1 & 0 \\ 0 & 0 & 18 & \vdots & -1 & 1 & 2 \end{pmatrix}$$

Now B has been diagonalized.   Set $P = \begin{pmatrix} 1 & 3 & -1 \\ 0 & 1 & 1 \\ 0 & 0 & 2 \end{pmatrix}$;   then $P^T B P = \begin{pmatrix} 1 & 0 & 0 \\ 0 & -2 & 0 \\ 0 & 0 & 18 \end{pmatrix}$.

19.56   Let $A = \begin{pmatrix} 0 & 1 & 1 \\ 1 & -2 & 2 \\ 1 & 2 & -1 \end{pmatrix}$,   a symmetric matrix.   Find a nonsingular matrix P such that $P^T A P$ is diagonal and find the diagonal matrix $P^T A P$.

▌   First form the block matrix (A, I):

$$(A, I) = \begin{pmatrix} 0 & 1 & 1 & \vdots & 1 & 0 & 0 \\ 1 & -2 & 2 & \vdots & 0 & 1 & 0 \\ 1 & 2 & -1 & \vdots & 0 & 0 & 1 \end{pmatrix}$$

In order to bring the nonzero diagonal entry -1 into the first diagonal position, apply the row operation $R_1 \leftrightarrow R_3$ and then the corresponding column operation $C_1 \leftrightarrow C_3$ to obtain

$$\begin{pmatrix} 1 & 2 & -1 & \vdots & 0 & 0 & 1 \\ 1 & -2 & 2 & \vdots & 0 & 1 & 0 \\ 0 & 1 & 1 & \vdots & 1 & 0 & 0 \end{pmatrix} \quad \text{and then} \quad \begin{pmatrix} -1 & 2 & 1 & \vdots & 0 & 0 & 1 \\ 2 & -2 & 1 & \vdots & 0 & 1 & 0 \\ 1 & 1 & 0 & \vdots & 1 & 0 & 0 \end{pmatrix}$$

Apply the row operations $R_2 \to 2R_1 + R_2$ and $R_3 \to R_1 + R_3$ and then the corresponding column operators $C_2 \to 2C_1 + C_2$ and $C_3 \to C_1 + C_3$ to obtain

$$\begin{pmatrix} -1 & 2 & 1 & \vdots & 0 & 0 & 1 \\ 0 & 2 & 3 & \vdots & 0 & 1 & 2 \\ 0 & 3 & 1 & \vdots & 1 & 0 & 1 \end{pmatrix} \quad \text{and then} \quad \begin{pmatrix} -1 & 0 & 0 & \vdots & 0 & 0 & 1 \\ 0 & 2 & 3 & \vdots & 0 & 1 & 2 \\ 0 & 3 & 1 & \vdots & 1 & 0 & 1 \end{pmatrix}$$

Apply the row operation $R_3 \to -3R_2 + 2R_3$ and then the corresponding column operation $C_3 \to -3C_2 + 2C_3$ to obtain

$$\begin{pmatrix} -1 & 0 & 0 & \vdots & 0 & 0 & 1 \\ 0 & 2 & 3 & \vdots & 0 & 1 & 2 \\ 0 & 0 & -7 & \vdots & 2 & -3 & -4 \end{pmatrix} \quad \text{and then} \quad \begin{pmatrix} -1 & 0 & 0 & \vdots & 0 & 0 & 1 \\ 0 & 2 & 0 & \vdots & 0 & 1 & 2 \\ 0 & 0 & -14 & \vdots & 2 & -3 & -4 \end{pmatrix}$$

Now A has been diagonalized. Set $P = \begin{pmatrix} 0 & 0 & 2 \\ 0 & 1 & -3 \\ 1 & 2 & -4 \end{pmatrix}$; then $P^T A P = \begin{pmatrix} -1 & 0 & 0 \\ 0 & 2 & 0 \\ 0 & 0 & -14 \end{pmatrix}$.

19.57 Prove Theorem 19.4.

▮ **Method 1.** If $f = 0$ or if $\dim V = 1$, then the theorem clearly holds. Hence we can suppose $f \neq 0$ and $\dim V = n > 1$. If $q(v) = f(v, v) = 0$ for every $v \in V$, then the polar form of f (see Problem 19.64) implies that $f = 0$. Hence we can assume there is a vector $v_1 \in V$ such that $f(v_1, v_1) \neq 0$. Let U be the subspace spanned by v_1 and let W consist of those vectors $v \in V$ for which $f(v_1, v) = 0$. We claim that $V = U \oplus W$.

(i) *Proof that* $U \cap W = \{0\}$: Suppose $u \in U \cap W$. Since $u \in U$, $u = kv_1$ for some scalar $k \in K$. Since $u \in W$, $0 = f(u, u) = f(kv_1, kv_1) = k^2 f(v_1, v_1)$. But $f(v_1, v_1) \neq 0$; hence $k = 0$ and therefore $u = kv_1 = 0$. Thus $U \cap W = \{0\}$.

(ii) *Proof that* $V = U + W$: Let $v \in V$. Set

$$w = v - \frac{f(v_1, v)}{f(v_1, v_1)} v_1 \qquad (1)$$

Then $\qquad\qquad f(v_1, w) = f(v_1, v) - \dfrac{f(v_1, v)}{f(v_1, v_1)} f(v_1, v_1) = 0$

Thus $w \in W$. By (1), v is the sum of an element of U and an element of W. Thus $V = U + W$. Accordingly, by (i) and (ii), we have $V = U \oplus W$.

Now f restricted to W is a symmetric bilinear form on W. But $\dim W = n - 1$; hence by induction there is a basis $\{v_2, \ldots, v_n\}$ of W such that $f(v_i, v_j) = 0$ for $i \neq j$ and $2 \leq i, j \leq n$. But by the very definition of W, $f(v_1, v_j) = 0$ for $j = 2, \ldots, n$. Therefore the basis $\{v_1, \ldots, v_n\}$ of V has the required property that $f(v_i, v_j) = 0$ for $i \neq j$.

Method 2. The Diagonalization Algorithm in Problem 19.52 shows that every symmetric matrix over K is congruent to a diagonal matrix. This is equivalent to the statement that f has a diagonal matrix representation.

19.58 Show that any bilinear form f on V is the sum of a symmetric bilinear form and a skew symmetric bilinear form.

▮ Set $g(u, v) = \frac{1}{2}[f(u, v) + f(v, u)]$ and $h(u, v) = \frac{1}{2}[f(u, v) - f(v, u)]$. Then g is symmetric because $g(u, v) = \frac{1}{2}[f(u, v) + f(v, u)] = \frac{1}{2}[f(v, u) + f(u, v)] = g(v, u)$ and h is skew symmetric because $h(u, v) = \frac{1}{2}[f(u, v) - f(v, u)] = -\frac{1}{2}[f(v, u) - f(u, v)] = -h(v, u)$. Furthermore, $f = g + h$.

19.5 QUADRATIC FORMS

19.59 Define a quadratic form.

▮ A mapping $q: V \to K$ is called a quadratic form if $q(v) = f(v, v)$ for some bilinear form f on V. Alternatively, a quadratic form is a polynomial $q(X) = X^T A X$ where $X^T = (x_1, \ldots, x_n)$ and A is a symmetric matrix. Thus

$$q(X) = (x_1, \ldots, x_n) \begin{pmatrix} a_{11} & a_{22} & \cdots & a_{1n} \\ a_{21} & a_{22} & \cdots & a_{2n} \\ \cdots\cdots\cdots\cdots\cdots\cdots \\ a_{n1} & a_{n2} & \cdots & a_{nn} \end{pmatrix} \begin{pmatrix} x_1 \\ x_2 \\ \vdots \\ x_n \end{pmatrix}$$

$$= \sum_{i, j} a_{ij} x_i x_j = a_{11} x_1^2 + a_{22} x_2^2 + \cdots + a_{nn} x_n^2 + 2 \sum_{i < j} a_{ij} x_i x_j$$

[Note $q(X)$ is a polynomial in which every term has degree two.]

Remark: Observe that if the above matrix A is diagonal, then the corresponding quadratic form q has the *diagonal representation* $q(X) = X^T A X = a_{11} x_1^2 + a_{22} x_2^2 + \cdots + a_{nn} x_n^2$, i.e., the quadratic polynomial representing q will contain no "cross product" terms. By Theorem 19.4, every quadratic form has such a representation [when $1 + 1 \neq 0$].

19.60 Find the quadratic form $q(x, y)$ corresponding to the symmetric matrix $A = \begin{pmatrix} 5 & -3 \\ -3 & 8 \end{pmatrix}$.

⬛
$$q(x, y) = (x, y)\begin{pmatrix} 5 & -3 \\ -3 & 8 \end{pmatrix}\begin{pmatrix} x \\ y \end{pmatrix} = (5x - 3y, -3x + 8y)\begin{pmatrix} x \\ y \end{pmatrix}$$
$$= 5x^2 - 3xy - 3xy + 8y^2 = 5x^2 - 6xy + 8y^2$$

Problems 19.61–19.63 refer to the following symmetric matrices:

$$A = \begin{pmatrix} 1 & 2 & -4 \\ 2 & 3 & 5 \\ -4 & 5 & -7 \end{pmatrix} \qquad B = \begin{pmatrix} 3 & & \\ & -4 & \\ & & 6 \end{pmatrix} \qquad C = \begin{pmatrix} 2 & -5 & 1 \\ -5 & -6 & -7 \\ 1 & -7 & 9 \end{pmatrix}$$

19.61 Find the quadratic form $q(x_1, x_2, x_3)$ corresponding to the symmetric matrix A.

⬛ The coefficient of x_i^2 is a_{ii} and the coefficient of $x_i x_j$ is $a_{ij} + a_{ji} = 2a_{ij}$. Thus $q(x_1, x_2, x_3) = x_1^2 + 4x_1 x_2 + 3x_2^2 - 8x_1 x_3 + 10x_2 x_3 - 7x_3^2$.

19.62 Find the quadratic form $q(x, y, z)$ corresponds to the diagonal matrix B.

⬛ Here $q(x, y, z) = 3x^2 - 4y^2 + 6z^2$. [There are no cross-product terms.]

19.63 Find the quadratic form $q(x, y, z)$ corresponding to the symmetric matrix C.

⬛ $q(x, y, z) = 2x^2 - 10xy - 6y^2 + 2xz - 14yz + 9z^2$. [As usual, we assume that x, y, z are the first, second, and third variable, respectively.]

19.64 Let q be the quadratic form associated with the symmetric bilinear form f. Show that f can be obtained from q by the following polar form of f: $f(u, v) = \frac{1}{2}(q(u + v) - q(u) - q(v))$. [Assume $1 + 1 \neq 0$ in K.]

⬛ $q(u + v) - q(u) - q(v) = f(u + v, u + v) - f(u, u) - f(v, v)$
  $= f(u, u) + f(u, v) + f(v, u) + f(v, v) - f(u, u) - f(v, v) = 2f(u, v)$

If $1 + 1 \neq 0$, we can divide by 2 to obtain the required identity.

19.65 Find the symmetric matrix A which corresponds to the quadratic form $q(x, y, z) = 3x^2 + 4xy - y^2 + 8xz - 6yz + z^2$.

⬛ The symmetric matrix $A = (a_{ij})$ representing $q(x_1, \ldots, x_n)$ has the diagonal entry a_{ii} equal to the coefficient of x_i^2 and has the entries a_{ij} and a_{ji} each equal to half the coefficient of $x_i x_j$. Thus

$$A = \begin{pmatrix} 3 & 2 & 4 \\ 2 & -1 & -3 \\ 4 & -3 & 1 \end{pmatrix}$$

19.66 Find the symmetric matrix B which corresponds to $q(x, y) = 4x^2 + 5xy - 7y^2$.

⬛ Here $B = \begin{pmatrix} 4 & \frac{5}{2} \\ \frac{5}{2} & -7 \end{pmatrix}$. [Division by 2 may introduce fractions even though the coefficients in q are integers.]

19.67 Find the symmetric matrix C which corresponds to $q(x, y, z) = 4xy + 5y^2$.

⬛ Even though only x and y appear in the polynomial, the expression $q(x, y, z)$ indicates that there are three variables. In other words, $q(x, y, z) = 0x^2 + 4xy + 5y^2 + 0xz + 0yz + 0z^2$. Thus

$$C = \begin{pmatrix} 0 & 2 & 0 \\ 2 & 5 & 0 \\ 0 & 0 & 0 \end{pmatrix}$$

19.68 Find the symmetric matrix D which corresponds to $q(x, y, z) = x^2 - 2yz + xz$.

▮ Here $D = \begin{pmatrix} 1 & 0 & \frac{1}{2} \\ 0 & 0 & -1 \\ \frac{1}{2} & -1 & 0 \end{pmatrix}$.

Problems 19.69–19.72 refer to the quadratic form $q(x, y) = 3x^2 + 2xy - y^2$ and the linear substitution, $x = s - 3t$, $y = 2s + t$.

19.69 Find $q(s, t)$.

▮ Substitute for x and y in q to obtain

$$q(s, t) = 3(s - 3t)^2 + 2(s - 3t)(2s + t) - (2s + t)^2$$
$$= 3(s^2 - 6st + 9t^2) + 2(2s^2 - 5st - 3t^2) - (s^2 + 4st + t^2) = 3s^2 - 32st + 20t^2$$

19.70 Find the matrix A which corresponds to the quadratic form $q(x, y)$ and rewrite the quadratic form in matrix notation.

▮ We have $A = \begin{pmatrix} 3 & 1 \\ 1 & -1 \end{pmatrix}$ and $q(X) = X^T A X$ where $X = (x, y)^T$.

19.71 Find the matrix P which corresponds to the linear substitution, and rewrite the linear substitution using matrix notation.

▮ We have $\begin{pmatrix} x \\ y \end{pmatrix} = \begin{pmatrix} 1 & -3 \\ 2 & 1 \end{pmatrix}\begin{pmatrix} s \\ t \end{pmatrix}$. Thus $P = \begin{pmatrix} 1 & -3 \\ 2 & 1 \end{pmatrix}$ and $X = PY$ where $X = (x, y)^T$ and $Y = (s, t)^T$.

19.72 Find $q(s, t)$ using the above matrix notation.

▮ We have $q(X) = X^T A X$ and $X = PY$. Thus $X^T = Y^T P^T$. Therefore,

$$q(s, t) = q(Y) = Y^T P^T A P Y = (s, t)\begin{pmatrix} 1 & 2 \\ -3 & 1 \end{pmatrix}\begin{pmatrix} 3 & 1 \\ 1 & -1 \end{pmatrix}\begin{pmatrix} 1 & -3 \\ 2 & 1 \end{pmatrix}\begin{pmatrix} s \\ t \end{pmatrix}$$

$$= (s, t)\begin{pmatrix} 3 & -16 \\ -16 & 20 \end{pmatrix}\begin{pmatrix} s \\ t \end{pmatrix} = 3s^2 - 32st + 20t^2$$

19.73 Let L be a linear substitution $X = PY$, as above. When is L nonsingular? Orthogonal?

▮ L is said to be nonsingular or orthogonal according as the matrix P representing the substitution is nonsingular or orthogonal.

19.74 Is the linear substitution in Problems 19.69–19.72 nonsingular?

▮ Yes, since the matrix $P = \begin{pmatrix} 1 & -3 \\ 2 & 1 \end{pmatrix}$ corresponding to the substitution is nonsingular.

19.75 Consider the quadratic form $q(x, y, z) = x^2 + 4xy + 3y^2 - 6xz + 10yz + 7z^2$. Find a nonsingular linear substitution expressing the variables x, y, z in terms of variables r, s, t such that $q(r, s, t)$ is diagonal.

▮ First find the matrix A which corresponds to the quadratic form. Here

$$A = \begin{pmatrix} 1 & 2 & -3 \\ 2 & 3 & 5 \\ -3 & 5 & 7 \end{pmatrix}$$

Then find a nonsingular matrix P such that $P^T A P$ is diagonal. Form the block matrix (A, I):

$$(A, I) = \begin{pmatrix} 1 & 2 & -3 & \vdots & 1 & 0 & 0 \\ 2 & 3 & 5 & \vdots & 0 & 1 & 0 \\ -3 & 5 & 7 & \vdots & 0 & 0 & 1 \end{pmatrix}$$

Apply the row operations $R_2 \to -2R_1 + R_2$ and $R_3 \to 3R_1 + R_3$ to (A, I) and then the corresponding column operations $C_2 \to -2C_1 + C_2$ and $C_3 \to 3C_1 + C_3$ to A to obtain

$$\begin{pmatrix} 1 & 2 & -3 & \vdots & 1 & 0 & 0 \\ 0 & -1 & 11 & \vdots & -2 & 1 & 0 \\ 0 & 11 & -2 & \vdots & 3 & 0 & 1 \end{pmatrix} \quad \text{and then} \quad \begin{pmatrix} 1 & 0 & 0 & \vdots & 1 & 0 & 0 \\ 0 & -1 & 11 & \vdots & -2 & 1 & 0 \\ 0 & 11 & -2 & \vdots & 3 & 0 & 1 \end{pmatrix}$$

Next apply the row operation $R_3 \to 11R_2 + R_3$ and then the corresponding column operation $C_3 \to 11C_2 + C_3$ to finally obtain

$$\begin{pmatrix} 1 & 0 & 0 & \vdots & 1 & 1 & 0 \\ 0 & -1 & 0 & \vdots & -2 & 1 & 0 \\ 0 & 0 & 119 & \vdots & -19 & 11 & 1 \end{pmatrix}$$

Thus

$$P = \begin{pmatrix} 1 & -2 & -19 \\ 0 & 1 & 11 \\ 0 & 0 & 1 \end{pmatrix} \quad \text{and} \quad P^T A P = \begin{pmatrix} 1 & & \\ & -1 & \\ & & 119 \end{pmatrix}$$

Thus the linear substitution $x = r - 2s - 19t$, $y = s + 11t$, $z = t$ will yield the quadratic form $q(r, s, t) = r^2 - s^2 + 119t^2$.

19.76 Let $q(x, y, z) = x^2 + 4xy + 3y^2 - 8xz - 12yz + 9z^2$. Find a nonsingular linear substitution expressing the variables x, y, z in terms of the variables r, s, t so that $q(r, s, t)$ is diagonal.

❚ Form the block matrix (A, I) where A is the matrix which corresponds to the quadratic form:

$$(A, I) = \begin{pmatrix} 1 & 2 & -4 & \vdots & 1 & 0 & 0 \\ 2 & 3 & -6 & \vdots & 0 & 1 & 0 \\ -4 & -6 & 9 & \vdots & 0 & 0 & 1 \end{pmatrix}$$

Apply $R_2 \to -2R_1 + R_2$ and $R_3 \to 4R_1 + R_4$ and the corresponding column operations, and then $R_3 \to 2R_2 + R_3$ and the corresponding column operation to obtain

$$\begin{pmatrix} 1 & 0 & 0 & \vdots & 1 & 0 & 0 \\ 0 & -1 & 2 & \vdots & -2 & 1 & 0 \\ 0 & 2 & -7 & \vdots & 4 & 0 & 1 \end{pmatrix} \quad \text{and then} \quad \begin{pmatrix} 1 & 0 & 0 & \vdots & 1 & 0 & 0 \\ 0 & -1 & 0 & \vdots & -2 & 1 & 0 \\ 0 & 0 & -3 & \vdots & 0 & 2 & 1 \end{pmatrix}$$

Thus the linear substitution $x = r - 2s$, $y = s + 2t$, $z = t$ will yield the quadratic form $q(r, s, t) = r^2 - s^2 - 3t^2$.

19.77 Let $q(x, y) = 2x^2 - 12xy + 5y^2$. Diagonalize q by the method known as "completing the square."

❚ First factor out the coefficient of x^2 from the x^2 term and the xy term to get $q(x, y) = 2(x^2 - 6xy \quad)$ $+ 5y^2$. Next complete the square inside the parentheses by adding an appropriate multiple of y^2 and then subtract the corresponding amount outside the parentheses to get $q(x, y) = 2(x^2 - 6xy + 9y^2) + 5y^2 - 18y^2 = 2(x - 3y)^2 - 13y^2$. [The -18 comes from the fact that the $9y^2$ inside the parentheses is multiplied by 2.] Let $s = x - 3y$, $t = y$. Then $x = s + 3t$, $y = t$. This linear substitution yields the quadratic form $q(s, t) = 2s^2 - 13t^2$.

19.78 Diagonalize $q(x, y) = 3x^2 - 12xy + 7y^2$ by completing the square.

❚ We have

$$q(x, y) = 3x^2 - 12xy + 7y^2 = 3(x^2 - 4xy \quad) + 7y^2$$
$$= 3(x^2 - 4xy + 4y^2) + 7y^2 - 12y^2 = 3(x - 2y)^2 - 5y^2$$

Let $s = x - 2y$, $t = y$. Then $x = s + 2t$, $y = t$. This linear substitution yields $q(s, t) = 3s^2 - 5t^2$.

Problems 19.79–19.80 refer to a diagonal matrix $A = \begin{pmatrix} a_1 & & & \\ & a_2 & & \\ & & \ddots & \\ & & & a_n \end{pmatrix}$ over a field K.

19.79   Show that for any nonzero scalars $k_1, \ldots, k_n \in K$, A is congruent to a diagonal matrix with diagonal entries $a_i k_i^2$.

▌ Let P be the diagonal matrix with diagonal entries k_i. Then

$$P^T A P = \begin{pmatrix} k_1 & & & \\ & k_2 & \cdots & \\ & & & k_n \end{pmatrix}\begin{pmatrix} a_1 & & & \\ & a_2 & \cdots & \\ & & & a_n \end{pmatrix}\begin{pmatrix} k_1 & & & \\ & k_2 & \cdots & \\ & & & k_n \end{pmatrix} = \begin{pmatrix} a_1 k_1^2 & & & \\ & a_2 k_2^2 & \cdots & \\ & & & a_n k_n^2 \end{pmatrix}$$

19.80   Show that if K is the real field $\mathbf{R}$, then A is congruent to a diagonal matrix with only 1s, -1s, and 0s as diagonal entries.

▌ Let P be the diagonal matrix with diagonal entries

$$b_i = \begin{cases} 1/\sqrt{|a_i|} & \text{if } a_i \neq 0 \\ 1 & \text{if } a_i = 0 \end{cases}$$

Then $P^T A P$ has the required form.

19.81   Show that $q(0) = 0$ for any quadratic form q on V.

▌ We have $q(0) = f(0, 0) = f(0v, 0) = 0f(v, 0) = 0$.

19.82   Suppose $q(u) = 0$ for a quadratic form q on V. Show that $q(ku) = 0$ for any $k \in K$.

▌ We have $q(ku) = f(ku, ku) = k^2 f(u, u) = k^2 q(u) = k^2 \cdot 0 = 0$.

19.83   Give an example of a quadratic form q on $\mathbf{R}^2$ such that $q(u) = 0$ and $q(v) = 0$ for some $u, v \in \mathbf{R}^2$ but $q(u + v) \neq 0$.

▌ Let $q(x, y) = x^2 - y^2$ and $u = (1, 1)$ and $v = (1, -1)$. Then $f(u) = 0$ and $f(v) = 0$ but $f(u + v) = f(2, 0) = 4 \neq 0$.

19.6   REAL SYMMETRIC BILINEAR AND QUADRATIC FORMS, LAW OF INERTIA

This section tests symmetric bilinear forms and quadratic forms on vector spaces over the real field $\mathbf{R}$. These forms appear in many branches of mathematics and physics. The special nature of $\mathbf{R}$ permits an independent theory.

The main content of this section is the following theorem, proved in Problem 19.96, and its corollary which follows.

Theorem 19.6:   Let f be a symmetric bilinear form on V over $\mathbf{R}$. Then there is a basis of V in which f is represented by a diagonal matrix; every other diagonal representation has the same number P of positive entries and the same number N of negative entries.

The following result for real quadratic forms is sometimes referred to as the Law of Inertia or Sylvester's theorem.

Corollary 19.7:   Any real quadratic form q has a unique representation in the form $q(x_1, \ldots, x_n) = x_1^2 + \cdots + x_s^2 - x_{s+1}^2 - \cdots - x_r^2$.

Remark: Throughout this section, unless otherwise stated or implied, f will denote a real symmetric bilinear form and q will denote the corresponding real quadratic form.

19.84   Define the signature of f and of q, denoted by $\text{Sig}(f)$ and $\text{Sig}(q)$, respectively.

▌ $\text{Sig}(f) = \text{Sig}(q) = P - N$ where P is the number of positive entries and N the number of negative entries in any diagonal representation of f and q. [By Theorem 19.6, P and N are unique for a given f and q.]

19.85 Show that $\text{rank}(f) = \text{rank}(q) = P + N$.

▮ Let D be a diagonal matrix representation of f and q. Then $\text{rank}(D)$ is equal to the number of nonzero entries on the diagonal of D, which is $P + N$. Thus $\text{rank}(f) = \text{rank}(q) = \text{rank}(D) = P + N$.

19.86 Find the signature of the quadratic form $q(x, y, z)$ in Problem 19.75.

▮ The equivalent diagonal quadratic form $q(r, s, t) = r^2 - s^2 + 119t^2$ has $P = 2$ positive entries on the diagonal and $N = 1$ negative entry on the diagonal. Thus $\text{Sig}(q) = P - N = 2 - 1 = 1$.

19.87 Find the signature of the quadratic form $q(x, y, z)$ in Problem 19.6.

▮ The equialent diagonal quadratic form is $q(r, s, t) = r^2 - s^2 - 3t^2$. Thus $P = 1$ and $N = 2$;   hence $\text{Sig}(q) = 1 - 2 = -1$.

19.88 Define a positive definite quadratic form.

▮ A quadratic form q is said to be *positive definite* if $q(v) = f(v, v) > 0$ for every vector $v \neq 0$. This is true if and only if any diagonal representation D of q only contains positive entries on the diagonal, i.e., if $\text{Sig}(q) = \dim V$.

19.89 Define a nonnegative semidefinite quadratic form.

▮ A quadratic form q is said to be nonnegative semidefinite if $q(v) = f(v, v) \geq 0$ for every vector v. Analogously, this is true if and only if any diagonal representation D of q only contains nonnegative entries on the diagonal, i.e., if $\text{Sig}(q) = \text{rank}(q)$.

19.90 Let $q(x, y, z) = x^2 + 2y^2 - 4xz - 4yz + 7z^2$. Is q positive definite?

▮ Diagonalize [under congruence] the symmetric matrix A corresponding to q [by applying $R_3 \to 2R_1 + R_3$ and $C_3 \to 2C_1 + C_3$,   and then $R_3 \to R_2 + R_3$ and $C_3 \to C_2 + C_3$]:

$$A = \begin{pmatrix} 1 & 0 & -2 \\ 0 & 2 & -2 \\ -2 & -2 & 7 \end{pmatrix} \to \begin{pmatrix} 1 & 0 & 0 \\ 0 & 2 & -2 \\ 0 & -2 & 3 \end{pmatrix} \to \begin{pmatrix} 1 & 0 & 0 \\ 0 & 2 & 0 \\ 0 & 0 & 1 \end{pmatrix}$$

The diagonal representation of q only contains positive entries, 1, 2, and 1, on the diagonal; hence q is positive definite.

19.91 Let $q(x, y, z) = x^2 + y^2 + 2xz + 4yz + 3z^2$. Is q positive definite?

▮ Diagonalize (under congruence) the symmetric matrix A corresponding to q:

$$A = \begin{pmatrix} 1 & 0 & 1 \\ 0 & 1 & 2 \\ 1 & 2 & 3 \end{pmatrix} \to \begin{pmatrix} 1 & 0 & 0 \\ 0 & 1 & 2 \\ 0 & 2 & 2 \end{pmatrix} \to \begin{pmatrix} 1 & 0 & 0 \\ 0 & 1 & 0 \\ 0 & 0 & -2 \end{pmatrix}$$

There is a negative entry -2 in the diagonal representation of q; hence q is not positive definite.

19.92 Show that $q(x, y) = ax^2 + bxy + cy^2$ is positive definite if and only if the discriminate $D = b^2 - 4ac < 0$.

▮ Suppose $v = (x, y) \neq 0$,   say $y \neq 0$. Let $t = x/y$. Then $q(v) = y^2[a(x/y)^2 + b(x/y) + c] = y^2[at^2 + bt + c]$. However, $s = at^2 + bt + c$ lies above the t axis, i.e., is positive for every value of t if and only if $D = b^2 - 4ac < 0$. Thus q is positive definite if and only if $D < 0$.

19.93 Let $q(x, y) = x^2 - 4xy + 5y^2$. Is q positive definite?

▮ **Method 1.** Diagonalize by completing the square: $q(x, y) = x^2 - 4xy + 4y^2 + 5y^2 - 4y^2 = (x - 2y)^2 + y^2 = s^2 + t^2$,   where $s = x - 2y$, $t = y$. Thus q is positive definite.

Method 2. Compute the discriminant $D = b^2 - 4ac = 16 - 20 = -4$. Since $D < 0$, q is positive definite.

19.94 Let $q(x, y) = x^2 + 6xy + 3y^2$. Is q positive definite?

▮ **Method 1.** Diagonalize by completing the square: $q(x, y) = x^2 + 6xy + 9y^2 + 3y^2 - 9y^2 = (x + 3y)^2 - 6y^2 = s^2 - 6t^2$,   where $s = x + 3y$, $t = y$. Since -6 is negative, q is not positive definite.

Method 2. Compute $D = b^2 - 4ac = 36 - 12 = 24$. Since $D > 0$, q is not positive definite.

19.95 Let f be the dot product on $\mathbf{R}^n$; that is, $f(u, v) = u \cdot v = a_1 b_1 + a_2 b_2 + \cdots + a_n b_n$, where $u = (a_i)$ and $v = (b_i)$. Is f positive definite?

▮ Note that f is symmetric since $f(u, v) = u \cdot v = v \cdot u = f(v, u)$. Furthermore, f is positive definite because $f(u, u) = a_1^2 + a_2^2 + \cdots + a_n^2 > 0$ when $u \neq 0$.

19.96 Prove Theorem 19.6.

▮ By Theorem 19.4, there is a basis $\{u_1, \ldots, u_n\}$ of V in which f is represented by a diagonal matrix, say, with P positive and N negative entries. Now suppose $\{w_1, \ldots, w_n\}$ is another basis of V in which f is represented by a diagonal matrix, say, with P' positive and N' negative entries. We can assume without loss of generality that the positive entries in each matrix appear first. Since $\text{rank}(f) = P + N = P' + N'$, it suffices to prove that $P = P'$.

Let U be the linear span of $u_1, \ldots, u_P$ and let W be the linear span of $w_{P'+1}, \ldots, w_n$. Then $f(v, v) > 0$ for every nonzero $v \in U$, and $f(v, v) \leq 0$ for every nonzero $v \in W$. Hence $U \cap W = \{0\}$. Note that $\dim U = P$ and $\dim W = n - P'$. Thus $\dim(U + W) = \dim U + \dim W - \dim(U \cap W) = P + (n - P') - 0 = P - P' + n$. But $\dim(U + W) \leq \dim V = n$; hence $P - P' + n \leq n$ or $P \leq P'$. Similarly, $P' \leq P$ and therefore $P = P'$, as required.

Remark: The above theorem and proof depend only on the concept of positivity. Thus the theorem is true for any subfield K of the real field $\mathbf{R}$.

19.97 An $n \times n$ real symmetric matrix A is said to be *positive definite* if $X^T A X > 0$ for every nonzero [column] vector $X \in \mathbf{R}^n$, i.e., if A is positive definite viewed as a bilinear form. Let B be any real nonsingular matrix. Show that (a) $B^T B$ is symmetric and (b) $B^T B$ is positive definite.
(a) $(B^T B)^T = B^T B^{TT} = B^T B$; hence $B^T B$ is symmetric.
(b) Since B is nonsingular, $BX \neq 0$ for any nonzero $X \in \mathbf{R}^n$. Hence the dot product of BX with itself, $BX \cdot BX = (BX)^T(BX)$, is positive. Thus $X^T(B^T B)X = (X^T B^T)(BX) = (BX)^T(BX) > 0$ as required.

19.7 ORTHOGONAL DIAGONALIZATION OF REAL QUADRATIC FORMS

Let q be a quadratic form on Euclidean space $\mathbf{R}^n$ and let A be the corresponding real symmetric matrix. Recall that a nonsingular matrix P is orthogonal if $P^T = P^{-1}$. The following theorem, proved in Chapter 20, shows that q may be diagonalized by means of an orthogonal change of coordinates.

Theorem 19.8: Let A be a real symmetric matrix. Then there exists an orthogonal matrix P such that $B = P^{-1}AP = P^T AP$ is diagonal.

19.98 Describe the algorithm which diagonalizes a quadratic form $q(X)$ on $\mathbf{R}^n$ by means of an orthogonal change of coordinates $X = PY$.

▮ *Orthogonal Diagonalization Algorithm*
Step 1. Find the symmetric matrix A which represents q and find its characteristic polynomial $\Delta(t)$.
Step 2. Find the eigenvalues of A which are the roots of $\Delta(t)$.
Step 3. For each eigenvalue λ of A in Step 2, find an orthogonal basis of its eigenspace.
Step 4. Normalize all eigenvectors in Step 3 which then forms an orthonormal basis of $\mathbf{R}^n$.
Step 5. Let P be the matrix whose columns are the normalized eigenvectors in Step 4.
Then $X = PY$ is the required orthogonal change of coordinates, and the diagonal entries of $P^T AP$ will be the eigenvalues $\lambda_1, \ldots, \lambda_n$ which correspond to the columns of P.

Remark: Theorem 20.6 guarantees that the eigenvectors belonging to distinct eigenvalues are orthogonal.

19.99 Find an orthogonal change of coordinates which diagonalizes the real quadratic form $q(x, y) = 2x^2 - 4xy + 5y^2$.

▮ First find the symmetric matrix A representing q and then its characteristic polynomial $\Delta(t)$:

$$A = \begin{pmatrix} 2 & -2 \\ -2 & 5 \end{pmatrix} \quad \text{and} \quad \Delta(t) = |tI - A| = \begin{vmatrix} t-2 & 2 \\ 2 & t-5 \end{vmatrix} = (t-6)(t-1)$$

The eigenvalues of A are 6 and 1. Substitute $t = 6$ into the matrix $tI - A$ to obtain the corresponding homogeneous system of linear equations $4x + 2y = 0$, $2x + y = 0$. A nonzero solution is $v_1 = (1, -2)$. Next substitute $t = 1$ into the matrix $tI - A$ to find the corresponding homogeneous system $-x + 2y = 0$, $2x - 4y = 0$. A nonzero solution is $v_2 = (2, 1)$. Normalize v_1 and v_2 to obtain the orthonormal basis $\{u_1 = (1/\sqrt{5}, -2/\sqrt{5}), u_2 = (2/\sqrt{5}, 1/\sqrt{5})\}$. Finally let P be the matrix whose columns are u_1 and u_2, respectively. Then

$$P = \begin{pmatrix} 1/\sqrt{5} & 2/\sqrt{5} \\ -2/\sqrt{5} & 1/\sqrt{5} \end{pmatrix} \quad \text{and} \quad P'AP = \begin{pmatrix} 6 & 0 \\ 0 & 1 \end{pmatrix}$$

Thus the required orthogonal change of coordinates is

$$\begin{pmatrix} x \\ y \end{pmatrix} = P \begin{pmatrix} x' \\ y' \end{pmatrix} \quad \text{that is,} \quad \begin{array}{l} x = \dfrac{x'}{\sqrt{5}} + \dfrac{2y'}{\sqrt{5}} \\[2mm] y = \dfrac{-2x'}{\sqrt{5}} + \dfrac{y'}{\sqrt{5}} \end{array}$$

Under this change of coordinates q is transformed into the diagonal form $q(x', y') = 6x'^2 + y'^2$. Note that the diagonal entries of q are the eigenvalues of A.

19.100 Find the signature of the above quadratic form q.

$\blacksquare$ Since both diagonal entries are positive, $P = 2$, $N = 0$; hence $\text{Sig}(q) = 2 - 0 = 2$.

19.101 Let C be the quadratic curve $2x^2 - 4xy + 5y^2 = 6$. Plot C in the coordinate plane $\mathbf{R}^2$. What kind of conic section is C?

$\blacksquare$ The change-of-basis matrix P in Problem 19.99 determines a new coordinate system for $\mathbf{R}^2$ with the new axis x' in the direction of the eigenvector $u_1 = (1/\sqrt{5}, -2/\sqrt{5})$ [or $v_1 = (1, -2)$] and the new axis y' in the direction of the eigenvector $u_2 = (2/\sqrt{5}, 1/\sqrt{5})$ [or $v_2 = (2, 1)$]. The equation of C with respect to the new coordinate system is $6x'^2 + y'^2 = 6$. The graph is an ellipse intersecting the x' axis at ± 1 and intersecting the y axis at $\pm\sqrt{6} \simeq \pm 2.3$, as in Fig. 19-1.

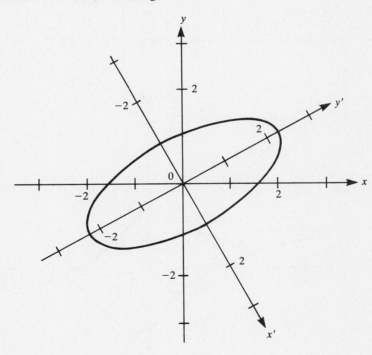

Fig. 19-1

19.102 Let $q(x, y) = x^2 + 4xy + y^2$. Find an orthogonal change of coordinates which diagonalizes q.

$\blacksquare$ First find the symmetric matrix A representing q and then its characteristic polynomial $\Delta(t)$:

$$A = \begin{pmatrix} 1 & 2 \\ 2 & 1 \end{pmatrix} \quad \text{and} \quad \Delta(t) = |tI - A| = \begin{vmatrix} t-1 & -2 \\ -2 & t-1 \end{vmatrix} = t^2 - 2t - 3 = (t-3)(t+1)$$

Thus the eigenvalues of A are 3 and -1. Substitute $t = 3$ into the matrix $tI - A$ to obtain the corresponding homogeneous system of linear equations $2x - 2y = 0$, $-2x + 2y = 0$. A nonzero solution is $v_1 = (1, 1)$.

Next substitute $t = -1$ into the matrix $tI - A$ to obtain the corresponding homogeneous system of linear equations $-2x - 2y = 0$, $-2x - 2y = 0$. A nonzero solution is $v_2 = (1, -1)$.

Normalize v_1 and v_2 to obtain the orthonormal basis $\{u_1 = (1/\sqrt{2}, 1/\sqrt{2}), u_2 = (-1/\sqrt{2}, 1/\sqrt{2})\}$. Finally let P be the matrix whose columns are u_1 and u_2, respectively; then

$$P = \begin{pmatrix} 1/\sqrt{2} & -1/\sqrt{2} \\ 1/\sqrt{2} & 1/\sqrt{2} \end{pmatrix} \quad \text{and} \quad P^T A P = \begin{pmatrix} 3 & 0 \\ 0 & -1 \end{pmatrix}$$

Thus the required orthogonal change of coordinates is

$$\begin{pmatrix} x \\ y \end{pmatrix} = P \begin{pmatrix} x' \\ y' \end{pmatrix} \quad \text{or} \quad \begin{aligned} x &= \frac{x'}{\sqrt{2}} - \frac{y'}{\sqrt{2}} \\ y &= \frac{x'}{\sqrt{2}} + \frac{y'}{\sqrt{2}} \end{aligned}$$

Under this change of coordinates, q is transformed into the diagonal form $q(x', y') = 3x'^2 - y'^2$. [Note that the diagonal entries of q are the eigenvalues of A.]

19.103 Find the signature of the above quadratic form q.

❙ Since one diagonal entry is positive and one is negative, $P = 1$ and $N = 1$. Thus $\mathrm{Sig}(q) = P - N = 1 - 1 = 0$.

19.104 Let C be the curve $x^2 + 4xy + y^2 = 3$. Plot C in the coordinate plane $\mathbf{R}^2$. What kind of conic section is C?

❙ Plot the transformed equation $3x'^2 - y'^2 = 3$ in the plane $\mathbf{R}^2$ with respect to a new axis x' in the direction of the eigenvector $u_1 = (1/\sqrt{2}, 1/\sqrt{2})$ [or $v_1 = (1, 1)$] and a new axis y' in the direction of the eigenvector $u_2 = (-1/\sqrt{2}, 1/\sqrt{2})$ [or $v_2 = (-1, 1)$]. The graph is a hyperbola with vertices on the x' axis at $x' = \pm 1$, as pictured in Fig. 19-2. [The asymptotes are $y' = \pm\sqrt{6}x'$.]

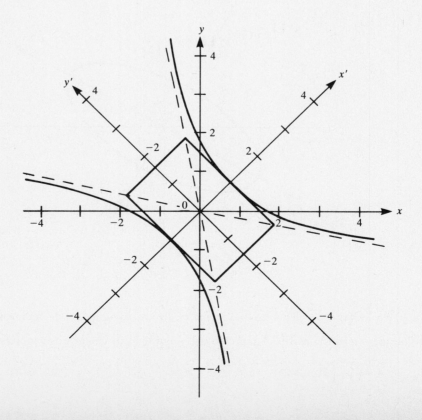

Fig. 19-2

19.105 Let $q(x, y) = 3x^2 - 6xy + 11y^2$. Find an orthogonal change of coordinates which diagonalizes q.

$\blacksquare$ Find the symmetric matrix A representing q and its characteristic polynomial $\Delta(t)$:

$$A = \begin{pmatrix} 3 & -3 \\ -3 & 11 \end{pmatrix} \quad \text{and} \quad \Delta(t) = \begin{vmatrix} t-3 & 3 \\ 3 & t-11 \end{vmatrix} = t^2 - 14t + 24 = (t-2)(t-12)$$

The eigenvalues are 2 and 12; hence a diagonal form of q is $q(x', y') = 2x'^2 + 12y'^2$. The corresponding change of coordinates is obtained by finding a corresponding set of eigenvectors of A.

Set $t = 2$ into the matrix $tI - A$ to obtain the homogeneous system $-x + 3y = 0$, $3x - 9y = 0$. A nonzero solution is $v_1 = (3, 1)$. Next substitute $t = 12$ into the matrix $tI - A$ to obtain the homogeneous system $9x + 3y = 0$, $3x + y = 0$. A nonzero solution is $v_2 = (-1, 3)$. Normalize v_1 and v_2 to obtain the orthonormal basis $u_1 = (3/\sqrt{10}, 1/\sqrt{10})$, $u_2 = (-1/\sqrt{10}, 3/\sqrt{10})$. The change-of-basis matrix P and the required change of coordinates follow:

$$P = \begin{pmatrix} 3/\sqrt{10} & -1/\sqrt{10} \\ 1/\sqrt{10} & 3/\sqrt{10} \end{pmatrix} \quad \text{and} \quad \begin{pmatrix} x \\ y \end{pmatrix} = P \begin{pmatrix} x' \\ y' \end{pmatrix} \quad \text{or} \quad \begin{aligned} x &= \frac{3x' - y'}{\sqrt{10}} \\ y &= \frac{x' + 3y'}{\sqrt{10}} \end{aligned}$$

One can also express x' and y' in terms of x and y by using $P^{-1} = P^T$, that is,

$$x' = \frac{3x + y}{\sqrt{10}} \qquad y' = \frac{-x - 3y}{\sqrt{10}}$$

Problems 19.106–19.112 refer to and orthogonally diagonalize the quadratic form $q(x, y, z) = 3x^2 + 2xy + 3y^2 + 2xz + 2yz + 3z^2$.

19.106 Find the symmetric matrix A which represents q and its characteristic polynomial $\Delta(t)$.

$\blacksquare$
$$A = \begin{pmatrix} 3 & 1 & 1 \\ 1 & 3 & 1 \\ 1 & 1 & 3 \end{pmatrix} \quad \text{and} \quad \Delta(t) = \begin{vmatrix} t-3 & -1 & -1 \\ -1 & t-3 & -1 \\ -1 & -1 & t-3 \end{vmatrix} = t^3 - 9t^2 + 24t - 20$$

19.107 Find the eigenvalues of A or, in other words, the roots of $\Delta(t)$.

$\blacksquare$ If $\Delta(t)$ has a rational root it must divide the constant 20, i.e., it must be among ± 1, ± 2, ± 4, ± 10, ± 20. Testing $t = 2$ we get

$$\begin{array}{r|rrrr} 2 & 1 & -9 & 24 & -20 \\ & & 2 & -14 & 20 \\ \hline & 1 & -7 & 10 & 0 \end{array}$$

Thus $\Delta(t) = (t-2)(t^2 - 7t + 10) = (t-2)^2(t-5)$. Hence the eigenvalues of A are 2 [with multiplicity two] and 5 [with multiplicity one].

19.108 Find an orthogonal basis of the eigenspace E_2 of the eigenvalue $\lambda = 2$.

$\blacksquare$ Subtract $t = 2$ down the diagonal of A to obtain the corresponding homogeneous system $x + y + z = 0$, $x + y + z = 0$, $x + y + z = 0$. That is, $x + y + z = 0$. The system has two independent solutions. One such solution is $v_1 = (0, 1, -1)$. We seek a second solution $v_2 = (a, b, c)$ which is orthogonal to v_1; that is, such that $a + b + c = 0$ and also $b - c = 0$. For example, $v_2 = (2, -1, -1)$. Thus $v_1 = (0, 1, -1)$, $v_2 = (2, -1, -1)$ is an orthogonal basis of E_2.

19.109 Find an eigenvector v_3 belonging to the eigenvalue $\lambda = 5$.

$\blacksquare$ Subtract $t = 5$ down the diagonal of A to obtain the corresponding homogeneous system $-2x + y + z = 0$, $x - 2y + z = 0$, $x + y - 2z = 0$. This system yields a nonzero solution $v_3 = (1, 1, 1)$.

[*Remark*: As expected from Theorem 20.6. v_3 is orthogonal to both v_1 and v_2; hence $\{v_1, v_2, v_3\}$ is an orthogonal basis of $\mathbf{R}^3$.]

19.110 Find an orthogonal change of coordinates which diagonalizes q.

▌ Normalize v_1, v_2, v_3 to obtain the orthonormal basis: $u_1 = (0, 1/\sqrt{2}, -1/\sqrt{2})$, $u_2 = (2/\sqrt{6}, -1/\sqrt{6}, -1/\sqrt{6})$, $u_3 = (1/\sqrt{3}, 1/\sqrt{3}, 1/\sqrt{3})$. Let P be the matrix whose columns are u_1, u_2, u_3. Then

$$P = \begin{pmatrix} 0 & 2/\sqrt{6} & 1/\sqrt{3} \\ 1/\sqrt{2} & -1/\sqrt{6} & 1/\sqrt{3} \\ -1/\sqrt{2} & -1/\sqrt{6} & 1/\sqrt{3} \end{pmatrix} \quad \text{and} \quad P^T A P = \begin{pmatrix} 2 & & \\ & 2 & \\ & & 5 \end{pmatrix}$$

Thus the required orthogonal change of coordinates is

$$x = \frac{2y'}{\sqrt{6}} + \frac{z'}{\sqrt{3}}$$

$$y = \frac{x'}{\sqrt{2}} - \frac{y'}{\sqrt{6}} + \frac{z'}{\sqrt{3}}$$

$$z = -\frac{x'}{\sqrt{2}} - \frac{y'}{\sqrt{6}} + \frac{z'}{\sqrt{3}}$$

Under this change of coordinates, q is transformed into the diagonal form $q(x', y', z') = 2x'^2 + 2y'^2 + 5z'^2$.

19.111 Find the signature of q.

▌ Since there are three positive diagonal entries and no negative diagonal entries, $P = 3$, $N = 0$. Thus $\text{Sig}(q) = P - N = 3 - 0 = 3$.

19.112 Describe the surface $3x^2 + 2xy + 3y^2 + 2xy + 2yz + 3z^2 = 1$.

▌ Under the above change of coordinates the equation of the surface is $2x'^2 + 2y'^2 + 5z'^2 = 1$. Thus the surface is an ellipsoid.

19.8 HERMITIAN FORMS

This section assumes that V is a vector space over the complex field $\mathbf{C}$. [As usual, $\bar{k}$ denotes the complex conjugate of $k \in \mathbf{C}$.]

Remark: If $A = (a_{ij})$ is an $n \times n$ matrix over $\mathbf{C}$, then we write $\bar{A}$ for the matrix obtained by taking the complex conjugate of every entry of A, that is, $\bar{A} = (\bar{a}_{ij})$. We also write A^* for $\bar{A}^T = \overline{A^T}$. That is, A^* is the conjugate transpose of A.

19.113 Consider the following matrices:

$$A = \begin{pmatrix} 2+3i & 5-4i \\ 6+7i & 1+9i \end{pmatrix} \qquad B = \begin{pmatrix} 6-2i & 7i \\ 16 & 2-5i \end{pmatrix} \qquad C = \begin{pmatrix} 3 & 5 & -4 \\ 2 & 7 & -5 \\ -5 & 6 & 8 \end{pmatrix}$$

Find A^*, B^*, and C^*.

▌ In each case, take the transpose of the matrix and then the conjugate of each element or, equivalently, take the conjugate of each element and then the transpose of the matrix. This yields

$$A^* = \begin{pmatrix} 2-3i & 6-7i \\ 5+4i & 1-9i \end{pmatrix} \qquad B^* = \begin{pmatrix} 6+2i & 16 \\ -7i & 2+5i \end{pmatrix} \qquad C^* = \begin{pmatrix} 3 & 2 & -5 \\ 5 & 7 & 6 \\ -4 & -5 & 8 \end{pmatrix}$$

[Observe that if a matrix M is real, then M^* is simply the transpose of M.]

19.114 Define a Hermitian matrix.

▌ A matrix H is Hermitian if $H^* = H$, i.e., if H is equal to its conjugate transpose. [This property is analogous to a matrix being symmetric in the real case.]

Problems 19.115–19.117 refer to the following matrices:

$$A = \begin{pmatrix} 2 & 2+3i & 4-5i \\ 2-3i & 5 & 6+2i \\ 4+5i & 6-2i & -7 \end{pmatrix} \qquad B = \begin{pmatrix} 3 & 2-i & 4+i \\ 2-i & 6 & i \\ 4+i & i & 3 \end{pmatrix} \qquad C = \begin{pmatrix} 4 & -3 & 5 \\ -3 & 2 & 1 \\ 5 & 1 & -6 \end{pmatrix}$$

19.115 Is A Hermitian?

▌ A is Hermitian since it is equal to its conjugate transpose.

19.116 Is B Hermitian?

▌ B is not Hermitian, even though it is symmetric.

19.117 Is C Hermitian?

▌ C is Hermitian. In fact, a real matrix is Hermitian if and only if it is symmetric.

19.118 Define a Hermitian form on a vector space V over the complex field **C**.

▌ A Hermitian form on V is a mapping $f: V \times V \rightarrow \mathbf{C}$ which satisfies
(i) $f(au_1 + bu_2, v) = af(u_1, v) + bf(u_2, v)$
(ii) $f(u, v) = \overline{f(v, u)}$
where $a, b \in \mathbf{C}$ and $u_i, v \in V$.

19.119 Suppose f is a Hermitian form on V. Show that
(iii) $f(u, av_1 + bv_2) = \bar{a}f(u, v_1) + \bar{b}f(u, v_2)$

▌ We have

$$f(u, av_1 + bv_2) = \overline{f(av_1 + bv_2, u)} = \overline{af(v_1, u) + bf(v_2, u)} = \bar{a}\,\overline{f(v_1, u)} + \bar{b}\,\overline{f(v_2, u)} = \bar{a}f(u, v_1) + \bar{b}f(u, v_2)$$

Remark: As before, we express condition (i) by saying f is linear in the first variable. On the other hand, we express condition (iii) by saying f is *conjugate linear* in the second variable.

19.120 Suppose f is a Hermitian form on V. Show that $f(v, v)$ is real for any $v \in V$.

▌ By condition (ii), $f(v, v) = \overline{f(v, v)}$. Thus $f(v, v)$ is real.

19.121 Let A be a Hermitian matrix. Show that f is a Hermitian form on $\mathbf{C}^n$ where f is defined by
$f(X, Y) = X^T A \bar{Y}$.

▌ For all $a, b \in \mathbf{C}$ and all $X_1, X_2, X_2, Y \in \mathbf{C}^n$, $f(aX_1 + bX_2, Y) = (aX_1 + bX_2)^T A \bar{Y} = (aX_1^T + bX_2^T)A\bar{Y} = aX_1^T A\bar{Y} + bX_2^T A\bar{Y} = af(X_1, Y) + bf(X_2, Y)$. Hence f is linear in the first variable. Also, $\overline{f(X, Y)} = \overline{X^T A \bar{Y}} = (\overline{X^T A \bar{Y}})^T = \bar{Y}^T A^T X = Y^T A^* \bar{X} = Y^T A \bar{X} = f(Y, X)$. Hence f is a Hermitian form on $\mathbf{C}^n$. [*Remark:* We use the fact that $X^T A \bar{Y}$ is a scalar and so it is equal to its transpose.]

19.122 Define a Hermitian quadratic form.

▌ Let f be a Hermitian form on V. The mapping $q: V \rightarrow \mathbf{R}$ defined by $q(v) = f(v, v)$ is called the *Hermitian quadratic form* or *complex quadratic form* associated with the Hermitian form f. Moreover, one can obtain f from q according to the following identity called the *polar form* of f: $f(u, v) = \frac{1}{4}(q(u + v) - q(u - v)) + \frac{1}{4}(q(u + iv) - q(u - iv))$.

19.123 Define a nonnegative semidefinite and a positive definite Hermitian form.

▌ A Hermitian form f and its quadratic form q are said to be *nonnegative semidefinite* if $q(v) = f(v, v) \geq 0$ for every $v \in V$, and are said to be *positive definite* if $q(v) = f(v, v) > 0$ for every $v \neq 0$.

19.124 Let f be the dot product on $\mathbf{C}^n$; that is, for $u = (z_i)$, $v = (w_i) \in \mathbf{C}^n$, let $f(u, v) = u \cdot v = z_1 \bar{w}_1 + z_2 \bar{w}_2 + \cdots + z_n \bar{w}_n$. Is f a Hermitian form? Is f positive definite?

▌ The mapping f is a Hermitian form on $\mathbf{C}^n$ since it satisfies properties (i) and (ii) for a Hermitian form. Moreover, f is positive definite since, for any $v \neq 0$, $f(u, u) = z_1 \bar{z}_1 + z_2 \bar{z}_2 + \cdots + z_n \bar{z}_n = |z_1|^2 + |z_2|^2 + \cdots + |z_n|^2 > 0$.

Remark: Every complex inner product on a vector space V over **C** is a positive definite Hermitian form and, conversely, any positive definite Hermitian form on V over **C** defines an inner product by $f(u, v) = \langle u, v \rangle$.

19.125 Define the matrix representation of a Hermitian form f on V relative to a basis $S = \{e_1, \ldots, e_n\}$ of V.

▐ The matrix $H = (h_{ij})$ where $h_{ij} = f(e_i, e_j)$ is called the *matrix representation* of f in the basis $\{e_i\}$. By (ii), $f(e_i, e_j) = \overline{f(e_j, e_i)}$; hence H is Hermitian and, in particular, the diagonal entries of H are real. Thus any diagonal representation of f contains only real entries.

19.126 Let f be a Hermitian form on V. Let H be the matrix of f in a basis $\{e_1, \ldots, e_n\}$ of V. Show that $f(u, v) = [u]^T H \overline{[v]}$ for all $u, v \in V$. [As usual, $[u]$ denotes the coordinate vector on u in the given basis.]

▐ Suppose $u = a_1 e_1 + a_2 e_2 + \cdots + a_n e_n$ and $v = b_1 e_1 + b_2 e_2 + \cdots + b_n e_n$. Then, as required,

$$f(u, v) = f(a_1 e_1 + \cdots + a_n e_n, b_1 e_1 + \cdots + b_n e_n)$$

$$= \sum_{i,j} a_i \overline{b_j} f(e_i, e_j) = (a_1, \ldots, a_n) H \begin{pmatrix} \overline{b_1} \\ \overline{b_2} \\ \vdots \\ \overline{b_n} \end{pmatrix} = [u]^T H \overline{[v]}$$

19.127 Let P be the change-of-basis matrix from a basis S of V to a new basis S'. Let H be the matrix of a Hermitian form f in the original basis S. Show that $B = P^T H \bar{P} = Q^* H Q$, where $Q = \bar{P}$, is the matrix of f in the new basis S'.

▐ Let $u, v \in V$. Since P is the change-of-basis matrix from S to S', we have $P[u]_{S'} = [u]_S$ and $P[v]_{S'} = [v]_S$; hence $[u]_S^T = [u]_{S'}^T P^T$ and $\overline{[v]}_S = \bar{P}\overline{[v]}_{S'}$. Thus, by Problem 19.126, $f(u, v) = [u]_S^T H \overline{[v]}_S = [u]_{S'}^T P^T H \bar{P} \overline{[v]}_{S'}$. But u and v are arbitrary elements of V; hence $P^T H \bar{P}$ is the matrix of f in the new basis S'.

Remark: The main structure theorem for Hermitian forms is the following theorem which is the complex analog of Theorem 19.4 on real symmetric bilinear forms.

Theorem 19.9: Let f be a Hermitian form on V. Then there exists a basis $\{e_1, \ldots, e_n\}$ of V in which f is represented by a diagonal matrix, i.e., $f(e_i, e_j) = 0$ for $i \neq j$. Moreover, every diagonal representation of f has the same number P of positive entries and the same number N of negative entries. The difference $S = P - N$ is called the *signature* of f.

19.128 The three elementary row operations and their corresponding column operations are as follows:

$$\begin{array}{llll}
[a_1] & R_i \leftrightarrow R_j & [a_2] & R_i \rightarrow k R_i, k \neq 0 & [a_3] & R_i \rightarrow k R_j + R_i \\
[b_1] & C_i \leftrightarrow C_j & [b_2] & C_i \rightarrow k C_i, k \neq 0 & [b_3] & C_i \rightarrow k C_j + C_i
\end{array}$$

Define the corresponding Hermitian column operations.

▐ Here the constant k is replaced by its conjugate $\bar{k}$; that is,

$$\begin{array}{lll}
[c_1] & C_i \leftrightarrow C_j & [c_2] & C_i \rightarrow \bar{k} C_i, \bar{k} \neq 0 & [c_3] & C_i \rightarrow \bar{k} C_j + C_i
\end{array}$$

19.129 Let $H = \begin{pmatrix} 1 & 1+i & 2i \\ 1-i & 4 & 2-3i \\ -2i & 2+3i & 7 \end{pmatrix}$, a Hermitian matrix. Find a nonsingular matrix P such that $P^T H \bar{P}$ is diagonal.

▐ First form the block matrix (H, I):

$$\begin{pmatrix} 1 & 1+i & 2i & \vdots & 1 & 0 & 0 \\ 1-i & 4 & 2-3i & \vdots & 0 & 1 & 0 \\ -2i & 2+3i & 7 & \vdots & 0 & 0 & 1 \end{pmatrix}$$

Apply the row operations $R_2 \rightarrow (-1 + i) R_1 + R_2$ and $R_3 \rightarrow 2i R_1 + R_3$ to (A, I) and then the corresponding "Hermitian column operations" [see Problem 19.128] $C_2 \to (-1-i)C_1 + C_2$ and $C_3 \to -2iC_1 + C_3$ to A to obtain

$$\begin{pmatrix} 1 & 1+i & 2i & \vdots & 1 & 0 & 0 \\ 0 & 2 & -5i & \vdots & -1+i & 1 & 0 \\ 0 & 5i & 3 & \vdots & 2i & 0 & 1 \end{pmatrix} \quad \text{and then} \quad \begin{pmatrix} 1 & 0 & 0 & \vdots & 1 & 0 & 0 \\ 0 & 2 & -5i & \vdots & -1+i & 1 & 0 \\ 0 & 5i & 3 & \vdots & 2i & 0 & 1 \end{pmatrix}$$

Next apply the row operation $R_3 \to -5iR_2 + 2R_3$ and the corresponding Hermitian column operation $C_3 \to 5iC_2 + 2C_3$ to obtain

$$\begin{pmatrix} 1 & 0 & 0 & \vdots & 1 & 0 & 0 \\ 0 & 2 & -5i & \vdots & -1+i & 1 & 0 \\ 0 & 0 & -19 & \vdots & 5+9i & -5i & 2 \end{pmatrix} \quad \text{and then} \quad \begin{pmatrix} 1 & 0 & 0 & \vdots & 1 & 0 & 0 \\ 0 & 2 & 0 & \vdots & -1+i & 1 & 0 \\ 0 & 0 & -38 & \vdots & 5+9i & -5i & 2 \end{pmatrix}$$

Now H has been diagonalized. Set

$$P = \begin{pmatrix} 1 & -1+i & 5+9i \\ 0 & 1 & -5i \\ 0 & 0 & 2 \end{pmatrix} \quad \text{and then} \quad P^T H \bar{P} = \begin{pmatrix} 1 & 0 & 0 \\ 0 & 2 & 0 \\ 0 & 0 & -38 \end{pmatrix}$$

19.130 Find the signature of the Hermitian matrix H in Problem 19.129.

▌ There are two positive entries, 1 and 2, and one negative entry, -38, on the diagonal representation of H; hence $P = 2$ and $N = 1$. Thus $\text{Sig}(H) = P - N = 2 - 1 = 1$.

19.9 MULTILINEARITY AND DETERMINANTS

19.131 Define a multilinear form and an alternating multilinear form on a vector space V over a field K.

▌ A mapping $f: \overbrace{V \times V \times \cdots \times V}^{m \text{ times}} \to K$ is called a *multilinear* [or *m-linear*] form on V if f is linear in each variable, i.e., for $i = 1, \ldots, m$, $f(\ldots, \widehat{au + bv}, \ldots) = af(\ldots, \hat{u}, \ldots) + bf(\ldots, \hat{v}, \ldots)$ where ^ denotes the ith component, and other components are held fixed. An m-linear form f is said to be *alternating* if $f(v_1, \ldots, v_m) = 0$ whenever $v_i = v_k$, $i \neq k$.

19.132 Let A be an n-square matrix over a field K. Then A may be viewed as an n-tuple of its row vectors, say, $A_1, A_2, \ldots, A_n$; that is, $A = (A_1, A_2, \ldots, A_n)$. Show that the determinant function is multilinear [with respect to the rows of A].

▌ Let D be the determinant function; i.e., $D(A) = D(A_1, A_2, \ldots, A_n) = |A|$. Suppose $A = (a_{ij})$ and suppose for a fixed i, $A_i = B_i + C_i$, where $B_i = (b_1, \ldots, b_n)$ and $C_i = (c_1, \ldots, c_n)$. Accordingly, $a_{i1} = b_1 + c_1$, $a_{i2} = b_2 + c_2, \ldots$, $a_{in} = b_n + c_n$. Expanding $D(A) = |A|$ by the ith row,

$$\begin{aligned} D(A) = D(A_1, \ldots, B_i + C_i, \ldots, A_n) &= a_{i1}A_{i1} + a_{i2}A_{i2} + \cdots + a_{in}A_{in} \\ &= (b_1 + c_1)A_{i1} + (b_2 + c_2)A_{i2} + \cdots + (b_n + c_n)A_{in} \\ &= (b_1 A_{i1} + b_2 A_{i2} + \cdots + b_n A_{in}) + (c_1 A_{i1} + c_2 A_{i2} + \cdots + c_n A_{in}) \end{aligned}$$

However, the two sums above are the determinants of the matrices obtained from A by replacing the ith row by B_i and C_i, respectively. That is, $D(A) = D(A_1, \ldots, B_i + C_i, \ldots, A_n) = D(A_1, \ldots, B_i, \ldots, A_n) + D(A_1, \ldots, C_i, \ldots, A_n)$. Furthermore, since multiplying a row by a scalar k multiplies the determinant by k, we have $D(A_1, \ldots, kA_i, \ldots, A_n) = kD(A_1, \ldots, A_i, \ldots, A_n)$. Thus D is multilinear.

19.133 Is the determinant an alternating form?

▌ Yes, since a matrix with two identical rows has zero determinant.

Theorem 19.10: Let $\mathscr{A}$ denote the set of n-square matrices over a field K. There exists a unique function $D: \mathscr{A} \to K$ such that (i) D is multilinear, (ii) D is alternating, (iii) $D(I) = 1$. This function D is none other than the determinant function; i.e., for any matrix $A \in \mathscr{A}$, $D(A) = |A|$.

19.134 Prove Theorem 19.10.

▌ By Problem 19.132 and results in Chapter 5, the determinant function does satisfy conditions (i), (ii), and (iii). Thus we need only prove the uniqueness of D.

Suppose D satisfies (i), (ii), and (iii). If $\{e_1, \ldots, e_n\}$ is the usual basis of K^n, then by (iii), $D(e_1, e_2, \ldots, e_n) = D(I) = 1$. Using (ii) we also have

$$D(e_{i_1}, e_{i_2}, \ldots, e_{i_n}) = \operatorname{sgn} \sigma \qquad \text{where } \sigma = i_1 i_2 \ldots i_n \tag{1}$$

Now suppose $A = (a_{ij})$. Observe that the kth row A_k of A is $A_k = (a_{k1}, a_{k2}, \ldots, a_{kn}) = a_{k1}e_1 + a_{k2}e_2 + \cdots + a_{kn}e_n$. Thus $D(A) = D(a_{11}e_1 + \cdots + a_{1n}e_n, a_{21}e_1 + \cdots + a_{2n}e_n, \ldots, a_{n1}e_1 + \cdots + a_{nn}e_n)$. Using the multilinearity of D, we can write $D(A)$ as a sum of terms of the form

$$D(A) = \sum D(a_{1i_1}e_{i_1}, a_{2i_2}e_{i_2}, \ldots, a_{ni_n}e_{i_n}) = \sum (a_{1i_1}a_{2i_2}\ldots a_{ni_n})D(e_{i_1}, e_{i_2}, \ldots, e_{i_n}) \tag{2}$$

where the sum is summed over all sequences $i_1 i_2 \ldots i_n$ where $i_k \in \{1, \ldots, n\}$. If two of the indices are equal, say $i_j = i_k$ but $j \neq k$, then by (ii), $D(e_{i_1}, e_{i_2}, \ldots, e_{i_n}) = 0$. Accordingly, the sum in (2) need only be summed over all permutations $\sigma = i_1 i_2 \ldots i_n$. Using (1), we finally have that

$$D(A) = \sum_\sigma (a_{1i_1}a_{2i_2}\ldots a_{ni_n})D(e_{i_1}, e_{i_2}, \ldots, e_{i_n})$$

$$= \sum_\sigma (\operatorname{sgn} \sigma)a_{1i_1}a_{2i_2}\ldots a_{ni_n} \qquad \text{where } \dot\sigma = i_1 i_2 \ldots i_n$$

Hence D is the determinant function and so the theorem is proved.

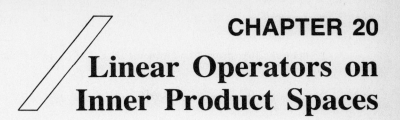

CHAPTER 20
Linear Operators on Inner Product Spaces

This chapter investigates the space $A(V)$ of linear operators T on an inner product space V. [See Chapter 14.] Thus the base field K is either the real field $\mathbf{R}$ or the complex field $\mathbf{C}$. In fact, different terminology will be used for the real case and for the complex case. We also use the fact that the inner product on Euclidean space $\mathbf{R}^n$ may be defined by $\langle u, v \rangle = u^T v$ and that the inner product on complex Euclidean space $\mathbf{C}^n$ may be defined by $\langle u, v \rangle = u^T \bar{v}$ where u and v are column vectors.

20.1 ADJOINT OPERATORS

20.1 Define the adjoint operator.

▮ A linear operator T on an inner product space V is said to have an *adjoint* operator T^* on V if $\langle T(u), v \rangle = \{u, T^*(v)\}$ for every $u, v \in V$.

20.2 Let A be a real n-square matrix viewed as a linear operator on $\mathbf{R}^n$. Show that A^T is the adjoint of A.

▮ For every $u, v \in \mathbf{R}^n$, $\langle Au, v \rangle = (Au)^T v = u^T A^T v = \langle u, A^T v \rangle$. Thus A^T is the adjoint of A.

20.3 Let B be a complex n-square matrix viewed as a linear operator on $\mathbf{C}^n$. Show that B^* is the adjoint of B [where B^* is the conjugate transpose of B].

▮ For every $u, v \in \mathbf{C}^n$, $\langle Bu, v \rangle = (Bu)^T v = u^T B^T v = u^T \overline{\bar{B}^T} \bar{v} = u^T \overline{B^* v} = \langle u, B^* v \rangle$. Thus B^* is the adjoint of B.

Remark: The notation B^* is used to denote the adjoint of B and, previously, to denote the conjugate transpose of B. Problem 20.3 shows that they both give the same result.

Problems 20.4–20.6 refer to the following matrices:

$$A = \begin{pmatrix} 2+3i & 5-4i \\ 6-9i & 2+7i \end{pmatrix} \qquad B = \begin{pmatrix} 3-7i & 18 & 4+1 \\ -7i & 6-i & 2-3i \\ 8+i & 7+9i & 6+3i \end{pmatrix} \qquad C = \begin{pmatrix} 1 & 2 & 3 \\ 4 & 5 & 6 \\ 7 & 7 & 7 \end{pmatrix}$$

20.4 Find the adjoint A^* of A.

▮ Take the conjugate transpose of A to get $A^* = \begin{pmatrix} 2-3i & 6+9i \\ 5+4i & 2-7i \end{pmatrix}$.

20.5 Find the adjoint B^* of B.

▮ The conjugate transpose gives us $B^* = \begin{pmatrix} 3+7i & 7i & 8-i \\ 18 & 6+i & 7-9i \\ 4-i & 2+3i & 6-3i \end{pmatrix}$.

20.6 Find the adjoint C^* of C.

▮ Since C is real, the adjoint C^* is simply the transpose of C. Thus $C^* = C^T = \begin{pmatrix} 1 & 4 & 7 \\ 2 & 5 & 7 \\ 3 & 6 & 7 \end{pmatrix}$.

Theorem 20.1: Let T be a linear operator on a finite dimensional inner product space V over K. Then
  (i) There exists a unique linear operator T^* on V such that $\langle T(u), v \rangle = \langle u, T^*(v) \rangle$ for every $u, v \in V$. [That is, T has an adjoint T^*.]
  (ii) If A is the matrix representation of T with respect to an orthonormal basis $S = \{e_i\}$ of V, then the matrix representation of T^* in the basis S is the conjugate transpose A^* of A [or the transpose A^T of A when K is real].

Theorem 20.1, proved in Problems 20.12–20.13, is the main result in this section.

20.7 Let T be the linear operator on $\mathbf{C}^3$ defined by $T(x, y, z) = (2x + iy, y - 5iz, x + (1 - i)y + 3z)$. Find $T^*(x, y, z)$.

▮ Find the matrix A representing T in the usual basis of $\mathbf{R}^3$:

$$A = \begin{pmatrix} 2 & i & 0 \\ 0 & 1 & -5i \\ 1 & 1-i & 3 \end{pmatrix}$$

Recall that the usual basis is orthonormal. Thus by Theorem 20.1, the matrix of T^* in this basis is the conjugate transpose A^* of A. Thus form

$$A^* = \begin{pmatrix} 2 & 0 & 1 \\ -i & 1 & 1+i \\ 0 & 5i & 3 \end{pmatrix}$$

Accordingly, $T^*(x, y, z) = (2x + z, -ix + y + (1 + i)z, 5iy + 3z)$.

20.8 Let $F: \mathbf{R}^3 \to \mathbf{R}^3$ defined by $F(x, y, z) = (3x + 4y - 5z, 2x - 6y + 7z, 5x - 9y + z)$. Find $F^*(x, y, z)$.

▮ First find the matrix A representing T in the usual basis of $\mathbf{R}^3$. [Recall the rows of A are the coefficients of x, y, z.] Thus

$$A = \begin{pmatrix} 3 & 4 & -5 \\ 2 & -6 & 7 \\ 5 & -9 & 1 \end{pmatrix}$$

Since the base field is $\mathbf{R}$, the adjoint F^* is represented by the transpose A^T of A. Thus form

$$A^T = \begin{pmatrix} 3 & 2 & 5 \\ 4 & -6 & -9 \\ -5 & 7 & 1 \end{pmatrix}$$

Then $F^*(x, y, z) = (3x + 2y + 5z, 4x - 6y - 9z, -5x + 7y + z)$.

20.9 Let T be the linear operator on $\mathbf{C}^3$ defined by $T(x, y, z) = (2x + (1 - i)y, (3 + 2i)x - 4iz, 2ix + (4 - 3i)y - 3z)$. Find $T^*(x, y, z)$.

▮ First find the matrix A representing T in the usual basis of $\mathbf{C}^3$:

$$A = \begin{pmatrix} 2 & 1-i & 0 \\ 3+2i & 0 & -4i \\ 2i & 4-3i & -3 \end{pmatrix}$$

Form the conjugate transpose A^* of A:

$$A^* = \begin{pmatrix} 2 & 3-2i & -2i \\ 1+i & 0 & 4+3i \\ 0 & 4i & -3 \end{pmatrix}$$

Thus $T^*(x, y, z) = (2x + (3 - 2i)y - 2iz, (1 + i)x + (4 + 3i)z, 4iy - 3z)$.

20.10 Let V be an inner product space. Each $u \in V$ determines a mapping $\hat{u}: V \to K$ defined by $\hat{u}(v) = \langle v, u \rangle$. Show that $\hat{u}$ is linear. [Thus u belongs to the dual space V^*.]

▮ For any $a, b \in K$ and any $v_1, v_2 \in V$, $\hat{u}(av_1 + bv_2) = \langle av_1 + bv_2, u \rangle = a\langle v_1, u \rangle + b\langle v_2, u \rangle = a\hat{u}(v_1) + b\hat{u}(v_2)$. Thus $\hat{u}$ is linear and, in other words, $\hat{u}$ is a linear functional on V.

Theorem 20.2: Let ϕ be a linear functional on a finite-dimensional inner product space V. Then there exists a unique vector $u \in V$ such that $\phi(v) = \langle v, u \rangle$ for every $v \in V$.

20.11 Prove Theorem 20.2 which is the converse of Problem 20.10 and which need not be true for vector spaces of infinite dimension.

▌ Let $\{e_1, \ldots, e_n\}$ be an orthonormal basis of V. Set $u = \sqrt{\phi(e_1)}e_1 + \overline{\phi(e_2)}e_2 + \cdots + \overline{\phi(e_n)}e_n$. Let $\hat{u}$ be the linear functional on V defined by $\hat{u}(v) = \langle v, u \rangle$, for every $v \in V$. Then, for $i = 1, \ldots, n$, $\hat{u}(e_i) = \langle e_i, u \rangle = \langle e_i, \overline{\phi(e_1)}e_1 + \cdots + \overline{\phi(e_n)}e_n \rangle = \phi(e_i)$. Since $\hat{u}$ and ϕ agree on each basis vector, $\hat{u} = \phi$.

Now suppose u' is another vector in V for which $\phi(v) = \langle v, u' \rangle$ for every $v \in V$. Then $\langle v, u \rangle = \langle v, u' \rangle$ or $\langle v, u - u' \rangle = 0$. In particular this is true for $v = u - u'$ and so $\langle u - u', u - u' \rangle = 0$. This yields $u - u' = 0$ and $u = u'$. Thus such a vector u is unique as claimed.

20.12 Prove (i) of Theorem 20.1.

▌ We first define the mapping T^*. Let v be an arbitrary but fixed element of V. The map $u \mapsto \langle T(u), v \rangle$ is a linear functional on V. Hence by Theorem 20.2 there exists a unique element $v' \in V$ such that $\langle T(u), v \rangle = \langle u, v' \rangle$ for every $u \in V$. We define $T^*V \to V$ by $T^*(v) = v'$. Then $\langle T(u), v \rangle = \langle u, T^*(v) \rangle$ for every $u, v \in V$.

We next show that T^* is linear. For any $u, v_i \in V$ and any $a, b \in K$, $\langle u, T^*(av_1 + bv_2) \rangle = \langle T(u), av_1 + bv_2 \rangle = \bar{a}\langle T(u), v_1 \rangle + \bar{b}\langle T(u), v_2 \rangle = \bar{a}\langle u, T^*(v_1) \rangle + \bar{b}\langle u, T^*(v_2) \rangle = \langle u, aT^*(v_1) + bT^*(v_2) \rangle$. But this is true for every $u \in V$; hence $T^*(av_1 + bv_2) = aT^*(v_1) + bT^*(v_2)$. Thus T^* is linear.

20.13 Prove (ii) of Theorem 20.1.

▌ The matrices $A = (a_{ij})$ and $B = (b_{ij})$ representing T and T^*, respectively, in the basis $\{e_i\}$ are given by $a_{ij} = \langle T(e_j), e_i \rangle$ and $b_{ij} = \langle T^*(e_j), e_j \rangle$. Hence $b_{ij} = \langle T^*(e_j), e_i \rangle = \overline{\langle e_i, T^*(e_j) \rangle} = \overline{\langle T(e_i), e_j \rangle} = \overline{a_{ij}}$. Thus $B = A^*$, as claimed.

Problems 20.14–20.17 prove Theorem 20.3 which summarizes some of the properties of the adjoint.

Theorem 20.3: Let S and T be linear operators on V and let $k \in K$. Then
        (i) $(S + T)^* = S^* + T^*$ (iii) $(ST)^* = T^*S^*$
        (ii) $(kT)^* = \bar{k}T^*$ (iv) $(T^*)^* = T$

20.14 Prove (i) of Theorem 20.3.

▌ For any $u, v \in V$, $\langle (S + T)(u), v \rangle = \langle S(u) + T(u), v \rangle = \langle S(u), v \rangle + \langle T(u), v \rangle = \langle u, S^*(v) \rangle + \langle u, T^*(v) \rangle = \langle u, S^*(v) + T^*(v) \rangle = \langle u, (S^* + T^*)(v) \rangle$. The uniqueness of the adjoint implies $(S + T)^* = S^* + T^*$.

20.15 Prove (ii) of Theorem 20.3.

▌ For any $u, v \in V$, $\langle (kT)(u), v \rangle = \langle kT(u), v \rangle = k\langle T(u), v \rangle = k\langle u, T^*(v) \rangle = \langle u, \bar{k}T^*(v) \rangle = \langle u, (\bar{k}T^*)(v) \rangle$. The uniqueness of the adjoint implies $(kT)^* = \bar{k}T^*$.

20.16 Prove (iii) of Theorem 20.3.

▌ For every $u, v \in V$, $\langle (ST)(u), v \rangle = \langle S(T(u)), v \rangle = \langle T(u), S^*(v) \rangle = \langle u, T^*(S^*(v)) \rangle = \langle u, (T^*S^*)(v) \rangle$. The uniqueness of the adjoint implies $(ST)^* = T^*S^*$.

20.17 Prove (iv) of Theorem 20.3.

▌ For every $u, v \in V$, $\langle T^*(u), v \rangle = \overline{\langle v, T^*(u) \rangle} = \overline{\langle T(v), u \rangle} = \langle u, T(v) \rangle$. The uniqueness of the adjoint implies $(T^*)^* = T$.

20.18 Let T be a linear operator on V, and let W be a T-invariant subspace of V. Show that $W^\perp$ is invariant under T^*.

▌ Let $u \in W^\perp$. If $w \in W$, then $T(w) \in W$ and so $\langle w, T^*(u) \rangle = \langle T(w), u \rangle = 0$. Thus $T^*(u) \in W^\perp$ since it is orthogonal to every $w \in W$. Hence $W^\perp$ is invariant under T^*.

20.19 Use the definition of the adjoint to show $I^* = I$.

▌ For every $u, v \in V$, $\langle I(u), v \rangle = \langle u, v \rangle = \langle u, I(v) \rangle$; hence $I^* = I$.

20.20 Use the definition of the adjoint to show $0^* = 0$.

▌ For every $u, v \in V$, $\langle 0(u), v \rangle = \langle 0, v \rangle = 0 = \langle u, 0 \rangle = \langle u, 0(v) \rangle$; hence $0^* = 0$.

20.21 Suppose T is invertible. Show that $(T^{-1})^* = (T^*)^{-1}$.

▌ $I = I^* = (TT^{-1})^* = (T^{-1})^* T^*$; hence $(T^{-1})^* = (T^*)^{-1}$.

20.22 Suppose $\langle T(u), v \rangle = 0$ for every $u, v \in V$. Show that $T = 0$.

▌ Set $v = T(u)$. Then $\langle T(u), T(u) \rangle = 0$ and hence $T(u) = 0$, for every $u \in V$. Accordingly, $T = 0$.

20.23 Suppose V is a complex inner product space and $\langle T(u), u \rangle = 0$ for every $u \in V$. Show that $T = 0$.

▌ By hypothesis, $\langle T(v + w), v + w \rangle = 0$ for any $v, w \in V$. Expanding and setting $\langle T(v), v \rangle = 0$ and $\langle T(w), w \rangle = 0$,

$$\langle T(v), w \rangle + \langle T(w), v \rangle = 0 \qquad (1)$$

Note w is arbitrary in (1). Substituting iw for w, and using $\langle T(v), iw \rangle = \bar{i} \langle T(v), w \rangle = -i \langle T(v), w \rangle$ and $\langle T(iw), v \rangle = \langle iT(w), v \rangle = i \langle T(w), v \rangle$, we get $-i \langle T(v), w \rangle + i \langle T(w), v \rangle = 0$. Dividing through by i and adding to (1), we obtain $\langle T(w), v \rangle = 0$ for any $v, w \in V$. Accordingly, $T = 0$.

20.24 Show that Problem 20.23 does not hold for a real space V, i.e., give an example of an operator T on a real space V for which $\langle T(u), u \rangle = 0$ for every $u \in V$ but $T \neq 0$.

▌ Let T be the linear operator on $\mathbf{R}^2$ defined by $T(x, y) = (y, -x)$. Then $\langle T(u), u \rangle = 0$ for every $u \in V$, but $T \neq 0$.

Remark: Let $A(V)$ denote the algebra of all linear operators on an inner product space V of finite dimension. The adjoint mapping $T \mapsto T^*$ on $A(V)$ is analogous to the conjugation mapping $z \mapsto \bar{z}$ on the complex field $\mathbf{C}$ as given in Table 20.1. Observe that the table identifies certain classes of operators $T \in A(V)$ whose behavior under the adjoint map imitates the behavior under conjugation of familiar classes of complex numbers. The next few sections investigate these classes of operators. In particular, the analogy between these classes of operators T and complex numbers z is reflected in Theorem 20.4 whose parts are proved later.

Table 20.1

Class of complex numbers	Behavior under conjugation	Class of operators in $A(V)$	Behavior under the adjoint map
Real axis	$\bar{z} = z$	Self-adjoint operators Also called: Symmetric (real case) Hermitian (complex case)	$T^* = T$
Unit circle ($\lvert z \rvert = 1$)	$\bar{z} = 1/z$	Orthogonal operators (real case) Unitary operators (complex case)	$T^* = T^{-1}$
Imaginary axis	$\bar{z} = -z$	Skew-adjoint operators Also called: Skew-symmetric (real case) Skew-Hermitian (complex case)	$T^* = -T$
Positive half axis $(0, \infty)$	$z = \bar{w}w, w \neq 0$	Positive definite operators	$T = S^*S$ with S nonsingular

Theorem 20.4: Let λ be an eigenvalue of a linear operator T on V.
    (i)   If $T^* = T$,   then λ is real.
    (ii)   If $T^* = T^{-1}$,   then $|\lambda| = 1$.
    (iii)   If $T^* = -T$,   then λ is pure imaginary.
    (iv)   If $T = S^*S$ with S nonsingular, then λ is real and positive.

20.2   SELF-ADJOINT OPERATORS, SYMMETRIC OPERATORS

20.25   Define a self-adjoint operator.

▮   An operator T on V is said to be self-adjoint if $T^* = T$.   The terms symmetric and Hermitian are also used for self-adjoint operators on V when the base fields are **R** and **C**, respectively.

Theorems 20.5–20.8 are the main content of this section.

Theorem 20.5:   Suppose T is a self-adjoint operator on V, i.e., suppose $T^* = T$.   Let λ be an eigenvalue of T.   Then λ is real.

Theorem 20.6:   Suppose T is self-adjoint, i.e., $T^* = T$.   Then eigenvectors of T belonging to distinct eigenvalues are orthogonal.

Theorem 20.7:   Let T be a symmetric [self-adjoint] operator on a real finite-dimensional inner product space V.   Then there exists an orthonormal basis of V consisting of eigenvectors of T; that is, T can be represented by a diagonal matrix relative to an orthonormal basis.

Theorem 20.8 [Alternate Form of Theorem 20.7]:   Let A be a real symmetric matrix.   Then there exists an orthogonal matrix P such that $B = P^{-1}AP = P^TAP$ is diagonal.

20.26   Prove Theorem 20.5.

▮   Let v be a nonzero eigenvector of T belonging to λ, that is, $T(v) = \lambda v$ with $v \neq 0$;   hence $\langle v, v \rangle$ is positive.   We show that $\lambda \langle v, v \rangle = \bar{\lambda} \langle v, v \rangle$:

$$\lambda \langle v, v \rangle = \langle \lambda v, v \rangle = \langle T(v), v \rangle = \langle v, T^*(v) \rangle = \langle v, T(v) \rangle = \langle v, \lambda v \rangle = \bar{\lambda} \langle v, v \rangle$$

But $\langle v, v \rangle \neq 0$;   hence $\lambda = \bar{\lambda}$ and so λ is real.

20.27   Prove Theorem 20.6.

▮   Suppose $T(v) = \lambda v$ and $T(w) = \mu w$ where $\lambda \neq \mu$.   We show that $\lambda \langle v, w \rangle = \mu \langle v, w \rangle$:

$$\lambda \langle v, w \rangle = \langle \lambda, w \rangle = \langle T(v), w \rangle = \langle v, T(w) \rangle = \langle v, \mu w \rangle = \bar{\mu} \langle v, w \rangle = \mu \langle v, w \rangle$$

[The last step uses the fact that μ is real by Theorem 20.5, so $\bar{\mu} = \mu$.]   But $\lambda \neq \mu$;   hence $\langle v, w \rangle = 0$ as claimed.

20.28   Let T be a symmetric operator on a real space V of finite dimension.   Show that (*a*) the characteristic polynomial $\Delta(t)$ of T is a product of linear factors [over **R**], (*b*) T has a nonzero eigenvector.

▮   (*a*)   Let A be a matrix representing T relative to an orthonormal basis of V; then $A = A^T$.   Let $\Delta(t)$ be the characteristic polynomial of A.   Viewing A as a complex self-adjoint operator, the matrix A has only real eigenvalues.   Thus $\Delta(t) = (t - \lambda_1)(t - \lambda_2) \cdots (t - \lambda_n)$ where the λ_i are all real.   In other words, $\Delta(t)$ is a product of linear polynomials over **R**.
    (*b*)   By (*a*), T has at least one [real] eigenvalue.   Hence T has a nonzero eigenvector.

20.29   Prove Theorem 20.7.

▮   The proof is by induction on the dimension of V.   If $\dim V = 1$,   the theorem trivially holds.   Now suppose $\dim V = n > 1$.   By the preceding problem, there exists a nonzero eigenvector v_1 of T.   Let W be the space spanned by v_1, and let u_1 be a unit vector in W, e.g., let $u_1 = v_1 / \|v_1\|$.

Since v_1 is an eigenvector of T, the subspace W of V is invariant under T. Hence $W^\perp$ is invariant under $T^* = T$. Thus the restriction $\hat{T}$ of T to $W^\perp$ is a symmetric operator.

We have dim $W^\perp = n - 1$ since dim $W = 1$. By induction, there exists an orthonormal basis $\{u, \ldots, u_n\}$ of $W^\perp$ consisting of eigenvectors of $\hat{T}$ and hence of T. But $\langle u_1, u_i \rangle = 0$ for $i = 2, \ldots, n$ because $u_i \in W^\perp$. Accordingly $\{u_1, u_2, \ldots, u_n\}$ is an orthonormal set and consists of eigenvectors of T. Thus the theorem is proved.

20.30 Let $A = \begin{pmatrix} 3 & 2 \\ 2 & 3 \end{pmatrix}$. Find a (real) orthogonal matrix P for which $P^T A P$ is diagonal.

▌ The characteristic polynomial $\Delta(t)$ of A is

$$\Delta(t) = |tI - A| = \begin{vmatrix} t-3 & -2 \\ -2 & t-3 \end{vmatrix} = t^2 - 6t + 5 = (t-5)(t-1)$$

and thus the eigenvalues of A are 5 and 1. Substitute $t = 5$ into the matrix $tI - A$ to obtain the corresponding homogeneous system of linear equations $2x - 2y = 0$, $-2x - 2y = 0$. A nonzero solution is $v_2 = (1, -1)$. Normalize v_1 to find the unit solution $u_1 = (1/\sqrt{2}, 1/\sqrt{2})$.

Next substitute $t = 1$ into the matrix $tI - A$ to obtain the corresponding homogeneous system of linear equations $-2x - 2y = 0$, $-2x\ 2y = 0$. A nonzero solution is $v_2 = (1, -1)$. Normalize v_2 to find the unit solution $u_2 = (1/\sqrt{2}, -1/\sqrt{2})$.

Finally let P be the matrix whose columns are u_1 and u_2, respectively; then

$$P = \begin{pmatrix} 1/\sqrt{2} & 1/\sqrt{2} \\ 1/\sqrt{2} & -1/\sqrt{2} \end{pmatrix} \quad \text{and} \quad P^T A P = \begin{pmatrix} 5 & 0 \\ 0 & 1 \end{pmatrix}$$

As expected, the diagonal entries of $P^T A P$ are the eigenvalues of A.

20.31 Let $B = \begin{pmatrix} 5 & 3 \\ 3 & -3 \end{pmatrix}$. Find a (real) orthogonal matrix P such that $P^T B P$ is diagonal.

▌ The characteristic polynomial $\Delta(t)$ of B is $\Delta(t)$ of B is $\Delta(t) = |tI - B| = t^2 - \text{tr}(B)t + |B| = t^2 - 2t - 24 = (t-6)(t+4)$. Thus the eigenvalues are $\lambda = 6$ and $\lambda = -4$. Hence

$$P^T B P = \begin{pmatrix} 6 & 0 \\ 0 & -4 \end{pmatrix}$$

To find the change-of-basis matrix P, we need to find the corresponding eigenvectors. Subtract $\lambda = 6$ down the diagonal of B to obtain the homogeneous system $-x + 3y = 0$, $3x - 9y = 0$. A nonzero solution is $v_1 = (3, 1)$. Subtract $\lambda = -4$ down the diagonal of B to obtain the homogeneous system $9x + 3y = 0$, $3x + y = 0$. A nonzero solution is $v_2 = (-1, 3)$. [As expected from Theorem 20.6, the vectors v_1 and v_2 are orthogonal.] Normalize v_1 and v_2 to obtain the orthonormal basis $u_1 = (3/\sqrt{10}, 1/\sqrt{10})$, $u_2 = (-1/\sqrt{10}, 3/\sqrt{10})$. Then

$$P = \begin{pmatrix} \dfrac{3}{\sqrt{10}} & \dfrac{-1}{\sqrt{10}} \\ \dfrac{1}{\sqrt{10}} & \dfrac{3}{\sqrt{10}} \end{pmatrix}$$

Problems 20.32–20.38 diagonalize the symmetric matrix $C = \begin{pmatrix} 11 & -8 & 4 \\ -8 & -1 & -2 \\ 4 & -2 & -4 \end{pmatrix}$.

20.32 Find the characteristic polynomial $\Delta(t)$ of C.

▌ $\Delta(t) = t^3 - \text{tr}(C)t^2 + (C_{11} + C_{22} + C_{33})t - |C| = t^3 - 6t^2 - 135t - 400$. [Here C_{ii} is the cofactor of c_{ii} in $C = (c_{ij})$.]

20.33 Find the eigenvalues of C or, in other words, the roots of $\Delta(t)$.

▌ If $\Delta(t)$ has a rational root it must divide 400. Testing $t = -5$ we get

$$
\begin{array}{r|rrrr}
-5 & 1 & -\ 6 & -135 & -400 \\
   &   & -\ 5 & +\ 55 & +400 \\
\hline
   & 1 & -11 & -\ 80 & +\ \ \ 0
\end{array}
$$

Thus $t + 5$ is a factor of $\Delta(t)$ and $\Delta(t) = (t + 5)(t^2 - 11t - 80) = (t + 5)^2(t - 16)$. Accordingly, the eigenvalues of C are $\lambda = -5$ [with multiplicity two] and $\lambda = 16$ [with multiplicity one].

20.34 Find orthogonal eigenvectors belonging to the eigenvalue $\lambda = -5$.

▮ Subtract $\lambda = -5$ down the diagonal of C to obtain the homogeneous system $16x - 8y + 4z = 0$, $-8x + 4y - 2z = 0$, $4x - 2y + z = 0$. That is, $4x - 2y + z = 0$. The system has two independent solutions. One solution is $v_1 = (0, 1, 2)$. We seek a second solution $v_2 = (a, b, c)$ which is orthogonal to v_1; i.e., such that $4a - 2b + c = 0$ and also $b - 2c = 0$. One such solution is $v_2 = (-5, -8, 4)$.

20.35 Find an eigenvector v_3 belonging to the eigenvalue $\lambda = 16$.

▮ Subtract $\lambda = 16$ down the diagonal of C to obtain the homogeneous system $-5x - 8y + 4z = 0$, $-8x - 17y - 2z = 0$, $4x - 2y - 20z = 0$. This system yields a nonzero solution $v_3 = (4, -2, 1)$. [As expected from Theorem 20.6b, the eigenvector v_3 is orthogonal to v_1 and v_2.]

20.36 Find an orthogonal matrix P such that $P^{-1}CP$ is diagonal.

▮ Normalize v_1, v_2, v_3 to obtain the orthonormal basis: $u_1 = (0, 1/\sqrt{5}, 2/\sqrt{5})$, $u_2 = (-5\sqrt{105}, 4/\sqrt{105})$, $u_3 = (4/\sqrt{21}, -2/\sqrt{21}, 1/\sqrt{21})$. Then P is the matrix whose columns are u_1, u_2, u_3. Thus

$$P = \begin{pmatrix} 0 & -5/\sqrt{105} & 4/\sqrt{21} \\ 1/\sqrt{5} & -8/\sqrt{105} & -2/\sqrt{21} \\ 2/\sqrt{5} & 4/\sqrt{105} & 1/\sqrt{21} \end{pmatrix} \quad \text{and} \quad P^T C P = \begin{pmatrix} -5 & & \\ & -5 & \\ & & 16 \end{pmatrix}$$

20.37 Consider the quadratic form $q(x, y, z) = 11x^2 - 16xy - y^2 + 8xz - 4yz - 4z^2$. Find an orthogonal change of coordinates which diagonalizes.

▮ Since C is the matrix which represents q, use the above matrix P to obtain the required change of coordinates:

$$x = -\frac{5y'}{\sqrt{105}} + \frac{4z'}{\sqrt{21}}$$

$$y = \frac{x'}{\sqrt{5}} - \frac{8y'}{\sqrt{105}} - \frac{2z'}{\sqrt{21}}$$

$$z = \frac{2x'}{\sqrt{5}} + \frac{4y'}{\sqrt{105}} + \frac{z'}{\sqrt{21}}$$

Under this change of coordinates, q is transformed into the diagonal form $q(x', y', ,z') = -5x' - 5y' + 16z'$.

20.38 Find the signature of q.

▮ Since there are two negative diagonal entries and one positive diagonal entry, $N = 2$, $P = 1$. Thus $\text{Sig}(q) = P - N = 1 - 2 = -1$.

20.39 Suppose T is self-adjoint and $\langle T(u), u \rangle = 0$ for every $u \in V$. Show that $T = 0$.

▮ By Problem 20.23, the result holds for the complex case; hence we need only consider the real case. Expanding $\langle T(v + w), v + w \rangle = 0$, we obtain

$$\langle T(v), w \rangle + \langle T(w), v \rangle = 0 \tag{1}$$

Since T is self-adjoint and since it is a real space, we have $\langle T(w), v \rangle = \langle w, T(v) \rangle = \langle T(v), w \rangle$. Substituting this into (1), we obtain $\langle T(v), w \rangle = 0$ for any $v, w \in V$. Thus $T = 0$.

20.40 Suppose T is self-adjoint and $T^2 = 0$. Show that $T = 0$.

▮ For any $v \in V$, we have $\|T(v)\|^2 = \langle T(v), T(v) \rangle = \langle v, T^2(v) \rangle = \langle v, 0v \rangle = \langle v, 0 \rangle = 0$. Thus $\|T(v)\| = 0$ and hence $T(v) = 0$. Since $T(v) = 0$ for every $v \in V$, we have $T = 0$.

20.41   Show that T^*T and TT^* are self-adjoint for any operator T on V.

$\blacksquare$ $(T^*T)^* = T^*T^{**} = T^*T$,   and hence T^*T is self-adjoint.   Also, $(TT^*)^* = T^{**}T^* = TT^*$,   and hence TT^* is self-adjoint.

20.42   Show that $T + T^*$ is self-adjoint for any operator T on V.

$\blacksquare$ $(T + T^*)^* = T^* + T^{**} = T^* + T = T + T^*$;   hence $T + T^*$ is self-adjoint.

20.43   Define a skew-adjoint operator.

$\blacksquare$ A linear operator T on V is said to be skew-adjoint if $T^* = -T$.

20.44   Suppose T is skew-adjoint, i.e., suppose $T^* = -T$.   Let λ be an eigenvalue of T.   Show that λ is pure imaginary, i.e., $\bar{\lambda} = -\lambda$.

$\blacksquare$ Let v be a nonzero eigenvector of T belonging to λ,   that is, $T(v) = v$ with $v \neq 0$.   Hence $\langle v, v \rangle \neq 0$.   We show that $\lambda \langle v, v \rangle = -\bar{\lambda} \langle v, v \rangle$: $\lambda \langle v, v \rangle = \langle \lambda v, v \rangle = \langle T(v), v \rangle = \langle v, T^*(v) \rangle = \langle v, -T(v) \rangle = \langle v, -\lambda v \rangle = -\bar{\lambda} \langle v, v \rangle$.   But $\langle v, v \rangle \neq 0$;   hence $\lambda = -\bar{\lambda}$ or $\bar{\lambda} = -\lambda$,   and so λ is pure imaginary.

20.45   Show that $T - T^*$ is skew-adjoint for any linear operator T on V.

$\blacksquare$ $(T - T^*)^* = T^* - T^{**} = T^* - T = -(T - T^*)$;   hence $T - T^*$ is skew-adjoint.

20.46   Show that any operator T is the sum of a self-adjoint operator and a skew-adjoint operator.

$\blacksquare$ Set $S = \frac{1}{2}(T + T^*)$ and $U = \frac{1}{2}(T - T^*)$.   Then $T = S + U$ where $S^* = (\frac{1}{2}(T + T^*))^* = \frac{1}{2}(T^* + T^{**}) = \frac{1}{2}(T^* + T) = S$ and $U^* = (\frac{1}{2}(T - T^*))^* = \frac{1}{2}(T^* - T) = -\frac{1}{2}(T - T^*) = -U$,   that is, S is self-adjoint and U is skew adjoint.

20.3   ORTHOGONAL AND UNITARY OPERATORS

20.47   Define a unitary and orthogonal operator.

$\blacksquare$ Let U be an invertible operator on V such that $U^* = U^{-1}$ or equivalently $UU^* = U^*U = I$.   Then U is said to be *orthogonal* or *unitary* according as the underlying field is real or complex.

Theorem 20.9, proved in Problem 20.55, gives alternative characterizations of these operators.

Theorem 20.9:   The following conditions on an operator U are equivalent
  (i) $U^* = U^{-1}$,   that is, $UU^* = U^*U = I$ [or U is unitary (orthogonal)].
  (ii)   *U preserves inner products,* i.e., for every $v, w \in V$, $\langle U(v), U(w) \rangle = \{v, w\}$.
  (iii)   *U preserves lengths,* i.e., for every $v \in V$, $\|U(v)\| = \|v\|$.

An orthogonal operator T need not be symmetric and so it may not be represented by a diagonal matrix relative to an orthonormal basis.   However, such an operator T does have a simple canonical representation, as described in Theorem 20.10 which is proved in Problem 20.58.

Theorem 20.10:   Let T be an orthogonal operator on a real inner product space V.   Then there exists an orthonormal basis B of V such that the matrix representation of T in the basis B has the form of the matrix in Fig. 20-1.

[The reader may recognize that the 2×2 diagonal blocks in Fig. 20-1 represent rotations in the corresponding two-dimensional subspaces.]

20.48   Suppose $T: \mathbf{R}^3 \to \mathbf{R}^3$ is the linear operator which rotates each vector v about the z axis by a fixed angle θ; i.e., suppose $T(x, y, z) = (x \cos \theta - y \sin \theta, x \sin \theta + y \cos \theta, z)$.   Is T orthogonal?

$\blacksquare$ As pictured in Fig. 20-2, the length (distance from the origin) of v is preserved under the rotation T. Thus T is an orthogonal operator.

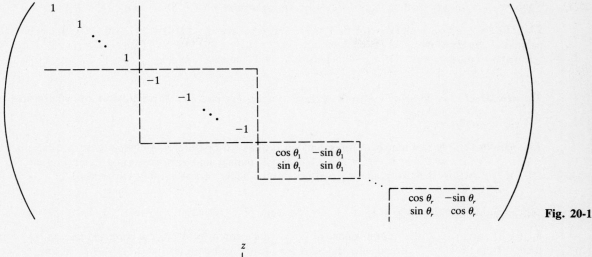

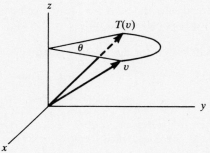

Fig. 20-1

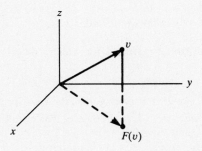

Fig. 20-2

20.49 Let $F: \mathbf{R}^3 \to \mathbf{R}^3$ be the linear operator which reflects each vector v through the xy plane; i.e., let $F(x, y, z) = (x, y, -z)$. Is F orthogonal?

▌ As pictured in Fig. 20-3, the length of v is preserved under the reflection F. Thus F is orthogonal.

Fig. 20-3

20.50 Given an example of a vector space V of infinite dimension and a linear map $T: V \to V$ for which Theorem 20.9 does not hold.

▌ Let V be the l_2-space of infinite sequences $v = (a_1, a_2, \ldots)$ or real numbers a_i such that $\sum a_i^2 < \infty$. Let $T: V \to V$ be the linear operator defined by $T(a_1, a_2, \ldots) = (0, a_1, a_2, \ldots)$. Clearly, T preserves inner products and lengths. However, T is not surjective since, for example, $(1, 0, 0, \ldots)$ does not belong to the image of T; hence T is not invertible.

20.51 Suppose U is unitary [orthogonal]. Show that u is an isometry on V.

▌ An isometry on V is a mapping which preserves distances. [Recall $d(v, w) = \|v - w\|$ is the distance between v and w.] Since U is unitary [orthogonal], $\|U(v) - U(w)\| = \|U(v - w)\| = \|v - w\|$. Thus U is an isometry.

20.52 Suppose T is unitary [orthogonal]. Let λ be an eigenvalue of T. Show that $|\lambda| = 1$.

∥ Let v be a nonzero eigenvector of T belonging to λ, that is, $T(v) = \lambda v$ with $v \neq 0$; hence $\langle v, v \rangle$ is positive. We show that $\lambda\bar{\lambda}\langle v, v \rangle = \langle v, v \rangle$: $\lambda\bar{\lambda}\langle v, v \rangle = \langle \lambda v, \lambda v \rangle = \langle T(v), T(v) \rangle = \langle v, T^*T(v) \rangle = \langle v, I(v) \rangle = \langle v, v \rangle$. But $\langle v, v \rangle \neq 0$; hence $\lambda\bar{\lambda} = 1$ and so $|\lambda| = 1$.

Theorem 20.11: A complex matrix A represents a unitary operator U [relative to an orthonormal basis] if and only if $A^* = A^{-1}$.

Theorem 20.12: A real matrix A represents an orthogonal operator U [relative to an orthonormal basis] if and only if $A^T = A^{-1}$. [That is, unitary and orthogonal matrices represent, respectively, unitary and orthogonal linear operators; and vice versa.]

20.53 Prove Theorems 20.11 and 20.12.

∥ By Theorem 20.1, the adjoint operator U^* is represented by A^* in the complex case and A^T in the real case. Thus $U^*U = I$ if and only if $A^*A = I$ in the complex case and $A^TA = I$ in the real case. In other words, T is unitary [orthogonal] if and only if $A^* = A^{-1}$ $(A^T = A^{-1})$.

20.54 Show that $T^*T - I$ is self-adjoint for any linear operator T.

∥ $(T^*T - I)^* = (T^*T)^* - I^* = T^*T^{**} - I = T^*T - I$. Thus $T^*T - I$ is self-adjoint.

20.55 Prove Theorem 20.9.

∥ Suppose (i) holds. Then, for every $v, w \in V$, $\langle U(v), U(w) \rangle = \langle v, U^*U(w) \rangle = \langle v, I(w) \rangle = \langle v, w \rangle$. Thus (i) implies (ii). Now if (ii) holds, then $\|U(v)\| = \sqrt{\langle U(v), U(v) \rangle} = \sqrt{\langle v, v \rangle} = \|v\|$. Hence (ii) implies (iii). It remains to show that (iii) implies (i).

Suppose (iii) holds. Then for every $v \in V$, $\langle U^*U(v), v \rangle = \langle U(v), U(v) \rangle = \langle v, v \rangle = \langle I(v), v \rangle$. Hence $\langle (U^*U - I)(v), v \rangle = 0$ for every $v \in V$. Since $U^*U - I$ is self-adjoint, we have $U^*U - I = 0$. Thus $U^*U = I$ and hence $U^* = U^{-1}$, as claimed.

20.56 Let U be a unitary [orthogonal] operator on V and let W be a subspace invariant under U. Show that $W^\perp$ is also invariant under U.

∥ Since U is nonsingular, $U(W) = W$; i.e., for any $w \in W$ there exists $w' \in W$ such that $U(w') = w$. Now let $v \in W^\perp$. Then for any $w \in W$, $\langle U(v), w \rangle = \langle U(v), U(w') \rangle = \langle v, w' \rangle = 0$. Thus $U(v)$ belongs to $W^\perp$. Therefore $W^\perp$ is invariant under U.

20.57 Show that every 2 by 2 orthogonal matrix A for which $\det(A) = 1$ is of the form $\begin{pmatrix} \cos\theta & -\sin\theta \\ \sin\theta & \cos\theta \end{pmatrix}$ for some real number θ.

∥ Suppose $A = \begin{pmatrix} a & b \\ c & d \end{pmatrix}$. Since A is orthogonal, its rows form an orthonormal set; hence $a^2 + b^2 = 1$, $c^2 + d^2 = 1$, $ac + bd = 0$, $ad - bc = 1$. The last equation follows from $\det(A) = 1$. We consider separately the cases $a = 0$ and $a \neq 0$.

If $a = 0$, the first equation gives $b^2 = 1$ and therefore $b = \pm 1$. Then the fourth equation gives $c = -b = \mp 1$, and the second equation yields $1 + d^2 = 1$ or $d = 0$. Thus

$$A = \begin{pmatrix} 0 & 1 \\ -1 & 0 \end{pmatrix} \quad \text{or} \quad \begin{pmatrix} 0 & -1 \\ 1 & 0 \end{pmatrix}$$

The first alternative has the required form with $\theta = -\pi/2$, and the second alternative has the required form with $\theta = \pi/2$.

If $a \neq 0$, the third equation can be solved to give $c = -bd/a$. Substituting this into the second equation, $b^2d^2/a^2 + d^2 = 1$ or $b^2d^2 + a^2d^2 = a^2$ or $(b^2 + a^2)d^2 = a^2$ or $a^2 = d^2$ and therefore $a = d$ or $a = -d$. If $a = -d$, then the third equation yields $c = b$ and so the fourth equation gives $-a^2 - c^2 = 1$ which is impossible. Thus $a = d$. But then the third equation gives $b = -c$ and so

$$A = \begin{pmatrix} a & -c \\ c & a \end{pmatrix}$$

Since $a^2 + c^2 = 1$, There is a real number θ such that $a = \cos\theta$, $c = \sin\theta$ and hence A has the required form in this case also.

20.58 Prove Theorem 20.10. Let T be an orthogonal operator on a real inner product space V. Then there exists an orthonormal basis B of V such that the matrix representation of T in the basis B is block diagonal with diagonal blocks consisting of 1s, -1s, and blocks of the form

$$\begin{pmatrix} \cos\theta_i & -\sin\theta_i \\ \sin\theta_i & \cos\theta_i \end{pmatrix}$$

(as in Fig. 20-1).

❚ Let $S = T + T^{-1} = T + T^*$. Then $S^* = (T + T^*)^* = T^* + T = S$. Thus S is a symmetric operator on V. By Theorem 20.7, there exists an orthonormal basis of V consisting of eigenvectors of S. If $\lambda_1, \ldots, \lambda_m$ denote the distinct eigenvalues of S, then V can be decomposed into the direct sum $V = V_1 \oplus V_2 \oplus \cdots \oplus V_m$ where the V_i consist of the eigenvectors of S belonging to λ_i. We claim that each V_i is invariant under T. For suppose $v \in V_i$; then $S(v) = \lambda_i v$ and $S(T(v)) = (T + T^{-1})T(v) = T(T + T^{-1})(v) = TS(v) = T(\lambda_i v) = \lambda_i T(v)$. That is, $T(v) \in V_i$. Hence V_i is invariant under T. Since the V_i are orthogonal to each other, we can restrict our investigation to the way that T acts on each individual V_i.

On a given V_i, $(T + T^{-1})v = S(v) = \lambda_i v$. Multiplying by T, $(T^2 - \lambda_i T + I)(v) = 0$. We consider the cases $\lambda_i = \pm 2$ and $\lambda_i \neq \pm 2$ separately. If $\lambda_i = \pm 2$, then $(T \pm I)^2(v) = 0$ which leads to $(T \pm I)(v) = 0$ or $T(v) = \pm v$. Thus T restricted to this V_i is either I or $-I$.

If $\lambda_i \neq \pm 2$, then T has no eigenvectors V_i since [Problem 20.52] the only eigenvalues of T are 1 or -1. Accordingly, for $v \neq 0$ the vectors v and $T(v)$ are linearly independent. Let W be the subspace spanned by v and $T(v)$. Then W is invariant under T, since $T(T(v)) = T^2(v) = \lambda_i T(v) - v$. We also have $V_i = W \oplus W^\perp$. Furthermore, by Problem 20.56, $W^\perp$ is also invariant under T. Thus we can decompose V_i into the direct sum of two-dimensional subspaces W_j where the W_j are orthogonal to each other and each W_j is invariant under T. Thus we can now restrict our investigation to the way T acts on each individual W_j.

Since $T^2 - \lambda_i T + I = 0$, the characteristic polynomial $\Delta(t)$ of T acting on W_j is $\Delta(t) = t^2 - \lambda_i t + 1$. Thus the determinant of T is 1, the constant term in $\Delta(t)$. By Problem 20.57, the matrix A representing T acting on W_j relative to any orthonormal basis of W_j must be of the form

$$\begin{pmatrix} \cos\theta & -\sin\theta \\ \sin\theta & \cos\theta \end{pmatrix}$$

The union of the basis of the W_j gives an orthonormal basis of V_i, and the union of the basis of the V_i gives an orthonormal basis of V in which the matrix representing T is of the desired form.

20.4 POSITIVE AND POSITIVE DEFINITE OPERATORS

20.59 Define a positive and positive definite operator.

❚ A linear operator P on an inner product space V is said to be *positive* [or *semidefinite*] if $P = S^*S$ for some operator S and is said to be *positive definite* if S is also nonsingular.

20.60 Show that a positive [positive definite] operator P is also self-adjoint.

❚ By definition, $P = S^*S$ for some operator S. Hence $P^* = (S^*S)^* = S^*S^{**} = S^*S = P$. Thus P is self-adjoint.

Theorems 20.13 and 20.14, proved in Problems 20.69 and 20.70, give alternative characterizations of these operators.

Theorem 20.13: The following conditions on an operator P are equivalent:
    (i) $P = T^2$ for some self-adjoint operator T.
    (ii) $P = S^*S$ for some operator S.
    (iii) P is self-adjoint and $\langle P(u), u \rangle \geq 0$ for every $u \in V$.

The corresponding theorem for positive definite operators is
Theorem 20.14: The following conditions on an operator P are equivalent:
    (i) $P = T^2$ for some nonsingular self-adjoint operator T.
    (ii) $P = S^*S$ for some nonsingular operator S.
    (iii) P is self-adjoint and $\langle P(u), u \rangle > 0$ for every $u \neq 0$ in V.

Theorem 20.15: A complex matrix $A = \begin{pmatrix} a & b \\ c & d \end{pmatrix}$ represents a positive [positive definite] operator if and only if A is self-adjoint [that is, $A^* = A$ in the complex case and $A^T = A$ in the real case] and a, d, and $|A| = ad - bc$ are nonnegative [positive] real numbers.

Problems 20.61–20.66 refer to Theorem 20.15 and the following matrices:

$$A = \begin{pmatrix} 1 & 1 \\ 1 & 1 \end{pmatrix} \qquad B = \begin{pmatrix} 3 & i \\ -i & 3 \end{pmatrix} \qquad C = \begin{pmatrix} 1 & 1 \\ 0 & 1 \end{pmatrix} \qquad D = \begin{pmatrix} 2 & 1 \\ 1 & 2 \end{pmatrix} \qquad E = \begin{pmatrix} 1 & i \\ -i & 1 \end{pmatrix} \qquad F = \begin{pmatrix} 1 & -2i \\ 2i & 1 \end{pmatrix}$$

20.61   Is A positive definite? Positive?

▮ Since $|A| = 0$, A is not positive definite. However, A is positive since $a = 1$, $d = 1$, and $|A| = 0$ are nonnegative.

20.62   Is B positive definite? Positive?

▮ Since $a = 3$, $d = 3$, and $|B| = 8$ are positive, B is positive definite [and hence positive].

20.63   Is C positive definite? Positive?

▮ Since C is not self-adjoint, i.e., $C^T \neq C$, C is neither positive definite nor positive.

20.64   Is D positive definite? Positive?

▮ Since $a = 2$, $d = 2$, and $|D| = 3$ are positive, D is positive definite [and hence positive].

20.65   Is E positive definite? Positive?

▮ Since $|E| = 0$, E is not positive definite. However, E is positive since $a = 1$, $d = 1$, and $|E| = 0$ are nonnegative.

20.66   Is F positive definite? Positive?

▮ Since $|F| = -3$, F is neither positive definite nor positive.

20.67   Suppose T is positive. Let λ be an eigenvalue of T. Show that λ is real and nonnegative.

▮ Since T is positive, T is self-adjoint; hence λ is real. Let v be a nonzero eigenvector of T belonging to λ, that is, $T(v) = \lambda v$ with $v \neq 0$; hence $\langle v, v \rangle$ is positive. Since T is positive, $T = S^*S$ for some operator S. We show $\lambda \langle v, v \rangle = \langle S(v), S(v) \rangle$: $\lambda \langle v, v \rangle = \langle \lambda v, v \rangle = \langle T(v), v \rangle = \langle S^*S(v), v \rangle = \langle S(v), S(v) \rangle$. Since $\langle S(v), S(v) \rangle$ is nonnegative and $\langle v, v \rangle$ is positive, we have λ is nonnegative, as required.

20.68   Suppose T is positive definite. Let λ be an eigenvalue of T. Show that λ is real and positive.

▮ Since T is positive definite, T is self-adjoint; hence λ is real. Let v be a nonzero eigenvector of T belonging to λ, that is, $T(v) = \lambda v$ with $v \neq 0$; hence $\langle v, v \rangle$ is positive. Since T is positive definite, $T = S^*S$ for some nonsingular operator S. Thus $S(v) \neq 0$, and hence $\langle S(v), S(v) \rangle$ is positive. We show that $\lambda \langle v, v \rangle = \langle S(v), S(v) \rangle$: $\lambda \langle v, v \rangle = \langle \lambda v, v \rangle = \langle T(v), v \rangle = \langle S^*S(v), v \rangle = \langle S(v), S(v) \rangle$. Since $\langle v, v \rangle$ and $\langle S(v), S(v) \rangle$ are both positive, we have λ is positive, as required.

20.69   Prove Theorem 20.13.

▮ Suppose (i) holds, i.e., $P = T^2$ where $T = T^*$. Then $P = TT = T^*T$ and so (i) implies (ii). Now suppose (ii) holds. Then $P^* = (S^*S)^* = S^*S^{**} = S^*S = P$ and so P is self-adjoint. Furthermore, $\langle P(u), u \rangle = \langle S^*S(u), u \rangle = \langle S(u), S(u) \rangle \geq 0$. Thus (ii) implies (iii), and so it remains to prove that (iii) implies (i).

Now suppose (iii) holds. Since P is self-adjoint, there exists an orthonormal basis $\{u, \ldots, u_n\}$ of V consisting of eigenvectors of P; say, $P(u_i) = \lambda_i u_i$. By Problem 20.67, the λ_i are nonegative real numbers. Thus $\sqrt{\lambda_i}$ is a real number. Let T be the linear operator defined by $T(u_i) = \sqrt{\lambda_i} u_i$, for $i = 1, \ldots, n$. Since T is represented by a real diagonal matrix relative to the orthonormal basis $\{u_i\}$, T is self-adjoint. Moreover, for each i, $T^2(u_i) = T(\sqrt{\lambda_i} u_i) = \sqrt{\lambda_i} T(u_i) = \sqrt{\lambda_i} \sqrt{\lambda_i} u_i = \lambda_i u_i = P(u_i)$. Since T^2 and P agree on a basis of V, $P = T^2$. Thus the theorem is proved.

20.70 Prove Theorem 20.14.

$\blacksquare$ Suppose (i) holds, i.e., $P = T^2$ where T is nonsingular and $T^* = T$. Then $P = TT = T^*T$ and hence (i) implies (ii). Now suppose (ii) holds. Then $P^* = (S^*S)^* = S^*S^{**} = S^*S = P$ and so P is self-adjoint. Suppose $u \neq 0$. Then $S(u) \neq 0$ since S is nonsingular and hence $\langle S(u), S(u) \rangle > 0$. Hence $\langle P(u), u \rangle = \langle S^*S(u), u \rangle = \langle S(u), S(u) \rangle > 0$. Thus (ii) implies (iii), and so it remains to prove that (iii) implies (i).

Now suppose (iii) holds. Since P is self-adjoint, there exists an orthonormal basis $\{u_1, \ldots, u_n\}$ of V consisting of eigenvectors of P; say, $P(u_i) = \lambda_i u_i$. By Problem 20.68, the λ_i are positive real numbers. Thus $\sqrt{\lambda_i}$ is a positive real number. Let T be the linear operator defined by $T(u_i) = \sqrt{\lambda_i} u_i$, for $i = 1, \ldots, n$. Since T is represented by a real diagonal matrix relative to the orthonormal basis $\{u_i\}$, T is self-adjoint, and since the diagonal entries are nonzero, T is nonsingular. Moreover, for each i, $T^2(u_i) = T(\sqrt{\lambda_i} u_i) = \sqrt{\lambda_i} T(u_i) = \sqrt{\lambda_i} \sqrt{\lambda_i} u_i = \lambda_i u_i = P(u_i)$. Since T^2 and P agree on a basis of V, $P = T^2$. Thus the theorem is proved.

Remark: The above operator T is the unique positive definite operator such that $P = T^2$; it is called the positive square root of P.

20.71 Suppose A is a diagonal matrix with positive real diagonal entries, say $\lambda_1, \lambda_2, \ldots, \lambda_n$. Show that A is positive definite.

$\blacksquare$ Let T be the diagonal matrix with diagonal entries $\sqrt{\lambda_1}, \sqrt{\lambda_2}, \ldots, \sqrt{\lambda_n}$. Then $A = T^2$ where T is nonsingular and T is self-adjoint [symmetric]. Hence A is positive definite.

20.72 Let A be a real positive definite matrix, and let Q be an orthogonal matrix. Show that $Q^T A Q = Q^{-1} A Q$ is also positive definite.

$\blacksquare$ Since A is a real positive definite matrix, $A = S^T S$ where S is nonsingular. Then $Q^T A Q = Q^T (S^T S) Q = (SQ)^T (SQ)$ where SQ is nonsingular. Thus $Q^T A Q$ is positive definite.

Problems 20.73–20.77 refer to the matrix $A = \begin{pmatrix} 5 & 1 \\ 1 & 5 \end{pmatrix}$.

20.73 Is A is positive definite?

$\blacksquare$ Since $a_{11} = 5$, $a_{22} = 5$, and $|A| = 24$ are positive, A is a positive definite matrix.

20.74 Find an orthogonal matrix Q such that $Q^T A Q$ is diagonal.

$\blacksquare$ The characteristic polynomial $\Delta(t)$ of A is

$$\Delta(t) = |tI - A| = \begin{vmatrix} t - 5 & -1 \\ -1 & t - 5 \end{vmatrix} = t^2 - 10^t + 24 = (t - 6)(t - 4)$$

Thus the eigenvalues are 6 and 4. Substitute $t = 6$ into the matrix $tI - A$ to obtain the corresponding homogeneous system of linear equations $x - y = 0$, $-x + y = 0$. A nonzero solution is $v_1 = (1, 1)$. Normalize v_1 to find the unit solution $u_1 = (1/\sqrt{2}, 1/\sqrt{2})$.

Next substitute $t = 4$ into the matrix $tI - A$ to obtain the corresponding homogeneous system of linear equations $-x - y = 0$, $-x - y = 0$. A nonzero solution is $v_2 = (1, -1)$. Normalize v_2 to find the unit solution $u_2 = (1/\sqrt{2}, -1/\sqrt{2})$.

Finally let Q be the matrix whose columns are u_1 and u_2, respectively; then

$$Q = \begin{pmatrix} \dfrac{1}{\sqrt{2}} & \dfrac{1}{\sqrt{2}} \\ \dfrac{1}{\sqrt{2}} & \dfrac{-1}{\sqrt{2}} \end{pmatrix} \quad \text{and} \quad Q^T A Q = \begin{pmatrix} 6 & 0 \\ 0 & 4 \end{pmatrix}$$

20.75 Find the square root S of $B = Q^T A Q = \begin{pmatrix} 6 & 0 \\ 0 & 4 \end{pmatrix}$.

$\blacksquare$ Take the square root of the diagonal entries to get $S = \begin{pmatrix} \sqrt{6} & 0 \\ 0 & 2 \end{pmatrix}$.

20.76   Show that $T = QSQ^T$ is the square root of A.

▎ We have $B = Q^T A Q = Q^{-1} A Q$;   hence $A = QBQ^{-1} = QBQ^T$.   Then $T^2 = (QSQ^T)(QSQ^T) = (QSQ^{-1})(QSQ^{-1}) = QS^2 Q^{-1} = QBQ^{-1} = A$.   Also, by Problem 20.71, T is positive definite.   Thus T is the positive square root of A.

20.77   Find T, the positive square root of A.

▎
$$T = QSQ^T = \begin{pmatrix} \dfrac{1}{\sqrt{2}} & \dfrac{1}{\sqrt{2}} \\ \dfrac{1}{\sqrt{2}} & \dfrac{-1}{\sqrt{2}} \end{pmatrix} \begin{pmatrix} \sqrt{6} & 0 \\ 0 & 2 \end{pmatrix} \begin{pmatrix} \dfrac{1}{\sqrt{2}} & \dfrac{1}{\sqrt{2}} \\ \dfrac{1}{\sqrt{2}} & \dfrac{-1}{\sqrt{2}} \end{pmatrix} = \frac{1}{2} \begin{pmatrix} \sqrt{6}+2 & \sqrt{6}-2 \\ \sqrt{6}-2 & \sqrt{6}+2 \end{pmatrix}$$

20.5   NORMAL OPERATORS

20.78   Define a normal operator.

▎ A linear operator T on an inner product space V is said to be normal if T commutes with its adjoint, i.e., if $TT^* = T^*T$.   [Analogously, a complex matrix A is normal if $AA^* = A^*A$,   and a real matrix A is normal if $AA^T = A^T A$.]

20.79   Show that self-adjoint and unitary [orthogonal] operators are normal

▎ Suppose $T^* = T$,   that is, T is self-adjoint. Then $TT^* = TT = T^*T$,   and hence T is normal. Suppose $T^* = T^{-1}$,   that is, T is unitary (orthogonal). Then $TT^* = I = T^*T$,   and hence T is normal.

Theorem 20.16:   Let T be a normal operator on a complex finite-dimensional inner product space V.   Then there exists an orthonormal basis of V consisting of eigenvectors of T; that is, T can be represented by a diagonal matrix relative to an orthonormal basis.

We give the corresponding statement for matrices.

Theorem 20.17 [Alternate Form of Theorem 20.5]:   Let A be a normal matrix.   Then there exists a unitary matrix P such that $B = P^{-1}AP = P^*AP$ is diagonal.

Problems 20.80–20.82 refer to the following matrices:

$$A = \begin{pmatrix} 1 & 1 \\ i & 3+2i \end{pmatrix} \qquad B = \begin{pmatrix} 1 & i \\ 0 & 1 \end{pmatrix} \qquad C = \begin{pmatrix} 1 & i \\ 1 & 2+i \end{pmatrix}$$

20.80   Is A normal?

▎ Compute

$$AA^* = \begin{pmatrix} 1 & 1 \\ i & 3+2i \end{pmatrix}\begin{pmatrix} 1 & -i \\ 1 & 3-2i \end{pmatrix} = \begin{pmatrix} 2 & 3-3i \\ 3+3i & 14 \end{pmatrix}$$

$$A^*A = \begin{pmatrix} 1 & -i \\ 1 & 3-2i \end{pmatrix}\begin{pmatrix} 1 & 1 \\ i & 3+2i \end{pmatrix} = \begin{pmatrix} 2 & 3-3i \\ 3+3i & 14 \end{pmatrix}$$

Since $AA^* = A^*A$,   the matrix A is normal.

20.81   Is B normal?

▎ Compute

$$BB^* = \begin{pmatrix} 1 & i \\ 0 & 1 \end{pmatrix}\begin{pmatrix} 1 & 0 \\ -i & 1 \end{pmatrix} = \begin{pmatrix} 2 & i \\ -i & 1 \end{pmatrix} \qquad B^*B = \begin{pmatrix} 1 & 0 \\ -i & 1 \end{pmatrix}\begin{pmatrix} 1 & i \\ 0 & 1 \end{pmatrix} = \begin{pmatrix} 1 & i \\ -i & 2 \end{pmatrix}$$

Since $BB^* \neq B^*B$,   the matrix B is not normal.

20.82   Is C normal?

❚ Compute

$$CC^* = \begin{pmatrix} 1 & i \\ 1 & 2+i \end{pmatrix}\begin{pmatrix} 1 & 1 \\ -i & 2-i \end{pmatrix} = \begin{pmatrix} 2 & 2+2i \\ 2-2i & 6 \end{pmatrix}$$

$$C^*C = \begin{pmatrix} 1 & 1 \\ -i & 2-i \end{pmatrix}\begin{pmatrix} 1 & i \\ 1 & 2+i \end{pmatrix} = \begin{pmatrix} 2 & 2+2i \\ 2-2i & 6 \end{pmatrix}$$

Since $CC^* = C^*C$, the matrix C is normal.

Problems 20.83–20.86 refer to a normal operator T.

20.83 Show that $T(v) = 0$ if and only if $T^*(v) = 0$.

❚ We show that $\langle T(v), T(v) \rangle = \langle T^*(v), T^*(v) \rangle$: $\langle T(v), T(v) \rangle = \langle v, T^*T(v) \rangle = \langle v, TT^*(v) \rangle = \langle T^*(v), T^*(v) \rangle$. Thus $T(v) = 0$ if and only if $T^*(v) = 0$.

20.84 Show that $T - \lambda I$ is normal.

❚ We show that $T - \lambda I$ commutes with its adjoint:

$$\begin{aligned}(T - \lambda I)(T - \lambda I)^* &= (T - \lambda I)(T^* - \bar{\lambda} I) = TT^* - \lambda T^* - \bar{\lambda} T + \lambda \bar{\lambda} I \\ &= T^*T - \bar{\lambda} T - \lambda T^* + \bar{\lambda} \lambda I = (T^* - \bar{\lambda} I)(T - \lambda I) \\ &= (T - \lambda I)^*(T - \lambda I)\end{aligned}$$

Thus $T - \lambda I$ is normal.

20.85 Show that if $T(v) = \lambda v$, then $T^*(v) = \bar{\lambda} v$; hence any eigenvector of T is also an eigenvector of T^*.

❚ If $T(v) = \lambda v$, then $(T - \lambda I)(v) = 0$. Since $T - \lambda I$ is normal, we have $(T - \lambda I)^*(v) = 0$. Thus $(T^* - \bar{\lambda} I)(v) = 0$; hence $T^*(v) = \bar{\lambda} v$.

20.86 Show that if $T(v) = \lambda_1 v$ and $T(w) = \lambda_2 w$ where $\lambda_1 \neq \lambda_2$, then $\langle v, w \rangle = 0$; i.e., eigenvectors and vectors of T belonging to distinct eigenvalues are orthogonal.

❚ We show that $\lambda_1 \langle v, w \rangle = \lambda_2 \langle v, W \rangle$: $\lambda_1 \langle v, w \rangle = \langle \lambda_1 v, w \rangle = \langle T(v), w \rangle = \langle v, T^*(w) \rangle = \langle v, \bar{\lambda}_2 w \rangle = \lambda_2 \langle v, w \rangle$. But $\lambda_1 \neq \lambda_2$; hence $\langle v, w \rangle = 0$.

20.87 Prove Theorem 20.16.

❚ The proof is by induction on the dimension of V. If $\dim V = 1$, then the theorem trivially holds. Now suppose $\dim V = n > 1$. Since V is a complex vector space, T has at least one eigenvalue and hence a nonzero eigenvector v. Let W be the subspace of V spanned by v and let u_1 be a unit vector in W.

Since v is an eigenvector of T, the subspace W is invariant under T. However, v is also an eigenvector of T^* by the preceding problem; hence W is also invariant under T^*. Therefore, $W^\perp$ is invariant under $T = T^{**}$. The restriction $\hat{T}$ of T to $W^\perp$ is a normal operator. Also, $\dim W^\perp = n - 1$ since $\dim W = 1$. By induction, there exists an orthonormal basis $\{u_2, \ldots, u_n\}$ of $W^\perp$ consisting of eigenvectors of $\hat{T}$ and hence of T. But $\langle u_1, u_i \rangle = 0$ for $i = 2, \ldots, n$ because $u_i \in W^\perp$. Accordingly $\{u_1, u_2, \ldots, u_n\}$ is an orthonormal set and consists of eigenvectors of T. Thus the theorem is proved.

20.6 SPECTRAL THEOREM

20.88 Define a diagonalizable operator.

❚ A linear operator T on an inner product space V is said to be diagonalizable if there exists operators $E_1, \ldots, E_r$ on V and scalars $\lambda_1, \ldots, \lambda_r$ such that
(i) $T = \lambda_1 E_1 + \lambda_2 E_2 + \cdots + \lambda_r E_r$ (iii) $E_1^2 = E_1, \ldots, E_r^2 = E_r$
(ii) $E_1 + E_2 + \cdots + E_r = I$ (iv) $E_i E_j = 0$ for $i \neq j$

20.89 Define an orthogonal projection.

▮ A linear operator E on an inner product space V is called an orthogonal projection if $E^2 = E$. [Thus the linear operators E_i in Problem 20.88 are orthogonal projections.]

20.90 Consider a diagonal matrix, say

$$A = \begin{pmatrix} 2 & & & \\ & 3 & & \\ & & 3 & \\ & & & 5 \end{pmatrix}$$

Show that A is diagonalizable [by the definition in Problem 20.88].

▮ Let

$$E_1 = \begin{pmatrix} 1 & & & \\ & 0 & & \\ & & 0 & \\ & & & 0 \end{pmatrix} \qquad E_2 = \begin{pmatrix} 0 & & & \\ & 1 & & \\ & & 1 & \\ & & & 0 \end{pmatrix} \qquad E_3 = \begin{pmatrix} 0 & & & \\ & 0 & & \\ & & 0 & \\ & & & 1 \end{pmatrix}$$

Then (i) $A = 2E_1 + 3E_2 + 5E_3$, (ii) $E_1 + E_2 + E_3 = I$, (iii) $E_i^2 = E_i$, and (iv) $E_i E_j = 0$ for $i \neq j$.

20.91 Previously an operator T was said to be diagonalizable if it could be represented by a diagonal matrix relative to some basis. What could be the reason for redefining diagonalizable operators in Problems 20.88?

▮ The definition in Problem 20.88 does not use the notion of matrices and hence can also apply to infinite-dimensional spaces V. The two definitions coincide when V has finite dimension as indicated in Problem 20.90.

20.92 Restate Theorems 20.7 and 20.16 using the definition in Problem 20.88 on diagonalizable operators.

Theorem 20.18 [**Spectral Theorem**]: Let T be a normal [symmetric] operator on a complex [real] finite-dimensional inner product space V. Then there exist orthogonal projections $E_1, \ldots, E_r$ on V and scalars $\lambda_1, \ldots, \lambda_r$ such that
  (i) $T = \lambda_1 E_1 + \lambda_2 E_2 + \cdots + \lambda_r E_r$ (iii) $E_1^2 = E_1, \ldots, E_r^2 = E_r$
  (ii) $E_1 + E_2 + \cdots + E_r = I$ (iv) $E_i E_j = 0$ for $i \neq j$

CHAPTER 21
Applications to Geometry and Calculus

21.1 VECTOR NOTATION IN $\mathbf{R}^3$

21.1 Define the **ijk** notation in $\mathbf{R}^3$.

    ▮ The notation $\mathbf{i} = (1, 0, 0)$, $\mathbf{j} = (0, 1, 0)$, $\mathbf{k} = (0, 0, 1)$ is used for the usual basis in $\mathbf{R}^3$.

21.2 Rewrite $u = (3, -5, 6)$ and $v = (1, 3, -2)$ in the **ijk** notation.

    ▮ Since $(a, b, c) = a(1, 0, 0) + b(0, 1, 0) + c(0, 0, 1)$, we have $u = 3\mathbf{i} - 5\mathbf{j} + 6\mathbf{k}$ and $v = \mathbf{i} + 3\mathbf{j} - 2\mathbf{k}$.

21.3 Find the dot products $\mathbf{i} \cdot \mathbf{i}$, $\mathbf{j} \cdot \mathbf{j}$, $\mathbf{k} \cdot \mathbf{k}$.

    ▮ Since $\mathbf{i}$, $\mathbf{j}$, $\mathbf{k}$ are unit vectors, $\mathbf{i} \cdot \mathbf{i} = 1$, $\mathbf{j} \cdot \mathbf{j} = 1$, and $\mathbf{k} \cdot \mathbf{k} = 1$.

21.4 Find the dot products $\mathbf{i} \cdot \mathbf{j}$, $\mathbf{i} \cdot \mathbf{k}$, and $\mathbf{j} \cdot \mathbf{k}$.

    ▮ Since $\mathbf{i}$, $\mathbf{j}$, $\mathbf{k}$ form an orthonormal basis, they are orthogonal; hence $\mathbf{i} \cdot \mathbf{j} = 0$, $\mathbf{i} \cdot \mathbf{k} = 0$, and $\mathbf{j} \cdot \mathbf{k} = 0$.

Problems 21.5–21.8 refer to vectors $u = a_1\mathbf{i} + a_2\mathbf{j} + a_3\mathbf{k}$ and $v = b_1\mathbf{i} + b_2\mathbf{j} + b_3\mathbf{k}$.

21.5 Give a formula for $u + v$ and cu for a scalar $c \in \mathbf{R}$.

    ▮ Using the fact that $\mathbf{i}$, $\mathbf{j}$, $\mathbf{k}$ form a basis of $\mathbf{R}^3$, $u + v = (a_1 + b_1)\mathbf{i} + (a_2 + b_2)\mathbf{j} + (a_3 + b_3)\mathbf{k}$ and $cu = ca_1\mathbf{i} + ca_2\mathbf{j} + ca_3\mathbf{k}$.

21.6 Give a formula for the dot product $u \cdot v$.

    ▮ Use the definition of the inner product in $\mathbf{R}^3$ to obtain $u \cdot v = a_1b_1 + a_2b_2 + a_3b_3$.

21.7 Give a formula for the cross product $u \times v$.

    ▮
$$u \times v = \begin{vmatrix} a_2 & a_3 \\ b_2 & b_3 \end{vmatrix}\mathbf{i} - \begin{vmatrix} a_1 & a_3 \\ b_1 & b_3 \end{vmatrix}\mathbf{j} + \begin{vmatrix} a_1 & a_2 \\ b_1 & b_2 \end{vmatrix}\mathbf{k}$$

or, equivalently

$$u \times v = \begin{vmatrix} \mathbf{i} & \mathbf{j} & \mathbf{k} \\ a_1 & a_2 & a_3 \\ b_1 & b_2 & b_3 \end{vmatrix}$$

21.8 Give a formula for the norm $\|u\|$.

    ▮
$$\|u\| = \sqrt{u \cdot u} = \sqrt{a_1^2 + a_2^2 + a_3^2}$$

21.9 Find the cross products $\mathbf{i} \times \mathbf{j}$, $\mathbf{j} \times \mathbf{k}$, $\mathbf{k} \times \mathbf{i}$, $\mathbf{j} \times \mathbf{i}$, $\mathbf{k} \times \mathbf{j}$, $\mathbf{i} \times \mathbf{k}$.

    ▮ Here $\mathbf{i} \times \mathbf{j} = \mathbf{k}$, $\mathbf{j} \times \mathbf{k} = \mathbf{i}$, $\mathbf{k} \times \mathbf{i} = \mathbf{j}$, and $\mathbf{j} \times \mathbf{i} = -\mathbf{k}$, $\mathbf{k} \times \mathbf{j} = -\mathbf{i}$, $\mathbf{i} \times \mathbf{k} = -\mathbf{j}$. In other words, if we view the triple $[\mathbf{i}, \mathbf{j}, \mathbf{k}]$ as a cyclic permutation, i.e., as arranged around a circle in the counterclockwise direction as in Fig. 21-1, then the product of two of them in the given direction is the third one, but the product of two of them in the opposite direction is the negative of the third one.

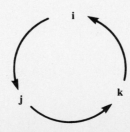

Fig. 21-1

Problems 21.10–21.32 refer to the vectors: $u = 2\mathbf{i} - 3\mathbf{j} + 4\mathbf{k}$, $v = 3\mathbf{i} + \mathbf{j} - 2\mathbf{k}$, $w = \mathbf{i} + 5\mathbf{j} + 3\mathbf{k}$.

21.10   Find $u + v$.

▌ Add corresponding components to get $u + v = 5\mathbf{i} - 2\mathbf{j} + 2\mathbf{k}$.

21.11   Find $2u - 3w$.

▌ First multiply the vectors by the scalars, and then add: $2u - 3w = (4\mathbf{i} - 6\mathbf{j} + 8\mathbf{k}) + (-3\mathbf{i} - 15\mathbf{j} - 9\mathbf{k}) = \mathbf{i} - 21\mathbf{j} - \mathbf{k}$.

21.12   Find $3u - 2v + 4w$.

▌ $3u - 2v + 4w = (6\mathbf{i} - 9\mathbf{j} + 12\mathbf{k}) + (-6\mathbf{i} - 2\mathbf{j} + 4\mathbf{k}) + (4\mathbf{i} + 20\mathbf{j} + 12\mathbf{k}) = 4\mathbf{i} + 9\mathbf{j} + 28\mathbf{k}$.

21.13   Find $u \cdot v$.

▌ Multiply corresponding components and then add to get $u \cdot v = 6 - 3 - 8 = -5$.

21.14   Find $u \cdot w$.

▌ $u \cdot w = 2 - 15 + 12 = -1$.

21.15   Find $v \cdot w$.

▌ $v \cdot w = 3 + 5 - 6 = 2$.

21.16   Find $\|u\|$.

▌ Square each component of u and then add to get $\|u\|^2$.   That is, $\|u\|^2 = 4 + 9 + 16 = 29$.   Thus $\|u\| = \sqrt{29}$.

21.17   Find $\|v\|$.

▌ $\|v\|^2 = 9 + 1 + 4 = 14$;   hence $\|v\| = \sqrt{14}$.

21.18   Find $\|w\|$.

▌ $\|w\|^2 = 1 + 25 + 9 = 35$;   hence $\|w\| = \sqrt{35}$.

21.19   Find $u \times v$.

▌
$$u \times v = \begin{vmatrix} \mathbf{i} & \mathbf{j} & \mathbf{k} \\ 2 & -3 & 4 \\ 3 & 1 & -2 \end{vmatrix} = (6 - 4)\mathbf{i} + (12 + 4)\mathbf{j} + (2 + 9)\mathbf{k} = 2\mathbf{i} + 16\mathbf{j} + 11\mathbf{k}$$

[*Remark*: Observe that the **j** component is obtained by taking the determinant "backwards." See Problem 21.7 on cross products.]

21.20   Find $u \times w$.

▌
$$u \times w = \begin{vmatrix} \mathbf{i} & \mathbf{j} & \mathbf{k} \\ 2 & -3 & 4 \\ 1 & 5 & 3 \end{vmatrix} = (-9 - 20)\mathbf{i} + (4 - 6)\mathbf{j} + (10 + 3)\mathbf{k} = -29\mathbf{i} - 2\mathbf{j} + 13\mathbf{k}$$

21.21   Find $v \times w$.

▌
$$v \times w = \begin{vmatrix} \mathbf{i} & \mathbf{j} & \mathbf{k} \\ 3 & 1 & -2 \\ 1 & 5 & 3 \end{vmatrix} = (3 + 10)\mathbf{i} + (-2 - 9)\mathbf{j} + (15 - 1)\mathbf{k} = 13\mathbf{i} - 11\mathbf{j} + 14\mathbf{k}$$

21.22   Find $v \times u$.

▌ $v \times u = -(u \times v) = -(2\mathbf{i} + 16\mathbf{j} + 11\mathbf{k}) = -2\mathbf{i} - 16\mathbf{j} - 11\mathbf{k}$.

21.23   Find $w \times v$.

▌ $w \times v = -(v \times w) = -(13\mathbf{i} - 11\mathbf{j} + 14\mathbf{k}) = -13\mathbf{i} + 11\mathbf{j} - 14\mathbf{j}$.

21.24 Find $w \times u$.

▌ $w \times u = -(u \times w) = -(-29\mathbf{i} - 2\mathbf{j} + 13\mathbf{k}) = 29\mathbf{i} + 2\mathbf{j} - 13\mathbf{k}$.

21.25 Find $\cos\theta$ where θ is the angle between u and v.

▌
$$\cos\theta = \frac{u \cdot v}{\|u\|\,\|v\|} = \frac{-5}{\sqrt{29}\sqrt{14}}$$

21.26 Find $\cos\theta$ where θ is the angle between v and w.

▌
$$\cos\theta = \frac{v \cdot w}{\|v\|\,\|w\|} = \frac{2}{\sqrt{14}\sqrt{35}} = \frac{2}{7\sqrt{10}}$$

21.27 Find $u \cdot v \times w$.

▌
$$u \cdot v \times w = \begin{vmatrix} 2 & -3 & 4 \\ 3 & 1 & -2 \\ 1 & 5 & 3 \end{vmatrix} = 6 + 6 + 60 - 4 + 20 + 27 = 115$$

[Note $u \cdot v \times w = \det(A)$ where u, v, w are the rows of A.]

21.28 Give a geometrical interpretation of $u \cdot v \times w$.

▌ The absolute value of $u \cdot v \times w$ represents the volume of the parallelopiped formed by the vectors u, v, and w as pictured in Fig. 21-2.

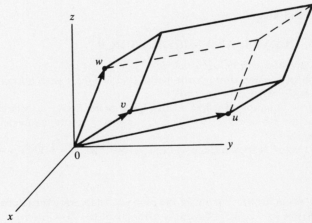

Fig. 21-2

21.29 Find $w \cdot v \times u$.

▌ Since $[w, v, u]$ is an odd permutation of $[u, v, w]$, we have $w \cdot v \times u = -(u \cdot v \times w) = -115$.

21.30 Find $w \cdot u \times v$.

▌ Since $[w, u, v]$ is an even permutation of $[u, v, w]$, we have $w \cdot v \times w = u \cdot v \times w = 115$.

21.31 Find $(u \cdot v) \times w$.

▌ By Problems 21.19, $u \times v = 2\mathbf{i} + 16\mathbf{j} + 11\mathbf{k}$. Thus

$$(u \times v) \times w = \begin{vmatrix} \mathbf{i} & \mathbf{j} & \mathbf{k} \\ 2 & 16 & 11 \\ 1 & 5 & 3 \end{vmatrix} = (48 - 55)\mathbf{i} + (11 - 6)\mathbf{j} + (10 - 16)\mathbf{k} = -7\mathbf{i} + 5\mathbf{j} - 6\mathbf{k}$$

21.32 Find $u \times (v \times w)$.

▌ By Problem 21.21, $v \times w = 13\mathbf{i} - 11\mathbf{j} + 14\mathbf{k}$. Thus

$$u \times (v \times w) = \begin{vmatrix} \mathbf{i} & \mathbf{j} & \mathbf{k} \\ 2 & -3 & 4 \\ 13 & -11 & 14 \end{vmatrix} = (-42 + 44)\mathbf{i} + (52 - 28)\mathbf{j} + (-22 + 39)\mathbf{k} = 2\mathbf{i} + 24\mathbf{j} + 17\mathbf{k}$$

[*Remark*: Observe that the cross product does not satisfy the associative law, i.e., $(u \times v) \times w \neq u \times (v \times w)$.]

21.33 Find a unit vector orthogonal to $u_1 = 4\mathbf{i} - 6\mathbf{j} + \mathbf{k}$ and $u_2 = 2\mathbf{i} + \mathbf{j} - 3\mathbf{k}$.

▌ First find $v = u_1 \times u_2$ which gives a vector orthogonal to both u_1 and u_2:

$$v = u_1 \times u_2 = \begin{vmatrix} \mathbf{i} & \mathbf{j} & \mathbf{k} \\ 4 & -6 & 1 \\ 2 & 1 & -3 \end{vmatrix} = (18 - 1)\mathbf{i} + (2 + 12)\mathbf{j} + (4 + 12)\mathbf{k} = 17\mathbf{i} + 14\mathbf{j} + 16\mathbf{k}$$

Normalize v by first finding $\|v\|$. We have $\|v\|^2 = 289 + 196 + 256 = 741$. Thus

$$\hat{v} = \frac{1}{\|v\|} v = \frac{1}{\sqrt{741}} (17\mathbf{i} + 14\mathbf{j} + 16\mathbf{k})$$

is the desired vector.

21.34 Find c so that $u = 4\mathbf{i} + 3\mathbf{j} + c\mathbf{k}$ is in the plane W spanned by $v_1 = \mathbf{i} + 2\mathbf{j} - 3\mathbf{k}$ and $v_2 = 2\mathbf{i} - \mathbf{j} + 4\mathbf{k}$.

▌ Note that u is in W if $u \cdot v_1 \times v_2 = 0$, i.e., if $\det(A) = 0$ where u, v_1, v_2 are the rows of A. Set

$$0 = \begin{vmatrix} 4 & 3 & c \\ 1 & 2 & -3 \\ 2 & -1 & 4 \end{vmatrix} = 32 - 18 - c - 4c - 12 - 12 = -5c - 10$$

Thus $5c = -10$ and hence $c = -2$.

21.2 PLANES, LINES, CURVES, AND SURFACES IN $\mathbf{R}^3$

The following formulas will be used below:

(a) The equation of a plane through the point $P_0(x_0, y_0, z_0)$ with normal direction $\mathbf{N} = a\mathbf{i} + b\mathbf{j} + c\mathbf{k}$ is $a(x - x_0) + b(y - y_0) + c(x - x_0) = 0$.

(b) The parametric equation of a line L through a point $P_0(x_0, y_0, z_0)$ in the direction of the vector $v = a\mathbf{i} + b\mathbf{j} + c\mathbf{k}$ is $x = at + x_0$, $y = bt + y_0$, $z = ct + z_0$ or, equivalently, $L(t) = (at + x_0)\mathbf{i} + (bt + y_0)\mathbf{j} + (ct + z_0)\mathbf{k}$.

(c) The normal vector $\mathbf{N}$ to a surface $F(x, y, z) = 0$ is $\mathbf{N} = F_x\mathbf{i} + F_y\mathbf{j} + F_z\mathbf{k}$.

21.35 Derive formula (a).

▌ Let $P(x, y, z)$ be an arbitrary point in the plane. The vector v from P_0 to P is $v = P - P_0 = (x - x_0)\mathbf{i} + (y - y_0)\mathbf{j} + (z - z_0)\mathbf{k}$. Since v is orthogonal to $\mathbf{N} = a\mathbf{i} + b\mathbf{j} + c\mathbf{k}$ [Fig. 21-3], we get $a(x - x_0) + b(y - y_0) + c(z - z_0) = 0$ as claimed.

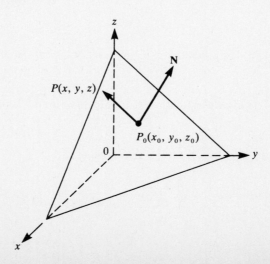

Fig. 21-3

21.36 Find the equation of the plane with normal direction $N = 5i - 6j + 7k$ and containing the point $P(3, 4, -2)$.

▮ Substitute P and N in formula (a) to get $5(x - 3) - 6(y - 4) + 7(z + 2) = 0$ or $5x - 6y + 7z = -23$.

21.37 Find a normal vector N to the plane $4x + 7y - 12z = 3$.

▮ The coefficients of x, y, z give a normal direction; hence $N = 4i + 7j - 12k$. [Any multiple of N also is normal to the plane.]

21.38 Find the plane H parallel to $4x + 7y - 12z = 3$ and containing the point $P(2, 3, -1)$.

▮ H and the given plane have the same normal direction; i.e., $N = 4i + 7j - 12k$ is normal to H. Substitute P and N in formula (a) to get $4(x - 2) + 7(y - 3) - 12(z + 1) = 0$ or $4x + 7y - 12z = 41$.

21.39 Let H and K be, respectively, the planes $x + 2y - 4z = 5$ and $2x - y + 3z = 7$. Find $\cos \theta$ where θ is the angle between the planes H and K.

▮ The angle θ between H and K is the same as the angle between the normal N of H and the normal N' of K. We have $N = i + 2j - 4k$ and $N' = 2i - j + 3k$. Then $N \cdot N' = 2 - 2 - 12 = -12$, $\|N\|^2 = 1 + 4 + 16 = 21$, $\|N'\|^2 = 4 + 1 + 9 = 14$. Thus

$$\cos \theta = \frac{N \cdot N'}{\|N\| \, \|N'\|} = -\frac{12}{\sqrt{21}\sqrt{14}} = -\frac{12}{7\sqrt{6}}$$

21.40 Derive formula (b).

▮ Let $P(x, y, z)$ be an arbitrary point on the line L. The vector w from P_0 to P is

$$w = P - P_0 = (x - x_0)i + (y - y_0)j + (z - z_0)k \qquad (1)$$

Since w and v have the same direction [Fig. 21-4],

$$w = tv = t(ai + bj + ck) = ati + btj + ctk \qquad (2)$$

Equations (1) and (2) give us our result.

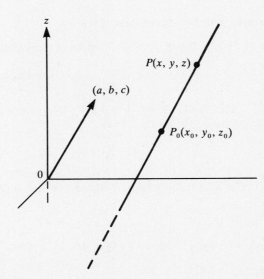

Fig. 21-4

21.41 Find the [parametric] equation of the line L through the point $P(3, 4, -2)$ and in the direction of $v = 5i - j + 3k$.

▮ Substitute in formula (b) to get $L(t) = (5t + 3)i + (-t + 4)j + (3t - 2)k$.

21.42 Find the equation of the line L through points $P(1, 3, 2)$ and $Q(2, 5, -6)$.

▮ First find the vector v from P to Q: $v = Q - P = \mathbf{i} + 2\mathbf{j} - 8\mathbf{k}$. Then use formula (b) with v and one of the given points, say P, to get $L(t) = (t + 1)\mathbf{i} + (2t + 3)\mathbf{j} + (-8t + 2)\mathbf{k}$.

21.43 Let H be the plane $3x + 5y + 7z = 15$. Find the equation of the line L perpendicular to H and containing the point $P(1, -2, 4)$.

▮ Since L is perpendicular to H, L must be in the same direction as the normal $\mathbf{N} = 3\mathbf{i} + 5\mathbf{j} + 7\mathbf{k}$ to H. Thus use formula (b) with $\mathbf{N}$ and P to get $L(t) = (3t + 1)\mathbf{i} + (5t - 2)\mathbf{j} + (7t + 4)\mathbf{k}$.

21.44 Define a curve in $\mathbf{R}^3$.

▮ Let D be an interval (finite or infinite) in the real line $\mathbf{R}$. A continuous function $F: D \rightarrow \mathbf{R}^3$ is a curve in $\mathbf{R}^3$. Thus to each $t \in D$ there is assigned the point $F(t) = F_1(t)\mathbf{i} + F_2(t)\mathbf{j} + F_3(t)\mathbf{k}$ in $\mathbf{R}^3$.

Remark: Suppose the above function $F(t)$ represents the position of a moving body B at time t. Then $V(t) = dF(t)/dt$ denotes the velocity of B and $A(t) = dV(t)/dt$ denotes the acceleration of B.

Problems 21.45–21.49 refer to the following curve where $0 \le t \le 5$: $F(t) = t^2\mathbf{i} + (3t + 4)\mathbf{j} + t^3\mathbf{k}$.

21.45 Find $F(t)$ when $t = 2$.

▮ Substitute $t = 2$ into $F(t)$ to obtain $F(2) = 4\mathbf{i} + 10\mathbf{j} + 8\mathbf{k}$.

21.46 Find $F(t)$ when $t = 4$.

▮ $F(t) = F(4) = 16\mathbf{i} + (12 + 4)\mathbf{j} + 64\mathbf{k} = 16\mathbf{i} + 16\mathbf{j} + 64\mathbf{k}$.

21.47 Find $F(t)$ when $t = 6$.

▮ $F(t)$ is not defined when $t = 6$ since the domain of F is the interval $0 \le t \le 5$.

21.48 Find the endpoints of the curve.

▮ The endpoints of the domain are $t = 0$ and $t = 5$. Hence the endpoints of the curve are $F(0) = 4\mathbf{j}$ and $F(5) = 25\mathbf{i} + 19\mathbf{j} + 125\mathbf{k}$.

21.49 Find the unit tangent vector $\mathbf{T}$ to the curve when $t = 2$.

▮ Take the derivative of $F(t)$ to obtain a vector V which is tangent to the curve:

$$V(t) = \frac{dF(t)}{dt} = 2t\mathbf{i} + 3\mathbf{j} + 3t^2\mathbf{k}$$

Next find V when $t = 2$. This yields $V = 4i + 3j + 12k$. Normalize V to get the unit tangent vector $\mathbf{T}$ to the curve when $t = 2$. We have $\|V\| = 13$. Thus

$$\mathbf{T} = \tfrac{4}{13}\mathbf{i} + \tfrac{3}{13}\mathbf{j} + \tfrac{12}{13}\mathbf{k}$$

Problems 21.50–21.53 refer to a moving body B whose position at time t is given by $R(t) = t^3\mathbf{i} + 2t^2\mathbf{j} + 3t\mathbf{k}$.

21.50 Find the position of B when $t = 1$.

▮ Substitute $t = 1$ into $R(t)$ to get $R(1) = \mathbf{i} + 2\mathbf{j} + 3\mathbf{k}$.

21.51 Find the velocity v of B when $t = 1$.

▮ Take the derivative of $R(t)$ to get

$$V(t) = \frac{dR(t)}{dt} = 3t^2\mathbf{i} + 4t\mathbf{j} + 3\mathbf{k}$$

Substitute $t = 1$ in $V(t)$ to get $v = V(1) = 3\mathbf{i} + 4\mathbf{j} + 3\mathbf{k}$.

21.52   Find the speed s of B when $t = 1$.

▮   The speed s is the magnitude of the velocity v. Thus $s^2 = \|v\|^2 = 9 + 16 + 3 = 34$ and hence $s = \sqrt{34}$.

21.53   Find the acceleration a of B when $t = 1$.

▮   Take the second derivative of $R(t)$ or, in other words, the derivative of $V(t)$ to get

$$A(t) = \frac{dV(t)}{dt} = 6t\mathbf{i} + 4\mathbf{j}$$

Substitute $t = 1$ in $A(t)$ to get $a = A(1) = 6\mathbf{i} + 4\mathbf{j}$.

Problems 21.54–21.55 refer to the following surface: $xy^2 + 2yz = 16$.

21.54   Find the normal vector $\mathbf{N}(x, y, z)$ to the surface.

▮   Find the partial derivatives F_x, F_y, F_z where $F(x, y, z) = xy^2 + 2yz - 16$. We have $F_x = y^2$, $F_y = 2xy + 2z$, $F_z = 2y$. Thus $\mathbf{N}(x, y, z) = y^2\mathbf{i} + (2xy + 2z)\mathbf{j} + 2y\mathbf{k}$.

21.55   Find the tangent plane H to the surface at the point $P(1, 2, 3)$.

▮   The normal to the surface at the point P is $\mathbf{N}(P) = \mathbf{N}(1, 2, 3) = 4\mathbf{i} + 10\mathbf{j} + 4\mathbf{k}$. Thus $\mathbf{N} = 2\mathbf{i} + 5\mathbf{j} + 2\mathbf{k}$ is also a normal vector at P. Substitute P and $\mathbf{N}$ into formula (a) to get $2(x - 1) + 5(y - 2) + 2(z - 3) = 0$ or $2x + 5y + 2z = 18$.

21.56   Consider the ellipsoid $x^2 + 2y^2 + 3z^2 = 15$. Find the tangent plane H at the point $P(2, 2, 1)$.

▮   First find the normal vector $\mathbf{N}(x, y, z) = F_x\mathbf{i} + F_y\mathbf{j} + F_z\mathbf{k} = 2x\mathbf{i} + 4y\mathbf{j} + 6z\mathbf{k}$. Evaluate the normal vector $\mathbf{N}(x, y, z)$ at P to get $\mathbf{N}(P) = \mathbf{N}(2, 2, 1) = 4\mathbf{i} + 8\mathbf{j} + 6\mathbf{k}$. Thus $\mathbf{N} = 2\mathbf{i} + 4\mathbf{j} + 3\mathbf{k}$ is normal to the ellipsoid at P. Substitute P and $\mathbf{N}$ into formula (a) to obtain H: $2(x - 2) + 4(y - 2) + 3(z - 1) = 0$ or $2x + 4y + 3z = 15$.

Problems 21.57–21.58 refer to the function $f(x, y) = x^2 + y^2$ whose solution set $z = x^2 + y^2$ represents a surface S in $\mathbf{R}^3$.

21.57   Find the normal vector $\mathbf{N}$ to the surface S when $x = 2$, $y = 3$.

▮ $\mathbf{N} = [f_x, f_y, -1] = 2x\mathbf{i} + 2y\mathbf{j} - \mathbf{k} = 4\mathbf{i} + 6\mathbf{j} - \mathbf{k}$. [*Remark*: We use the fact that when $F(x, y, z) = f(x, y) - z$, we have $F_x = f_x$, $F_y = f_y$, and $F_z = -1$.]

21.58   Find the tangent plane H to the surface S when $x = 2$, $y = 3$.

▮   If $x = 2$, $y = 3$, then $z = 4 + 9 = 13$; hence $P(2, 3, 13)$ is the point on the surface S. Substitute P and $\mathbf{N} = 4\mathbf{i} + 6\mathbf{j} - \mathbf{k}$ into formula (a) to obtain H: $4(x - 2) + 6(y - 3) - (z - 13) = 0$ or $4x + 6y - z = 13$.

21.3   SCALAR AND VECTOR FIELDS

21.59   Define a scalar field.

▮   A function $f: \mathbf{R}^3 \to \mathbf{R}$ is called a scalar field in $\mathbf{R}^3$. In other words, a scalar field f assigns a scalar $f(x, y, z)$ to each point $P(x, y, z)$ in $\mathbf{R}^3$. [Analogously, a function $f: \mathbf{R}^n \to \mathbf{R}$ is called a scalar field in $\mathbf{R}^n$.]

21.60   Define a vector field.

▮   A function $F: \mathbf{R}^3 \to \mathbf{R}^3$ is called a vector field in $\mathbf{R}^3$. In other words, a vector field F assigns a vector $F_1(x, y, z)\mathbf{i} + F_2(x, y, z)\mathbf{j} + F_3(x, y, z)\mathbf{k}$ to each point $P(x, y, z)$ in $\mathbf{R}^3$. [Analogously, a function $F: \mathbf{R}^n \to \mathbf{R}^n$ is called a vector field in $\mathbf{R}^n$.]

Remark: Frequently, the domain of a scalar field or vector field is a subset D of $\mathbf{R}^n$, rather than $\mathbf{R}^n$ itself.

Problems 21.61–21.64 refer to the following fields on some domain D:
(a) The temperature at a point
(b) The velocity of the wind at a point
(c) The height above sea level of a point
(d) The magnetic field

21.61 Is (a) a scalar or vector field?

❚ Since temperature is a scalar, (a) is a scalar field.

21.62 Is (b) a scalar or vector field?

❚ The velocity of the wind at a point is a vector with magnitude and direction; hence (b) is a vector field.

21.63 Is (c) a scalar or vector field?

❚ Since the height of a point is a scalar quantity; (c) is a scalar field.

21.64 Is (d) a scalar or vector field?

❚ The magnetic field is a vector field since there is a magnetic force with magnitude and direction at each point.

Problems 21.65–21.67 refer to the scalar field $f(x, y, z) = x^2 + yz$.

21.65 Find $f(P_1)$ for the point $P_1(1, 2, -4)$.

❚ $f(P_1) = f(1, 2, -4) = 1 - 8 = -7$.

21.66 Find $f(P_2)$ for the point $P_2(2, -3, 5)$.

❚ $f(P_2) = f(2, -3, 5) = 4 - 15 = -11$.

21.67 Find $f(P_3)$ for the point $P_3(3, 1, -2)$.

❚ $f(P_3) = f(3, 1, -2) = 9 - 2 = 7$.

21.68 Consider the scalar field $g(x, y) = x^2 + 2y^2$ in $\mathbf{R}^2$. Describe and plot the level curves of g.

❚ For each scalar $c \in \mathbf{R}$ there is the level curve $g(x, y) = x^2 + 2y^2 = c$. These curves are ellipses with centers at the origin as pictured in Fig. 21-5.

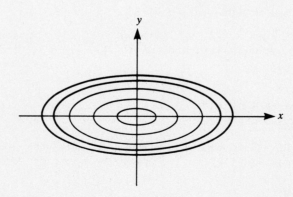

Fig. 21-5

Problems 21.69–21.71 refer to the following vector field in $\mathbf{R}^3$: $F(x, y, z) = xyz\mathbf{i} + (x^2 + y^2 + z^2)\mathbf{j} + (x^2 - yz)\mathbf{k}$.

21.69 Find $F(P_1)$ for the point $P_1(1, 2, -4)$.

▮ $F(P_1) = F(1, 2, -4) = -8\mathbf{i} + (1 + 4 + 16)\mathbf{j} + (1 + 8)\mathbf{k} = -8\mathbf{i} + 21\mathbf{j} + 9\mathbf{k}$.

21.70   Find $F(P_2)$ for the point $P_2(2, -3, 5)$.

▮ $F(P_2) = F(2, -3, 5) = -30\mathbf{i} + (4 + 9 + 25)\mathbf{j} + (4 + 15)\mathbf{k} = -30\mathbf{i} + 38\mathbf{j} + 19\mathbf{k}$.

21.71   Find $F(P_3)$ for the point $P_3(3, 1, -2)$.

▮ $F(P_3) = F(3, 1, -2) = -6\mathbf{i} + (9 + 1 + 4)\mathbf{j} + (9 + 2)\mathbf{k} = -6\mathbf{i} + 14\mathbf{j} + 11\mathbf{k}$.

21.72   Consider the vector field $F(x, y) = \frac{1}{2}x\mathbf{i} - \frac{1}{2}y\mathbf{j}$ in $\mathbf{R}^2$.   Describe the field.

▮   At each point $P(x, y)$ in the plane there is a vector $F(P)$ which has $\frac{1}{2}$ the length from the origin 0 to P and which is perpendicular to the vector from 0 to P as pictured in Fig. 21-6.

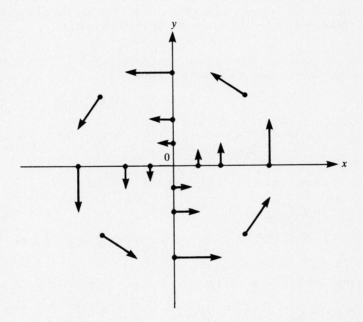

Fig. 21-6

21.4   DEL OPERATOR ∇ AND GRADIENT, DIVERGENCE AND CURL

21.73   Define the del operator.

▮   The vector differential operator del, denoted by ∇, is defined by $\nabla \equiv D_x\mathbf{i} + D_y\mathbf{j} + D_z\mathbf{k} = \mathbf{i}D_x + \mathbf{j}D_y + \mathbf{k}D_z$ where D_x, D_y, and D_z denote the partial derivatives with respect to x, y, and z, respectively.   That is,

$$\nabla \equiv \frac{\partial}{\partial x}\mathbf{i} + \frac{\partial}{\partial y}\mathbf{j} + \frac{\partial}{\partial z}\mathbf{k} \equiv \mathbf{i}\frac{\partial}{\partial x} + \mathbf{j}\frac{\partial}{\partial y} + \mathbf{k}\frac{\partial}{\partial z}$$

More generally, the del operator ∇ is defined for any set of variables $x_1, x_2, \ldots, x_n$ by $\nabla \equiv [D_1, D_2, \ldots, D_n]$ where D_i denotes the partial derivative with respect to the ith variable x_i.

21.74   Define the gradient of a scalar field.

▮   Let $f(x, y, z)$ be a differentiable scalar field.   The gradient of f, denoted by ∇f or grad f, is defined by $\nabla f = (D_x\mathbf{i} + D_y\mathbf{j} + D_z\mathbf{k})(f) = D_x(f)\mathbf{i} + D_y(f)\mathbf{j} + D_z(f)\mathbf{k}$.   Note that ∇f is a vector field.

21.75   Define the divergence of a vector field.

▮   Let $F(x, y, z) = F_1\mathbf{i} + F_2\mathbf{j} + F_3\mathbf{k}$ be a differentiable vector field.   Then the divergence of F, written $\nabla \cdot F$ or div F, is defined by $\nabla \cdot F = (D_x\mathbf{i} + D_y\mathbf{j} + D_z\mathbf{k}) \cdot (F_1\mathbf{i} + F_2\mathbf{j} + F_3\mathbf{k}) = D_x(F_1) + D_y(F_2) + D_z(F_3)$. Note that $\nabla \cdot F$ is a scalar field.

21.76   Define the curl of a vector field.

▌ Let $F(x, y, z) = F_1\mathbf{i} - F_2\mathbf{j} + F_3\mathbf{k}$ be a differentiable vector field.   Then the curl or rotation of F, written $\nabla \times F$ or curl F or rot F, is defined by

$$\nabla \times F = (D_x\mathbf{i} + D_y\mathbf{j} + D_z\mathbf{k}) \times (F_1\mathbf{i} + F_2\mathbf{j} + F_3\mathbf{k}) = \begin{vmatrix} \mathbf{i} & \mathbf{j} & \mathbf{k} \\ D_x & D_y & D_z \\ F_1 & F_2 & F_3 \end{vmatrix}$$

$$= [D_y(F_3) - D_z(F_2)]\mathbf{i} + [D_z(F_1) - D_x(F_3)]\mathbf{j} + [D_x(F_2) - D_y(F_1)]\mathbf{k}$$

Note that $\nabla \times F$ is again a vector field.

21.77   Is the gradient, divergence, and curl defined for a function in $\mathbf{R}^n$?

▌ If $f(x_1, \ldots, x_n)$ is a scalar field in $\mathbf{R}^n$, then   div $f = \nabla f = [D_1(f), D_2(f), \ldots, D_n(f)]$.   If $F = [F_1, F_2, \ldots, F_n]$ is a vector field in $\mathbf{R}^n$, then   div $F = \nabla \cdot F = D_1(F_1) + D(F_2) + \cdots + D_n(F_n)$.   The curl of a vector field F in $\mathbf{R}^n$ is not defined except when $n = 3$,   just like the cross product of vectors is not defined except in $\mathbf{R}^3$.   [Observe that we cannot use the **ijk** notation in $\mathbf{R}^n$.]

Theorem 21.4:   Suppose f is a differentiable scalar field.   Let $D_u(f)$ denote the directional derivative of f in the direction of a vector u.   Then

$$D_u(f)(P) = (\nabla f)(P) \cdot \frac{u}{\|u\|} = \frac{(\nabla f)(P) \cdot u}{\|u\|}$$

at a point P in the domain.   [That is, the component of the gradient of f in the direction of a vector u is equal to the directional derivative of f in the direction of u.]

Problems 21.78–21.80 refer to the scalar field $f(x, y, z) = x^2y^2 + z^3$.

21.78   Find the gradient of f, that is, ∇f.

▌ $\nabla f = (D_x\mathbf{i} + D_y\mathbf{j} + D_z\mathbf{k})(x^2y^2 + z^3) = D_x(x^2y^2 + z^3)\mathbf{i} + D_y(x^2y^2 + z^3)\mathbf{j} + D_z(x^2y^2 + z^3)\mathbf{k} = 2xy^2\mathbf{i} + 2x^2y\mathbf{j} + 3z^2\mathbf{k}$.

21.79   Find the directional derivative of f at the point $P(3, 2, 1)$ in the direction of $u = 2\mathbf{i} - \mathbf{j} + 2\mathbf{k}$.

▌ Evaluate ∇f at the point P: $(\nabla f)(P) = (\nabla f)(3, 2, 1) = 24\mathbf{i} + 36\mathbf{j} + 3\mathbf{k}$.   Take the dot [inner] product of $(\nabla f)(P)$ with u and divide by $\|u\| = \sqrt{4 + 1 + 4} = \sqrt{9} = 3$ to get $D_u(f)(P)$:

$$D_u(f)(P) = \frac{(24\mathbf{i} + 36\mathbf{j} + 3\mathbf{k}) \cdot (2\mathbf{i} - \mathbf{j} + 2\mathbf{k})}{3} = \frac{18}{3} = 6$$

21.80   Find the directional derivative of f at the point $Q(1, -2, 3)$ in the direction $v = \mathbf{i} + 3\mathbf{j} - 2\mathbf{k}$.

▌ Evaluate the gradient ∇f at the point Q: $(\nabla f)(Q) = (\nabla f)(1, -2, 3) = 8\mathbf{i} - 4\mathbf{j} + 27\mathbf{k}$.   Then

$$D_v(f)(Q) = \frac{(\nabla f)(Q) \cdot v}{\|v\|} = \frac{(8\mathbf{i} - 4\mathbf{j} + 27\mathbf{k}) \cdot (\mathbf{i} + 3\mathbf{j} - 2\mathbf{k})}{\sqrt{1 + 4 + 9}} = -\frac{58}{\sqrt{14}}$$

Problems 21.81–21.83 refer to the following scalar field in $\mathbf{R}^4$: $f(x, y, z, t) = xt^3 + yz^3$.

21.81   Find the gradient ∇f.

▌ $\nabla f = [D_x, D_y, D_z, D_t]f = [D_x(f), D_y(f), D_z(f), D_t(f)] = [t^3, z^3, 3yz^2, 3xt^2]$.

21.82   Find the directional derivative of f at the point $P(1, 2, -2, 1)$ in the direction $u = [2, -1, 3, -2]$.

▌ Evaluate the gradiant ∇f at the point P: $(\nabla f)(P) = (\nabla f)[1, 2, -2, 1] = [1, -8, 24, 3]$.   Then find $(\nabla f)(P) \cdot u = [1, -8, 24, 3] \cdot [2, -1, 3, -2] = 2 + 8 + 72 - 6 = 76$.   Next compute $\|u\| = \sqrt{4 + 1 + 9 + 4} = \sqrt{18} = 3\sqrt{2}$.   Then

$$D_u(f)(P) = \frac{76}{3\sqrt{2}}$$

21.83 Find the directional derivative of f at the point $Q(1, -3, 2, -1)$ in the direction of $v = [3, -2, 1, 4]$.

▮ We have $(\nabla f)(Q) = (\nabla f)[1, -3, 2, -1] = [1, 8, -36, 3]$, $(\nabla f)(Q) \cdot v = [1, 8, -36, 3] \cdot [3, -2, 1, 4] = 3 - 16 - 36 + 12 = -37$, $\|v\| = \sqrt{9 + 4 + 1 + 16} = \sqrt{30}$. Thus $D_v(f)(Q) = -37/\sqrt{30}$.

Problems 21.84–21.89 refer to the following vector field: $F(x, y, z) = xz^2\mathbf{i} + xy^2\mathbf{j} + xyz\mathbf{k}$.

21.84 Find the divergence $\nabla \cdot F$.

▮ $\nabla \cdot F = (D_x\mathbf{i} + D_y\mathbf{j} + D_z\mathbf{k}) \cdot (xz^2\mathbf{i} + xy^2\mathbf{j} + xyz\mathbf{k}) = D_x(xz^2) + D_y(xy^2) + D_z(xyz) = z^2 + 2xy + xy = z^2 + 3xy$.

21.85 Find $\nabla \cdot F$ at $P(3, 2, 1)$.

▮ $(\nabla \cdot F)(P) = (\nabla \cdot F)(3, 2, 1) = 1 + 18 = 19$.

21.86 Find $\nabla \cdot F$ at $Q(1, -2, 3)$.

▮ $(\nabla \cdot F)(Q) = (\nabla \cdot F)(1, -2, 3) = 9 - 6 = 3$.

21.87 Find the curl $\nabla \times F$.

▮

$$\nabla \times F = \begin{vmatrix} \mathbf{i} & \mathbf{j} & \mathbf{k} \\ D_x & D_y & D_z \\ xz^2 & xy^2 & xyz \end{vmatrix}$$

$$= [D_y(xyz) - D_z(xy^2)]\mathbf{i} + [D_z(xz^2) - D_x(xyz)]\mathbf{j} + [D_x(xy^2) - D_y(xz^2)]\mathbf{k}$$

$$= (xz - 0)\mathbf{i} + (2xz - yz)\mathbf{j} + (y^2 - 0)\mathbf{k} = xz\mathbf{i} + (2xz - yz)\mathbf{j} + y^2\mathbf{k}$$

21.88 Find $\nabla \times F$ at $P(3, 2, 1)$.

▮ $(\nabla \times F)(P) = (\nabla \times F)(3, 2, 1) = 3\mathbf{i} + 4\mathbf{j} + 4\mathbf{k}$.

21.89 Find $\nabla \times F$ at $Q(1, -2, 3)$.

▮ $(\nabla \times F)(Q) = (\nabla \times F)(1, -2, 3) = -3\mathbf{i} + 12\mathbf{j} + 4\mathbf{k}$.

Problems 21.90–21.93 refer to the following vector field: $F(x, y, z) = xyz\mathbf{i} + (x^2 + y^2 + z^2)\mathbf{j} + (x^2 - yz)\mathbf{k}$.

21.90 Find the divergence $\nabla \cdot F$.

▮ $\nabla \cdot F = (D_x\mathbf{i} + D_y\mathbf{j} + D_z\mathbf{k})[xyz\mathbf{i} + (x^2 + y^2 + z^2)\mathbf{j} + (x^2 - yz)\mathbf{k}] = D_x(xyz) + D_y(x^2 + y^2 + z^2) + D_z(x^2 - yz)$
$= yz + 2y - y = yz + y$.

21.91 Find the curl $\nabla \times F$.

▮

$$F = \begin{vmatrix} \mathbf{i} & \mathbf{j} & \mathbf{k} \\ D_x & D_y & D_z \\ xyz & x^2 + y^2 + z^2 & x^2 - yz \end{vmatrix}$$

$$= [D_y(x^2 - yz) - D_z(x^2 + y^2 + z^2)]\mathbf{i} + [D_z(xyz) - D_x(x^2 - yz)]\mathbf{j}$$

$$+ [D_x(x^2 + y^2 + z^2) - D_y(xyz)]\mathbf{k}$$

$$= (-z - 2z)\mathbf{i} + (xy - 2x)\mathbf{j} + (2x - xz)\mathbf{k} = -3z\mathbf{i} + (xy - 2x)\mathbf{j} + (2x - xz)\mathbf{k}$$

21.92 Find $\nabla(\nabla \cdot F)$.

▮ $\nabla(\nabla \cdot F) = (D_x\mathbf{i} + D_y\mathbf{j} + D_z\mathbf{k})(yz + y) = D_x(yz + y)\mathbf{i} + D_y(yz + y)\mathbf{j} + D_z(yz + y)\mathbf{k} = (z + 1)\mathbf{j} + y\mathbf{k}$.

21.93 Find $\nabla \times (\nabla \times F)$.

$$\nabla \times (\nabla \times F) = \begin{vmatrix} \mathbf{i} & \mathbf{j} & \mathbf{k} \\ D_x & D_y & D_z \\ -3z & xy - 2x & 2x - xz \end{vmatrix}$$

$$= (0 - 0)\mathbf{i} + (-3 - 2 + z)\mathbf{j} + (y - 2 - 0)\mathbf{k} = (z - 5)\mathbf{j} + (y - 2)\mathbf{k}.$$

21.94   Show that $\nabla \cdot (\nabla \times V) = 0$ for any vector field V.

I   Suppose $V = F\mathbf{i} + G\mathbf{j} + H\mathbf{k}$.   Then

$$\nabla \times V = \begin{vmatrix} \mathbf{i} & \mathbf{j} & \mathbf{k} \\ D_x & D_y & D_z \\ F & G & H \end{vmatrix} = (H_y - G_z)\mathbf{i} + (F_z - H_x)\mathbf{j} + (G_x - F_y)\mathbf{k}$$

Thus, $\nabla \cdot (\nabla \times V) = D_x(H_y - G_z) + D_y(F_z - H_x) + D_z(G_x - F_y) = H_{xy} - G_{xz} + F_{yz} - H_{xy} + G_{xz} - F_{yz} = 0.$

21.5   DIFFERENTIAL EQUATIONS

21.95   Rewrite the following system of differential equations in matrix form:

$$\frac{dx}{dt} = 4x - y$$

$$\frac{dy}{dt} = 2x + y$$

I   Let $X = \begin{pmatrix} x \\ y \end{pmatrix}$ and $A = \begin{pmatrix} 4 & -1 \\ 2 & 1 \end{pmatrix}$.   Then the system is equivalent to the matrix differential equation $dX/dt = AX$.

21.96   Consider a linear matrix differential equation

$$\frac{d}{dt}(X) = AX \tag{1}$$

Suppose $X = PY$ is a nonsingular change of variables.   Show that the transformed system has the form

$$\frac{d}{dt}(Y) = P^{-1}APY$$

I   Substitute $X = PY$ in (1) to get $d(PY)/dt = APY$.   Since the differential operator is linear, it commutes with the matrix P; that is, $d(PY)/dt = P[dY/dt]$.   Thus we get $P[dY/dt] = APY$.   Multiplying by P^{-1} yields (1).

21.97   Consider the matrix $A = \begin{pmatrix} 4 & -1 \\ 2 & 1 \end{pmatrix}$ in Problem 21.95.   Find a nonsingular matrix P such that $B = P^{-1}AP$ is diagonal.

I   The characteristic polynomial $\Delta(t)$ of A is

$$\Delta(t) = |tI = A| = \begin{vmatrix} t - 4 & 1 \\ -2 & t - 1 \end{vmatrix} = t^2 - 5t + 6 = (6 - 3)(t - 2)$$

Thus the eigenvalues of A are 3 and 2.   Substitute $t = 3$ into the matrix $tI - A$ to obtain the corresponding homogeneous system $-x + y = 0$, $-2x + 2y = 0$.   A nonzero solution is $v_1 = (1, 1)$. Substitute $t = 2$ into the matrix $tI - A$ to obtain the homogeneous system $-2x + y = 0$, $-2x + y = 0$. A nonzero solution is $v_2 = (1, 2)$.   Let P be the matrix whose columns are v_1 and v_2, respectively.   Then

$$P = \begin{pmatrix} 1 & 1 \\ 1 & 2 \end{pmatrix} \quad \text{and} \quad B = P^{-1}AP = \begin{pmatrix} 3 & 0 \\ 0 & 2 \end{pmatrix}$$

21.98   Solve the system of differential equations in Problem 21.95.

I   Diagonalize the system by a change of variables using the matrix P in Problem 21.97 as follows:

$$\begin{pmatrix} x \\ y \end{pmatrix} = P \begin{pmatrix} r \\ s \end{pmatrix} \quad \text{or} \quad \begin{matrix} x = r + s \\ y = r + 2s \end{matrix} \tag{1}$$

By Problems 21.96 and 21.97, the system with this change of variables now has the diagonal form:

$$\frac{dr}{dt} = 3r$$

$$\frac{ds}{dt} = \quad 2s$$

The solution of this diagonal system is $r = ae^{3t}$, $s = be^{2t}$ where a, b are parameters. Substitute in *(1)* to obtain the required solution:

$$x = ae^{3t} + be^{2t}$$
$$y = ae^{3t} + 2be^{2t}$$

Problems 21.99–21.104 refer to the following system of differential equations:

$$\frac{dx}{dt} = \quad 4x + 2y + \quad z$$

$$\frac{dy}{dt} = \quad 2x + 5y + 2z$$

$$\frac{dz}{dt} = -2x - 4y - \quad z$$

21.99 Rewrite the system in matrix form.

▌ Let

$$X = \begin{pmatrix} x \\ y \\ z \end{pmatrix} \quad \text{and} \quad A = \begin{pmatrix} 4 & 2 & 1 \\ 2 & 5 & 2 \\ -2 & -4 & -1 \end{pmatrix}$$

Then the system is equivalent to the matrix equation $dX/dt = AX$.

21.100 Find the characteristic polynomial $\Delta(t)$ of A.

▌ $\Delta(t) = t^3 - \text{tr}(A)t^2 + (A_{11} + A_{22} + A_{33})t - \det(A) = t^3 - 8t^2 + 17t - 10$. [Here $\text{tr}(A)$ is the trace of A and A_{ii} is the cofactor of the diagonal element a_{ii}.]

21.101 Find the eigenvalues of A or, in other words, the roots of $\Delta(t)$.

▌ If $\Delta(t)$ has a rational root it must divide 10. Testing $t = 1$ we get

$$
\begin{array}{r|rrrr}
1 & 1 - 8 + 17 - 10 \\
  & \ 1 - \ 7 + 10 \\
\hline
  & 1 - 7 + 10
\end{array}
$$

Thus $t - 1$ is a factor of $\Delta(t)$ and $\Delta(t) = (t - 1)(t^2 - 7t + 10) = (t - 1)(t - 2)(t - 5)$. Accordingly, the eigenvalues of A are $\lambda = 1$, $\lambda = 2$, $\lambda = 5$.

21.102 Find a nonzero eigenvector for each of the eigenvalues of A.

▌ Subtract $\lambda = 1$ down the diagonal of A to obtain the corresponding system $3x + 2y + z = 0$, $2x + 4y + 2z = 0$, $-2x - 4y - 2z = 0$. A nonzero solution is $v_1 = (0, 1, -2)$. Subtract $\lambda = 2$ down the diagonal of A to obtain the system $2x + 2y + z = 0$, $2x + 3y + 2z = 0$, $-2x - 4y - 3z = 0$. A nonzero solution is $v_2 = (1, -2, 2)$. Subtract $\lambda = 5$ down the diagonal of A to obtain the homogeneous system $-x + 2y + z = 0$, $2x - 2z = 0$, $-2x - 4y - 6z = 0$. A nonzero solution is $v_3 = (1, 1, -1)$.

21.103 Find a nonsingular matrix P such that $B = P^{-1}AP$ is diagonal.

▌ Let P be the matrix whose columns are v_1, v_2, v_3, respectively. Then

$$P = \begin{pmatrix} 0 & 1 & 1 \\ 1 & -2 & 1 \\ -2 & 2 & -1 \end{pmatrix} \quad \text{and} \quad B = P^{-1}AP = \begin{pmatrix} 1 & & \\ & 2 & \\ & & 5 \end{pmatrix}$$

21.104 Solve the system of differential equations.

▎ Diagonalize the system by a change of variables using the above matrix P as follows:

$$\begin{pmatrix} x \\ y \\ z \end{pmatrix} = P \begin{pmatrix} x' \\ y' \\ z' \end{pmatrix} \qquad \text{or} \qquad \begin{aligned} x &= & y' + z' \\ y &= & x' - 2y' + z' \\ z &= & -2x' + 2y' - z' \end{aligned} \qquad (1)$$

With this change of variables, the system now has the diagonal form

$$\frac{dx'}{dt} = x' \qquad \frac{dy'}{dt} = 2y' \qquad \frac{dz'}{dt} = 5z'$$

The solution of the diagonal system is $x' = ae^t$, $y' = be^{2t}$, $z' = ce^{5t}$ where a, b, c are parameters. Substitute in (1) to obtain the required solution:

$$\begin{aligned} x &= & be^{2t} + ce^{5t} \\ y &= & ae^t - 2be^{2t} + ce^{5t} \\ z &= & -2ae^t + 2be^{2t} - ce^{5t} \end{aligned}$$